Dynamics: The Geometry of Behavior

Studies in Nonlinearity

Series Editor: Robert L. Devaney

Robert L. Devaney, *A First Course in Chaotic Dynamical Systems: Theory and Experiment* (1992)

Gerald Edgar, *Classics in Fractals (1992)*

Nicholas Tufillaro, Tyler Abbott, and Jeremiah Reilly, *An Experimental Approach to Nonlinear Dynamics and Chaos* (1992)

James E. Georges, Delbert L. Johnson, and Robert L. Devaney, *Dynamical Systems Software* (1992)

DYNAMICS
THE GEOMETRY OF BEHAVIOR

Second Edition

Ralph H. Abraham

and

Christopher D. Shaw

ADDISON-WESLEY PUBLISHING COMPANY
The Advanced Book Program
Redwood City, California • Menlo Park, California •
Reading, Massachusetts • New York •
Don Mills, Ontario • Wokingham, United Kingdom •
Amsterdam • Bonn • Sydney • Singapore •
Tokyo • Madrid • San Juan

Addison-Wesley Publishing Company
Advanced Book Program
350 Bridge Parkway
Redwood City, California 94065

Mathematics Editor: Barbara Holland
Editorial Assistant: Diana Tejo
Production Manager: Pam Suwinsky
Production Assistant: Karl Matsumoto
Copyeditor: Andrew Alden
Cover designer: Nancy Brescia
Studies in Nonlinearity logo: Marek Antoniak
Typesetting: TypaGraphix

Library of Congress Cataloging-in-Publication Data

Abraham, Ralph.
Dynamics—the geometry of behavior / Ralph H. Abraham and Christopher D. Shaw — 2nd ed.
p. cm. — (Studies in nonlinearity)
Includes bibliographical references and index.
ISBN 0-201-56716-4 (HB). — ISBN 0-201-56717-2 (PB)
1. Dynamics. I. Shaw, Christopher D. II. Title. III. Series: Addison-Wesley studies in nonlinearity.
QA845.A24 1992
531'.11—dc20 91-30958
CIP

Published by arrangement with Aerial Press, Inc.

Printed in the United States of America

1 2 3 4 5 6 7 8 9 10-XX-96 95 94 93 92

Contents

Foreword

During the Renaissance, algebra was resumed from Near Eastern sources, and geometry from the Greek. Scholars of the time became familiar with classical mathematics. When calculus was born in 1665, the new ideas spread quickly through the intellectual circles of Europe. Our history shows the importance of the diffusion of these mathematical ideas, and their effects upon the subsequent development of the sciences and technology.

Today, there is a cultural resistance to mathematical ideas. Due to the widespread impression that mathematics is difficult to understand, or to a structural flaw in our educational system, or perhaps to other mechanisms, mathematics has become an esoteric subject. Intellectuals of all sorts now carry on their discourse in nearly total ignorance of mathematical ideas. We cannot help thinking that this is a critical situation, as we hold the view that mathematical ideas are essential for the future evolution of our society.

The absence of visual representations in the curriculum may be part of the problem, contributing to mathematical illiteracy and the math-avoidance reflex. This book is based on the idea that mathematical concepts may be communicated easily in a format that combines visual, verbal, and symbolic representations in tight coordination. It aims to attack math ignorance with an abundance of visual representations.

In sum, the purpose of this book is to encourage the diffusion of mathematical ideas by presenting them *visually.*

Preface

Dynamics is a field emerging somewhere between mathematics and the sciences. In our view, it is the most exciting event on the concept horizon for many years. The new concepts appearing in dynamics extend the conceptual power of our civilization and provide new understanding in many fields.

We discovered, while working together on the illustrations for a book in 1978,[1]* that we could explain mathematical ideas visually, within an easy and pleasant working partnership. In 1980, we wrote an expository article on dynamics and bifurcations,[2] using hand-animation to emulate the *dynamic picture technique* universally used by mathematicians in talking among themselves: a picture is drawn slowly, line by line, along with a spoken narrative—the dynamic picture and the narrative tightly coordinated.

Our efforts inevitably exploded into four volumes, now combined into this book. The dynamic picture technique, evolved through our work together, and in five years of computer graphic experience with the *Visual Math Project* at the University of California at Santa Cruz, is the basis of this work. The majority of the book is devoted to visual representations, in which four colors are used according to a strict code.

Math symbols have been kept to a minimum. In fact, they are almost completely suppressed. Our purpose is to make the book work for readers who are not practiced in symbolic representations. We rely exclusively on visual representations, with brief verbal explanations. Some formulas are shown with the applications, as part of the graphics, but are not essential. However, *this strategy is exclusively pedagogic.* We do not want anyone to think that we consider symbolic representations unimportant in mathematics. On the contrary, this field evolved primarily in the symbolic realm throughout the classical period. Even now, a full understanding of our subject demands a full measure of formulas, logical expressions, and technical intricacies from all branches of mathematics. A brief introductions to these is included in the Appendix.

We have created this book as a short-cut to the research frontier of dynamical systems: theory, experiments, and applications. It is our goal—we know we may fail to reach it—to provide any interested person with an acquaintance with the basic concepts:

*Footnotes refer to the Notes, which follow the Appendix.

- state spaces: manifolds—geometric models for the virtual states of a system
- attractors: static, periodic, and chaotic—geometric models for its local asymptotic behavior
- separatrices: repellors, saddles, insets, tangles—defining the boundaries of regions (basins) dominated by different behaviors (attractors), and characterizing the global behavior of a system
- bifurcations: subtle and catastrophic—geometric models for the controlled change of one system into another.

The ideas included are selected from the literature of dynamics: Part One, "Periodic Behavior," covers the classical period from 1600 to 1950. Part Two, "Chaotic Behavior," is devoted to recent developments, 1950 to the present, on the chaotic behavior observed in experiments. Part Three, "Global Behavior," describes the concept of structural stability, discovered in 1937, and the important generic properties discovered since 1959, relating to the tangled insets and outsets of a dynamical system. These are fundamental to Part Four, "Bifurcation Behavior." In fact, the presentation in Part Four of an atlas of bifurcations in dynamical schemes with one control parameter was the original and primary goal of this whole book, and all of the topics in the first three parts have been selected for their importance to the understanding of these bifurcations. For we regard the *response diagram*, a molecular arrangement of the atomic bifurcation events described here, as the most useful dynamical model available to a scientist.

We assume nothing in the way of prior mathematical training, beyond vectors in three dimensions, and complex numbers. Nevertheless, it will be tough going without a basic understanding of the simplest concepts of calculus.

Our first attempt at the pictorial style used here evolved in the first draft of *Dynamics: A Visual Introduction*, during the summer of 1980. Our next effort, the preliminary draft of Part Two of this book, was circulated among friends in the summer of 1981. Extensive feedback from them has been very influential in the evolution of this volume, and we are grateful to them:

Fred Abraham
Ethan Akin
Michael Arbib
Jim Crutchfield
Larry Cuba
Richard Cushman
Larry Domash
Jean-Pierre Eckman
Len Fellman
George Francis
Alan Garfinkel
John Guckenheimer
Moe Hirsch
Phil Holmes
Dan Joseph
Jean-Michel Kantor
Bob Lansdon
Arnold Mandell
Jerry Marsden
Nelson Max
Jim McGill
Kent Morrison
Charles Musès
Norman Packard
Tim Poston
Otto Rossler
Lee Rudolph
Katie Scott
Rob Shaw
Mike Shub
Steve Smale
Joel Smoller
Jim Swift
Bob Williams
Art Winfree
Marianne Wolpert
Gene Yates
Chris Zeeman

We are especially grateful to Tim Poston and Fred Abraham for their careful reading of the manuscript; to the Dynamics Guild (J. Crutchfield, D. Farmer, N. Packard, and R. Shaw) for their computer plots used in many places in this book; to Richard Cushman for history lessons; to Phyllis Wright and Claire Moore of TypaGraphix for their care in typesetting and production; to Lauro Lato of Aerial Press for her expert assistance in the production process; to Diane Rigoli for her splendid final drawings based on our rough sketches for Part Four; and to Rob Shaw for providing photos for

Section 16.3 and computer plots for Section 17.3. The generosity and goodwill of many dynamicists has been crucial in the preparation of this book; we thank them all. We are grateful to Tom Jones, André Leroi-Gourhan, Preston James, Goeffrey Martin, and their publishers for permitting the reproduction of their illustrations. Finally, it is a pleasure to thank the National Science Foundation for financial support.

Ralph H. Abraham
Christopher D. Shaw
Santa Cruz, California
October, 1991

PART 1

Periodic Behavior

Dedicated to Lord Rayleigh

DYNAMICS HALL OF FAME

Dynamics has evolved into three disciplines: applied, mathematical, and experimental. Applied dynamics is the oldest. Originally regarded as a branch of natural philosophy, or physics, it goes back to Galileo at least. It deals with the concept of change, rate of change, rate of rate of change, and so on, as they occur in natural phenomena. We take these concepts for granted, but they emerged into our consciousness only in the fourteenth century.[1]

Mathematical dynamics began with Newton and has become a large and active branch of pure mathematics. This includes the theory of ordinary differential equations, now a classical subject. But since Poincaré, the newer methods of topology and geometry have dominated the field.

Experimental dynamics is an increasingly important branch of the subject. Founded by Galileo, it showed little activity until Rayleigh, Duffing, and Van der Pol. Experimental techniques have been revolutionized with each new development of technology. Analog and digital computers are now accelerating the advance of the research frontier, making experimental work more significant than ever.

This chapter presents a few words of description for some of the leading figures of the history of dynamics. Their positions in a two-dimensional tableau — date versus specialty (applied, mathematical, or experimental dynamics) — are shown in Table 1.1. Those included are not more important than numerous others, but limitations of space and knowledge prevent us from giving a more complete museum here.

TABLE 1.1–THE HISTORY OF DYNAMICS

Date	APPLIED DYNAMICS	MATHEMATICAL DYNAMICS	EXPERIMENTAL DYNAMICS
1600	Kepler		Galileo
1650			
	Huyghens	Newton Leibniz	
1700			
		Euler	
1750			
		Lagrange	
1800			
1850			
	Helmholtz Rayleigh	Poincaré Lie Liapounov	Rayleigh
1900			
	Lotka Volterra Rashevsky	Birkhoff Andronov Cartwright	Duffing Van der Pol Hayashi
1950			

Galileo Galilei, 1564–1642. One of the first to deal thoroughly with the concept of acceleration, Galileo founded dynamics as a branch of natural philosophy. The close interplay of theory and experiment, characteristic of this subject, was founded by him.

photo courtesy of D.J. Struik, A Concise History of Mathematics, Dover Publications, New York (1948)

Johannes Kepler, 1571–1630. The outstanding and original exponent of applied dynamics. Kepler made use of extensive interaction between theory and observation to understand the planetary motions.

courtesy of Kepler, Gesammelte Werke, Beck, München (1960)

Isaac Newton, 1642–1727. Mathematical dynamics, as well as the calculus on which it is based, was founded by Newton at age 23. Applications and experiments were basic to his ideas, which were dominated by the doctrine of determinism. His methods were geometric.

photo courtesy of the Trustees of the British Museum

Gottfried Wilhelm Leibniz, 1646–1716. The concepts of calculus, mathematical dynamics, and their implications for natural philosophy, occurred independently to Leibniz. His methods were more symbolic than geometric.

photo courtesy of the Trustees of the British Museum

Leonhard Euler, 1707–1783. Primarily known for his voluminous contributions to algebra, Euler developed the techniques of analysis which were to dominate mathematical dynamics throughout its classical period.

photo courtesy of E. T. Bell, Men of Mathematics, Simon and Schuster, New York (1937)

Joseph-Louis Lagrange, 1736–1813. A disciple of Euler, Lagrange developed the analytical method to extremes, and boasted that his definitive text on the subject contained not a single illustration.

photo courtesy of the Bibliothèque Nationale, Paris, France.

Marius Sophus Lie, 1842–1899. In combining the ideas of symmetry and dynamics, Lie built the foundations for a far-reaching extension of dynamics, the theory of groups of transformations.

photo courtesy of Minkowski, H., Briefe an David Hilbert, Mit Beiträgen und herausgegeben von L. Rüdenberg, H. Zassenhaus; Springer-Verlag, Heidelberg (1973)

John William Strutt, Baron Rayleigh, 1842–1919. In a career of exceptional length and breadth, spanning applied mathematics, physics, and chemistry, Rayleigh dwelled at length on acoustical physics. In this context, he revived the experimental tradition of Galileo in dynamics, laying the foundations for the theory of nonlinear oscillations. His text on acoustics, published in 1877, remains to this day the best account of this subject.

photo courtesy of Applied Mech. Rev. 26 (1973)

Jules Henri Poincaré, 1854–1912. Known for his contributions to many branches of pure mathematics, Poincaré devoted the majority of his efforts to mathematical dynamics. Among the first to accept the fact that the classical analytical methods of Euler and Lagrange had serious limitations, he revived geometrical methods. The results were revolutionary for dynamics, and gave birth to topology and global analysis as well. These branches of pure mathematics are very active yet.

photo courtesy of the Library of Congress, Washington, D.C.

Aleksandr Mikhailovich Liapounov, 1857–1918. Another pioneer of geometric methods in mathematical dynamics, Liapounov contributed basic ideas of stability.

photo courtesy of Akademija Nauk, SSR (1954)

Georg Duffing, 1861–1944. A serious experimentalist, Duffing studied mechanical devices to discover geometric properties of dynamical systems. The theory of oscillations was his explicit goal.

photo courtesy of Mrs. Monika Murasch and Prof. Dr.-Ing. R. Gasch, Berlin.

George David Birkhoff, 1884–1944. The first dynamicist in the New World, Birkhoff picked up where Poincaré left off. Although a geometer at heart, he discovered new symbolic methods. He saw beyond the theory of oscillations, created a rigorous theory of ergodic behavior, and foresaw dynamical models for chaos.

photo courtesy of G. D. Birkhoff, Collected Mathematical Papers, American Mathematical Society, New York (1950)

Balthasar van der Pol, 1889–1959. The first radio transmitter became, in the hands of this outstanding experimentalist, a high-speed laboratory of dynamics. Many of the basic ideas of modern experimental dynamics came out of this laboratory.

photo courtesy of Balthasar van der Pol, Selected Scientific Papers, Vol. 1, H. Bremmer and C. J. Boukamp (eds.), North Holland, Amsterdam (1960)

Nicholas Rashevsky, 1899–1972. From antiquity until the 1920's, applied dynamics meant physics. At last, the important applications to the biological and social sciences came into view, in the visionary minds of the general scientists Lotka, Volterra, and Rashevsky.

photo courtesy of Bull. Math. Biophys. 34 (1972)

Mary Lucy Cartwright, 1900– . Dame Cartwright, together with J. E. Littlewood, revived dynamics in England, during World War II. Inspired by the work of Van der Pol, they obtained important results on the ultraharmonics of forced electronic oscillations, using analytical and topological methods.

photo courtesy of Math. Gazette 36 (1952)

Chihiro Hayashi, 1911–1986. The experiments of dynamicists were restricted to a few simple systems (Duffing's system, Van der Pol's system, etc.) until the appearance of the general purpose analog computer. One of the creators of this type of machine, and the first to fully exploit one as a laboratory of dynamics, Hayashi contributed much to our knowledge of oscillations.

photo courtesy of Ch. Hayashi, Selected Papers on Nonlinear Oscillators, Kyoto (1975)

1

Basic Concepts of Dynamics

The key to the geometric theory of dynamical systems created by Poincaré is the *phase portrait* of a dynamical system. The first step in drawing this portrait is the creation of a geometric model for the set of all possible states of the system. This is called the *state space*. On this geometric model, the dynamics determine a cellular structure of *basins* enclosed by *separatrices*. Within each cell or basin is a nucleus called the *attractor*. The states that will actually be observed in this system are the attractors. Thus, the portrait of the dynamical system, showing the basins and attractors, is of primary importance in applications. This chapter introduces these basic concepts.

1.1. State Spaces

The strategies for making mathematical models for observed phenomena have been evolving since ancient times. An organism—physical, biological, or social—is observed in different states. This *observed system* is the target of the modeling activity. Its states cannot really be described by only a few observable parameters, but we pretend that they can. This is the first step in the process of "mathematical idealization" and leads to a geometric model for the set of all idealized states: the *state space* of the model. Different models may begin with different state spaces. The relationship between the actual states of the real organism and the points of the geometric model is a fiction maintained for the sake of discussion, theory, thought, and so on: this is known as the *conventional interpretation.* This section describes some examples of this modeling process.

The simplest scheme is the one-parameter model. The early history of science used this scheme extensively.

1.1.1. The actual state of this waffle iron cannot be described completely by a single observable parameter, such as the temperature. But usually we find it convenient to pretend that it can. This pretense is an agreement, the *conventional interpretation*, within the modeling process. It is justified by its usefulness in describing the behavior of the device.

1.1.2. The correlation between the internal state of a complex system, such as a mammal, and a single observed parameter may be very good or very bad, depending on the context. In the case of George Washington, the oral temperature correlates better with his health than his honesty.

1.1.3. In these examples, the geometric model for the set of all (mathematically idealized) states is the real number line. This is one of the simplest state spaces.

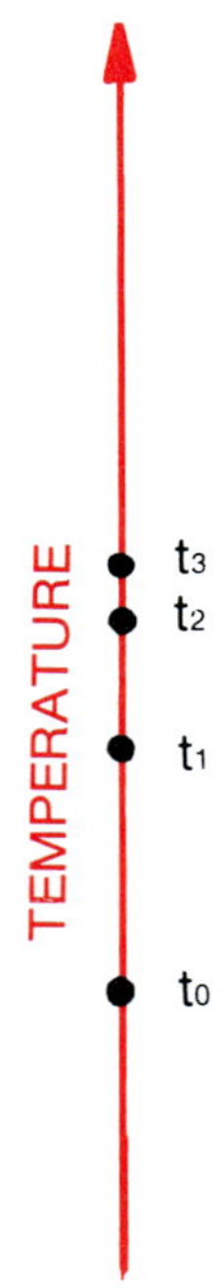

1.1.4. Observing the parameter for a while, it will probably change. The different values observed may be labeled by the time of their observation: the states observed at four different times are shown here.

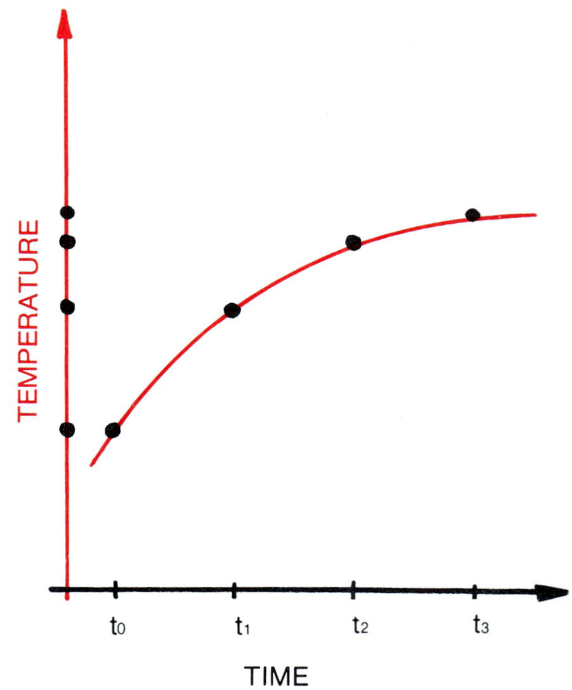

1.1.5. These data compose a *time series* of observations, and are shown here as a graph. The vertical line represents the state space, and the horizontal line axis indicates time.

Closer observation may suggest two parameters for the description of a given state of the actual organism.

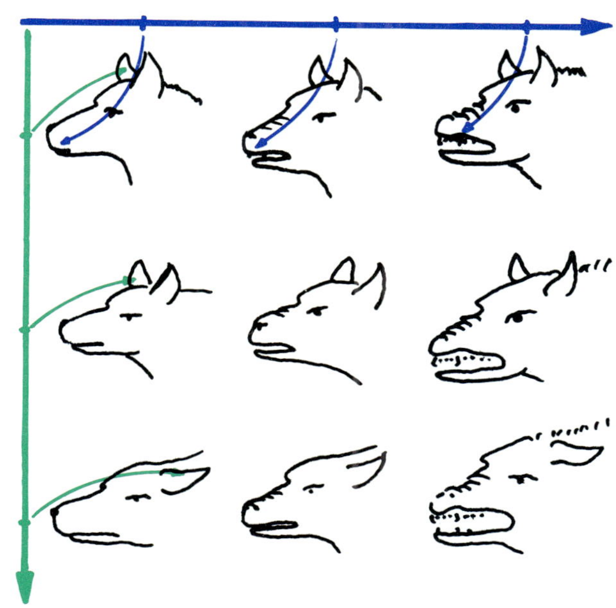

1.1.6. In this modeling scheme of Konrad Lorenz and Christopher Zeeman,[1] two parameters are used for the emotional state of a dog. The two observed parameters are *ear attitude*, which correlates with the emotional state of fear, and *fang exposure*, corresponding to the degree of rage.

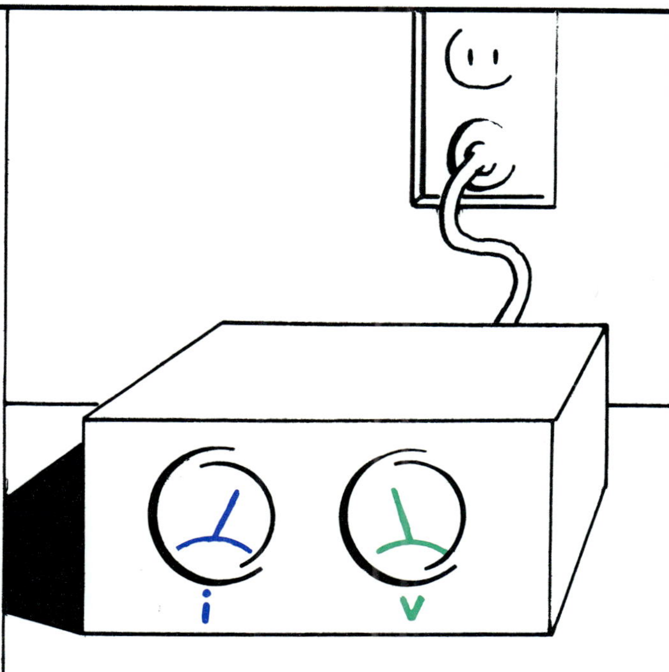

1.1.7. Electronic devices are simple to model, as the observations are explicitly numerical. This electronic "black box" is provided with panel meters, which indicate the instantaneous values of voltage and current at specific points of the electronic network within the box.

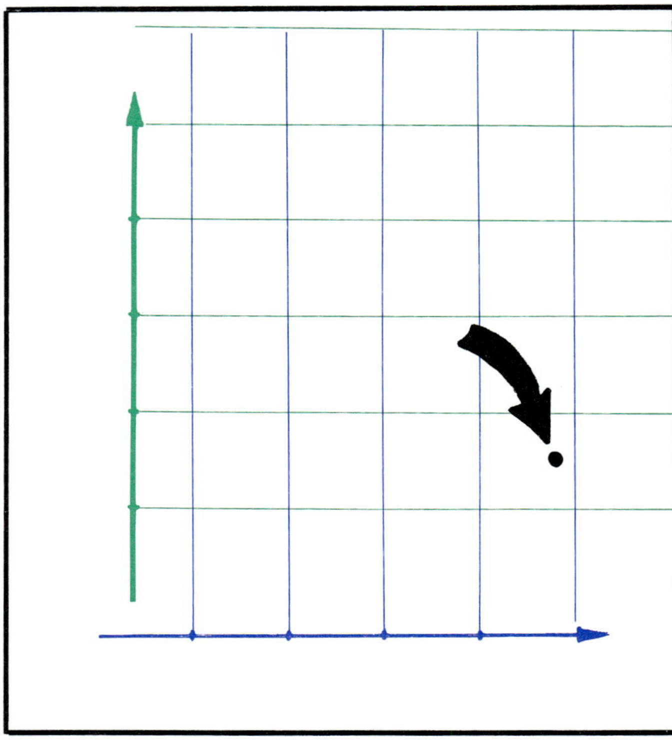

1.1.8. The values of two numerical parameters (of a model such as those of the preceding two examples) may be represented by a single point in this two-dimensional state space, the *plane* of Euclidean geometry.

Changes in the actual state of the system are observed and are represented as a curve in the state space. Each point on this curve carries (implicity at least) a label recording the time of observation. This is called a *trajectory* of the model.

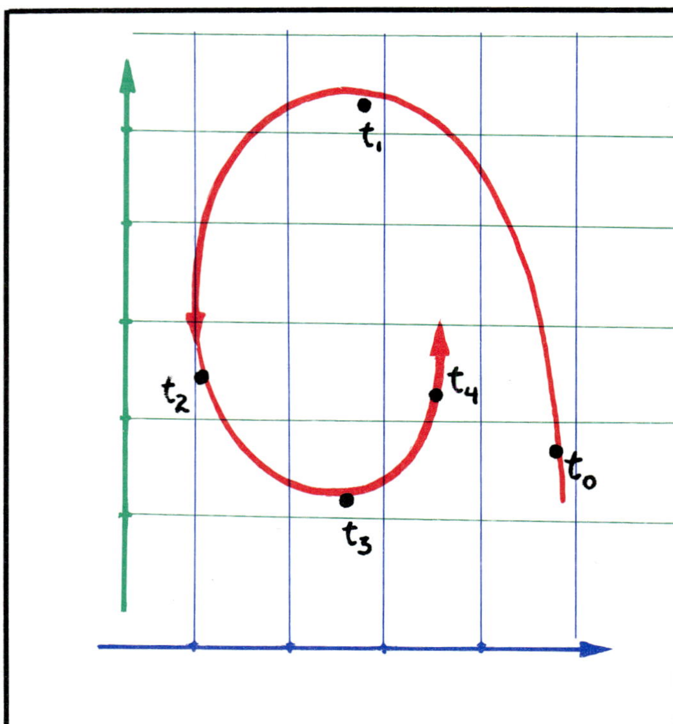

1.1.9. For example, if the two parameters representing the emotional state of a dog, or the internal state of an electronic black box, are observed at successive times and recorded in the plane with labels, a trajectory of the model is obtained.

Another style of representing the changing data is by its *time series,* which means the *graph* of a trajectory. We have already seen a time series, in a one dimensional context. But this style of data representation may also be used in higher dimensions.

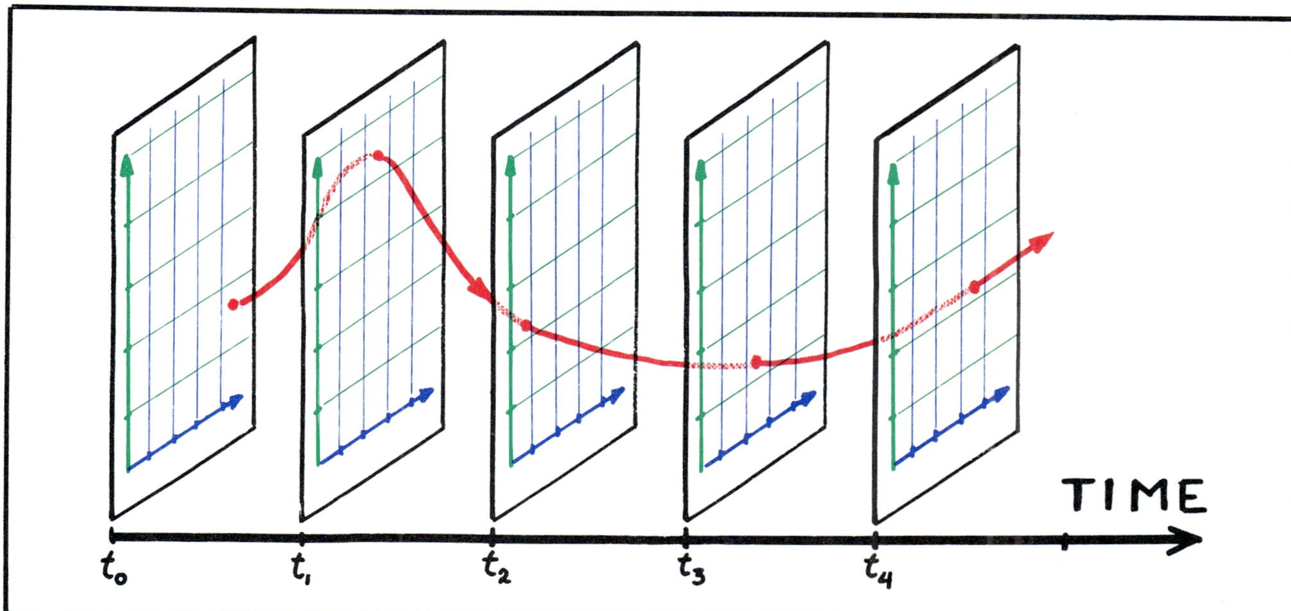

1.1.10. Here the vertical plane represents the state space, and the horizontal axis represents the time of observation. The parameters observed at a given time are plotted in the vertical plane passing through the appropriate point on the time axis.

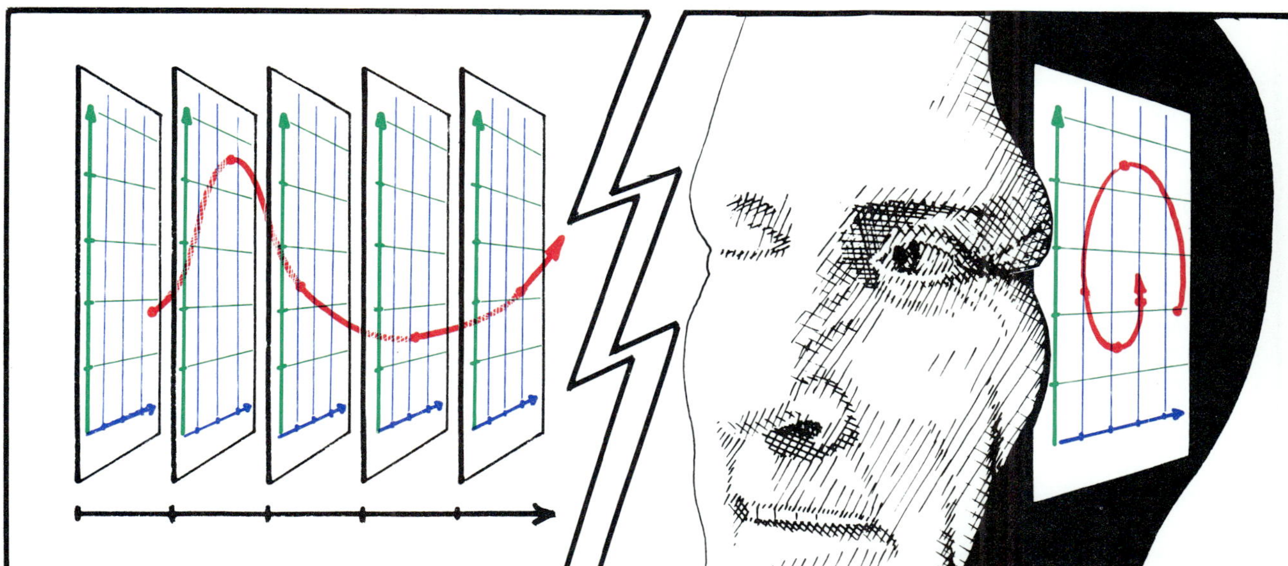

1.1.11. The trajectory may be obtained from the time series, by simply viewing it from the right angle—straight down the time axis from the end, infinitely far away.

Observing more parameters leads to models of higher dimensions.

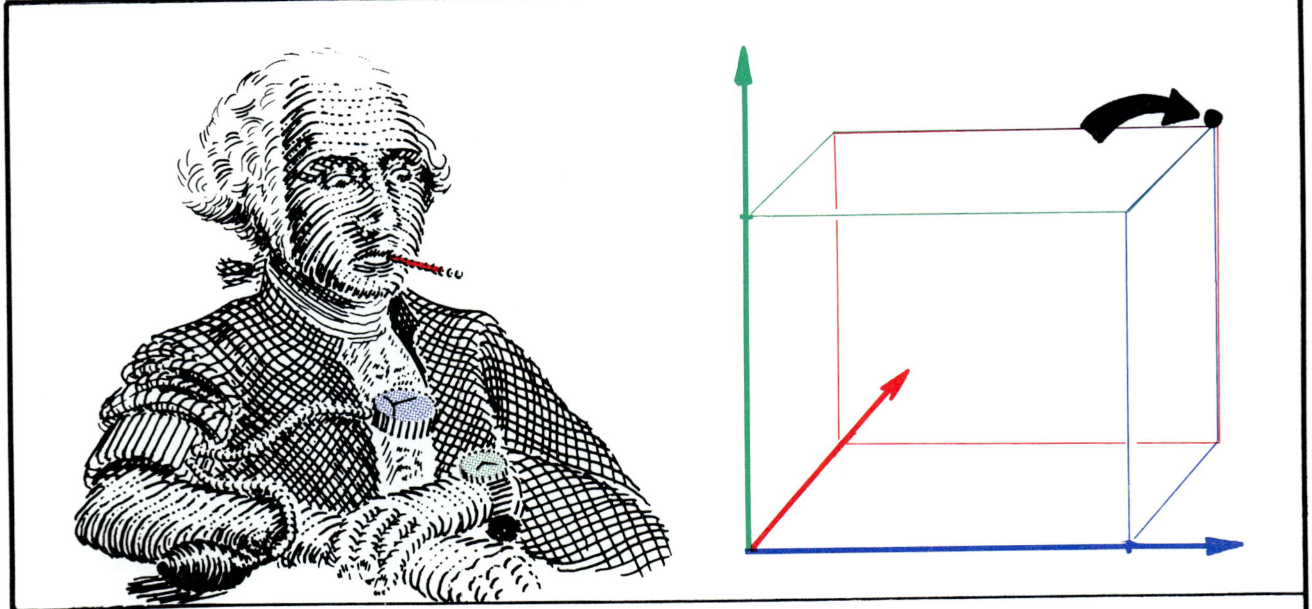

1.1.12. Suppose that at 7 am, this athlete observes three of his body parameters (say temperature, blood pressure, and pulse rate), records these three data as a point in three-dimensional state space.

Many phenomena require geometric models that are not simply coordinate spaces. In dynamical systems theory, the geometric models used are *manifolds.*

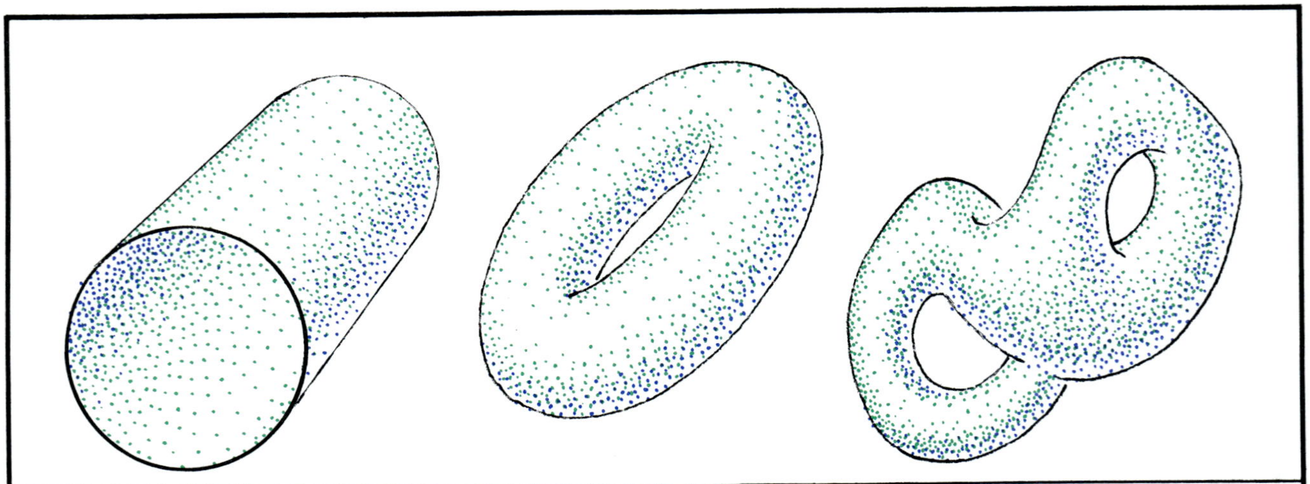

1.1.13. Here are some examples of manifolds. Other examples will arise in later chapters. They are made of pieces of flat spaces, bent and glued together.

1.2. Dynamical Systems

At this point, the history of a real system has been represented graphically, as a *trajectory* in a geometric *state space.* An alternative representation is the *time series,* or *graph,* of the trajectory. The dynamical concepts of the Middle Ages included these kinds of representation. But in the 1660's, something new was added—*the instantaneous velocity, or derivative, of vector calculus*—by Newton. As dynamical systems theory evolved, the *velocity vectorfield* emerged as one of the basic concepts. Trajectories determine velocity vectors by the *differentiation* process of calculus. Conversely, velocity vectors determine trajectories by the *integration* process of calculus.

This is the differentiation process, which determines the velocity vectorfield from the trajectories.

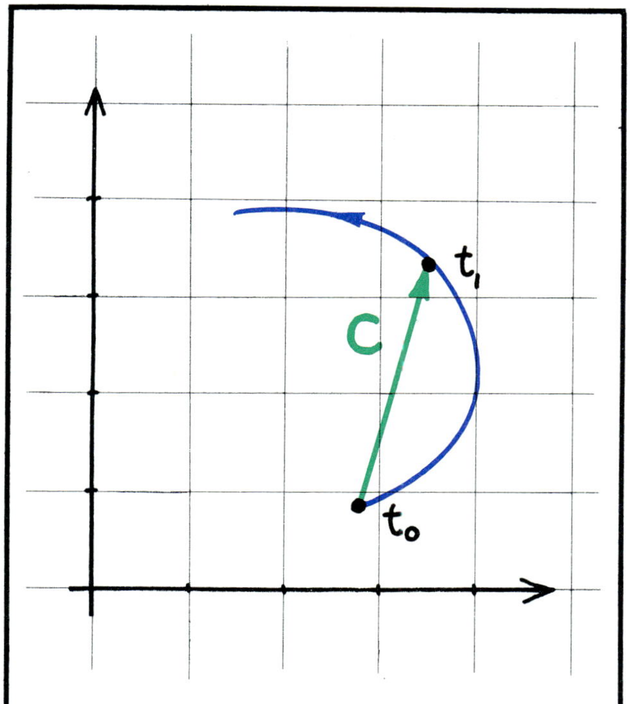

1.2.1. On this trajectory, the states obseved at two different times, t_0 and t_1, are connected by a *bound vector,* represented here by a line segment pointed on one end. Let C denote this bound vector.

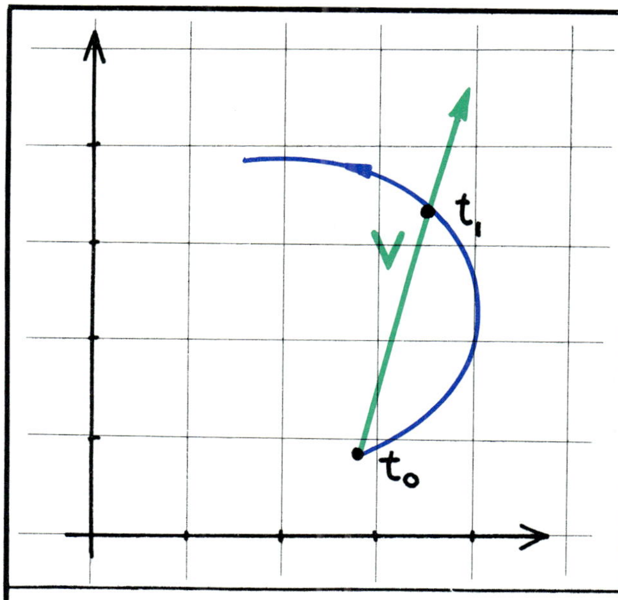

1.2.2. The *average velocity* of the change of state, C, is the vector starting at the point labeled t_0 on the curve, and directed along the vector of change of state, C, but divided by T, the time elapsed between t_0 and t_1. Let V denote this vector, V=C/T. It represents the average speed and direction of the change of state.

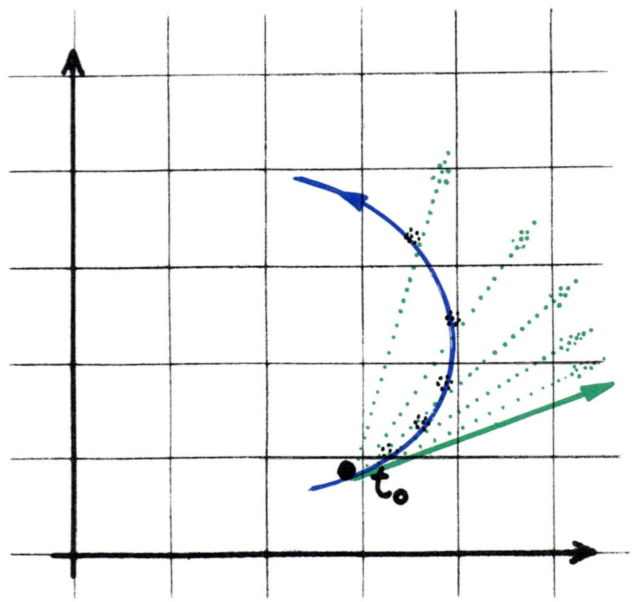

1.2.3. The *instantaneous velocity* of the trajectory at the time t_0 is the bound vector that V tends to as the elapsed time T shrinks smaller and smaller. This limiting vector, denoted here by $C\uparrow$, is also known as the *tangent vector.* The construction of this velocity, or tangent vector, from the curve is called *differentiation* in vector calculus.

The modeling process begins with the choice of a particular state space in which to represent observations of the system. Prolonged observations lead to many trajectories within the state space. At any point on any of these curves, a velocity vector may be derived. This is the new dynamical concept of Newton and Leibniz. It is useful in describing an inherent tendency of the system to move with a habitual velocity, at particular points in the state space.

The prescription of a velocity vector at each point in the state space is called a *vectorfield.*

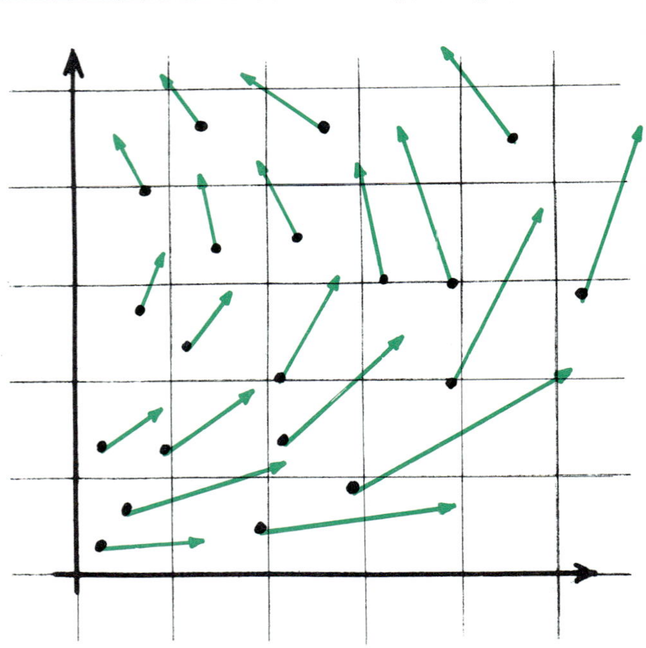

1.2.4. A *vectorfield* is a field of bound vectors, one defined at (and bound to) each and every point of the state space. Here only a few of the vectors are drawn, to suggest the full field.

The state space, filled with trajectories, is called the *phase portrait* of the dynamical system. The velocity vectorfield has been derived from the phase portrait by *differentiation*.

We regard this vectorfield as the model for the system under study. In fact, the phrase *dynamical system* will specifically denote this vectorfield.

In the practice of this modeling art, the choice of a vectorfield is a difficult and critical step. Extensive observations of the organism being modeled, over a long period of time, will usually reveal tendencies (to a dynamicist, at least) that can be represented as a dynamical system. The history of applied dynamics provides excellent examples of this process. Several of these are described in the next four chapters. The usefulness of this kind of model depends on the followig fundamental hypotheses.

Hypothesis 1. The observation of the organism over time, represented as a trajectory in the state space, will have this property, at each of its points: its velocity vector is exactly the same as the vector specified by the dynamical system.

Henceforward, the word *trajectory* will always carry this assumption. That is, the trajectories of the phase portrait have the specified velocity vectors, and further, they will be assumed to represent the behavior of the system being modeled. Further, for technical reasons we also assume:

Hypothesis 2. The vectorfield of the model is smooth.

Smoothness, in this context, is most easily seen in the one-dimensional case. On a one-dimensional state space, a vectorfield is specified by a graph in the plane. In this context, the graph is *smooth* if it is continuous as well: no jumps, no sharp corners.

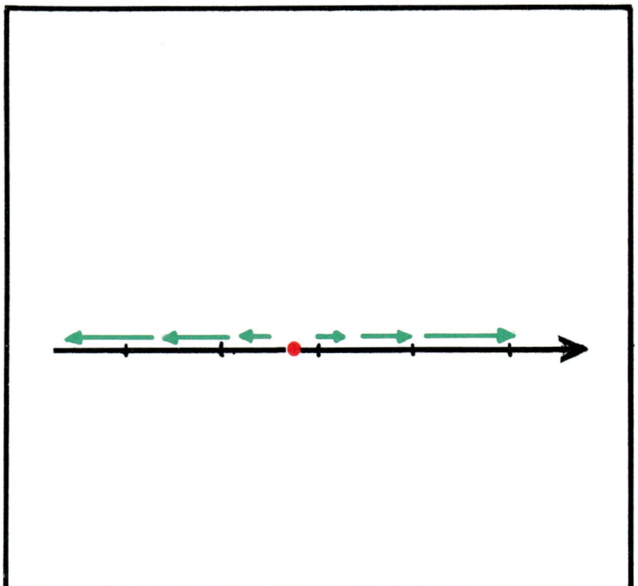

1.2.5. For example, here is a vectorfield (green) on a one-dimensional state space (black). The vector at the rest point (red) is the "zero vector"; its length is zero.

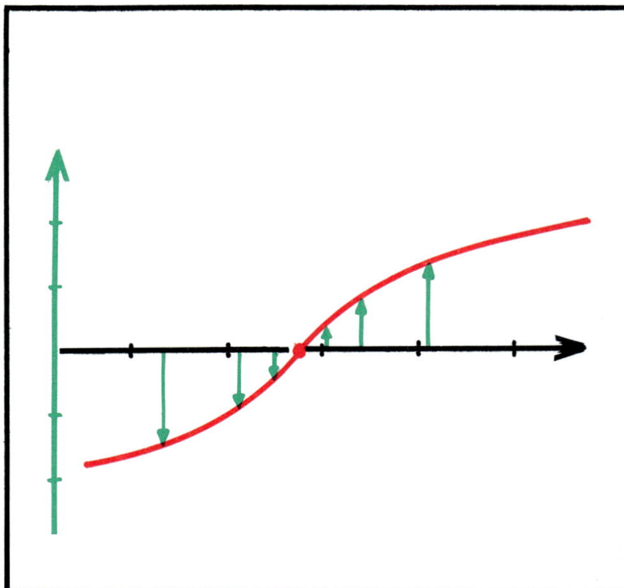

1.2.6. Stand up each green vector by rotating each one counterclockwise by a right angle. The arrowheads (green) trace out a curve (red), which is the graph of a function. The vectorfield is completely described by this function.

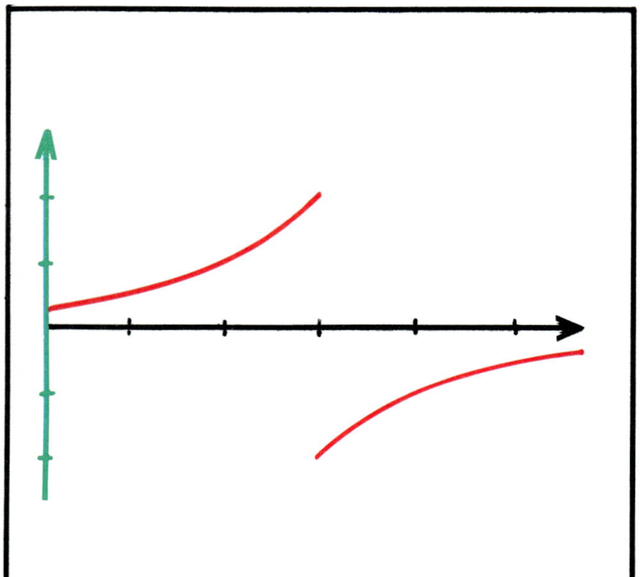

1.2.7. Another vectorfield is described by this function. This function is not continuous, so the vectorfield is not smooth.

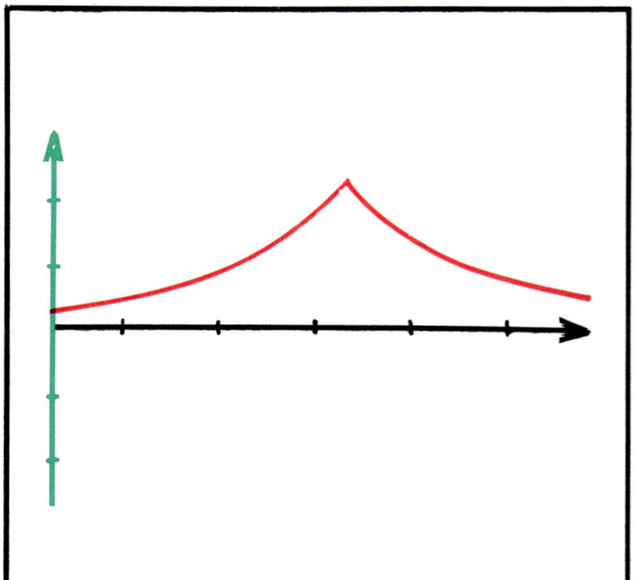

1.2.8. Yet another vectorfield is described by this function. This function is continuous but has a sharp corner. This vectorfield is not smooth either.

We suppose now that a dynamical system has been chosen as a model for a system. Given this vectorfield, how can we deduce the trajectories, thus the phase portrait, and the behavior of the system?

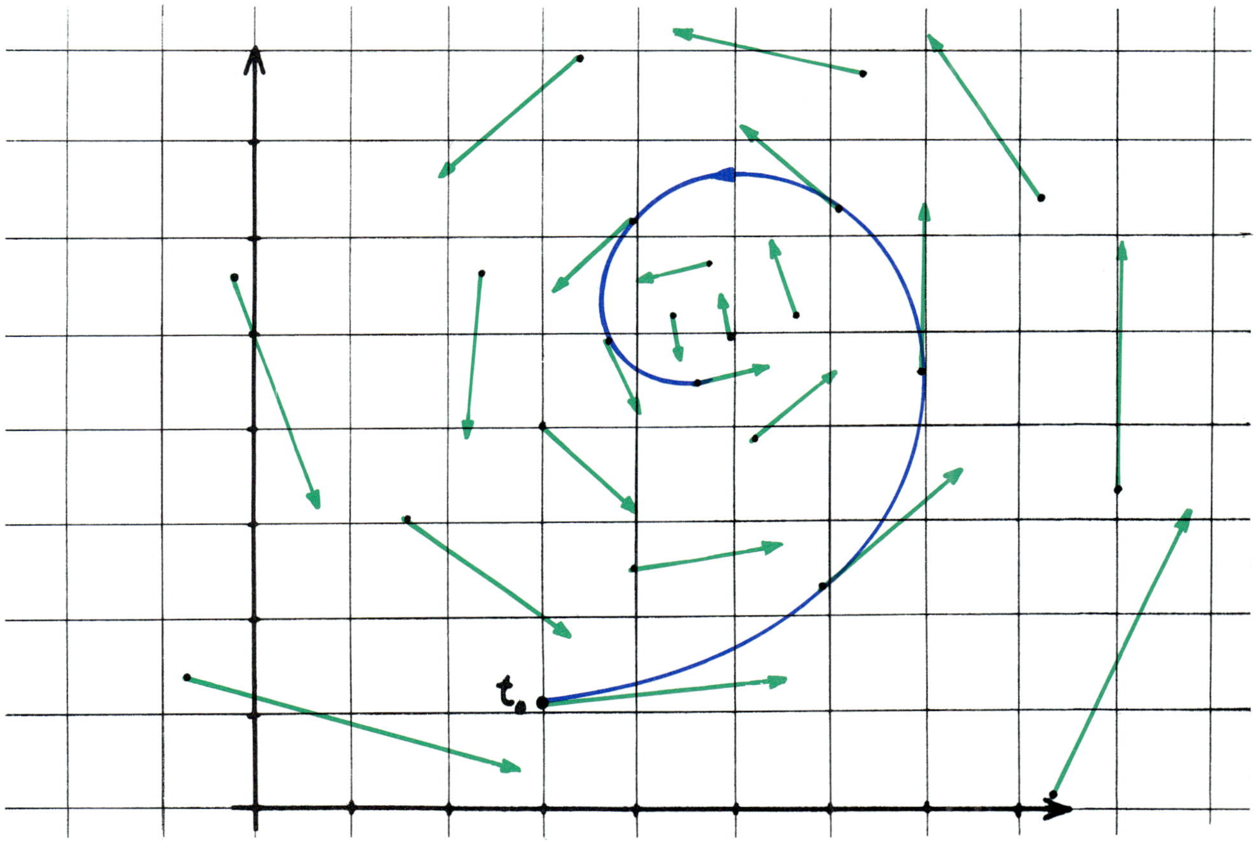

1.2.9. Given a state space and a dynamical system (smooth vectorfield), a curve in the state space is a *trajectory, or integral curve*, of the dynamical system if its velocity vector agrees with the vectorfield at each point along the curve. This means the curve must evolve so as to be tangent to the vectorfield at each point, as shown here. The point on the trajectory corresponding to elapsed time zero, t_0, is the *initial state* of the trajectory.

Given a dynamical system (a smooth vectorfield on a manifold), how can we find its trajectories? Analysis, the mathematical theory which as evolved since Newton and Leibniz, has established that from each initial point, there is a single trajectory of the system. Finding it requires the construction called *integration* in vector calculus. Thus, trajectories are sometimes called *integral curves*.

A graphical construction that approximates the integration of a trajectory, or integral curve, was discovered by Euler.

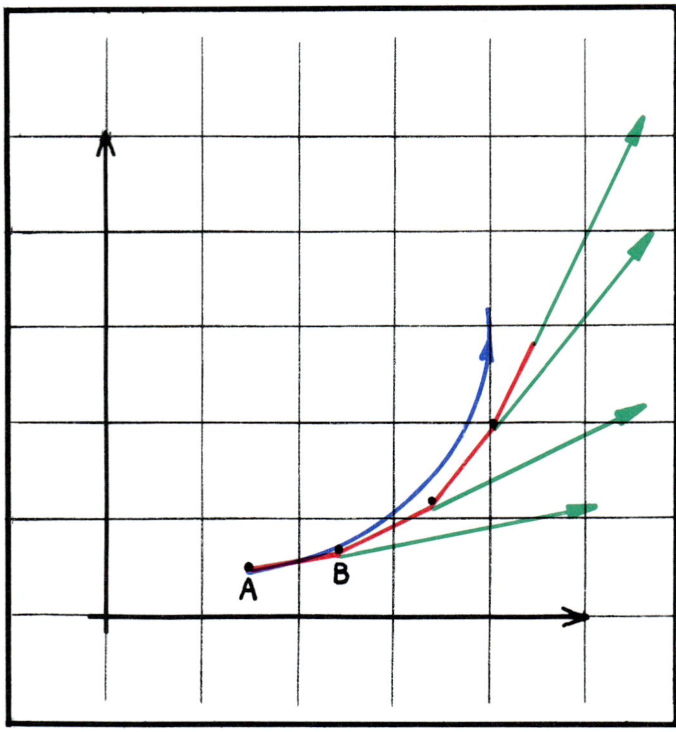

1.2.10. Euler's method approximates an integral curve by a polygon. Starting from the initial point, *A*, a straight line is drawn along the vector of the dynamical system attached to that point, *V*(*A*). The length of this straight line is a small proportion of the length of this vector, say one-tenth. At point *B*, at the far end of this line segment, the construction is repeated, using the vector, *V*(*B*), attached to this point by the dynamical system. This construction is repeated as many times as necessary, to draw the polygonal, approximate trajectory.

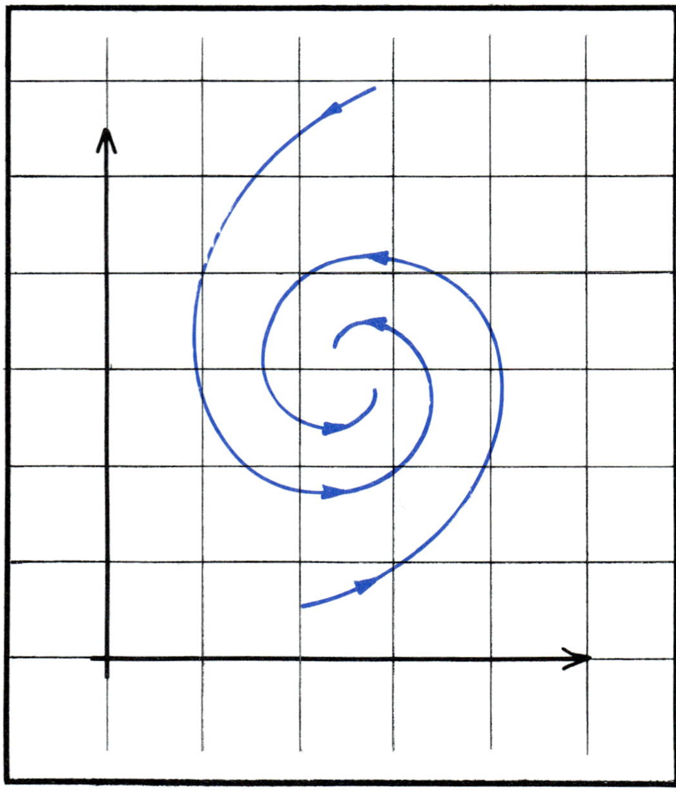

1.2.11. The state space is filled with trajectories, completely determined by the dynamical system. The display of the state space, decomposed into these curves, is the ***phase portrait*** of the system. Furthermore, the space of states may be imagined to flow, as a fluid, around in itself, with each point following one of these curves. This motion of the space within itself is called the ***flow*** of the dynamical system.

In the next chapter, simple examples will show that flat, Euclidean spaces will not suffice for all of our geometric models. In some cases, curved spaces (that is, *manifolds*) will be necessary. In global analysis, the calculus of Newton and Leibniz is generalized to the context of manifolds. This generalization provides the basic tools of mathematical dynamics.

The trajectory and velocity vector concepts fit nicely into the context of manifolds. Here we illustrate these ideas on a two-dimensional manifold, which is simply a curved surface.

1.2.12. Here the instantaneous velocity vector is obtained as a limit of average velocity vectors, as in an earlier illustration. But in this case, the state space is *curved.* The velocity vector does not lie in the curved surface. It sticks out into the ambient three-dimensional space. It is *tangent* to the surface.

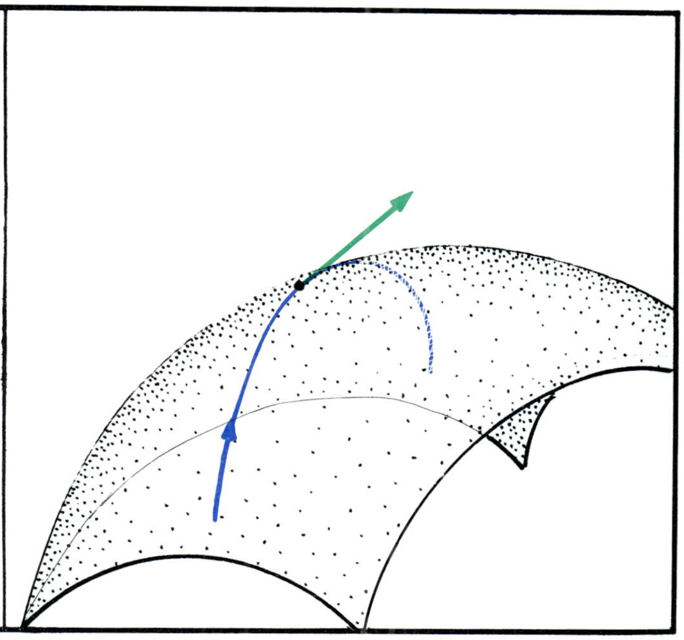

1.2.13. Now repeat this construction many times, with different curves lying in the surface, all passing through the same point. All the vectors lie in the same plane, tangent to the curved surface at a point. This plane is the *tangent space* of the space of states at that point.

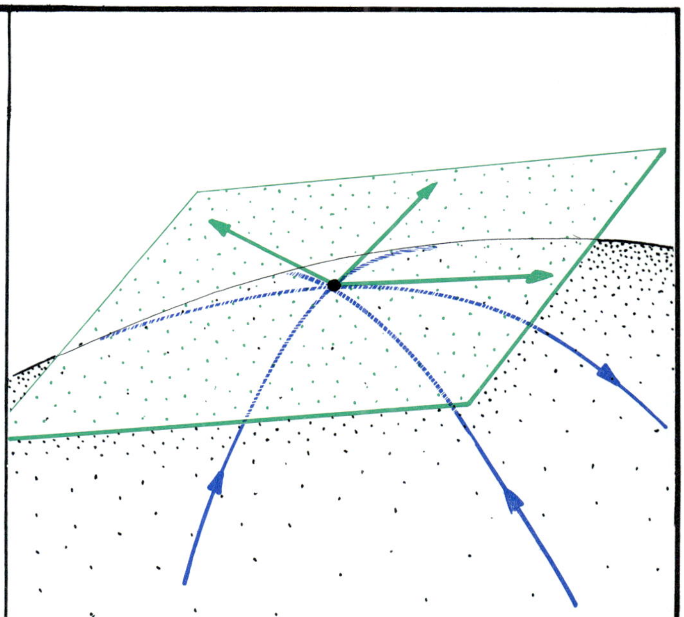

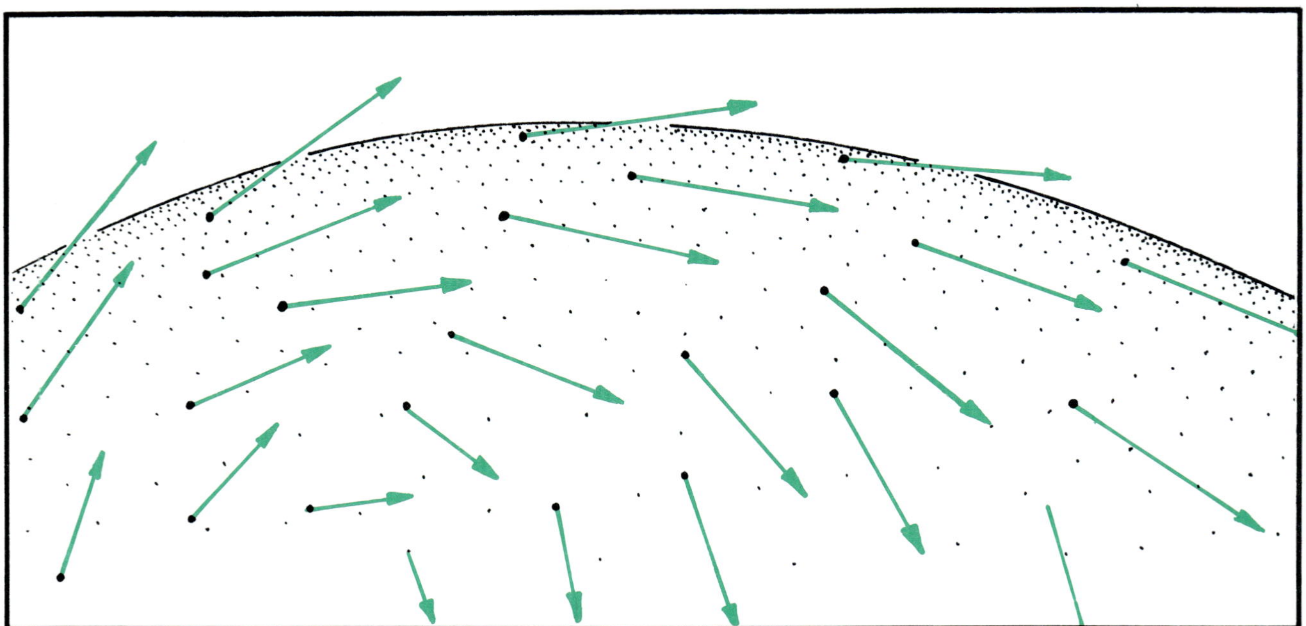

1.2.14. A vectorfield, in this context, means the assignment of a tangent vector to every point of the curved surface.

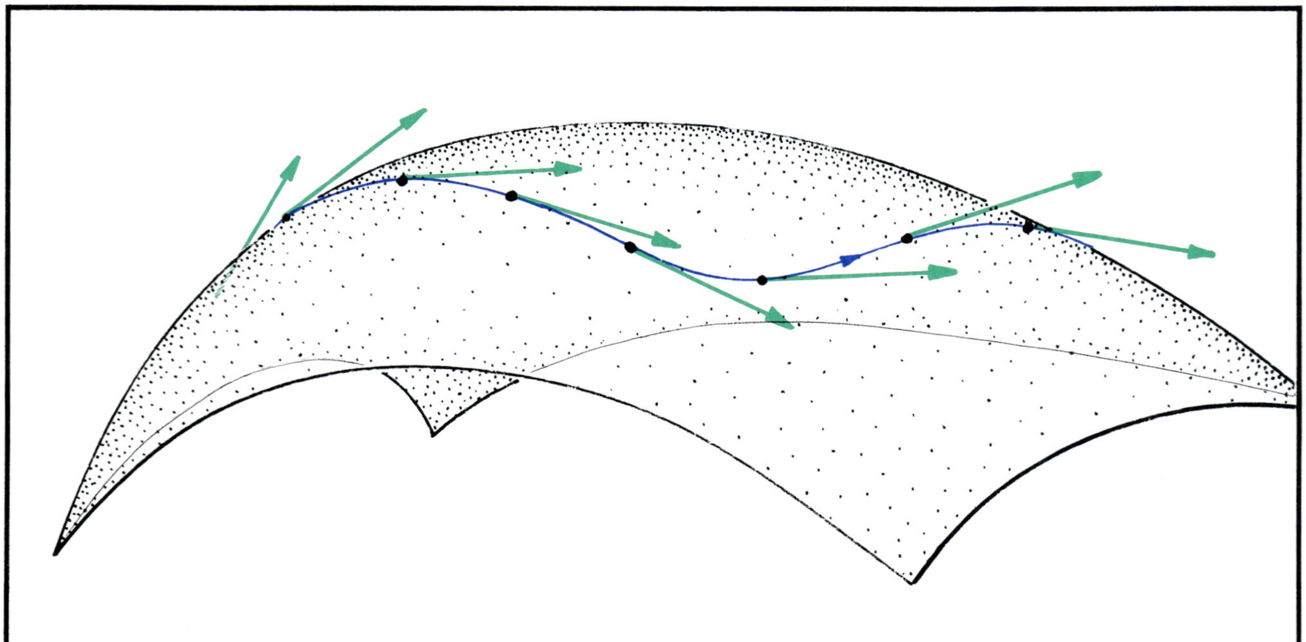

1.2.15. This is a trajectory, or integral curve, of a vectorfield (dynamical system) on a curved space of state. The tangent vectors at each point project off the surface, yet the integral curve stays within it.

This completes the introduction of the basic concept of a dynamical system, and the modeling process, which we may summarize as follows.

Suppose a dynamical model has been proposed for some experimental situation. This situation may be a laboratory device, an organism, a social group, or whatever. The model consists of a manifold and a vectorfield. The manifold is a geometrical model for the observed states of the experimental situation and is called the *state space* of the model. The vectorfield is a model for the habitual tendencies of the situation to evolve from one state to another and is called the *dynamic* of the model. Now mathematics can be brought out of the toolbox and used to draw many trajectories of the dynamical system, creating its *phase portrait*. These basic concepts of dynamical systems theory have been illustrated in the preceding two sections.

Now you may ask: SO WHAT? Well, according to our *conventional interpretation*, the agreed rules of the game, these trajectories are supposed to describe the behavior of the system as observed over an interval of time. Either they do this, with an accuracy sufficient to impress you and be useful for predicting the behavior of the experimental situation, or they do not. In many examples of this art, called *applied dynamics,* they do. These models succeed remarkably well and have been used by many satisfied customers over the years. Some of these examples are presented in the next four chapters.

But some obstinate readers may still exclaim: SO WHAT? Well, dynamical systems theory has yet more to offer: PREDICTION FOREVER. Sophisticated techniques from the research frontier of pure mathematics have been employed to yield *qualitative predictions of the asymptotic behavior of the system in the long run, or even forever.* Although qualitative predictions are not as precise as quantitative ones, they are a whole lot better than no predictions at all. *And for most problems of applied dynamics, quantitative predictions are impossible.* So, the remaining sections of this chapter are devoted to illustrating the concepts of *asymptotic behavior.*

1.3. Special Trajectories

The first step in the quest for qualitative predictions of asymptotic behavior is the examination of the phase portrait for special types of trajectories. Here we illustrate some of these special trajectories.

The simplest special trajectory is a point. Let's consider this in a one-dimensional context first.

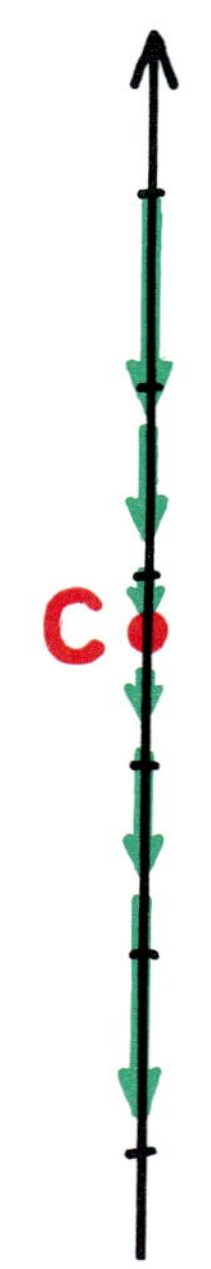

1.3.1. Here is a vectorfield on a one-dimensional state space. At one point in the state space, marked C, the associated velocity vector is the *zero vector.* This vector has length zero. The point marked C is called a *critical point*, or an *equilibrium point*, of the vectorfield. Because we assumed at the start that the vectorfield is continuous, the velocity vectors attached to points near the critical point are very short.

1.3.2. This is the phase portrait of the dynamical system to the left. Three trajectories are shown, starting at the points, A, B, and C. Tick marks along the trajectories indicate the positions at successive seconds. Note that they are closer together near the critical point. One trajectory is piled up on the critical point. The velocity of this trajectory is zero at all times. It does not move. It is called a *constant trajectory.*

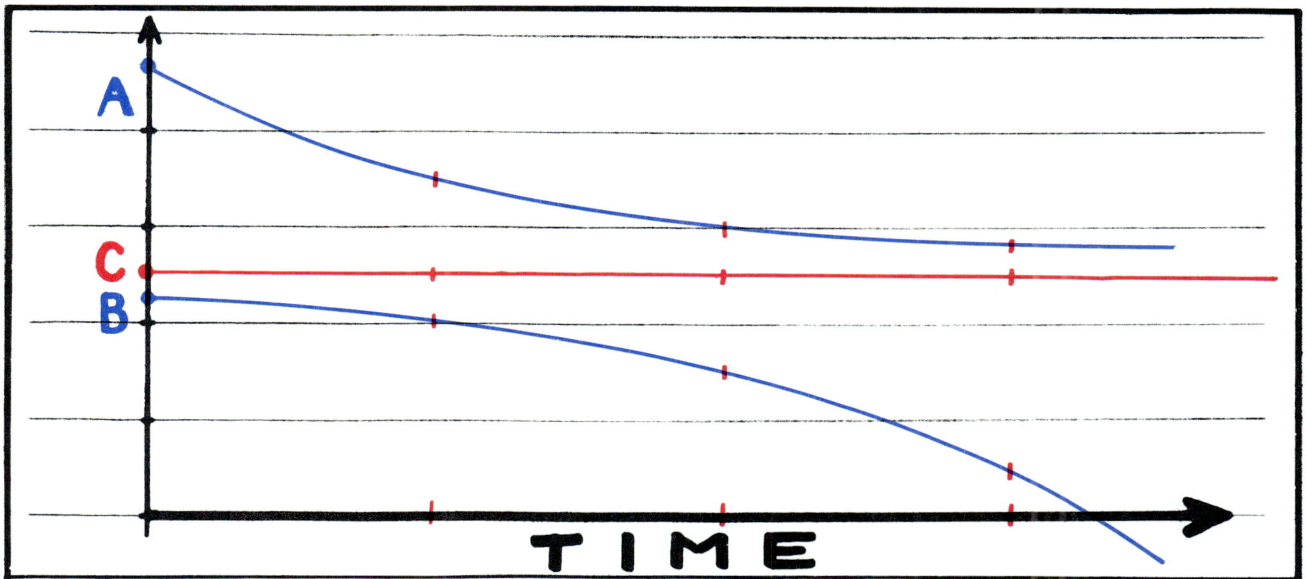

1.3.3. These are the *time series* corresponding to the three trajectories of the preceding phase portrait. The graph (time series) of the trajectory of *C* is a horizontal line, a constant function of time. This represents the constant trajectory of the critical point.

Now let's look at the same idea in a two-dimensional context.

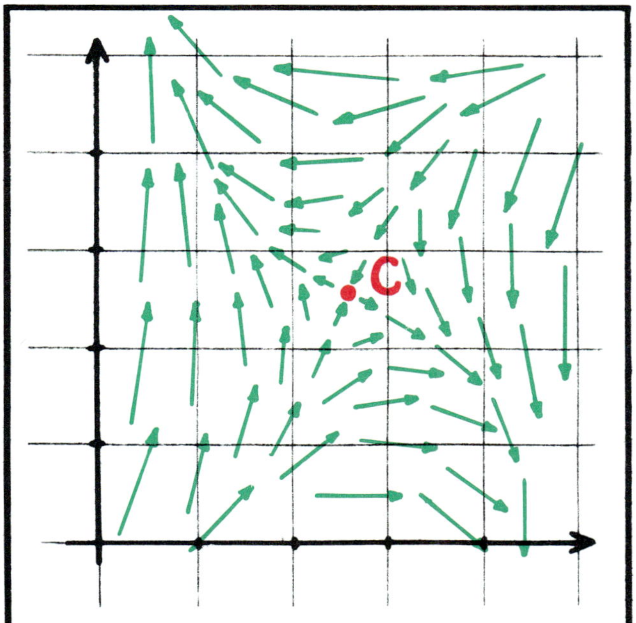

1.3.4. This is a garden-variety vectorfield in the plane. The zero vector appears once, as the velocity vector of the critical point *C*. Nearby, the vectors are short.

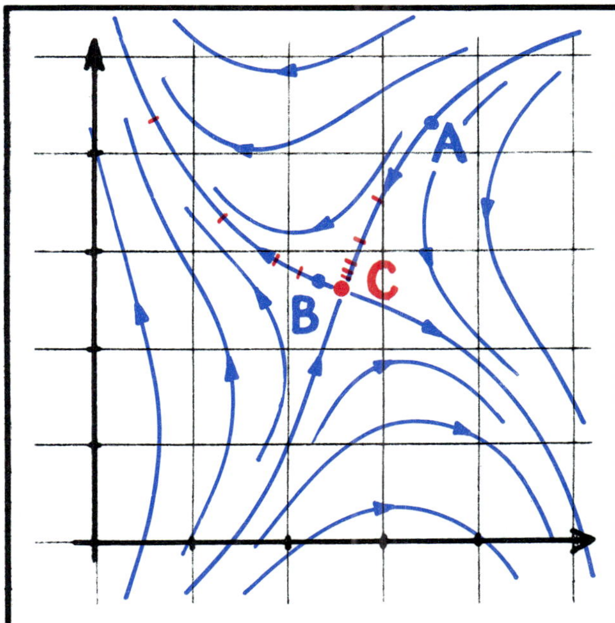

1.3.5. This is the phase portrait of the vectorfield to the left. The trajectory of the critical point is again piled up on the critical point. It is a constant trajectory.

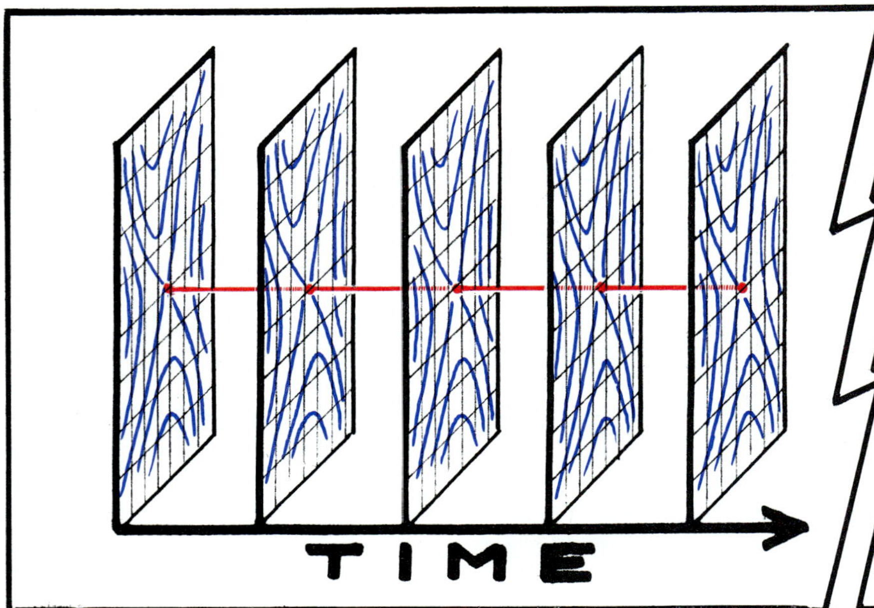

1.3.6. This is the time series of the constant trajectory in the two-dimensional context. Once again, it is a constant function of the time parameter.

In two dimensions or more, other types of special trajectories frequently occur. Here is a very important one, the *cycle*.

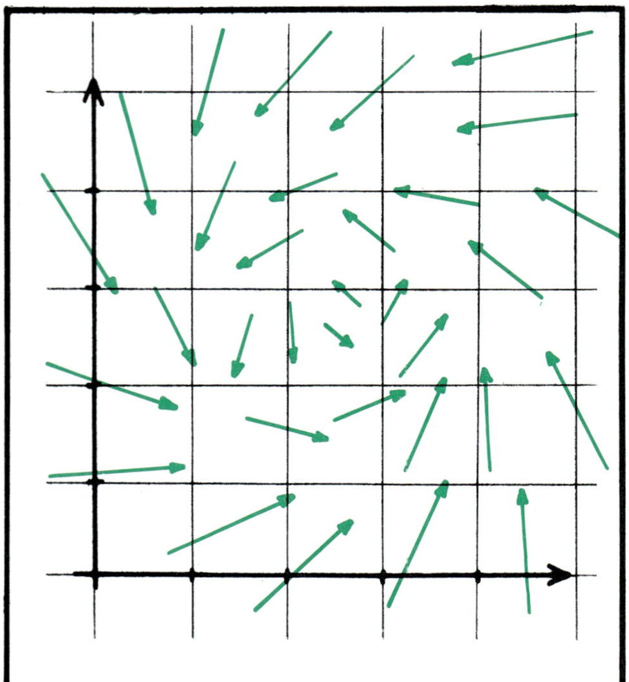

1.3.7. This planar vectorfield has an *eddy*. It seems as if the flow must somehow circle about a point.

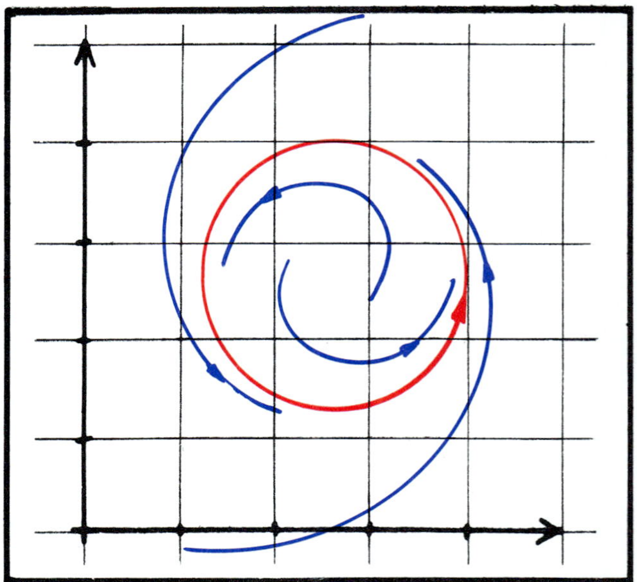

1.3.8. This is the phase portrait of the preceding dynamical system. Indeed, here we find a trajectory that is wrapped around and around the same curve. This is called a *closed trajectory*. It is also known as a *closed orbit*, *periodic trajectory*, *cycle*, or *oscillation*.

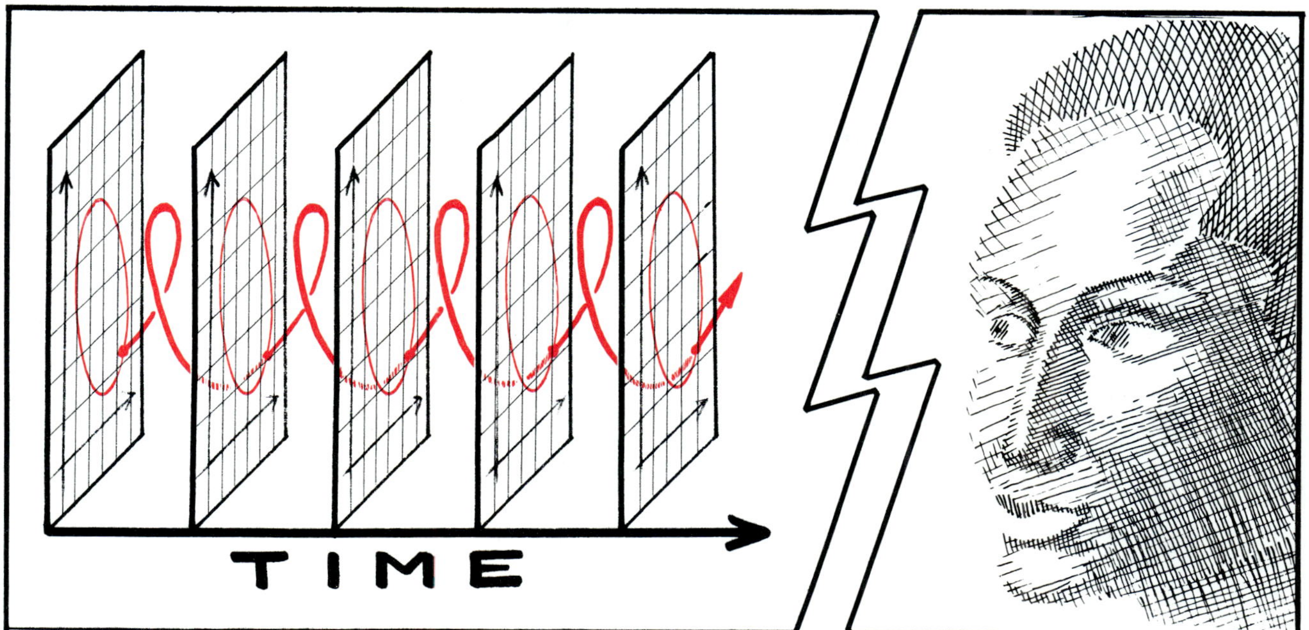

1.3.9. This is the time series representation of a closed trajectory. The graph wraps around a horizontal cylinder. The same interval of time is required to complete each wrap. This time interval is called the *period* of the closed trajectory.

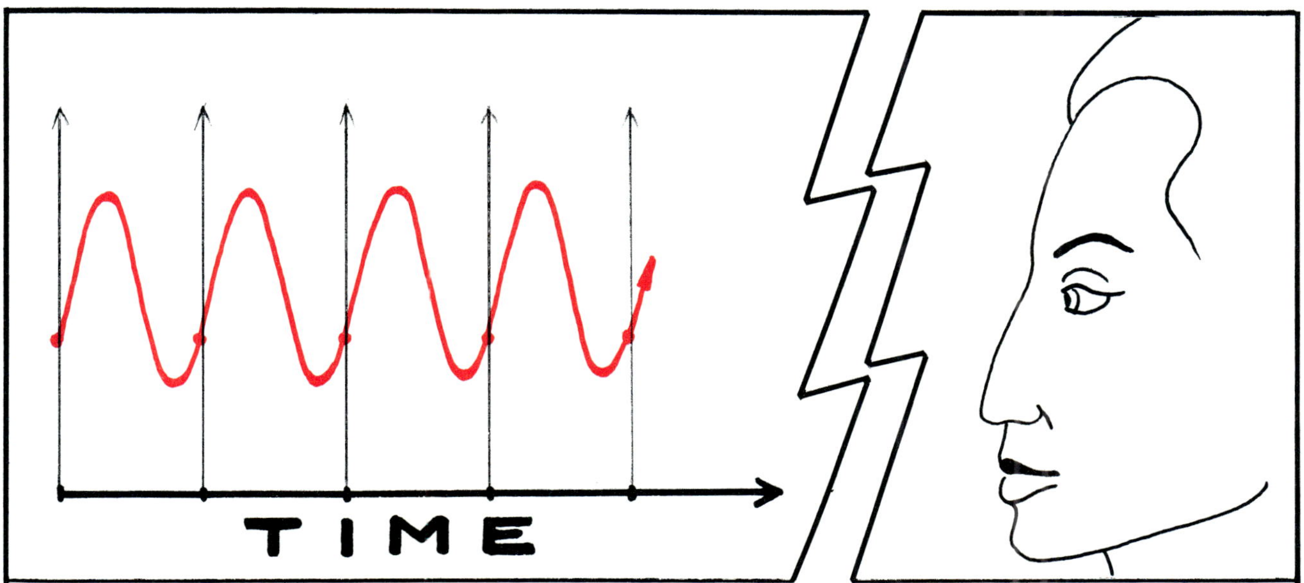

1.3.10. If a single parameter (of the two coordinates in the plane) is chosen, and the other data are forgotten, the time series of the chosen data may be plotted in the plane. The result, the *time series of the preferred parameter*, is a *periodic function*. This means that in every vertical strip corresponding to one period (or wrap, or cycle) of the closed trajectory, the graph exactly repeats itself.

These special types of trajectories, constant and periodic trajectories, also occur in phase portraits of dimension three or more.

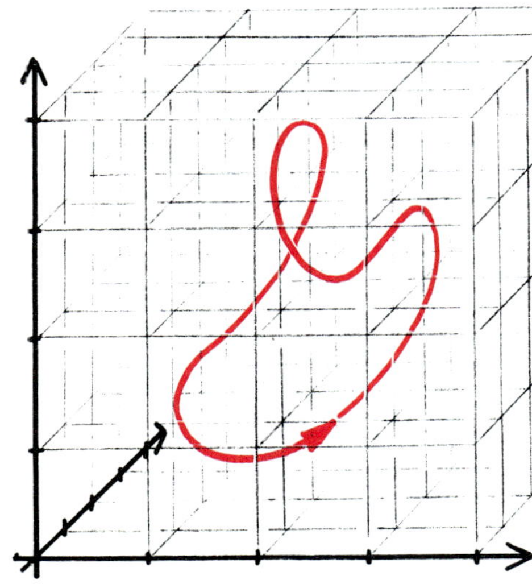

1.3.11. This is what a periodic trajectory looks like in 3D.

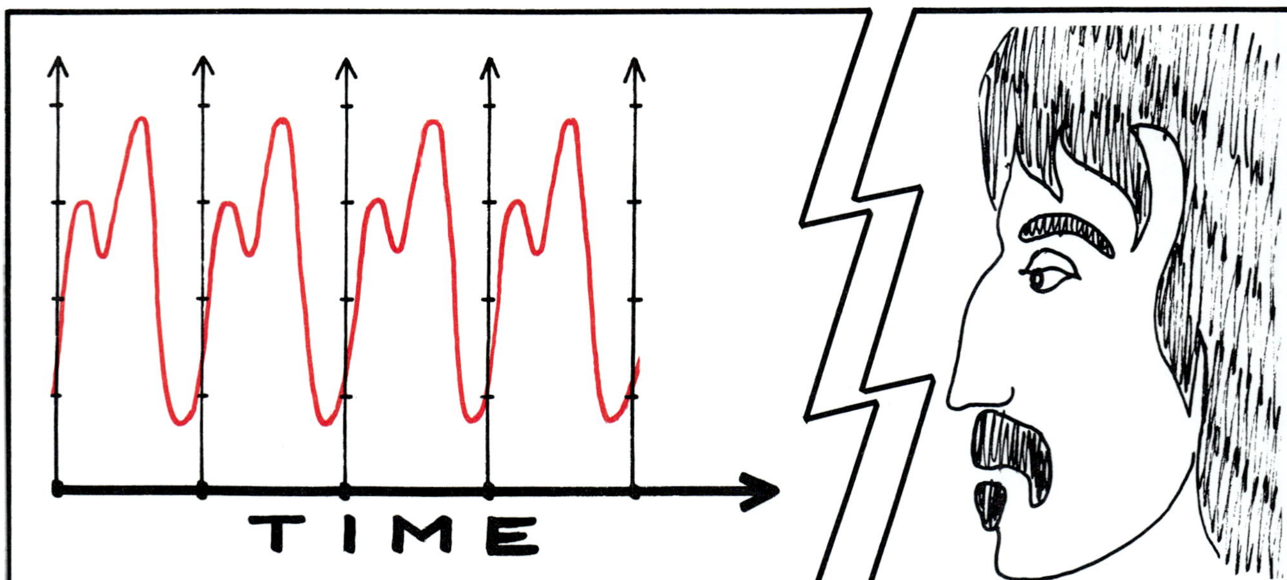

1.3.12. Choosing one parameter from the three coordinates, this preferred parameter may be recorded along the periodic trajectory. The time series of these data is again a periodic function.

But in higher dimensional phase portraits, other special trajectories may be found.

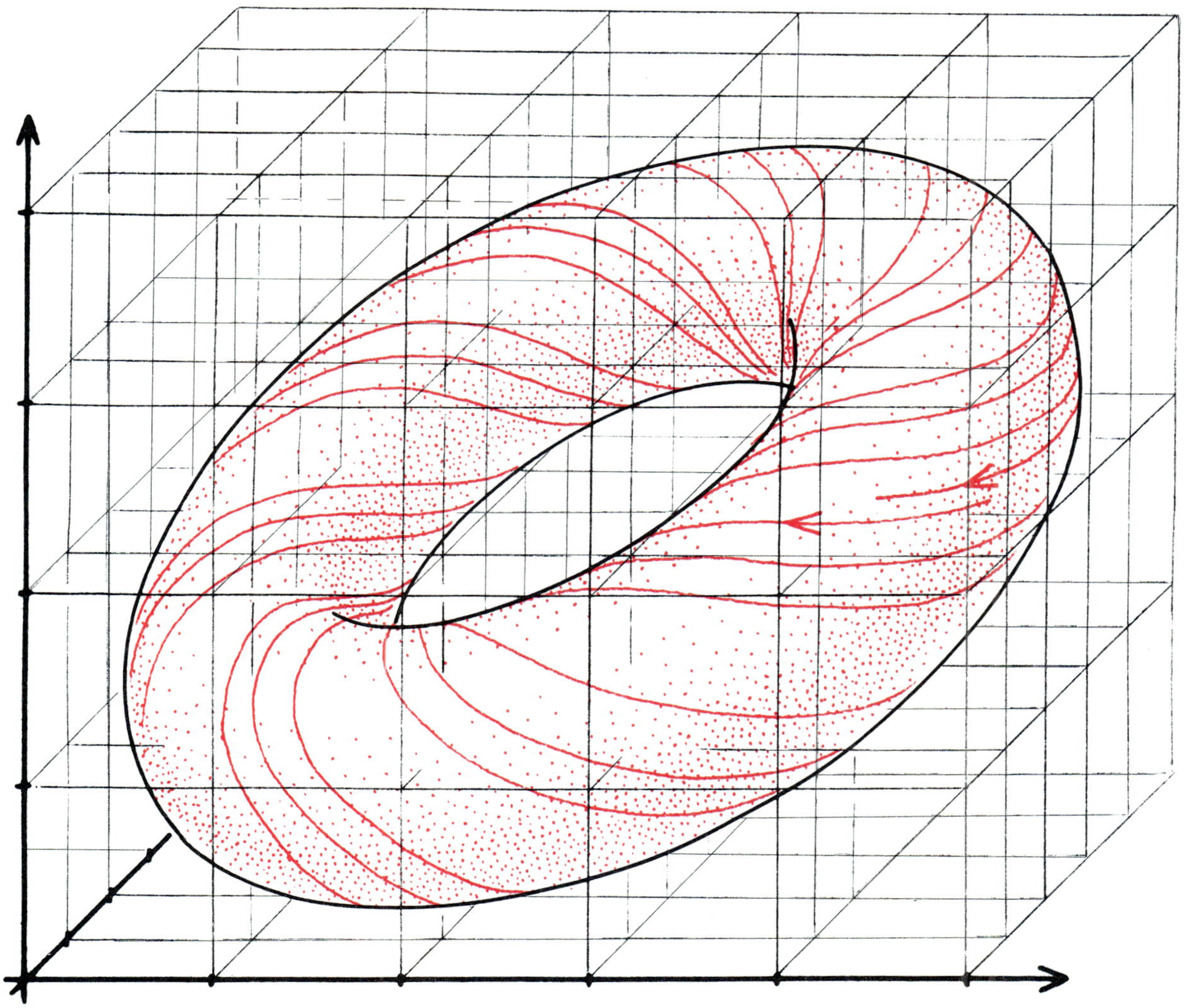

1.3.13. In the three-dimensional state space, imagine a torus (doughnut-like surface) with an infinitely long coil of wire wrapped endlessly around it, but never crossing itself or piling up. This can occur as a trajectory of a dynamical system, as we whall see in Section 4.4. It is called a *solenoidal* or *almost periodic trajectory.* An application is discussed in Chapter 5.

This finishes our list of special types of trajectory. Their significance in applications depends on the fact they occur as limit sets, as we describe in the next section.

1.4. Asymptotic Approach to Limit Sets

The second step in the dynamical systems quest for qualitative predictions of asymptotic behavior is the examination of the phase portrait for *asymptotic limit sets*. Let's see what this means.

We reconsider critical points in one dimension first, to start with the simplest case.

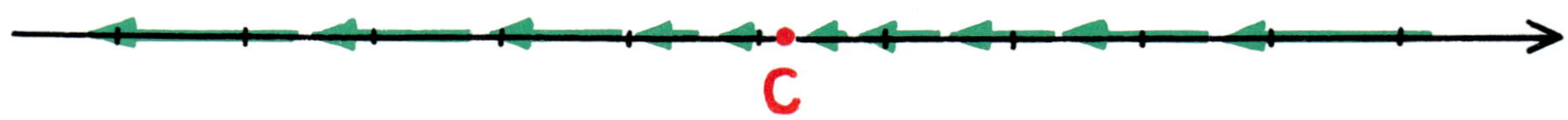

1.4.1. Here is the vectorfield on a one-dimensional state space. Recall that the point marked C is a critical point of the vectorfield. Because the vectorfield is smooth, the velocity vectors attached to points near the critical point are very short.

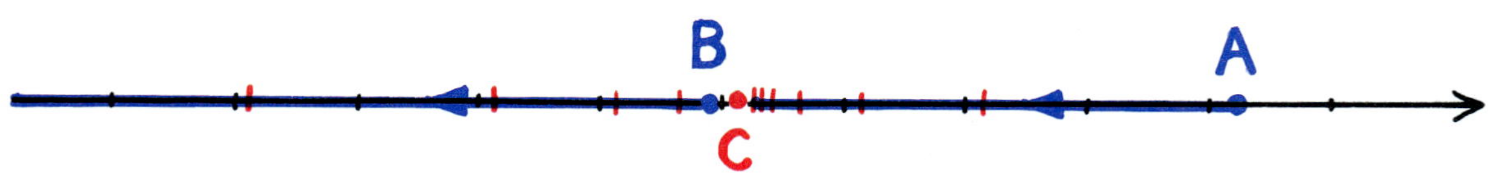

1.4.2. This is the phase portrait of the dynamical system above. Two trajectories are shown, starting at the points A and C. Tick marks along the trajectories indicate the positions at successive seconds. Note, again, that they are closer together near the critical point. The trajectory of C is a constant trajectory, piled up on the critical point. As time marches on, the trajectory of A gets ever closer to the point C. As it gets closer, it slows down. It gets closer and slower indefinitely and approaches the critical point *asymptotically*. That is, it takes forever to reach C. We say that C is the *limit point* of the trajectory through A.

This trajectory approaches its limit point asymptotically.

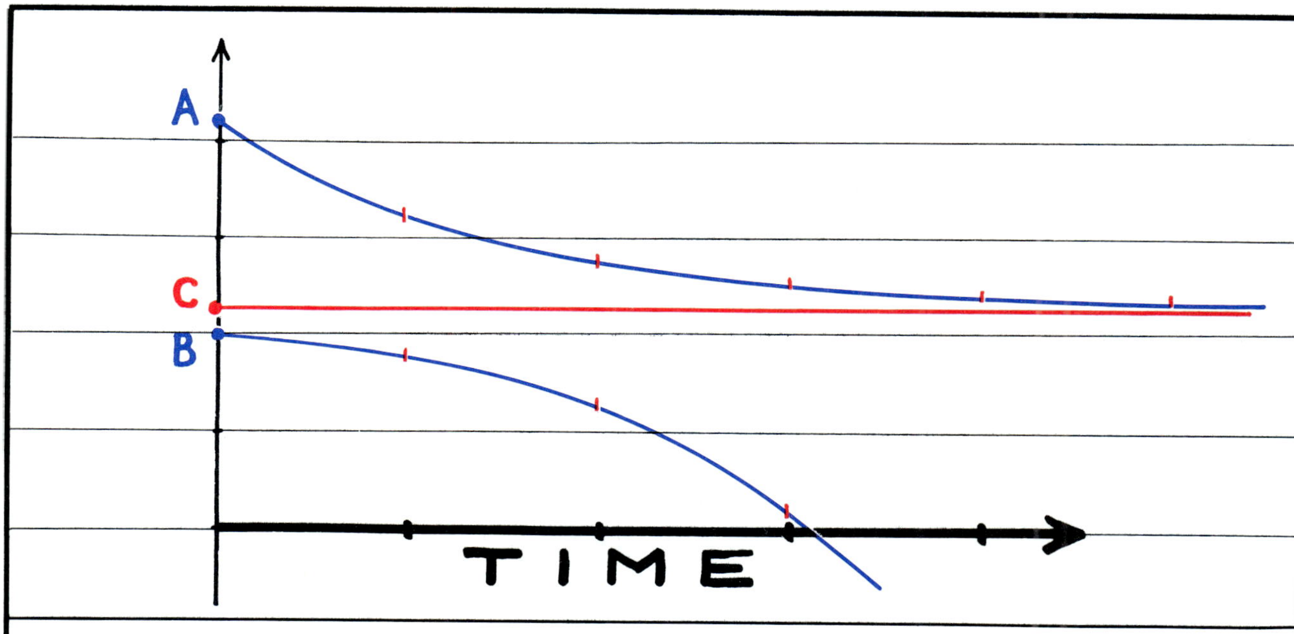

1.4.3. These are the time series corresponding to the three trajectories of the preceding phase portrait. The graph (time series) of the trajectory of *C* is a horizontal line, a constant function of time. This represents the constant trajectory of the critical point. The graph of the trajectory of *A* is descending to the right toward the horizontal line. It approaches this line *asymptotically*, as time increases to the right. The horizontal line is the *asymptote* of the graph of the trajectory of point *A*.

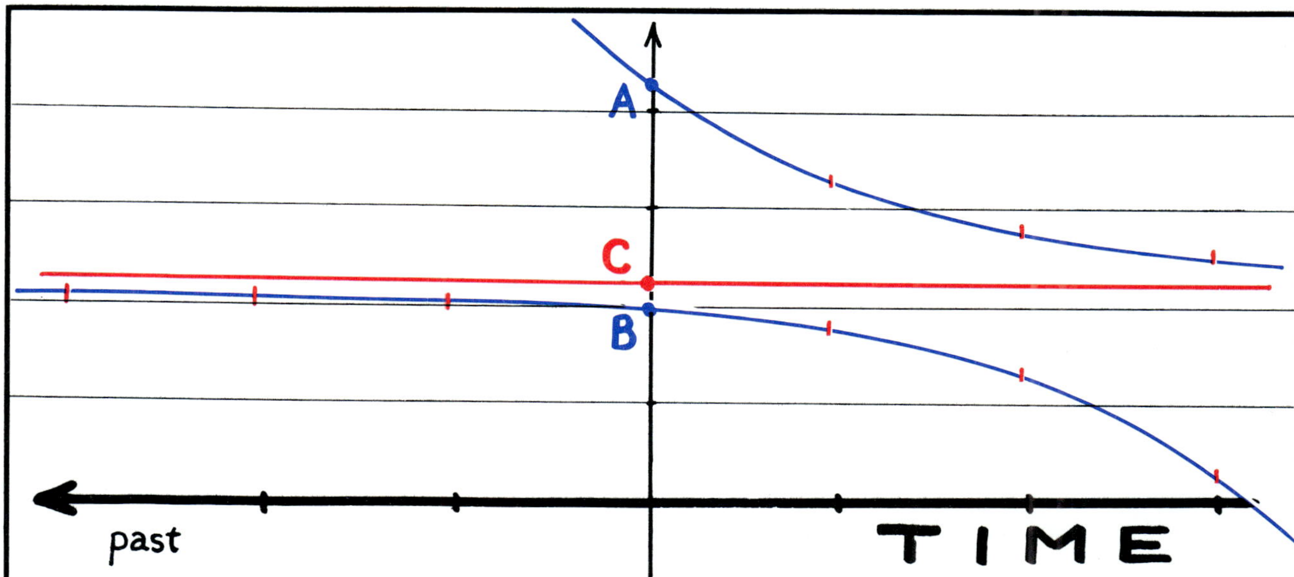

1.4.4. The graph of the trajectory of *B* similarly approaches the horizontal line asymptotically, but going *backwards in time*.

Now let's look at these ideas in a two-dimensional context.

1.4.5. This is a garden-variety vectorfield in the plane. As in the linear example in the preceding section, the zero vector appears once, at the point *C*. Nearby, the vectors are short. But this vectorfield is different. The vectors spiral around the critical point. It is a critical point of *spiral type*.

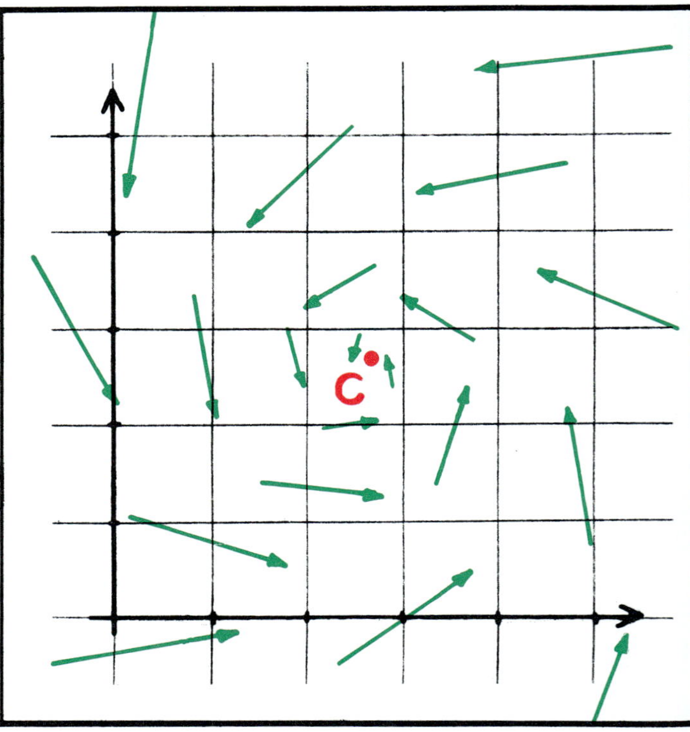

1.4.6. This is the phase portrait of the vectorfield. The trajectory of the critical point is again piled up on the critical point. It is a constant trajectory. The other neaby trajectories, such as the one through the point marked *A*, spiral around the critical point, getting closer and closer. They approach this point *asymptotically,* slowing down as they close in. We say that the critical point *C* is a *limit point* of the trajectory through the point *A*.

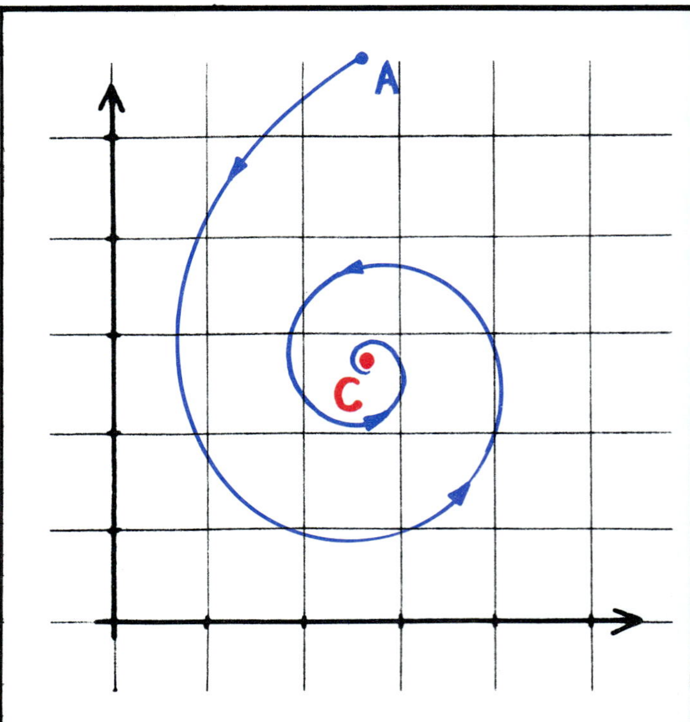

This trajectory approaches its limit point asymptotically.

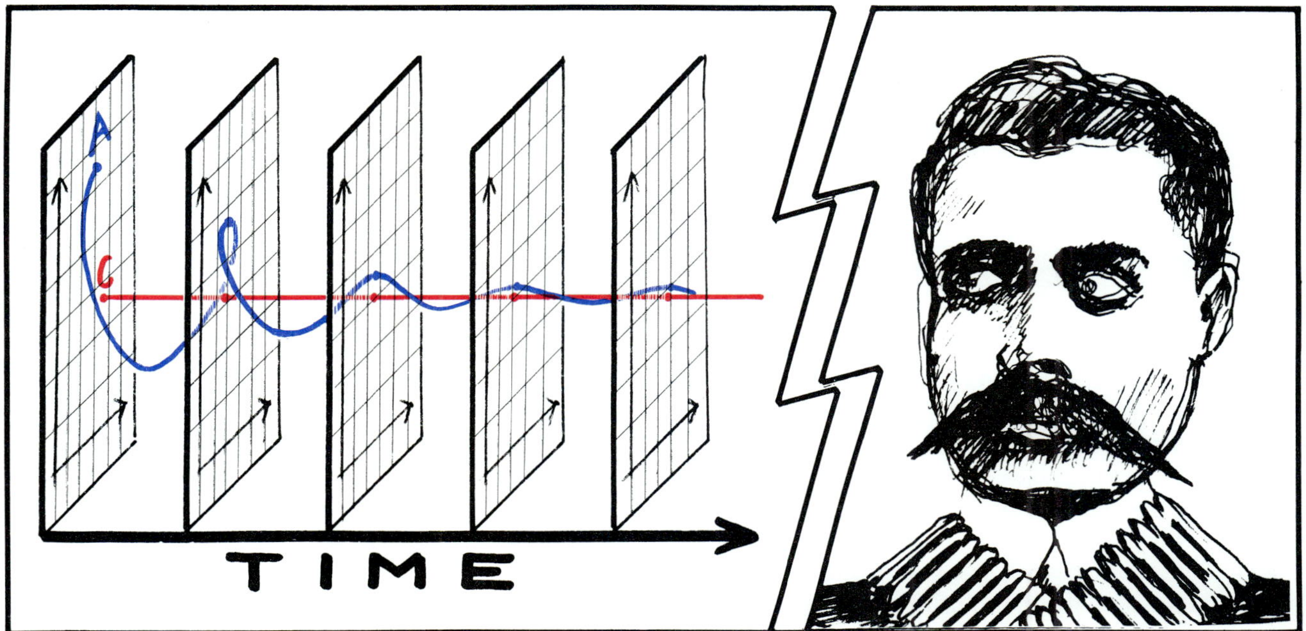

1.4.7. Here are the time series of two trajectories of this phase portrait. The time series of the constant trajectory, piled up on the critical point C, is the graph of a constant (vector-valued) function. This graph is a horizontal straight line. The time series of the nearby point A spirals around this straight line, approaching closer and closer as time moves to the right.

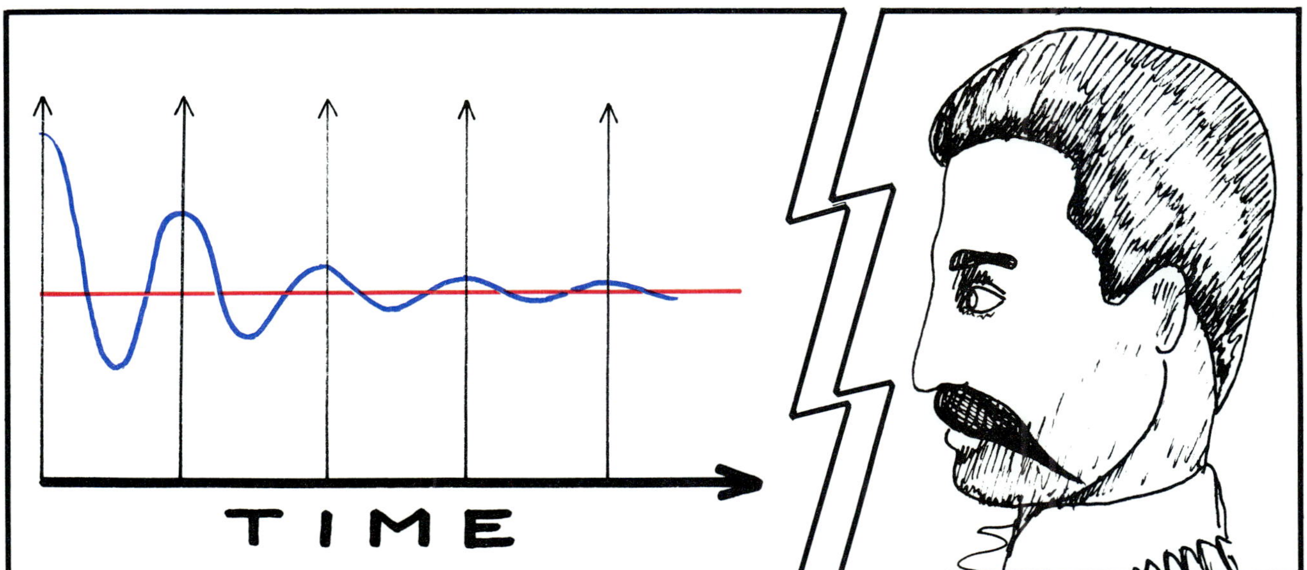

1.4.8. Choosing one of the two coordinates of the plane as a preferred parameter, the time series of this parameter along the two trajectories looks like this. The time series of the trajectory through A (wavy curve) approaches asymptotically toward the time series of the constant trajectory (horizontal straight line) as time increases to the right.

In two dimensions or more, other types of special trajectories frequently occur. One of these, as we have seen in the previous section, is the *cycle.* A cycle may be the asymptotic limit set for a trajectory.

1.4.9. This is the phase portrait of a planar vectorfield with a cycle, marked C. A point on the cycle is marked B. The trajectory through B is a closed trajectory, winding around and around this cycle. Another trajectory is shown, through the point marked A. This trajectory spirals around the cycle, getting closer and closer as time goes on. We say that C is a *limit cycle.* It is the *limit set* for the trajectory through the point A.

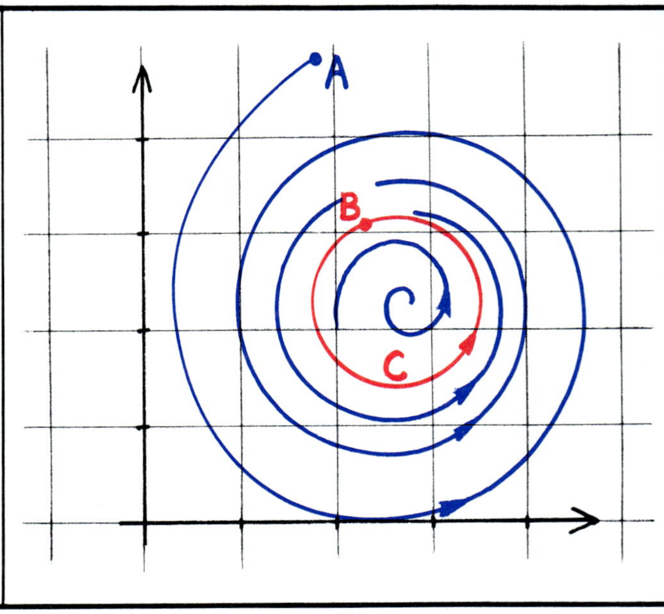

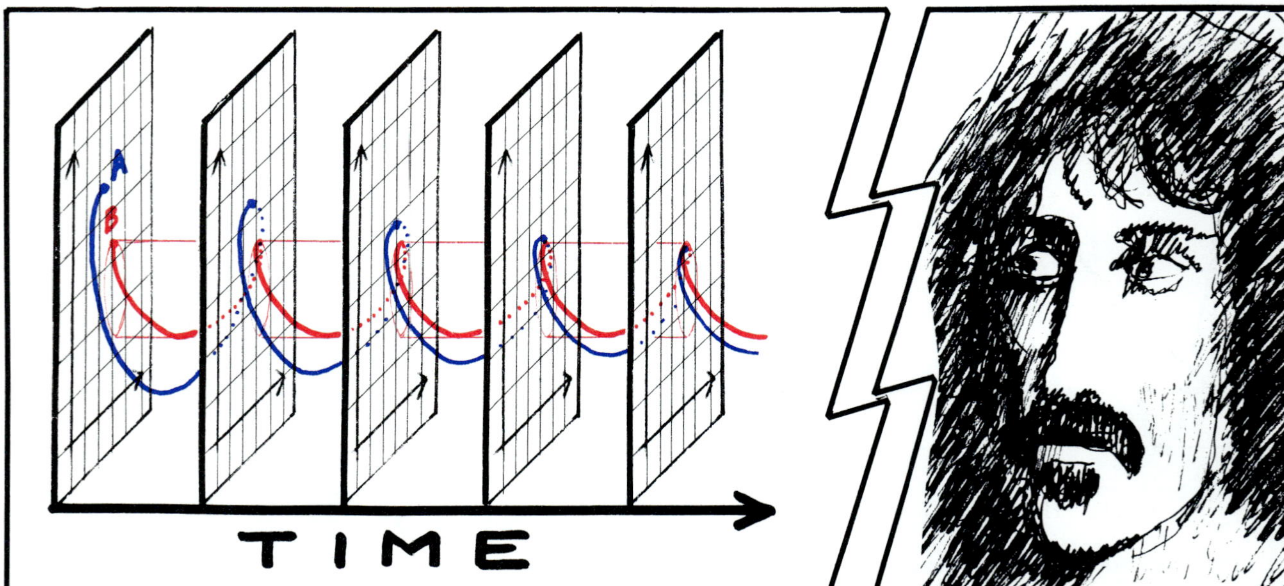

1.4.10. This is the time series representation of the two trajectories. A horizontal cylinder is shown, extending to the right from the cycle C. The time series for the spiraling trajectory, A, is wound loosely around the cylinder, and gets tighter and tighter as time increases to the right.

Limit points and limit cycles also occur in phase portraits of higher dimensions. Further, in dimensions greater than two, other limit sets may turn up. For example, a torus can occur as a limit set in a three-dimensional system. The solenoid, described in the preceding section, is a case in point.

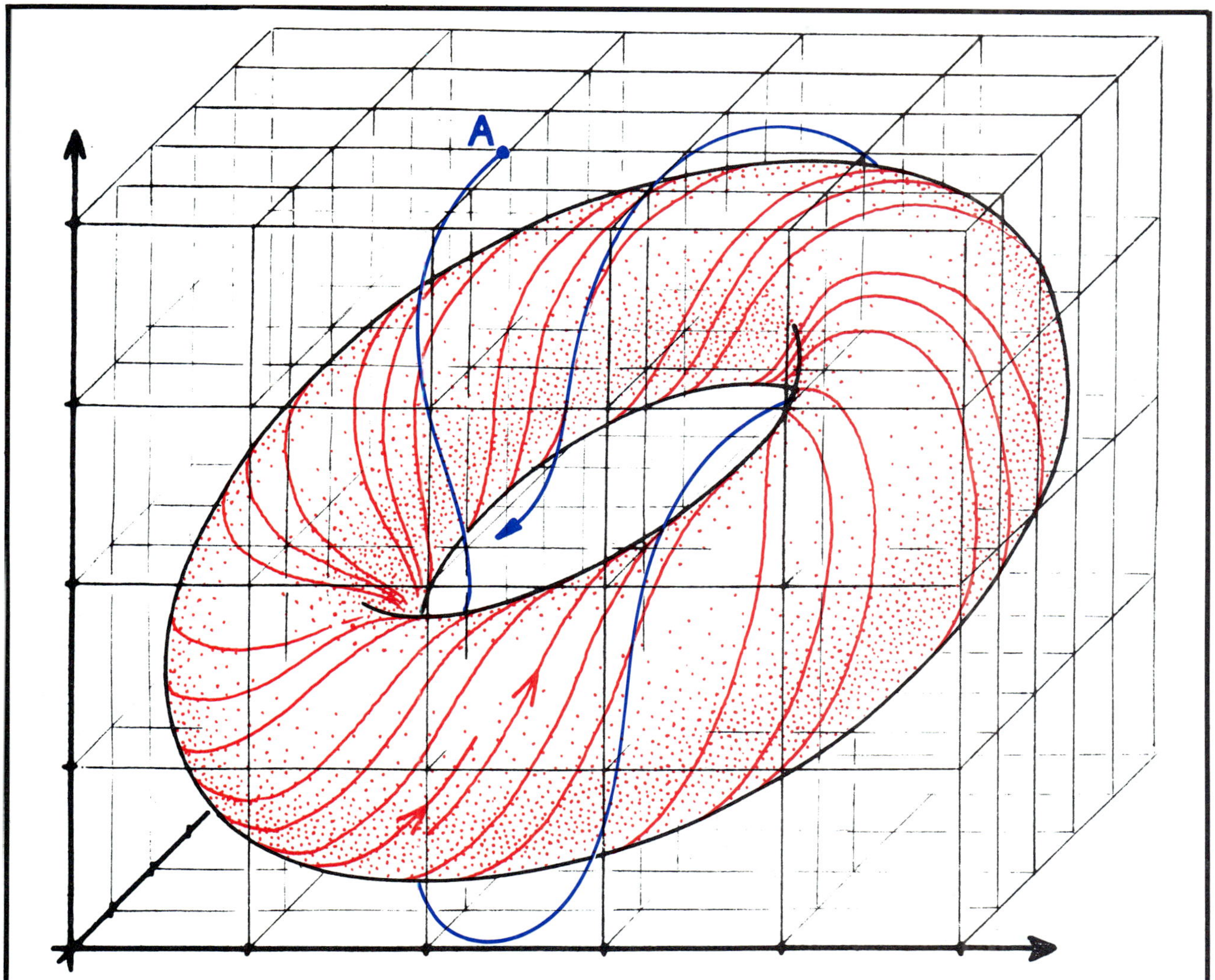

1.4.11. Here, the trajectory through the point marked *A* is wound, like a loose solenoid, around the torus. As time goes on, it winds around tighter and tighter. It approaches its limit set, the torus, asymptotically.

There are many more limit sets. Some of the more exotic ones will be shown in Part Two. But we have not yet made our case for the importance of the geometric theory of dynamical systems. To explain what we mean by *prediction forever*, a further concept is needed; that of an *attractor*. This is an outstanding type of limit set. It represents the behavior of a system in *dynamical equilibrium, after transients die away*. So let's go on.

1.5. Attractors, Basins, and Separatrices

If an organism is dropped into a prepared environment, or an experimental deviced is prepared in an initial state and then turned on, we expect to see a brief settling-in period before it settles down to an observable behavior. The erratic behavior during the initial settling-in period is called the *start-up transient*. The settled-in, eventual observable behavior is the *equilibrium state* of the experiment.. *Warning: Equilibrium, as used here, does not imply a static equilibrium, nor a steady state.*

In a dynamical system, modeling this experimental situation, a trajectory will model the start-up transient, while its limit set models the equilibrium state that follows. The asymptotic approach of the trajectory to its limit set models the dying away of the transient, as the system settles to its dynamic equilibrium.

For probability reasons to be explained shortly, the only equilibrium states which may be observed experimentally are those modeled by the limit sets that receive most of the trajectories. These are called *attractors*.

Here is the attractor concept, illustrated in two dimensions. The same ideas apply in all dimensions. First, we consider limit points, the simplest limit sets.

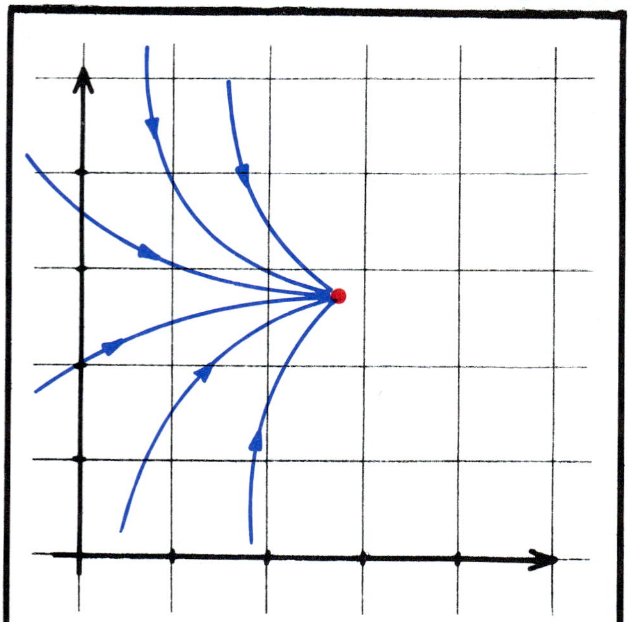

1.5.1. Suppose a dynamical system in the plane has a critical point. And let's suppose further that this critical point is the limit set of some trajectories in the phase portrait.

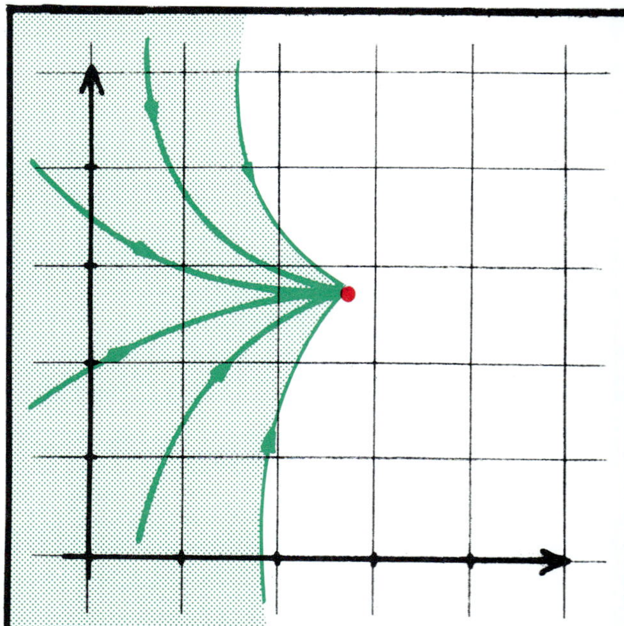

1.5.2. Now, find every single trajectory which approaches this limit point asymptotically, and color it green. The green portion of the plane is the *inset* of the limit set (that is, the critical point).

The inset of a limit set represents, in a dynamical model, all the initial states that end up in the same equilibrium state, after the start-up transient dies away.

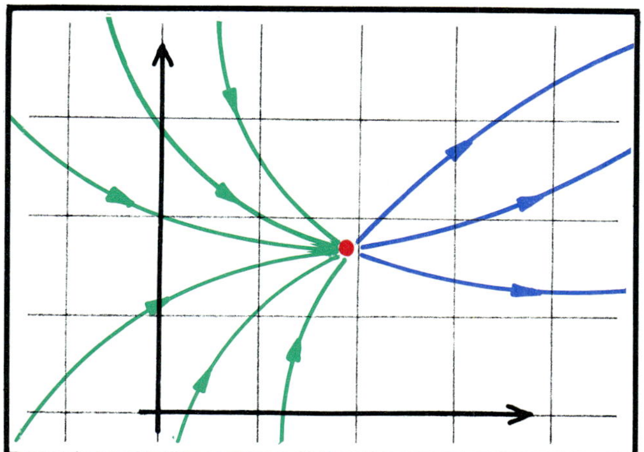

1.5.3. Other trajectories depart from the limit point. That is, *if the direction of time were reversed, these trajectories would approach asymptotically to this limit point.* Restoring the direction of time to normal again, we say these departing trajectories have the critical point as their *alpha-limit set*.

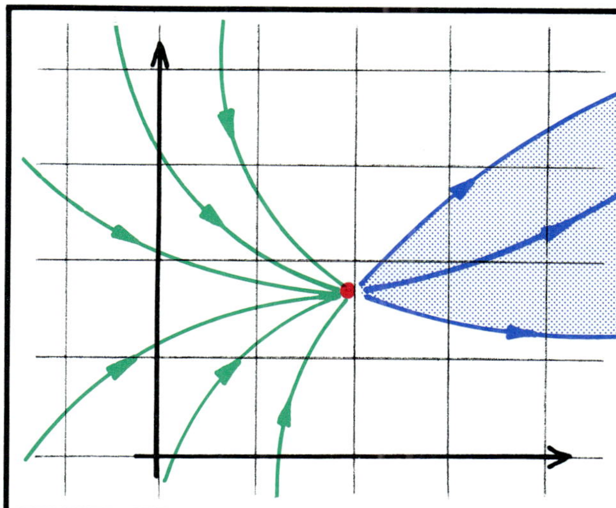

1.5.4. Now find every trajectory that has this critical point as its alpha-limit set, and color it blue. The blue portion of the plane is the *outset* of the limit set (that is, the critical point).

Sometimes we say *omega-limit* in place of just plain limit. The omega-limit set refers to the future asymptotic behavior, while alpha-limit set refers to the past. The trajectory goes from alpha-limit to omega-limit.

Warning: Some trajectories neither arrive nor depart at a critical point, although they may pass close by!

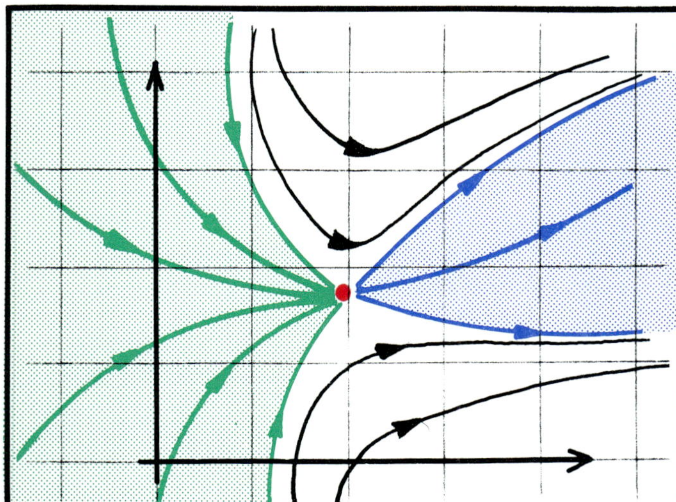

1.5.5. Here we see the same dynamical system in the plane. Both the inset and the outset are colored in. The black trajectories start near the inset and go toward the limit set for a while. When they get near the limit point, they feel the influence of the outset, and turn aside. Following the outset, they disappear into the distance.

When a trajectory flies by, it is on its way somewhere. Either it has an omega-limit set elsewhere in the phase portrait, or it departs from the state space, never to be seen again. Similarly, it came from somewhere else. Thus, every trajectory may belong simultaneously to the outset of one limit set (its alpha-limit set) and to the inset of another (its omega-limit set). In two dimensions, cycles may be limit sets. Limit cycles have insets and outsets, too. (So do the other limit sets, in higher dimensions.)

What if all nearby trajectories are arriving?

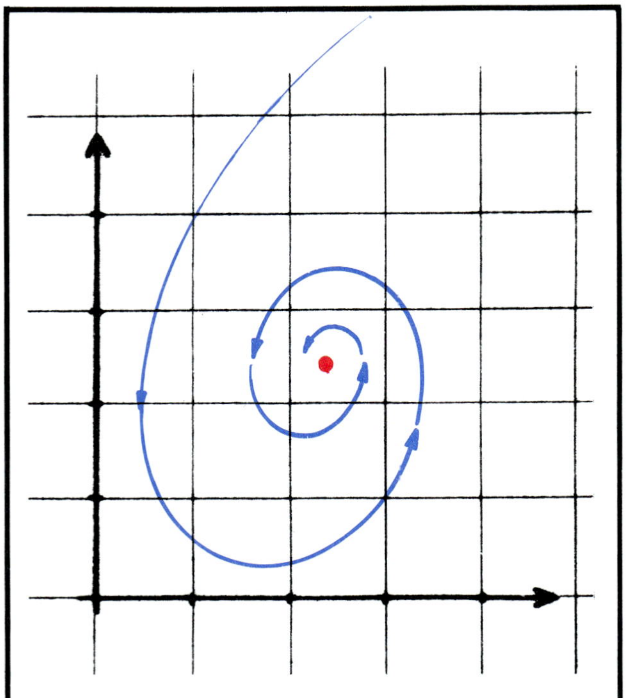

1.5.6. Here is a limit point in two dimensions. Every nearby trajectory is arriving. The inset contains an open disk around the limit point. Every initial point in this disk is captured by the limit point, when its transient dies away. If an initial point is chosen at random from all the points in the state space, the probability of it asymptotically approaching this limit point, instead of some other, is positive. (In fact, it is 100% in this particular example.) This limit point is an *attractor.* As it is a point, and represents static equilibrium, it is also called a *static attractor.*

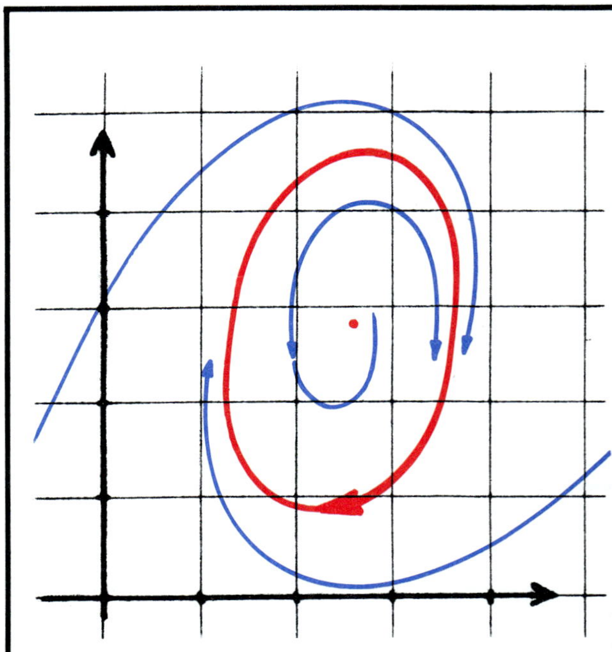

1.5.7. This is the phase portrait of a different dynamical system in the plane. It has a limit cycle. Again, the inset of this limit set is as large as possible. Except for the solitary red point in the center, which is a critical point, every single initial state evolves to the same limit set. The inset includes an open annulus (ring) around the limit cycle. The probability that an initial state, chosen at random from among all the initial states in the state space, will end up at this limit cycle is positive. (In fact, it is 100% in this particular case.) This limit cycle is an attractor. As it is a cycle, and represents a periodic equilibrium, it is also called a *periodic attractor.*

An *attractor* is a limit set with an *open inset*. That is, there is an open neighborhood of the limit set within its inset.

Of all limit sets that represent possible dynamical equilibria of the system, the attractors are the most prominent, experimentally. This is because the probability of an initial state of the experiment to evolve asymptotically to a limit set is proportional to the volume of its inset. We will say that a limit set is *probable* if the volume of its inset is a positive number instead of zero. Open sets have positive volume, although not every set of positive volume is an open set. Attractors have open insets, so they are probable. They are experimentally discoverable. Other limit sets may be probable, without being attractors in the strict sense of the preceding paragraph. They are called *vague attractors*. This is synonymous with *probable limit set*. These are also experimentally discoverable. Limit sets that have *thin insets* (that is, that have probability zero) are *non-attractors*. They are experimentally insignificant. These are called *exceptional limit sets*, or synonymously, *improbable limit sets*[2].

The inset of an attractor is called its *basin*. In a typical phase portrait, there will be more than one attractor. The phase portrait will be divided into their different basins. The dividing boundaries (or regions) are called *separatrices*. In fact, any point that is not in the basin of an attractor belongs to a separatrix, by definition.

Here are some examples of attractors, basins, and separatrices, in two dimensions. The same concepts apply in three or more dimensions, but are harder to visualize.

COLOR CONVENTION:

Attractors = red
Basins = blue
Separatrices = green

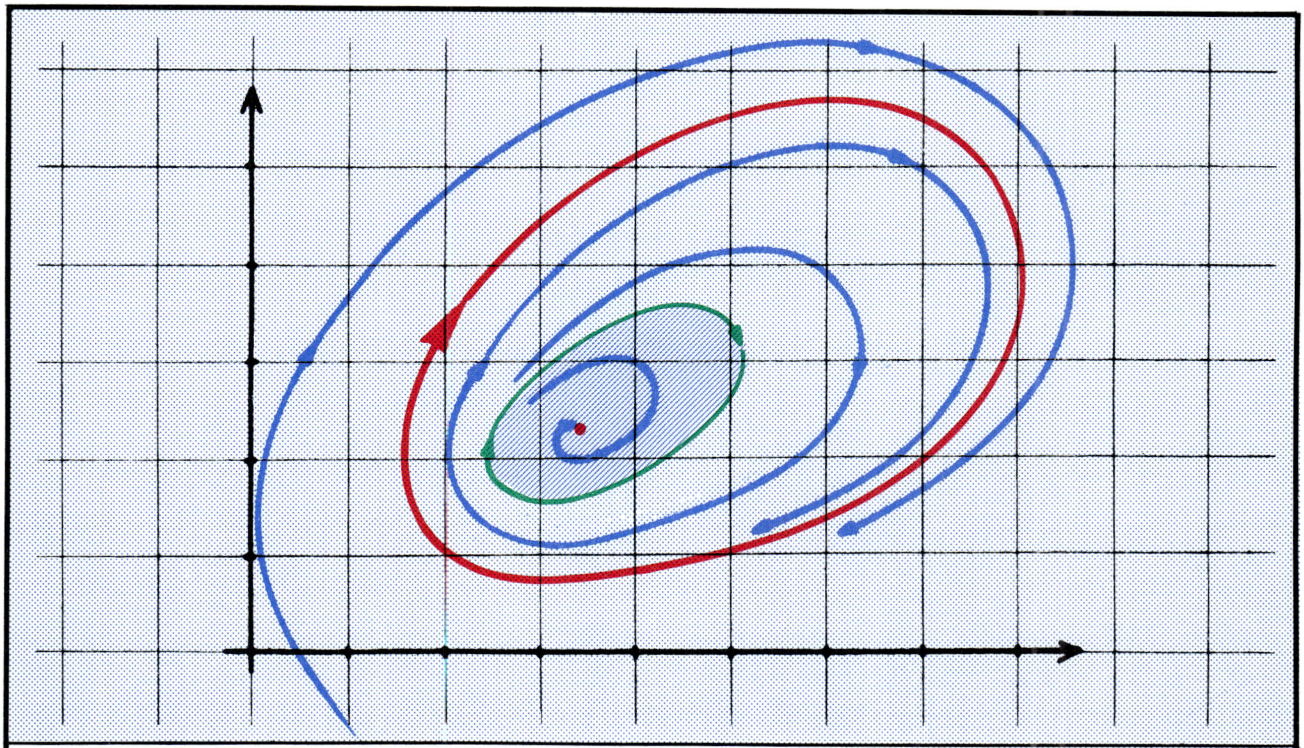

1.5.8. In this example, there are two attractors: a point and a cycle. A third limit set, a cycle, comprises the separatrix.

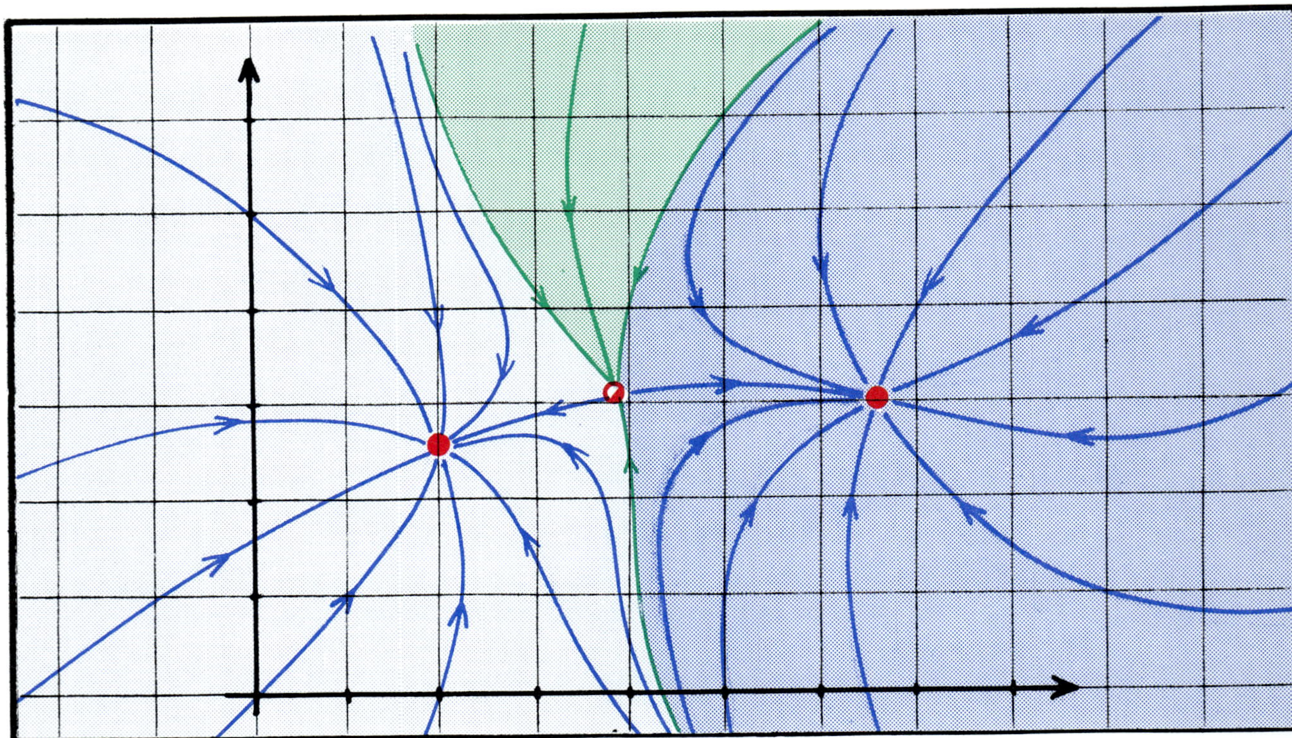

1.5.9. Here, also, there are two attractors, both points. A third limit set, also a point, is not an attractor. It is a vague limit set. The inset of this limit set comprises the separatrix.

This reveals the pattern of the general case: the separatrix consists of all points, not in a basin. Every point tends to a limit set. If its limit set is an attractor, it belongs to a basin. So, if it belongs to the separatrix (and therefore not to a basin), it must tend to a non-attractor. Thus, the *separatrix consists of insets of exceptional limit sets.*

The preceding examples are artificial, made up to illustrate the concepts. But we are overdue for some more meaningful examples. So, at this point, let's turn to gradient systems—a rich source of simple examples, based on a geometrical construction.

1.6. Gradient Systems

The *gradient* operation of vector calculus provides dynamical systems (vectorfields) of an especially simple type called a *gradient system*. In these, there is an auxiliary function, called the *potential function*. The velocity vectorfield is simply the *gradient vectorfield* of this potential function.

This means that, at each point in the state space, the velocity vector indicates the direction of most rapid increase of the function, and the magnitude (slope) of this increase.

This section develops a typical example of the gradient dynamics in the plane.

1.6.1. The state space, in this example, is the plane. The potential function is a function from the state space to the real number line. To each point in the state space, it assigns a real number. This number is the *potential* of the corresponding state. In an application of this scheme, this potential would presumably be observable, or deducible from observations.

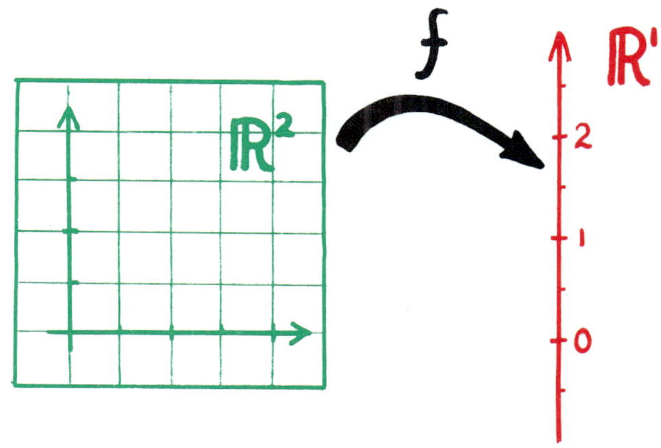

1.6.2. Represent this function as a graph in three-dimensional space. The state space is the horizontal coordinate plane. From each point in this plane, move vertically a distance equal to the potential of that point—up if the potential is a positive real number, down if it is negative.

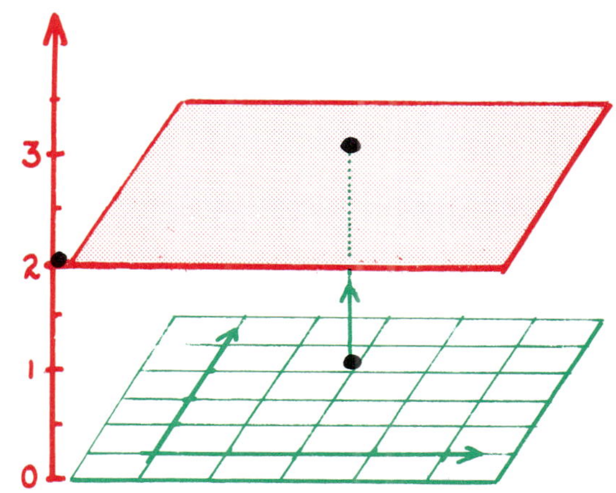

The graph of a potential function on a planar state space is a surface in three-dimensional space, called the *potential surface.* We may think of this as a landscape.

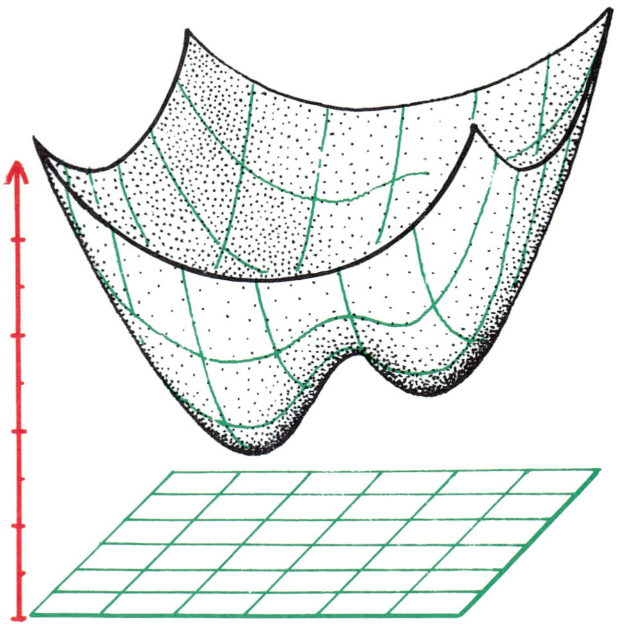

1.6.3. For the sake of definiteness, let's choose a particular potential function, and visualize it as a potential surface. This one, for example, has two valleys, with a saddle ridge in between.

An alternative representation of a function is its *contour map.* Let us represent our exemplary potential this way.

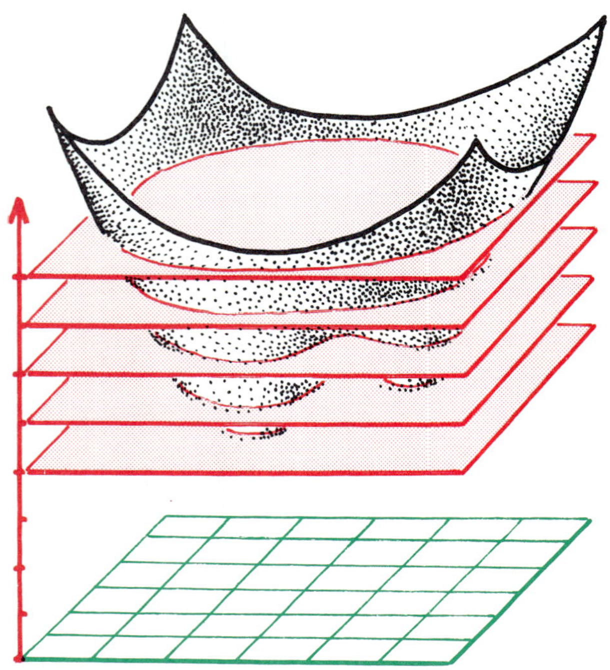

1.6.4. At regular intervals along the vertical axis, draw horizontal cutting planes. Mark the potential surface with a red curve where each cutting plane cuts through the surface. These are called the *level curves* of the surface.

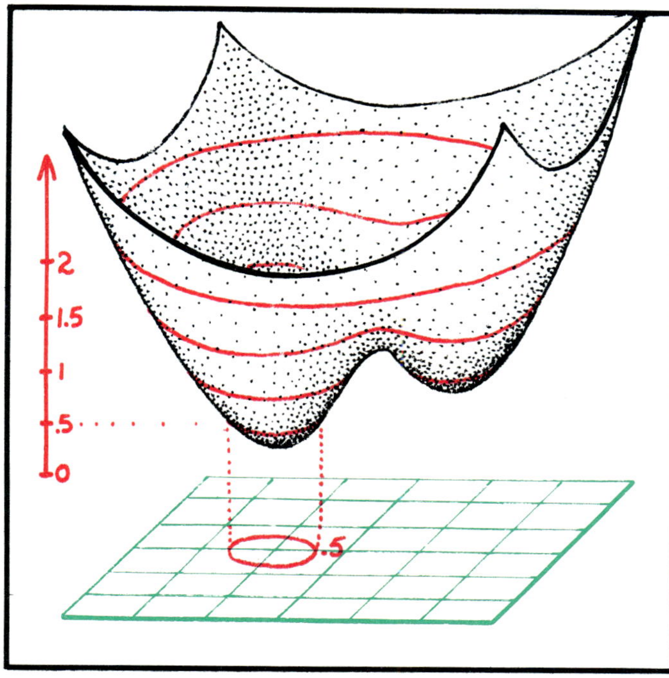

1.6.5. Next, each level curve is projected onto the horizontal coordinate plane. This curve in the state space, also called a *contour curve*, contains every state with the same value of the potential function. Over this curve, the potential surface has a constant height. Label this contour with its common value of the potential.

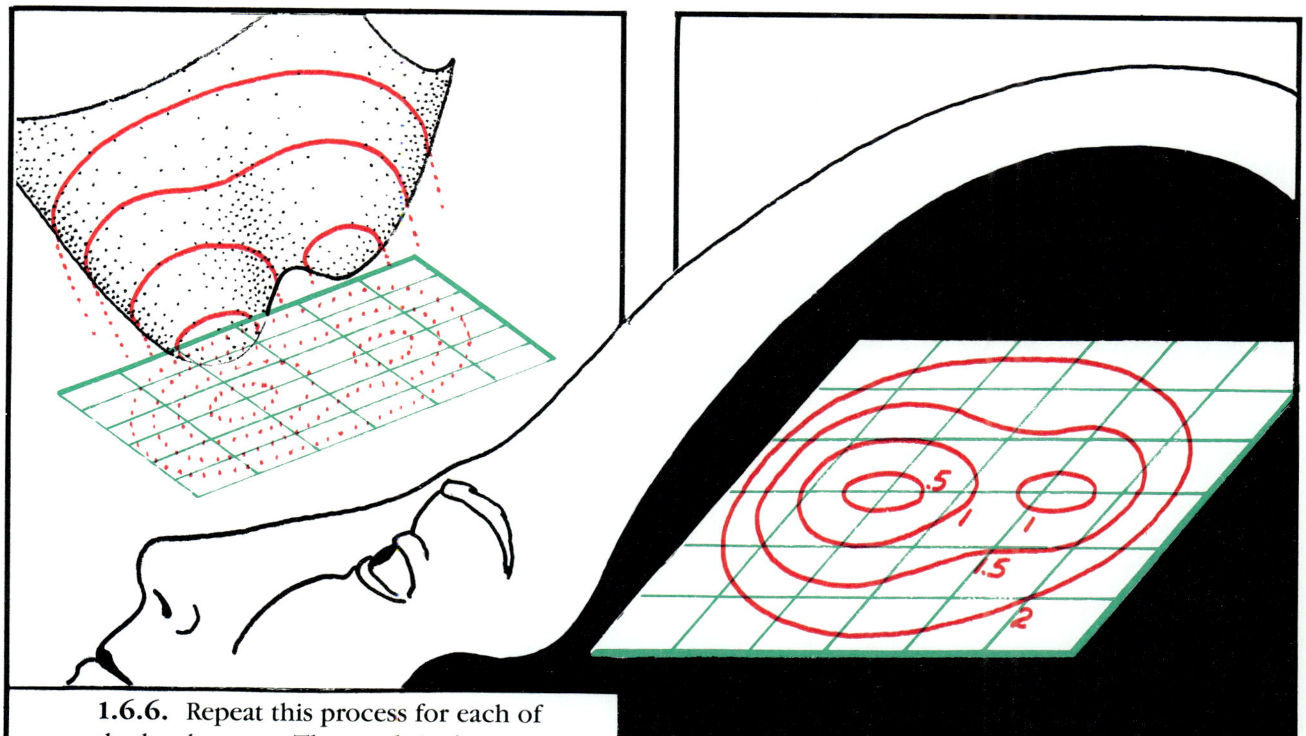

1.6.6. Repeat this process for each of the level curves. The result is the contour map of the potential surface, drawn in the horizontal coordinate plane. This is essentially what you would see, looking up from far below at the potential surface, with level curves drawn upon it.

1.6.7. Finally, extract the state space, with the contour map drawn within it, from the three-dimensional context of the graph. This is the alternate representation.

The gradient dynamical system for this particular potential function is derived as follows.

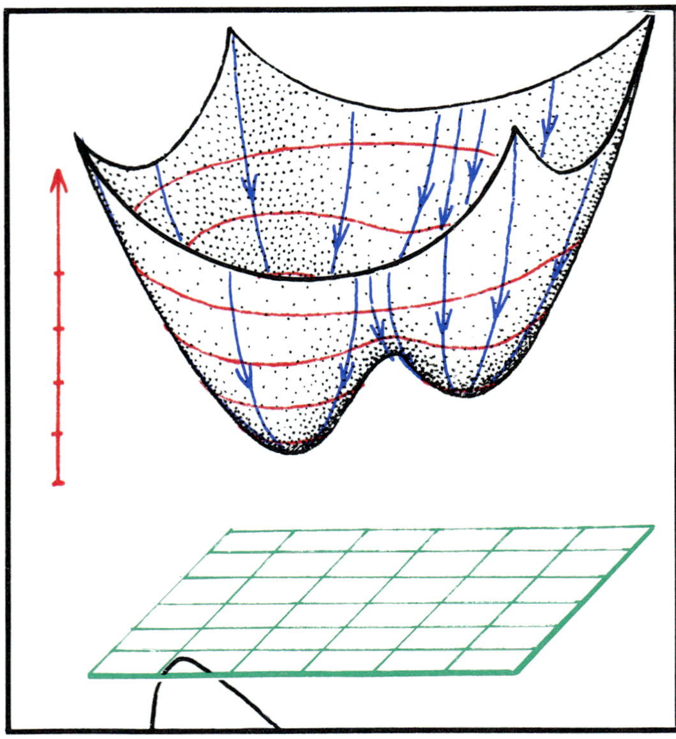

1.6.8. Sprinkle the potential surface with a fine mist of blue ink. Droplets will run down the fall lines, that is, the routes of steepest descent. Suppose that the speed of a droplet is exactly the steepness of the slope.

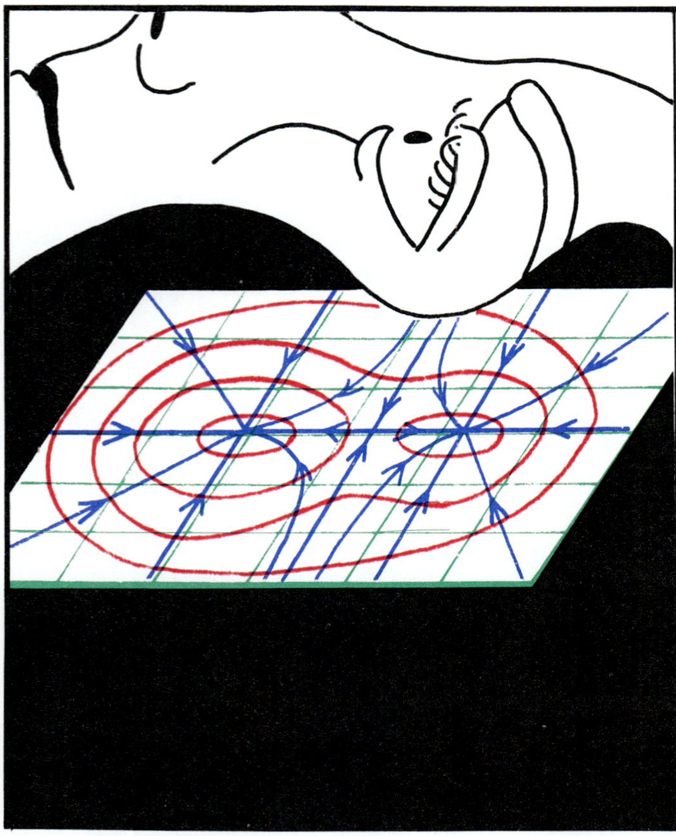

1.6.9. Viewed from far below, the blue droplets appear to move over the contour map, at right angles to the contours.

The blue curves, perpendicular to the contours, together with the parameter of time along each, comprise the phase portrait of the gradient dynamical system.

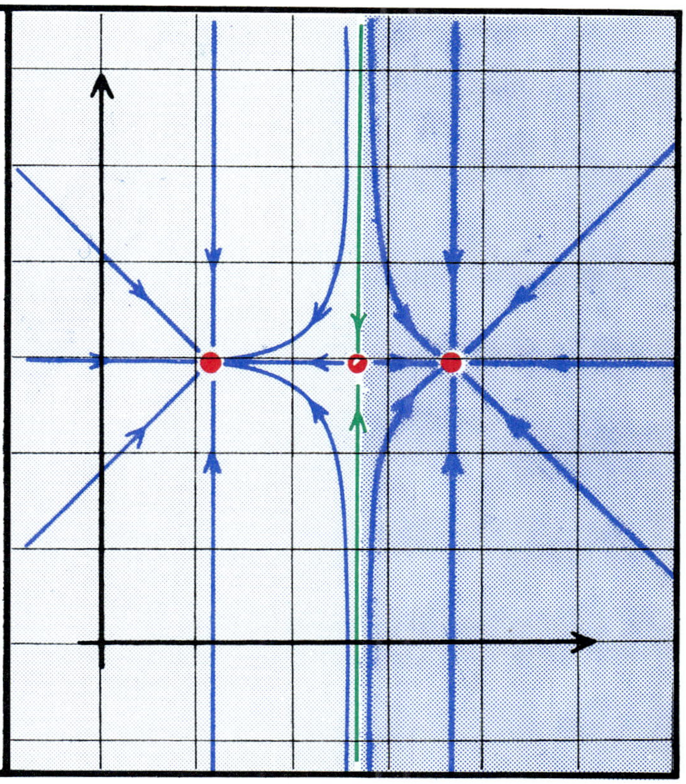

1.6.10. In this example, the gradient phase portrait has two basins, with a point attractor in each. Between them is a limit point of saddle type, or *saddle point,* corresponding to the saddle on the ridge between the two valleys in the landscape. The inset of the saddle point consists of the two green trajectories. This inset is also the separatrix, dividing the state space into the two basins.

Gradient systems, generally, are much like this example. Their limit sets are generally equilibrium points. A limit cycle is impossible in a gradient system, as you cannot go steadily downhill and still return to your starting point (except in an Escher print). Although gradient systems are useful in some elementary physics problems, their usefulness in general applications is severely limited by the lack of limit cycles. The next chapter will show why limit cycles are so important.

2

Classical Applications:

Limit Points in 2D from Newton to Rayleigh

The early work in applied dynamics, before Rayleigh and Poincaré, was devoted primarily to the motions of the planets around the sun. This application, called *celestial mechanics*, comprises an enormous and important field. It was the main subject of Poincaré's research, and is still very active. Under the name *conservative mechanics*, it has been enlarged to include all non-dissipative, that is, frictionless, mechanical systems.

General dynamical systems in nature—whether physical, chemical, biological, or social—are not conservative, but *dissipative*. The theory and applications of these dynamical systems have been elaborated mostly in this century, after the lead of Euler, Rayleigh, Poincaré, and Liapounov.

The classical examples of dissipative dynamical systems—primarily those included in Rayleigh's text of 1877—are described in this chapter. Some of the analytical descriptions, as systems of ordinary differential equations of the first order written classically, are included here without discussion. Explanations of these expressions, not essential for understanding the rest of this text, may be found in the Appendix.

2.1. Pendula

The pendulum may be the most classical example of the dynamical modeling process. It has a two-dimensional state space, and a dynamical system established by Newton.

This model assumes that the rod is very light, but rigid. The hinge at the top is perfectly frictionless. The weight at the lower end is heavy, but very small. It moves in a vacuum. The force of gravity always pulls it straight down.

These idealizations describe the modeling assumptions in this example, called the *simple pendulum*.

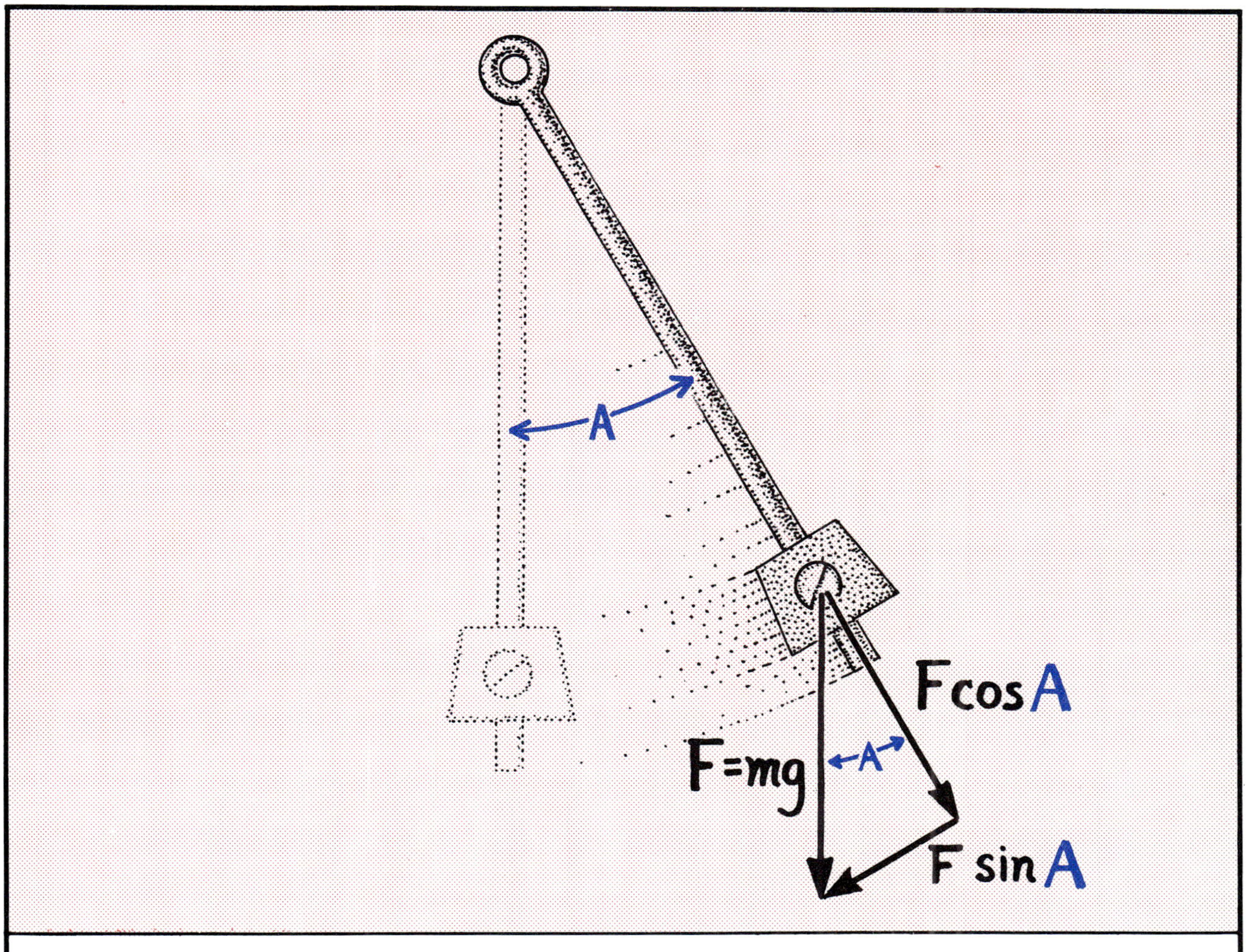

2.1.1. If A denotes the angle of elevation of the pendulum and F the force of gravity, then $F \cos A$ is the pull along the rod, and $F \sin A$ the force turning it, as shown here.

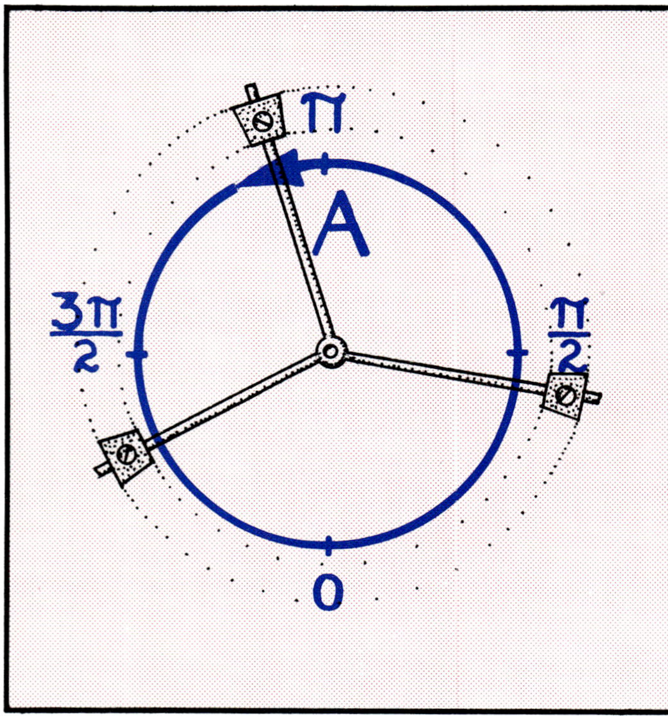

2.1.2. The angle of elevation, A, parameterizes a circle. That is, values of A can be any real number, but $A = 0$ and $A = 2\pi$ denote the same angle. The angle A represents a point of the circle. It is called an *angular variable.*

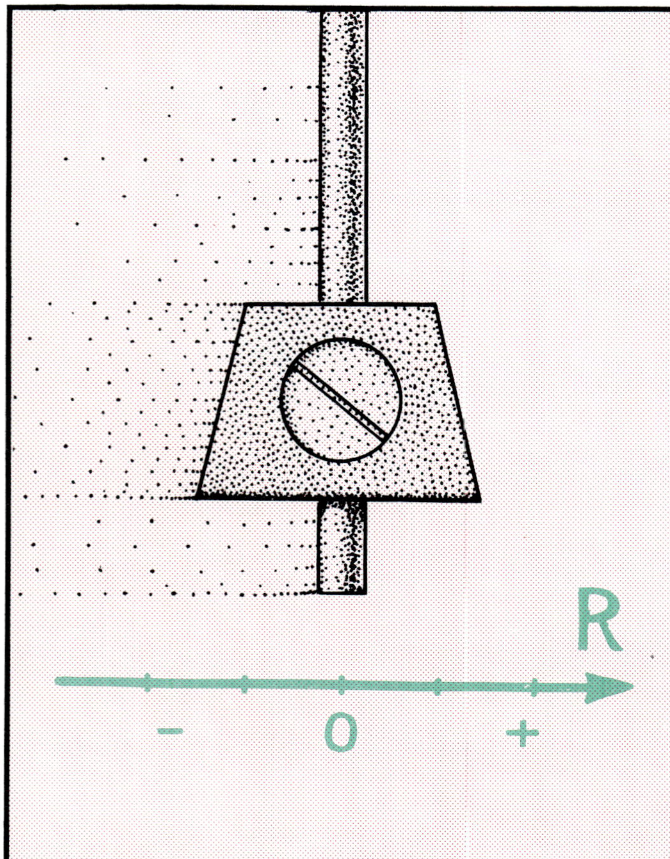

2.1.3. Let R denote the rate of rotation of the rod at a given moment. This rate is also observable, by radar for example. In Newton's model, this parameter is included, along with the angle A, as a descriptor of the state of the pendulum. The rate of rotation, R, may have any real number as its value. It represents a point of the real number line.

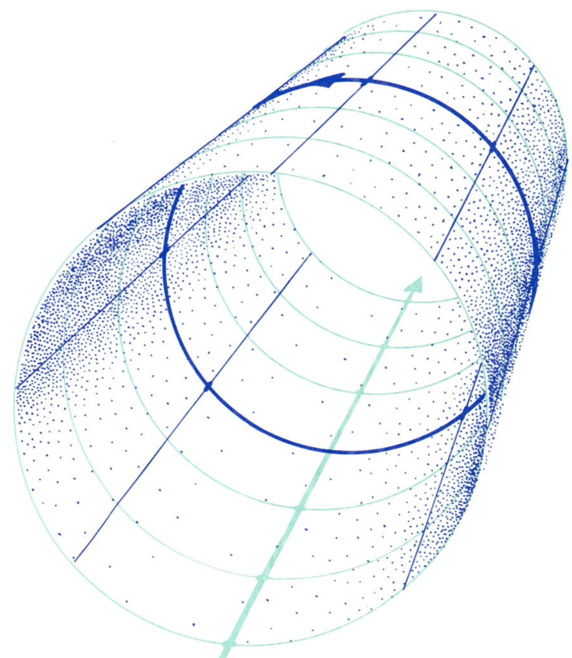

2.1.4. The two parameters, A and R, together locate a point on a circular cylinder. This is the state space of Newton's model. The vertical circle in the center of this cylinder denotes the states of zero angular velocity, $R = 0$. The straight line from front to back, at the bottom of the cylinder, is the axis of zero inclination, $A = 0$, where the pendulum is lowest.

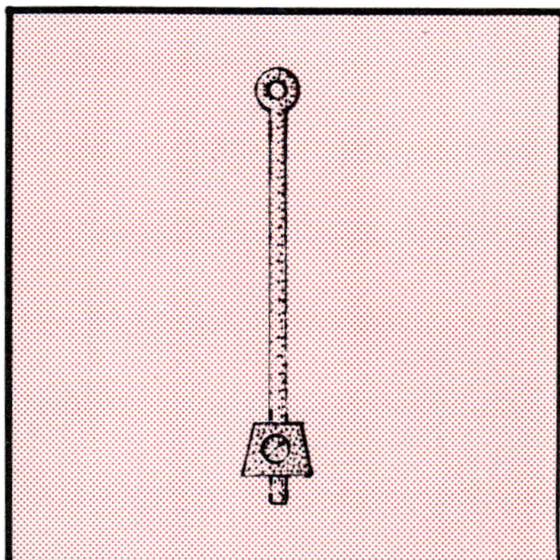

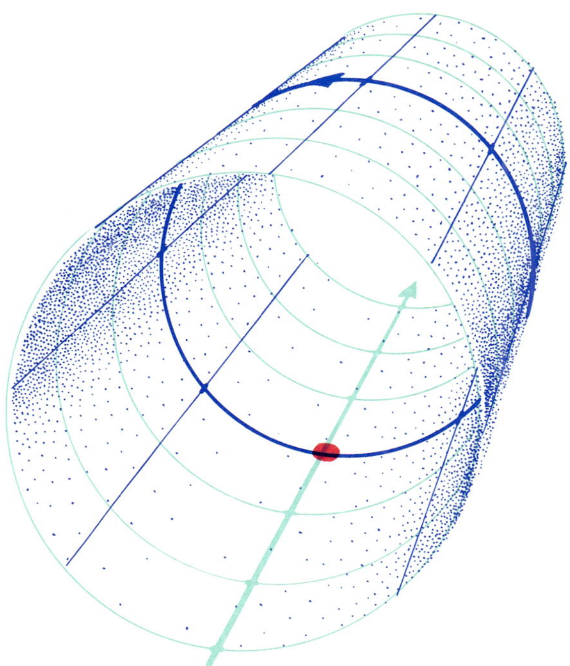

2.1.5. At the *origin*, defined by $(A,R) = (0,0)$, the pendulum is at rest at its lowest position.

Moving the pendulum a little to the left and then letting it go with no shove causes it to swing indefinitely. Remember, there is no friction in the hinge and no air in the way. The representation of this motion as a trajectory in Newton's model is shown here in four steps.

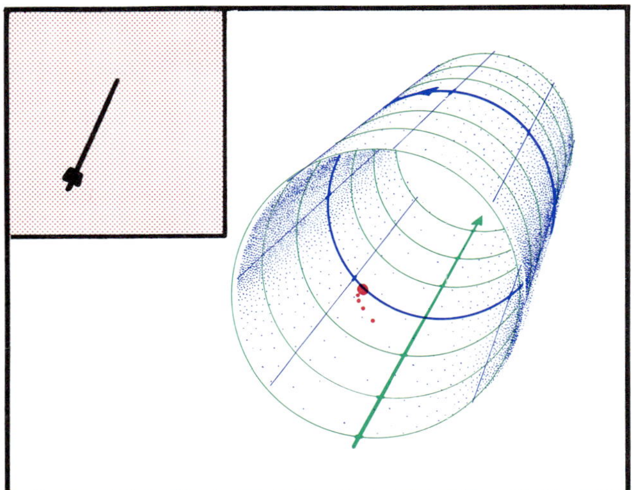

2.1.6. Step 1. Immediately after the pendulum is released, the representative point is on the circle of $R = 0$ to the left of the origin, moving away from us as the rate R increases.

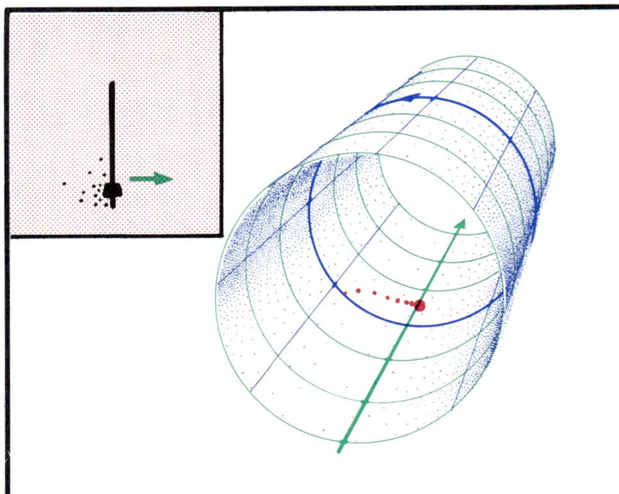

2.1.7. Step 2. It also moves to the right as the inclination increases. Here it has just reached the axis, $A = 0$, as the pendulum goes by its bottom point.

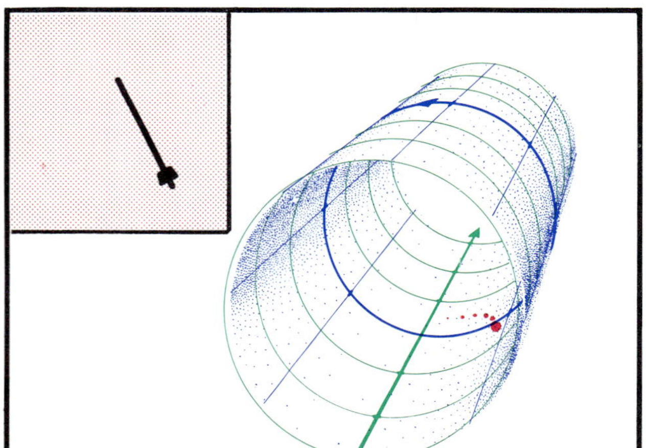

2.1.8. Step 3. It continues to move to the right, moving towards us rather than away, as R decreases. It reaches the circle of $R = 0$, when the pendulum attains its maximum swing to the right, and turns to fall again toward its bottom, $A = 0$.

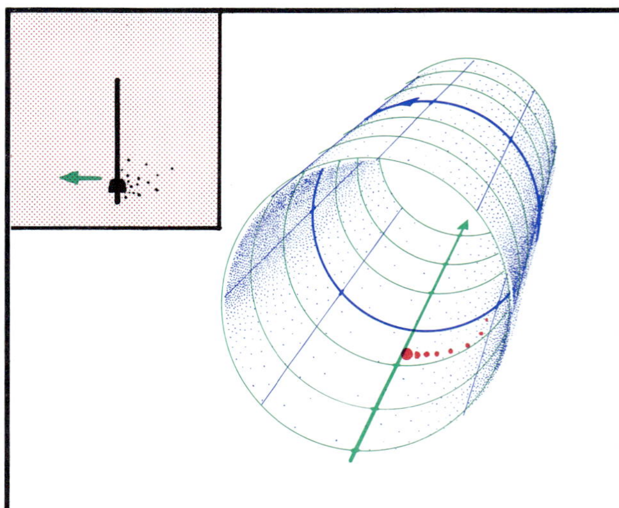

2.1.9. Step 4. It approaches us and moves to the left, as the pendulum falls. It crosses the axis, $A = 0$, as the pendulum swings through bottom.

Then the cycle begins again. The full trajectory in the state space, corresponding to this oscillating motion of the pendulum, is a cycle, or closed loop.

The next sequence shows the trajectory in Newton's cylindrical model representing the motion of the pendulum dislodged from a precarious balance at the top.

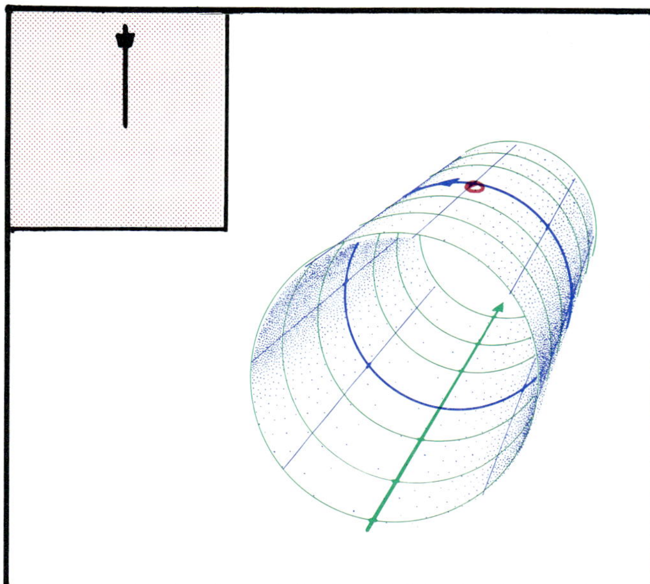

2.1.10. Step 1. The pendulum can be balanced at the top, in *unstable equilibrium.*

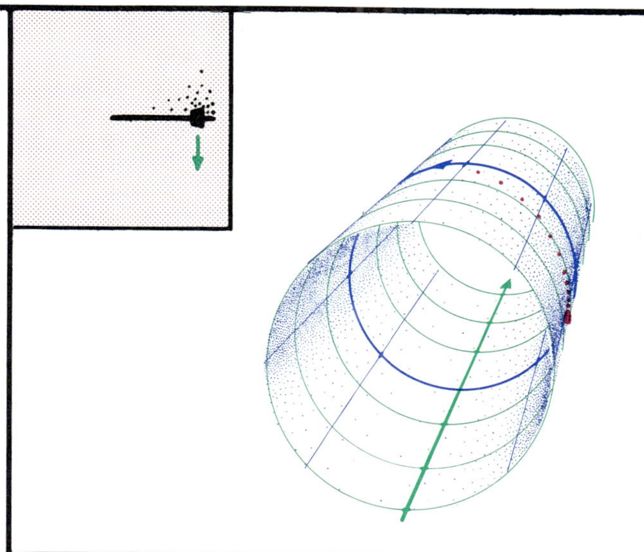

2.1.11. Step 2. An infinitesimal touch on the left will start it falling to the right. Picking up speed, its states trace the red trajectory shown here.

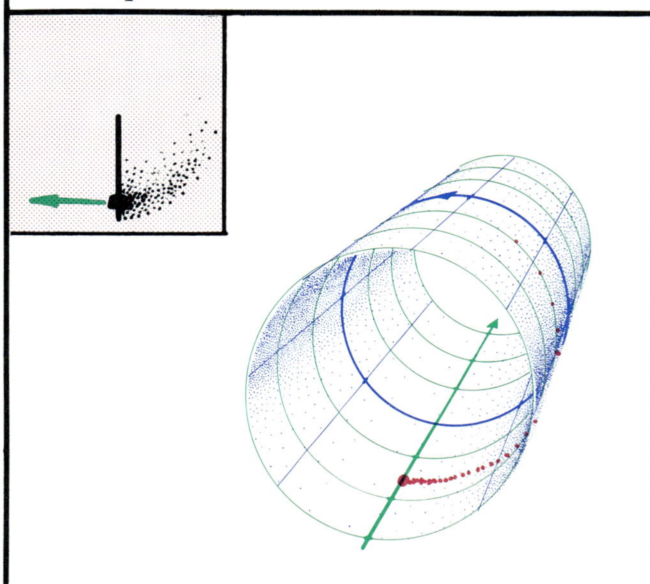

2.1.12. Step 3. As the pendulum passes the bottom of its swing, the rate of rotation reaches its maximum.

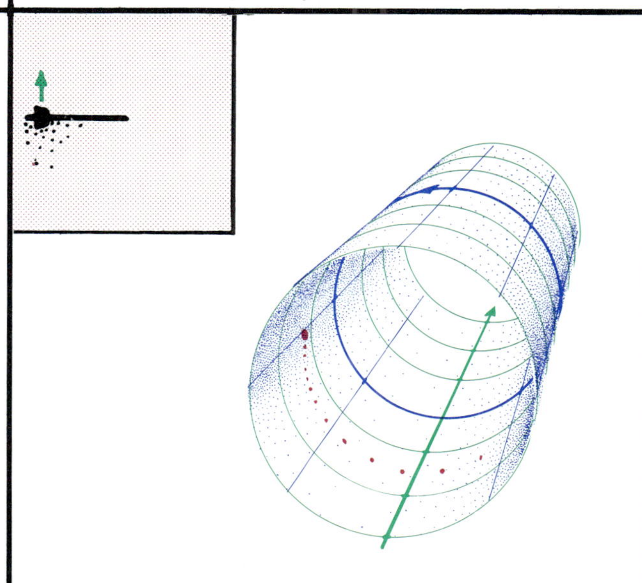

2.1.13. Step 4. On the way back up to the top again, the pendulum is slowing down.

As this trajectory approaches the position of balance at the top (its omega-limit point, an unstable equilibrium) it moves slower and slower. The pendulum actually balances at the top again, at the end of the motion, but this motion takes forever.

***Warning.* This trajectory is not a cycle, because the point at the top, the omega-limit point, does not belong to the trajectory.**

The next sequence is a slight modification of the preceding—the pendulum is balanced at the top, then shoved hard to the right.

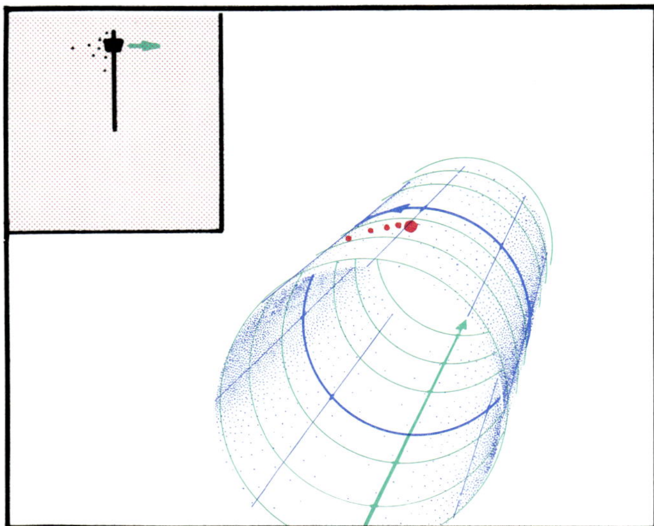

2.1.14. Step 1. The pendulum is just to the right of the top, but its rate is large. Thus, the point in the geometric model corresponding to this state—shown as a red dot here—is closer to us than it was in the preceding sequence.

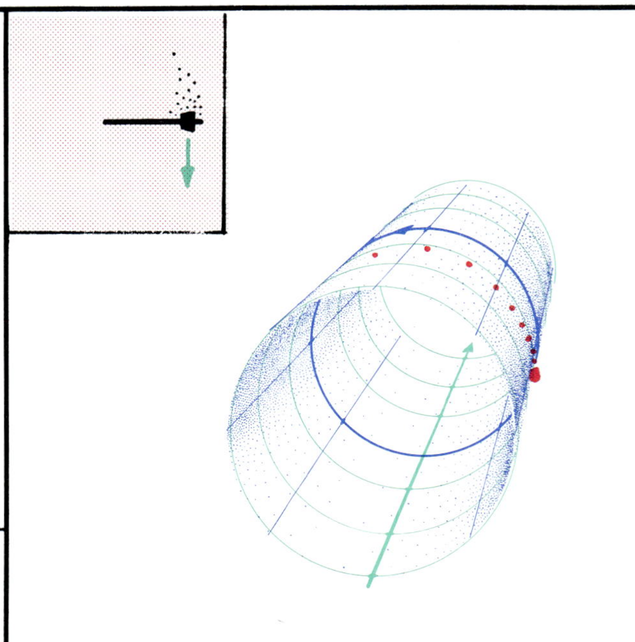

2.1.15. Step 2. The pendulum accelerates as it falls down to the right.

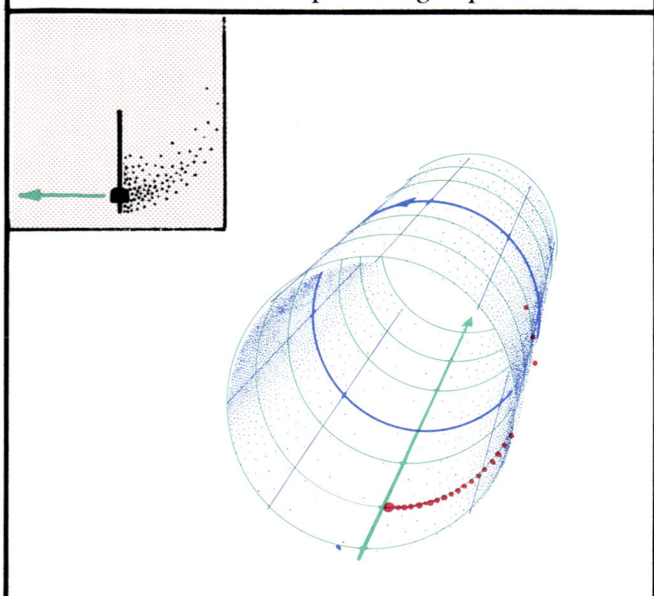

2.1.16. Step 3. It passes the bottom, moving fast, attains its maximum speed, and keeps on going.

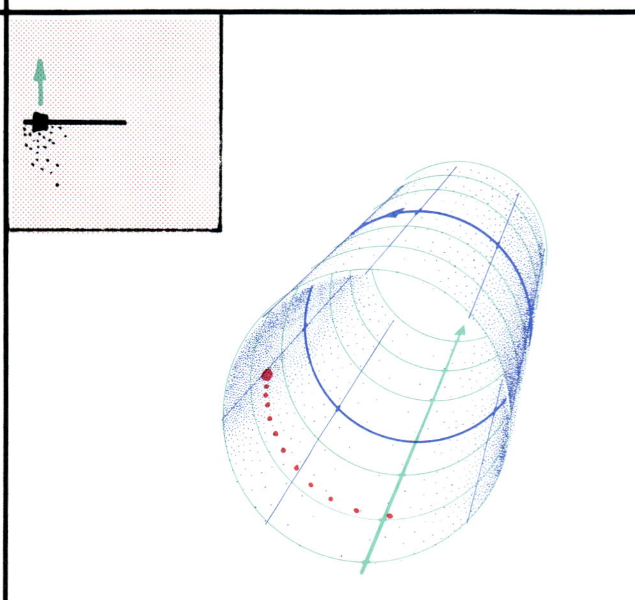

2.1.17. Step 4. Moving back up again, it slows down, but not enough to come to rest at the top.

In this motion, the pendulum rotates clockwise indefinitely. The corresponding (red) trajectory in the cylindrical state space closes at the top. It is a cycle. But unlike the slow oscillation described above, this fast cycle goes around the cylinder.

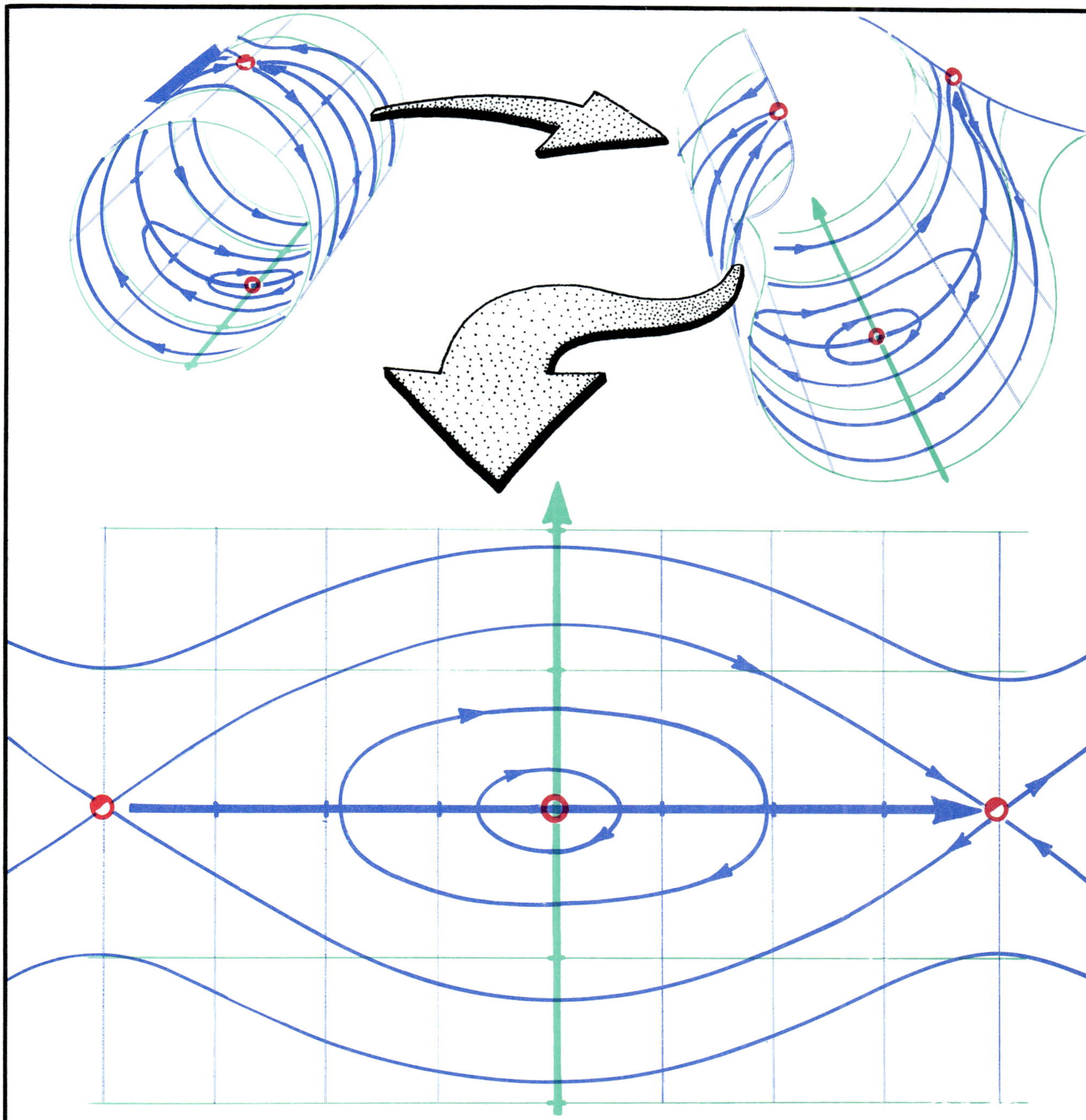

2.1.18. Performing many such experiments with a pendulum, or emulating one with an analog or digital computer, would reveal the phase portrait of Newton's model. This cylinder full of trajectories is easier to see if it is cut down the top and flattened, as shown here. Notice that there are two equilibrium points. One at the top is a *saddle*, a type we have already seen in the gradient dynamical system in the preceding chapter. The other, at the origin, is another type called a *center*, or *vortex point*. This type will recur throughout the rest of this book. The critical point in the center is *not a limit point* of the nearby trajectories.

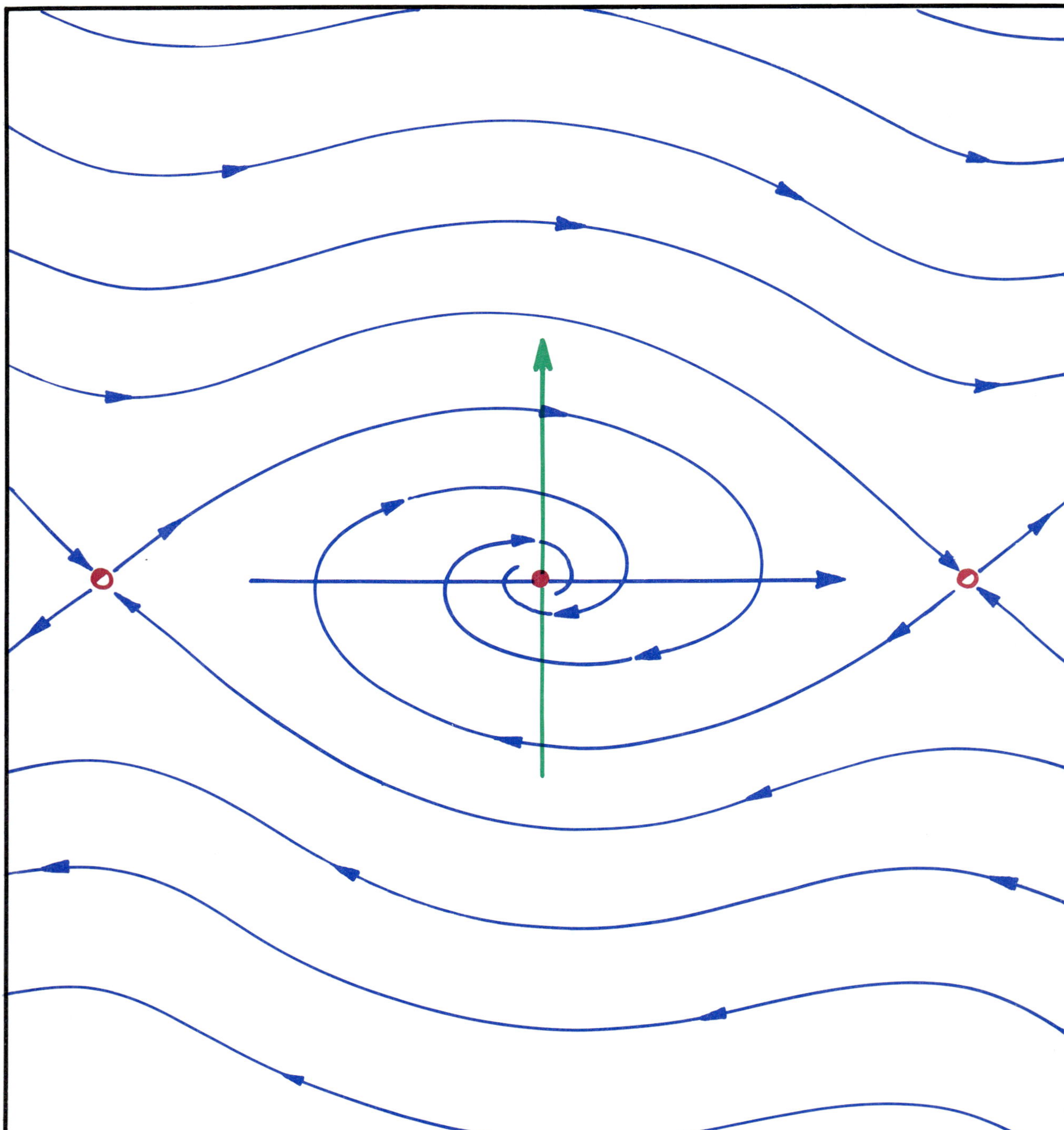

2.1.19. A more realistic example may be obtained by including the effects of friction in the hinge, and air resistance. Here is the phase portrait which results. Notice that it is very similar to the preceding portrait, but the equilibrium point at the origin is no longer a *center.* It has become an *attractor.* This is because any nearby trajectory, representing a slow motion of the pendulum near the bottom, will die away because of friction, and the pendulum will *come to rest*. This spiraling type of point attractor is sometimes called a *focal point*.

An instructive variation of this example may be produced by adding small forces that break the symmetry.

2.1.20. Suppose the pendulum is magnetic, and two magnets are added near the bottom of the arc. The one on the left is stronger.

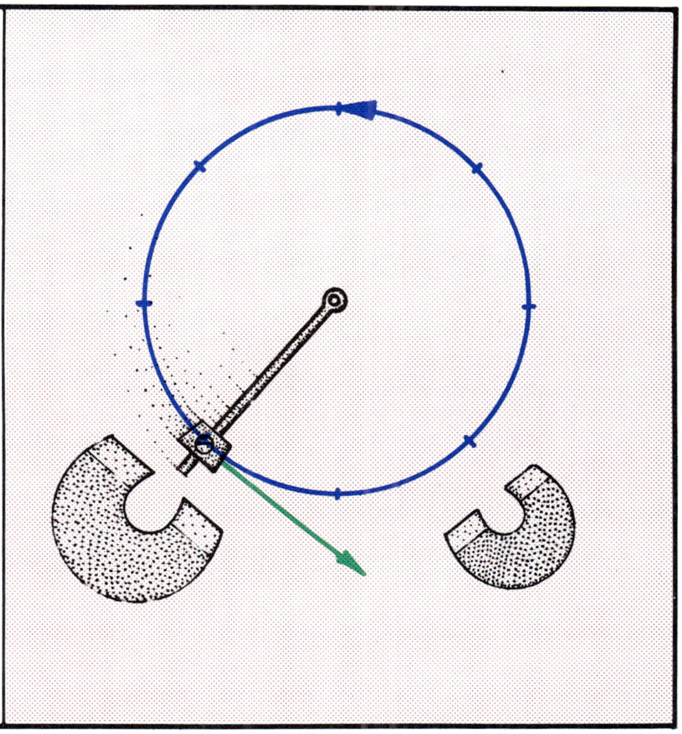

2.1.21. As the pendulum bob could be stopped near either magnet and held slightly aloft by its attraction, the dynamical system modeling this device will have two basins, with a point attractor in each.

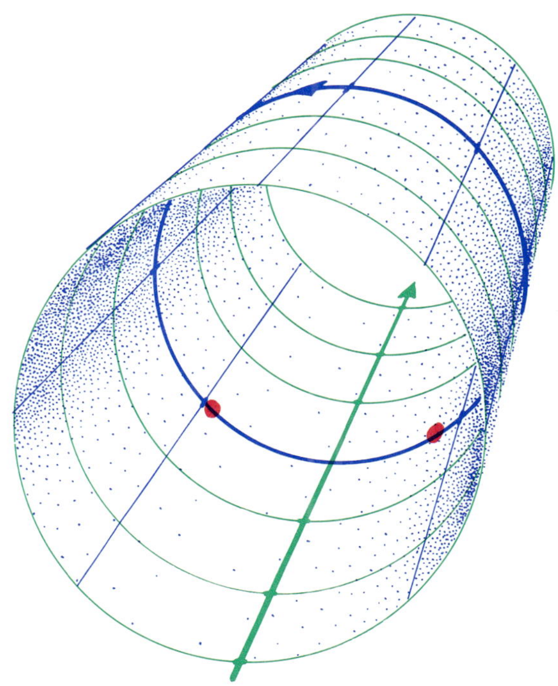

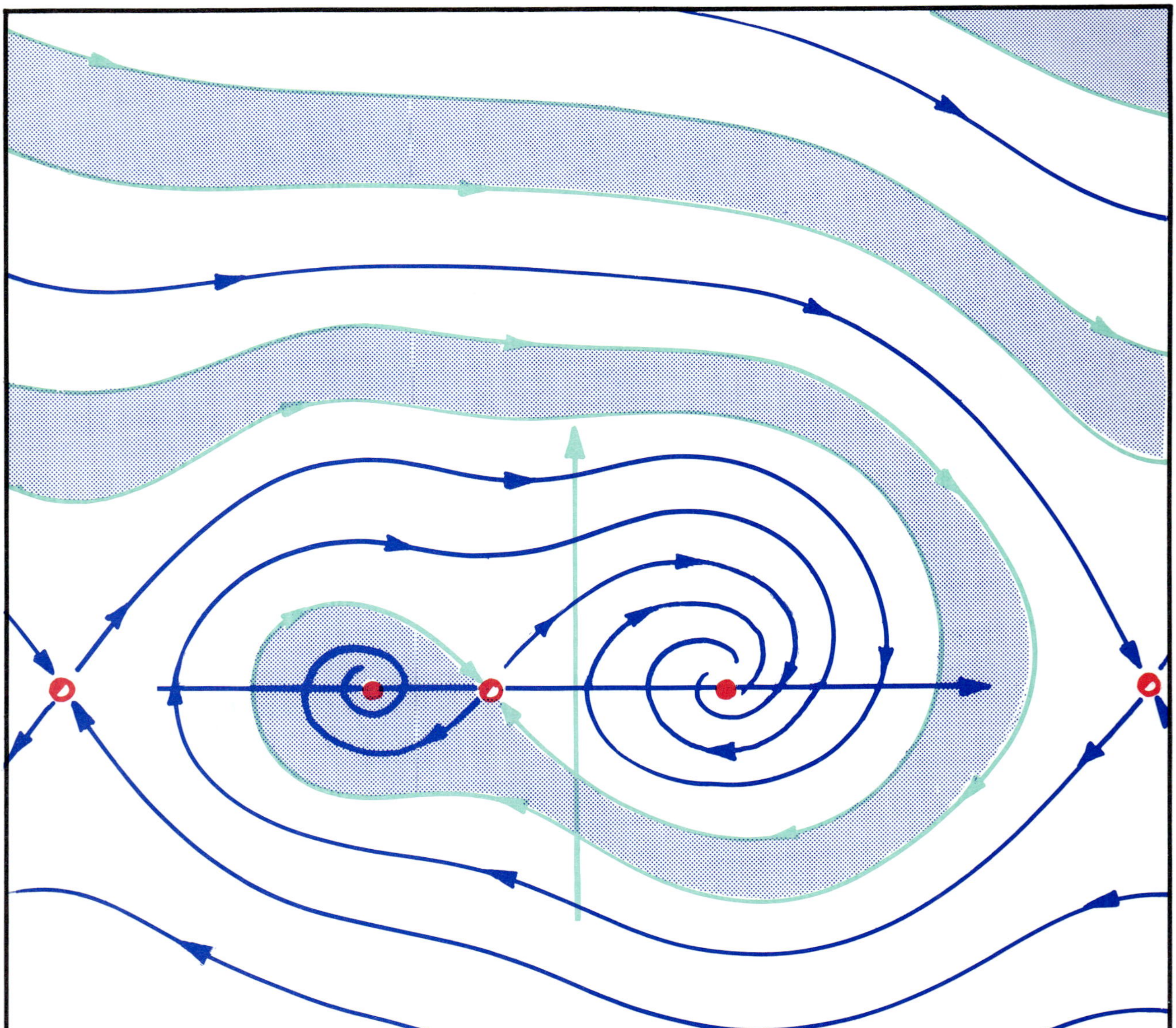

2.1.22. Here is the phase portrait of the magnetic pendulum, unrolled. The basin of the rest point near the smaller magnet, shaded, is smaller. Thus, the probability of an initial state (angular position plus angular velocity) evolving asymptotically to the rest point adjacent to the smaller magnet is less than 50%.

Note. In this particular case, the shaded basin extends upward only. Thus, if the bob is swinging rapidly counterclockwise, it cannot come to rest at the smaller magnet.

This is not solely an armchair experiment. It has actually been studied.[1]

This is the most classical application. Although pendulum theory may not turn everyone on, it has certainly been fundamental to the growth of dynamics. The writings of Lord Rayleigh are held together with this common thread, and he saw pendula everywhere. The spirit of his work lives on in many dynamics laboratories, even today.

2.2. Buckling Columns

The deformations of elastic solid material provided the context for some of the early applications of dynamics. The buckling of a vertical, elastic column—under a weight balanced on its top—was studied by Euler. A particularly simple analysis of this experimental situation has been made by Stoker in 1950. His dynamical model, illustrated in this section, is closely related to the model for the simple pendulum, shown in the preceding section.

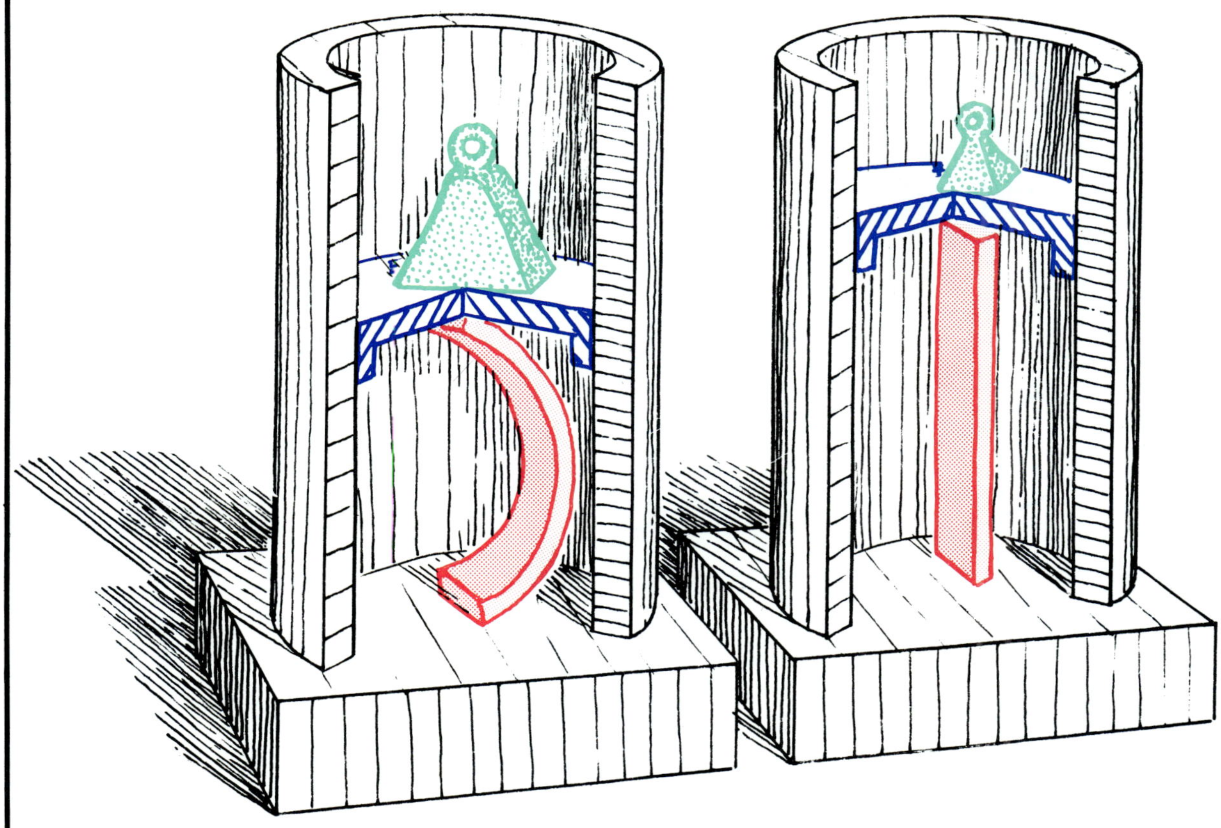

2.2.1. An *elastic column*—a thin metal bar balanced on end—is slightly compressed by a light weight on top. If the center is pushed to one side and released, it oscillates back and forth. Under a heavier weight, it buckles to one side or the other.

This physical system may be crudely modeled by a dynamical system in the plane.

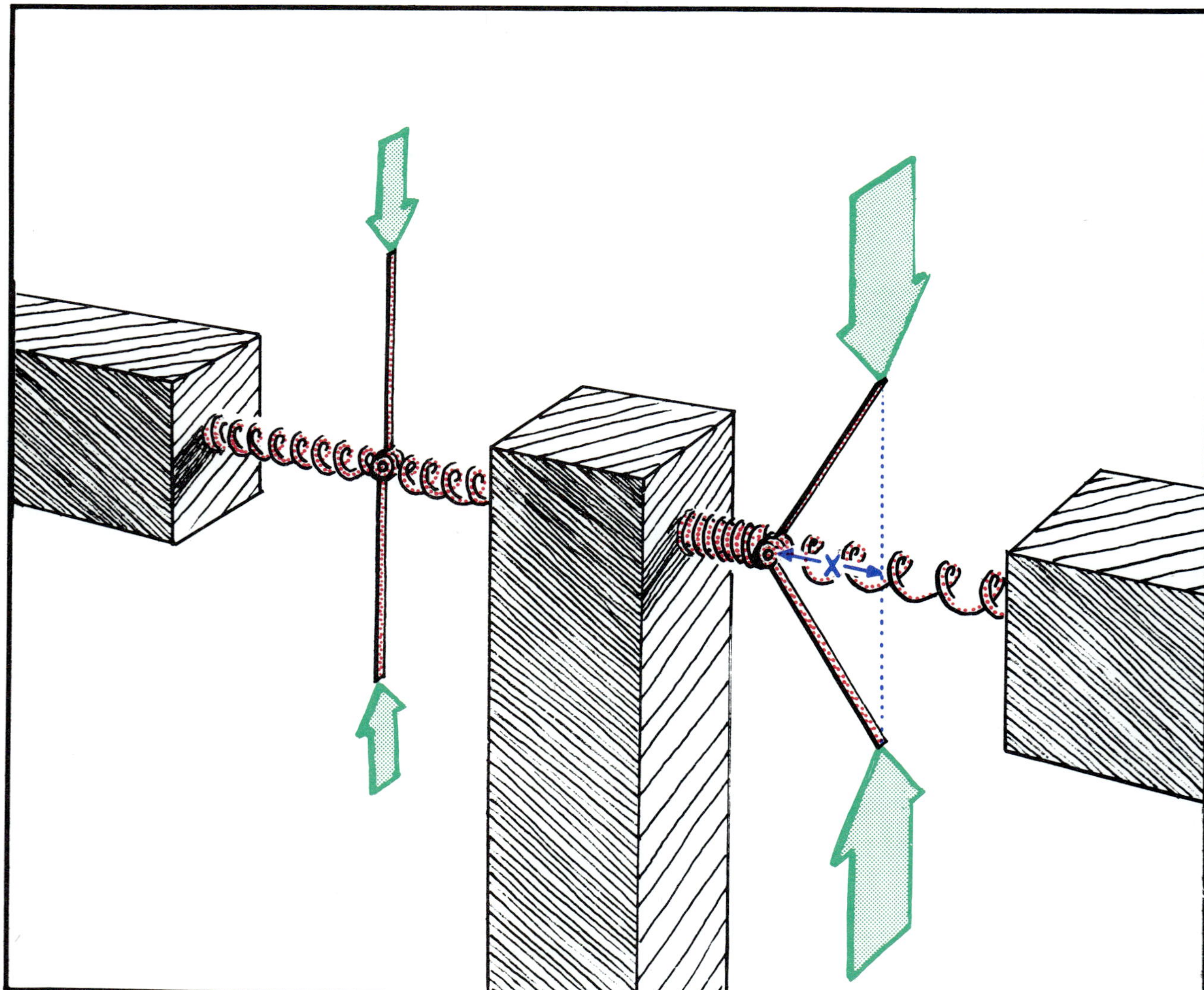

2.2.2. Stoker's model idealizes the elastic column as a hinged rod, restrained by two springs. When the compressive force is less than a critical value, the rods return to vertical alignment. Under a larger force, the hinge moves to one side or the other, compressing one of the springs and stretching the other. This corresponds to the buckling of the column.

The geometrical model for the states of the hinged rod is the plane. The two observed parameters are the displacement of the hinge to the right of vertical, recorded on the horizontal axis: X, and the velocity of the hinge on the vertical axis: $V = X'$.

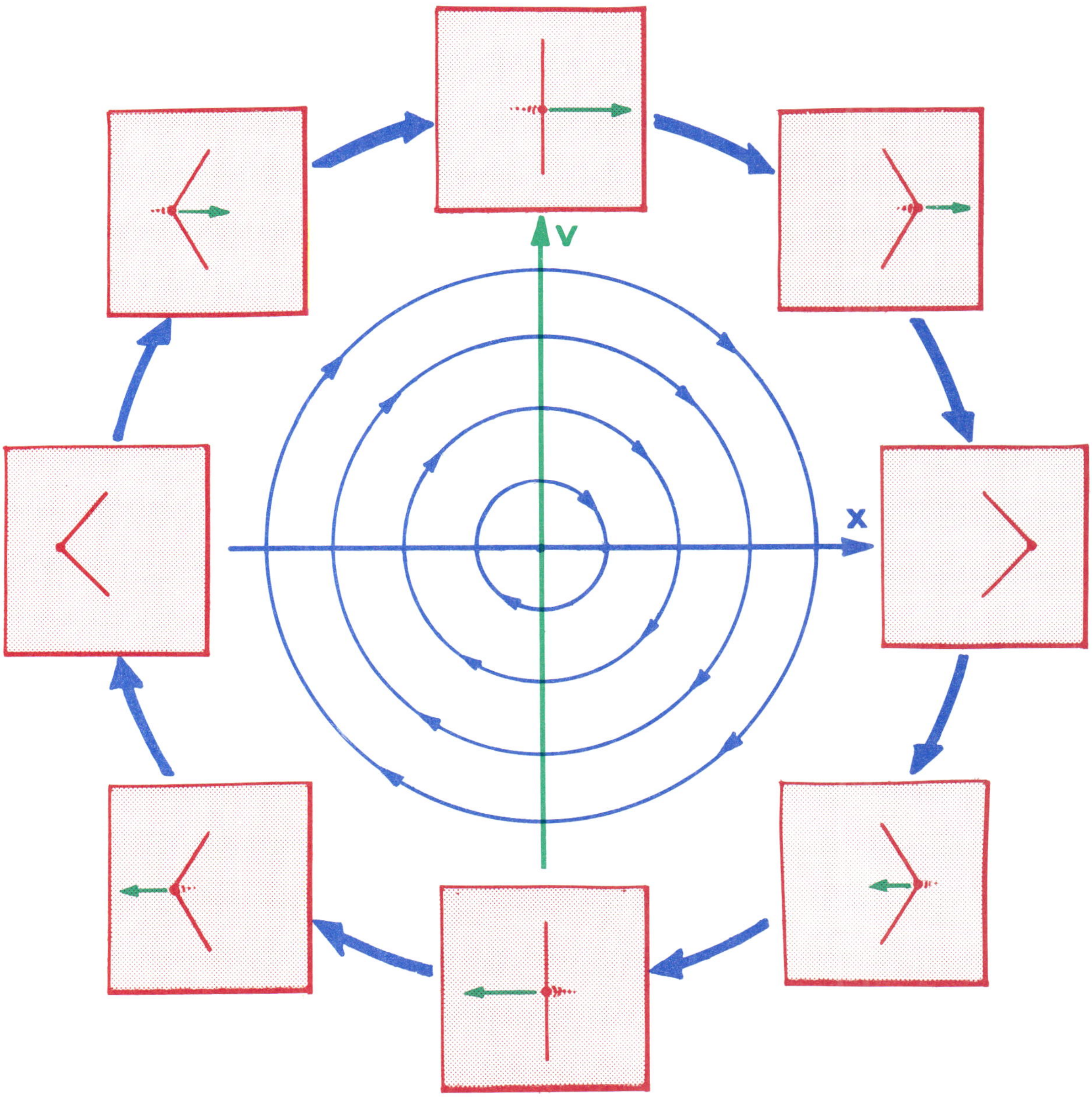

2.2.3. This tableau shows a succession of states, and their corresponding representations in the state space, as the hinge vibrates back and forth. Assume the weight is lighter than the critical value for buckling and there is no friction in the system. Then the phase portrait is a *center*, as in the preceding example. That is, it consists of concentric closed trajectories. These are periodic trajectories, representing oscillations. The breadth of a closed trajectory represents the *amplitude* of the oscillation.

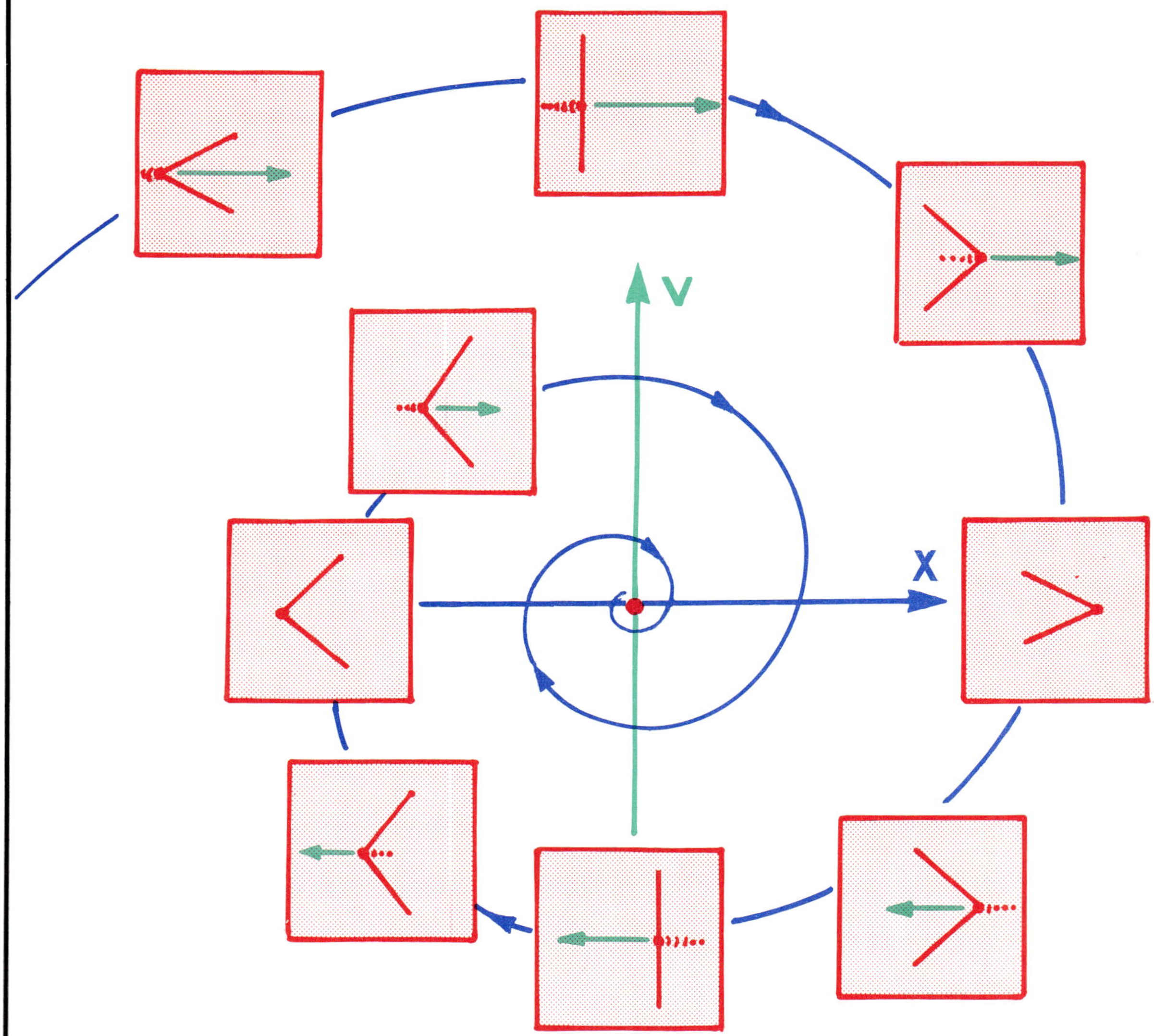

2.2.4. Adding friction, the concentric loops are replaced by spiraling trajectories. As in the case of the pendulum with friction, they approach the origin as the amplitude of oscillation decreases to zero. This is a typical phase portrait, with one basin surrounding its unique *attractor*—in this case a *focal point*. Recall that a point attractor, or *rest point*, is a critical point (equilibrium point, limit point) that attracts all nearby initial states. Thus, all the state parameters describing these nearby initial states evolve asymptotically, as time increases, to constant values. Their omega-limit state is at rest, at the rest point.

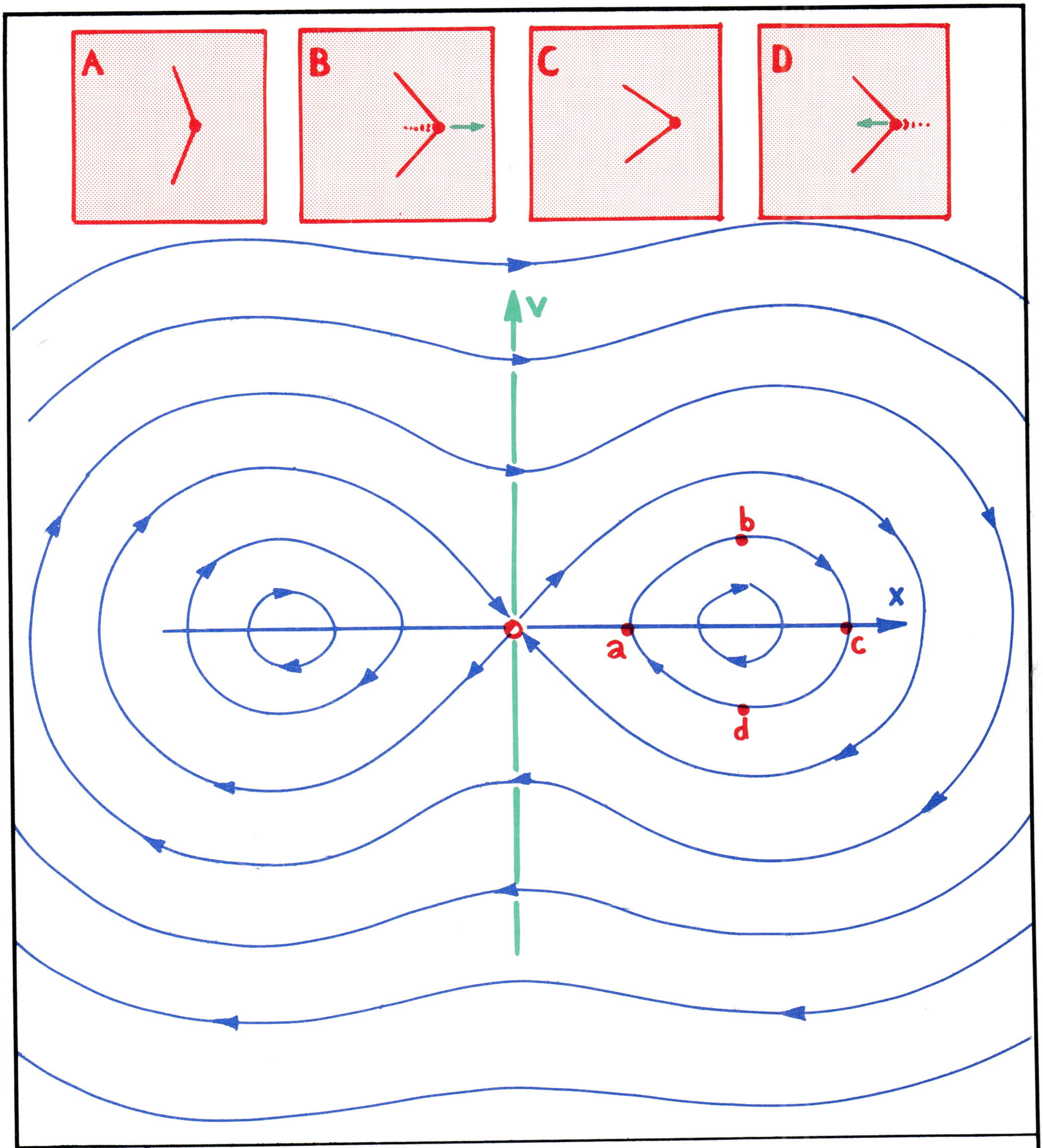

2.2.5. This is the phase portrait for the frictionless system, with the heavier weight. The sequence of states—A,B,C,D,A—describes an oscillation around an average displacement to the right. The *equilibrium point* at the origin is a *saddle*.

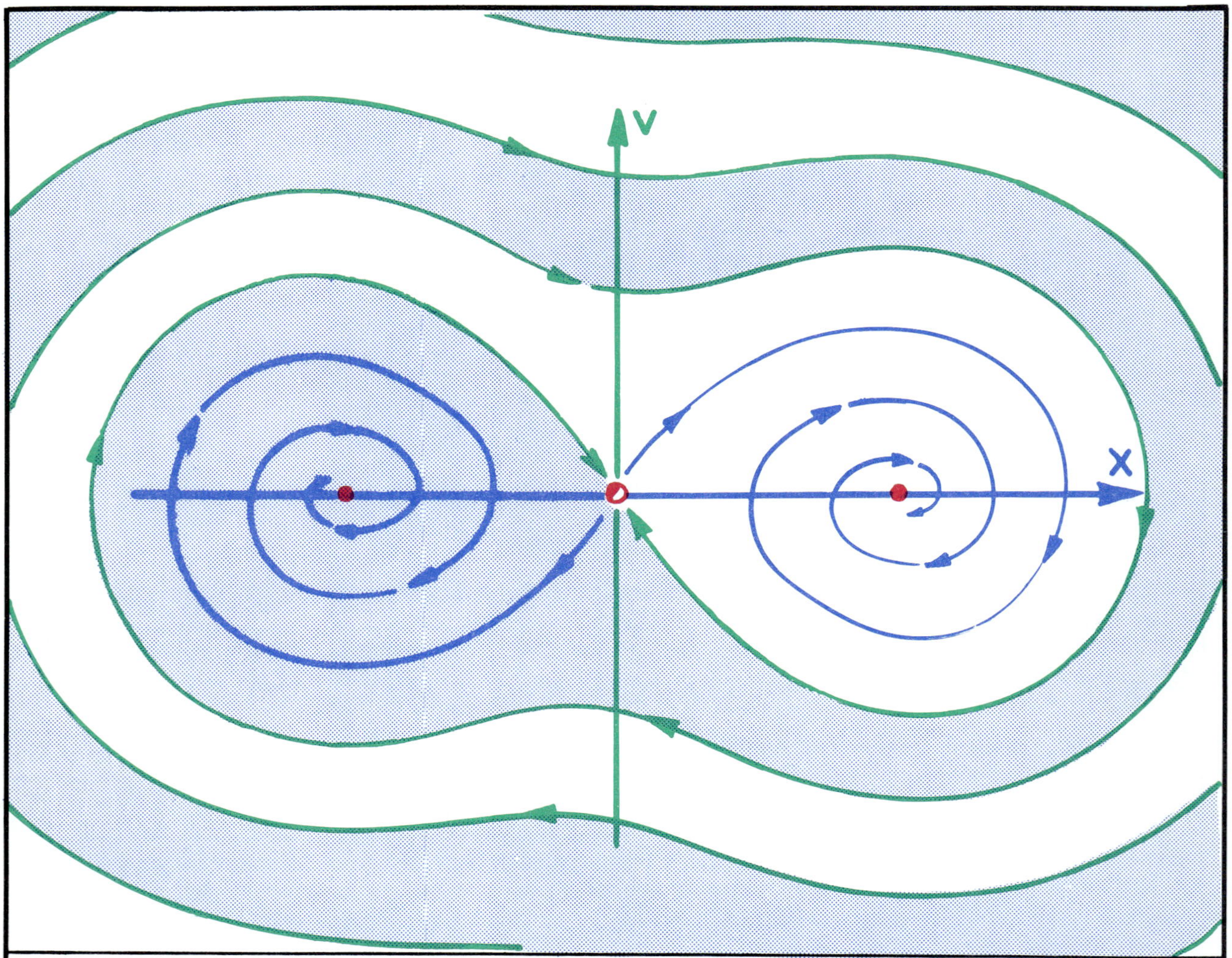

2.2.6. Adding friction, the concentric circles are again replaced by spirals. This is a typical portrait with two *basins*, a *focal point attractor* in each. Note that the insets of the saddle at the origin—shown here in green—do not belong to either basin. Thus, they comprise the *separatrix*, defining the boundaries of the two basins, one of which is shaded here. The relative area of a basin determines the *probability* of its attractor. That is, the chance of choosing an initial state that evolves to the attractor on the left is proportional to the area of the shaded basin. In this example, the two point attractors are equally probable.

This example is like the gradient systems described earlier, in that the limit sets are all points. Thanks to friction, closed trajectories are impossible. But unlike those of gradient systems, the trajectories approach the limit point in spirals, rather than radially. The point attractor is a focal point. The spirals correspond to oscillations of diminishing amplitude—*damped oscillations*. The point attractor that occurs in gradient systems is *radial*. That is, the approaching trajectories do not spiral. The radial type of point attractor is sometimes called a *star point*, or *node*.

2.3. Percussion Instruments

In *The Theory of Sound*, Lord Rayleigh studied separately the production, propagation, and reception of sound. His efforts to explain the production of sound by musical instruments became the theory of nonlinear oscillations.[2] From the point of view of dynamics, musical instruments may be divided in two classes:

percussion instruments, such as drums, guitars and pianos, to be modeled by damped oscillations (focal points, or point attractors of spiral type), and

sustained instruments, such as bowed strings and winds, which are to be modeled by self-sustaining oscillations (periodic attractors).

This section describes the classical models for the percussion instruments. The sustained instruments will be treated in the next section.

2.3.1. The percussion instruments all produce musical tones which decay (die out) in time. We hear the transient response of the system. The aymptotic limit of the audible transient is silence. Although in principle it takes a very long time for the note to die away, it actually becomes inaudible a short time after being struck.

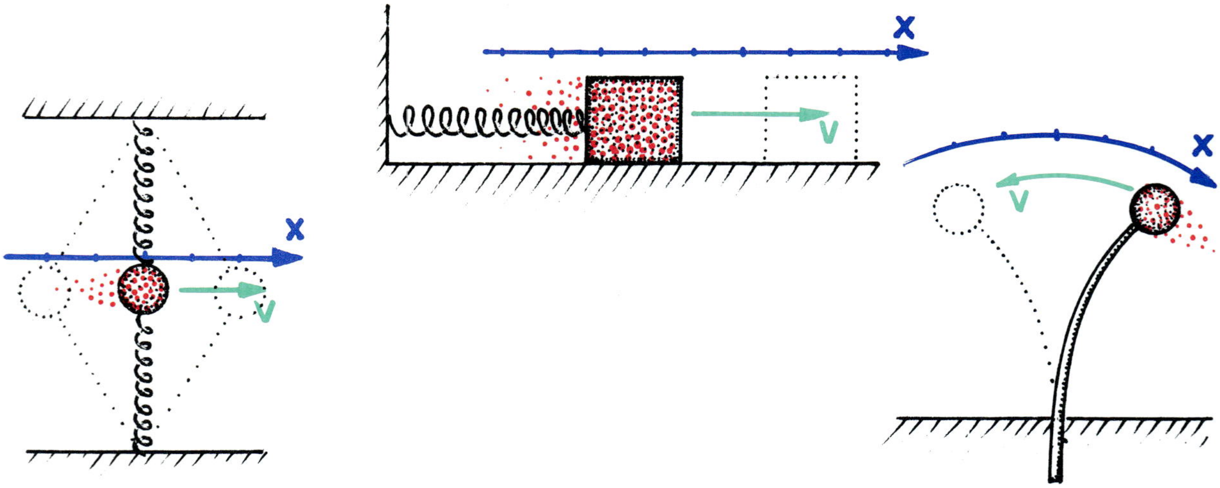

2.3.2. Simple mechanical models for the most sophisticated instruments look like an elementary physics lab. Different configurations of springs and weights behave, very approximately, like the instruments. As in the case of the buckling column, discussed in the preceding section, the resulting model is surprisingly useful.

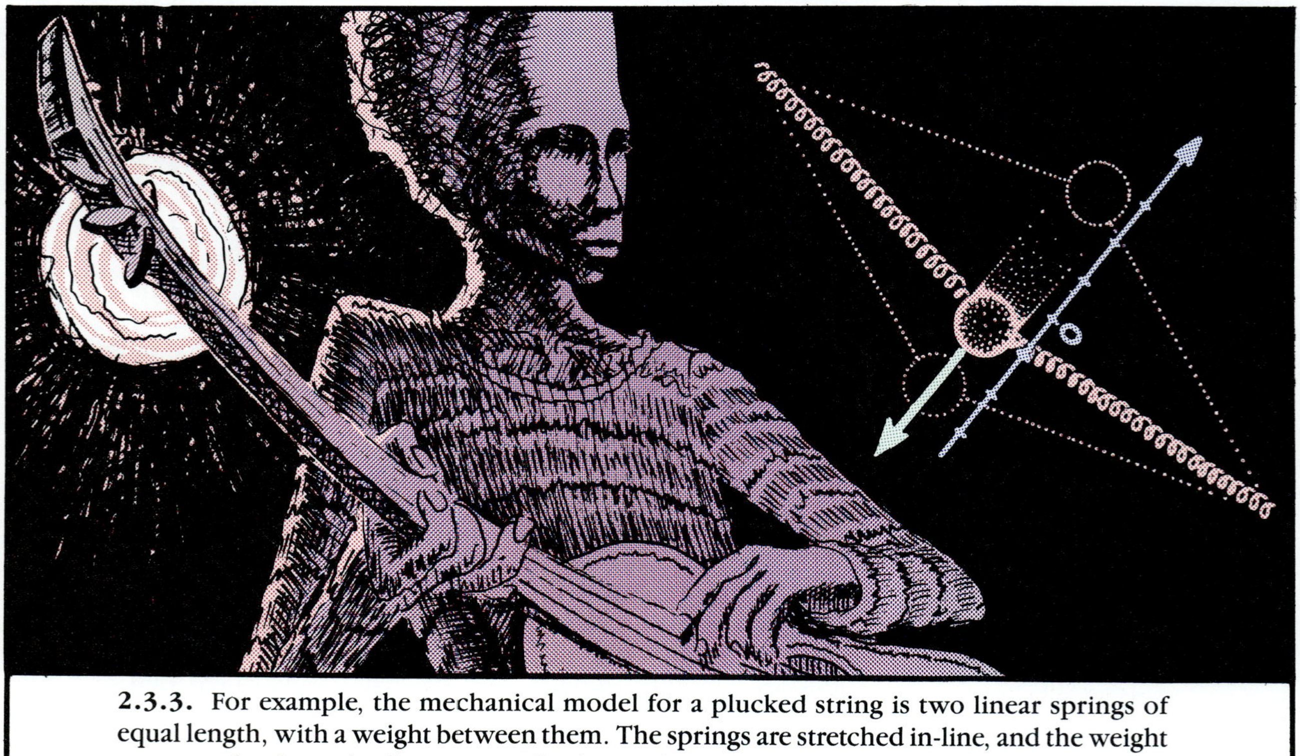

2.3.3. For example, the mechanical model for a plucked string is two linear springs of equal length, with a weight between them. The springs are stretched in-line, and the weight moves only along the line perpendicular to the springs.

2.3.4. As in the two preceding applications, the geometric model for the state space of the mechanical system is the plane. The parameters are the displacement of the weight to the right of equilibrium, and the velocity of its motion. If there is no friction in the mechanical model, the phase portrait of its dynamical model (Newton's Law of Motion) is a *center.* This is very much like the frictionless pendulum described at the beginning of this chapter.

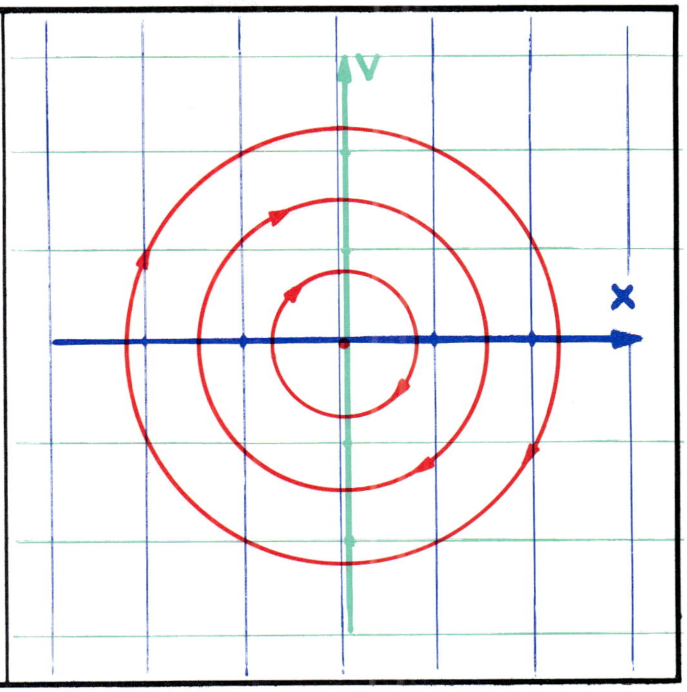

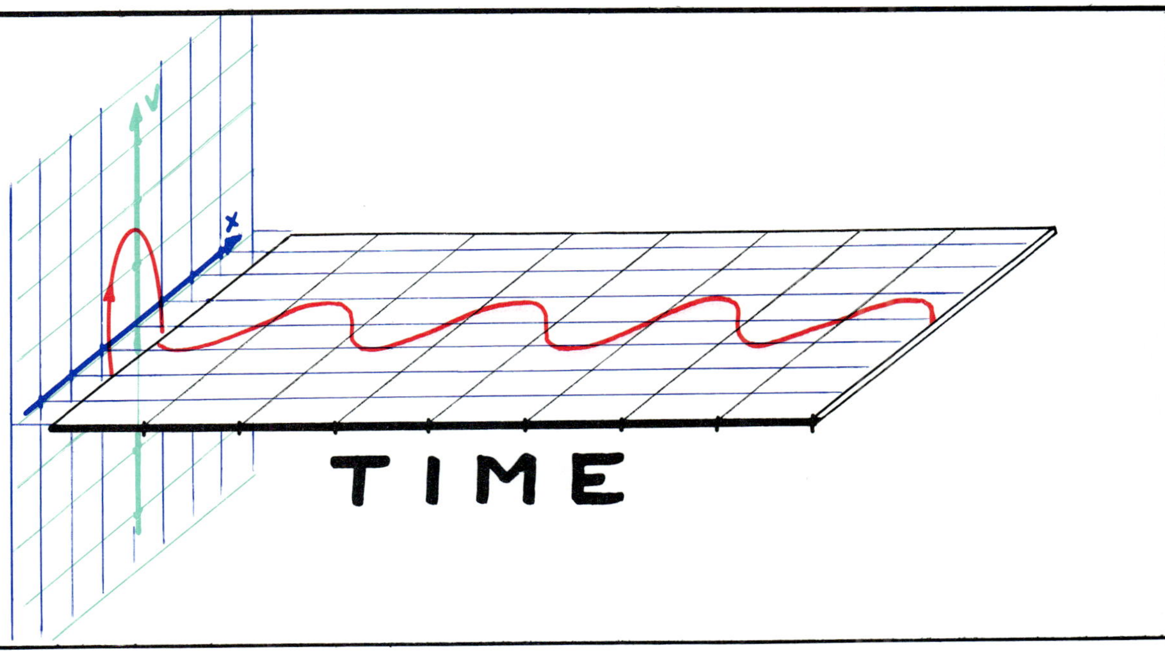

2.3.5. Each trajectory of the center is closed. As described in Chapter 1, the time series of a preferred parameter (for example, the displacement of the weight concentrated at the center of the string) is a periodic function. The weight oscillates back and forth, periodically. In fact, under simplifying assumptions, this motion is sinusoidal. It corresponds to a *pure tone.*

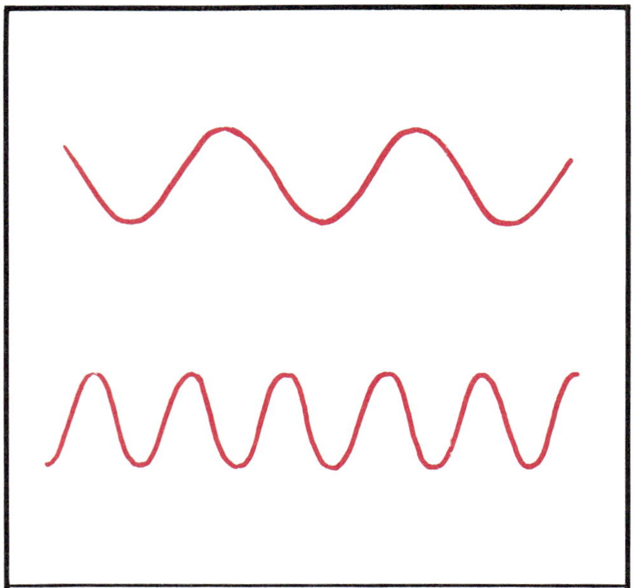

2.3.6. More oscillations per second correspond to higher frequencies, or tones.

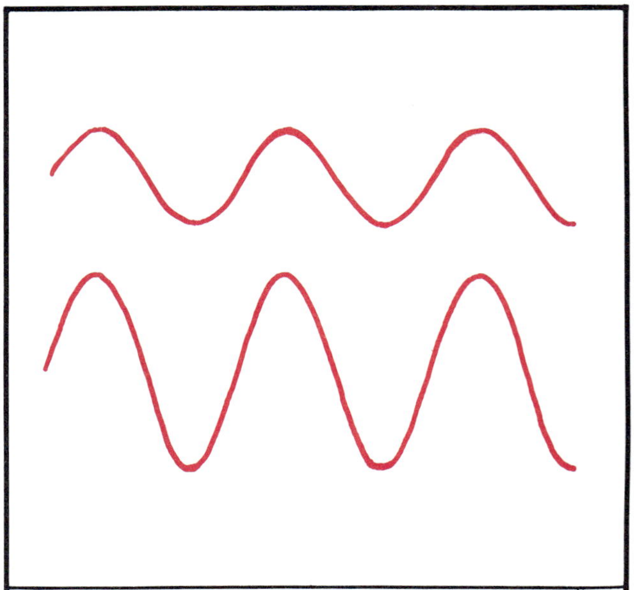

2.3.7. The vertical displacement of the time series, or *amplitude*, corresponds to the loudness of the note.

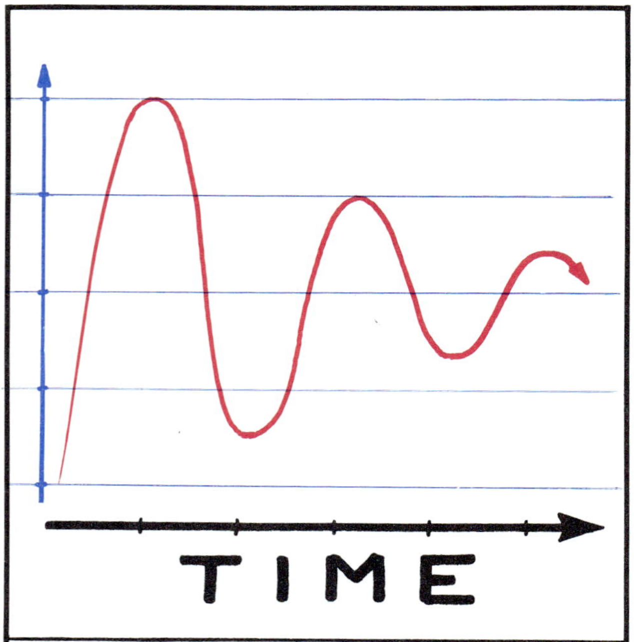

2.3.8. The time series of a plucked guitar or struck piano string is a function that decays, as shown here.

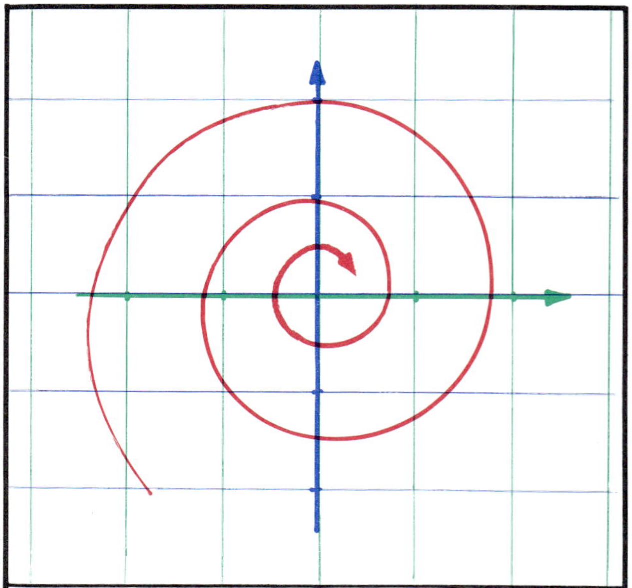

2.3.9. A trajectory in the state space of the mechanical model must spiral asymptotically toward the origin, in order to have the function on the left as its time series.

A model for decaying tones must include friction. To further simplify the discussion, we now replace the two springs (perpendicular to the motion of the weight) with a single spring (along the line of motion).

2.3.10. The dynamical model (Newton's Law) for this mechanical system is similar to that of the preceding system, with the two colinear springs. The difference is insignificant for small displacements.

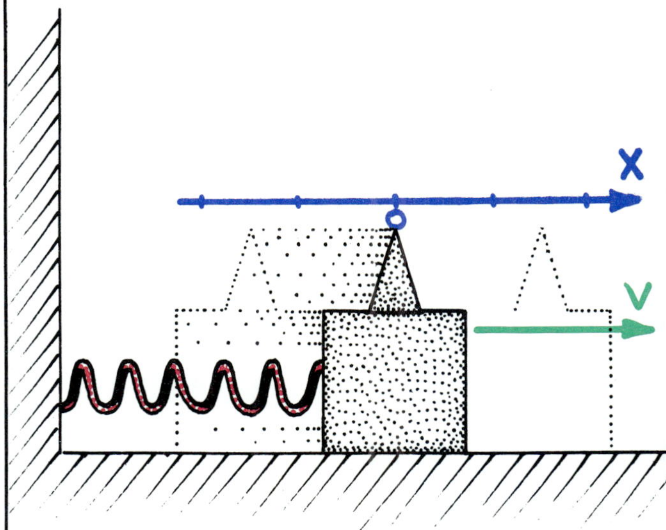

2.3.11. Next, in the single spring system, we assume the spring is linear. That is, we assume *Hooke's Law*: The force required to extend the spring a certain distance is a constant times that distance. Here is a graph of force versus extension, under this assumption.

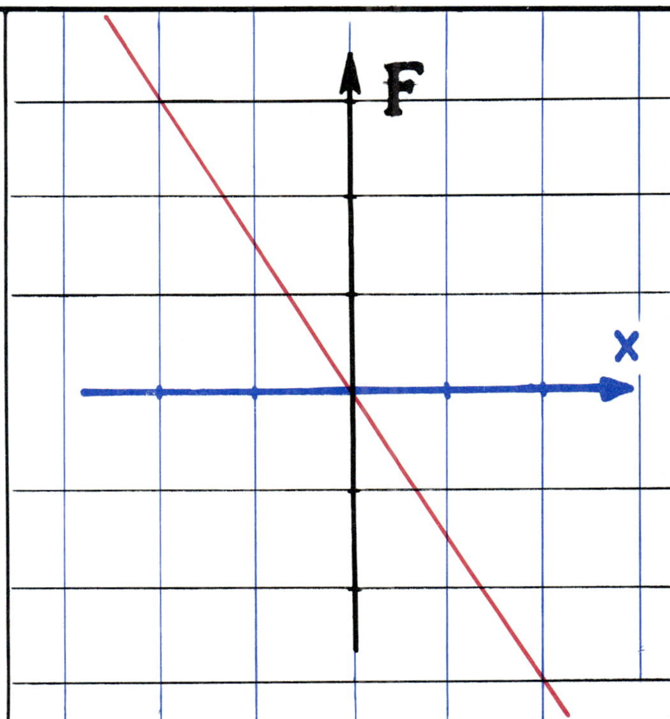

The mechanical system described here, with a linear spring, is called the *harmonic oscillator.* Without going through the mathematical analysis of this system, which is classical, we simply present the results.

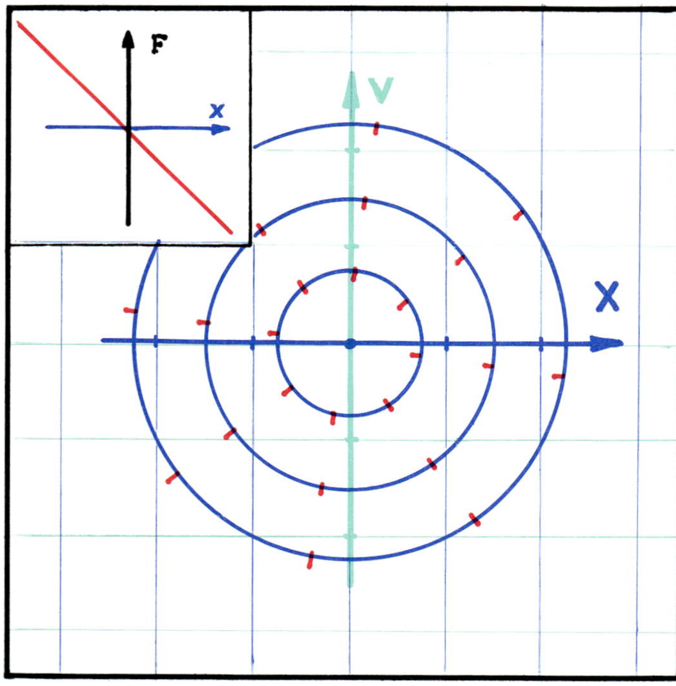

2.3.12. If after all there is no friction between the weight and the surface it moves upon, the phase portrait is a center. Further, all the concentric trajectories are periodic, with *the same period*. This is shown by the tick marks on the trajectories, in this illustration. Thus, no matter how hard you pluck the string, the note will have the same pitch. The spring characteristic (force versus extension) is shown in the inset.

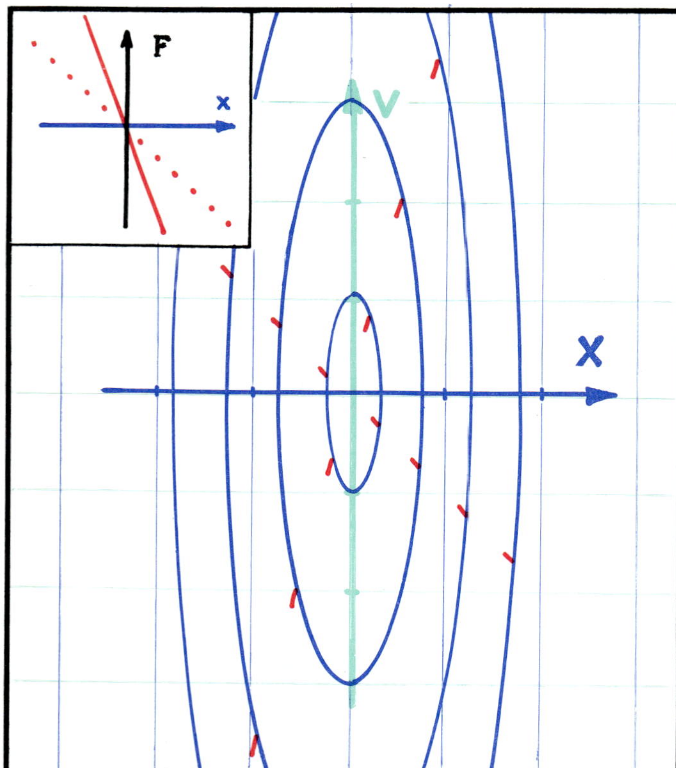

2.3.13. Tightening the guitar string corresponds to increasing the slope of the spring force versus extension graph. The steeper spring characteristic is again shown in the inset. The phase portrait is still a center, with concentric (more eccentric) elliptical trajectories. All these periodic trajectories still have the same period. But it is shorter than before. That is, the frequency of the oscillation is greater, and the pitch is higher.

The harmonic oscillator may be a poor model for a guitar string, for two reasons: (1) a guitar string is not linear, and (2) it is not frictionless either. Let's remove these objections one at a time.

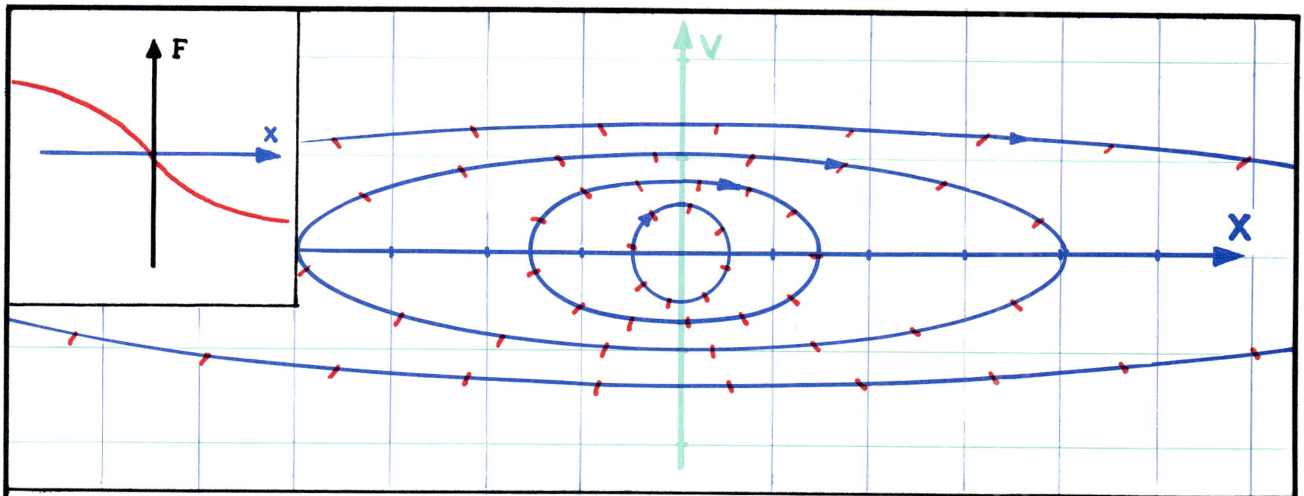

2.3.14. Suppose the string is not linear. The graph of force versus extension will not be a line. But we may still suppose that its deviation from linearity is symmetric. The simplest such deviation from linearity is a *cubic* one. Two cases have been extensively studied. In one of these, called the *soft spring*, the force is less than linear, as shown in the inset. The phase portrait in this case is still a center. But in this case, the eccentricity (and thus also the frequency) of the periodic trajectories depends upon the amplitude. The larger trajectories have a lower frequency. Thus, the pitch of the plucked note will be lower for louder notes than for softer ones—such a string may make a poor guitar.

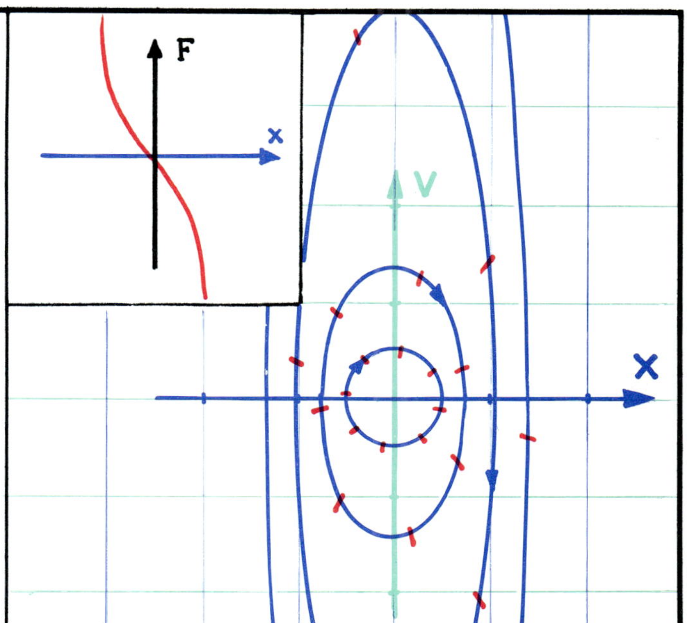

2.3.15. The other well-studied case of a nonlinear spring is the *hard spring*. Here, the spring force is a cubic function that is *more* than linear, as shown in the inset. The larger trajectories have a higher frequency in this case.

Now we deal with objection 2 by introducing friction between the weight and its supporting surface.

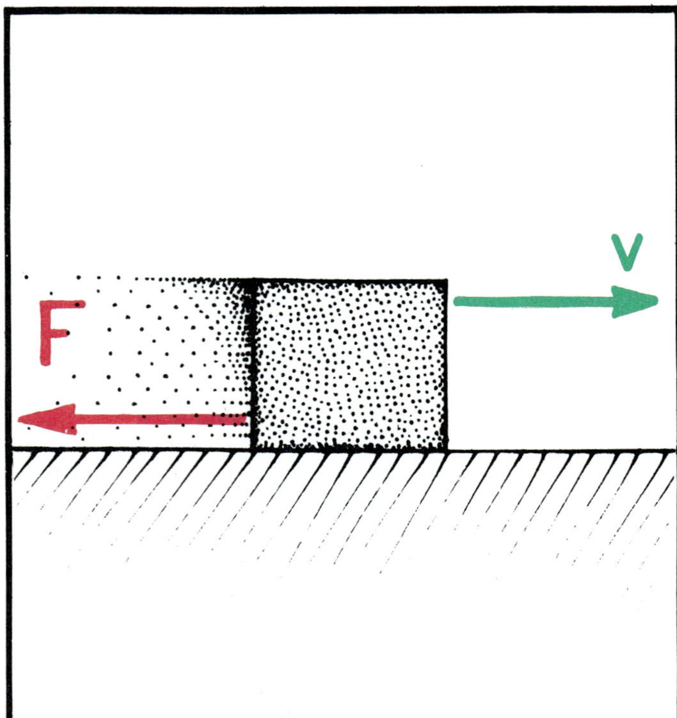

2.3.16. Friction, unlike the spring force, does not depend upon the extension of the spring. But it does depend on the velocity, and it works against the motion. We assume the table top is equally rough all over.

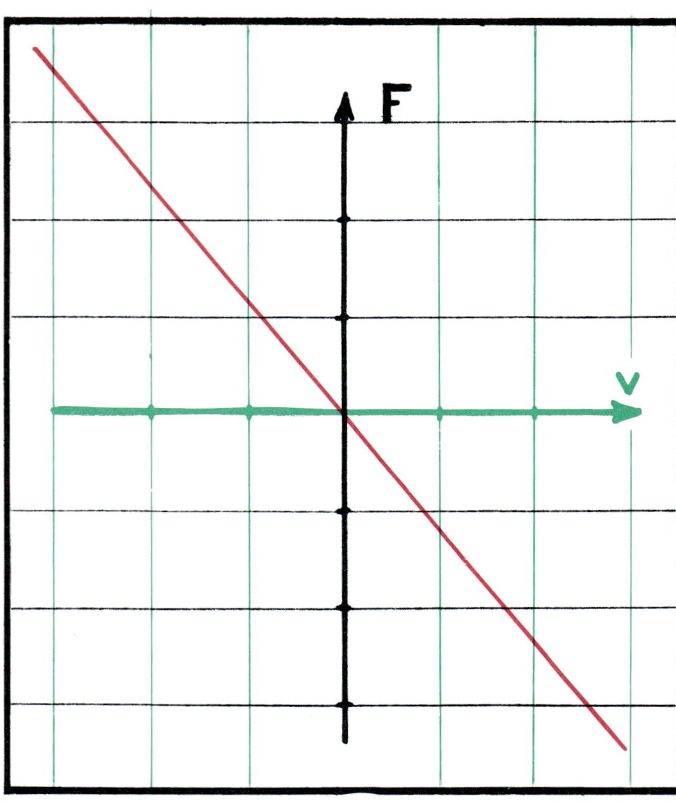

2.3.17. The force of friction, as a function of the velocity, need not be linear. But for the present, we assume it is. Thus its graph, as shown here, is a straight line. This system is called *the harmonic oscillator with linear damping.*

The mathematical analysis of the damped harmonic oscillator, likewise, is classical. Here again, we simply present the results of this analysis, as it was known to Newton. We return to the case of a linear spring, but add linear friction.

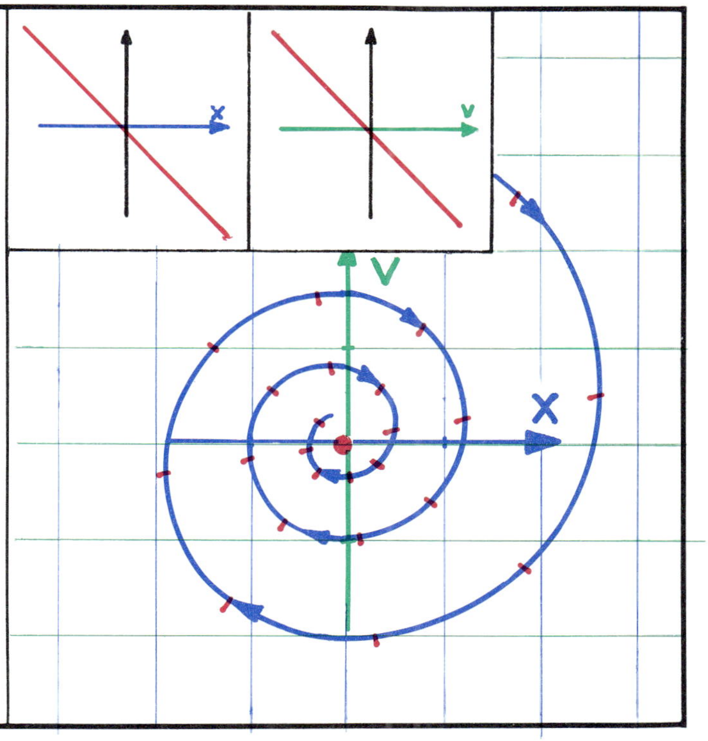

2.3.18. The phase portrait of the damped linear oscillator has a point attractor (focal point) at the origin. This is very like the damped pendulum. The linear spring and friction functions are shown in the insets. The damped oscillation has a constant period (hence also constant frequency or pitch) which is the same as the undamped system. The amplitude decays exponentially.

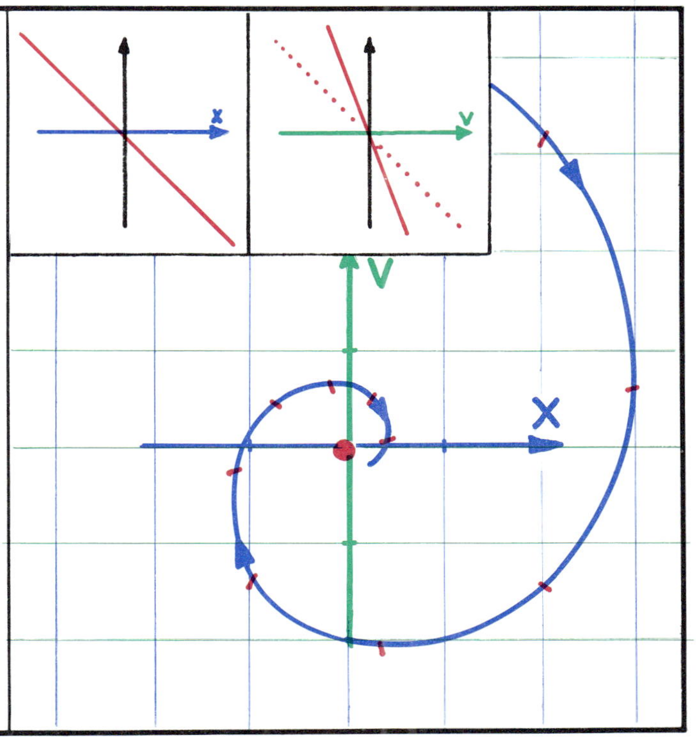

2.3.19. In the case of greater friction, the same spring will exhibit damped oscillation of the same period. But in this case, the amplitude decays faster. The spiraling trajectory approaches the focal point more quickly.

Now let's consider the spring model with both objections eliminated.

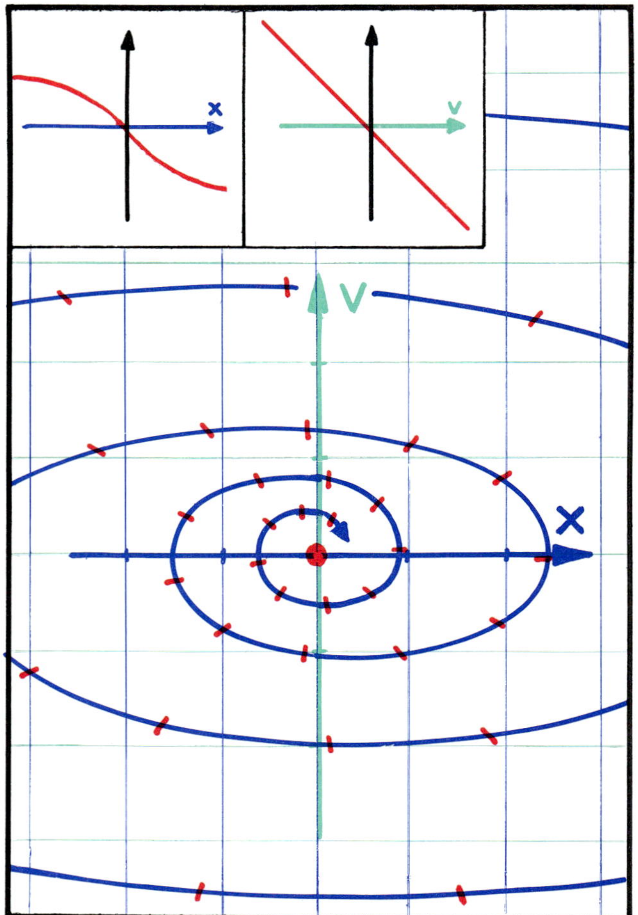

2.3.20. Here is the soft spring with linear damping. The spring and friction functions are shown in the insets. The pitch of a note rises as the tone dies away.

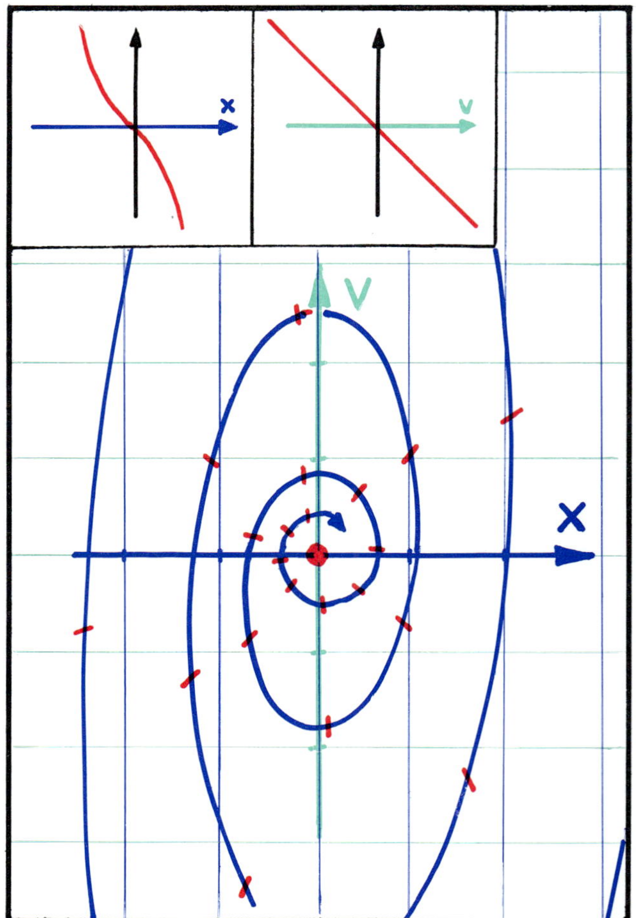

2.3.21. Here is the hard spring with linear damping. The pitch of a note falls as the tone dies away. Another poor guitar!

All the possible combinations and deviations from these simplifying assumptions have been explored, but by now you get the idea: A damped nonlinear oscillator is a reasonable model for a percussion instrument.

As a general rule, *attractors* model the observed states of the system.

Our first three applications violate this general rule. For in this case the transient response models the tone heard, the attractor models the silence that follows. And in the preceding examples the transient response modeled the damped oscillation of the pendulum or column, the attractor modeled the stillness that follows.

Finally, let's consider an extreme variation: what if the friction force were reversed?

2.3.22. Previously, we considered friction that is normal. That is, it drags, or pulls against the motion. The dashed line here represents an *inverse friction*: it aids the motion. And the faster the motion, the more the inverse friction aids it.

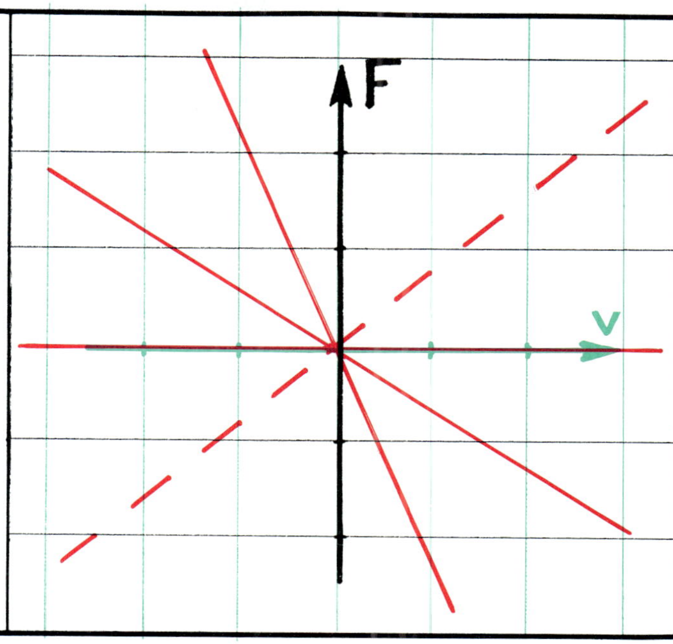

2.3.23. Here is the classical phase portrait for the harmonic oscillator with inverse friction. Again, the spring and friction graphs are shown in the insets. Here we have a *point repellor* at the origin. For any initial state, the alpha-limit set (asymptotic limit for the infinitely distant past) is the origin (no motion). The time series for any of these trajectories is an oscillation that grows (exponentially) with time. Presumably the spring breaks after a while.

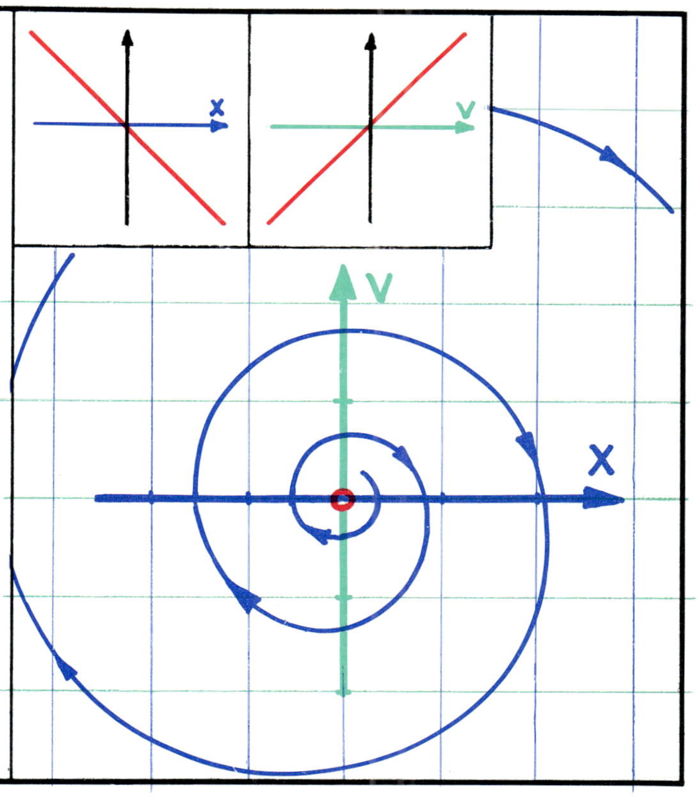

In the next chapter, we will find this unusual model useful.

2.4. Predators and Prey

In this section, we illustrate an ecological application. This is a 1925 classic, due to Lotka and Volterra, the early pioneers of mathematical biology.[3] Consider a fictitious ecosystem, containing substantial populations of two species only — say big fish and small fry — along with a large supply of food for small fry. The choice of a state space for this application is easy.

The number of small fry and the number of big fish, respectively, are represented as coordinates in the plane.

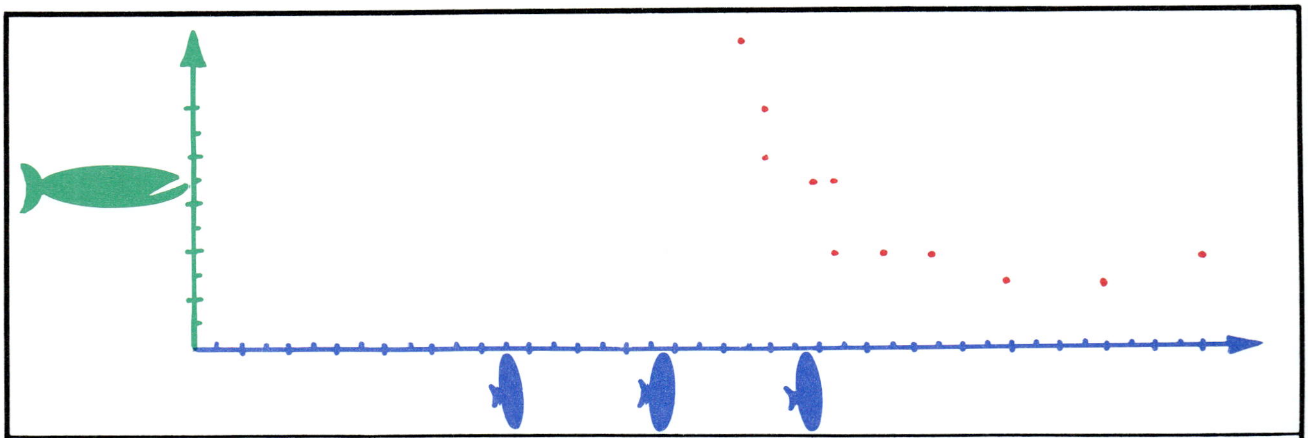

2.4.1. To apply the modeling concepts of dynamics, the dotted lines must be idealized into continuous curves by interpolation.

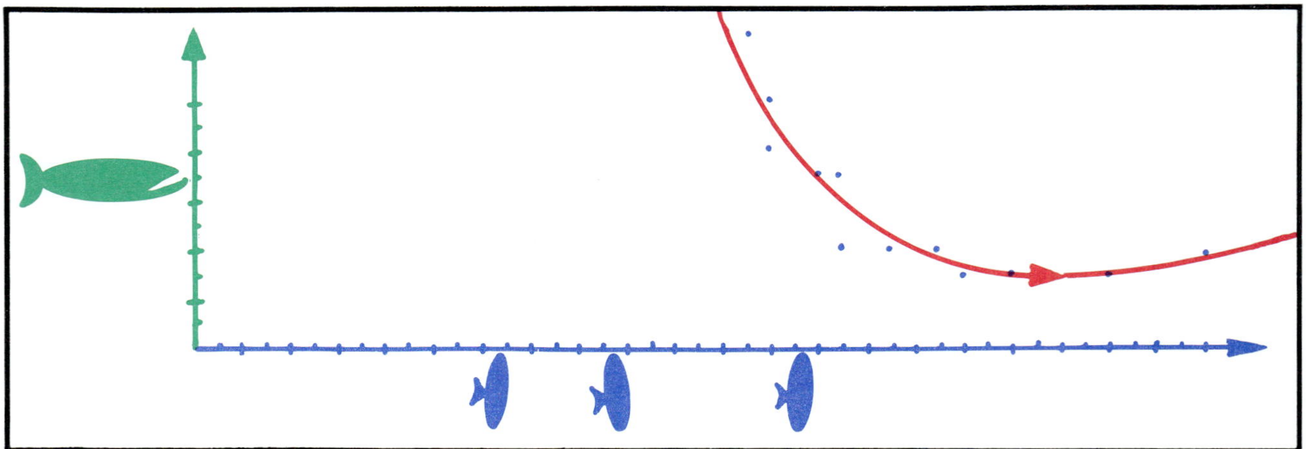

2.4.2. The observations, over time, of the two populations describe a dotted line in the plane. Births and deaths change the coordinates by integers, a few at a time.

The dynamical system for this model, the *Lotka-Volterra vectorfield,* can be roughly described in four regions.

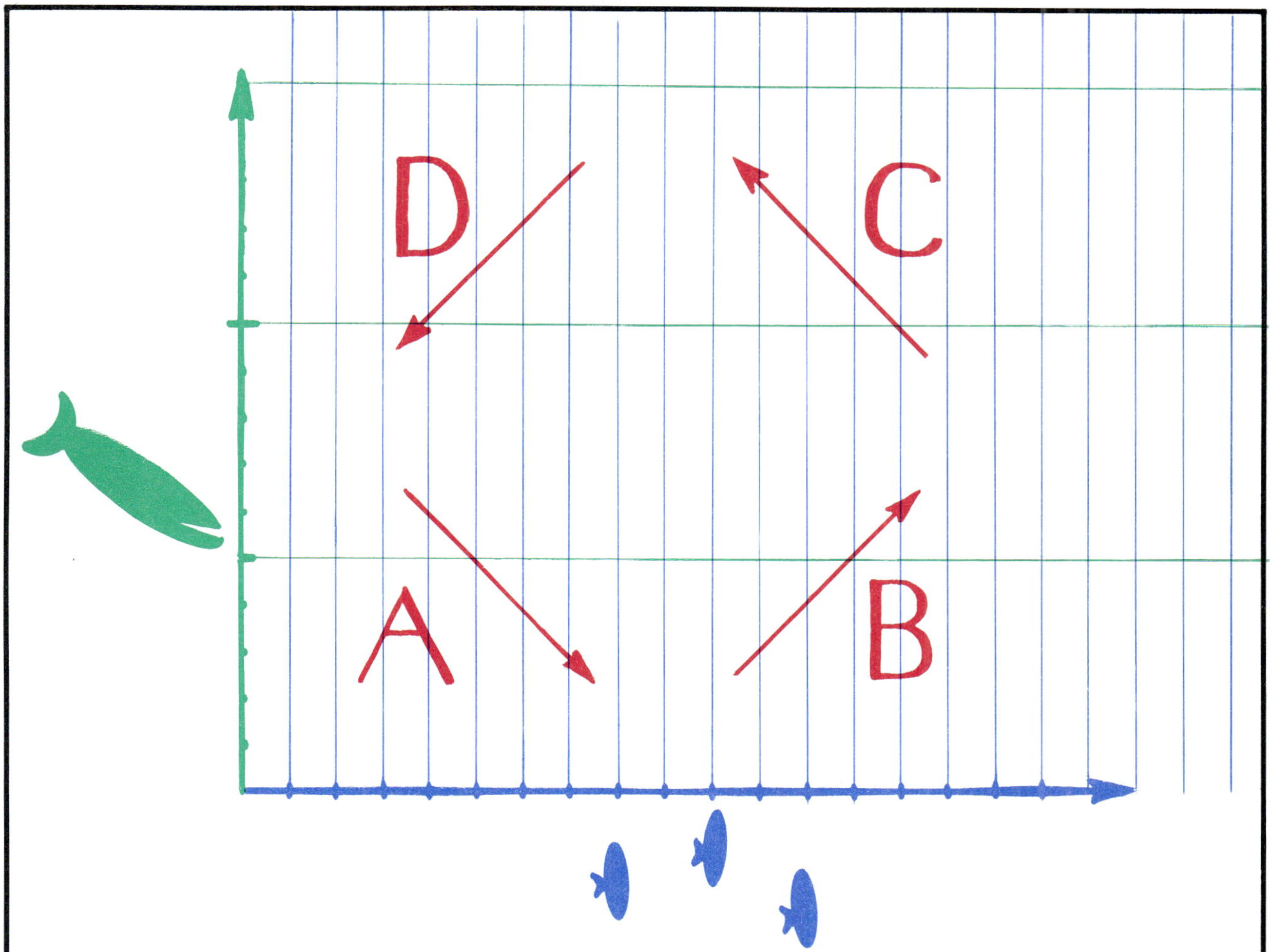

2.4.3. Region A. In this part of the state space, both populations are relatively low. When both populations are low, big fish decrease for lack of food (small fish) while small fry increase thanks to less predation. This is the habitual tendency for states in this region. The interpretation of this tendency as a bound velocity vector is shown in region A.

Region B. In this region, there are many small fry but relatively few predators. But when there are many small fry and few big fish, both populations increase. This is interpreted by the direction of the vector shown in region B.

Region C. Here both populations are relatively large. The big fish are well fed and multiply, while the small fry population declines drastically. This tendency is shown by the vector in region C.

Region D. In this part of the state space, there are few small fry but many big fish. Both populations decline. This tendency is shown by the vector inn region D.

The Lotka-Volterra vectorfield is not just *some* vectorfield with these features, it is a *particular* one, which seemed the simplest choice at the time.

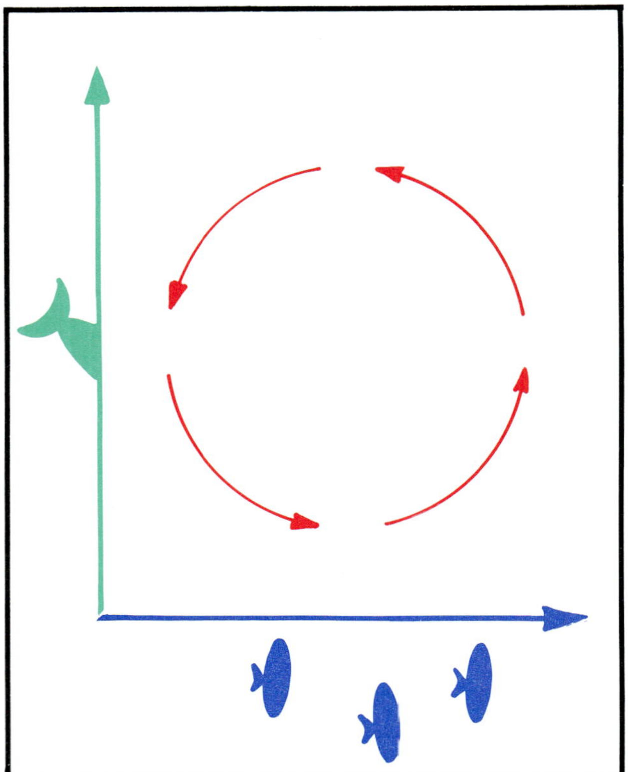

2.4.4. The phase portrait of this system can be visualized, in part, from these features: the flow tends to circulate counterclockwise. The ecologist would like to know what happens to the two populations *in the long run*.

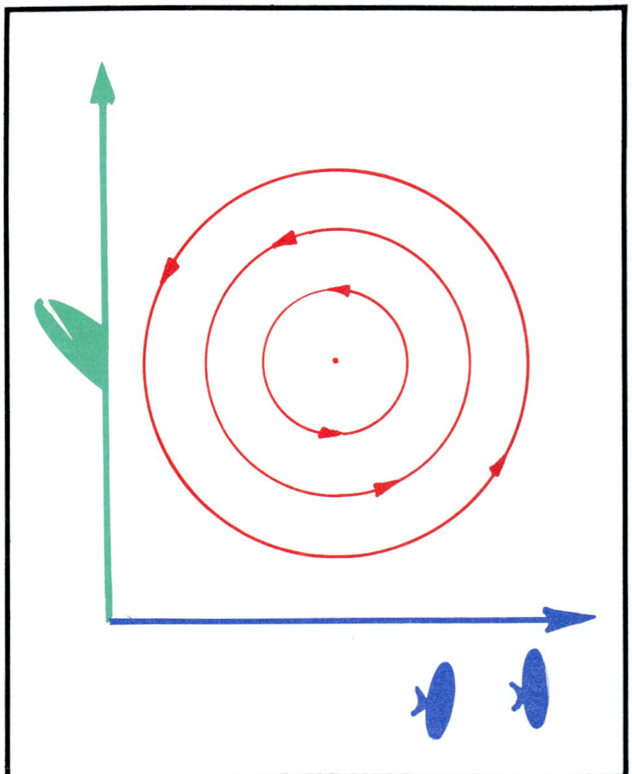

2.4.5. The answer, in this case, is a prediction of the mathematical model. The phase portrait is a *center*: a nest of closed trajectories around a central equilibrium point.[4]

Conclusion: every trajectory is periodic. Each initial population of big fish and small fry will recur periodically.

Now that the modeling process had been described, we may return to the question: why bother? This question has an exceptionally convincing answer, which accounts for the numerous examples of the process now proliferating in the literature of applied dynamics: *dynamical systems theory tells what to expect in the long run.* In this case, the two populations persistently oscillate. The same cycle of population numbers, for both species, will recur indefinitely, each time with the same elapsed time, or *period*. This is an example of a *prediction forever.* The periodicity of fish populations in the Adriatic Sea actually inspired Volterra to make his model.

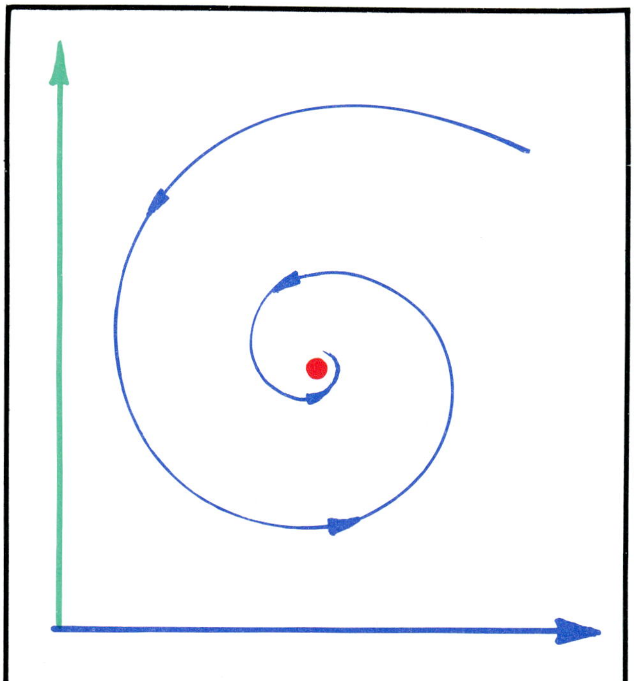

2.4.6. If some kind of ecological friction were added to the model, the center would become a focal point attractor. This would be a reasonable model for an ecological system in static equilibrium.

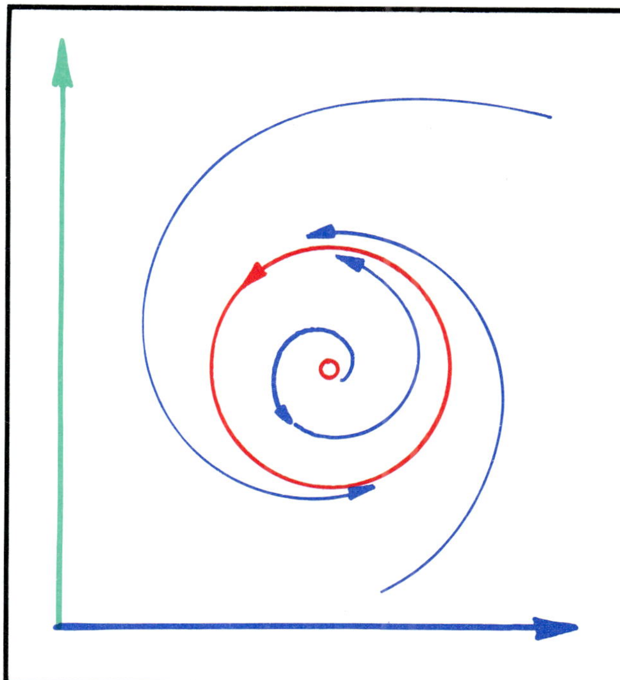

2.4.7. A more subtle modification of the model could result in a phase portrait like this, with only one periodic trajectory. This would be a more satisfying model for the observed periodicity of the fish in the Adriatic.

These improvements to the Volterra and Lotka model have been made recently.[5] But the type of phase portrait on the right, with the limit cycle, was well-known to Lord Rayleigh, as we shall see in the next chapter.

3

Vibrations:
Limit Cycles in 2D from Rayleigh to Rashevsky

The history of dynamics, from Pythagoras to the present, has been enlivened by music. Until around 1800, this history was dominated by the limit point concept. Then, Chladni's experiments with musical instruments attracted Napoleon's attention. And the limit cycle idea began to grow in the consciousness of the scientific community. This is an abstract analog of the discovery of the wheel. In this chapter, we present the key dynamical steps of this bifurcation in the history of science.

89

The damping depends on the string only. It is not changed by the bow. The violinist uses rosin on the bow, making it sticky. This changes the shape of the curve describing the friction as a function of the velocity. To get the idea, we consider a weight moving on a sticky tabletop.

3.1. Wind Instruments

Following his analysis of the percussion instruments, Lord Rayleigh went further. In his attempt to explain all aspects of sound, he created a successful model for the sustained instruments. He managed to combine inverse friction to small motions with normal friction for large motions in a single dynamical system. The result is a simple example of ***self-sustained oscillation***. The same model turned out, 45 years later, to be useful in the field of radio frequency electronics. This later application is described in the last chapter. Here, we resume our story in 1877.

3.1.1. What do these instruments have in common with a radio transmitter? As long as you don't run out of juice, they keep on playing.

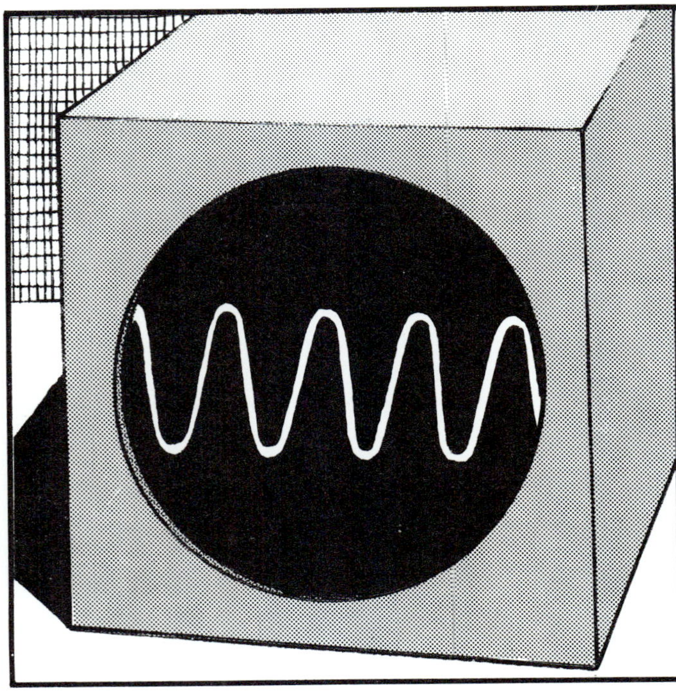

3.1.2. The sound of a sustained instrument, portrayed as a time series (amplitude of air motion versus time) by an oscilloscope for example, is a periodic function that does not decay in time. As long as the player puts energy into the instrument, the oscillation may be sustained at the same loudness (amplitude).

The dynamical model must have a closed trajectory, or periodic attractor, with this function as the time series for a preferred parameter. For the sake of definiteness, let's choose a clarinet reed as the target of the model.

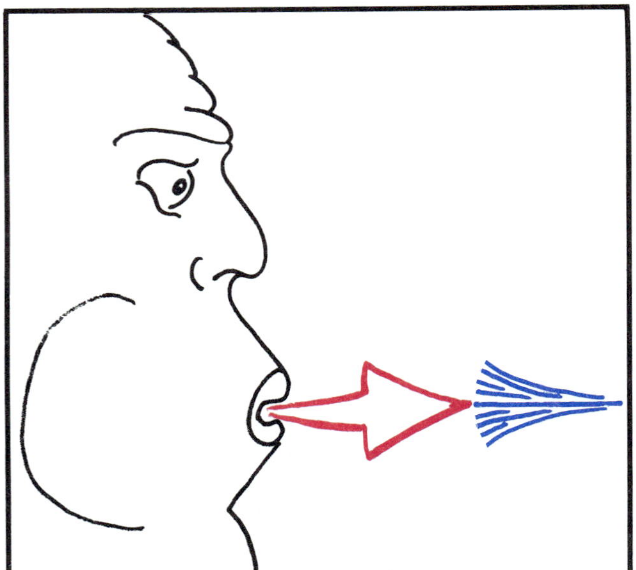

3.1.3. By blowing along the reed, the clarinetist adds energy to the system, sustaining the vibration.

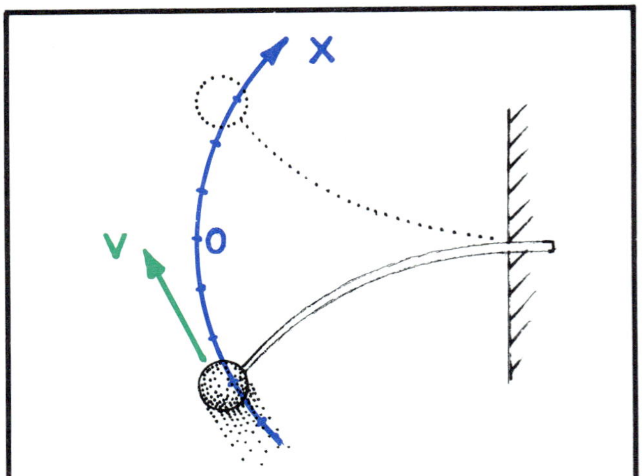

3.1.4. The 1870's style model for the reed is a flexible wand, with a concentrated small weight at the end. Somehow, we agree on a way to measure the amount the reed is bent and its velocity (rate of change of this amount).

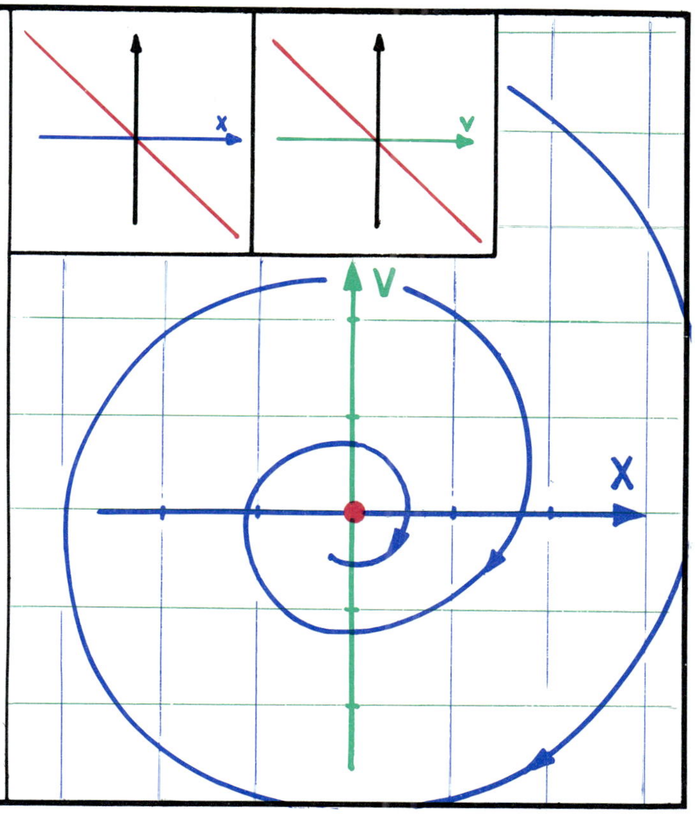

3.1.5. With no blowing, the wand is a type of pendulum. A reasonable dynamical model will look like this. A point attractor of spiral type is located at the origin. This is the spring model from the preceding section. The characteristic functions describing the damping and the spring are displayed in the insets.

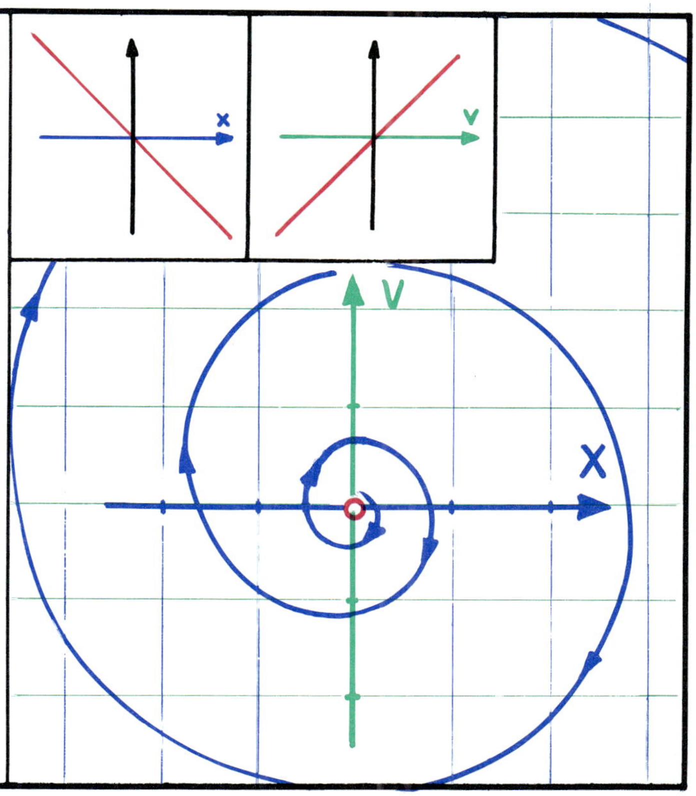

3.1.6. Rayleigh modifies the spring model to include the clarinetist by replacing the behavior near the origin by inverse friction. As described at the end of the preceding section, the origin becomes a point repellor. The behavior far from the origin is not changed.

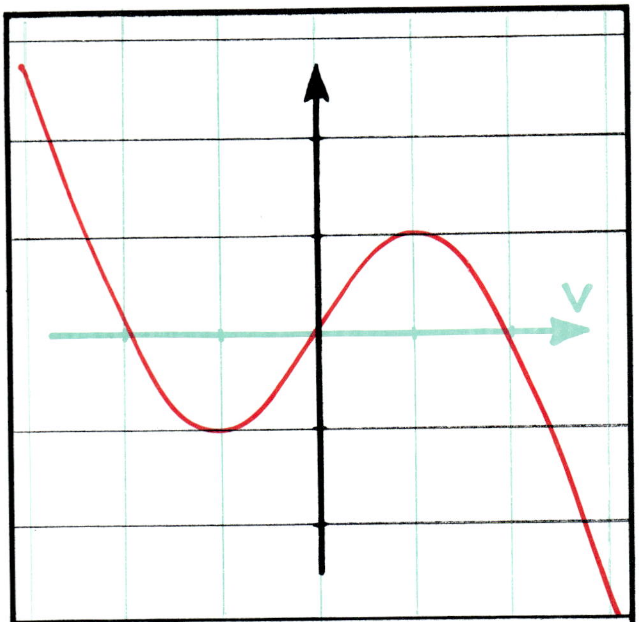

3.1.7. A simple way to mate normal friction for large motions with inverse friction for small motions is with this characteristic curve, a cubic (polynomial of degree three). This is the simplest curve with negative slope near the origin, and positive slope far away.

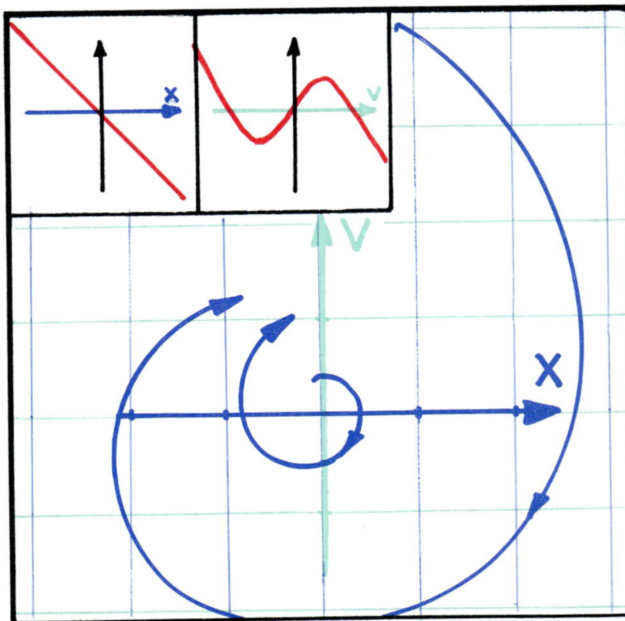

3.1.8. The resulting phase portrait has a poin repellor at the origin. Yet far from the origin, all the trajectories are spiraling in.

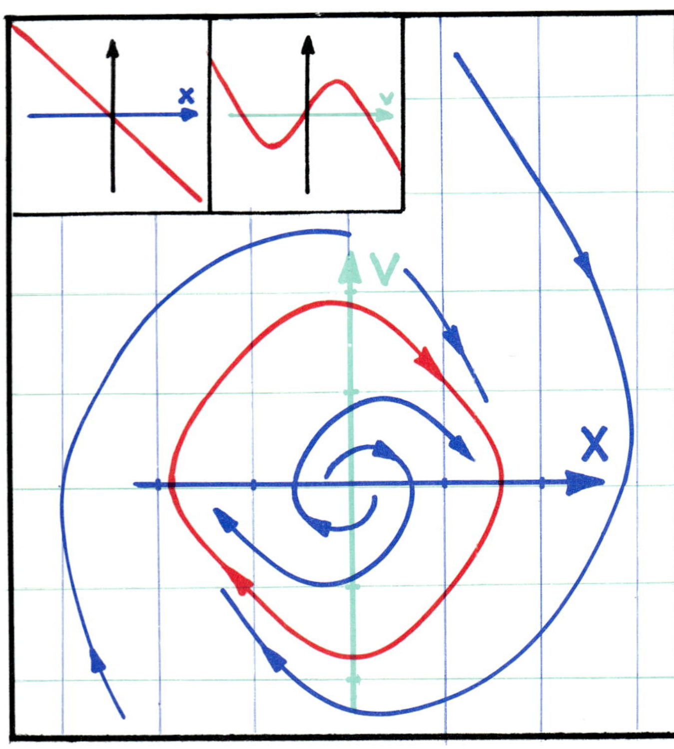

3.1.9. Between the distant spiraling in and the central spiraling out, a periodic trajectory (limit cycle) is trapped. Although this was obvious to Lord Rayleigh, mathematicians succeeded in proving it to their own satisfaction only about 50 years later.

This limit cycle is the dynamic model for the sustained oscillation of the blown clarinet reed. What is the relationship between the parameters in the model and the sound of the clarinet?

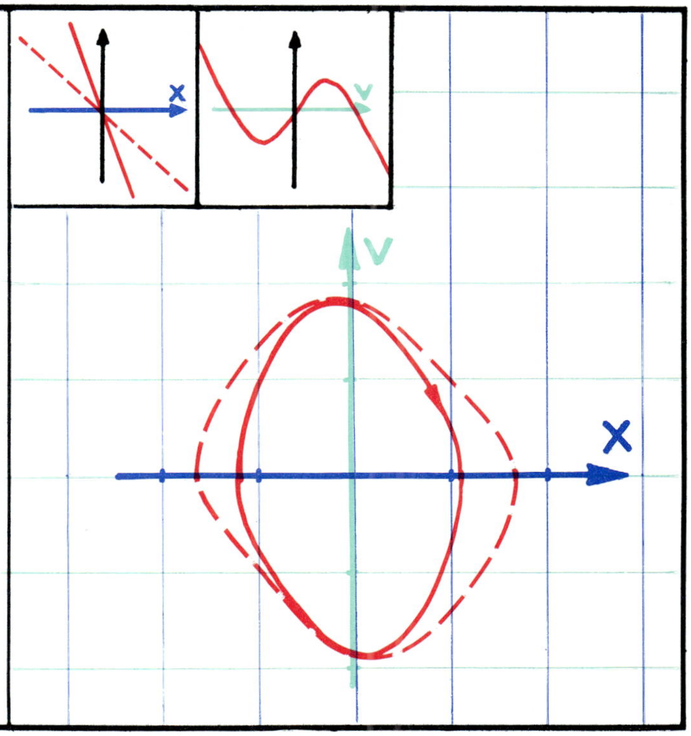

3.1.10. A stiffer reed is modeled by a stronger spring. The characteristic function of the spring is a steeper line, as shown in the inset on the left. The limit cycle has a different shape, and the tone (timbre) of the clarinet is richer.

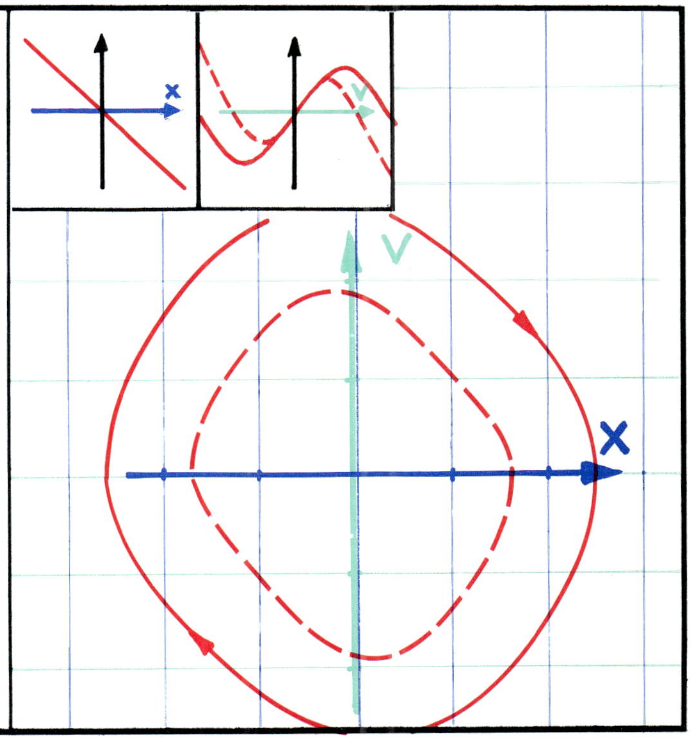

3.1.11. Blowing harder is modeled by a broader friction characteristic, as shown in the inset on the right. The limit cycle is larger, and the tone of the clarinet is louder.

This example of Lord Rayleigh's turned out to the be most important single item in the dynamics field for a century. To get more familiar with it, let's start again, this time with a violin.

3.2. Bowed Instruments

Many musical instruments produce sustained tones with the bowing mechanism. Besides the violin, cello, bass, and so on, Lord Rayleigh mentions the wine goblet, bowed by a finger on the rim. His great precursor, Chladni, applied his experimental bow to plates of glass, in hopes of creating new instruments.

To choose one example, let's consider the violin.

3.2.1. Ignoring the bow, the string of the violin is modeled by this nineteenth-century gadget. This is identical to the guitar string model of the preceding section. Again, we use x to denote values of the displacement of the spring (that is, the string at the point of bowing) and y for the velocity, that is, the rate of change of the displacement.

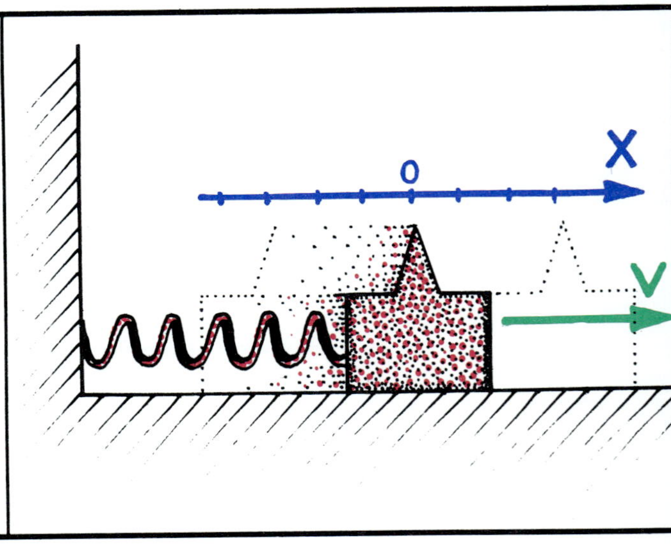

3.2.2. The phase portrait has, again, a *focal point attractor* at the origin. The *frequency* (rate of spiraling) and the rate of decay are determined by the characteristic functions of the spring (left inset) and the friction (right inset). Both of these are aspects of the violin string itself and of its tightness.

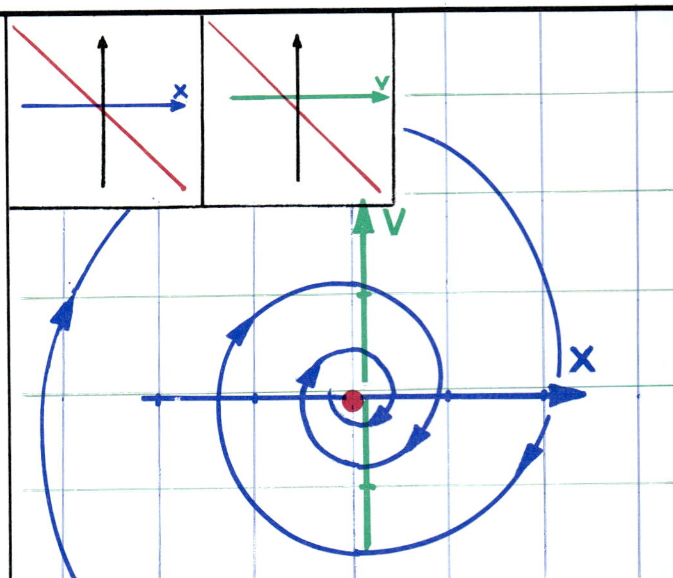

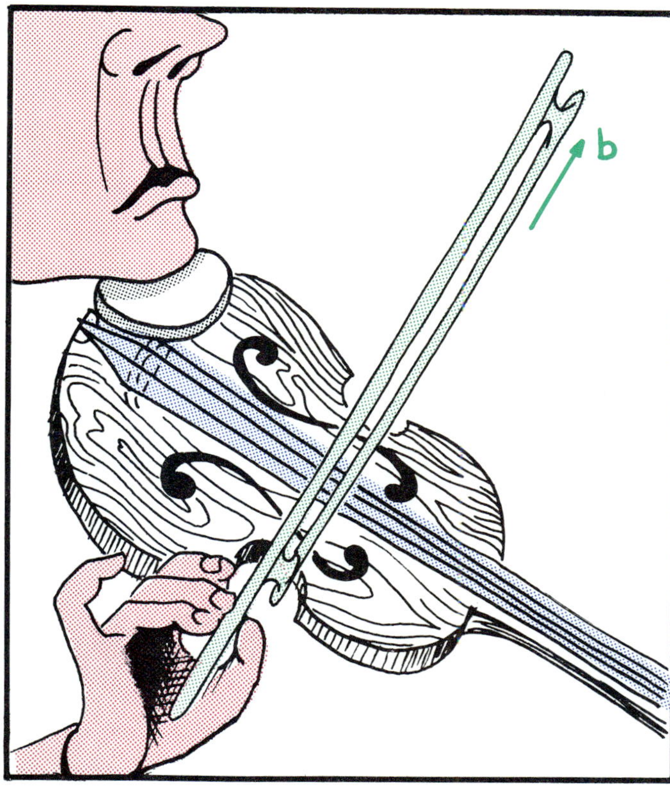

3.2.3. The violinist sustains the vibration by putting energy into the string with the bow. The friction of the bow on the string depends on the rate of bowing. We introduce a new symbol, b, to denote the rate of drawing the bow across the string.

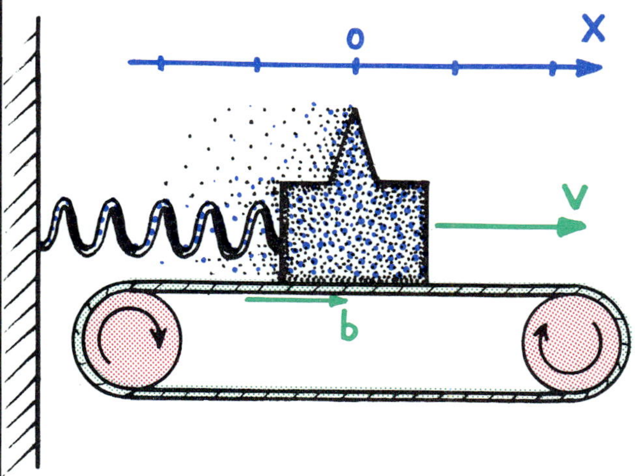

3.2.4. The spring model may be simply modified to include the action of the bow. Replace the tabletop on which the spring slides by a conveyor belt. This represents the bow. The weight, as before, represents the violin string.

The damping depends on the string only. It is not changed by the bow. The violinist uses rosin on the bow, making it sticky. This changes the shape of the curve describing the friction as a function of the velocity. To get the idea, we consider a weight moving on a sticky tabletop.

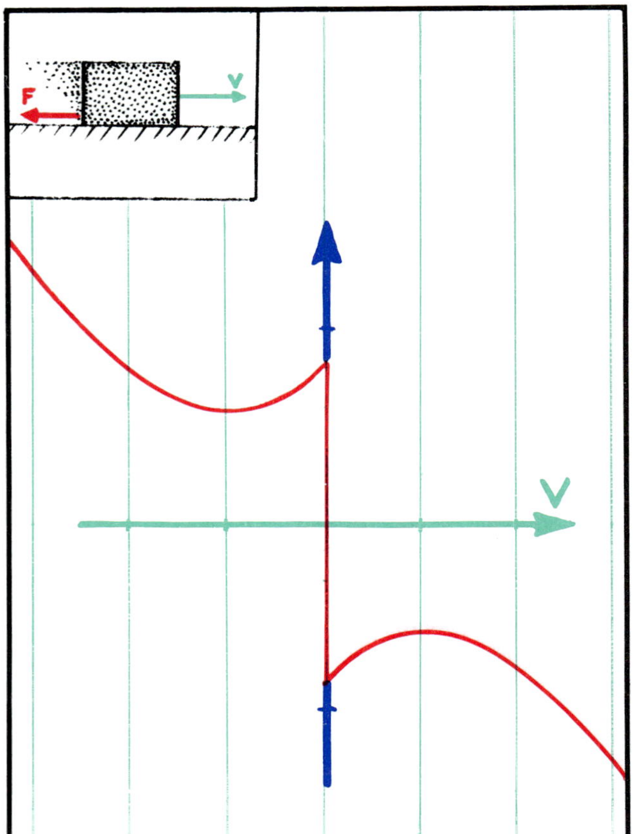

3.2.5. Further, suppose the tabletop is very sticky with rosin, and we just begin to push the weight lightly to the right. The speed is zero. But the force of friction is building up as we push. Suddenly it slides, as the force reaches a critical value. The same thing happens if we pull instead of push. This experimental situation is represented in the mechanical model by the shape of the characteristic function of friction shown here.

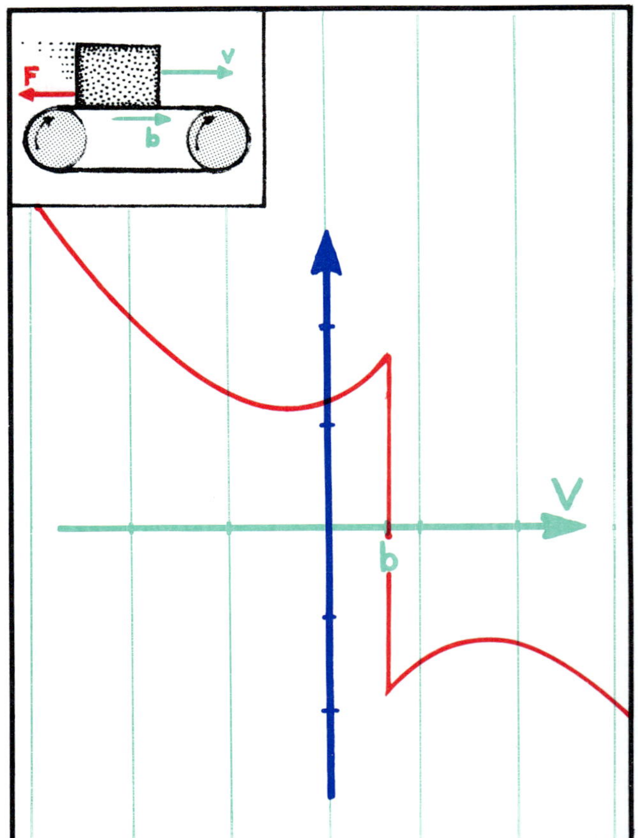

3.2.6. In the preceding discussion, we assumed the weight had hardly begun to move. Now, let's imagine the same experience—sticking, pulling up to critical force, then slipping—while the tabletop is moving relative to the weight with speed b. This situation is modeled by this friction function, obtained from the preceding example by sliding the graph to the right. You see, the tabletop is moving at speed b. So when the weight sticks to the tabletop, it must be moving at the same speed, $v = b$. Thus, the vertical segment of the graph must be located at the common velocity.

Identifying the tabletop with the bow, and the spring-weight system with the violin string, we now have a dynamical model for the bowed string, The dynamical model corresponds to a 19-century mechanical model: a weight, fixed to a linear damped spring, oscillating on a conveyor belt. This dynamical model is the same as that for the clarinet reed, except that the smooth (cubic) friction function for the blown reed is replaced by this one with a glitch. The glitch is located at the speed of the bow. We have only drawn the function for one bowing speed. This happens to be positive. That is, the violinist is pushing toward positive deflection. We do not have, in this case, a precise function in mind. We just assume that it is shaped something like this.

Assuming the friction function characterizing the bowed violin string looks something like this, what can we deduce about the phase portrait of the dynamical system? (If you just want a quick answer to this question, you could skip to the end of the section.)

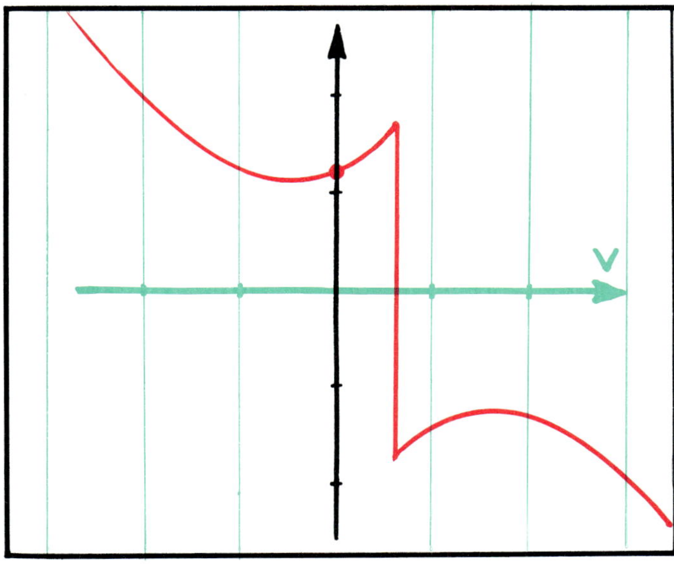

3.2.7. When the motion pauses, the velocity is zero. The force of friction, read from the graph here, is then some number, $F(0)$. This friction is independent of the displacement of the spring. It depends only on the velocity.

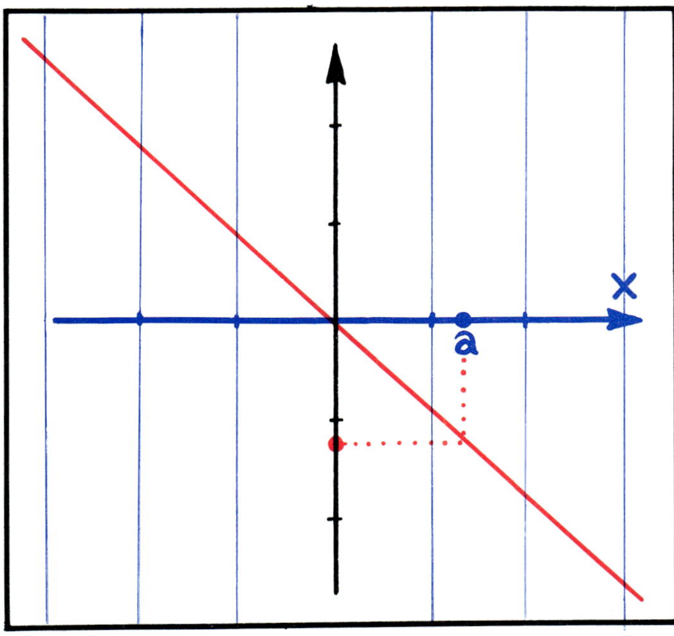

3.2.8. Meanwhile, the force of the spring depends linearly on the deflection. It is independent of the velocity. For some certain deflection, $x = a$, the spring force will be $-F(0)$.

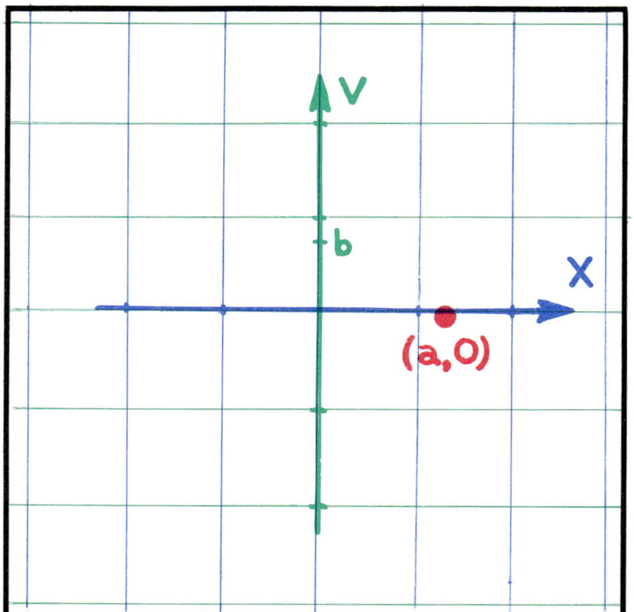

3.2.9. At this special deflection, and at zero velocity, the friction force exactly balances the spring force. The phase portrait has a critical point at $(a,0)$.

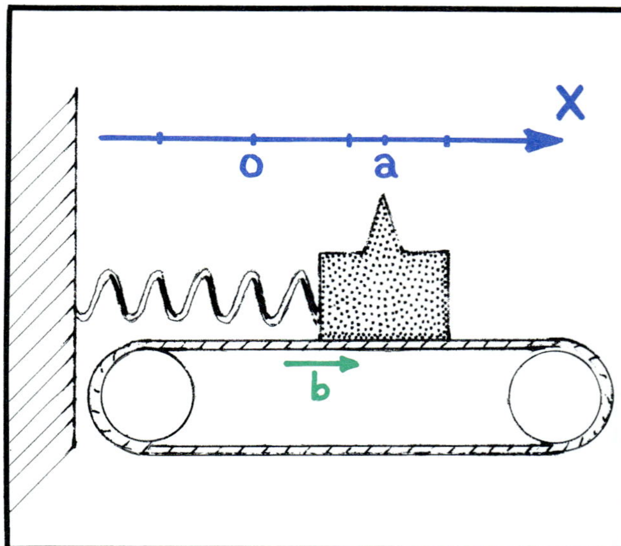

3.2.10. Here is the mechanical model at the critical point. The bow is moving to the right at its fixed speed b. The weight has paused ($v = 0$) at the critical displacement ($x = a$) where the friction force is balanced by the spring force.

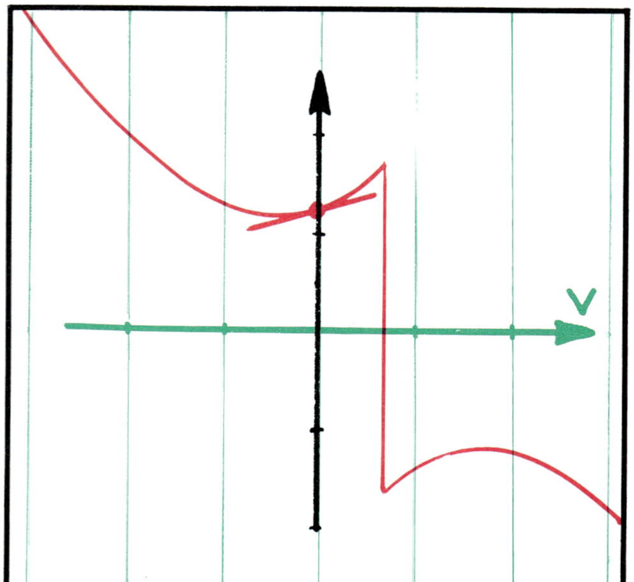

3.2.11. At this critical point, the inclination of the friction function is positive. This is the situation called *inverse friction* at the end of Section 2.3.

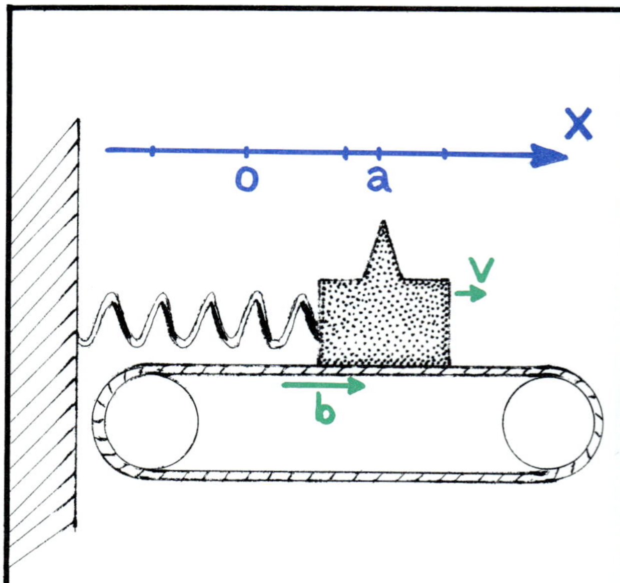

3.2.12. This means that a small motion of the weight to either side will create a runaway oscillation, as explained at the end of Section 2.3.

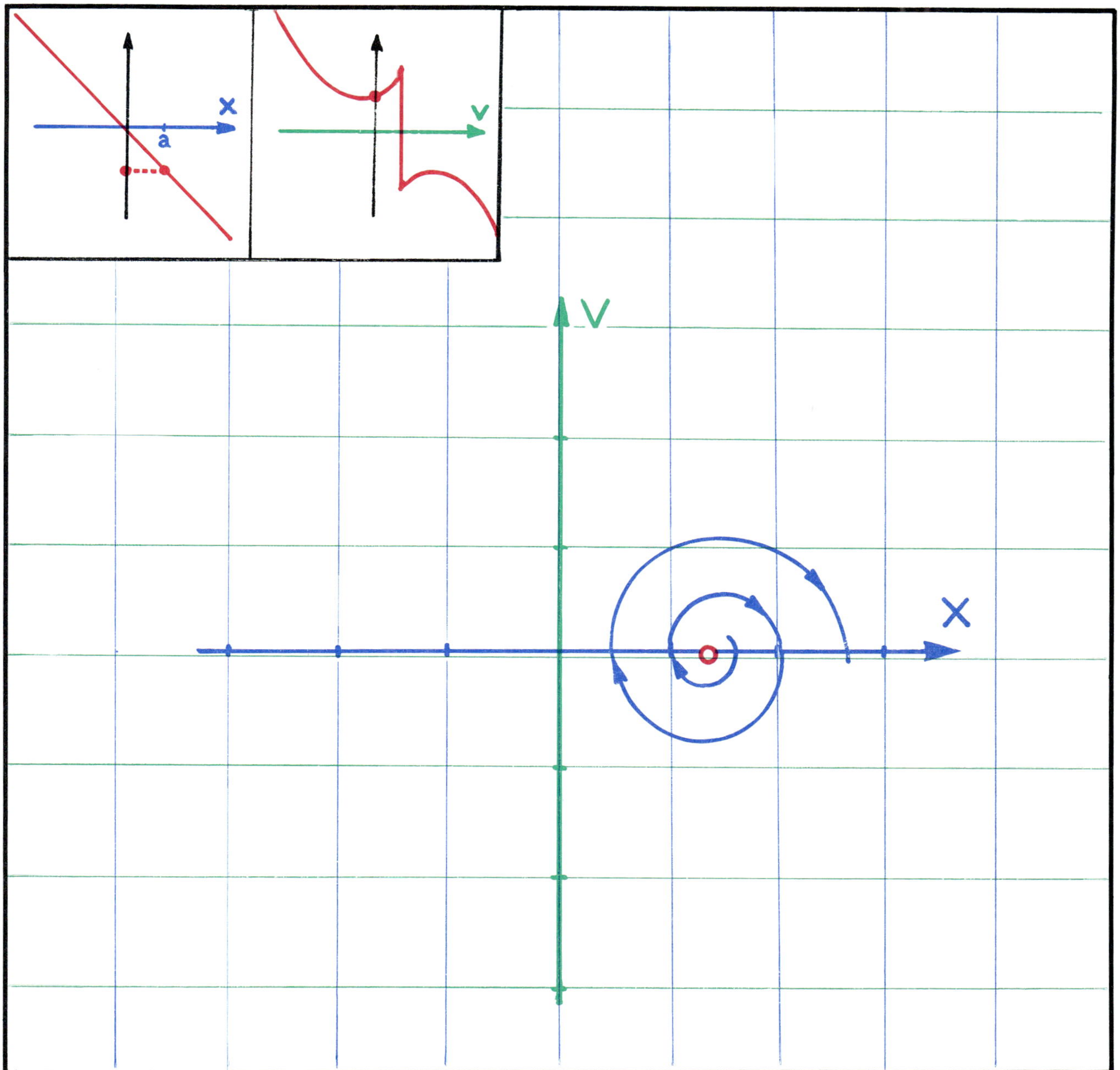

3.2.13. So far, we have figured out that there is a critical point in the phase portrait of our dynamical model for the bowed violin string, and that is a *focal point repellor.*

As the small motions about the critical point grow, and the large motions decay, it seems plausible that there is a limit cycle in the phase portrait for the bowed violin string. This would be just like the model for the blown clarinet reed. Unfortunately, this cannot be proved without assuming more about the shape of the friction function.

Assuming that there is a limit cycle in the phase portrait, what does this mean the violin string is doing when it is bowed? It means the endless cycle of states shown here: 1,2,3,4,1,2,3,4, and so on. *Note:* The spring force graphs are drawn upside down here for easier comparison with the adjacent friction force graphs in the inserts.

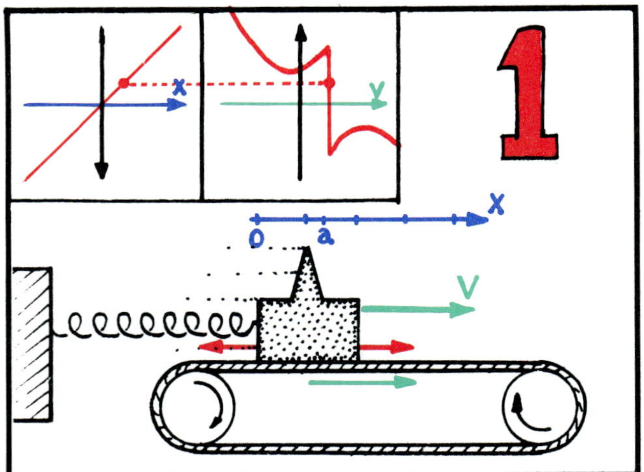

3.2.14. 1. STUCK. The friction and spring forces are balanced and the weight is stuck to the belt. The weight is a little to the the right of zero (equilibrium of the spring) but not as far as the critical point.

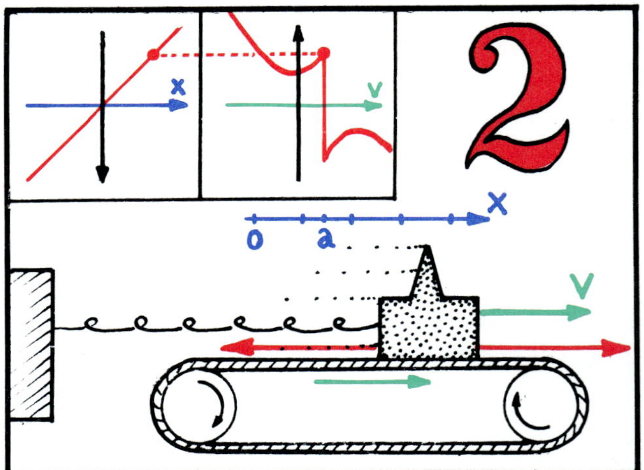

3.2.15. 2. BEGINNING TO SLIP. The friction and spring forces are balanced, but larger, as the displacement increases. At the critical force for friction, slipping begins.

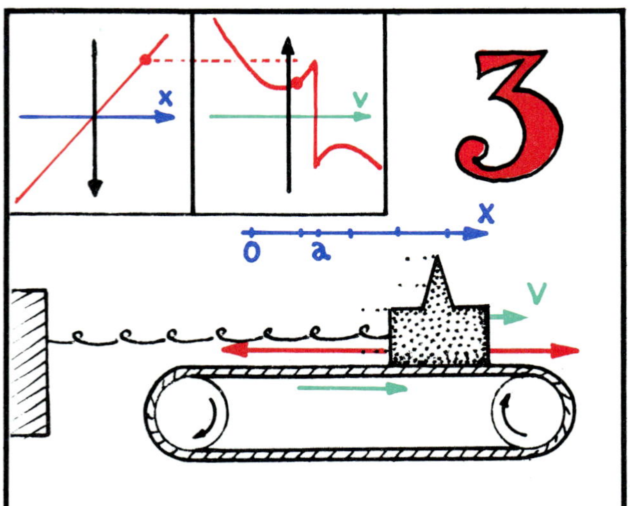

3.2.16. 3. SLIPPING. When the velocity begins decreasing, while x is still increasing, then the sudden drop in the friction yields the tug of war to the spring. The acceleration is negative. Rightward motion slows to a halt, and the weight begins to move back to the left.

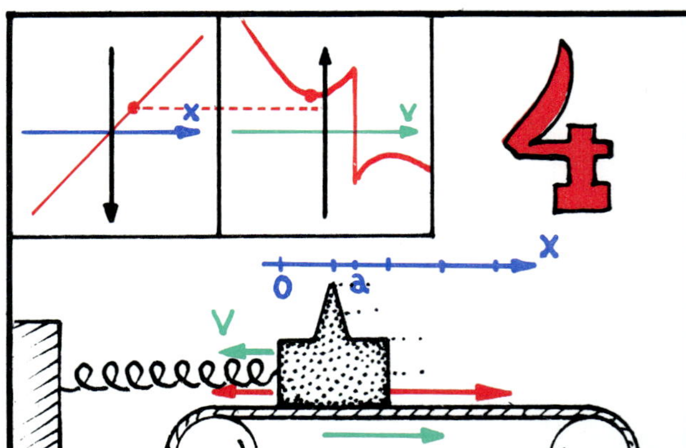

3.2.17. 4. GRABBING. When the leftward motion has decreased the spring force to a value smaller than the slipping friction, the tug of the belt wins once more. Motion to the left slows, and the weight turns and begins once again to move to the right. When the velocity reaches the critical value (the red dot on the glitch) in the friction function, slippage is going to happen again (return to 1).

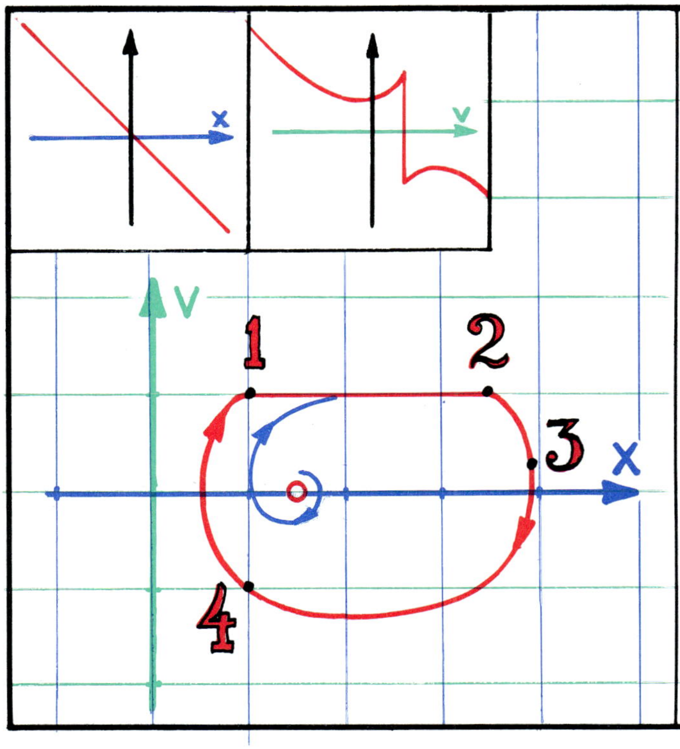

3.2.18. Here are the four stages in the cycle, located in the phase portrait. The rest of the limit cycle has been interpolated. The flat part at the top corresponds to the stuck phase. The repelling critical point is inside the cycle. Initial states near this repellor will spiral outward, clockwise, approaching the self-sustaining oscillation.

What happens outside the cycle? Let's make a change of scale, so the cycle is only about one-tenth its former size.

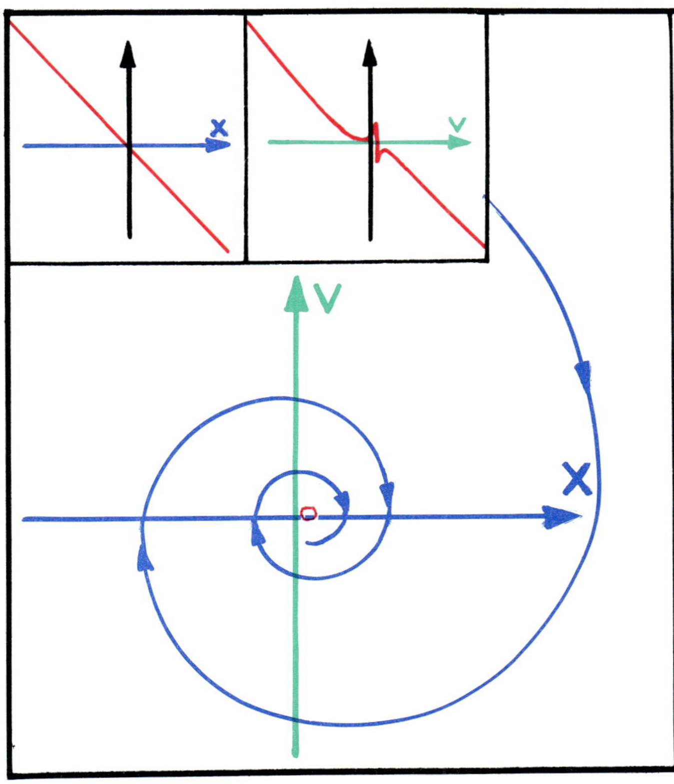

3.2.19. Now the inverse friction region is smaller than grape seed and is almost at the origin. The friction function looks essentially linear and normally dissipative. The phase portrait has, roughly, an attractive point near the origin. Actually, it is not a point. It is an attractive region about the size of a grape seed or so, containing a limit cycle. The initial states in this picture, corresponding to very large scale motions of the weight on the conveyor belt, will decay to the vicinity of the limit cycle (the seed). So essentially, the closed trajectory is attractive (a limit cycle), as we have assumed all along.

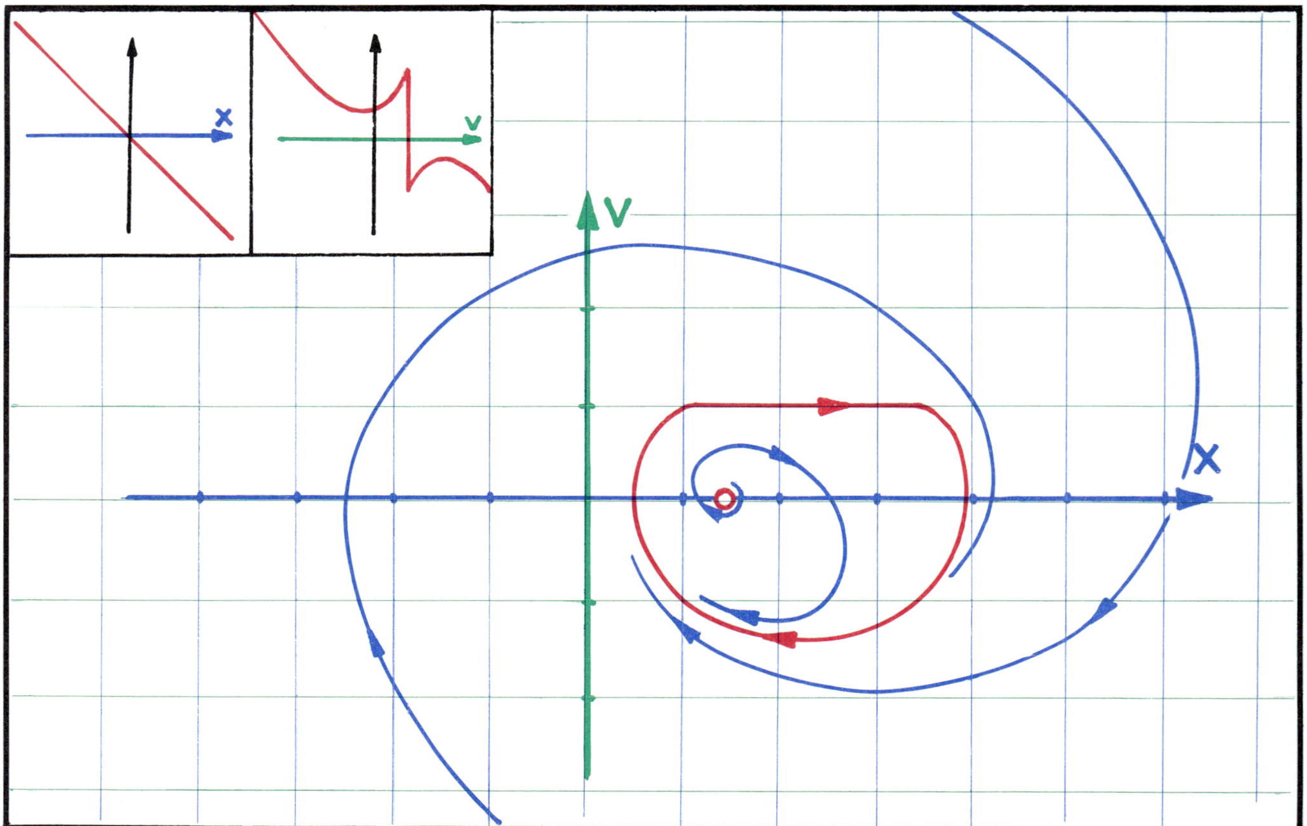

3.2.20. Returning to normal scale, here is the complete phase portrait for the dynamic model for the bowed violin string.

All this was child's play for Lord Rayleigh. He managed to make mechanical models very simply, learn from them, translate back and forth to the symbolic expressions of differential equations, and relate them equally well to electrical models. We will refer to the dynamical system used by him as a model for self-sustained oscillations—equally for wind or bowed instruments, or electrical oscillator of Helmholtz—as *Rayleigh's system.* It comes up again in the next section, as *Van der Pol's model for electronic oscillations,* and again in Chapter 5.

3.3 Radio Transmitters

Rayleigh had already observed that his model for self-systained mechanical oscillations applied equally well to an electrical oscillator suggested a few years earlier by Helmholtz. Among Rayleigh's followers the early experimentalists, Duffing and Van der Pol were particularly influential. Duffing was especially interested in mechanical vibrations, while Van der Pol worked with the first electronic oscillators based on vacuum tubes. In the next two chapters, we will describe the main results of these two experimentalists. At this point, the work of Van der Pol provides a second example of the representation of an oscillating physical system by a dynamical model with a periodic attractor.

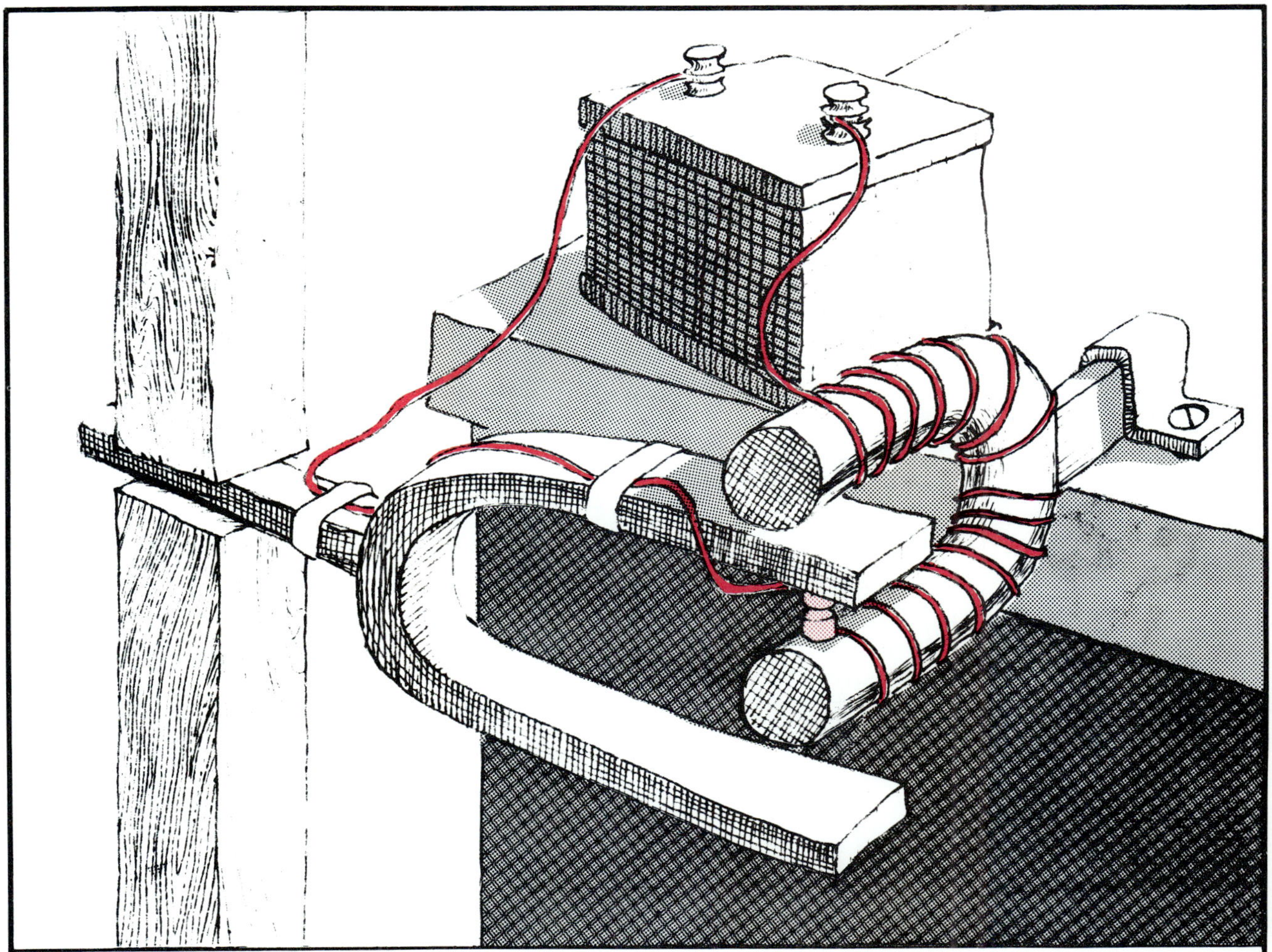

3.3.1. Here is the scheme of Helmholtz's electrical vibrator, the *tuning fork interrupter.* This device lives on, even today, as an alarm-bell ringer, or doorbell buzzer.

The invention of the triode vacuum tube made possible the realization of Helmholtz's scheme at very high frequencies, and so radio transmission was born. But to Van der Pol, this device became an extremely manageable laboratory instrument for experimental dynamics.

3.3.2. The physical system consists of the original radio transmitter. The chassis contains power supplies, a triode vacuum tube, a *tank circuit* consisting of an inductive coil and a variable capacitor in parallel, load resistors, and a feedback coil from the plate tank circuit to the grid of the tube, to induce oscillations. The two dials on the front of the chassis monitor the radio frequency current and voltage at the plate of the tube.

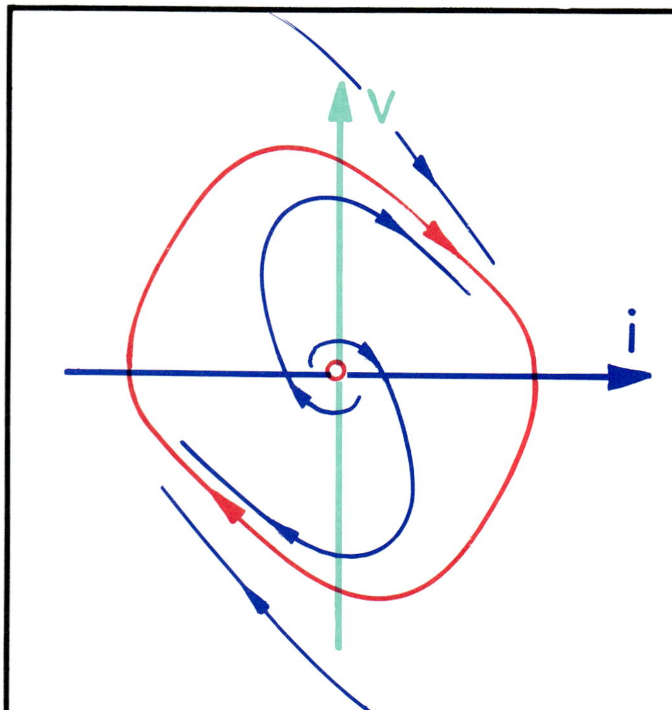

3.3.3. The observed parameters of this system are voltage and current, shown by the panel meters. Thus, the appropriate state space is plane. Here is the phase portrait of the vectorfield deduced by Van der Pol for this system. This is based upon electronic circuit theory, now standard. Its chief features are a repelling equilibrium point at the origin, and a periodic attractor around the origin. The mathematical proof of these facts is arduous compared to the ease of their discovery by experiments.

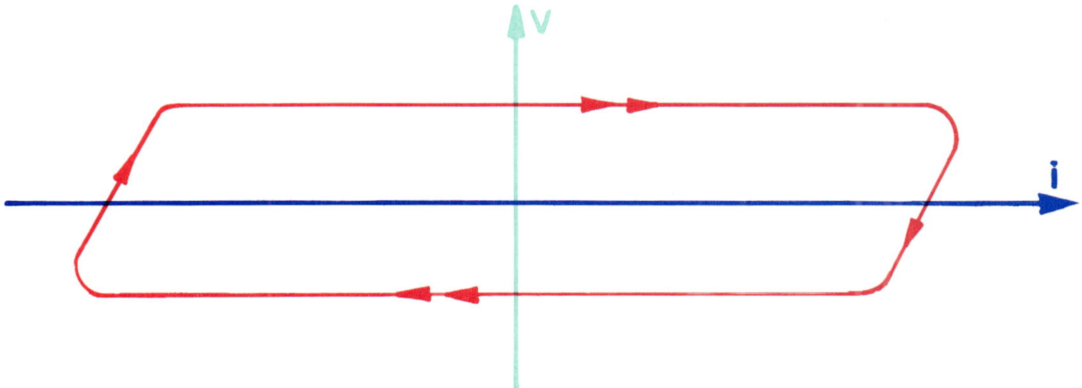

3.3.4. A simple modification of the preceding dynamical model for the triode oscillator yields this portrait, called a *relaxation oscillator.* The speed along the periodic attractor in this case is relatively slow on the near-vertical segments, and fast on the longer horizontal segments. Thus, the equilibrium oscillation lingers long at the minimum voltage (horizontal axis), snaps over to the maximum voltage, lingers there, then snaps back. Van der Pol proposed this as a model for the heartbeat.

3.4. Biological Morphogenesis

Many novel and exciting applications of dynamics to topics in biology and social theory were envisioned by Rashevsky. The best known of these, a model for biological morphogenesis, was rediscovered by Turing, and later studied by others. In this section, we will illustrate our interpretation of Rashevsky's model in the context of *phyllotaxis,* the morphogenesis of plant growth.[1]

The empirical domain for this application is an idealized vine.

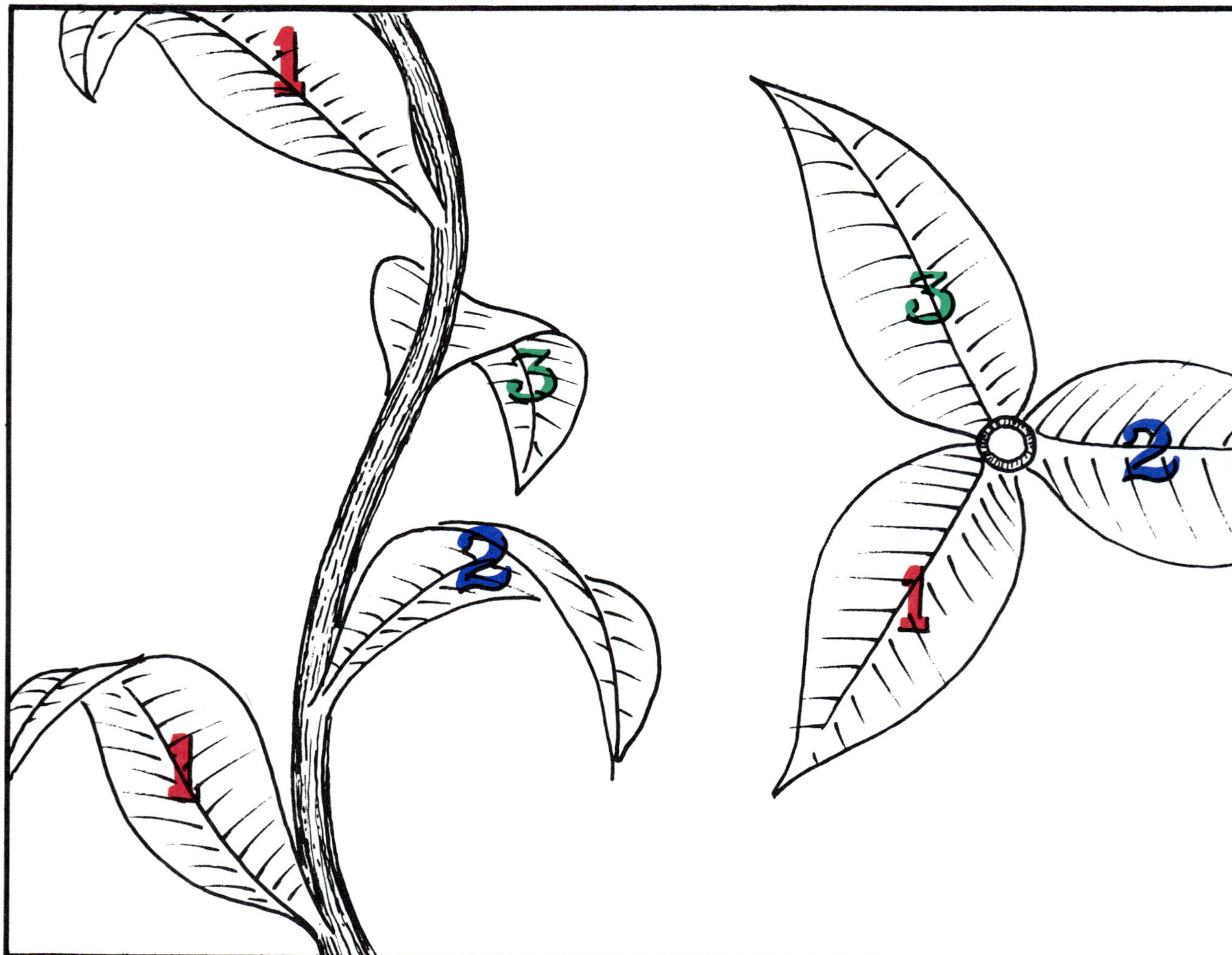

3.4.1. Have you noticed how some plant stalks sprout branchlets symmetrically? This vine, for example, sprouts one branchlet at a time. The direction of these branchlets rotate around the stalk with trihedral symmetry.

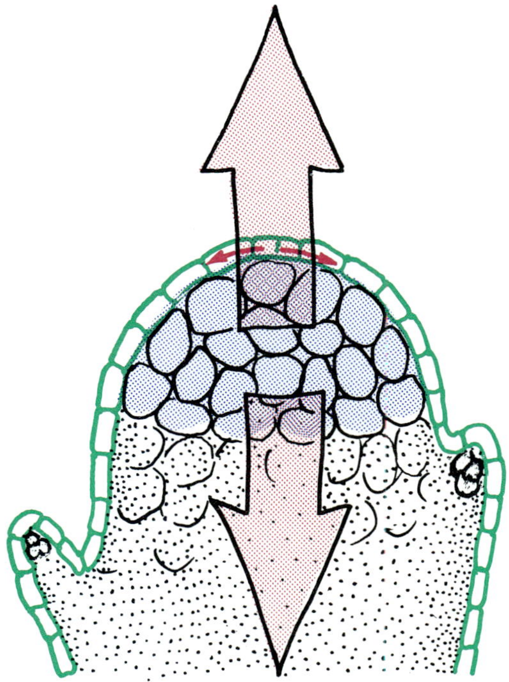

3.4.2. At the tip of the growing stalk is a growth bud. Beneath the epidermis is the *apical meristem,* a mass of undifferentiated, totipotent cells. Trailing in their wake, as they move upward on the growing stalk, are various derivative, cytologically differentiated, cells. Among these are the leaf bud cells, the branch cells, and so on. The question of morphogenesis, also called *phyllotaxis* in this context, is the formation of the *pattern* of these differentiated cells, and thus, of the leaf buds and branchlets.

The Rashevsky model for morphogenesis is based on a ring of growth cells, around the circumference of the stalk, near the growth bud at the top. Let's build up his model for this ring of cells, one cell at a time.

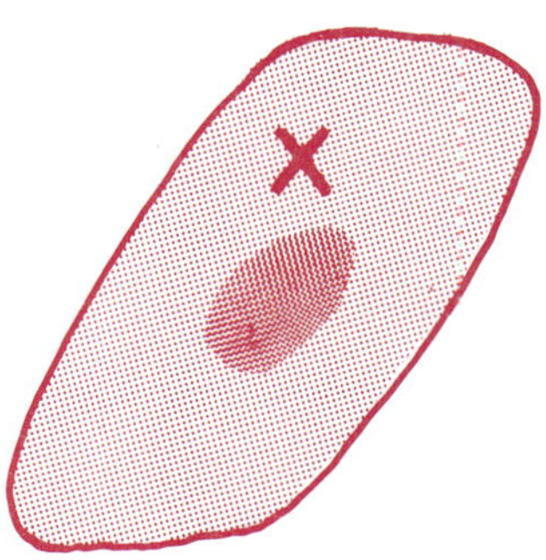

3.4.3. The cell is regarded as a bag of fluid. Convection stirs this fluid, so that its chemical composition is homogeneous, or well-stirred. One of the chemical constituents, called a *morphogen*, is a growth hormone. The relative concentration of this morphogen, x, is the observed parameter in the model for one cell. The state space is a line segment, as the parameter varies only between 0 and 1.

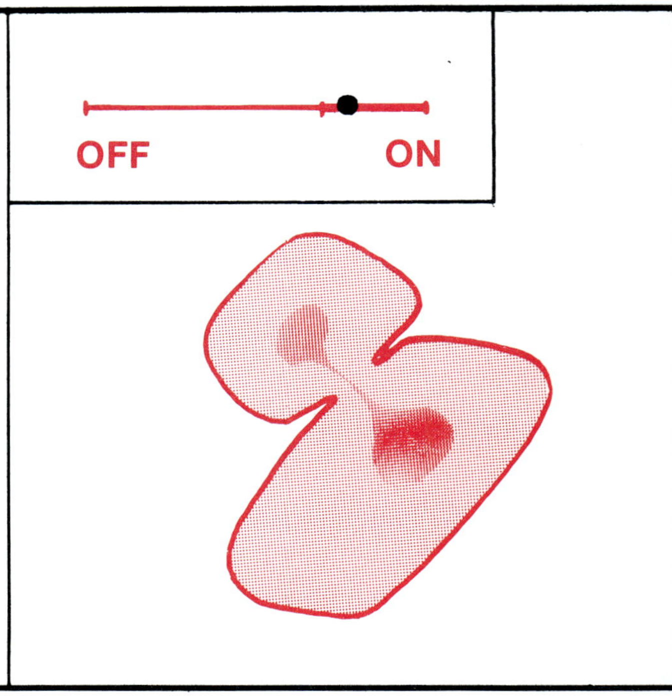

3.4.4. If the concentration of this morphogen in the fluid of the cell exceeds a certain critical value, the growth function of the cell is turned on. It divides, and a branchlet is born.

Next step: two cells, with one morphogen, in an *open system*. This means that the morphogen can come and go between the two-celled system and its environment.

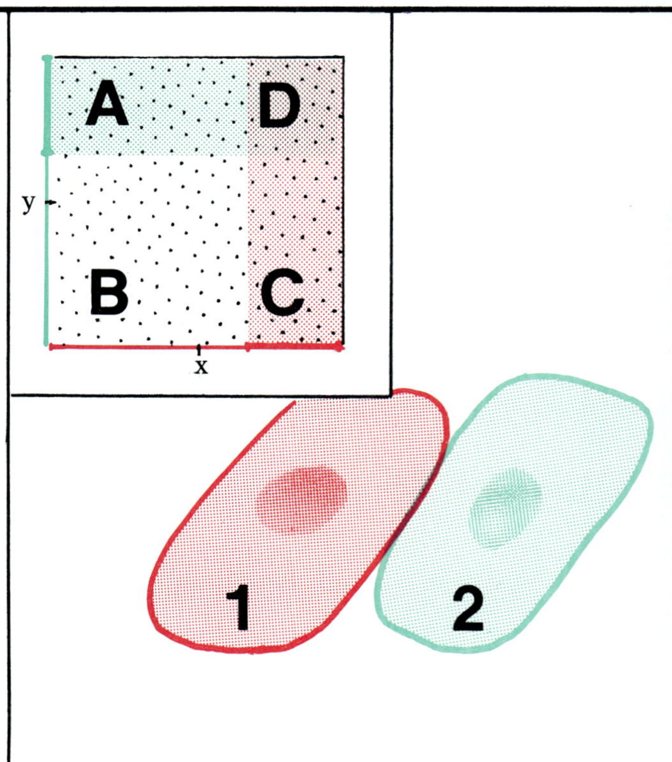

3.4.5. Observing the relative concentrations of morphogen in *each* cell determines a point in the unit square, (x,y). The square is the state space for the system of two cells, with one morphogen, as an open system, The state space is shown here, divided into four regions:

A. Cell 1 off, cell 2 growing.
B. Both cells' growth turned off.
C. Cell 1 growing, cell 2 off.
D. Both cells growing.

Now let's close the system. In the closed system of two cells, the morphogen cannot enter or leave. The total amount is constant.

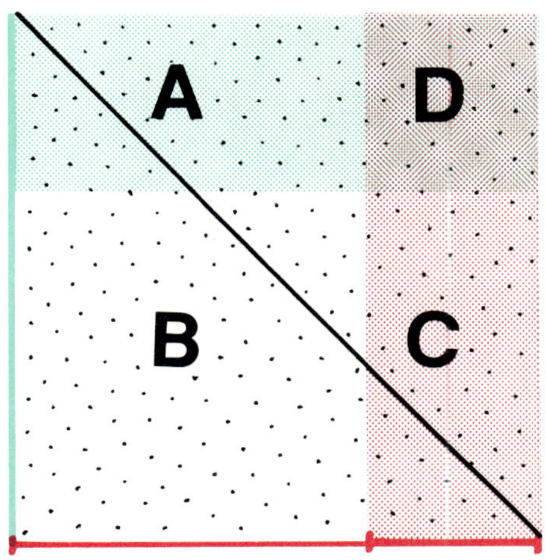

3.4.7. Extracting this line segment, we have the state space for the closed system. Notice that the line meets the zones *A*, *B*, and *C* of the square.

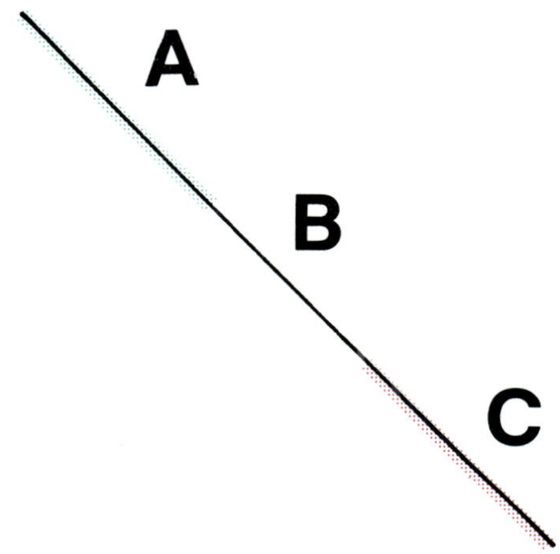

3.4.6. The state space for the closed system is a *subspace* of the square. Only the points of the square on the black line segment satisfy the constraint of the closed system: the sum of the concentrations is constant, or $x + y = 1$.

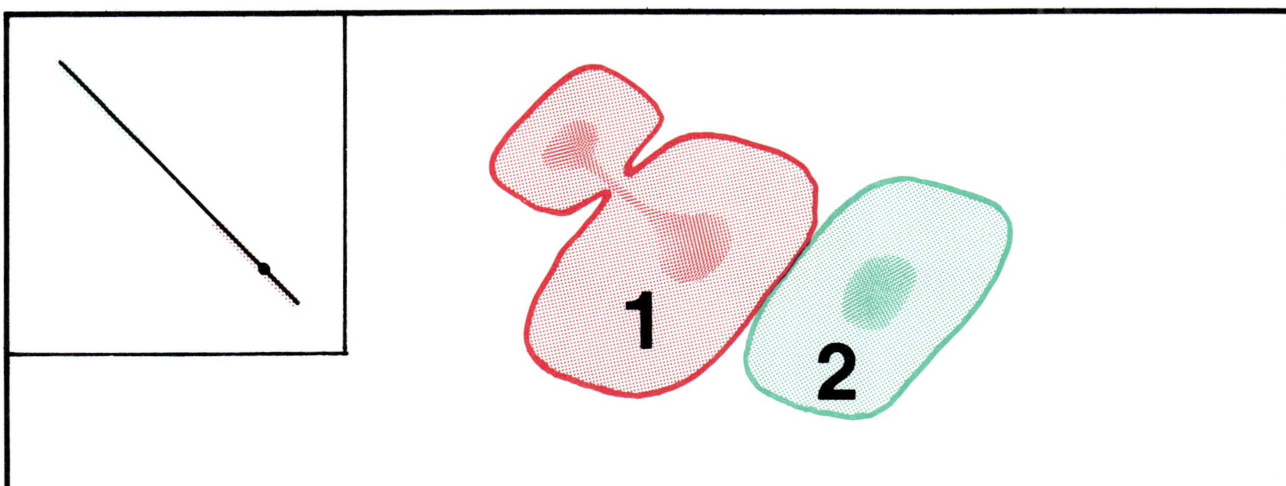

3.4.8. If the state of the closed system of two cells is in the line segment *A,* cell 1 is off and cell 2 is growing. In segment *B,* neither cell is growing, and in segment *C,* cell 1 is growing while cell 2 is off.

Final step: three cells, one morphogen, closed system. We think of them as a *ring* of cells.

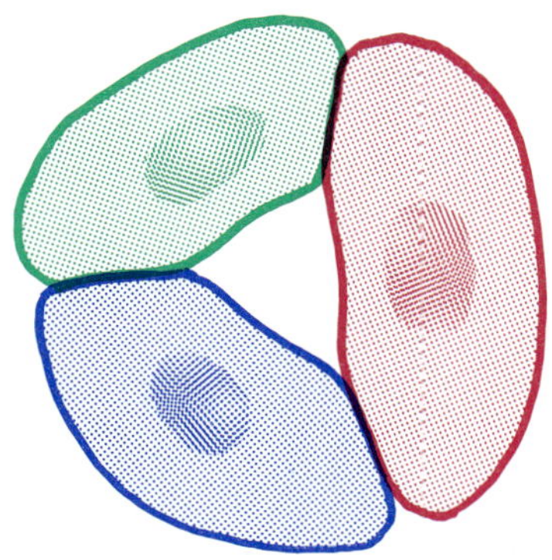

3.4.9. Here is the ring of three cells, with a uniform concentration of the morphogen in each. The point (x,y,z) in the unit cube of three-space represents a state of the system.

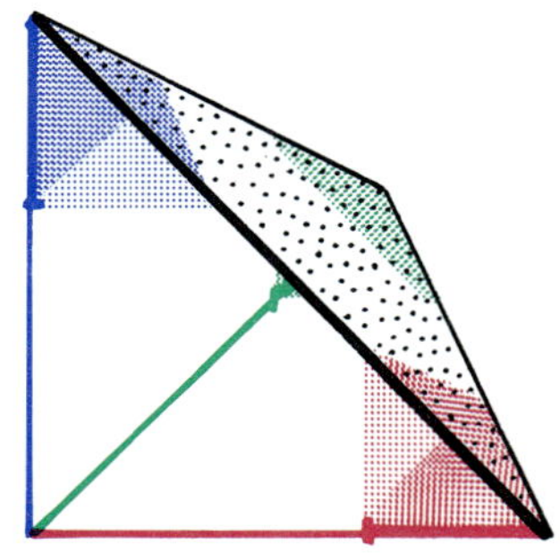

3.4.10. Plotting the state in three space, we find the assumption of a closed system, $x + y + z = 1$, constrains the state to lie on this triangle, called the *unit simplex* of three space. This planar, equilateral triangle is the state space for the closed system of three cells, with one morphogen.

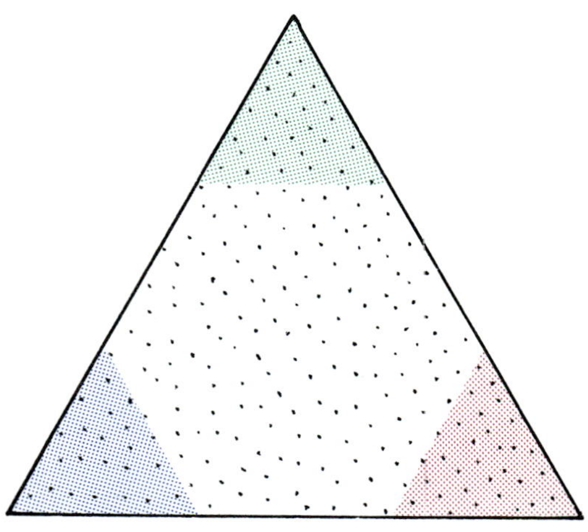

3.4.11. The three small triangles at the corner points of the state space correspond to a distribution of morphogen in which the amount in one of the cells exceeds the critical value for growth.

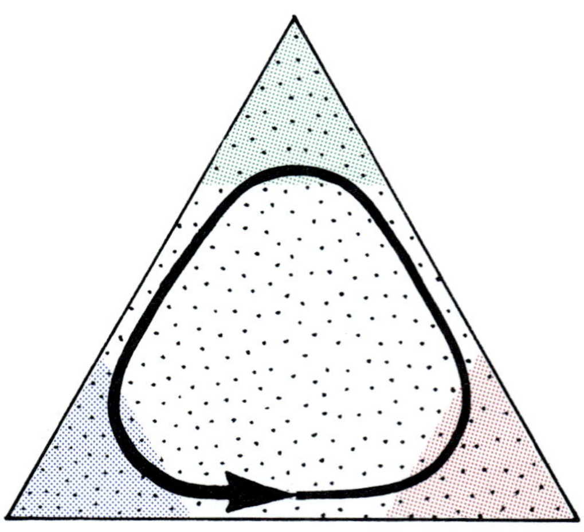

3.4.12. Suppose now that a dynamical system has been added to the model, and that it has a periodic attractor like this. Then periodically, one after another of the three cells is turned on, then off.

We may now connect with the bean stalk by imagining a *stack* of rings of cells, as a simplified model for the *stalk*.

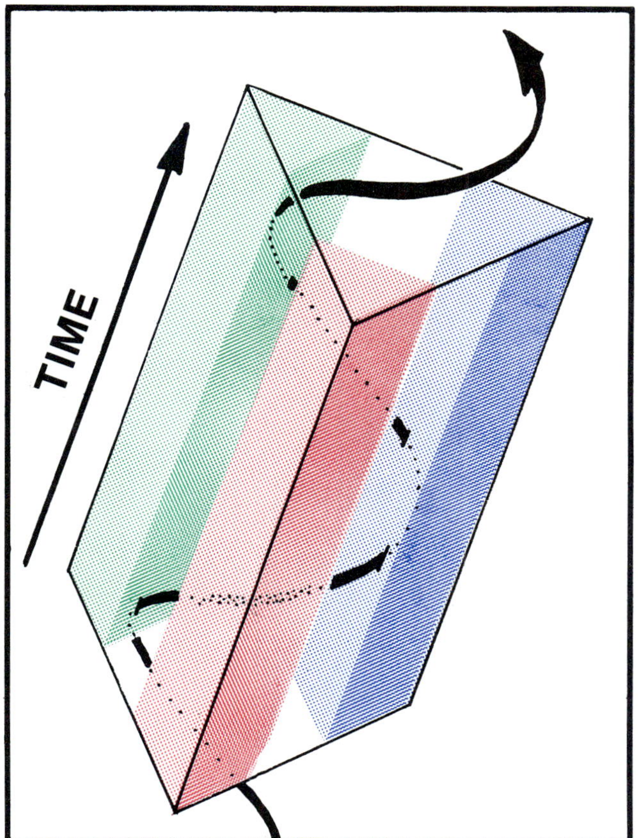

3.4.13. Here is the stack of rings of cells, each ring represented as an identical copy of the triangular model. Growth of the stalk upwards in time is represented by associating time with the upward direction. The periodic attractor of the preceding illustration thus gives rise to a *periodic time series* spiraling upwards in time.

3.4.14. Successively, the turned-on cell proceeds around the stalk, initiating branchlets. Many improvements to this model come immediately to mind, and no doubt they had occurred to Rashevsky already, in 1940.

The further biological and social applications of this scheme for morphogenesis await us in the near future.

4

Forced Vibrations:
Limit Cycles in 3D from Rayleigh to Duffing

Lord Rayleigh's study of musical instruments provided the early examples of limit points and limit cycles in the plane, discussed in the two preceding chapters. He went on to study *forced oscillations,* with musical applications in mind. Besides these applications, involving tuning forks and the determination of pitch, he envisioned further applications to tides and electrical motors. This led him into experimental work, progressive abstraction into theory, and the foundation of a new branch of dynamics: *forced vibration*.

Forced vibration is one of the most significant topics in dynamics, and its potential applications are manifold. We distinguish two separate cases:

(1) *A system which tends to rest is subject to a periodic force.* Classical example: effect of mechanical vibration on a pendulum. Biological example: effect of the seasons on big fish and small fry (Figure 2.4.6).

(2) *A system which tends to self-sustained oscillation is subject to a periodic force.* The preceding section on biological morphogenesis, for example, suggests the question: what happens if a biological oscillator is influenced by an external periodic force, such as sunlight?

In the next chapter, we will describe the results obtained for case 2 by Rayleigh and Van der Pol. In this chapter, we describe Rayleigh's work on case 1 with the double pendulum, and the related results obtained later by Duffing. We begin by constructing a three-dimensional model for the states of the forced system, the *ring model.*

4.1. The Ring Model For Forced Springs

To the early experimentalists, a self-sustained oscillator was hard to arrange, so they approximated one with a very large pendulum. The decay in the amplitude of its swing would be insignificant in a short experiment. As the source of the periodic force applied to the *driven system,* a smaller pendulum, it would be relatively unmoved by the motion of the driven system. Of course, we would properly consider this a *coupled system* of two swinging pendula. It only *approximates a forced vibration,* which is an unreal idealization. This was well understood by Rayleigh, who wrote:[1]

> As has already been stated, the distinction of forced and free vibrations is imporant; but it may be remarked that most of the forced vibrations which we shall have to consider as affecting a system, take their origin ultimately in the motion of a second system, which influences the first, and is influenced by it. A vibration may thus have to be reckoned as forced in its relation to a system whose limits are fixed arbitrarily, even when that system has a share in determining the period of the force which acts upon it. On a wider view of the matter embracing both the systems, the vibration in question will be recognized as free.

Our goal in this section is to turn this intuition into a geometric model in three dimensions. This is the state space for the coupled system. In it, a free vibration of the coupled system (equivalent for a forced oscillation of the driven system) is represented as a *periodic attractor.*

4.1.1. Here are the actual experimental devices of three early workers: Rayleigh, Duffing, and Ludeke.

4.1.2. Helmholtz and Rayleigh also analyzed mechanical systems subject to periodic electrical forces. These systems more closely approximate the ideal forced vibration, in that the forcing system (alternating current generator) is relatively indifferent to the motion of the driven system (tuning fork).

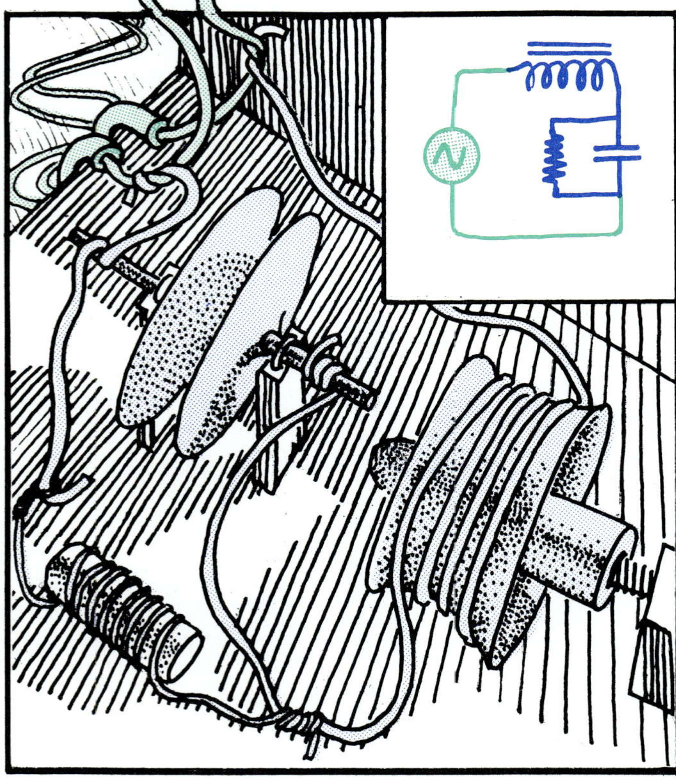

4.1.3. The early experimentalists also analyzed electrical systems subject to periodic electrical forces. For example, a parallel plate capacitor and an inductive coil in series was regarded as an electrical analog of the tuning fork in the driven system.

We now describe Duffing's results in the original mechanical context, the double pendulum. First, the driven system.

4.1.4. This pendulum will be the driven system in the mechanical apparatus for studying forced vibration in case 1. One observed parameter, *A*, is the angle of deflection of the bob from vertical. We also observe another parameter, the rate of change of the angle, *R*.

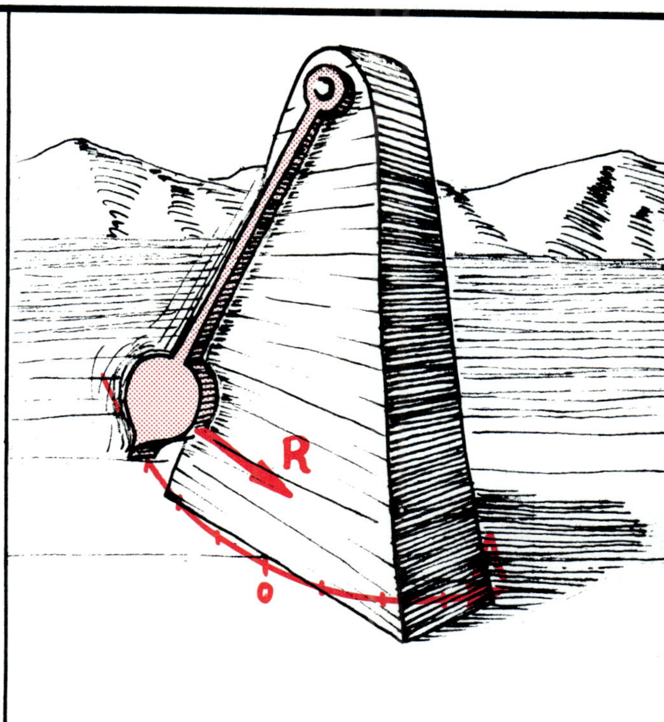

4.1.5. Recall that the state space for this device is a cylinder, as described in Section 2.1. But we may cut the cylinder and unroll it into a plane. This will be useful here, as we will consider only small motions of the bob, represented by points near the origin, $(A,R) = (0,0)$, of the state space. Recall also that the phase protrait for this system has a focal point attractor at the origin.

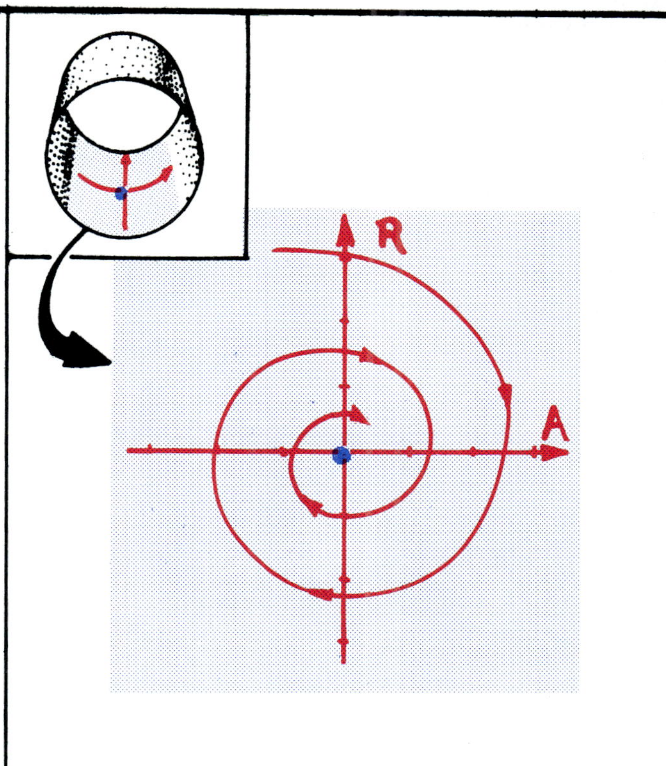

And now, the driving system, with twentieth century additions. . . .

4.1.6. This turntable motor has a sophisticated governor, which works hard to maintain a constant frequency of rotation. To its turntable is connected a push-rod and lever. The upper (pointed) end of this lever will eventually be the oscillating point of support for the driven pendulum.

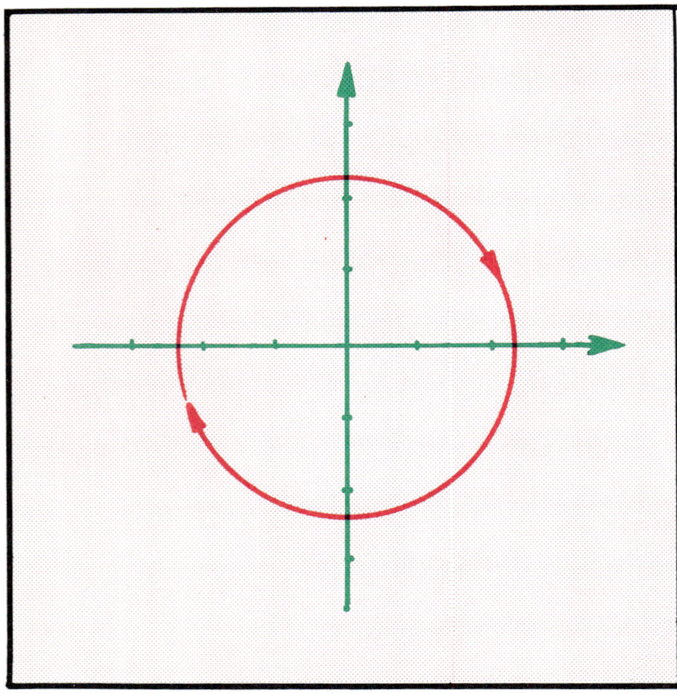

4.1.7. This motor is a replacement for the giant pendulum in Rayleigh's original scheme. It has a self-sustained oscillation. Like the clarinet reed and violin string of the preceding chapter, its dynamical model has an attractive limit cycle in a planar state space.

4.1.8. If we leave the motor running, we may forget start-up transients, and regard the limit cycle itself as the entire state space. Thus, there is only one observed parameter for this driving system: its phase, ϕ. The phase varies from 0 to 2π around the cycle of phases, which is the state space. The frequency of the cycle is supposed to be fixed by the governor. It is not a variable, but a constant in the model.

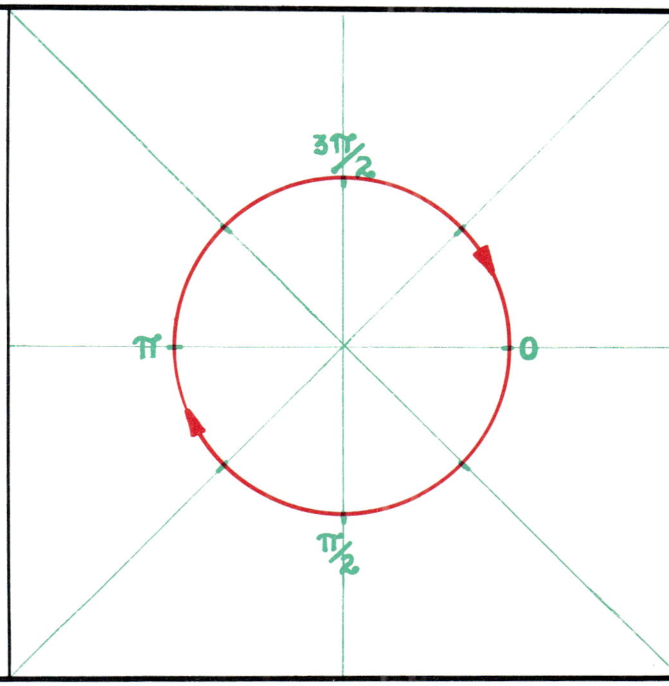

4.1.9. For ease of visualization, we now cut the cycle, and unroll it into a straight line segment. Cut on the right, at *phase zero*. Holding the upper end fast, bend the lower end down, to the left, and up, until straight. Having cut the cycle at phase zero, both ends of this line segment correspond to the beginning of the cycle.

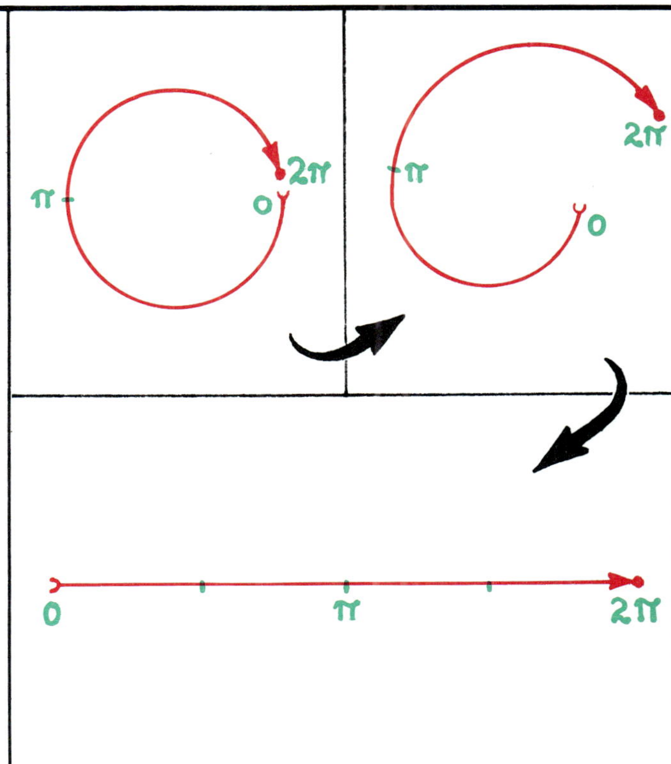

The dot marks *phase zero,* $\phi = 0$, the beginning of a cycle. This choice is somewhat arbitrary, but we show here the most common choice, called *the cosine convention*. In this convention, phase zero means the rod has just arrived full left, and is turning to go back. Thus, the pointed top of the lever is *full right at driving phase zero.*

Next, we combine the geometric models for the driven pendulum and the driving motor into a single, combined model. This model represents *the compound system of the two devices, uncoupled.*

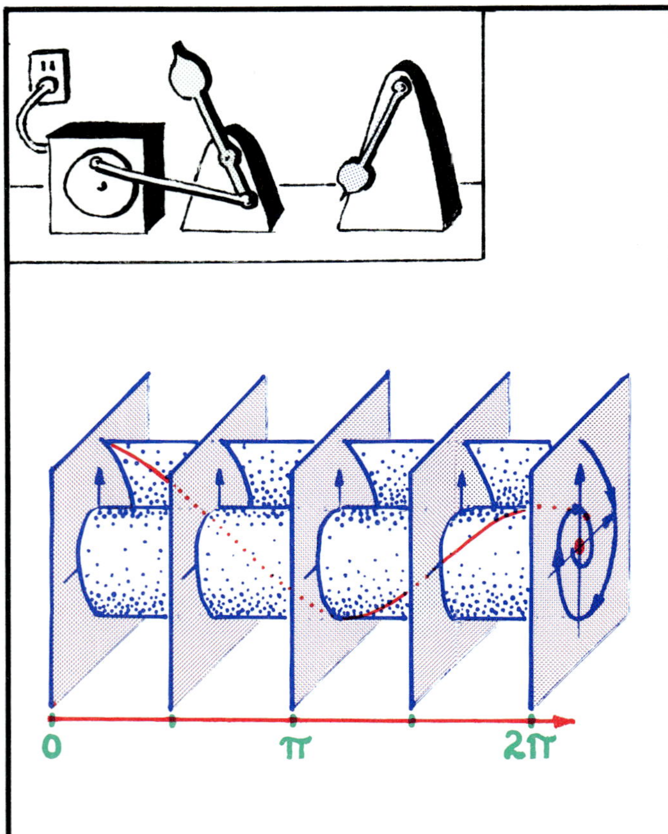

4.1.10. At each point of the unrolled cycle of phases of the driving motor, place an identical copy of the plane of states of the driven system. Orienting these planes vertically as shown, we may think of the scheme as a deck of cards on edge. Each card has the phase portrait of the damped pendulum printed on it. (This is an example of *Cartesian product construction.*) Every point in the resulting three-dimensional scheme represents, simultaneously, a state of the pendulum, (A,R), and a phase of the driver, ϕ. The three dimensions thus represent the observed parameters, (ϕ,A,R), of the combined (but uncoupled) system. Comparing with Figure 1.4.7, notice that the *phase* of the driving oscillation has replaced *time* as the parameter. Thus, the driving oscillation has become the clock. During one cycle of this clock, the state of the driven oscillation approaches its attractor, as shown by the exemplary trajectory (red) in this illustration.

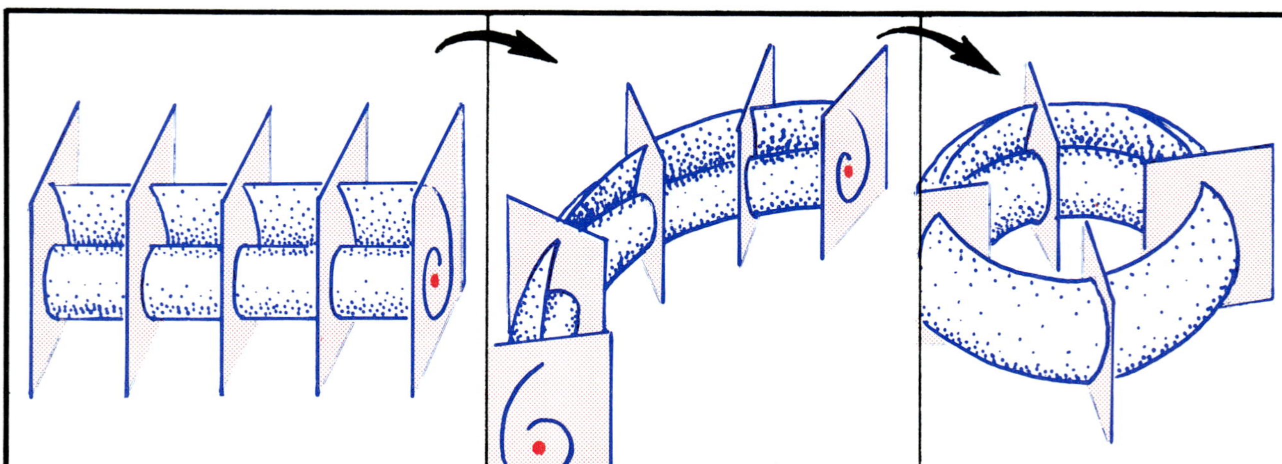

4.1.11. Finally, to get the correct model for the combined system, we roll up the cycle of phases again, carrying the deck of cards along with the cycle. Bend the left end down, to the right, and up again, until the two ends meet at the right. Glue the two end cards together. This *ring model* is the geometric model for the combined system.

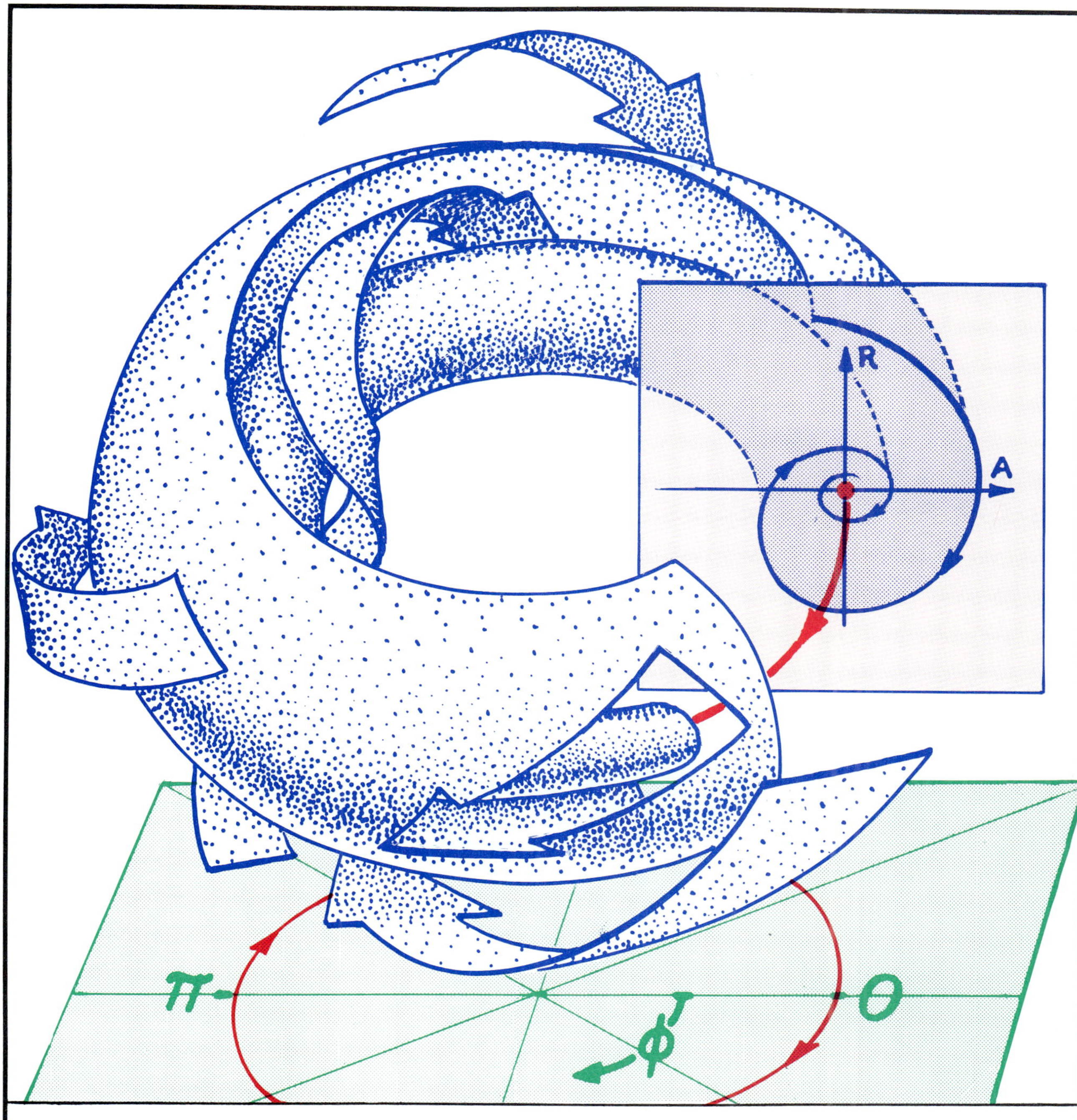

4.1.12. The phase portrait for the combined, uncoupled system looks like this. Scrolls within the ring contain the trajectories, which spiral around the red cycle in the middle of the ring. This is an attractive limit cycle for the combined dynamical system. It represents the pendulum bob coming to rest, as the driving motor keeps on running at its regulated frequency. Each scroll is actually a cylinder, rolled up like a pant leg. Also, it is an *invariant manifold* of the flow. This means, simply, that it is a collection of trajectories. No trajectory enters or leaves a scroll. A slice has been removed from the scroll at phase zero for better visibility.

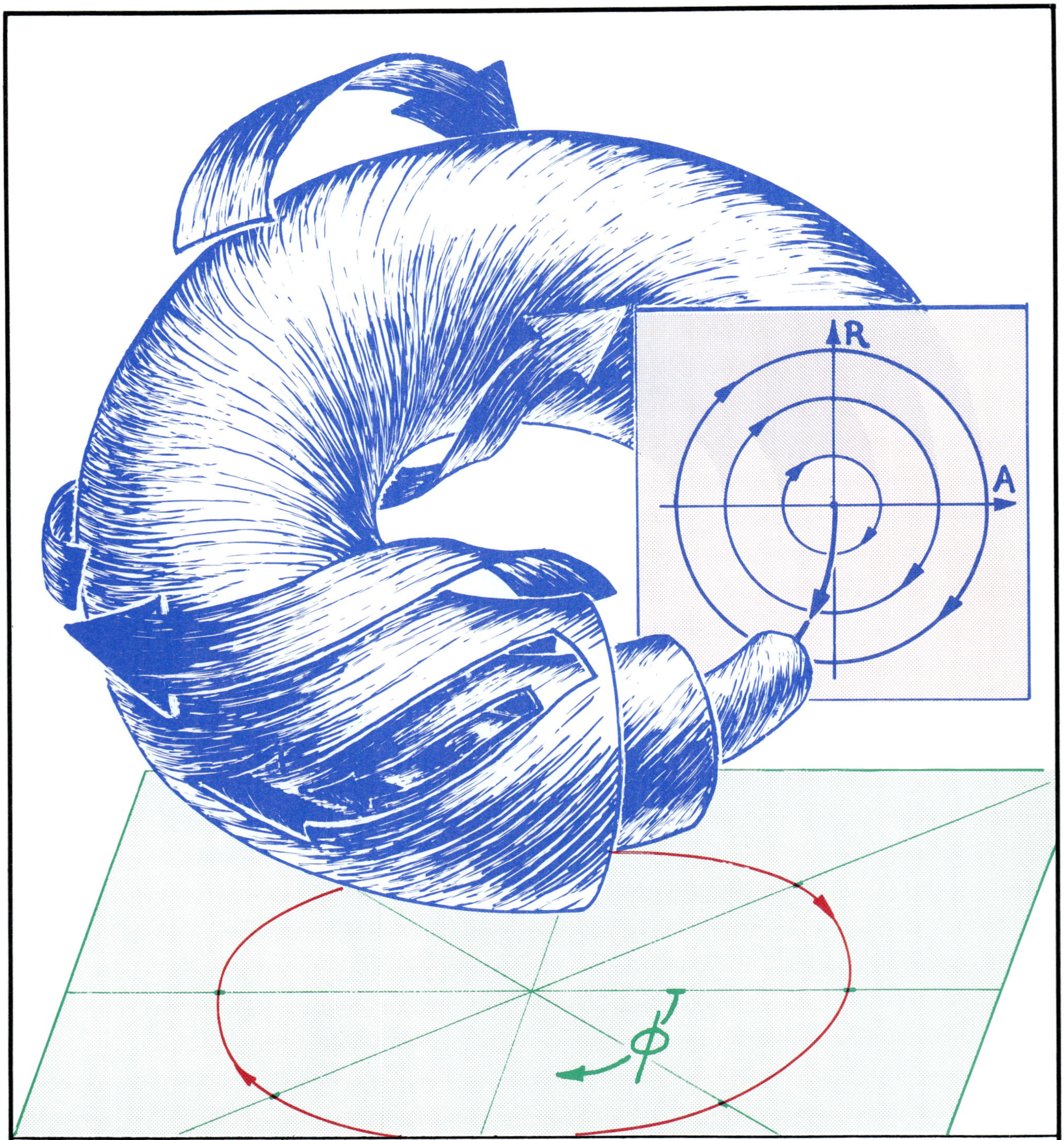

4.1.13. For comparison, here is the phase portrait for the combined dynamical system, in the case of the ***undamped*** pendulum, with the same driving motor. The scrolls are replaced by concentric *tori.* Each card of the deck is printed with concentric circles. The trajectories of the combined system spiral around these tori, which are invariant manifolds. The central cycle (red) is *not a limit cycle* for nearby trajectories. These concentric tori have also been cut through, at phase zero, for visibility.

The trick of cutting through the combined phase portrait at a fixed phase, for better visualization, came from the early experimentalists. It is now called stroboscopy: Rayleigh names Plateau (1836) as the inventor.[2]

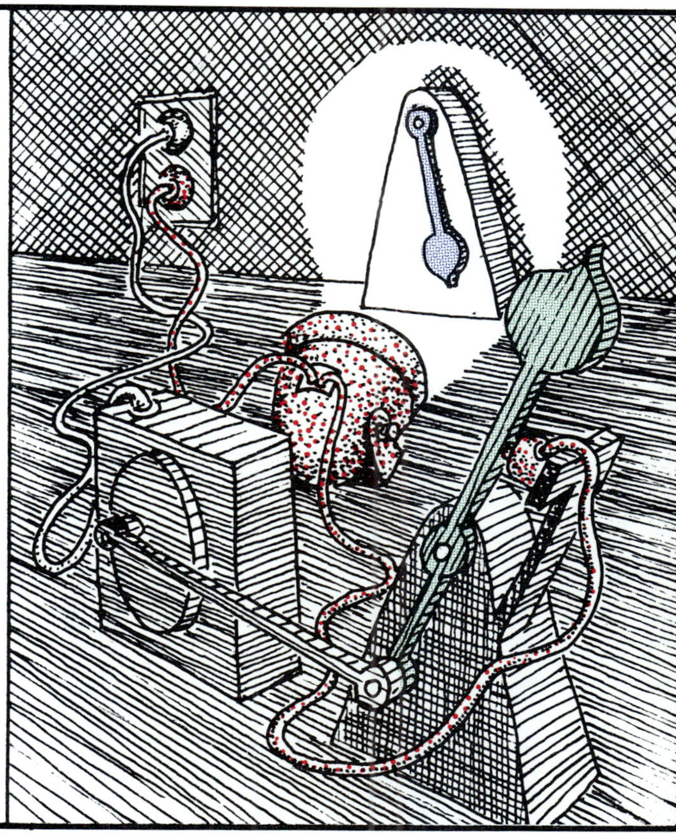

4.1.14. The strobe lamp is aimed at the driven bob. It is turned on momentarily, when the drive (blue) pendulum contacts the microswitch, at phase zero. Lord Rayleigh used the rays of the sun, interrupted by an electrical diaphragm.

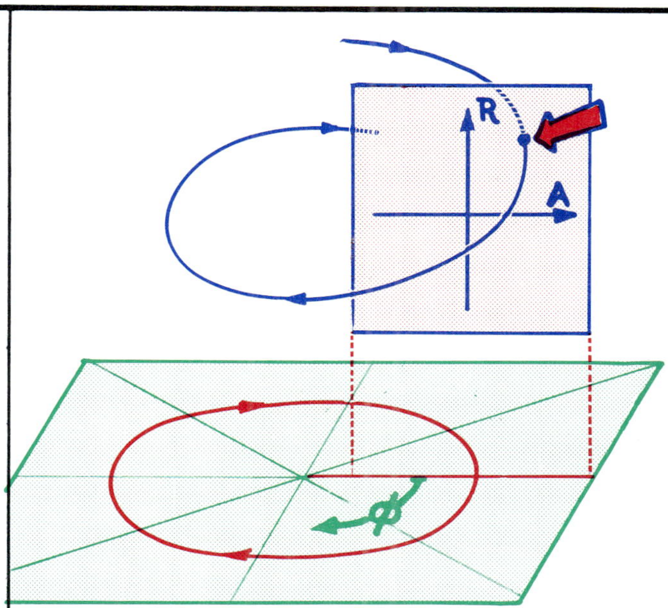

4.1.15. In the light of the stroboscope, the parameters of the driven bob may be observed at the fixed phase of the driving motor. The observed data define a point in the *strobe plane,* the card of the deck corresponding to this fixed phase. At this point, the trajectory of the combined system pierces the strobe plane.

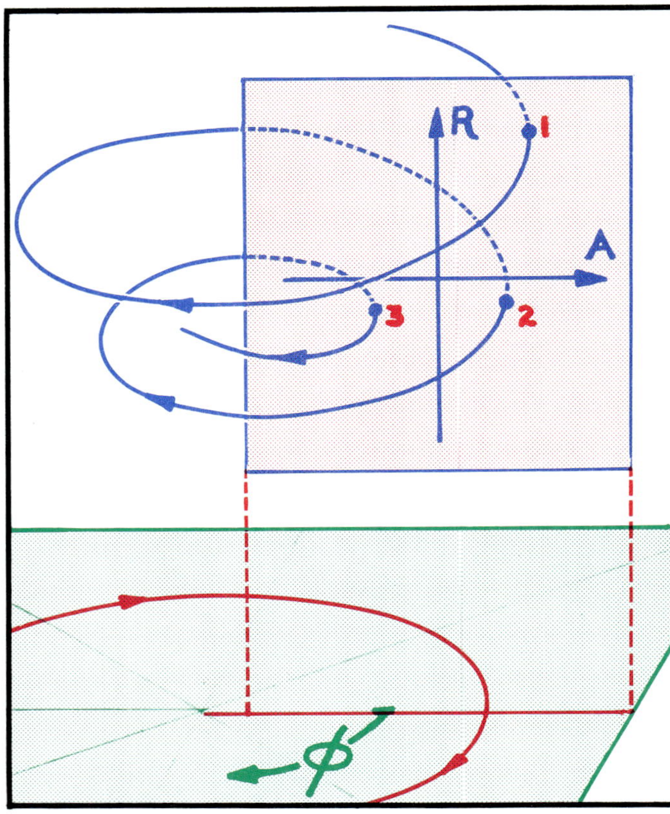

4.1.16. Recording the observations of successive flashes of the strobe light, a sequence of points in the strobe plane is obtained, instead of a continuous trajectory, as the record of motion of the bob. We may call this a *strobed trajectory.*

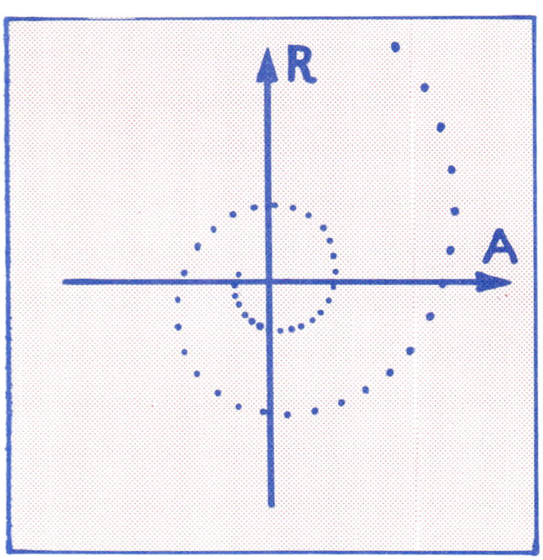

4.1.17. If the driving frequency is relatively fast with respect to the free (unforced) frequency of the driven pendulum, the strobed trajectory will appear to step along a spiral in the strobe plane. In other words, if the driving clock (see Figure 4.1.10) runs quickly, the free pendulum will seem slow. That is, it will run around the spiral almost like a continuous trajectory. Here the driving frequency is approximately 24 times the free frequency.

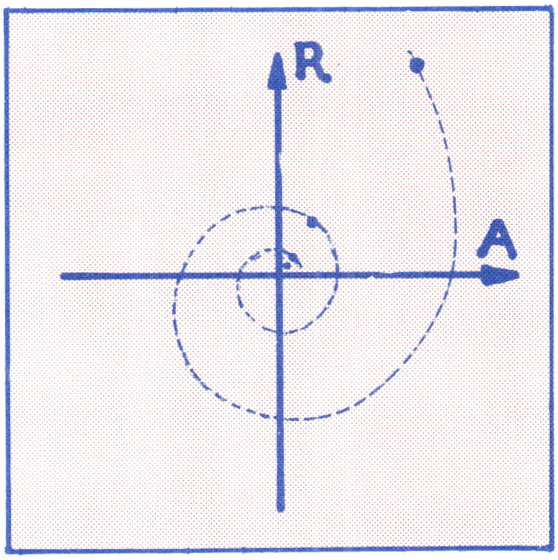

4.1.18 If the driving frequency is about the same as the free frequency of the driven system, the strobed trajectory may appear to walk directly toward the origin, as the pendulum comes to rest. In the strobe light of the experiment, the blue pendulum will seem to swing slowly to rest from one side.

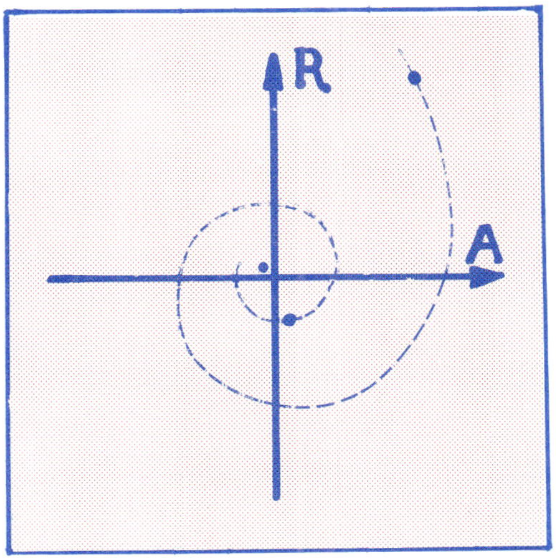

4.1.19. Finally, if the driving frequency is relatively slow with respect to the pendulum, the strobed trajectory will take giant steps, spiraling at a very slow pace toward rest. Here, the driving frequency is approximately ¾ of the free frequency.

All this regard for the uncoupled system has been for practice in 3D visualization and getting used to the ring model. In the next section, we will finally connect the two gadgets.

4.2. Forced Linear Springs

At last, let's pin the pendulum to the swinging end of the motor-driven lever.

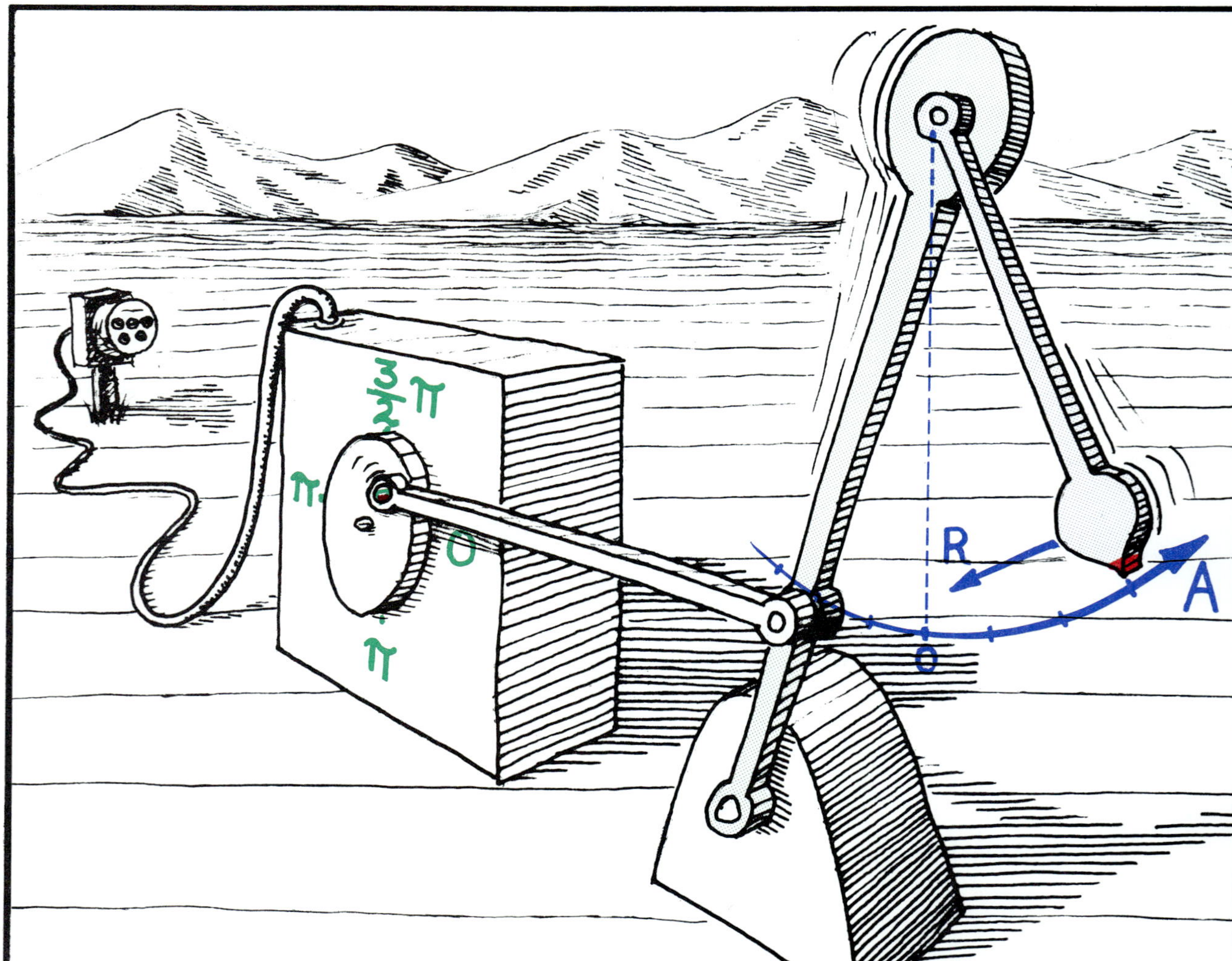

4.2.1. The coupled system has the same observed parameters as the uncoupled system: angle of the bob from vertical, rate of change of the angle, and phase of the driving motor.

The state space is still the ring. The dynamics, however, are changed. The vectorfield is quite different. So are the trajectories and the phase portrait.

To get the idea of the phase portrait for the coupled system (forced vibrations), a few armchair experiments will be helpful. We call this *the game of bob.*

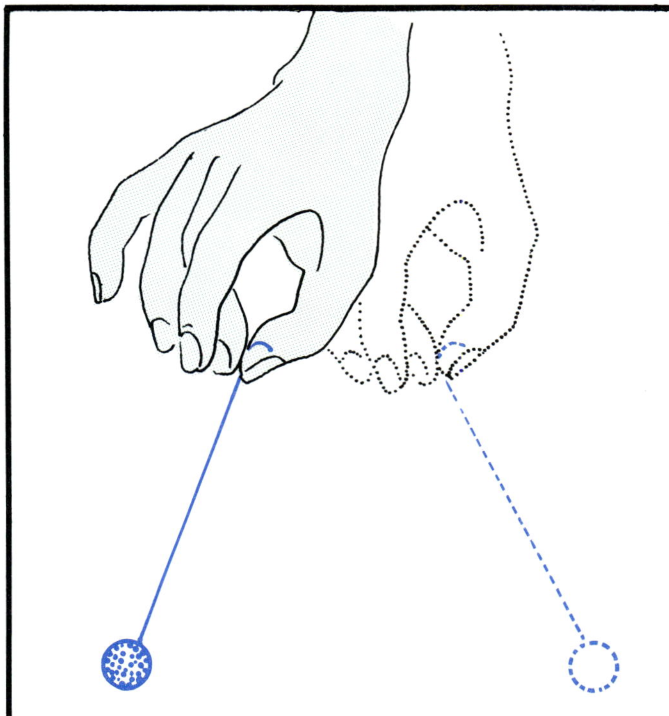

4.2.2. Give your hand a little jerk, then hold it still. The pendulum swings slowly to rest. Notice the frequency. This is the frequency of the free, unforced pendulum. Now, without changing the length of the string, oscillate your hand back and forth along a horizontal line. If the driving frequency of your hand is *slower* than the free frequency, the pendulum follows your hand. Instead of coming to rest, it is in *sustained oscillation.*

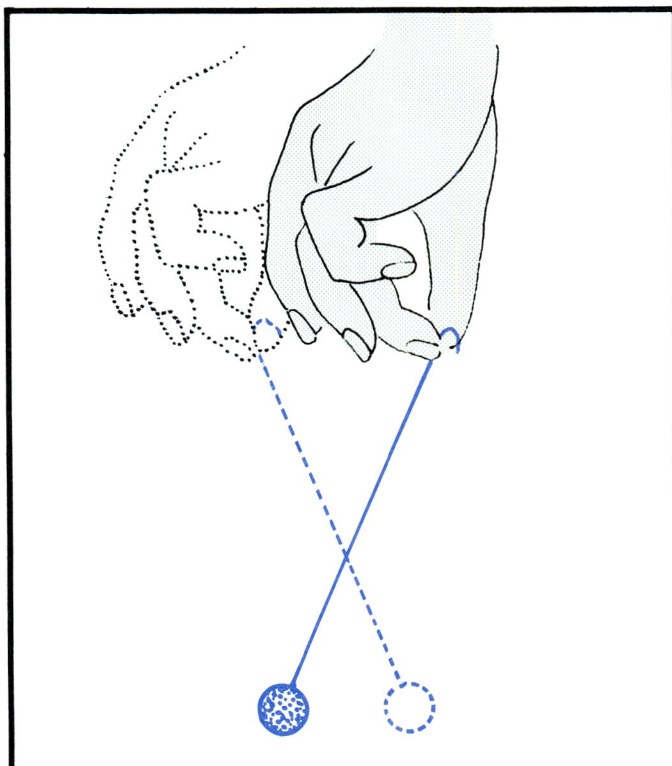

4.2.3. Start again with your hand and the bob at rest. This time, oscillate your hand *faster* than the free frequency. Notice that the bob swings *opposite* to the motion of your hand.

If you are in public that's enough for now. But if nobody is watching, here is another experiment to try.

4.2.4. If the driving frequency is relatively close to the free frequency, the amplitude of the swinging bob is greater.

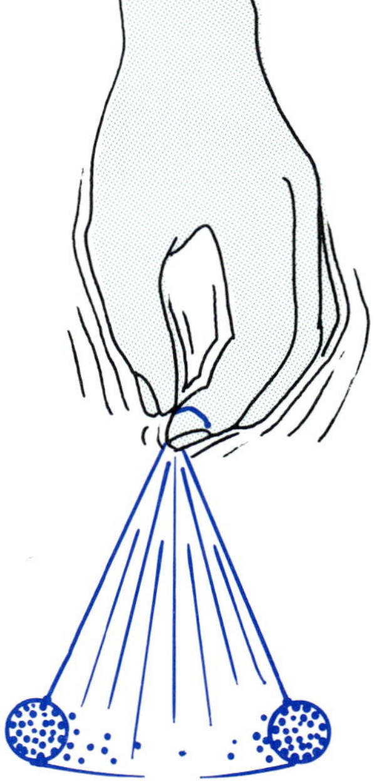

4.2.5. If the forcing frequency is much faster or slower than the free frequency, and the amplitude of the forcing motion is unchanged, the amplitude of the swinging bob is less.

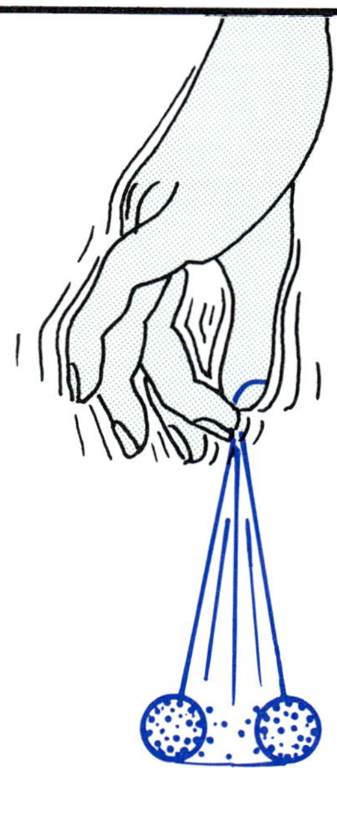

Now let's get on with Duffing's discoveries. First, we will improve the mechanical setup to simplify the analysis.

4.2.6. The driving motor will be the same as in the previous section, but the connecting rod is disconnected from the vertical lever, and attached to a weight. This weight is forced to move horizontally on the tabletop.

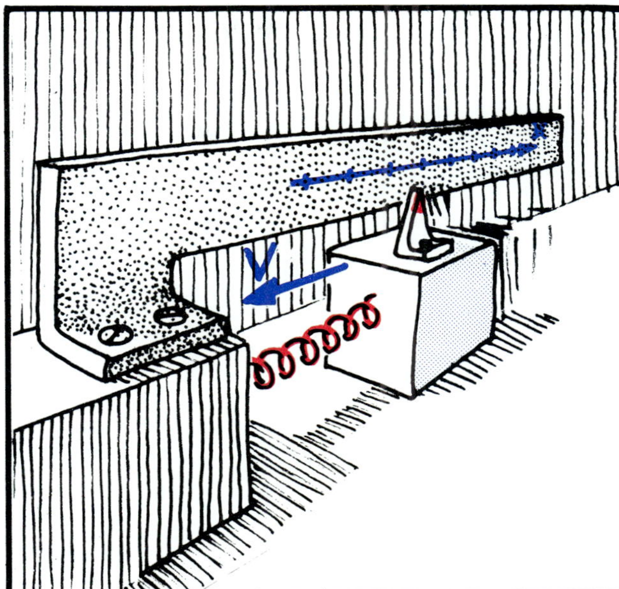

4.2.7. As in Chapter 2, we replace the pendulum with a spring. The spring is fixed to a support on the left, and to the weight on the right. The weight is free to move on the tabletop, which resists the motion with friction.

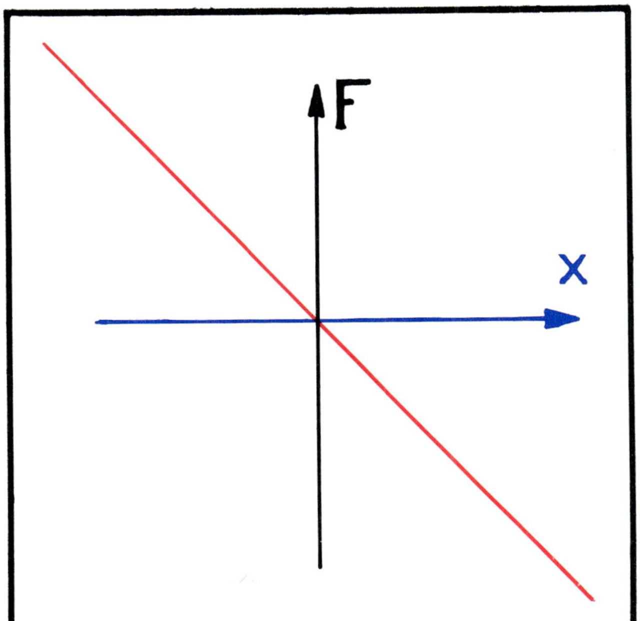

4.2.8. For a start, we assume the spring is linear. This restriction will be removed in the following section.

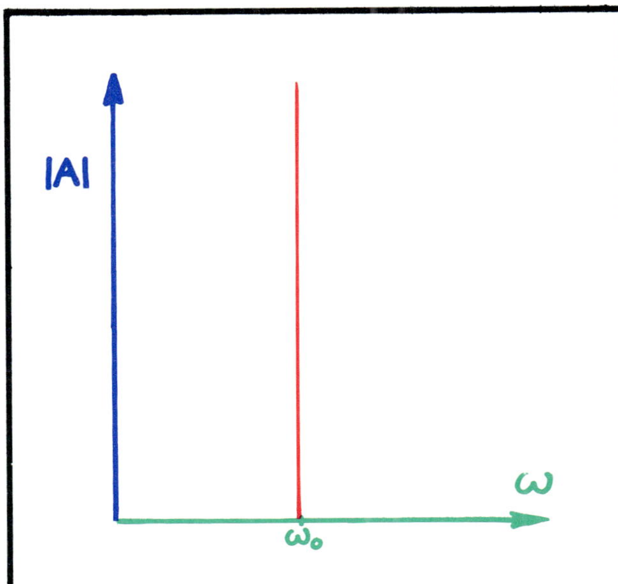

4.2.9 Recall that weights with linear springs have a natural frequency independendent of the amplitude of the oscillation. They make perfect guitar strings.

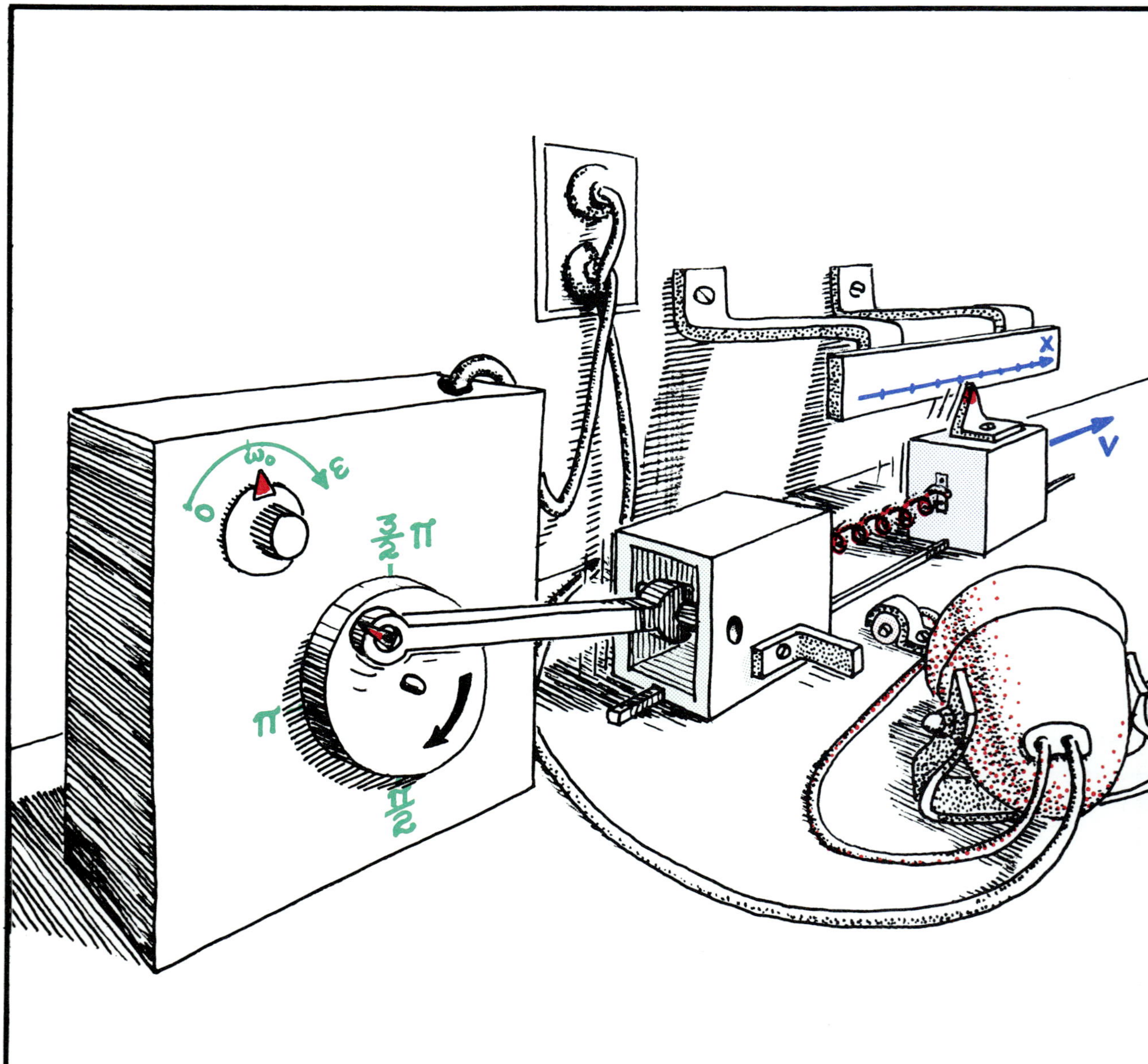

4.2.10. Here is the coupled system. The connecting rod of the driving motor moves the support of the spring, previously fixed to the tabletop, in a horizontal oscillation. For experimental purposes, we include a speed control for the forcing frequency, and a strobe light set for phase zero.

The state space for this forced vibration is again the ring, with observed parameters: phase of the forcing cycle, deflection of the spring, and velocity of the driving weight. As for the phase portrait, we have discovered by experiment that it contains an attractive limit cycle. Thus, in the coupled system, the drive (blue) weight is forced to oscillate by the driving (green) weight.

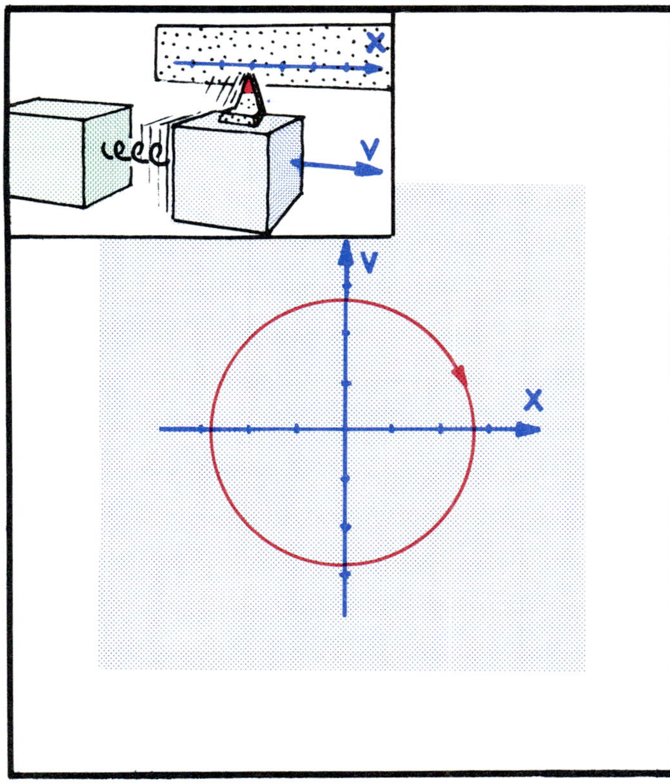

4.2.11. The bob oscillates. To represent this as a limit cycle in the ring, let's begin by recording a full cycle of the bob in its own, planar, state space. This is *not a trajectory* of a dynamical system in the plane, just a step in a graphical construction. Later we will erase it.

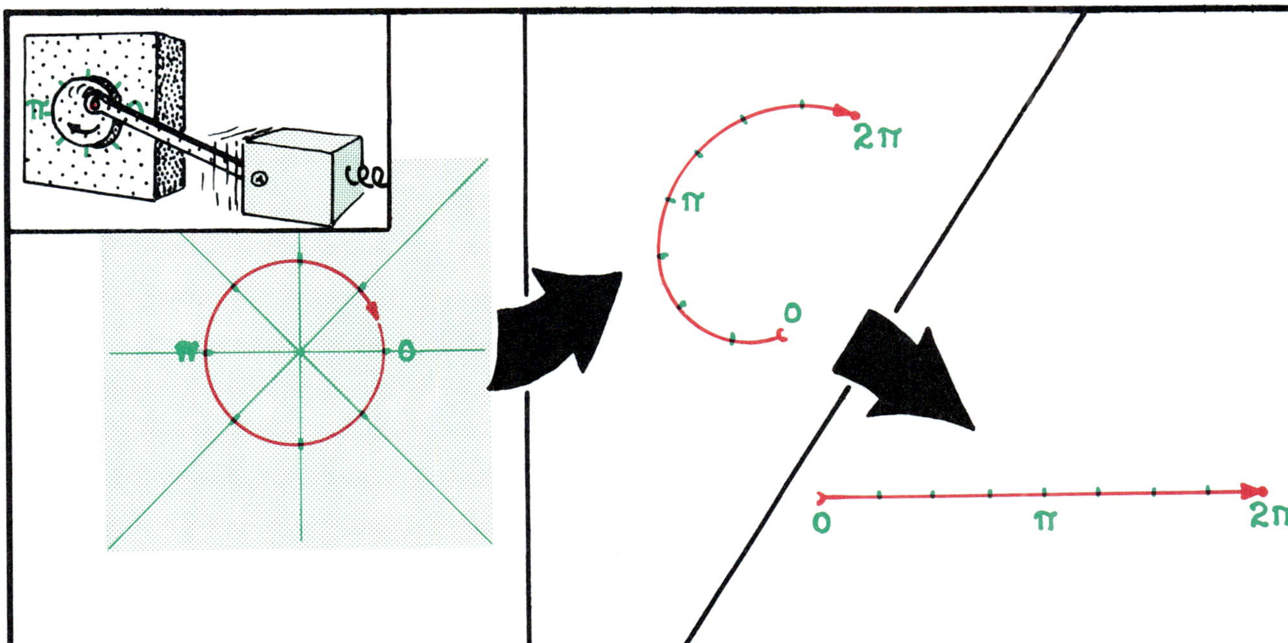

4.2.12. Recall that the state space for the driving motor is a cycle of phases. Earlier, the ring was constructed by cutting this cycle and straightening it out.

4.2.13. The deck of cards represents the ring, cut and straightened. On each card, we have drawn (in green) the observed cycle of the bob, as shown in the panel before last. This makes a green cylinder in the 3D model.

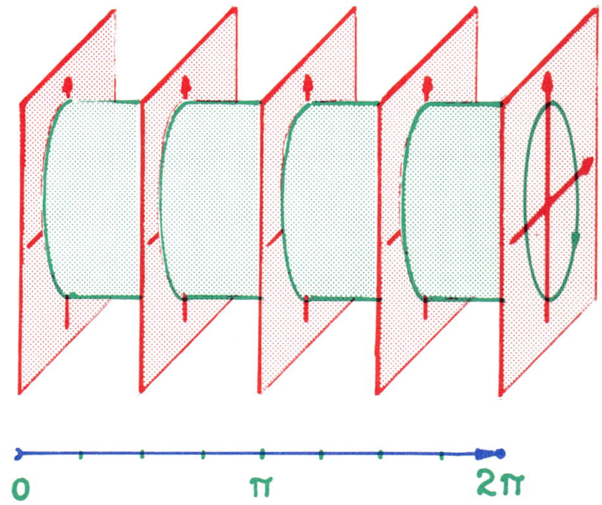

4.2.14. Now, record the observed motion of the combined system in this model as the bob oscillates and the driving motor traverses one full cycle. The red curve is this record. It is seen to stay on the green cylinder. It is a trajectory. Note that this illustration closely resembles Figure 1.3.9, with driving phase in place of time as the parameter.

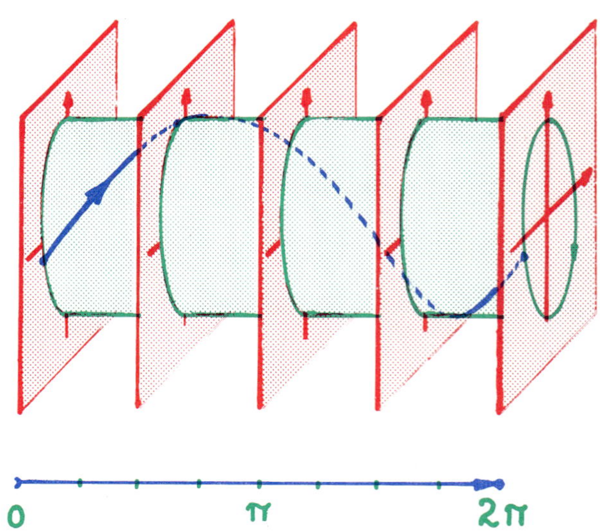

The green cylinder is not a feature of the phase portrait; it is just another step in our graphical construction. Later, we will erase this, also.

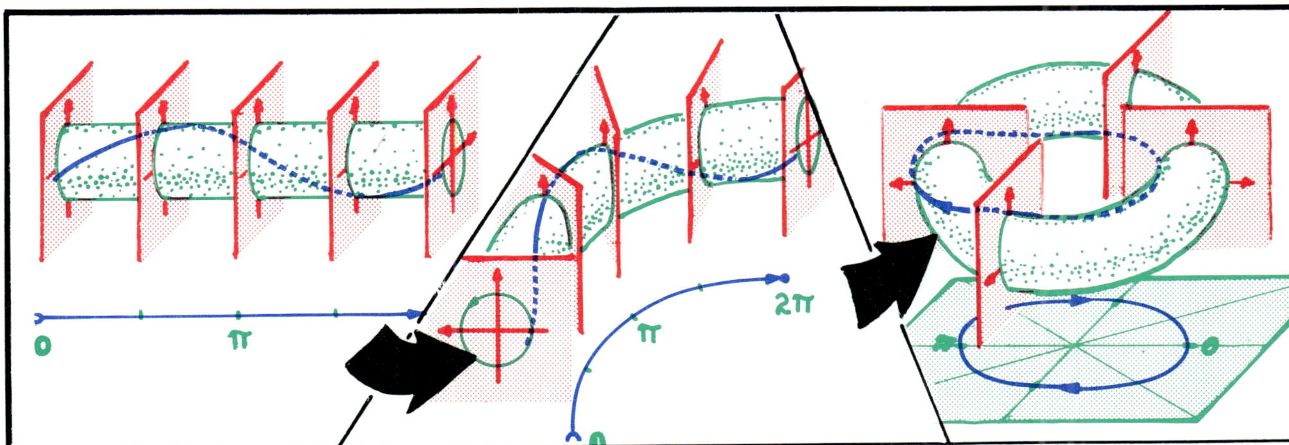

4.2.15. Next step, roll up the deck of cards again to make the ring. The green cylinder becomes a green torus. The red curve closes up, making a cycle on the green torus.

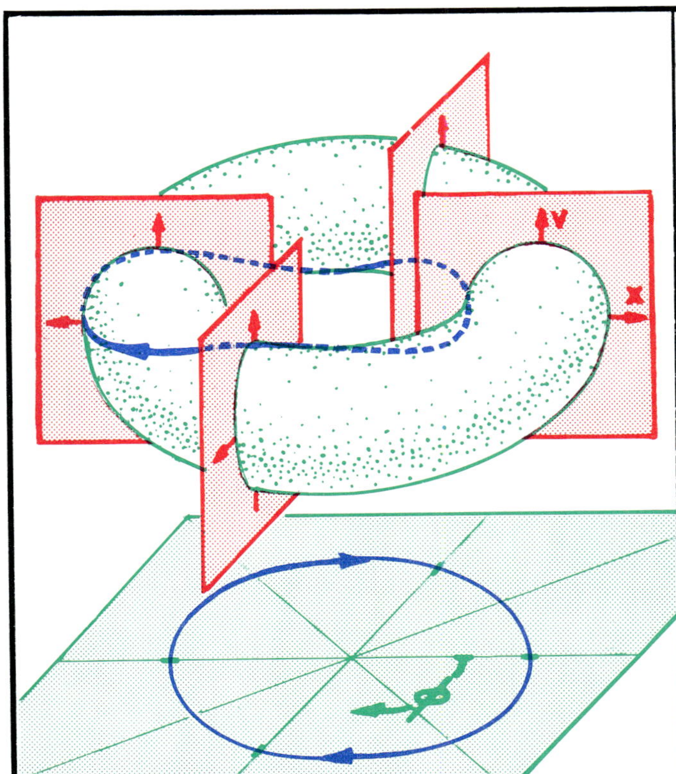

4.2.16. This is the attractive limit cycle known to Rayleigh, Duffing, and all those who have played the game of bob as we instructed. It is called an *isochronous harmonic,* as it completes one full cycle of the bob to each full cycle of the driving motor.

Now is a good time to erase the green torus. It is not an invariant manifold; most trajectories pass through it. It is just an aid for visualizing the isochronous harmonic, which is the only trajectory on it. Next, let's see what the nearby trajectories actually do.

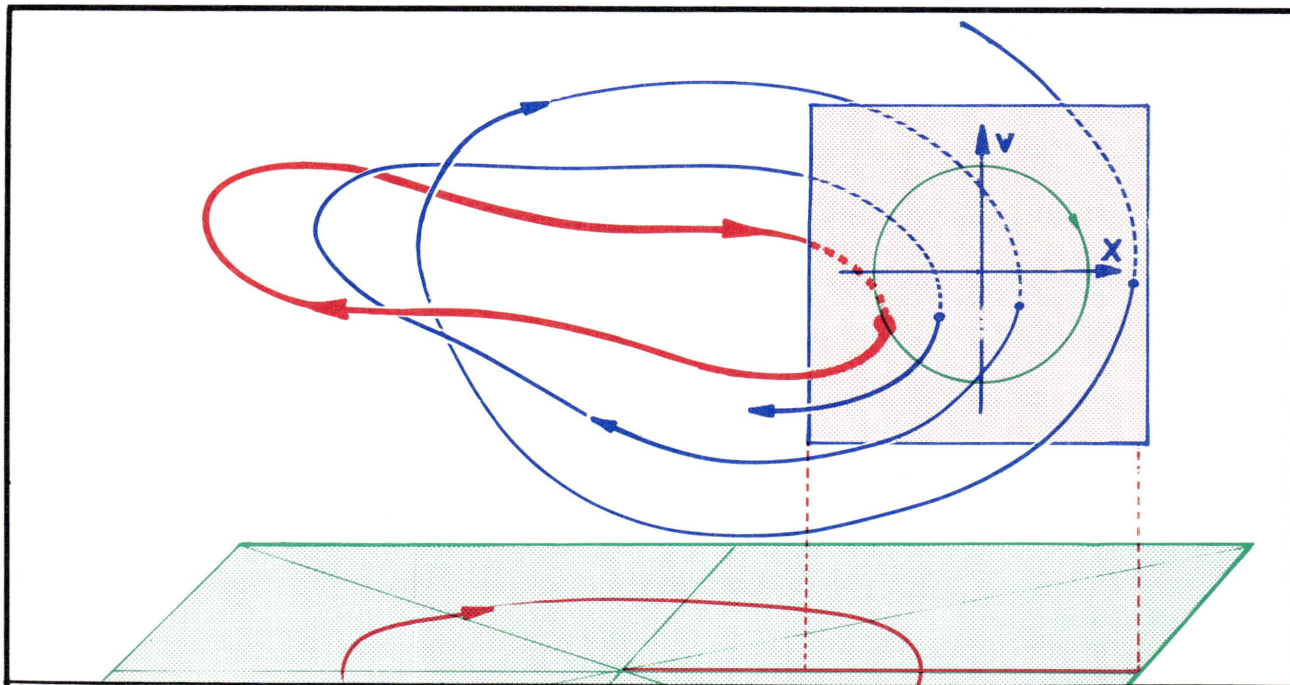

4.2.17. Here we have cut through the ring at phase zero for better seeing. The blue trajectory is a typical one. It spirals around the ring, getting closer to the red attractor with each cycle.

4.2.18. Observing this motion with the strobe light, we plot the tracks of several blue dots in the strobe plane at each flash. This is the strobe trajectory of this motion. The isochronous harmonic meets the strobe plane at a single point, the red one. The blue strobe trajectory approaches closer and closer to the red point. All nearby trajectories approach the red point like this, and may cross the green ring.

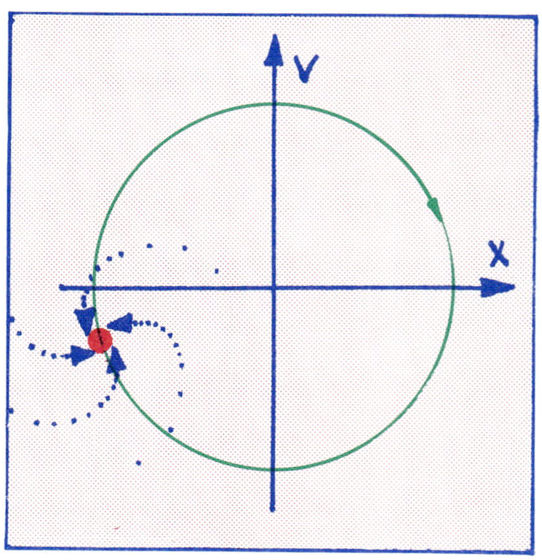

Retracing the steps of Duffing, let's play the game of bob in earnest with this apparatus. Suppose we change the speed of the driving motor.

4.2.19. If the forcing frequency is very slow, the driven (blue) weight will follow the motion of the driving (green) weight. The strobe at phase zero (according to the cosine convention, this means full right, and turning to go back) will flash on the blue weight at its phase zero position.

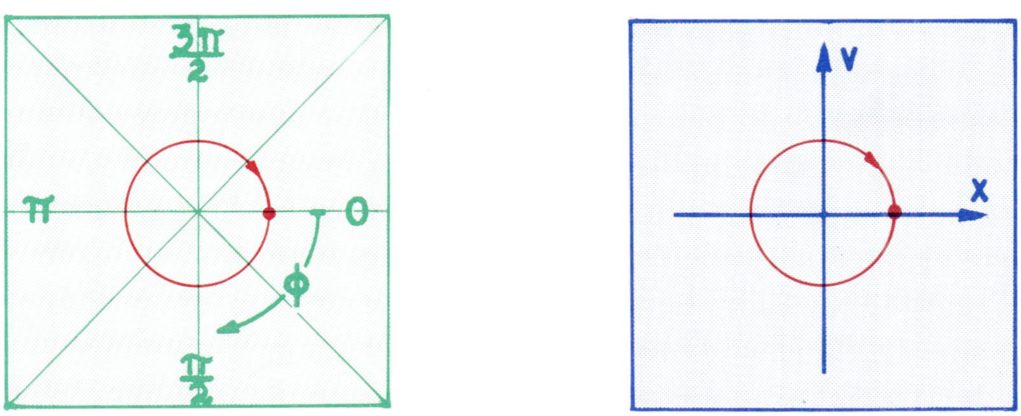

4.2.20. This means that the two separate trajectories are cycles that are *in phase.* That is, they both reach phase zero at the same time. They are both full right, and turning to go back, at this moment.

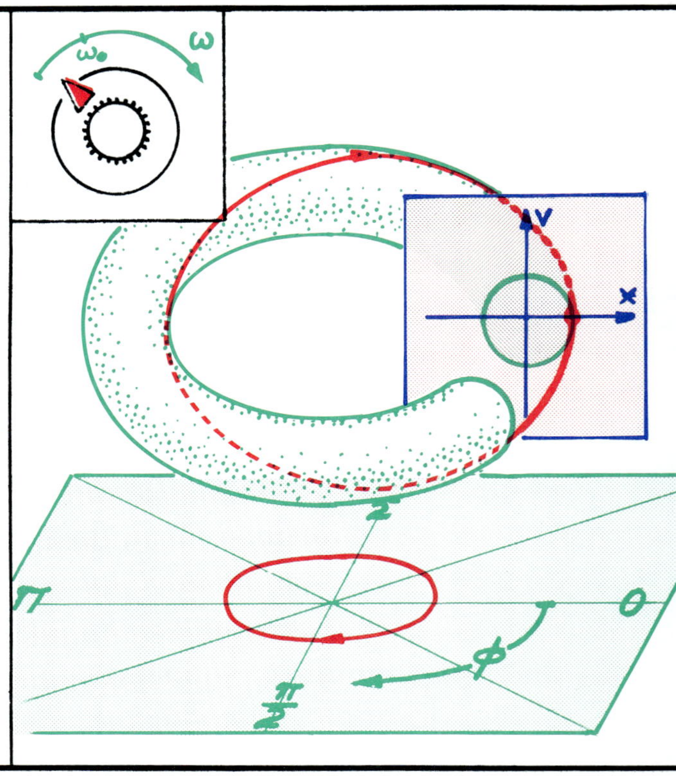

4.2.21. Thus, the combined trajectory, a periodic attractor, passes through the line of constant phase,. zero, in the state plane of the driven (blue) weight (the positive horizontal axis) at the same time the driving phase passes the same value. This occurs just as the strobe flashes and records the point shown here in the strobe plane.

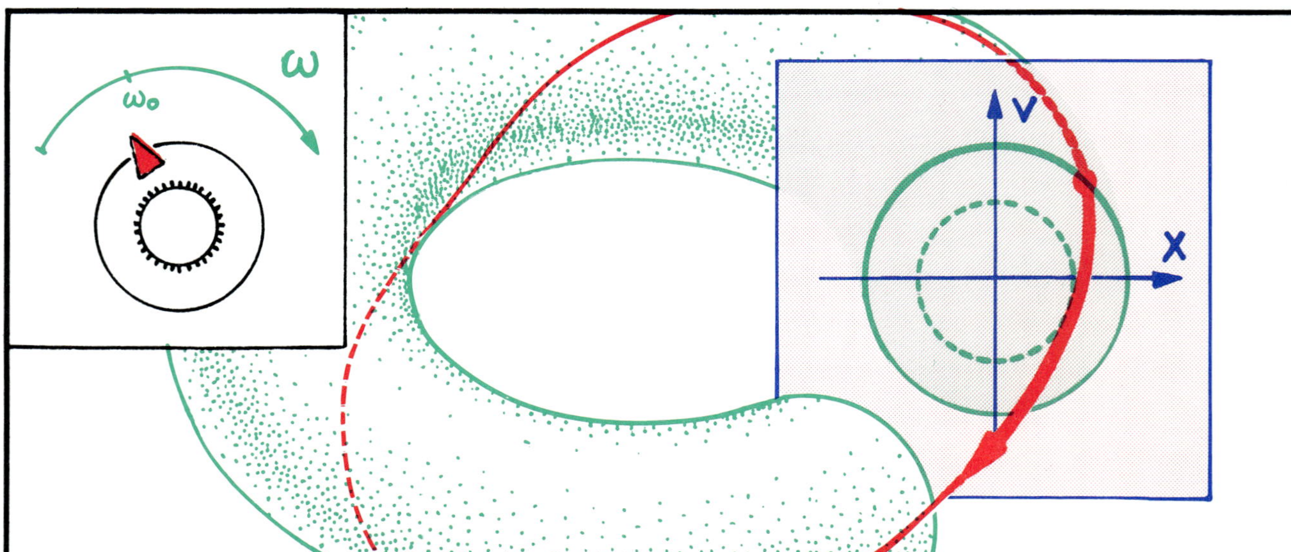

4.2.22. Increasing the driving frequency to a value close to the free frequency of the linear spring and drive (blue) weight, but still smaller, the attractive limit cycle is still an isochronous harmonic. But its phase has slipped. Here, the drive (blue) weight is *behind* the driving (green) weight in phase. This is represented by the red strobe point position *above* the horizontal axis in the strobe plane. Further, the amplitude of the driven (blue) weight's motion is larger than before. This is represented by drawing the red limit cycle on a larger imaginary torus.

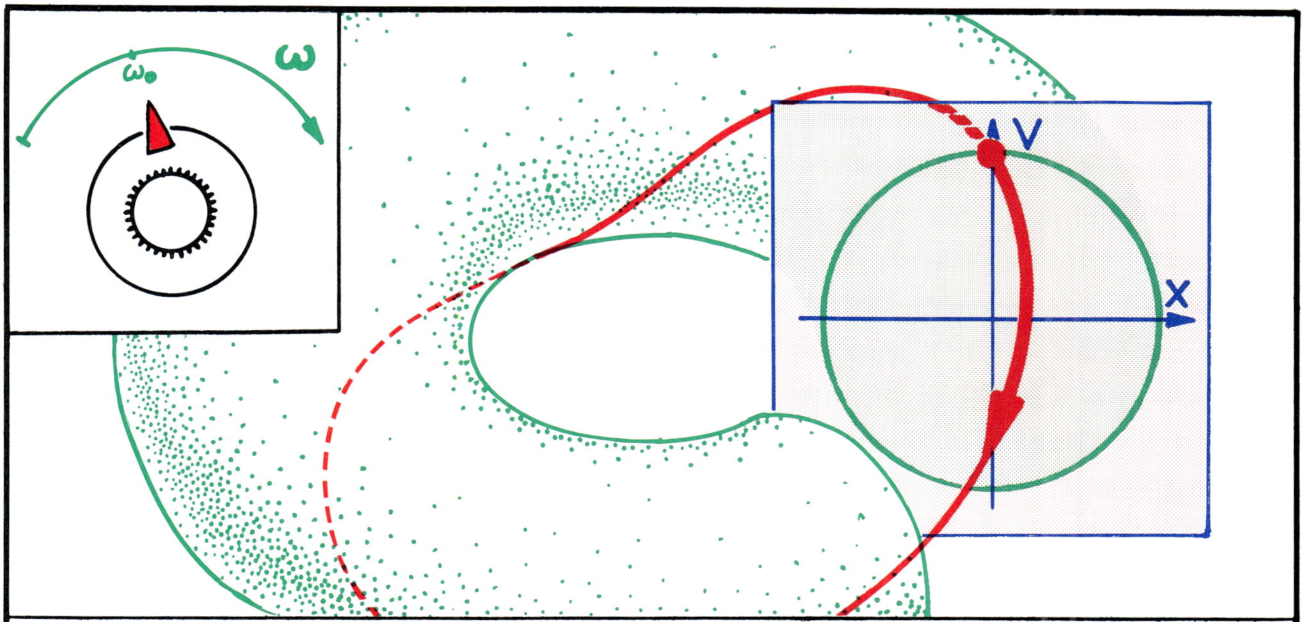

4.2.23. As you may have found in playing the game of bob (Figure 4.2.4), the forced oscillation attains its largest amplitude when the driving frequency is equal to the free frequency of the driven oscillator. But the strobe light on our machine reveals further that at this maximum amplitude, the phase of the forced (blue) weight is $\pi/2$ (a quarter cycle) behind the driving (green) phase.

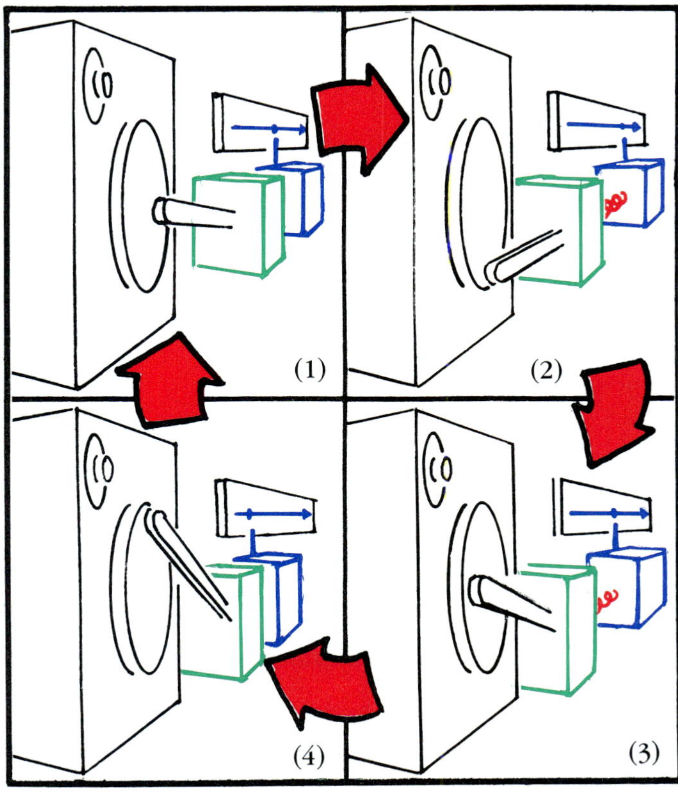

4.2.24. The mechanics of this phase delay is very well known to children who swing. (4) When the driven weight (yourself) goes through phase π (all the way up in back) the driving force goes through phase $-\pi/2$ (you begin to pump forward). (1) At the bottom of the swing forward, your pump has peaked. (2) At the top of your swing to the front, you begin to pump back. (3) At the bottom of your back swing, your back pump has peaked. (Be careful with the larger amplitudes.)

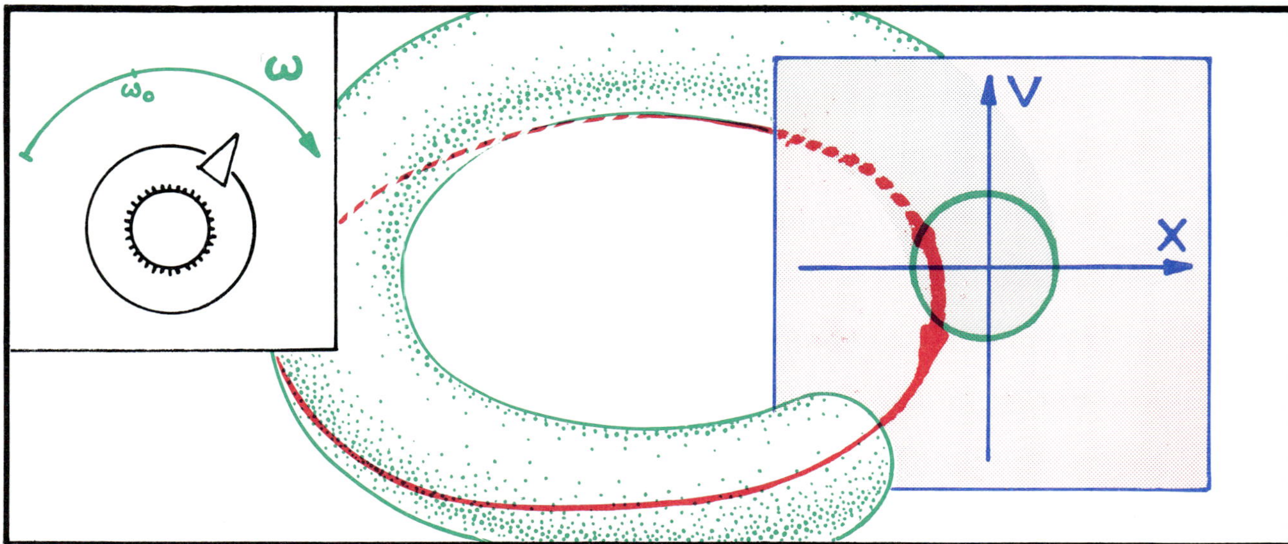

4.2.25. With the driving frequency much faster than the green frequency, the phase of the driven (blue) weight lags a half-cycle ($-\pi$) behind the forcing (green) weight. Also, the amplitude of the forced oscillation is smaller than it was, when the frequencies were the same. But we know this already, from the game of bob. The smaller amplitude is shown, in this illustration, but the smaller "waist" of the locating torus. Compare with Figure 4.2.23.

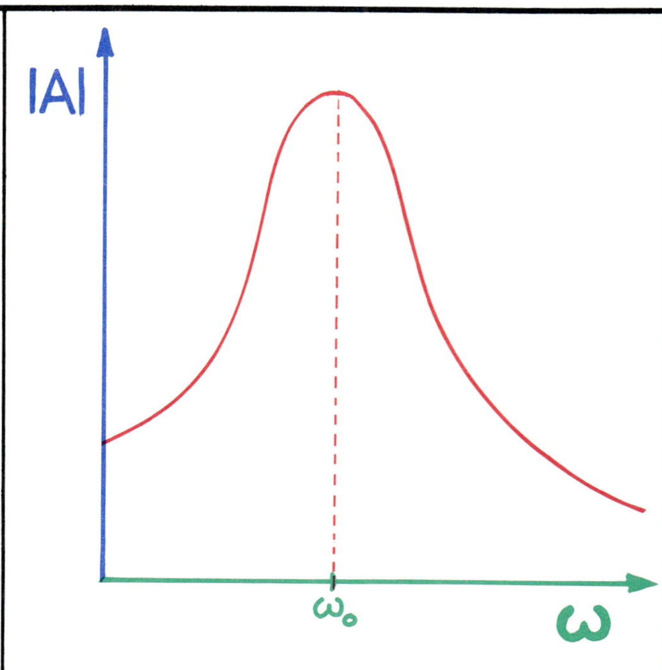

4.2.26. The *amplitude* of the driven bob's motion, as well as its phase, *depends on the driving frequency.* Here is the graph of this relationship, as revealed by Duffing's mathematical analysis, as well as by his experiments.

There are precious few linear springs around, so we really should do this all over again for hard and soft springs. And so, on to the next section for hard springs. Soft springs will be left to the reader as a test.

4.3.3. Forced Hard Springs

A hard spring is slightly more realistic than a linear spring. Recall that with a linear spring, if the force required to stretch it a distance x is $F(x)$, then the force required to stretch it twice as far is twice as great, or $F(2x) = 2F(x)$.

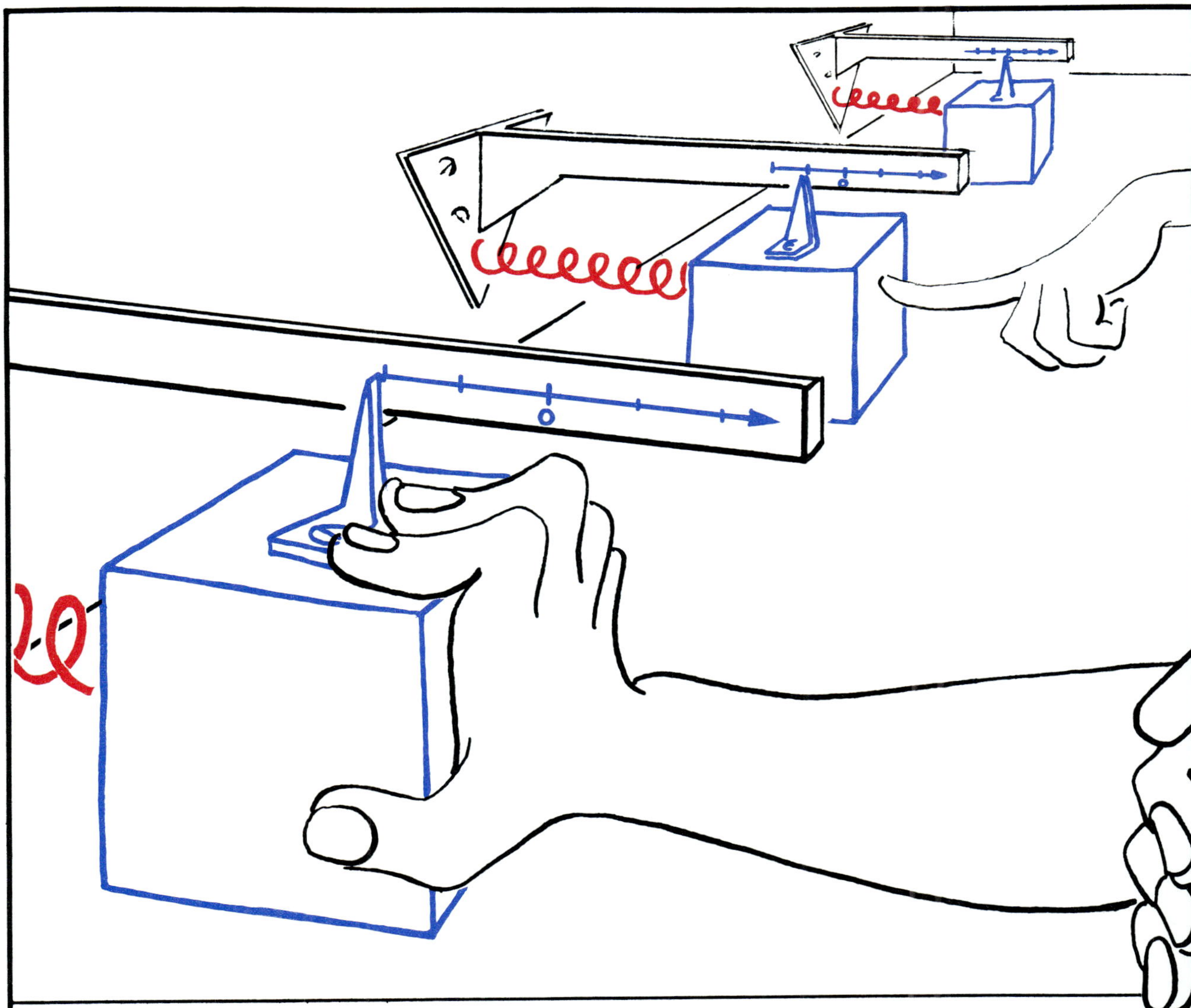

4.3.1. Most springs in real life are *hard.* This means that the force required to stretch the spring twice as far is *more* than twice as great, or $F(2x) > 2F(x)$.

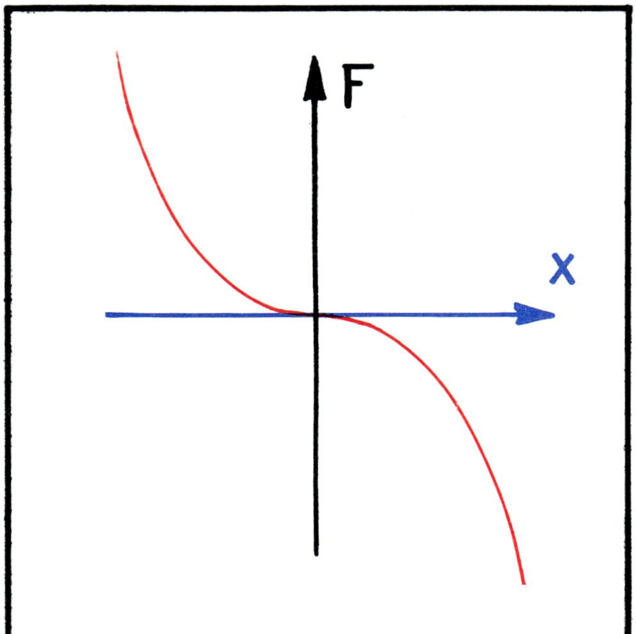

4.3.2. Here is an example of an ideal hard spring. The characteristic function of this spring is a cubic, as shown in this graph.

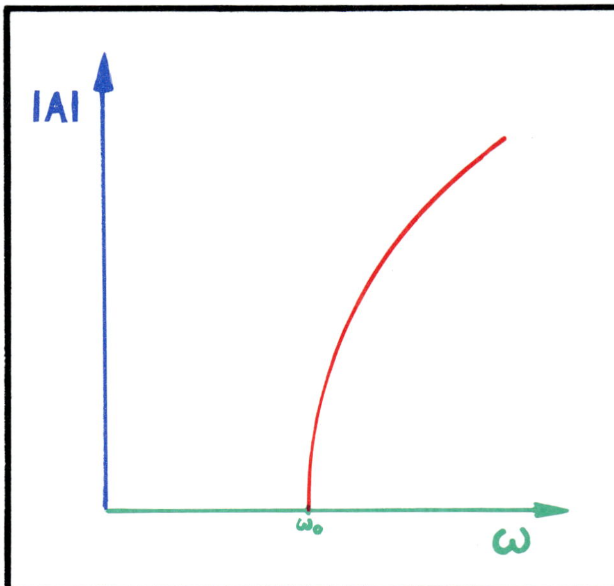

4.3.3. The dynamic consequence of this stiffness is this bend in the response curve: as the amplitude decreases, the frequency decreases—another poor guitar string. The next illustrations explain this bend.

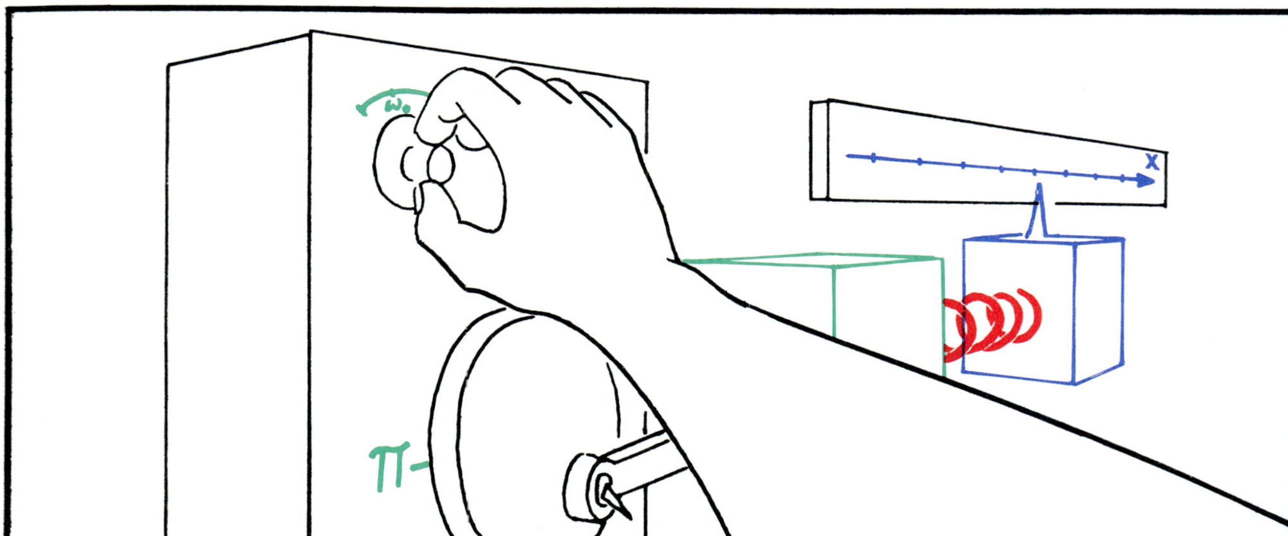

4.3.4. Here is our laboratory apparatus, left over from the experiments of the preceding section. Let's replace the linear spring with a hard one and observe a sequence of forced vibrations with different forcing frequencies. The amplitude of the forcing oscillation is not changed in this sequence.

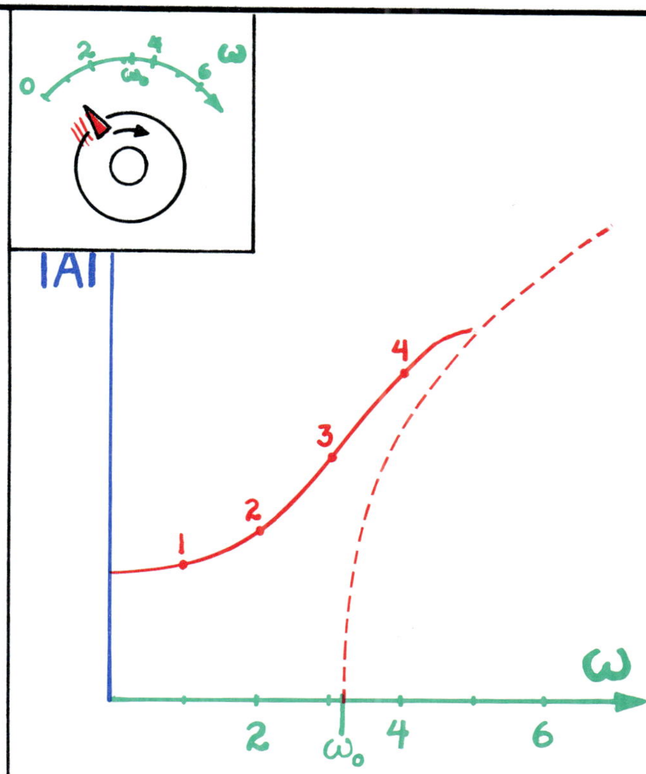

4.3.5. Starting with the speed control at position 1 (slow forcing) and then increasing it, we see a shift in phase, and an increase in amplitude, of the resulting oscillation of the driven (blue) weight. At positions 1 and 2, this is about the same as the linear case. At positions 3 and 4, the effect of stiffness can be observed. The amplitude keeps on increasing, even after the forcing frequency exceeds the free frequency.

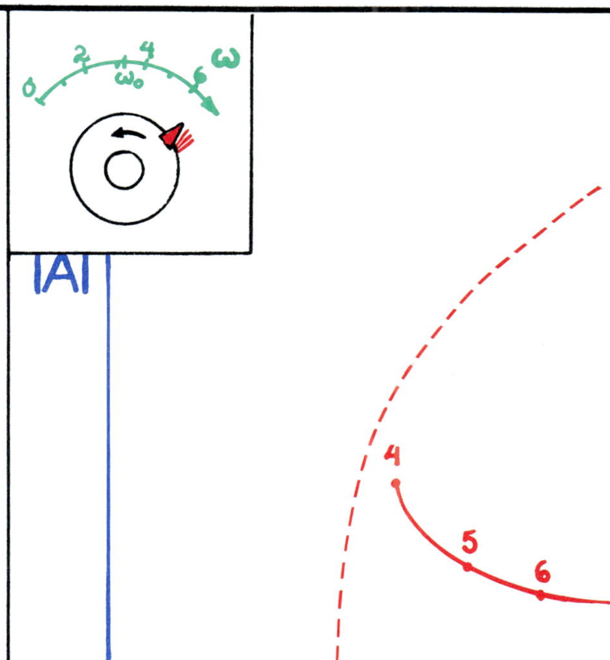

4.3.6. Now start a new sequence of experiments, with the driving motor running at top speed, and decreasing it for each test. Again, the response is much like the linear spring at positions 6 and 5. But at 4, the amplitude of the response vibration suddenly jumps to the higher value. This higher amplitude is the same one found in the first sequence of observations, after increasing the speed from position 3 to 4.

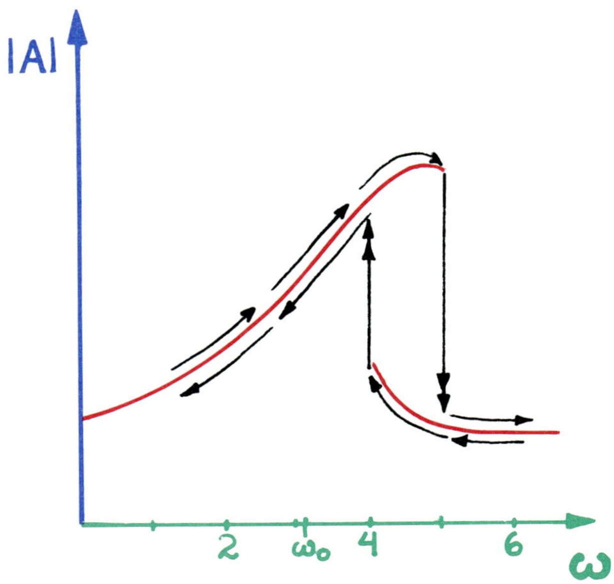

4.3.7. Moving the speed lever all the way to slowest, then back to fastest, back to slowest, and so on, we observe this *hysteresis loop of Duffing.* At 4 and decreasing, the amplitude suddenly increases. At 5 and increasing, it suddenly decreases. If you don't believe it, try it out. Because a pendulum behaves like a soft spring, you may observe this phenomenon in the game of bob.

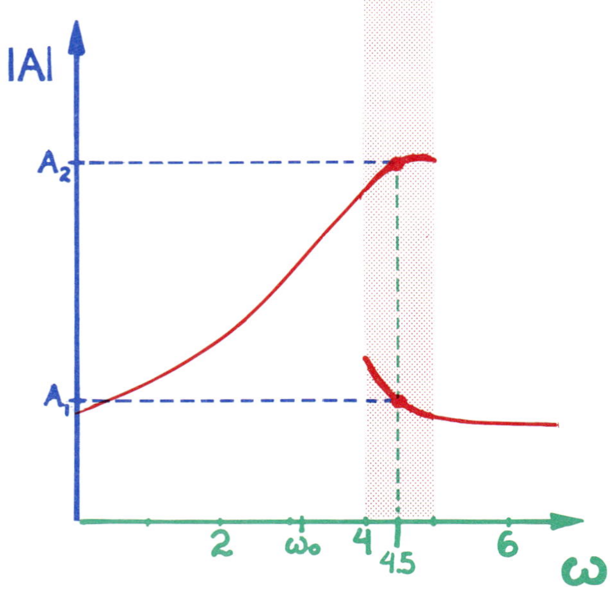

4.3.8. If you do believe it, then you agree that when the speed control of the driving motor is between positions 4 and 5, *there are two periodic attractors* in the phase portrait of the forced system.

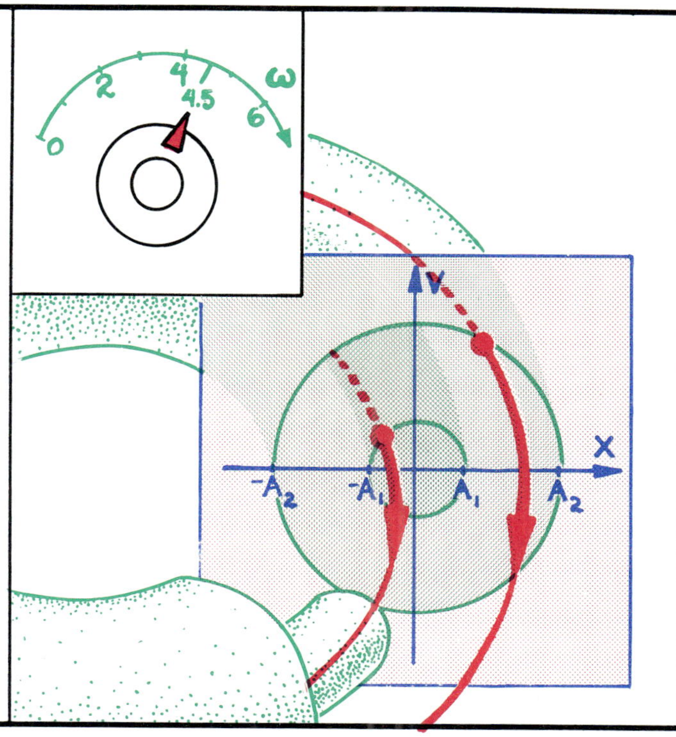

4.3.9. One of the attractive limit cycles, corresponding to the smaller vibration, is close to a half cycle out of phase with the driver. (See Figure 4.2.25.) It circles around a smaller (imaginary) torus. The other, corresponding to the larger vibration, is roughly in phase with the driver. (See Figure 4.2.22.) It circles around a larger (imaginary) torus.

Both of these limit cycles are isochronous harmonics. Neither of the tori is an invariant manifold—most trajectories go through them. Some nearby trajectories tend asymptotically to one of these attractors, some to the other. There are two basins, as in the examples of Sections 1.6 and 2.2.

The two basins are divided by a separatrix. Where is it located in the phase portrait?

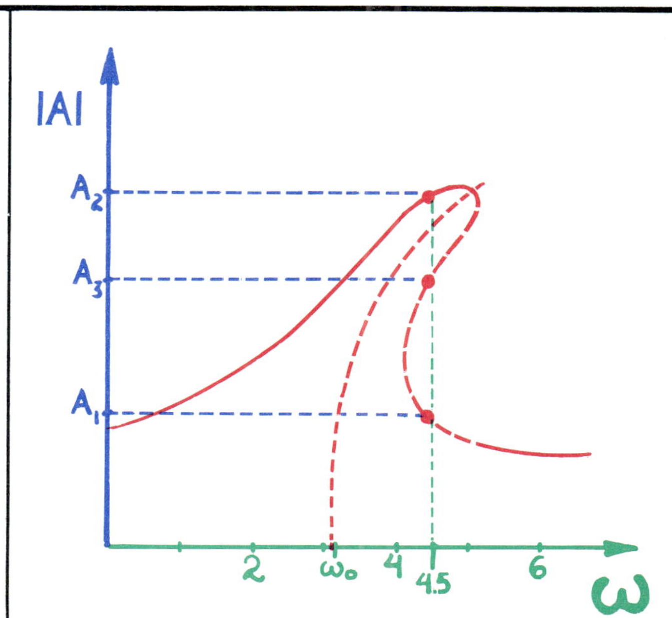

4.3.10. In the response diagram, the two branches of the response curve, observed in the experiments described above, are *connected* as a smooth curve. The connecting segment corresponds to another limit cycle in the phase portrait—one that is *experimentally invisible.*

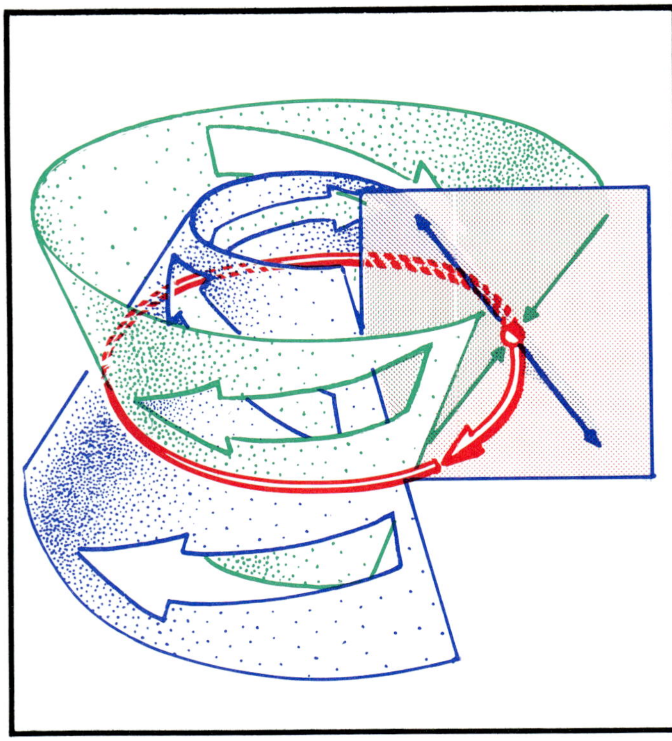

4.3.11. The invisible third limit cycle is a *saddle cycle.* These occur in state spaces of dimension three or more. Saddle cycles are the periodic analog of saddle points. That is, they are related to periodic attractors as saddle points are related to point attractors. In three dimensions, the *inset* of a saddle cycle is a surface, a deformed cylinder belted by the saddle cycle. The inset is shown in green in this illustration. It is a *separatrix*; it separates two basins. The *outset* of a saddle cycle is another deformed cylinder, which crosses through the inset where it is belted by the saddle cycle. The outset is shown in blue. Half of the outset is in one basin, half in the other. These halves are divided by the cycle itself, shown in red.

This green inset is an *invariant manifold.* It consists of trajectories which stay on it forever and tend asymptotically toward the red saddle cycle as time increases. The saddle cycle is the omega-limit set for the trajectories on the green inset, as explained in Section 1.5.

To better visualize the phase portrait, we may use the trick immortalized by Poincaré: extract the strobe plane.

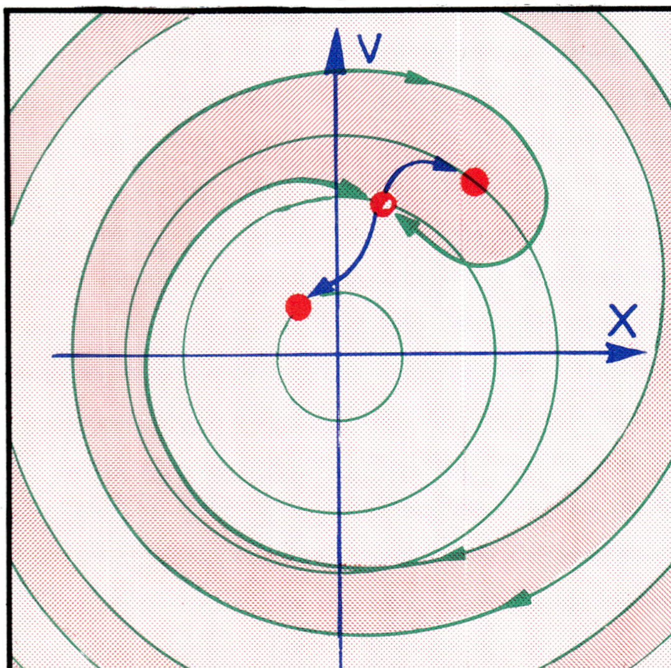

4.3.12. Here the strobe positions of the three limit cycles are shown as red points in the strobe plane. The solid red dots represent attractors. The small half-filled circle denotes the saddle cycle. The inset of the saddle, two green spirals, intersect the strobe plane in two curves, whon here in green. They divide the strobe plane into two regions: the dark-shaded teardrop shape, and the rest. These two regions comprise the *basins in cross-section,* that is, in the strobe plane. Note that the saddle point in the cross-section is *between* the two attractors, both in phase and in amplitude.

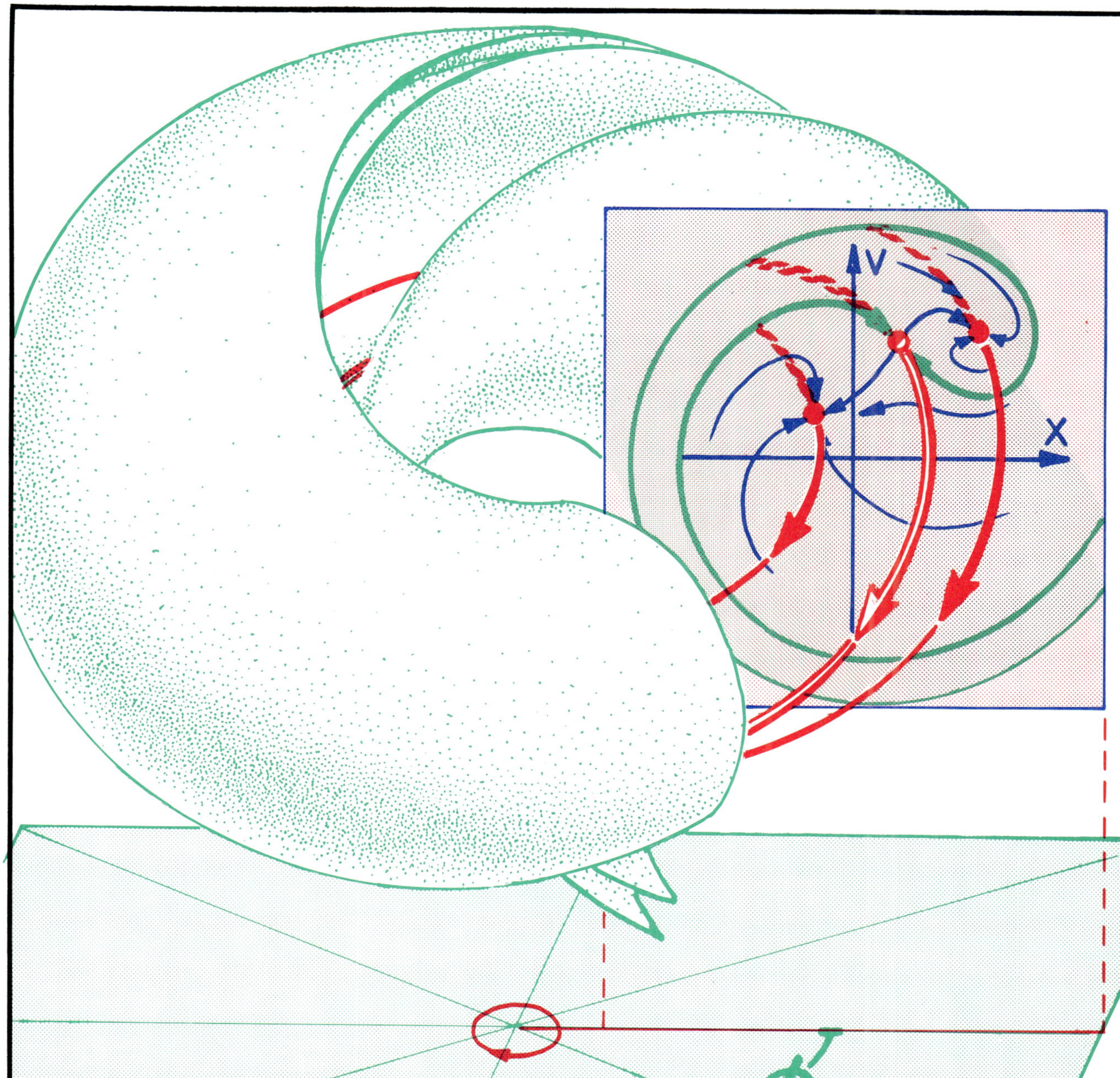

4.3.13. To visualize the basins in full, as three-dimensional regions, we repeat the strobe process at many successive phases of the driving motor. The full basins, revealed by this technique, look like this: yin-yang in 3D. The teardrop revolved once as the section progresses around the driving cycle. The three red cycles are the isochronous harmonics discovered by Duffing.

All this from experimenting with different forcing frequencies, at the same forcing amplitude. What if we changed the amplitude?

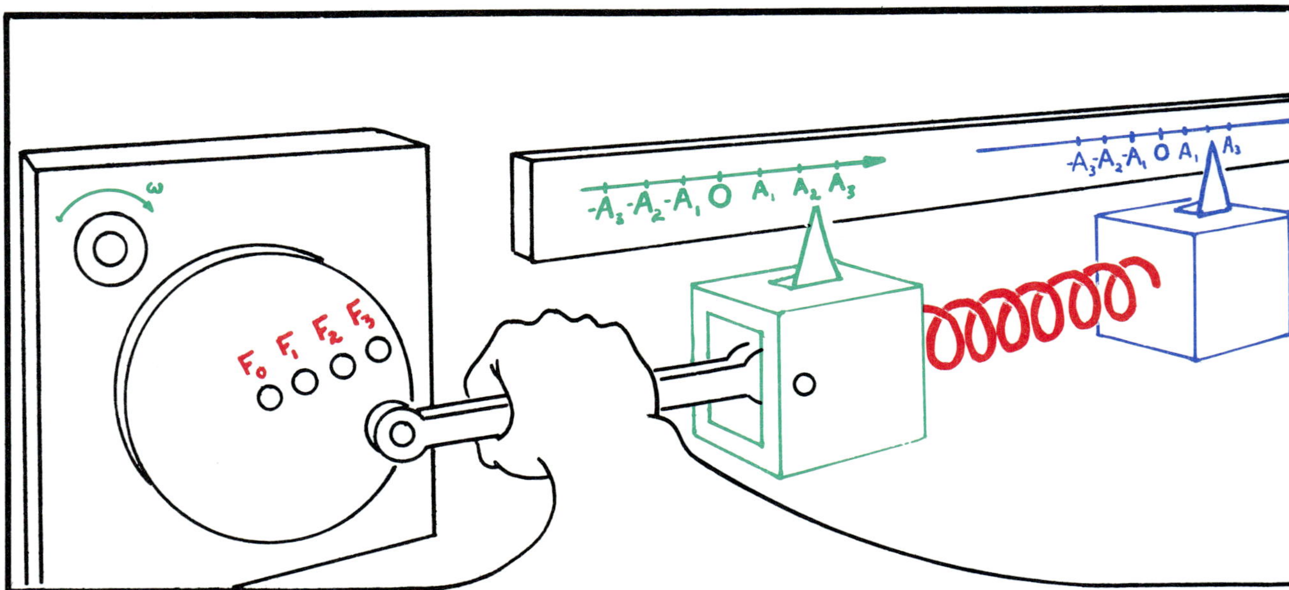

4.3.14. Drilling more holes in the turntable, we gain easy access to the amplitude of the forcing oscillation, without affecting its frequency.

Having changed the amplitude, we may repeat all the experiments described above, with the frequency changed variously.

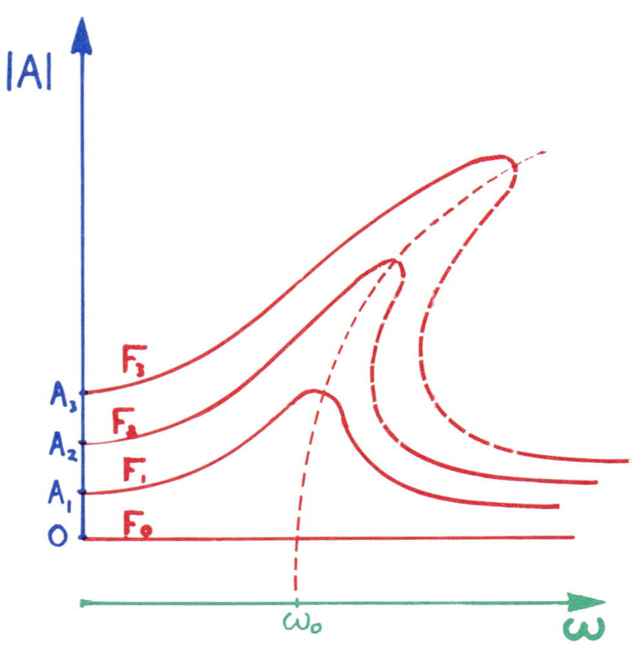

4.3.15. The results are always similar, except that the location of the response curve is shifted, kink and all. Here are the response curves for three different amplitudes. The larger the amplitude, the higher the response curve in this diagram. Notice that the lower curve (smallest forcing amplitude) looks rather like the response curve of a linear spring.

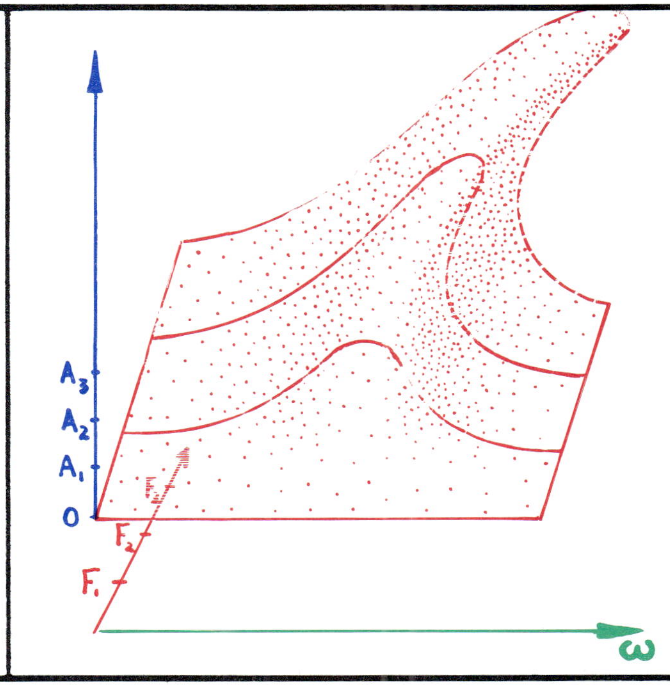

4.3.16. Regarding the response as a function of the two control parameters, frequency and amplitude of the forcing oscillation, we find that the curves on the right are successive slices of a kinked surface in a three-dimensional response diagram.

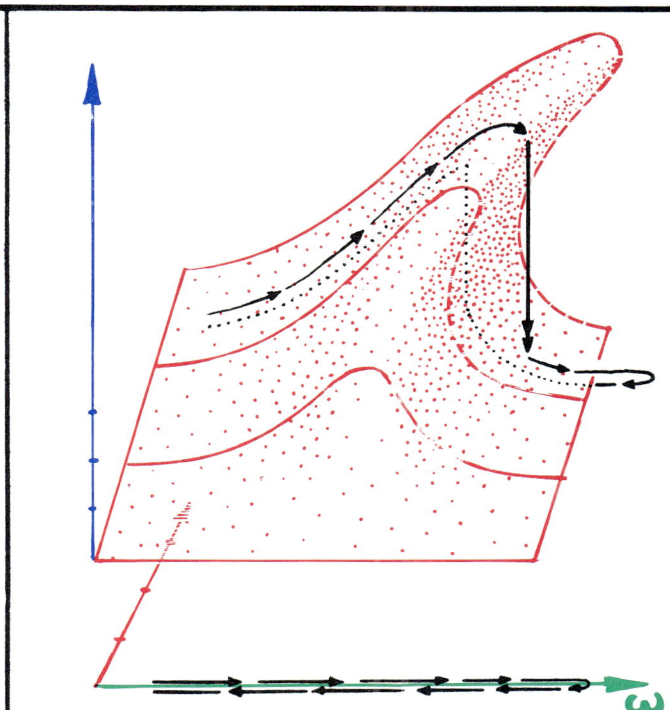

4.3.17. If you have already heard about catastrophe theory, you may recognize this as the *cusp catastrophe.*[3] Here, we have added the *histeresis loop* to the cusp figure.

Actually, since Duffing there has appeared an enormous literature on the forced pendulum. Much of this is devoted to additional attractors that have been found in Duffing's phase portrait, which are *non-isochronous harmonics.*

In the next section, we give a brief introduction to these other harmonics.

4.4. Harmonics

When the driving amplitude is fixed but the frequency is changed in a sequence of experiments, as described in the preceding section, unusual motions of the bob turn up when the driving frequency is twice the free frequency, or half of it, and so on. These are the *non-isochronous harmonics.* Here is the simplest example, sometimes called the *second harmonic.* Suppose that the driving motor is turning at about half of the free frequency of the spring-bob system. We find that the driven system responds with a sustained oscillation at about the free frequency.

Get out your game of bob again for this one.

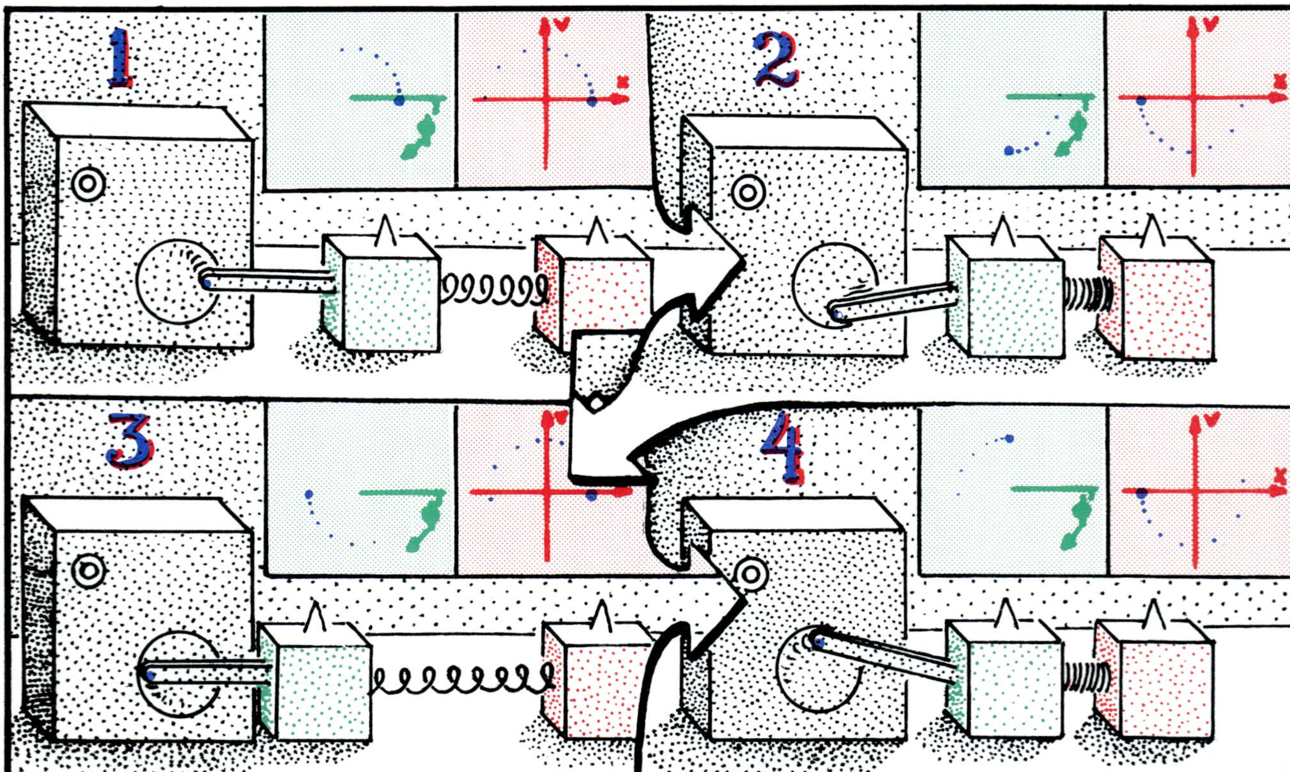

4.4.1. Here are four strobe views of the second harmonic motion.

1. Driver phase 0, driven phase 0. They are both at the beginning of a cycle, in phase.

2. Driver phase $\pi/2$, driven phase π.

3. Driver phase π, driven phase 0. Since being in phase in 1, the drive (blue) weight has completed a full cycle, while the driving turntable has completed only half a turn.

4. Driver phase $-\pi/2$, driven phase π.

1. Driver phase 0, driven phase 0. Since 1 above, the turntable has completed a full turn at last, while the driven weight has done two swings.

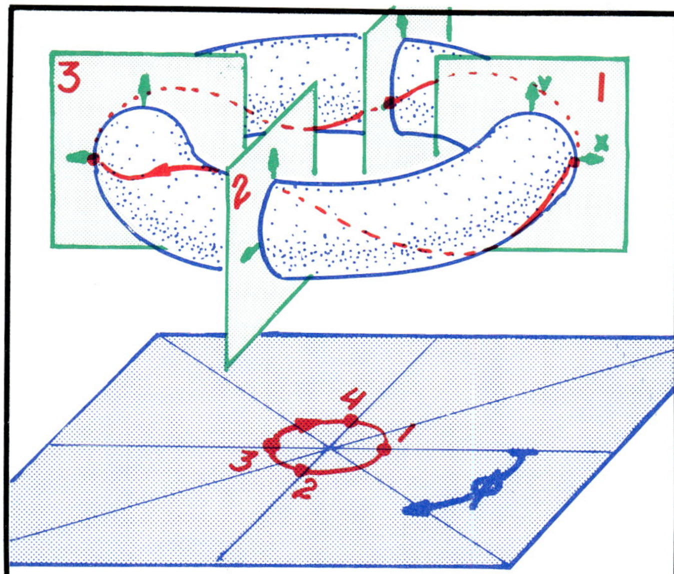

4.4.2. Let's now pick up the four cards (strobe planes) of the preceding panel, and place them in their proper places within the ring model. Interpolating the rest of the trajectory, we obtain this red trajectory. It is an attractive limit cycle, which winds twice around the waist of the imaginary green torus, while making one full circuit of the ring. It represents the second harmonic of the driven spring-bob.

In the preceding section, there was already an imaginary green torus or two in the ring model, representing the isochronous harmonics of Duffing. If this double-twisting attractor is also in the ring, how is it situated with respect to the isochronous harmonics, or the rotating teardrop?

The isochronous harmonics, the second harmonics, and all the other little harmonics, all live harmoniously together in the ring model.

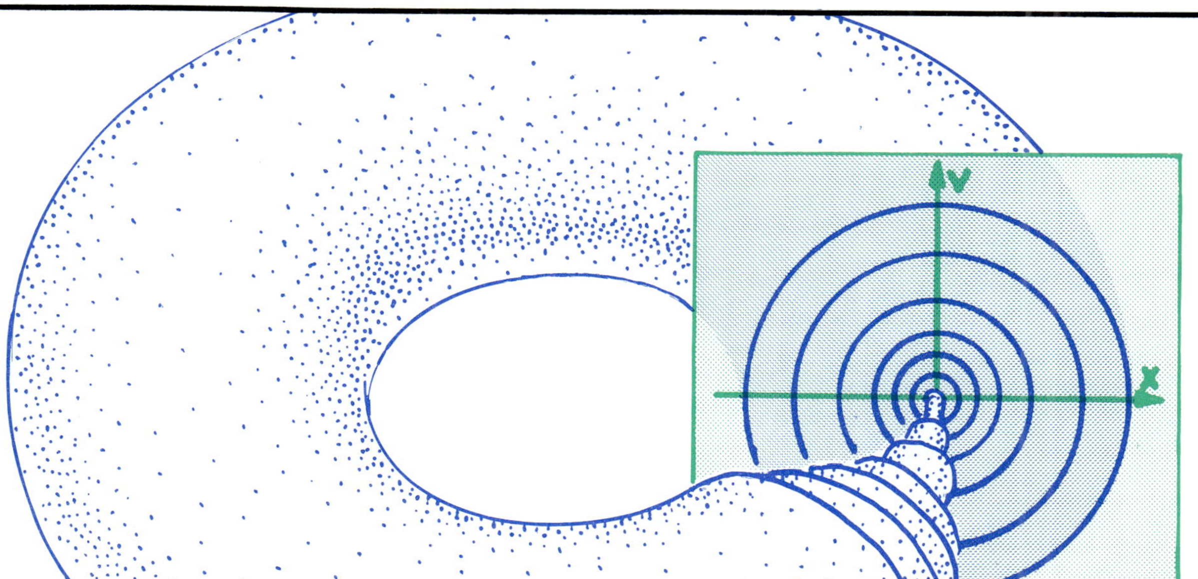

4.4.3. Here are several of the harmonics in the ring model. Each has its own home, in one of the concentric rings. When the forced oscillation creates a harmonic (periodic attractor) in the compound system with a non-isochronous frequency, *the amplitude of the harmonic is smaller.* Thus, it winds around a locating torus with a smaller waist. Thus, the outer rings in this illustration contain the osichornous harmonics. The smaller rings house the more exotic harmonics, just like the circles of Hell in the Inferno of Dante.

To get a clear picture of the harmonic vibrations of a forced pendulum or spring, we need a certain familiarity with trajectories on a torus.

We pause now for a short course in toral arrangement.

4.4.4. Earlier, in Figure 4.3.21, we introduced a phantom torus to depict the isochronous harmonic, in phase with the driving oscillation.

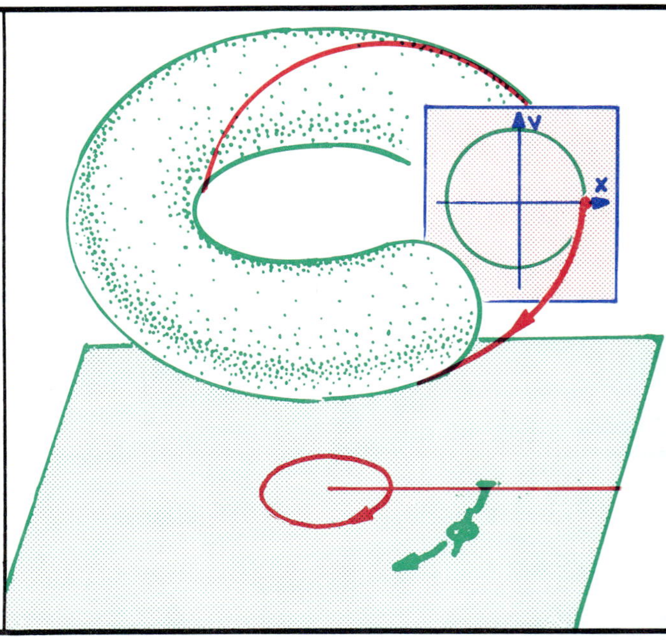

4.4.5. We may regard the state space as a cyclical deck of cards. Each card slices the phantom torus in a vertical (green) circle. Thus, the torus is made of green circles. Each of these corresponds to a particular phase of the driving oscillation of the green weight.

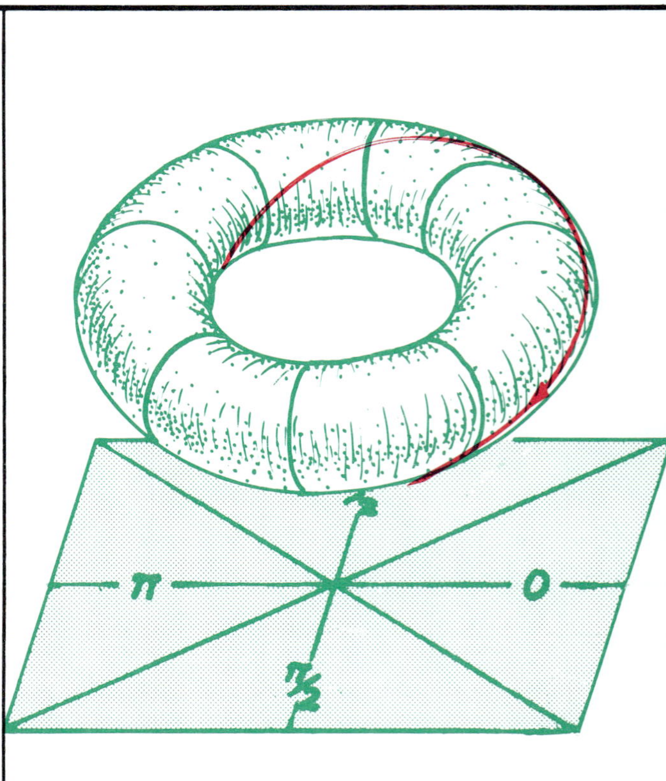

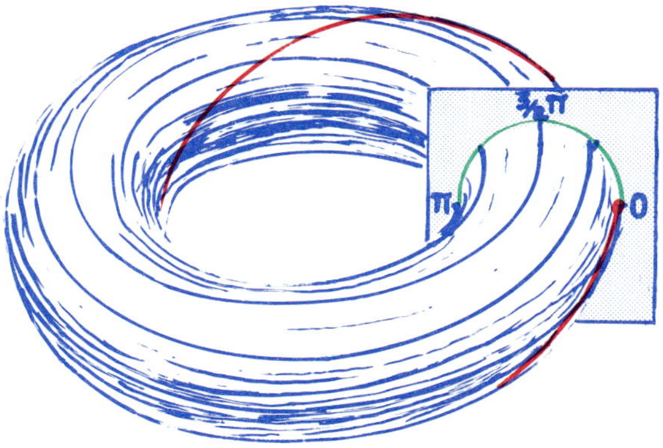

4.4.6 On the other hand, we may think of the torus as being made of these horizontal blue stripes. Each of these horizontal circles corresponds to a particular phase of the drive (blue) weight.

Let's take an imaginary green torus out of the ring. Slice vertically through it at the strobe plane on the extreme right, corresponding to green phase zero.

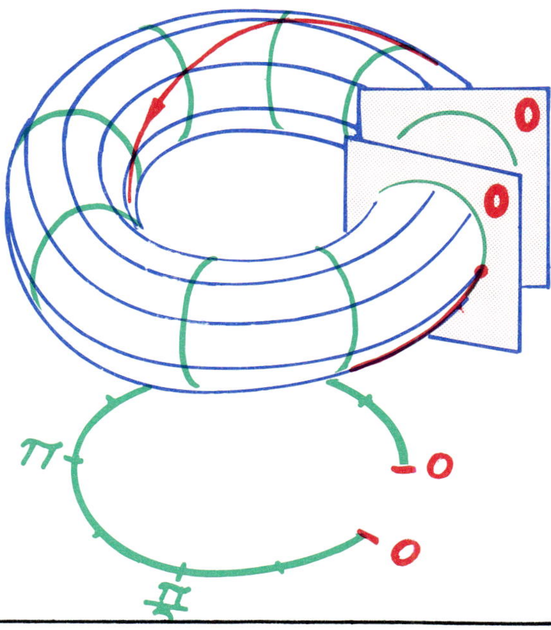

4.4.7. Grab the loose end closer to us. Pull it out and leftward and push it back, to straighten it out.

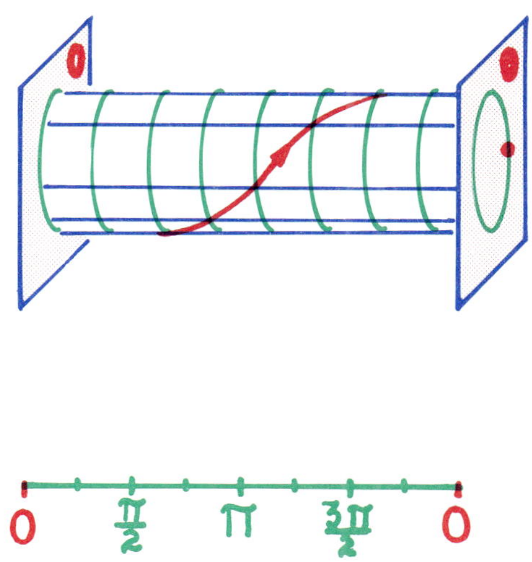

4.4.8 Straightening out the tube and reducing the scale of the green phase, we have a section of a cylinder. Both ends correspond to green phase zero.

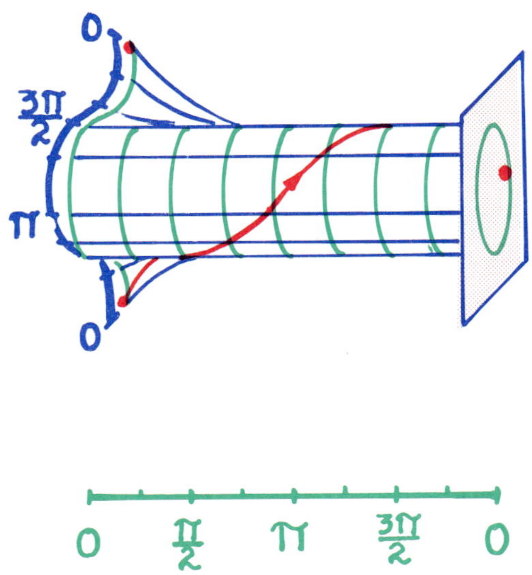

4.4.9 Now cut the back of the cylinder, along the line of zero phase of the blue cycle. Push the lower edge down, while pulling the upper edge up.

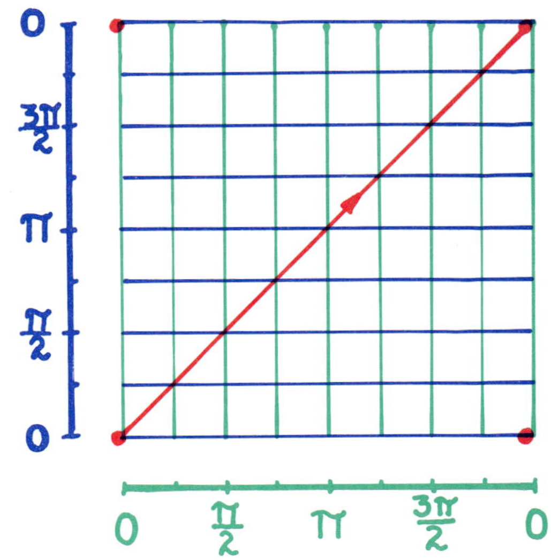

4.4.10. We obtain this flat square. Both horizontal edges correspond to blue (driven) phase zero while both vertical edges correspond to green (driver) phase zero.

The red trajectory turns out to be a straight line through two corners, in the case of a linear spring.

This red trajectory represents an isochronous harmonic, with both cycles in phase. Thus the driving (green) weight and the driven (blue) weight begin together, each at phase zero. After one cycle of each, they return to phase zero simultaneously. To be sure of this, move along the red trajectory, little by little. At each point, observe the turntable phase (by looking down to the lower edge) and the bob phase (by looking left, to the vertical edge) and see how they are changing.

Here is another translation exercise, from toral to flat representation.

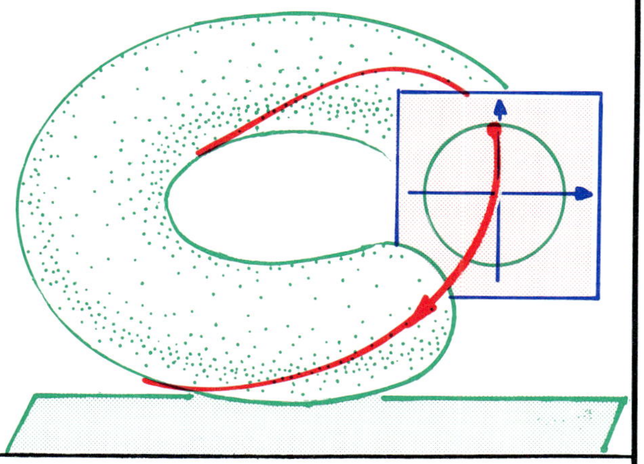

4.4.11. This is another isochronous harmonic on the torus. But the two oscillations are a quarter-cycle out of phase, as in Figure 4.3.23.

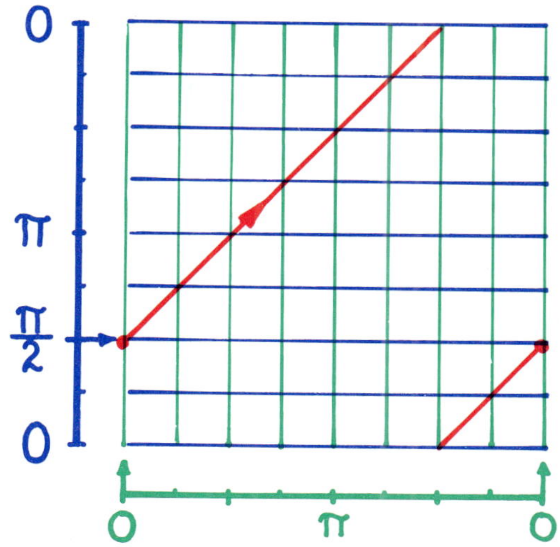

4.4.12. The phase relationship is easier to analyze in the flat representation. On the left edge, driving weight is at phase zero while driven weight is at phase $\pi/2$.

To step along this red trajectory, you must remember that when you fall off the top of the square, you reappear at the bottom. Likewise, when you run off the right edge, you reappear on the left. It's just like a TV screen.

That finishes the digression on toral arrangement. Now, back to harmonics. The ideas developed, originally, in the context of musical instruments.

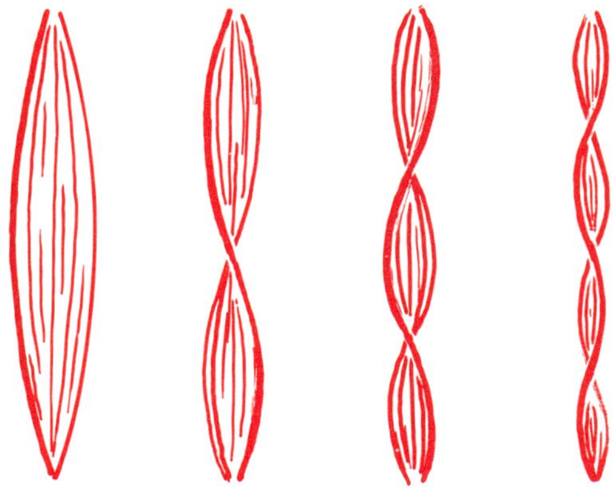

4.4.13. A guitar string vibrates in differnt *modes*. Explained mathematically by Euler, they were exhaustively studied in experiments by Pythagoras. The sound of each of these modes is a pure tone. The mode shown on the left is called the *fundamental*. The frequency of each mode is an integer multiple of the fundamental frequency.

Musical harmonics may be plotted on a torus. But instead, we will return to the context of the driving and the driven weights. In this context, the driving weight plays the role of the fundamental tone. We shall explain, in sequence, *ultraharmonics, subharmonics,* and *ultrasubharmonics.*

This is an example of an ultraharmonic, in toral and TV representations.

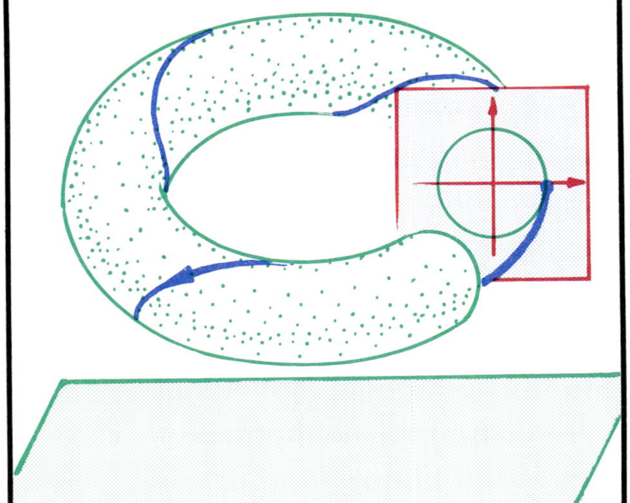

4.4.14. This red trajectory on the torus winds three times around the waist (driven cycle) while going only once around the hole (driving cycle).

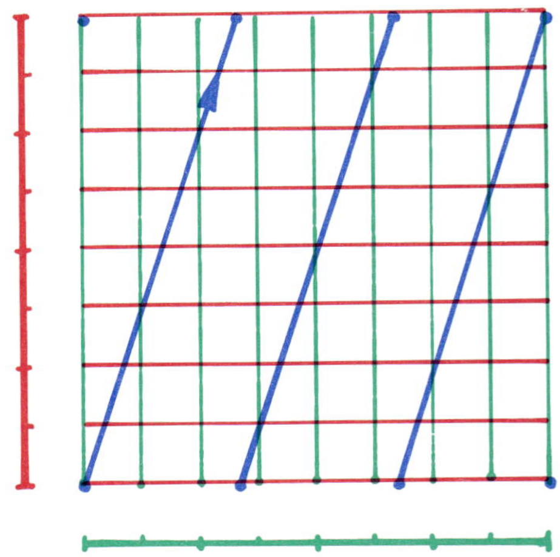

4.4.15. In the flat square, the red trajectory crosses the horizontal edge three times and the vertical edge only once.

This example is an untraharmonic of *harmonic ratio 3:1.* That is, 3 bob cycles occur during 1 turntable cycle. There are untraharmonics of all integer ratios.

4.4.16. Here is the general ultraharmonic. There are an arbitrary number of driven cycles, say *P*, for one turntable cycle. The harmonic ratio is *P*:1.

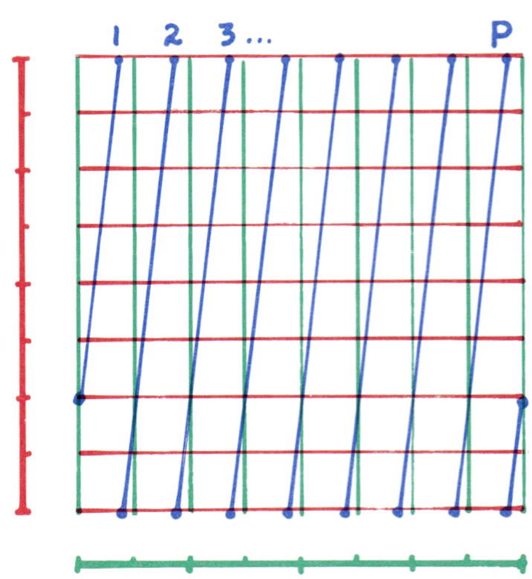

A subharmonic is characterized by slower motions of the driven weight. The drive must complete several cycles before the driven weight completes one.

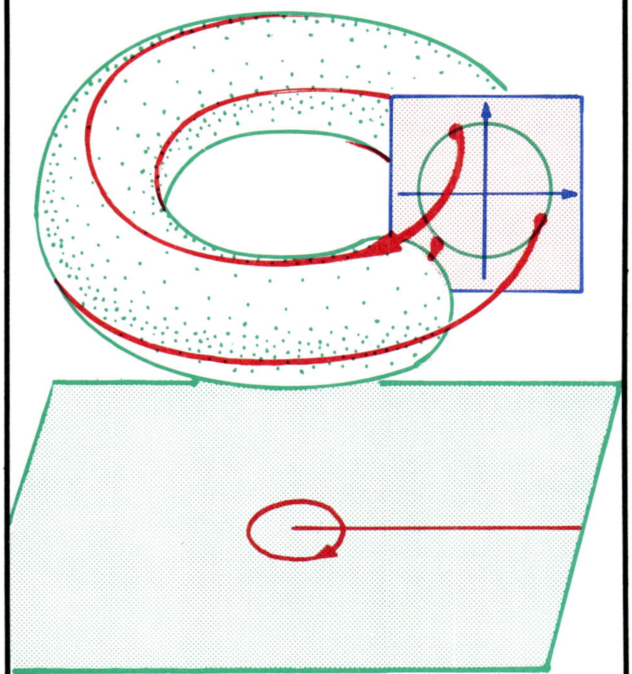

4.4.17. This is a subharmonic requiring 3 driven cycles, to complete 1 response cycle.

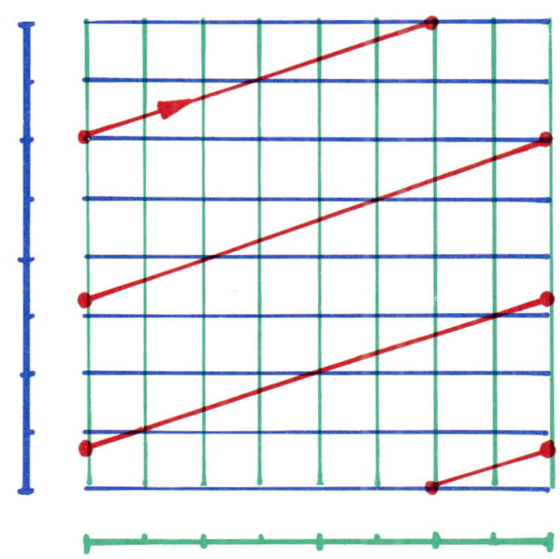

4.4.18. The flat representation of the red trajectory wraps around the vertical edge 3 times and the horizontal edge just once. The harmonic ratio is 1:3.

The general subharmonic requires some number of turntable cycles, say *Q*, for one cycle of the bob. The harmonic ratio in this case is 1:*Q*.

Isochronous harmonics have harmonic ratio 1:1. Isochronous, ultra, and subharmonics are special types of the general harmonic, also known as the *ultra-subharmonic*. These have harmonic ratios *P*:*Q*, with any integers for *P* and *Q*.

The next example gives rise to the musical interval do-sol (a fifth). The harmonic ratio is 3:2.

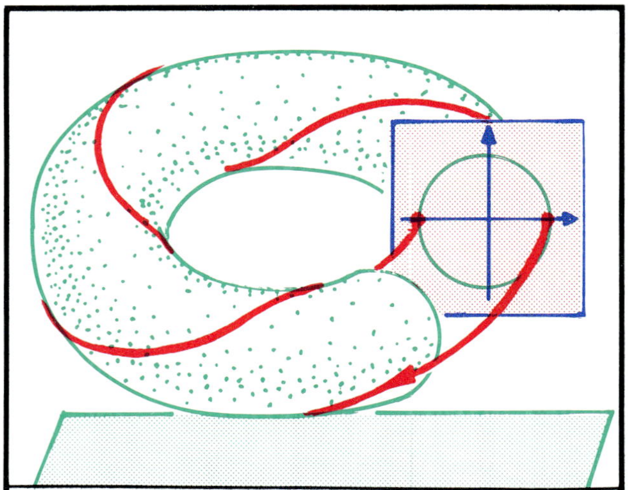

4.4.19. The red trajectory goes twice around the hole, each time making a turn and a half around the waist.

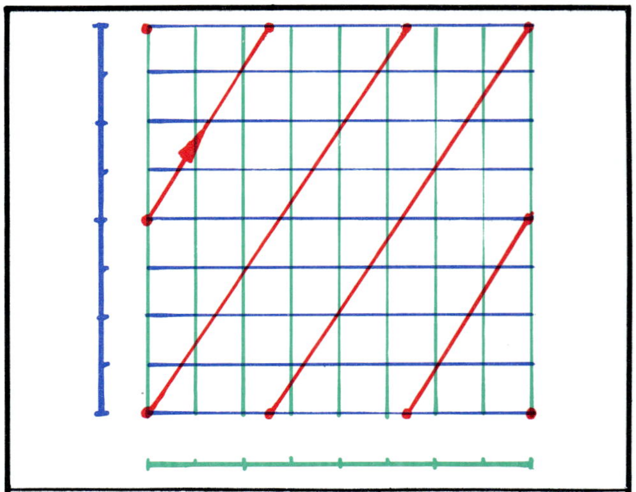

4.4.20. In the flat representation, it transits the horizontal 3 times and the vertical edge twice.

You see now how to draw a general harmonic, with any rational number *P*:*Q* as ratio. Beware: we have drawn the trajectories on the flat square more or less straight, for simplicity. But in actuality, they may be very curved.

How about a harmonic with an irrational ratio?

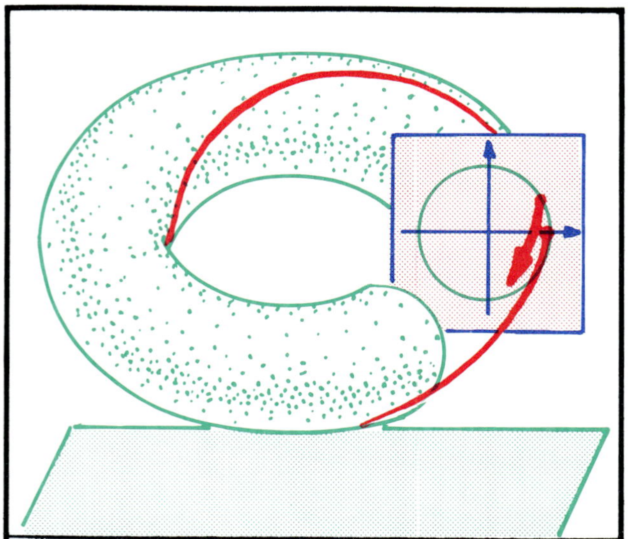

4.4.21. Here is a piece of a red trajectory. If we drew it all, the torus would be almost all red. It goes round indefinitely, without closing. This is the *solenoid* of Section 1.4.

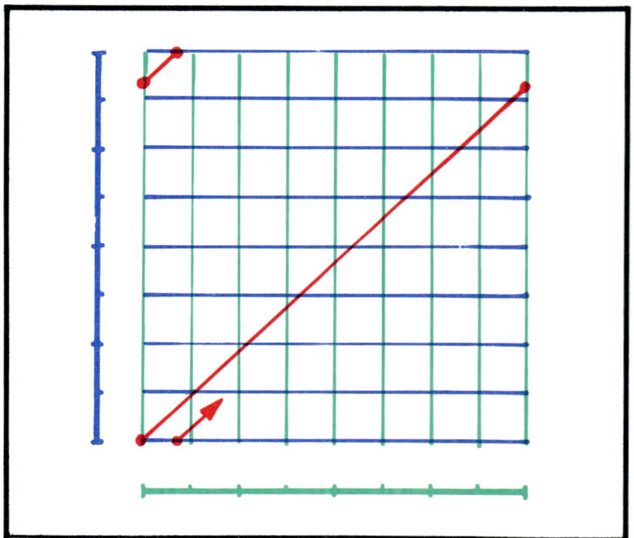

4.4.22. Again we have chosen the simplest situation to draw: the trajectory becomes a straight line in the flat square. But after each wrap-around, it hits a blank space upon re-entry. It never closes.

As we shall see in Part Two, this situation is very rare. Even in the case of actual harmonics (rational harmonic ratios) the more exotic ratios are relatively rare, as experimentalists have found.[4] They are restricted to very small rings isn the ring model. This means they occur with very small amplitudes.

There is much more to be learned from Duffing's game of bob, as we shall see in Part Two. But now let's move on to the penultimate topic in classical dynamics—compound oscillations.

5

Compound Oscillations
Invariant Tori in 3D from Huyghens to Hayashi

The considerations of Rayleigh, in the question of the production of sound by bowed strings in particular, evolved into a large branch of dynamics, called *forced oscillations,* or *nonlinear vibrations.*[1] The first results were obtained by Duffing, in 1918, for the forced pendulum, and have already been described in the preceding chapter. In the following decade, similar results were obtained by Appleton, Van der Pol, Andronov, and others for the periodic, forced perturbation of a self-sustained oscillator. In this chapter, we describe their results. These discoveries for forced oscillators are subtly different from Duffing's observations of forced pendula, described in the preceding chapter. There, we forced a point attractor. Here, we will force a periodic attractor.

160

Now the two oscillators are *coupled.* This means that the phase portrait is *perturbed* by the addition of small vectors at each point of the state space. This small vectorfield is added to the dynamical model representing the uncoupled system. Without saying exactly what this small perturbation is, one can conclude something about the coupled system anyway. This amazing conclusion is a geometric theorem of Peixoto, described at length in Part Three.

Here is the geometrical model for the same physical system, with coupling introduced between the mechanical oscillators.

5.1. The Torus Model for Two Oscillators

The first step toward understanding forced oscillations is to make a geometric model for the states. We will use the torus, now familiar from the previous chapter. Then, we describe the dynamics on the torus corresponding to two *uncoupled* oscillators. In the following section, we allow coupling between the two oscillators, describing the dynamics on the torus. In the third section. we introduce the *ring model* for forced oscillations.

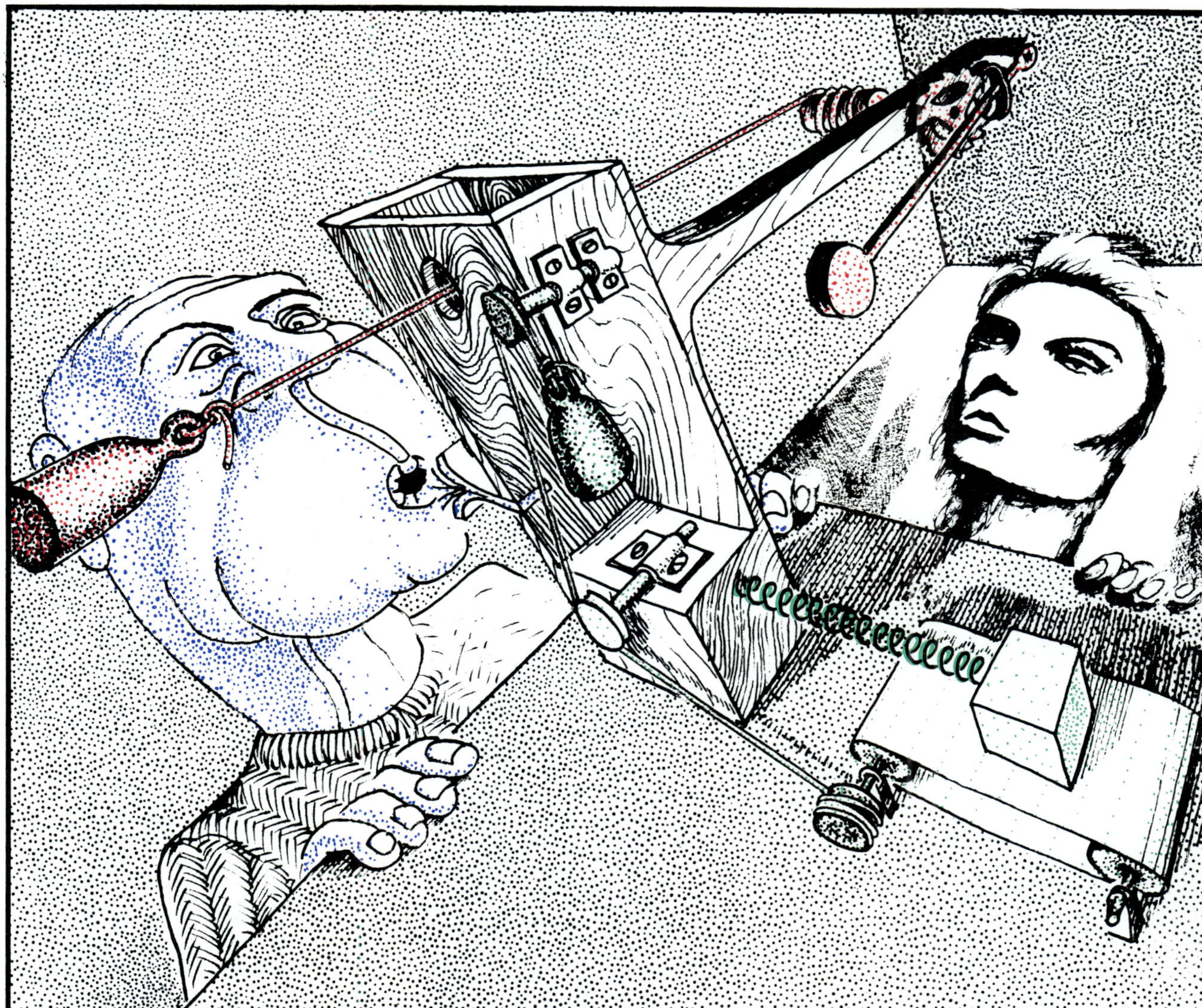

5.1.1. These mechanical systems have dynamical models, described in Chapter 3, with similar phase portraits. They are self-sustained *oscillators.*

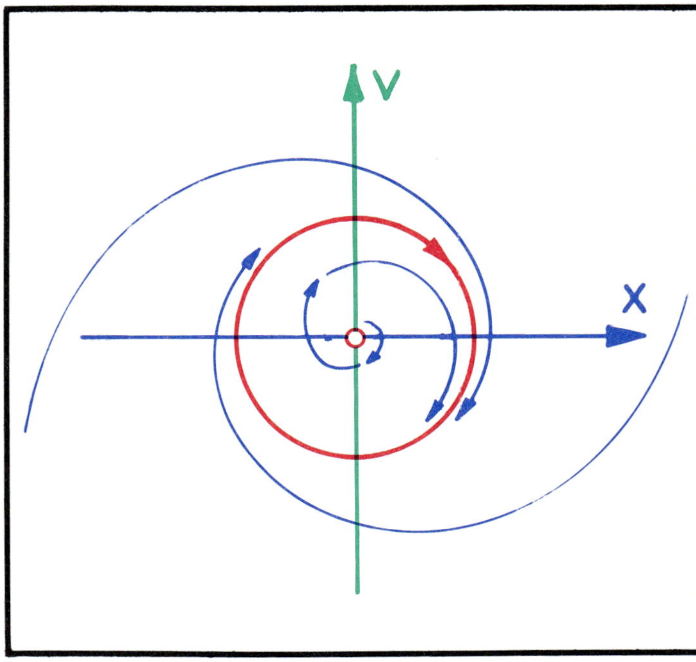

5.1.2. Their phase portraits contain attractive limit cycles, that is, periodic attractors. After the start-up transient dies away, the trajectory follows the attractive limit cycle.

For the sake of visualizing the asymptotic behavior of the oscillators, we may ignore the transient behavior in the dynamical model. Thus, the two-dimensional model (the plane with a limit cycle around the origin) may be replaced by the limit cycle itself, standing alone.

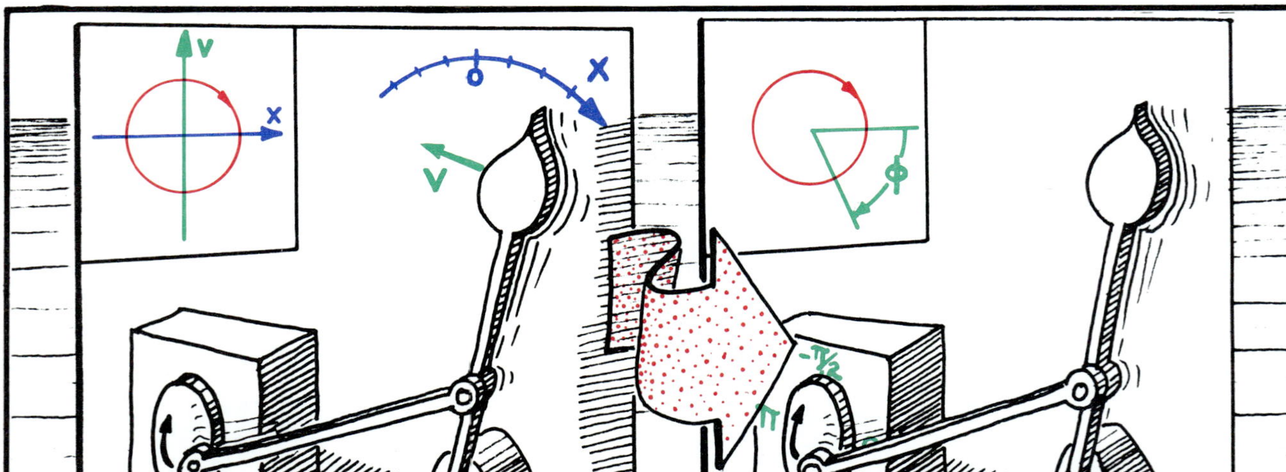

5.1.3. Recall that in the last chapter, we introduced the *reduced model* for the states of an oscillator. The state of the oscillator is represented by a point of this cycle, corresponding to its *phase*. The two-dimensional model for the states of one oscillator is thus replaced by a one-dimensional model. We allow the plane to fade away, while the cycle remains.

Next, we consider two different oscillators. If the two oscillators are physically separate, the motion of each is uninfluenced by the other. We say they are *uncoupled oscillators*.

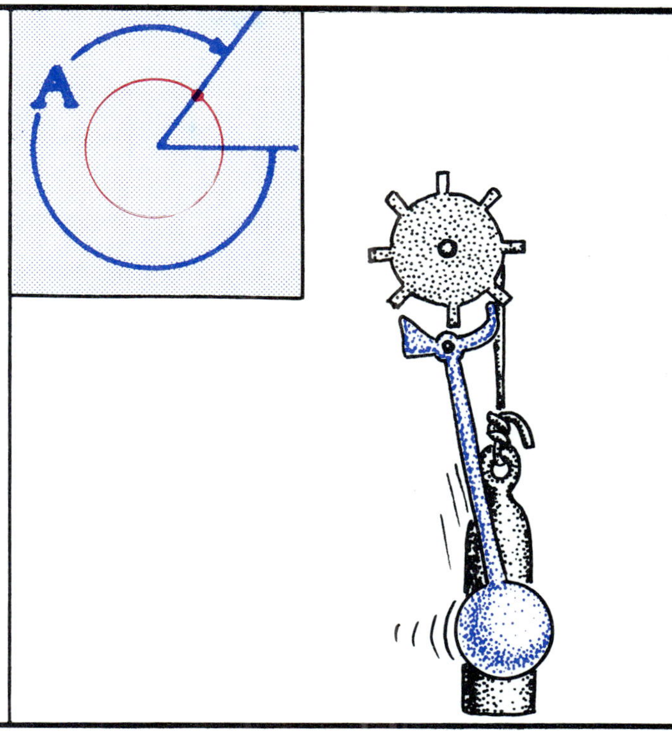

5.1.4. In the reduced model, the state space for one oscillator is a cycle. Although this cycle may belong to the plane of displacement and velocity, we will forget the plane and concentrate on this cycle. A state of this oscillator is specified by an angle, *A*, corresponding to its phase.

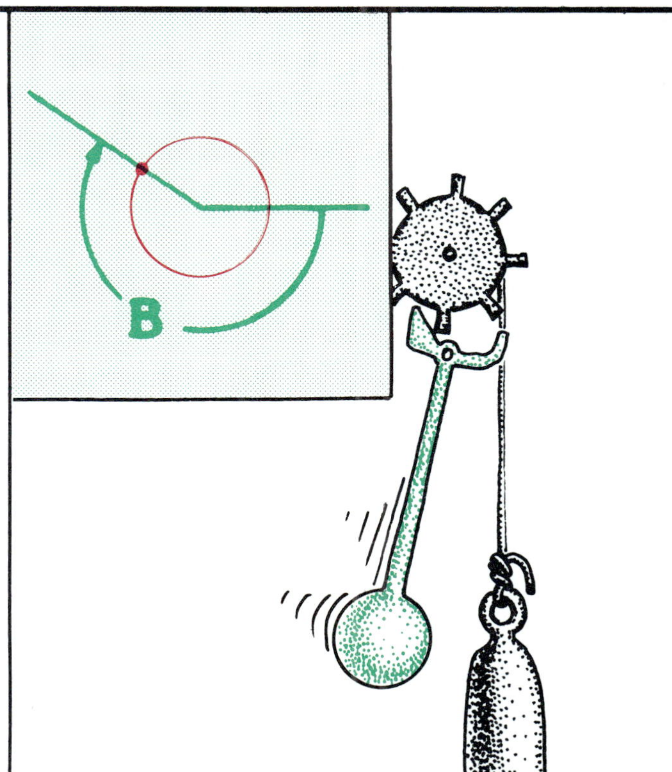

5.1.5. The second oscillator is represented by another cycle. We also forget the plane we have used to define this cycle. A state of this oscillator may be specified by another angle, *B*, representing its phase.

The two oscillators may be described simultaneously in a single state space as follows.

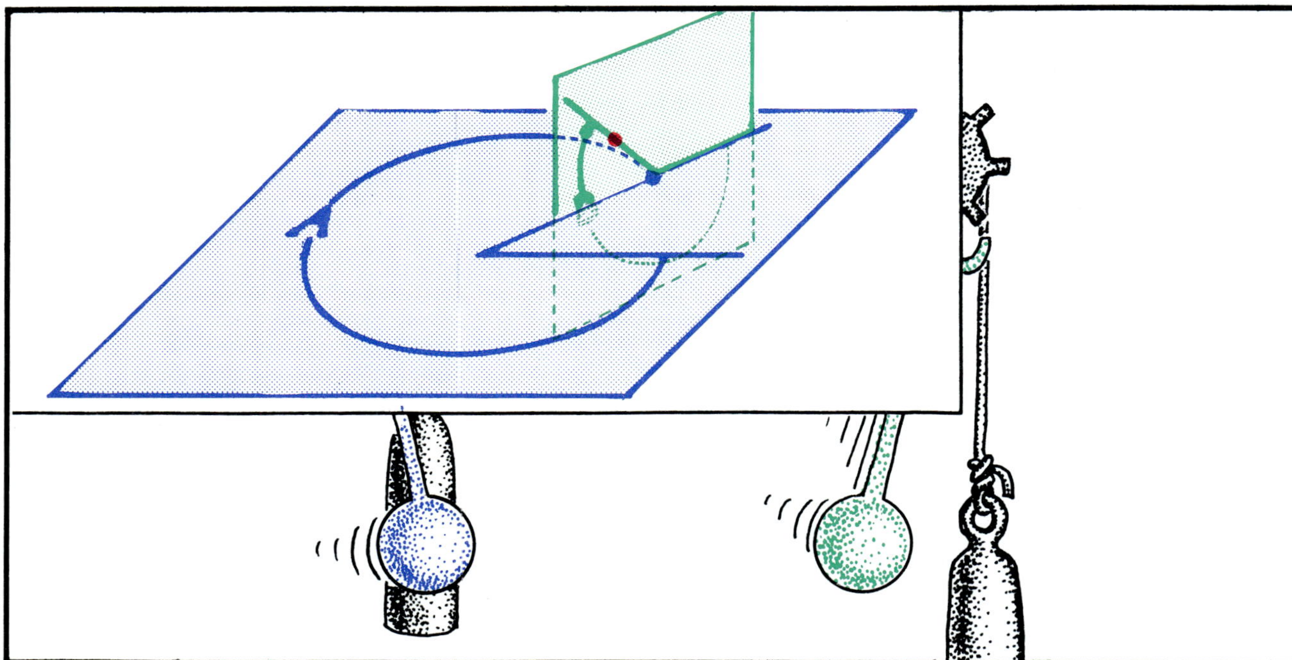

5.1.6. First, recline the planar model for the blue oscillator horizontally. As we say above, each point of the limit cycle in this plane, A, describes a unique phase of the blue oscillator. At this point, A, we erect a planar model of the green oscillator standing vertically. We imagine this plane to be *perpendicular* to the limit cycle of the blue oscillator. Within this vertical plane, we visualize the limit cycle of the green oscillator. The red point in this drawing is described exactly by the *two phases, A* and B. This pair of phases (A,B), describes the red point, and represents the state of the *combined system* consisting of the two oscillators.

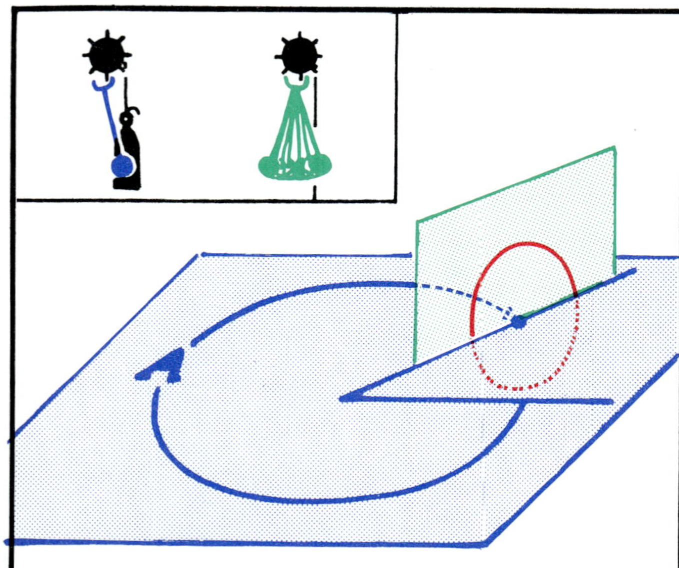

5.1.7. If the blue oscillator is stuck at phase A and the green oscillator moves through a full cycle, the red point describing the combined system traverses this red cycle.

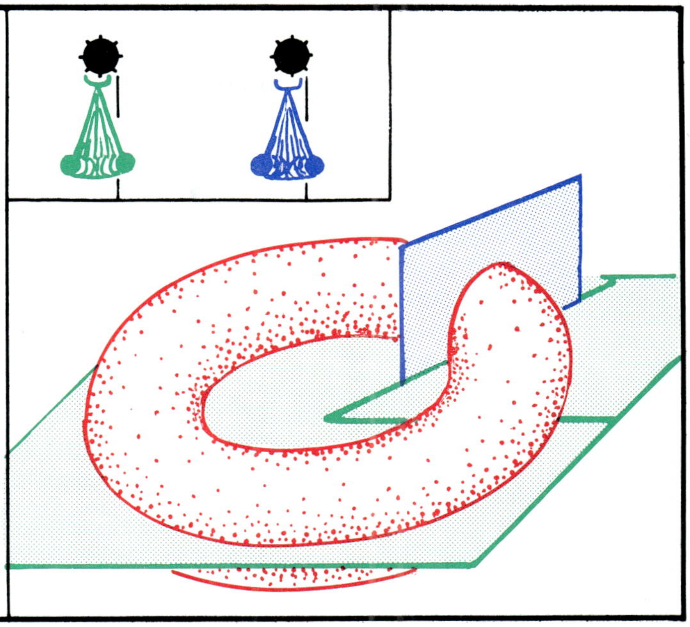

5.1.8. If the blue oscillator moves through a full cycle, the red cycle is pushed around the horizontal cycle, sweeping out this red torus.

Reducing both oscillator models from planes to cycles, the planes in the preceding panels fade away, leaving just the torus. The state space for the combined system of two oscillators is this torus, which is the Cartesian product of the two cycles. The full model for two oscillators is four-dimensional, but this doubly reduced model has only two dimensions. It is easier to visualize.

This all seems very much like the imaginary little green tori in the ring model for the forced pendulum of Duffing, constructed in the preceding chapter. Yet we took pains to point out repeatedly that the green tori were not *invariant manifolds.* Trajectories went right through them as if they were not even there. The red torus is different. It is the state space, there is a dynamical system on it modeling the two oscillators, and the trajectories must stay within the torus.

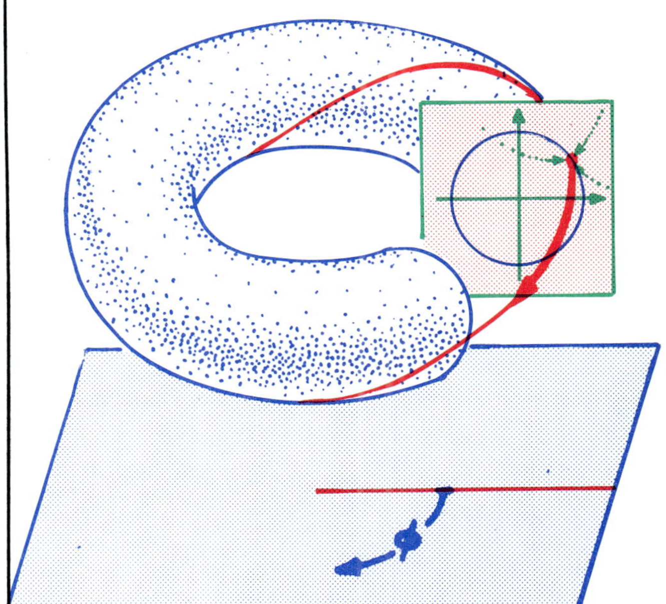

5.1.9. Here is a reminder about the green torus in the ring model for the forced pendulum, in the preceding chapter. It served, in a graphical construction, to locate an attractive limit cycle. Trajectories go right through it. We cannot reduce the ring model for the forced pendulum of the preceding chapter to the green torus.

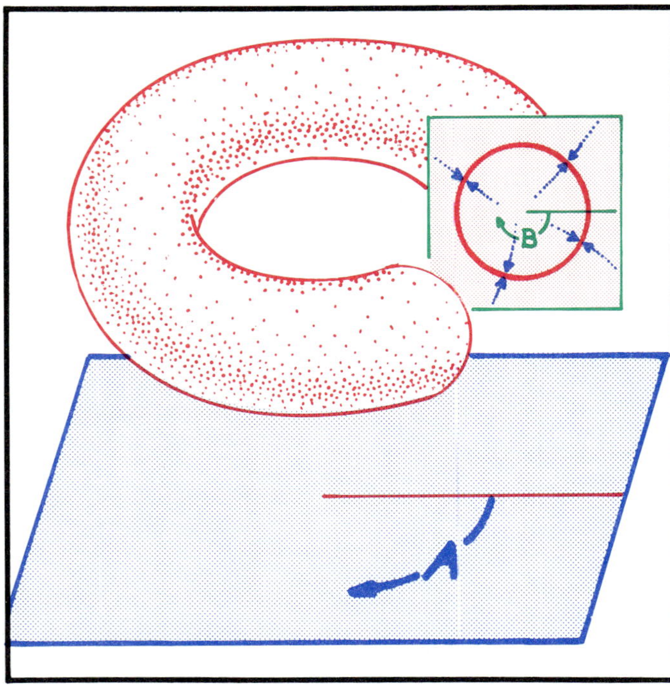

5.1.10. Here, for contrast, is the red torus. This is an attractive, invariant manifold for a dynamical system in another ring model. It is made of trajectories that stay within it. We are regarding this as the state space, ignoring the rest of the ring.

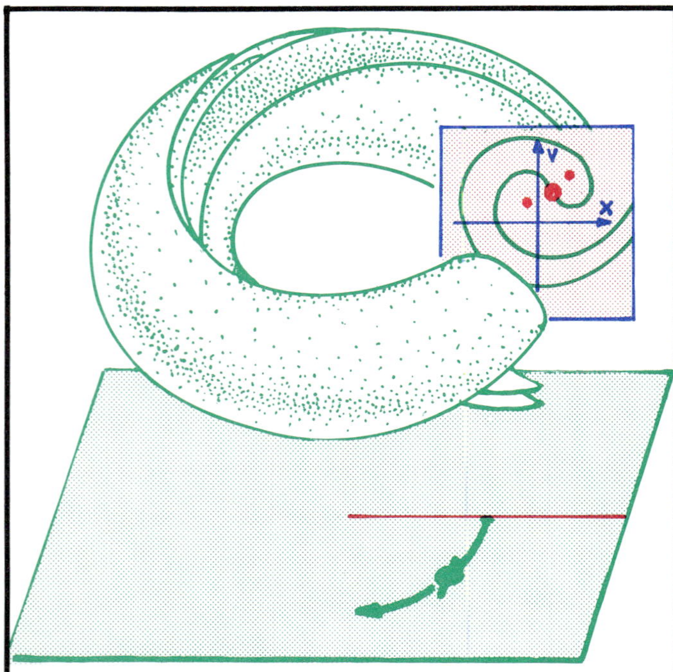

5.1.11. Although the red torus is not an analog of the green tori in the ring model, it does have a relative in the forced pendulum context. The separatrix dividing the basins of the two isochronous harmonics found by Duffing is also an invariant set.

Having constructed the state space for the two oscillators, we may now describe the dynamical system for compound oscillations, by *coupling* the two oscillators.

5.2. The Torus Model for Coupled Oscillators

Any two dynamical systems may be combined into a single system by the Cartesian product construction. A small perturbation of this combined system is called a *coupling* of the two systems. For example, the Cartesian product of two one-dimensional systems is a two-dimensional system, as described in the preceding section. In this section, we introduce a perturbation, or *coupling*, in the model for two oscillators. We will obtain a dynamical system on the torus as a geometric model for the behavior of the two coupled oscillators.

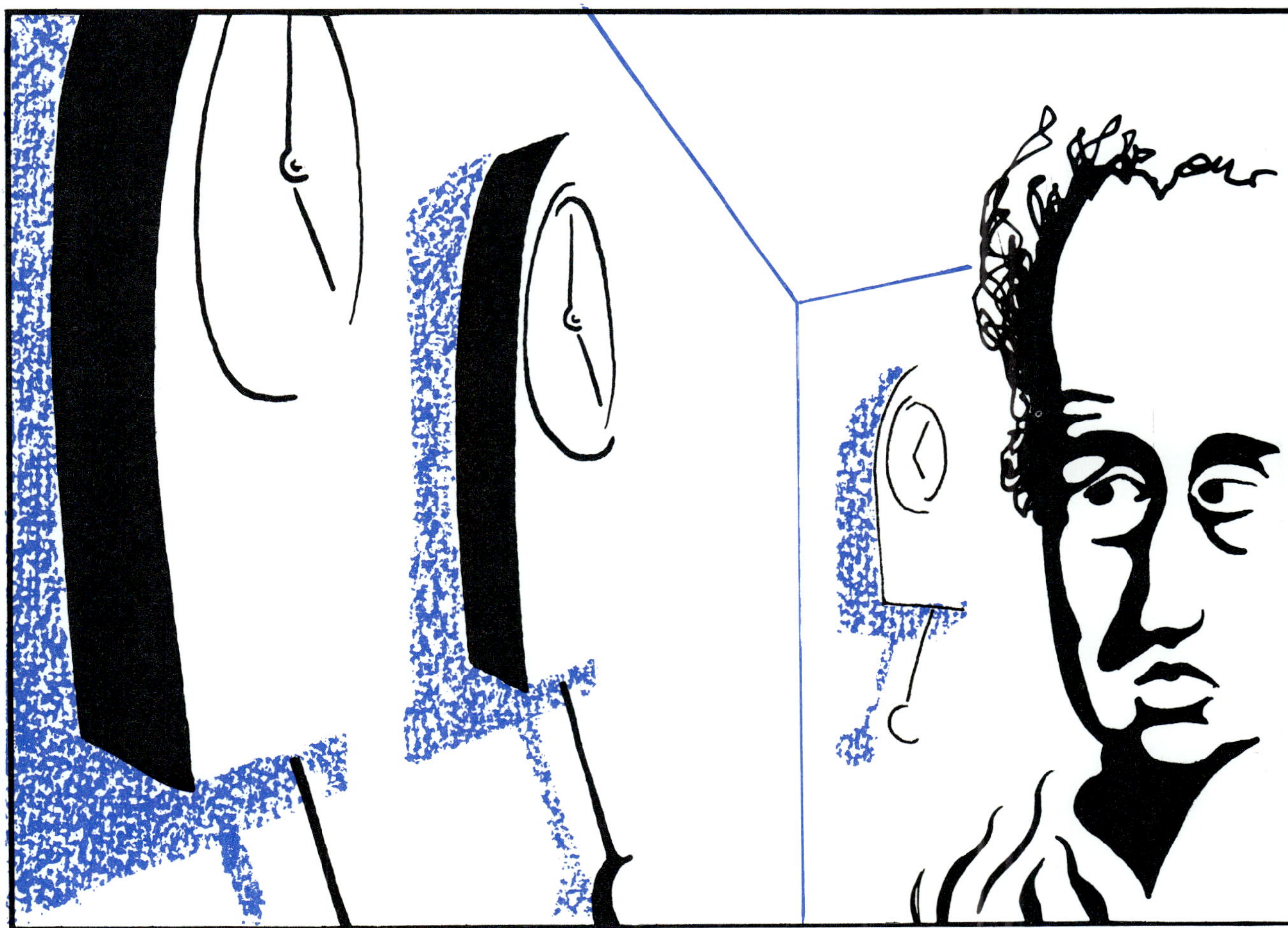

5.2.1. An example is provided by coupling two clocks. This particular system was observed by Christiaan Huyghens, an outstanding dynamicist of the seventeenth century. He noticed that two clocks hung on the same wall tend to synchronize and suspected that this *entrainment* phenomenon was caused by nonlinear coupling through the elasticity of the wall. The full explanation of entrainment, a recent result in dynamical systems theory, is due to Peixoto and is described in Part Two.

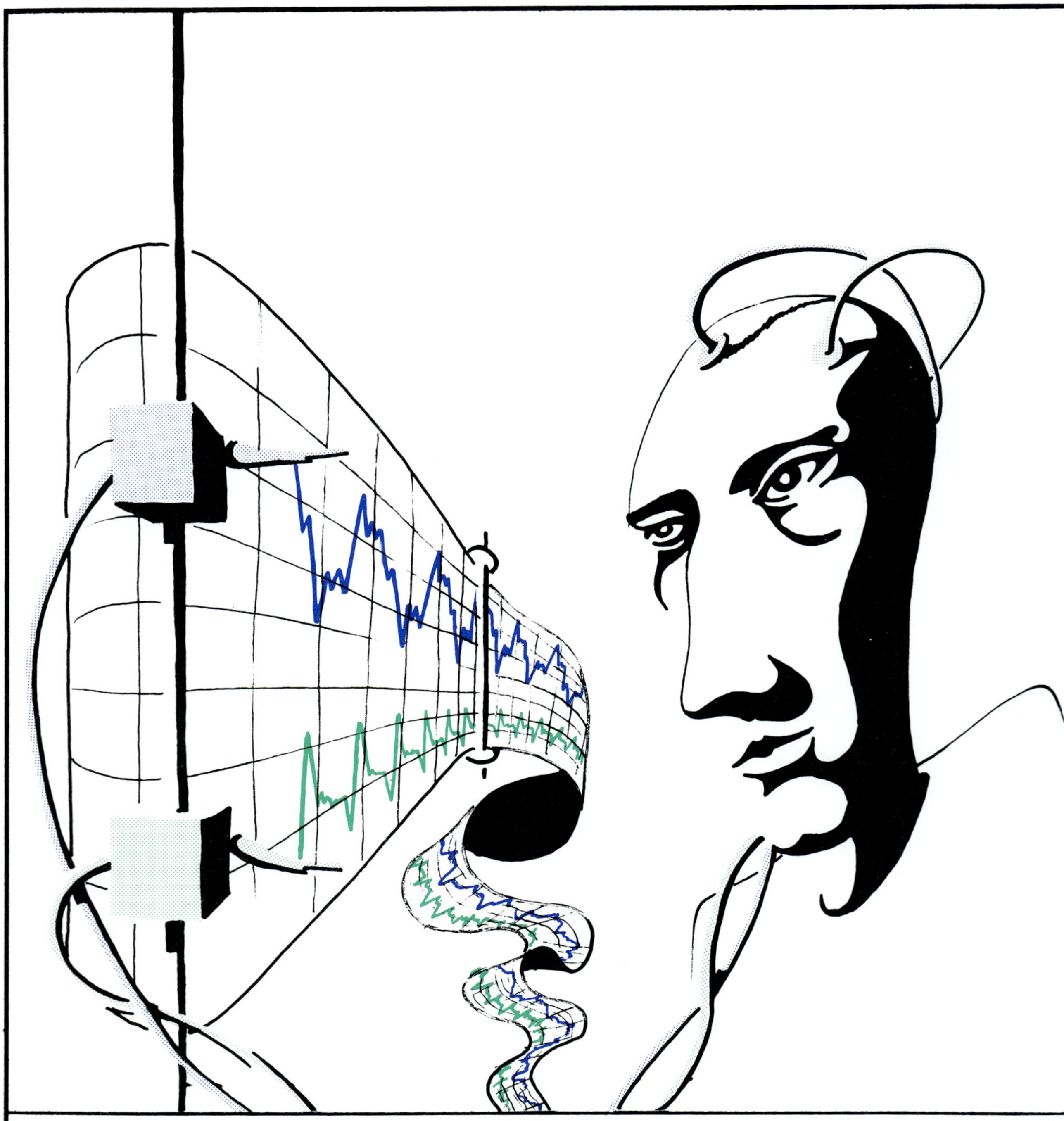

5.2.2. In many applications, two parameters are observed, and both are periodic. The torus will provide a geometric model for all of these empirical situations.

Let's choose a particular situation to model: to two clocks observed by Huyghens. To begin with consider *two uncoupled clocks.*

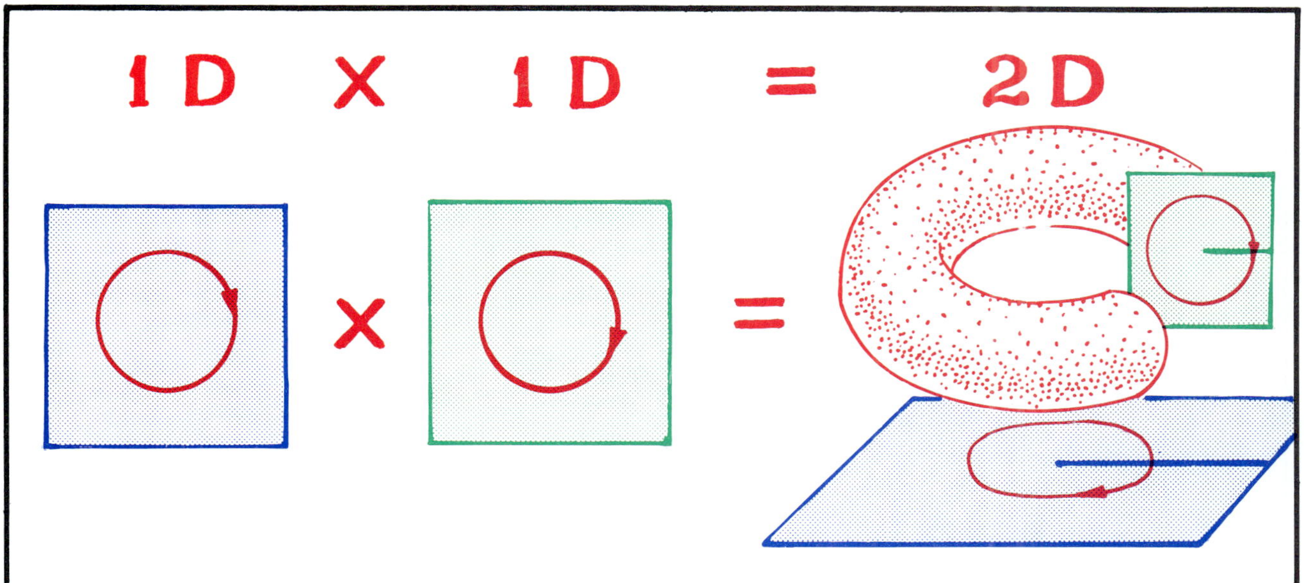

5.2.3. The geometric model for the states of this combined system is formed as follows: reduce the model for each clock (oscillator) to one dimension (a cycle), then take the Cartesian product of the two one-dimensional state spaces. The result is a two-dimensional torus, as explained in the preceding section.

In this *uncoupled* situation, there is *no entrainment*. Here's why.

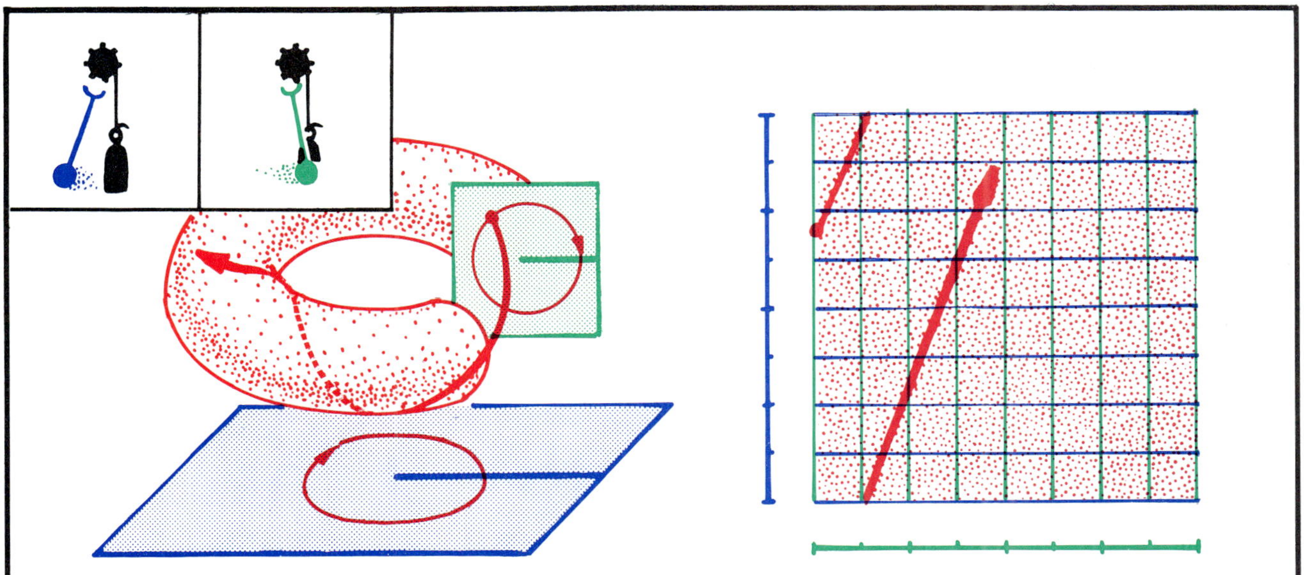

5.2.4. The trajectory of a point on the torus, corresponding to the time (phase) of each clock, winds around the torus. Assume the rate of each clock is constant. Then on the flat rectangular model of the torus, the trajectory is a straight line.

The slope of this line is the ratio: rate of second (green) clock to rate of first (blue) clock. This ratio, a contant, may be rational or not.

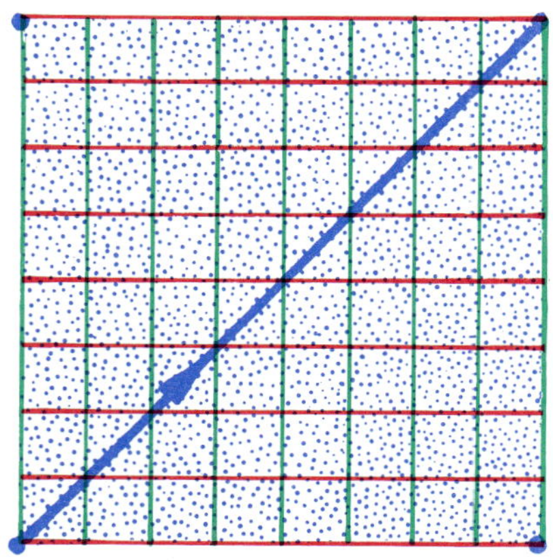

5.2.5. Ideally, if the two clocks run at the same rates, the ratio is one. Further, if they tell the same time, their phases are identical. Therefore, the trajectory on the flat torus is the diagonal line running from corner to corner.

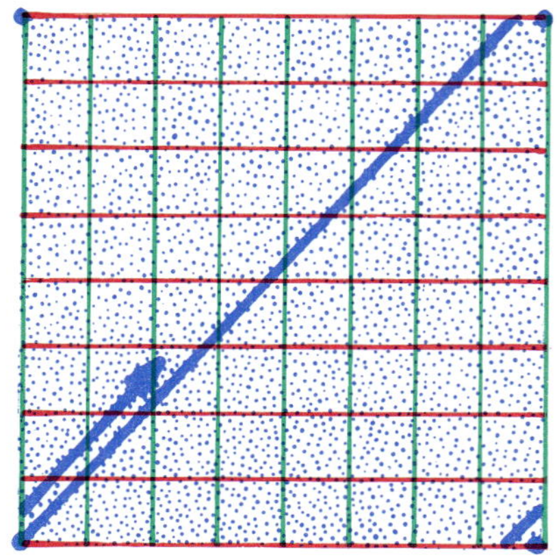

5.2.6. If the system is slightly changed, by a speck of dust in the works of one of the clocks for example, the ratio of the rates changes. The slope of the straight line on the flat model of the torus changes slightly. It is no longer exactly equal to one. And the trajectory on the torus changes from a periodic trajectory to a solenoid, perhaps, or to a periodic trajectory which winds many times around, instead of just once.

This is the situation called *non-entrainment*. This means that *a slight change in the system results in a slight change in the ratio of the rates (frequencies) of the oscillators.*

Now we may explain *entrainment*.

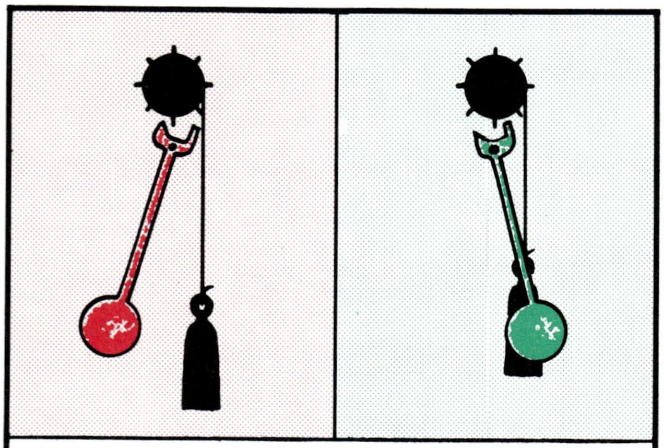

5.2.7. So far, we have assumed each clock is totally indifferent to the state of the other: the clocks are *uncoupled.*

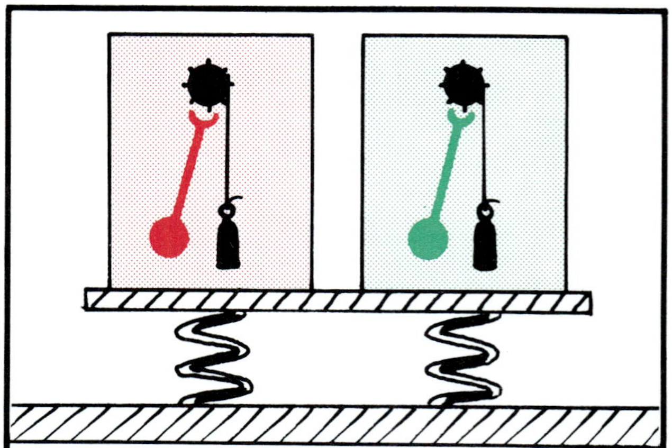

5.2.8. A slight mechanical connection between the two oscillators will create an interaction: the motion of each will influence the motion of the other.

Now the two oscillators are *coupled*. This means that the phase portrait is *perturbed* by the addition of small vectors at each point of the state space. This small vectorfield is added to the dynamical model representing the uncoupled system. Without saying exactly what this small perturbation is, one can conclude something about the coupled system anyway. This amazing conclusion is a geometric theorem of Peixoto, described at length in Part Three.

Here is the geometrical model for the same physical system, with coupling introduced between the mechanical oscillators.

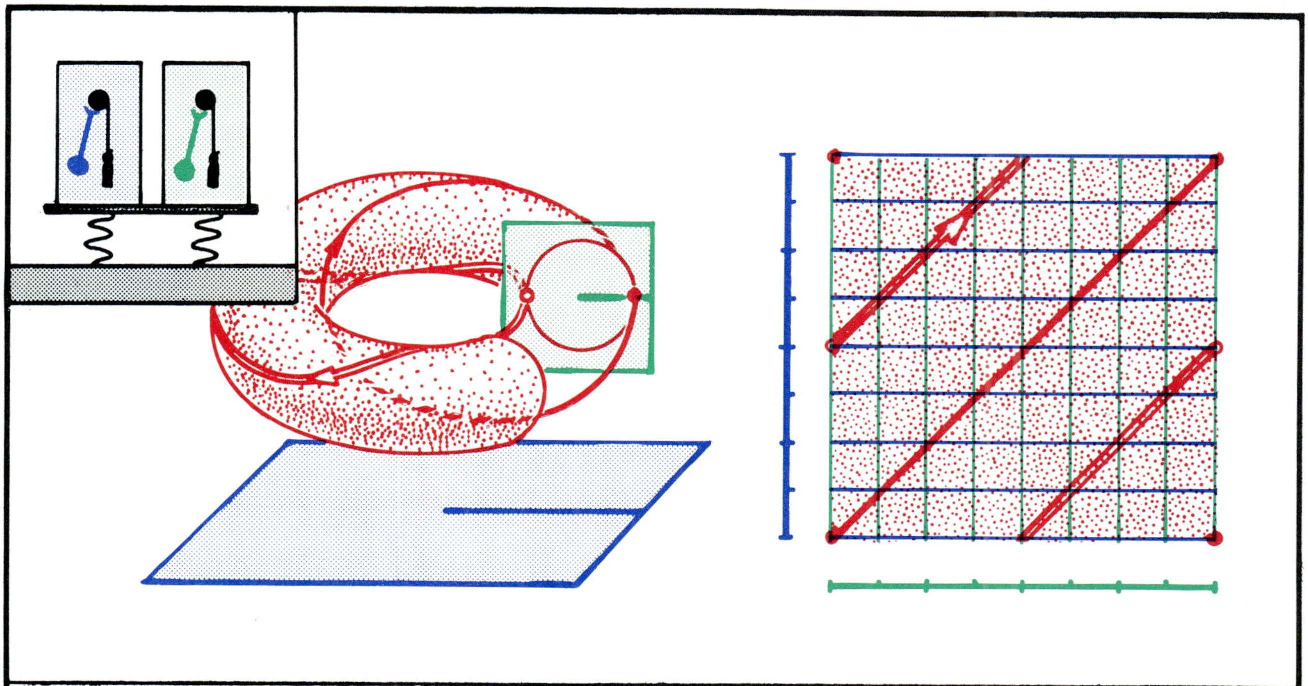

5.2.9. The dynamical system modeling this is a *perturbation of the uncoupled system described previously. The theory of Peixoto ensures that for a typical perturbation, the perturbed model looks like this. There is a finite, even number of closed trajectories. They all wind around the same number of times. Every other one is an attractor. The intermediate ones are repellors. There are no other limit sets. This kind of portrait is called a braid.*

Even more, Peixoto's theorem says that this braid is *structurally stable*. This means that a small perturbation (another speck of dust) will make *no significant change in the phase portrait*. For example, a braid with periodic trajectories winding once around (equal rates of the clocks) will still be a once-winding braid after a small perturbation. Thus, Peixoto's theory provides a mathematical rationale for the *frequency entrainment* phenomenon observed by Huyghens. *Warning: The phases need not be entrained, only the frequencies.*

How do braids arise? Although Peixoto's theory provides a mathematical basis for this phenomenon, the actual mechanism of it is not yet clear. We will demystify the mechanics of braids in Section 5.4. First, we need to enlarge the geometric model from two dimensions to three. So, on to the next section.

5.3. The Ring Model for Forced Oscillators

We are now ready to apply the abstract ideas of coupled oscillators to some concrete examples. In this section (and the next two), we consider the coupling of mechanical oscillators. This is analogous to the work of Duffing described in the preceding chapter. But there, the forcing oscillator (motor) was coupled to a system that tended to come to rest (damped pendulum). Here, we couple the forcing oscillator (motor) to a system that tends to a self-sustained oscillation (clockworks). This clockworks will be assumed to be *affected by the coupling,* while the turntable motor is *unaffected.*

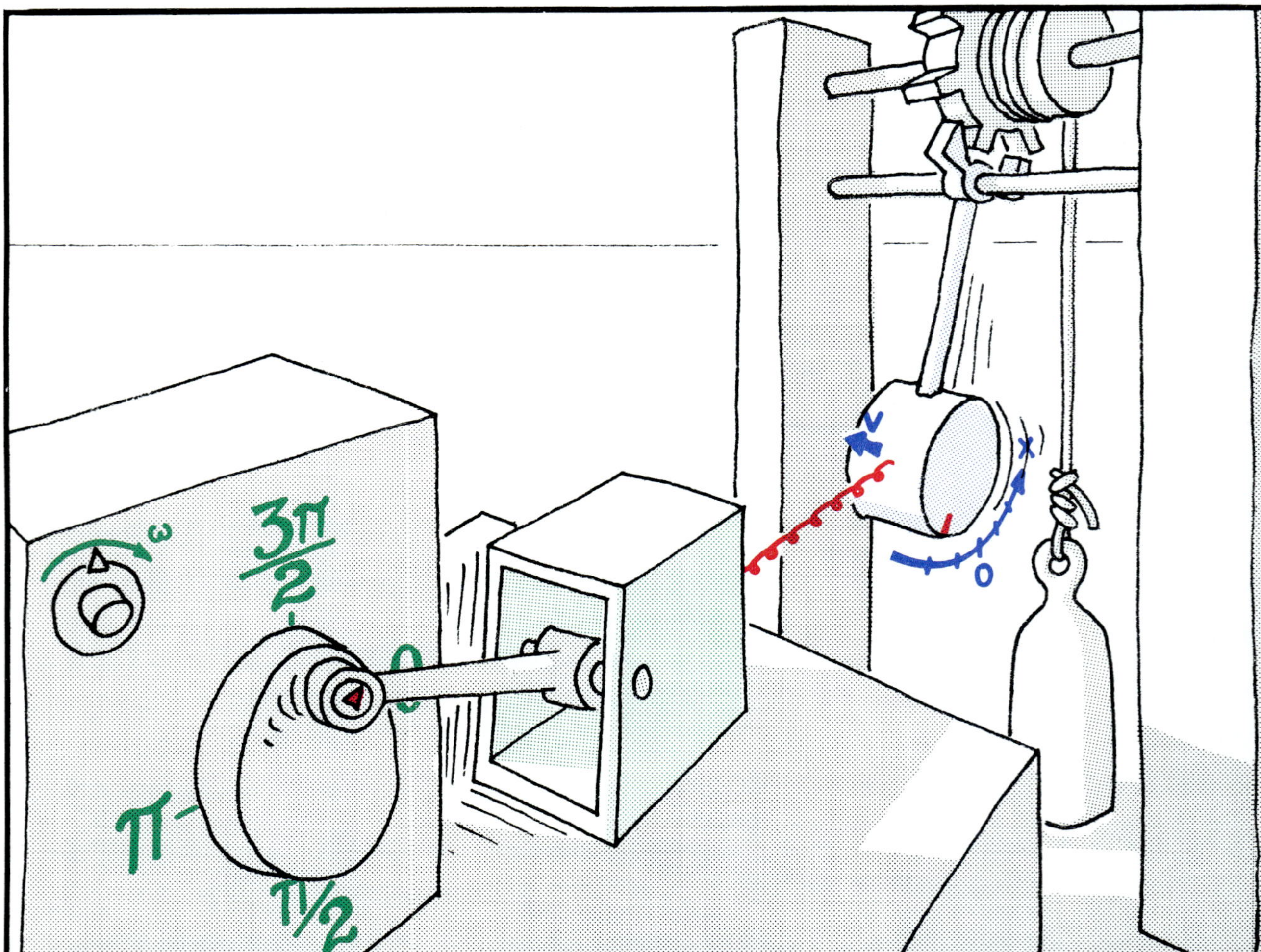

5.3.1. The turntable motor is so well regulated that its speed, once set with the control knob, is unaffected by the load. The forcing oscillation is coupled to the clock pendulum by a light spring. The stiffer the spring, the greater the effect of the driving oscillation on the periodic motion of the clock pendulum.

The most useful geometric model for the state space of this system is three-dimensional. It is essentially the same as the *ring model* of Section 4.1. We will place the torus in this three-dimensional context.

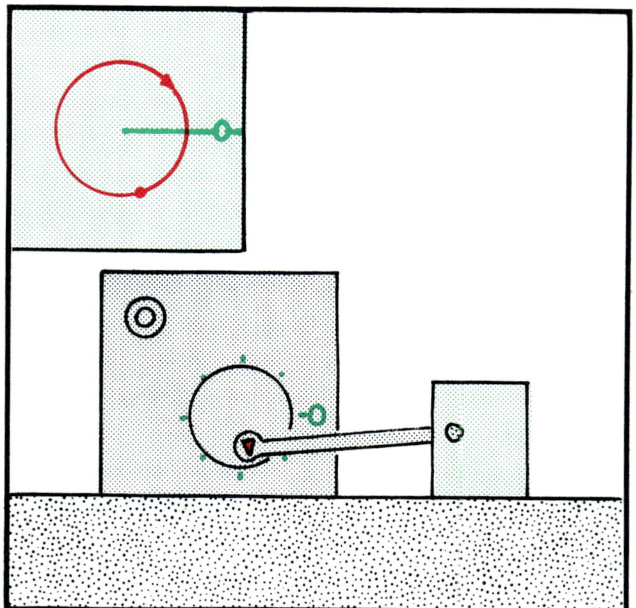

5.3.2. We use the reduced model for the driving motor, as this oscillation is unchanged by the coupling. Only its *phase* is important.

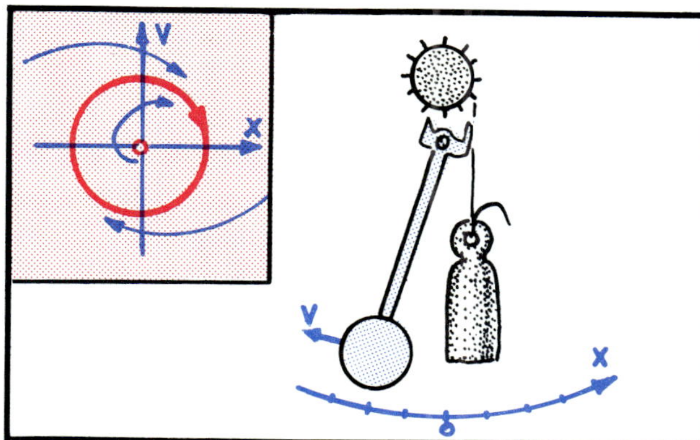

5.3.3. But the state space of the clockworks will be visualized as a two-dimensional phase plane. The phase portrait consists of a periodic attractor, representing the self-sustained oscillation of the pendulum. The width of this cycle then corresponds to the *emplitude, A*, of the oscillation of the pendulum of the clock. This amplitude will be affected by the force of the driving motor, communicated through the coupling spring.

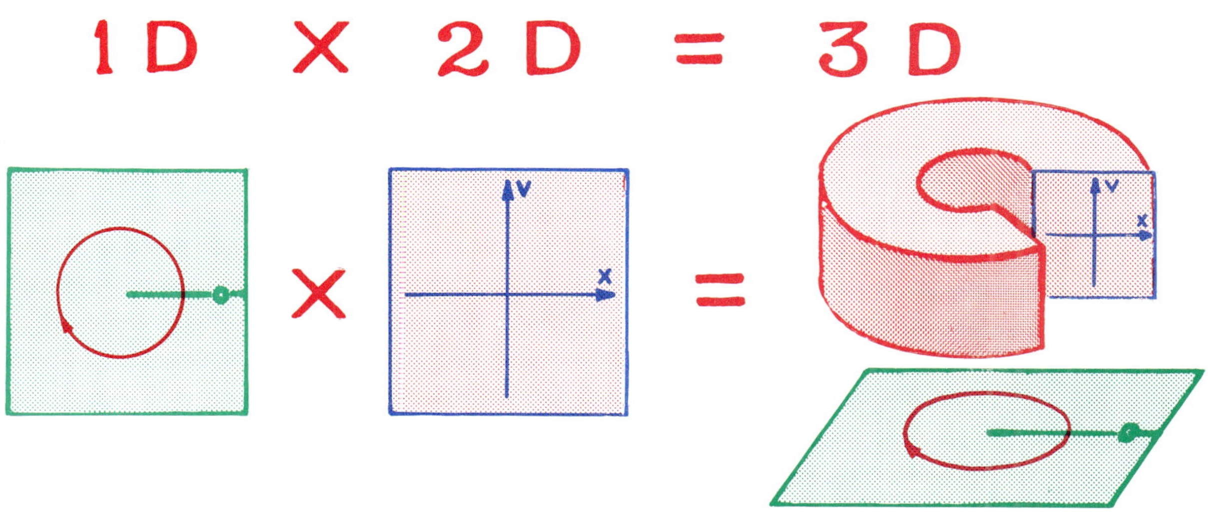

5.3.4. The Cartesian product of the clock plane and the motor cycle is the three-dimensional *ring model* for the combined system. (Compare with Figure 5.2.3.)

This scheme represents a special case of coupled systems, in which there is a dominant partner and a more flexible one. We have used the same strategy in the ring model for the forced pendulum (see Section 4.2). A more general scheme would use two dimensions for each oscillator, for a total of four dimensions.

The regulated motor maintains its rate, but the motion of the pendulum is affected by the periodic force of the spring. To see how it is affected, we may take out the coupling spring, then replace it.

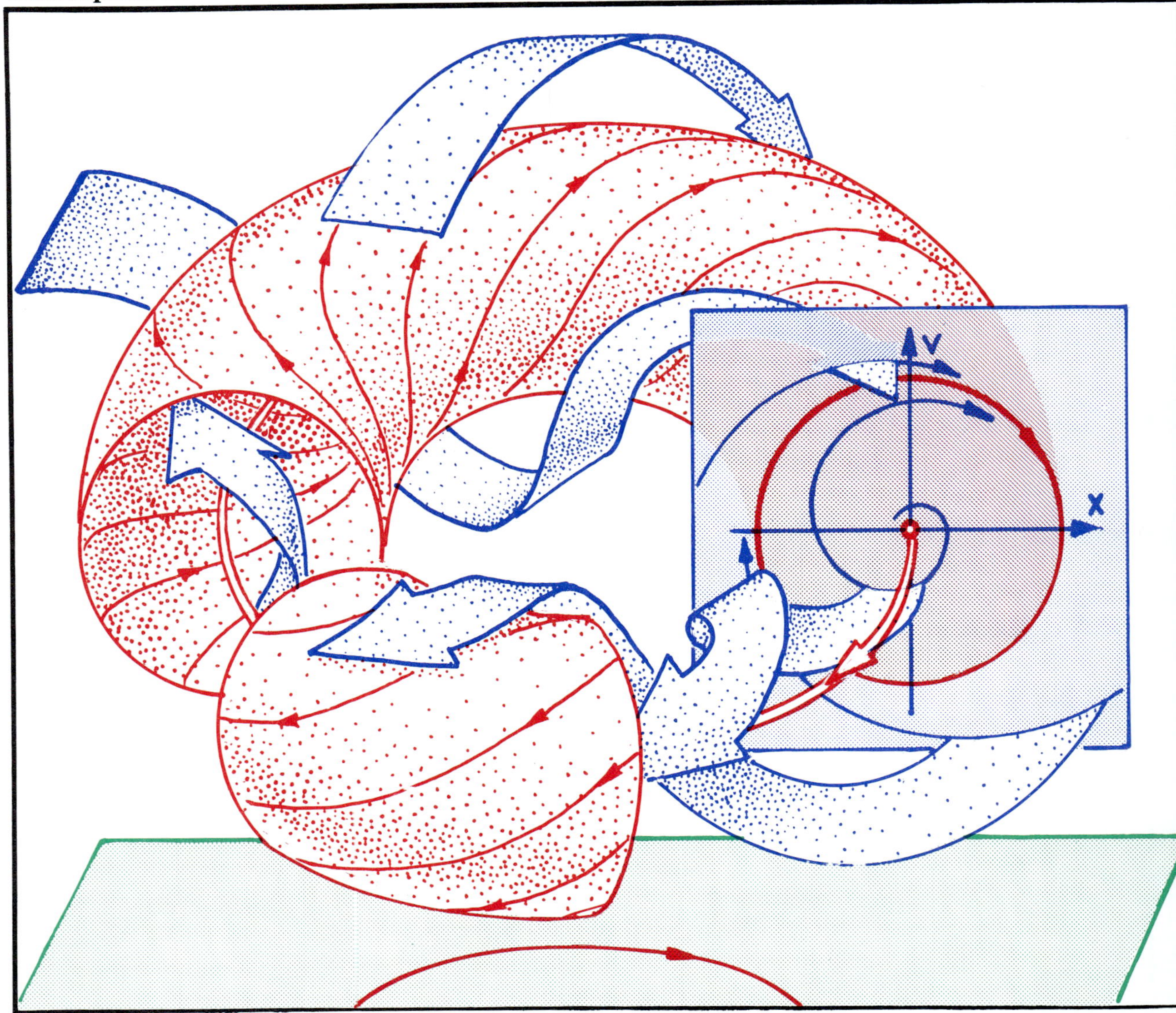

5.3.5. With no spring, the two oscillators are uncoupled. Once the transients die away, the motion may be recorded as a trajectory on a torus, as explained in Section 5.1. Unlike the green torus in the ring model of the forced pendulum in Chapter 4, this red torus is an *invariant manifold* of the dynamical system. This means that every trajectory that begins on the torus stays on the torus. In fact, the red torus is *attractive*, yet probably *not an attractor.* It is *attractive* in this sense: if we put the combined system in an initial state *off the torus* (for example, by giving the pendulum a shove to a larger amplitude) and let it go, the resulting trajectory will be attracted to the torus, as the amplitude decays to the original value.

And it may be *not an attractor* in the following sense: if we put the combined system in a typical initial state *on the torus* and let it go, the resulting trajectory may be a periodic trajectory on the torus. But when we say a set is an attractor, we mean not only that it is attractive, but also that it is *transitive:* that is, most trajectories on it wander all over it.

Thus, not all attractive tori are attractors.

When we reattach the spring, the two oscillators are coupled. The phase portrait is a perturbation of the picture described above for the uncoupled system. According to an important theorem of mathematical dynamics, the perturbed portrait still has an attractive, invariant torus. If the perturbation (coupling) were gradually turned on, the uncoupled torus would be gradually deformed into the coupled one. Therefore the theory of Peixoto applies, as described in the preceding section. The phase portrait contains a braid of periodic attractors on the torus. The invariant torus is attractive, yet not an attractor. The braided periodic trajectories within the torus are the actual attractors, in the coupled case.

Because of these braids, the clockworks and the motor are *entrained in frequency, but not in phase.* In the next section, we will examine the dynamics of these braids.

5.4. Braids: The Dynamics of Entrainment

In Section 5.2, we encountered the braided periodic attractors in the two-dimensional torus model for coupled oscillators. To explain the mechanics of frequency entrainment in our forced mechanical oscillator, we will reconsider the braids in the three-dimensional ring model of the preceding section.

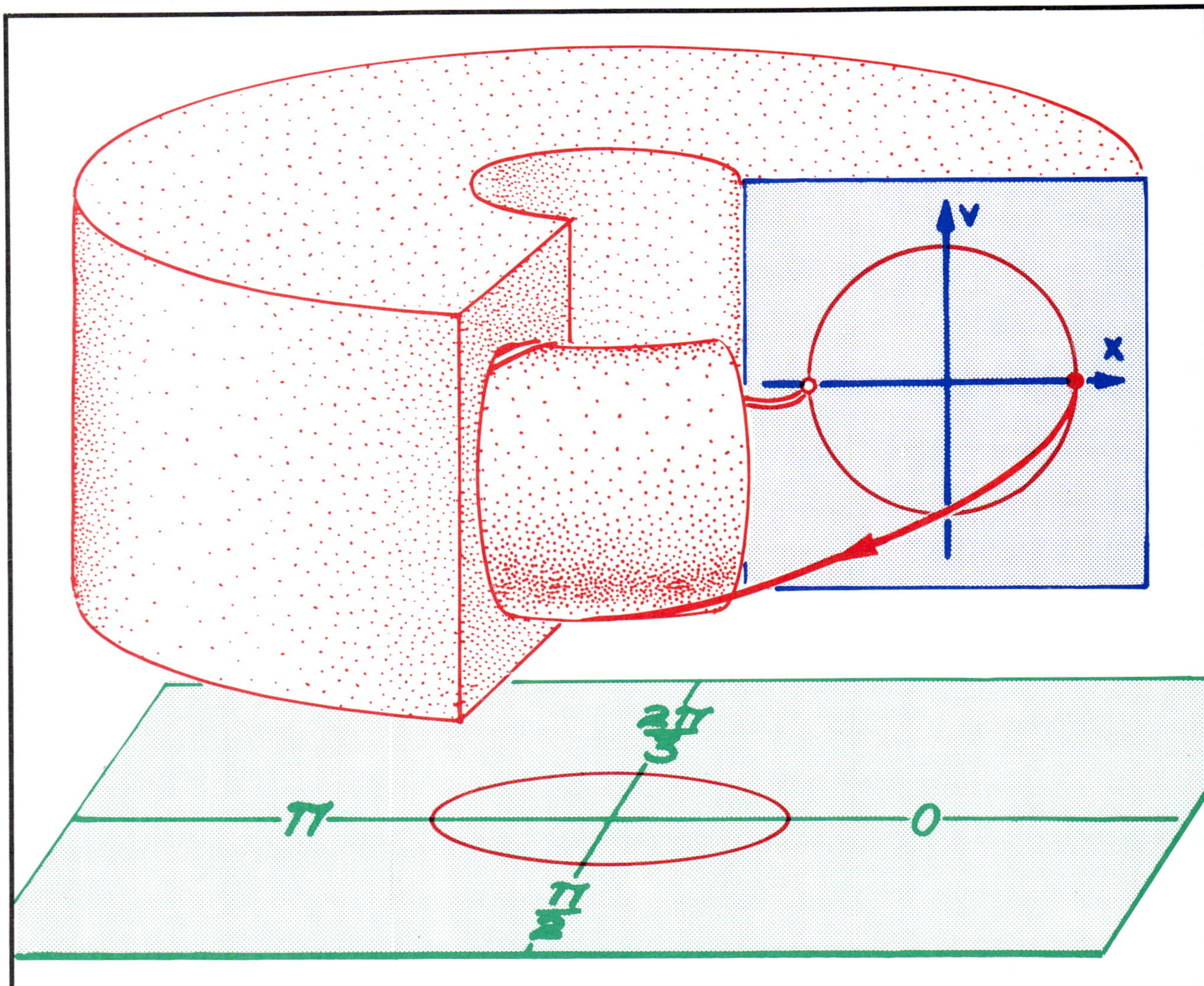

5.4.1. Recall that the ring model for forced oscillators has an invariant torus. This attractive torus corresponds to the torus (reduced) model of Section 5.2. Now we put the 2D torus back into the context of the 3D ring model. By performing armchair experiments with out mechanical device, we will see how braids arise on this torus.

In this series of experiments, we will couple the turntable motor to the pendulum of the clockworks with a very feeble spring. Thus, the phase portrait of the combined system will be a slight variation of the uncoupled system, in the ring model. The uncoupled system was shown in Figure 5.3.5. Also, we will begin by setting the speed of the driving motor to the natural frequency of the clock. This is the situation called weakly coupled, *isochoronous* oscillators.

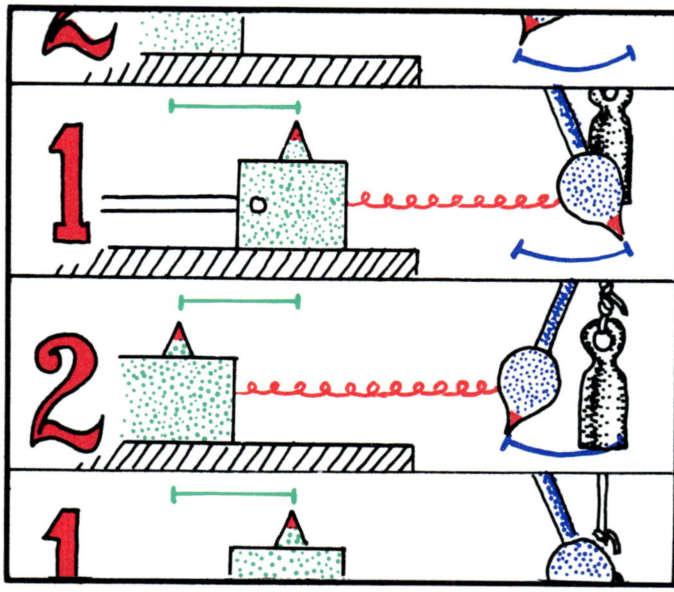

5.4.2. Start up the motor. After a brief start-up transient, we observe the driven weight and the pendulum swinging *in phase*. The motion of the pendulum has been influenced by the motion of the wght, through the intercession of the weak coupling spring. We suppose the weight has a bigger swing than the pendulum, as shown here. Then the only result of this influence, in contrast to the uncoupled situation, is a slight increase in the amplitude (width of swing) of the pendulum.

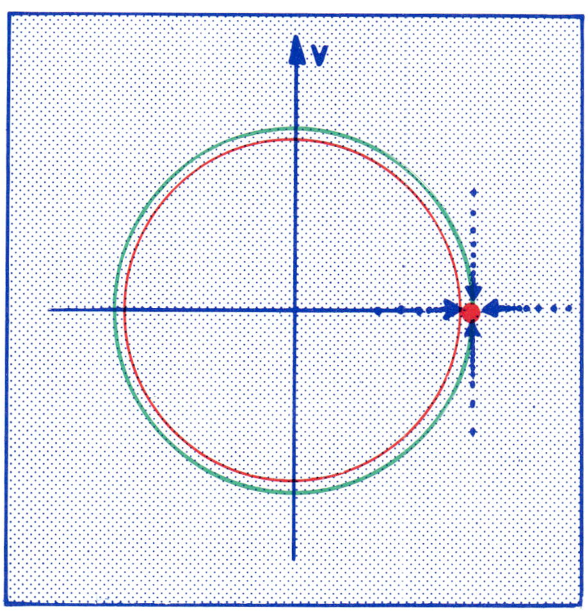

5.4.3. Now let's look at the trajectory of this motion in the ring model. First, we slice the ring in the strobe plane corresponding to phase zero of the driving oscillator. In this slice, the invariant torus of the uncoupled system shown in Figure 5.4.1 appears as a red circle. Because the amplitude of the pendulum has been slightly increased by the coupling, the green locating torus (not invariant) for this trajectory is slightly fatter than the uncoupled red torus. Thus, in this strobe plane, it appears as a slightly larger circle. The trajectory of the isochronous, in-phase, periodic motion observed in the preceding panel winds around a green locating torus, and appears in the strobe plane as a point, shown here as a *solid red dot*. The blue dots represent trajectories that are approaching the attractor.

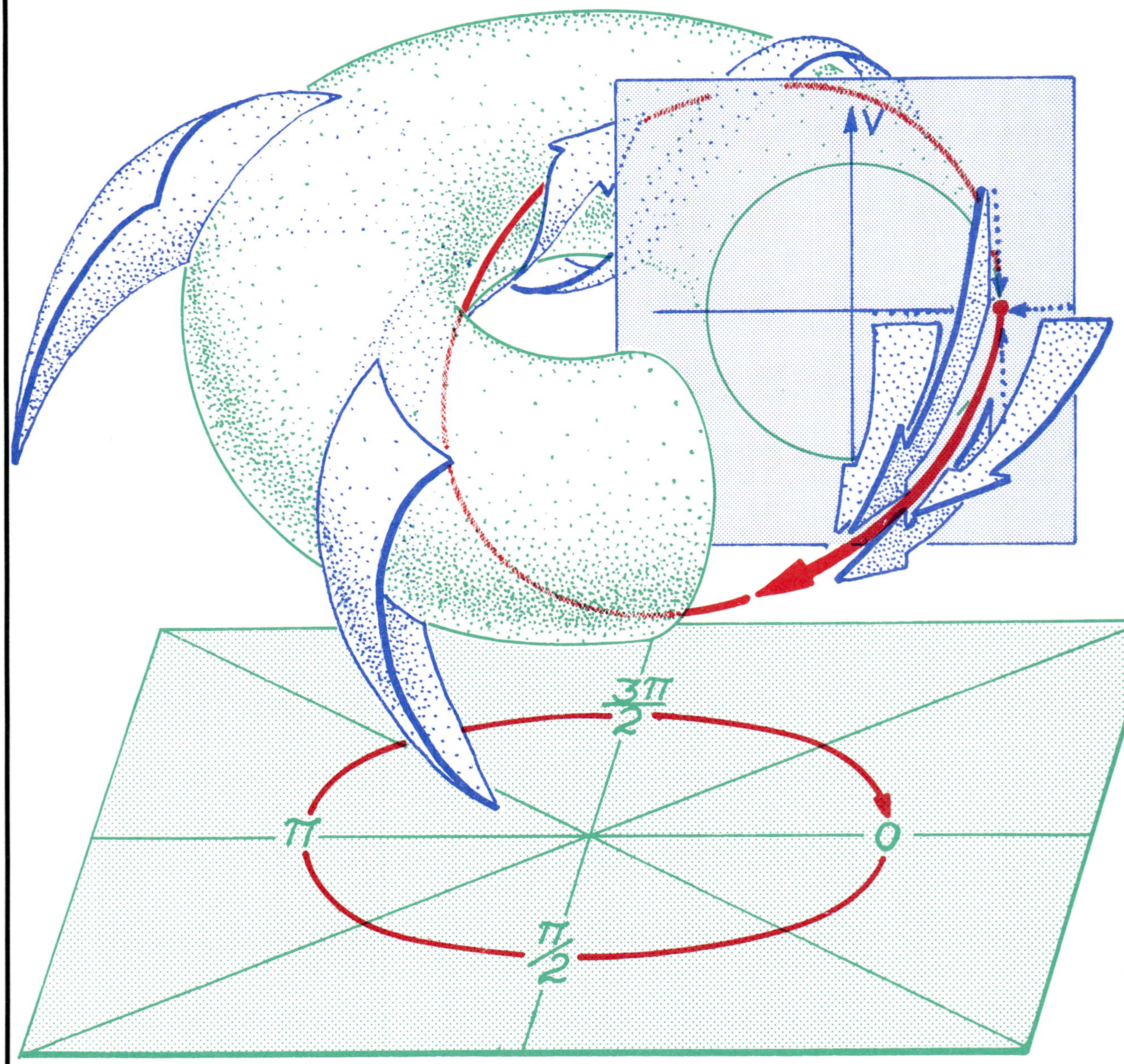

5.4.4. Here is the isochronous in-phase trajectory in the ring model. Note that the attractive trajectory goes through phase zero of the driving motor at roughly the same time it passes phase zero of the driven clock. It is *in phase.* Also, it is *isochronous,* as the frequencies are entrained to be equal. This periodic trajectory is actually an *attractor.*

Recall that in Section 5.2, Peixoto's theory predicted that in the torus model, we would find not only a periodic attractor, but a periodic repellor as well. Another experiment, with the motor and clock isochronous but out of phase, will locate this periodic motion in the ring model.

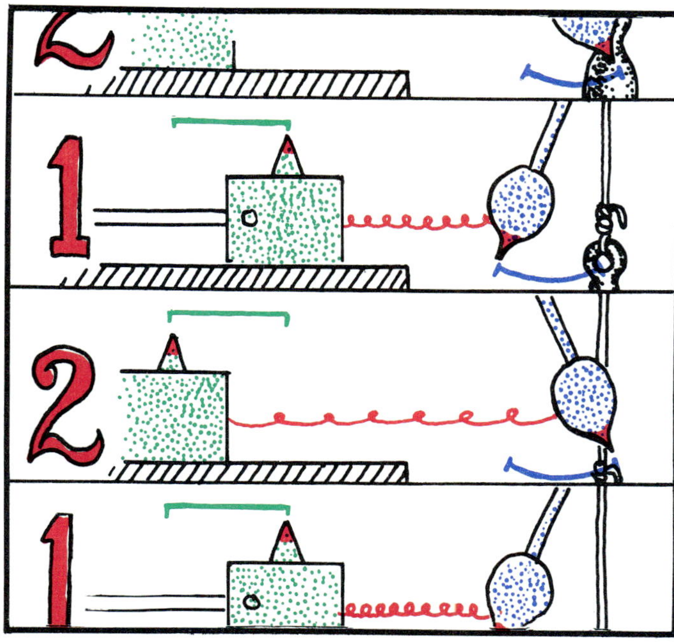

5.4.5. In this experiment, the clock pendulum is set at phase π (half-cycle, or full left) and the driving weight at phase zero (full right). They are out of phase when the experiment begins. They are as close together as they can get. We let them go, and the start-up transient dies away. After half a cycle, in step 2, they are as far apart as they can get. Here the clock's pendulum is opposed by the full force of the stretched coupling spring. After a full cycle, step 1 repeats, and they are closest together at the same time again. Once more, the clock's pendulum is opposed by the spring's greatest force. The effect of the coupling is to reduce the amplitude of the pendulum's motion.

But this motion does not persist. Its trajectory is not an attractor, for if the clock began slightly ahead of phase π, its phase would drift forward until it were in phase with the driving weight. Likewise, if it began slightly behind phase , its phase would drift backwards until it were in phase. In fact, in the two-dimensional torus model (Figure 5.2.9), this motion was a repellor, belonging to the separatrix. Let's plot this motion in the ring model, beginning with the driving-phase-zero strobe plane.

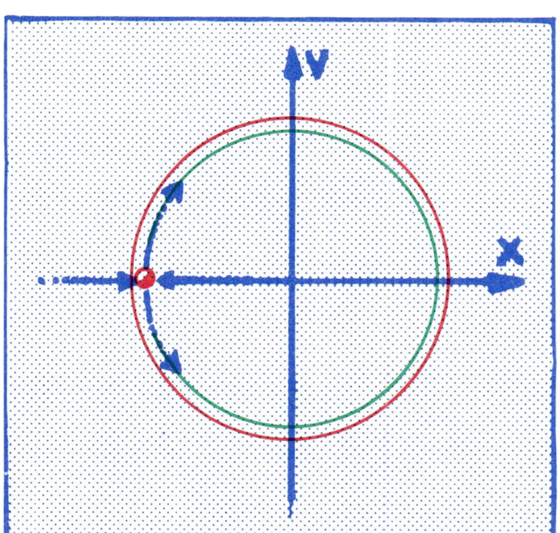

5.4.6. Again, the invariant torus of the uncoupled system cuts the strobe plane in a circle, shown here in red. Another smaller torus, not invariant but convenient for visualization, cuts the plane in the green circle. The periodic trajectory corresponding to the isochronous, out-of-phase oscillation winds once around the green, locating torus. It cuts the strobe lane in a single point, shown here as a half-filled red dot. Note that this trajectory *repels* in the direction along the torus and *attracts* in the perpendicular direction. The repellor in the torus model becomes a saddle in the ring model.

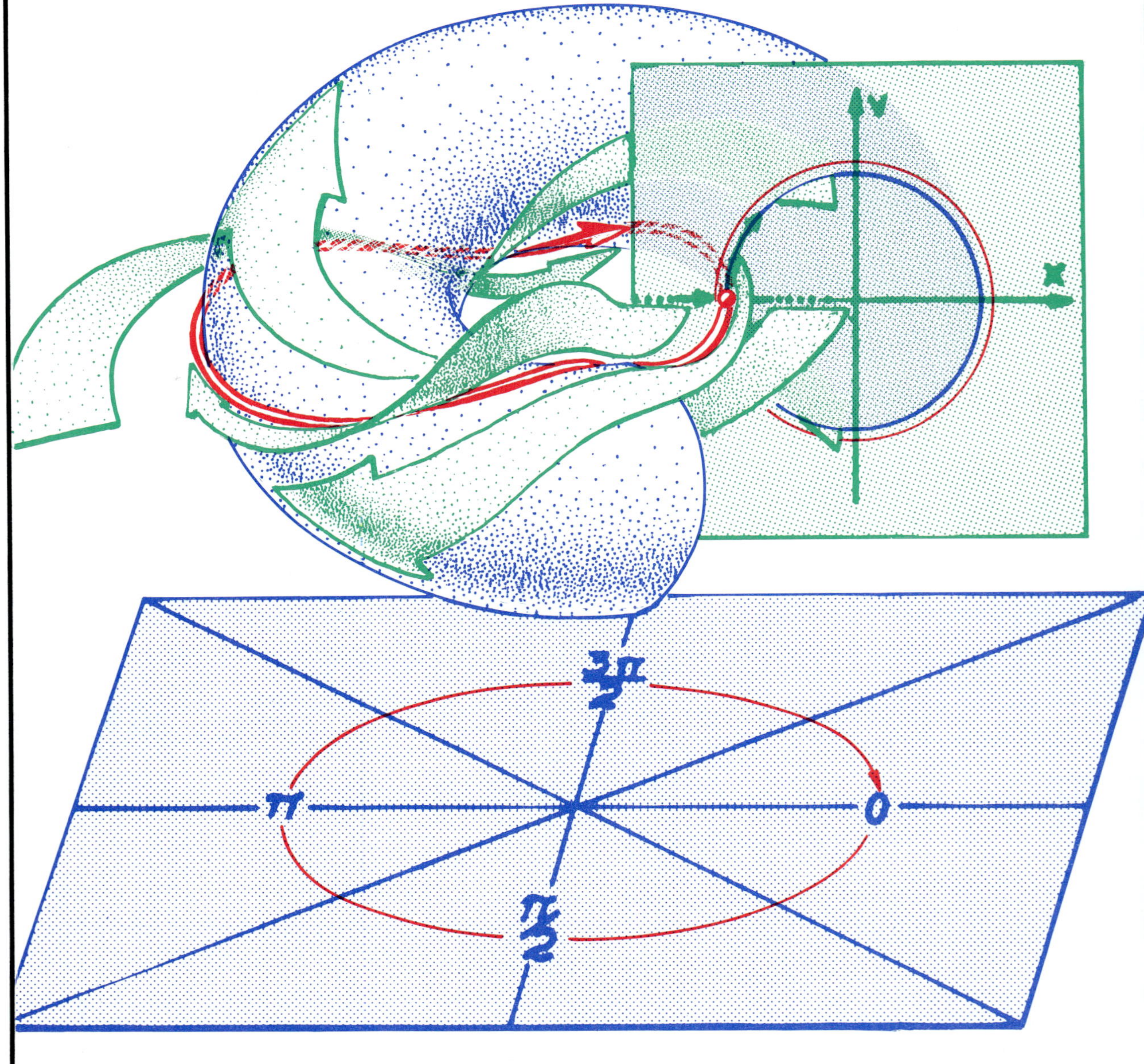

5.4.7. And here is the strobe plane, placed in the three-dimensional ring model. The red trajectory records the motion of the ideal experiment, with a perfectly out-of-phase, isochronous oscillation. This periodic trajectory is not actually an attractor, but a *periodic saddle.* It attracts amplitudes but repels phases, as shown by the ribbon arrows.

Now we have located the braided cycles, and we are ready to determine the shape of the invariant torus they are braided around.

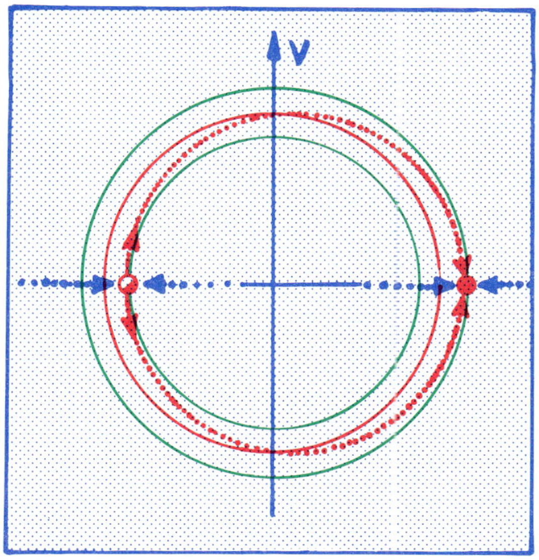

5.4.8. The invariant torus of the weakly coupled system is close to that of the uncoupled system. It is composed of the outset of the periodic saddle. Here, we shown this outset in red, as cut by the strobe plane. Note that it lies between the green locating tori used in the preceding constructions.

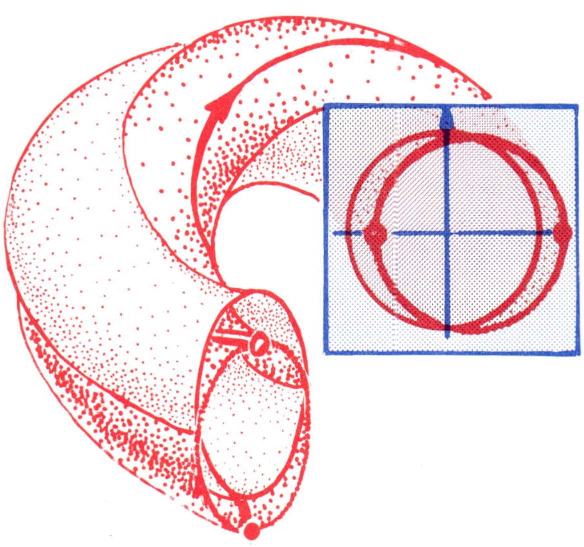

5.4.9. Here, in three dimensions, is the new invariant torus of the weakly coupled system (coarse shading). It is close to the old invariant torus of the uncoupled system (fine shading). The periodic attractor and the periodic saddle wind around the new torus. Both are isochronous. That is, they wind once around the waist of their invariant torus.

Another important feature of the phase portrait of the combined system is the *central repellor.* There is a repellent periodic trajectory near the center of the ring model. It corresponds to a small oscillation of the clock pendulum, out of phase with the driving motor. It is an *unstable equilibrium.* That is, it is possible in principle to balance the clock precariously in this mode of oscillation, just as it is possible to balance a pendulum at the top of its swing (see Section 2.1).

5.4.10. Again, we imagine a green, noninvariant, locating torus. This one is quite thin, and lies near the center of the invariant torus. Here, we show the strobe plane. It cuts the locating torus in the small green circle and the invariant torus in the dotted red circle. The isochronous central repellor winds once around the green locating torus, meeting the strobe plane in a single point, shown here as a *hollow red dot*.

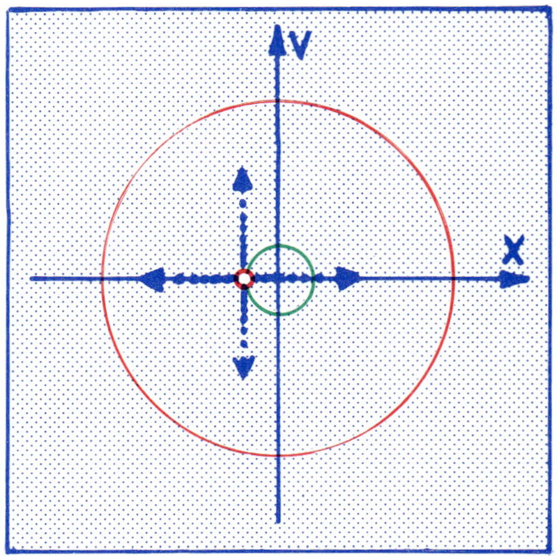

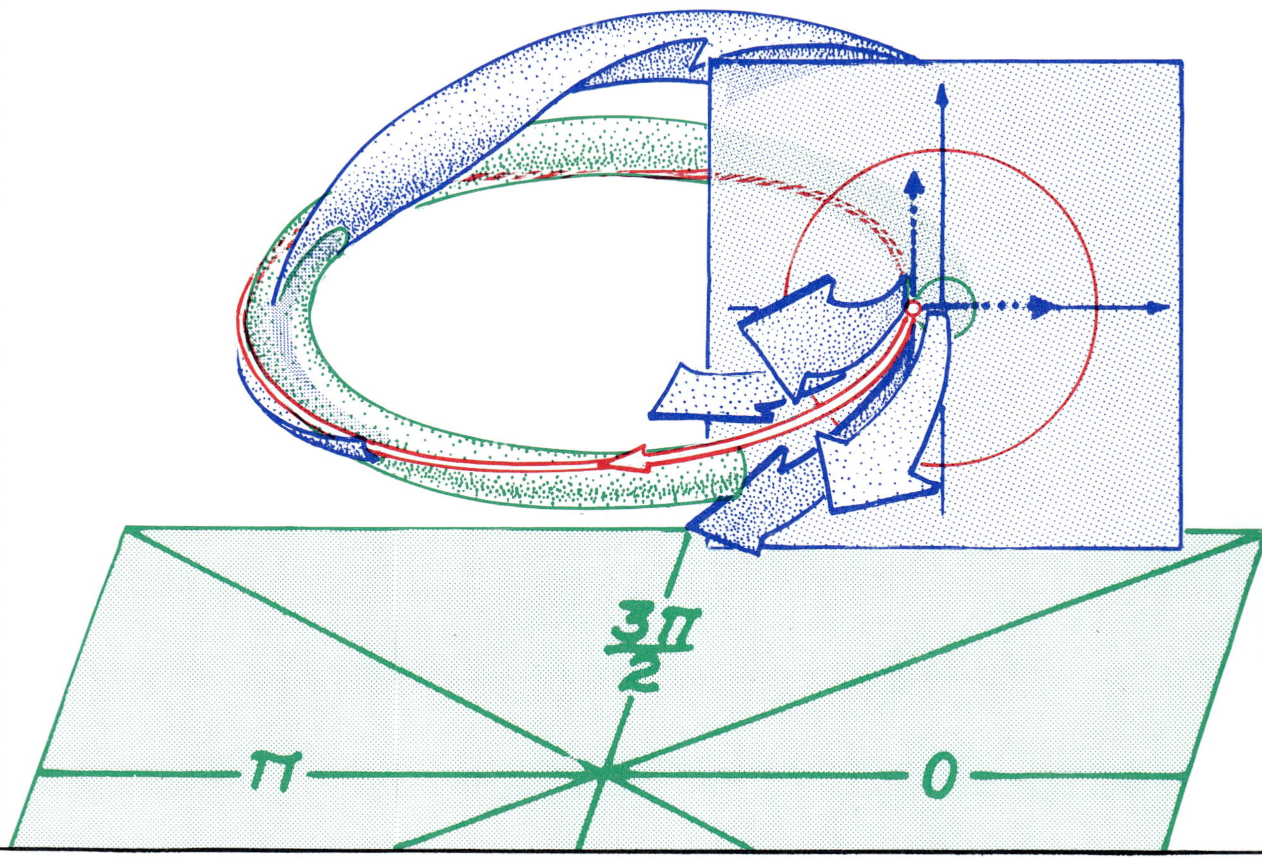

5.4.11. Here is the strobe plane, in the three-dimensional context of the ring model. The outset of the periodic saddle comprises the invariant torus. The outset of the central repellor comprises the central portion of the basin of the periodic attractor.

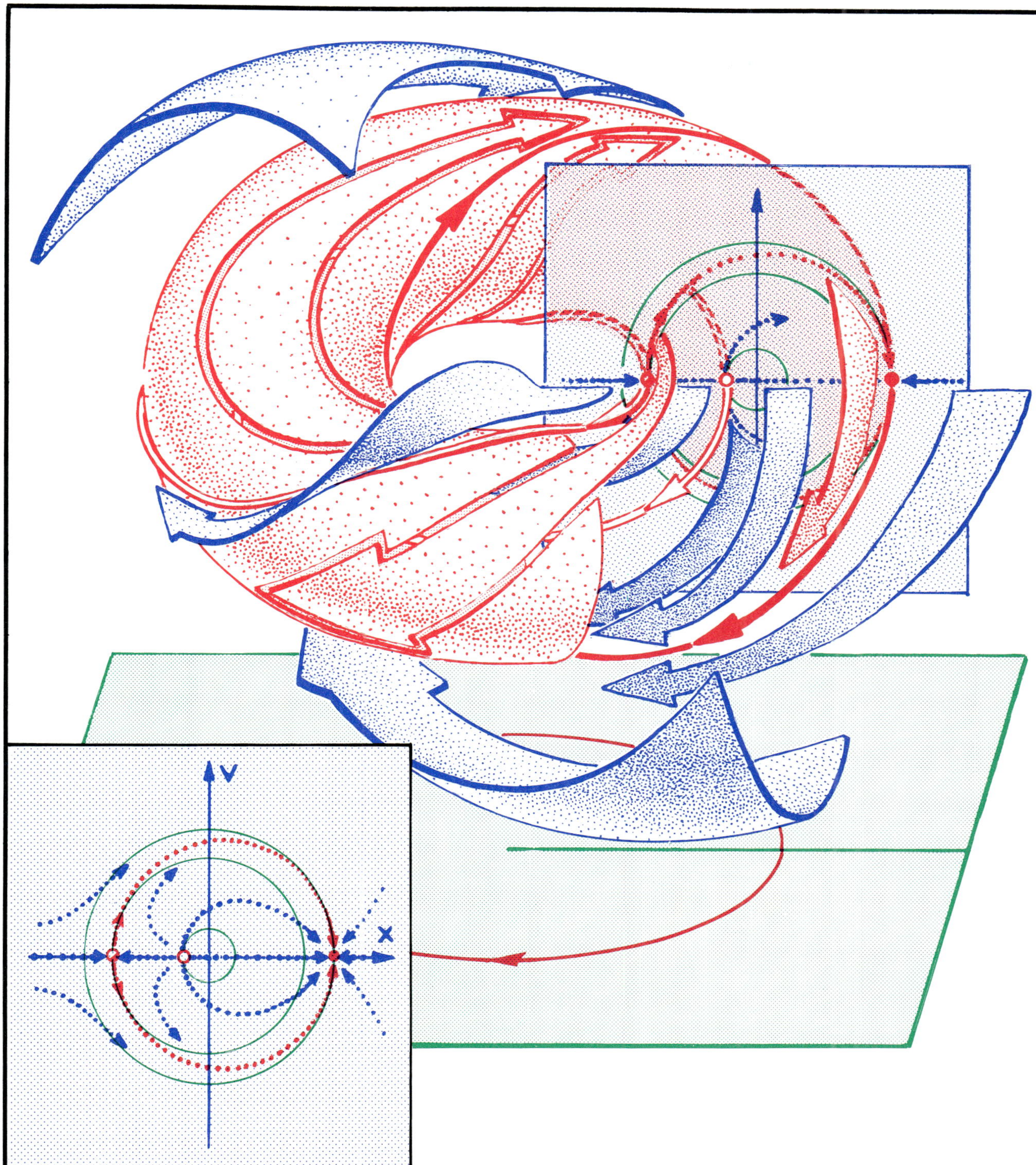

5.4.12. Putting all the pieces together, we see the two braided periodic trajectories on the invariant (red) torus, with the central repellor on its locating (green) torus within. The inset shows the driving-phase-zero strobe plane.

Now we see the braid: one periodic attractor and one periodic saddle, each isochronous, occupy the perturbed invariant torus. The attractor represents an in-phase oscillation of the coupled system. This situation is sometimes called *phase entrainment.* But unlike frequency entrainment, it is not *structurally stable.* This means that a further perturbation of the system (for example, a slight change in the speed of the driving motor) may shift the phase difference of the periodic attractor away from zero, while the clock frequency will still be entrained by the motor. Worse, for two general oscillators without an obvious zero phase, phase entrainment does not even make sense.

We have established that the phase portrait of the compound oscillator has at least three periodic trajectories: the braided saddle and attractor, and the central repellor. These three motions are isochronous. But we have only experimented with approximately equal frequencies. What happens if we really change the speed of the driving motor?

5.5. Response Curves for Frequency Changes

To see how this phase portrait depends on the speed of the driving motor, we must repeat the experiments of the preceding section many times, with different driving frequencies.

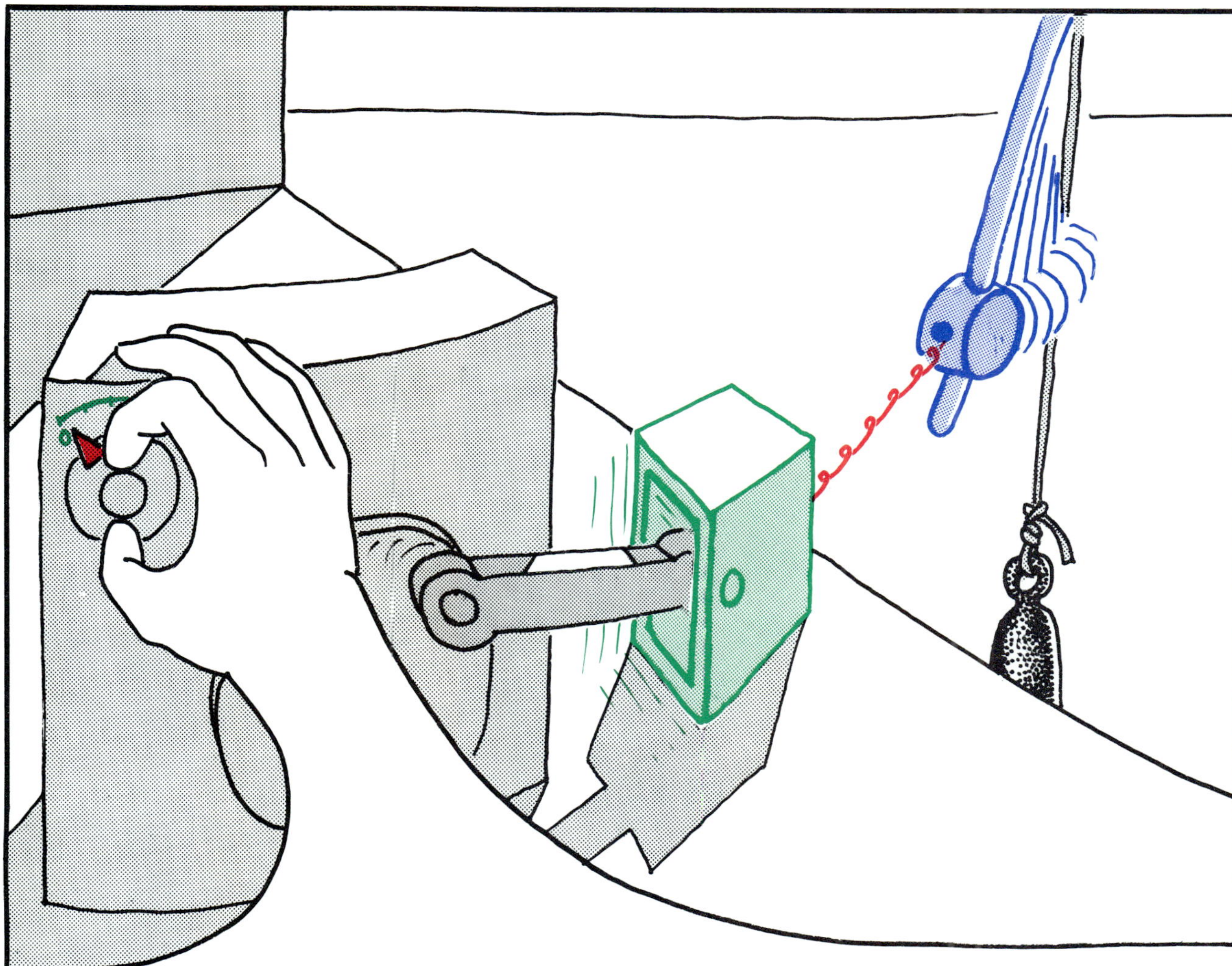

5.5.1. Changing the speed control of the driving motor a little and repeating the experiment, we will observe almost identical results: three isochronous periodic motions of the compound system. The pendulum is still isochronous, or entrained, with the driving frequency. Yet there are subtle differences in the *amplitude* of the three isochronous periodic motions. So, we must carefully measure these amplitudes in our experiments.

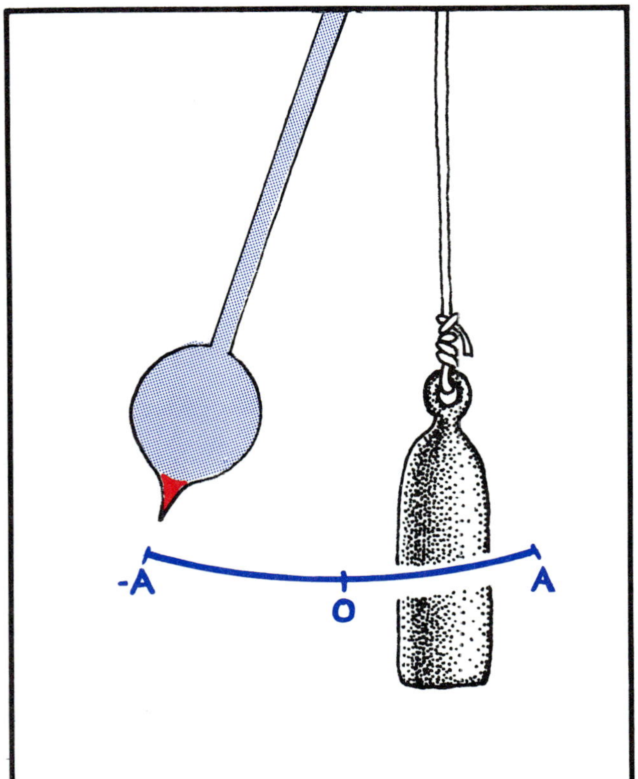

5.5.2. The amplitude is measured along the arc of the swinging pendulum. It is the maximum angle attained by the bob, in deviation from the vertical. It is a positive number.

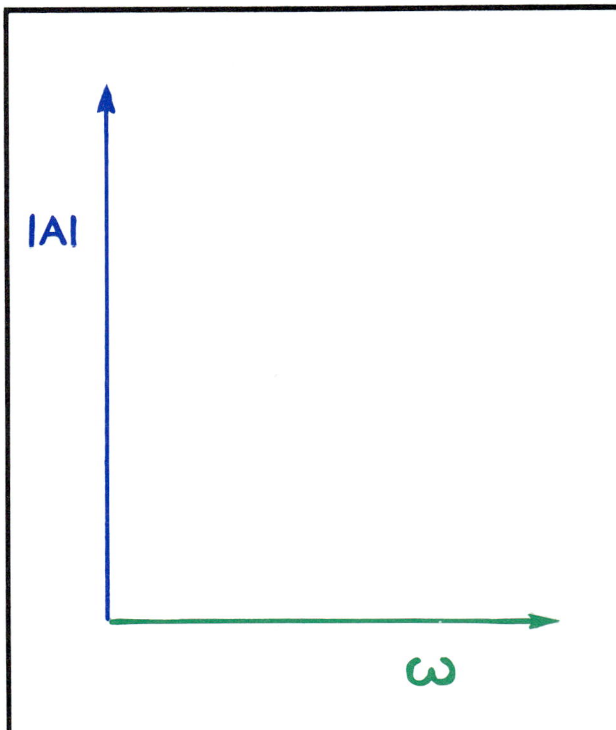

5.5.3. In each experiment with the forced oscillator, we will measure only the amplitude of the periodic motion found, called the *response amplitude,* and the frequency of the driving motor, or *driving frequency.* These two measurements may then be plotted in this *response plane* of response amplitude versus driving frequency.

After many experiments, we will obtain the graphs of these response amplitudes, regarded as functions of the driving frequency. These *response curves* comprise the *response diagram,* also called the *bifurcation diagram of the one-parameter system.*

We have already done the isochronous experiments, and may find the response amplitudes by looking back at the strobe plane illustrations of the preceding section. Recall that we use the following *strobe plane convention:*

solid dot	=	**attractor**
half-filled dot	=	**saddle**
hollow dot	=	**repellor**

for representing the three trajectories as points in the strobe plane.

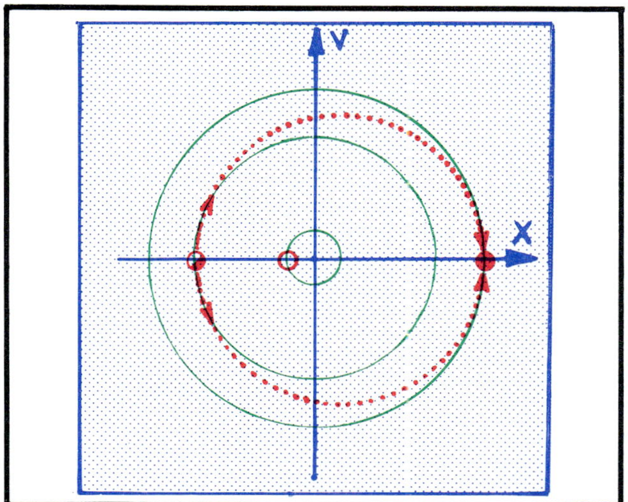

5.5.4. Here is the strobe plane portrait for the isochronous case. The diameter of the locating torus, represented as a green circle here for each of the three periodic trajectories, is the response amplitude.

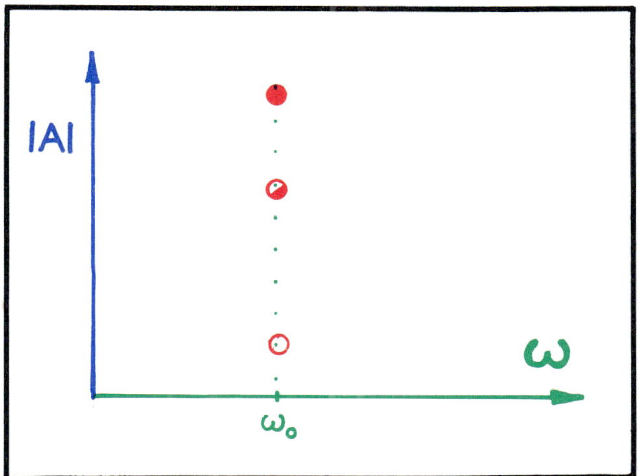

5.5.5. Here, the three response amplitudes are plotted above the point on the horizontal axis, w_0, representing the driving frequency. This corresponds to the speed of the motor, which in this case is isochronous with the clock.

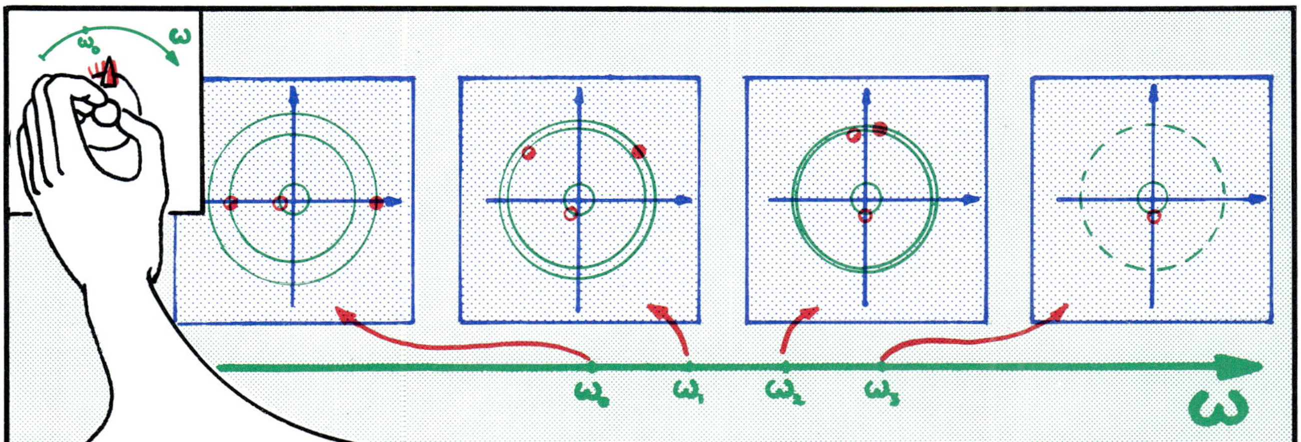

5.5.6. As we increase the frequency of the driving oscillation, w, the isochronous oscillation of the clock pendulum (periodic attractor, solid dot in the preceding plot) lags behind the driving oscillation. Simultaneously, the phase of the unstable mode (periodic saddle, half-filled dot in the preceding plot) advances ahead of the driver. As the driving frequency increases further, the amplitudes of these two periodic trajectories become closer to each other. Eventually, they coincide, cancel, and disappear. This is a catastrophic change in the phase portrait, which now has only one isochronous periodic trajectory, instead of three. And the remaining one is repelling! In this illustration, the amplitudes and phases of the periodic trajectories are shown in the cross-section of the ring model corresponding to phase zero of the driving oscillation. The sequence of sections, from left to right, corresponds to increasing driving frequencies, w_0, w_1, w_2, and so on.

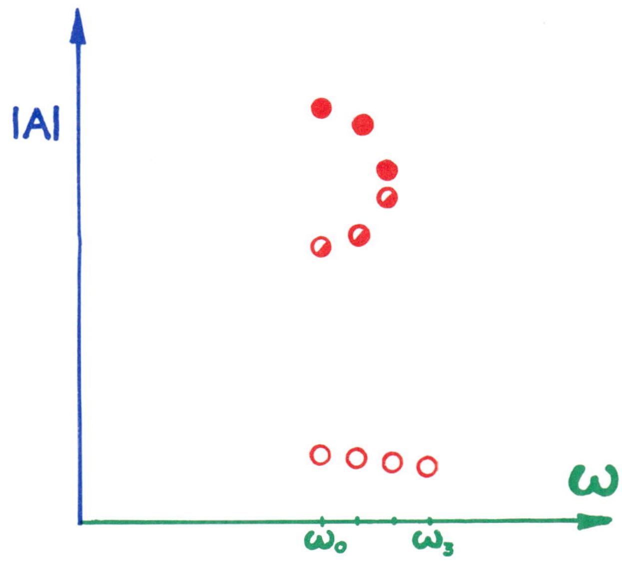

5.5.7. Measuring the response amplitudes of the periodic motions, seen as the diameters of the green circles in each of the strobe planes in the preceding panel, we record the results of the series of armchair experiments in the response plane. Over each of the chosen driving frequencies, we record the observed response amplitudes.

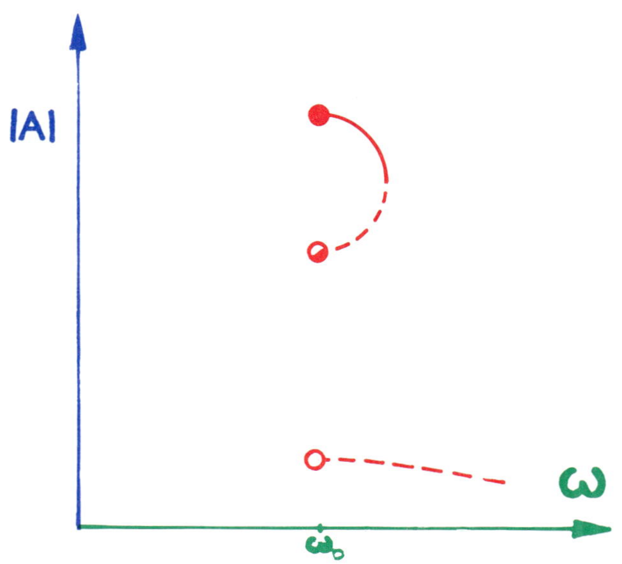

5.5.8. After more experiments if necessary, we can fill in the complete response curves for the isochronous motions. The change from three motions to only one turns out to be a smooth one, as two curves join in a circular shape.

Here we have used the *response plane convention:* solid curves are the tracks of attractors, dotted curves are the tracks of saddles or repellors.

Known as the *dynamic annihilation catastrophe,* this particular diagram is an example of *bifurcation behavior,* the subject of Part 4.

A similar event results from *decreasing* the speed of the driving motor, as we shall see in the next illustration.

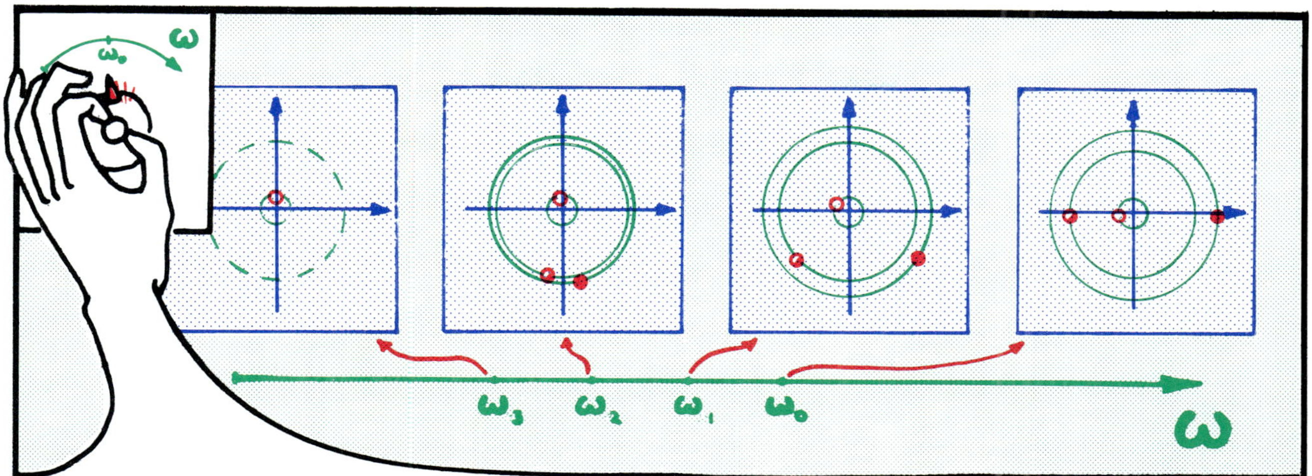

5.5.9. Decreasing the driving frequency results in a similar bifurcation event, as shown in this sequence of sections. The phase of the attractor advances, while the phase of the saddle lags behind. Eventually, they meet and wipe each other out. The sequence of sections shown here, *from right to left,* corresponds to decreasing driving frequencies, w_0, w_1, w_2, and so on.

5.5.10. Again, we plot the response amplitudes (observed above as the diameters of the green circles) as functions of the decreasing driving frequency, to obtain the rest of the response diagram for the compound system with the weak coupling spring. Here we have adjoined the new data to the response diagram for increased driving frequencies in Figure 5.5.8.

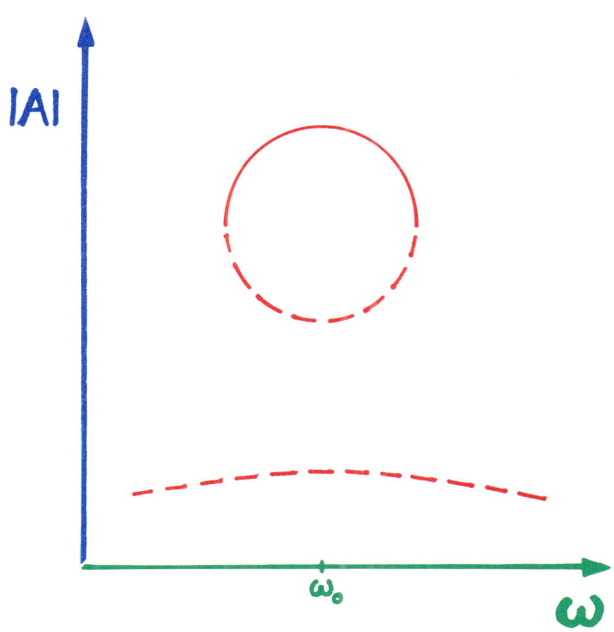

On the other hand, we could change the spring and repeat all of the experiments. For example . . .

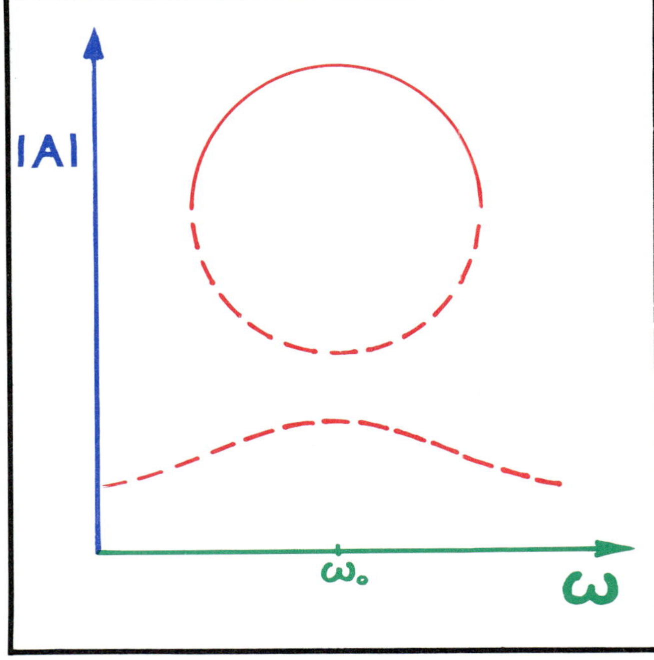

5.5.11. Here are the data produced by a stronger spring. The amplitudes change over a wider range under the influence of stronger coupling.

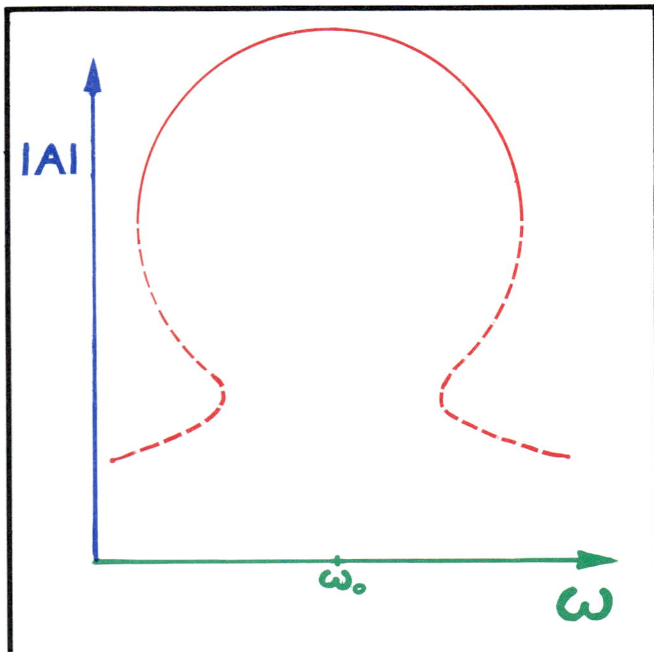

5.5.12. Here are the data produced by an even stronger spring. In this case, the saddle may cancel either the attractor *or the repellor.*

These response data record amplitudes of oscillation, as functions of *two control parameters:* driving frequency and coupling strength. Stacking up a number of response diagrams such as the two preceding ones, for stronger and stronger springs, would produce a three-dimensional plot. This is known as the *bifurcation diagram of the two-parameter system.* Many examples will be described in Part Four. A more compact representation of the same data may be made, just by superimposing the plots on the same planar diagram. This is called the *response* diagram of the two-parameter system.

Here is the response diagram for the two-parameter system. The parameters are coupling strength and driving frequency.

5.5.13. Here, the three different one-parameter response diagrams, described above, are superimposed. A regular structure begins to emerge.

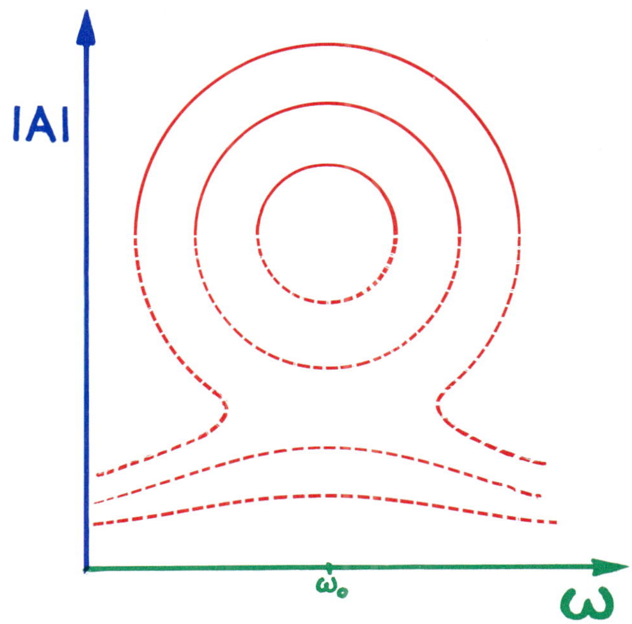

This diagram tells almost everything about the amplitudes of the isochronous oscillations of an ideal, self-sustained oscillator, when driven by another, regulated oscillator. Many more oscillations, called harmonics, are known to occur in such systems, however. Their discovery and analysis was greatly facilitated by the development of electrical oscillators, to which we now turn.

5.6. Forced Electrical Oscillators

Lord Rayleigh's model for self-sustained oscillations in organ pipes and violin strings (Chapter 3) applies very naturally to vibrations in an electrical context. In fact, Rayleigh himself carried out the application to an electrical vibrator invented by Helmholtz. Forty years later, the early electrical engineers found that Rayleigh's model worked very well for the vacuum tube oscillators used in the first radio transmitters. Further, Rayleigh had briefly studied a coupled system of two such oscillators — a large one forcing a small one. But real progress in understanding coupled oscillators awaited the further development of radio frequency electronics. In 1921 the time was right, and Van der Pol began this progress. In this section, we describe his program, which culminates Part One.

Electronic oscillators are easier for experimentalists to manipulate than clocks, but harder for onlookers to understand.

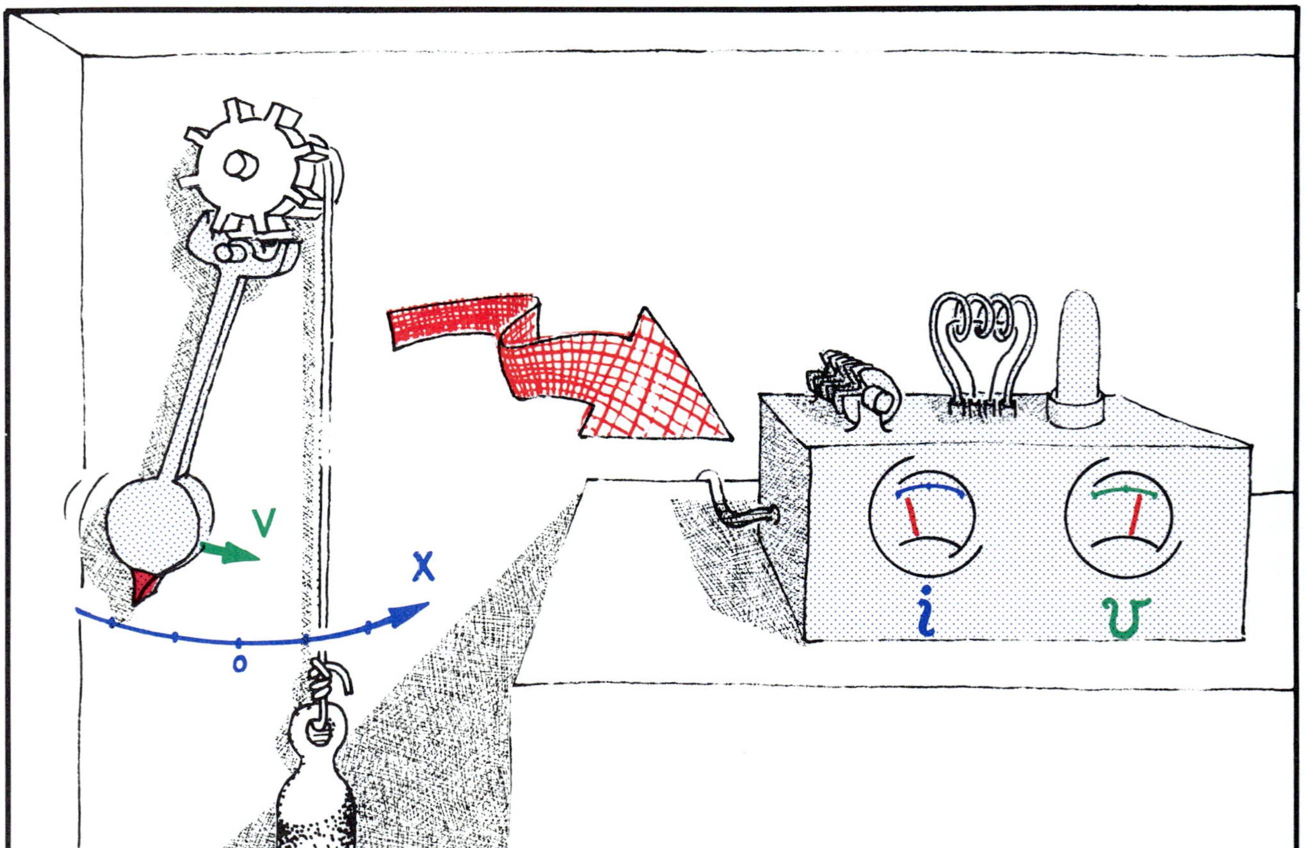

5.6.1. First, we replace the clockworks with an electric oscillator; not an electric clock, but a radio frequency oscillator running millions of times faster. (See Section 3.4.)

5.6.2. As the current and voltage fluctuations are much too rapid to be observed on the panel meters, we connect an oscilloscope to the device. Horizontal deflection measures current and vertical measures voltage of the output circuit of the oscillator.

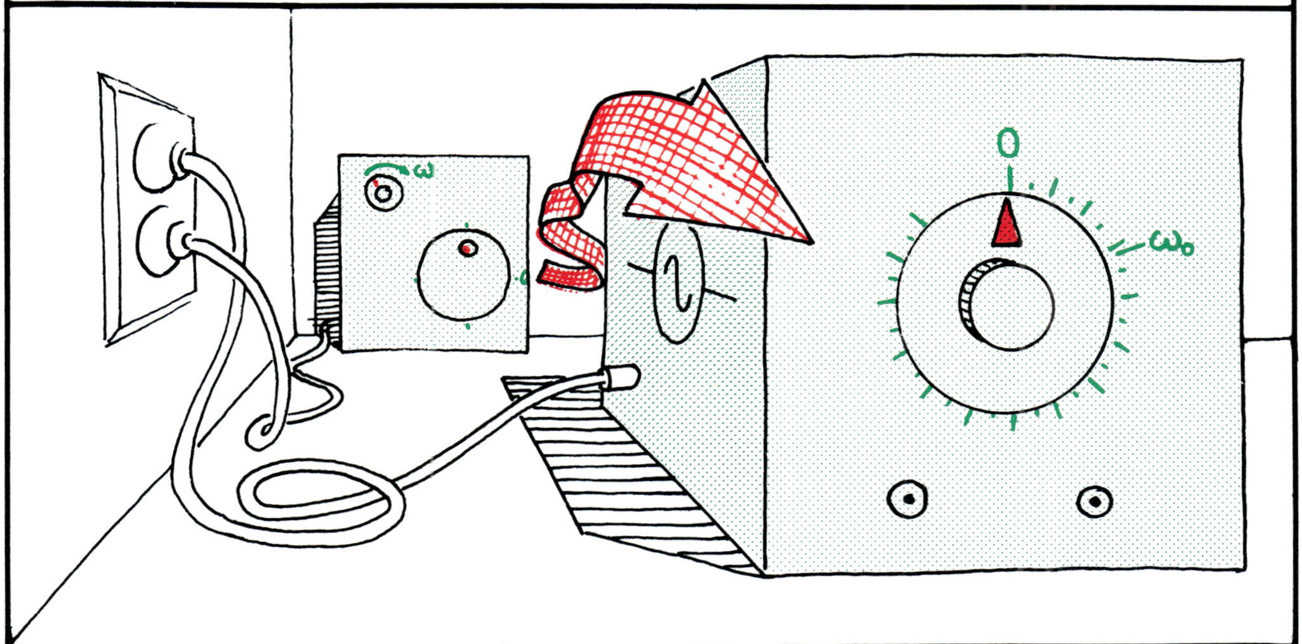

5.6.3. Next, we replace the regulated driving motor with another electrical oscillator. This is a VFO (variable frequency oscillator) a well-regulated sinusoidal oscillator with a frequency adjustment knob.

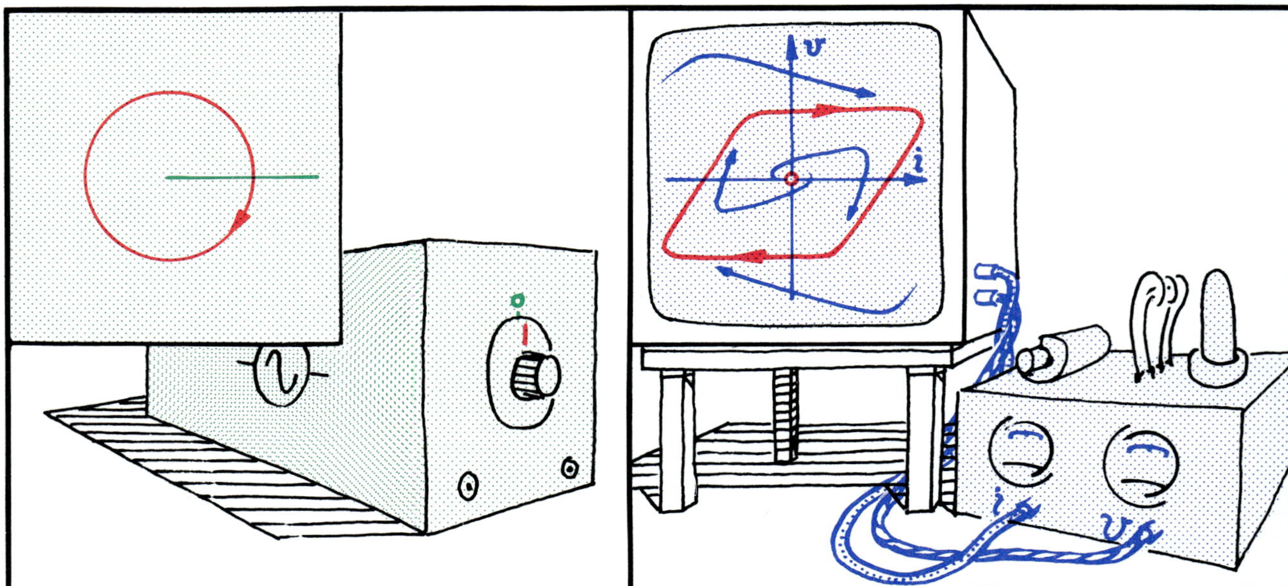

5.6.4. As in the preceding section, the state space for the driving oscillator is a circle. One dimension, the phase of the driving oscillation, suffices. The state space of the driven oscillator is a plane. Before coupling, the phase portrait consists of a single periodic attractor, cycling around a point repellor.

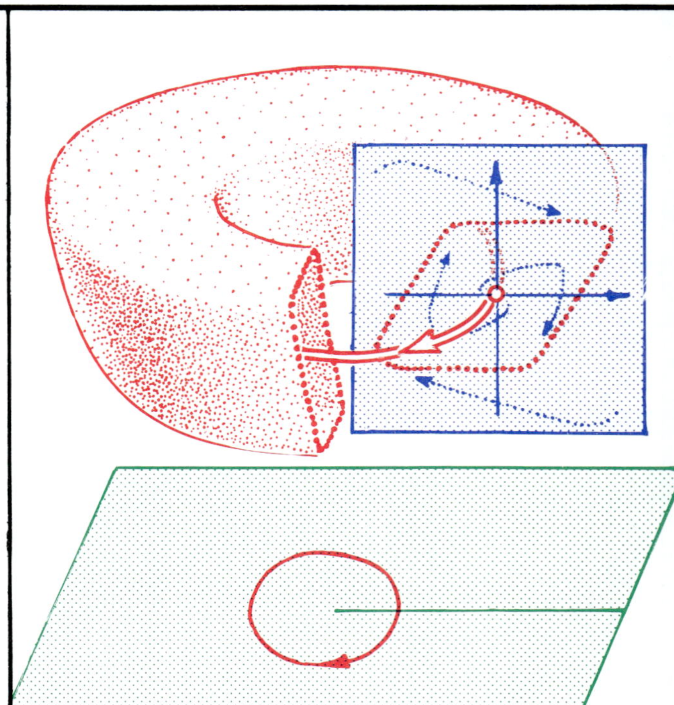

5.6.5. The combined state space is again a *ring model,* the Cartesian product of the circle and the plane. Before coupling the two oscillators, the combined phase portrait consists of an attractive, invariant torus around a periodic repellor.

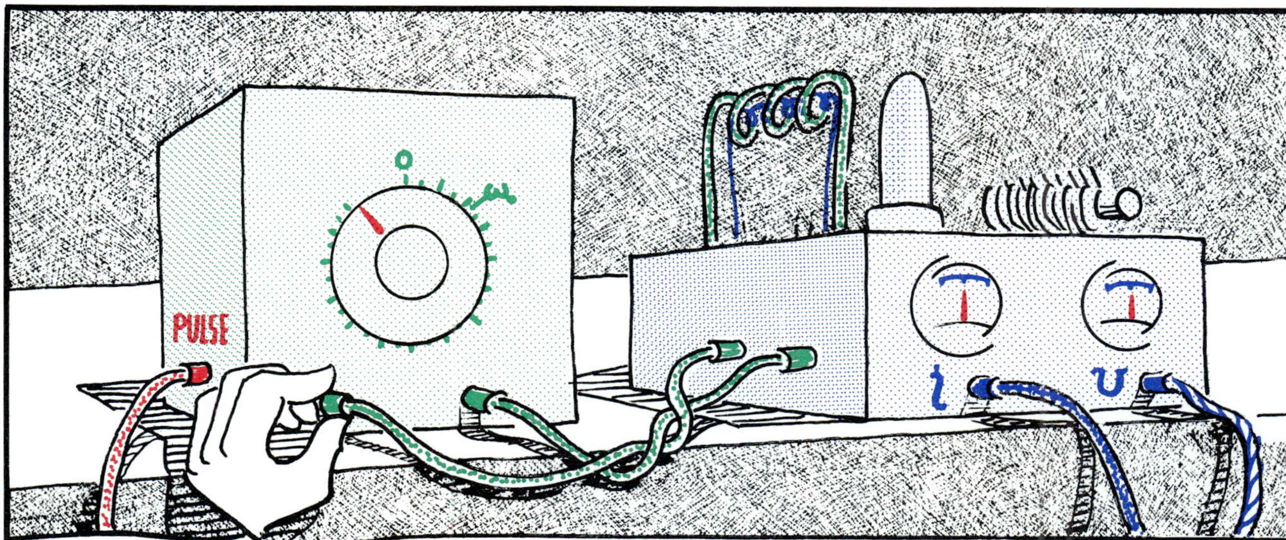

5.6.6. Next, the coupling. The output of the driving generator is connected to the input of the driven oscillator. The actual coupling is electromagnetic, through an auxiliary coil around the inductor of the driven oscillator.

5.6.7. Finally, we will utilize a pulse from the driving oscillator. This brief pulse is emitted each time the driving oscillation passes phase zero. It is connected to the high-voltage power supply of the oscilloscope, so that the beam is on only during the brief pulse. This scheme replaces the strobe lamp of the mechanical device in the preceding section. Thus, we see on the screen of the oscilloscope only the cross-section of the ring model corresponding to driving phase zero.

Let's try out this setup, which is essentially the same as the appratus in Van der Pol's laboratory. First, we unplug the coupling wire between the two oscillators. Then . . .

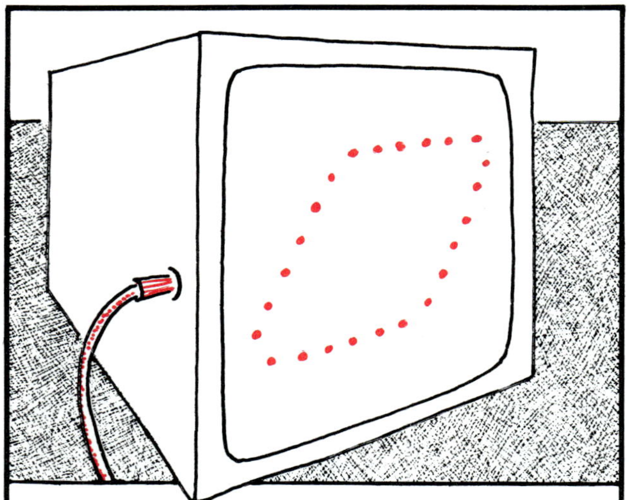

5.6.8. Turning on everything, we see a dotted outline of the periodic attractor of the driven oscillator on the oscilloscope screen. These dots are trajectories crossing driving-phase-zero strobe plane. The oscilloscope is strobed by the zero-phase pulses from the driving oscillator.

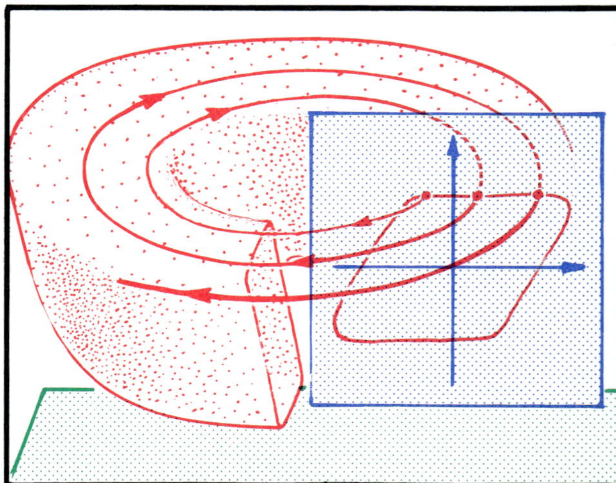

5.4.9. Prolonging this cross-section around the ring model, we see that the trajectory is a periodic attractor or a solenoid. The way it winds around the attractive (red) torus depends on the relationship between the two frequencies, as described in Section 5.1.

Before plugging the coupling wire back in, we can try changing the frequency of the driving oscillator.

5.6.10. Setting the driving frequency to exactly the same frequency as the driven oscillator would reduce the trajectory to a single dot, in the cross-section of the attractive torus. But it is impossible to set it exactly right, so the dot wanders around.

Now we plug in the coupling wire between the two oscillators.

5.6.11. Miraculously, the wandering dot settles down in a single spot. Further tiny movements of the driving frequency knob may move this spot back and forth on the driven (red) cycle, but at any one frequency it does not wander.

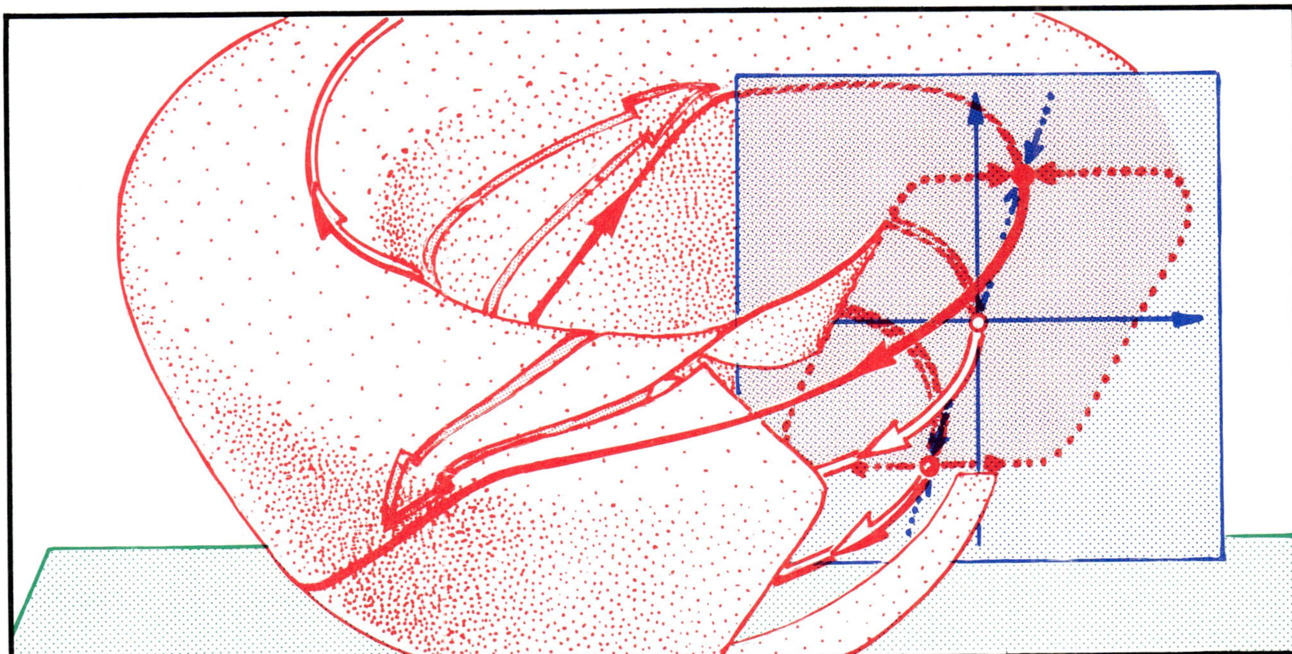

5.6.12. But we expect this, after the explanations of the preceding section. The *structural stability* of the coupled phase portrait implies *frequency entrainment* of the coupled oscillators.

For example, here is one called the third subharmonic (see Section 5.2). It may be observed with Van der Pol's original apparatus, by setting the driving frequency to three times the frequency of the driven oscillator.

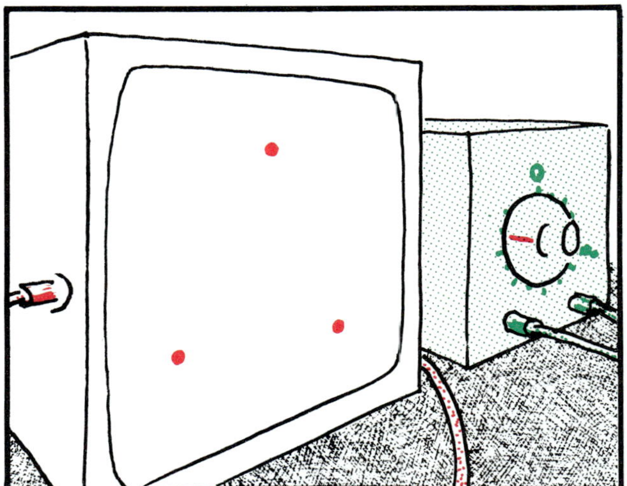

5.6.13. Set the frequencies, then turn on the oscillators, and connect the coupling wire. You will see three dots on the red cycle.

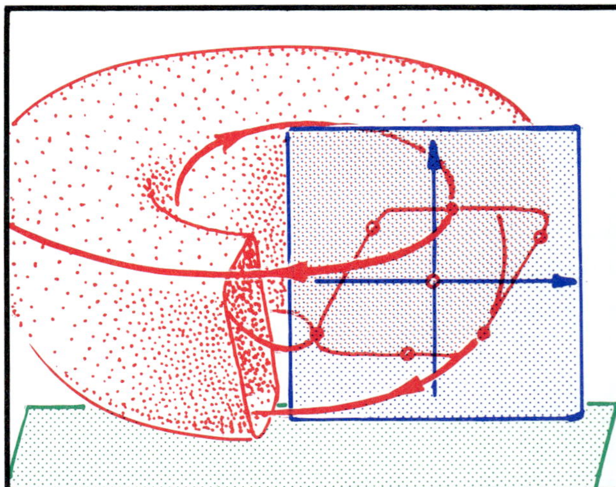

5.6.14. These three dots, prolonged around the attractive (red) torus, trace a periodic attractor winding thrice around.

A few zillion experiments like these will reveal further details of the response diagram, as we mentioned at the end of the preceding section.

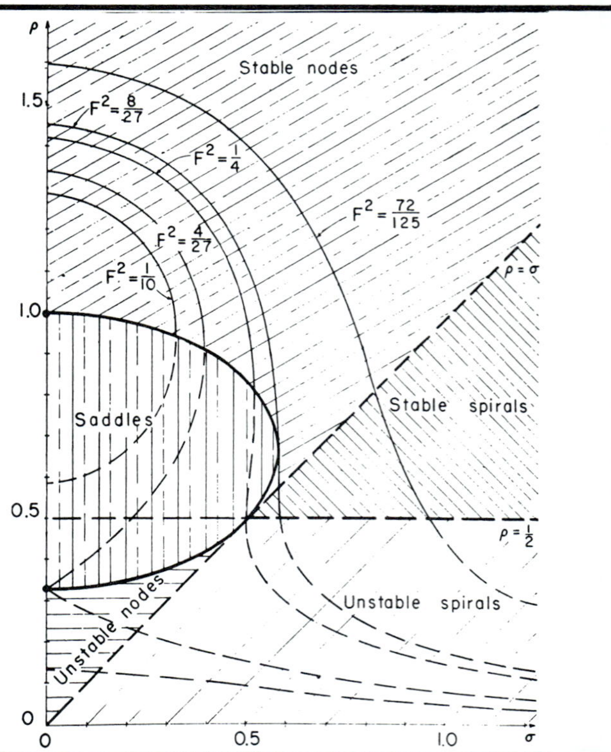

5.6.15. Here is the response diagram, as described by Stoker in 1950. Unexpected exotic details would be discovered a decade later.

And for these exotic details, of chaotic attractors and tangled separatrices, turn to Part Three.

CONCLUSION

The hardy reader, arriving at this point without extensive short cuts, has seen the dynamics of the classical period, 1600–1950. In fact, this subject had a full and happy life, and died of natural causes. The masterly text of Stoker (1950) is the eloquent epitaph of *classical dynamics.* Fortunately for the evolution of our species on planet Earth, this demise was foreseen in 1882 by Poincaré. His vision, nourished in secrecy by his faithful followers in Russia and America, emerged in the 1950's into the mainstream of mathematics, united with the outstanding applications of the classical period and acquired new electronic laboratory tools. Thus, *modern dynamics* was born.

Our goal has been not only to survey classical dynamics, but to do so from the viewpoint of the modern period. Thus, the groundwork for the following chapters has been laid. Part Two will report on the amazing discoveries of the modern period: *generic tangles* and *attractive chaotics.*

Basically, the classical period ended because the analytical techniques were exhausted. The moden period relies upon the geometric techniques pioneered by Poincaré. But simultaneously, experimental dynamics was revolutionized by electronic inventions. With fast analog and digital computers replacing the mechanical apparatus of the classical dynamics laboratory, amazing discoveries were rapidly made. The theory and experiements evolved together in the recent decades, and a new set of paradigms emerged to revolutionize the sciences. This story is the subject of Parts Two and Three.

PART 2

Chaotic Behavior

Dedicated to Henri Poincaré and George D. Birchoff

Courtesy of the Library of Congress

By permission of the American Mathematical Society

Chaotic Dynamics Hall of Fame

As described in Part One, experiments play an increasingly important role in dynamics. Helmholtz, Rayleigh, Duffing, Van der Pol, and Hayashi relied on experimental simulations to discover the main properties of periodic motions in nonlinear oscillations. Sincd 1950, digital simulations have become increasingly important, especially since the experimental discovery of chaotic attractors in 1962 by Lorenz.

But chaotic limit sets had been known to theoretical dynamics since Poincaré. The first chaotic attractor in a dynamical system (in the mathematical sense of a vector field on a state space) was discovered in 1932 by Birkhoff. It took many years before these objects emerged into the theoretical literature, in the works of Charpentier, Levinson, Cartwright, Littlewood, Smale, and others. Experimental studies accelerated this process.

But we must emphasize at once, and we cannot possibly do this as strongly as we would like, that *the connection between the chaotic attractors of theory and those of experiments is hypothetical at this time.*

In fact, many believe that the connection is fictitious. An account of this view may be found in the literature (Abraham, 1983). But in this Part, we will be very casual about this fundamental problem. We will speak of *chaotic attractor* in either the theoretical or experimental context.

This introduction presents a few words of description for some of the leading figures in the history of chaotic dynamics. Their positions in a two-dimensional tableau—date versus specialty (applied, mathematical, or experimental dynamics)—are shown in Table 2.1.

TABLE 2.1–THE HISTORY OF DYNAMICS

Date	THEORY	EXPERIMENT
1850		
	Helmholtz	
	Poincaré	Rayleigh
1900		
		Duffing
	Birkhoff	Van der Pol
	Charpentier	
	Levinson	Hayashi
1950	Cartwright & Littlewood	
		Lorenz, Stein & Ulam
	Smale	Rössler, Ueda
		Shaw
2000		

Here are some capsule histories. Further details may be found in Chapter 8.

Jules Henri Poincaré. 1854–1912. In his studies of celestial motion, Poincaré discovered ***homoclinic trajectories*** (see Section 3.1). Eventually, in 1962, Smale showed that these are chaotic (non-attractive) limit sets. Thus the first appearance of a chaotic limit set in the mathematical literature, as far as we know, was in Poincaré's three-volume work on celestial mechanics, in 1892.

George David Birkhoff, 1884–1944. The first chaotic attractor to appear in the mathematical literature, to our knowledge, is in Birkhoff's paper on remarkable curves, published in 1932. (See Section 3.2.) The actual discovery, in a context derived from celestial mechanics, occurred in 1916.

Marie Charpentier. Charpentier further developed the properties of Birkhoff's remarkable curves, in a series of papers in the 1930's.

Norman Levinson. In 1944, Levinson conjectured that Birkhoff's chaotic attractor might occur in the three-dimensional dynamics of forced oscillation. In 1948, he announced he had proved that it does.

Mary Lucy Cartwright and J.E. Littlewood. In a joint paper published in 1945, Cartwright and Littlewood announced a similar results. Their proof appeared in 1949. They were inspired by experimental observations of Van der Pol and Van der Mark, published in 1927.

Edward N. Lorenz. In one of the first digital simulations of a dynamical system, around 1961, Lorenz discovered the experimental object which has come to be known as a chaotic attractor, in the experimental sense, in a model for atmospheric air currents.

Stephen Smale. In 1962, Smale proved that Poincaré's homoclinic trajectories are chaotic limit sets, in the mathematical sense.

More recently, there has been an enormous explosion of research work on both mathematical and experimental chaos. Reluctantly, we close our history at this point, in 1962.

6

Static Limit Sets and Characteristic Exponents

In Part One, limit points and cycles in dimensions one, two, and three were introduced. In this chapter and the next, we review the geometry of these simple limit sets. Further, we introduce the *Liapounov characteristic exponents,* which completely characterize the geometry of the simple limit sets. This is prosaic, but important. In Chapter 8, more complicated limit sets will be introduced.

6.1. Limit Points In One Dimension

In one-dimensional state spaces, dynamical systems are naturally very constrained. For example, the typical limit set must be a point. As in Part One, we begin with this case for completenesss. In addition, this provides a simple context in which to describe a new idea, the *Liapounov characteristic exponent.*

Recall that one-parameter state spaces were used extensively in the early history of science.

For example, George's temperature may be represented as a point on a line. A simple dynamical model for this experimental domain (George's oral temperature) could be the vector field shown here. It has a single point attractor. Starting with the thermometer at an arbitrary initial state (room temperature, say), he inserts it under his tongue. In three minutes, the transient dies away, leading to a static attractor.

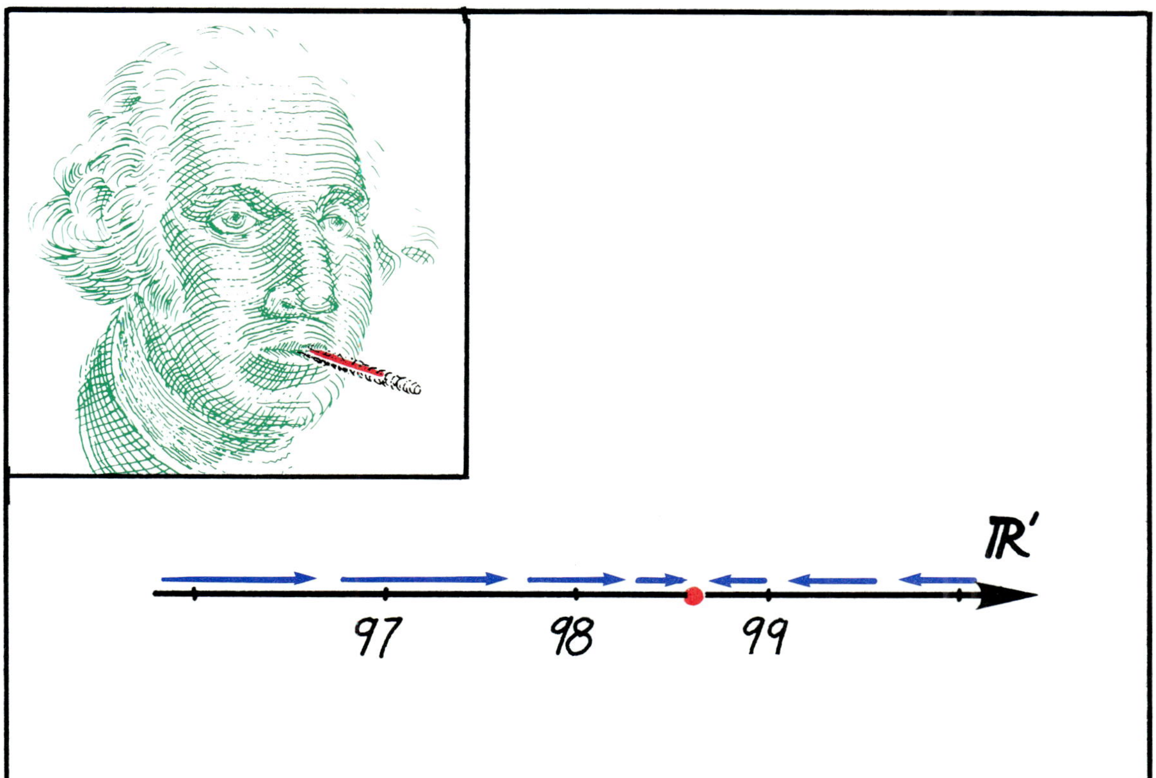

6.1.1. In this dynamical system, there is a single attractor, a limit point, at 98.6 degrees.

We now describe the *characteristic exponent or CE* of this limit point. Rotate each vector 90 degrees counterclockwise. Their ends delineate the *graph* of the vectorfield. In this context (one-dimensional linear state space) the graph is a curve in the two-dimensional plane. The graph goes through the horizontal axis at the limit point.

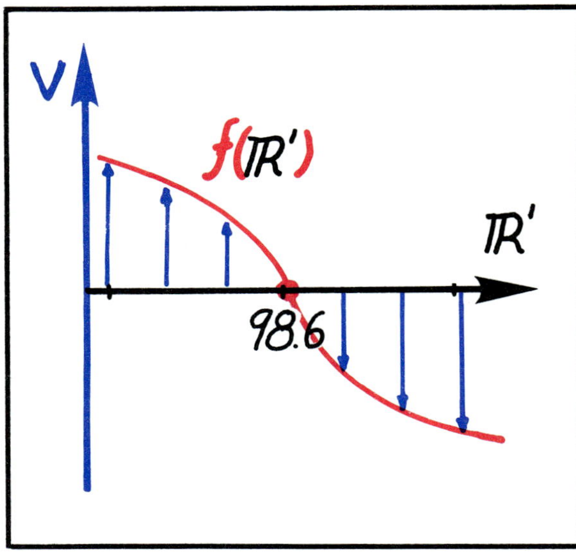

6.1.2. The slope of the graph is *negative* at the limit point, as the limit point is *attractive.* The steeper the slope, the faster the temperature of the thermometer approaches equilibrium.

There is only *one CE* because the dimension of the state space is *one.* In general, *the number of CEs is equal to the dimension of the state space.* Here, the dimension of the inset is one, which is maximal. The inset is an open subset of the state space. It is the *basin* of the point attractor. Meanwhile, the *outset* consists of a single point, the attractor. Thus, the *index* (dimension of the outset) is zero. The *magnitude of the CE* tells the *strength of the attractor,* that is, the rate of approach of the nearby trajectories, within the inset or basin.

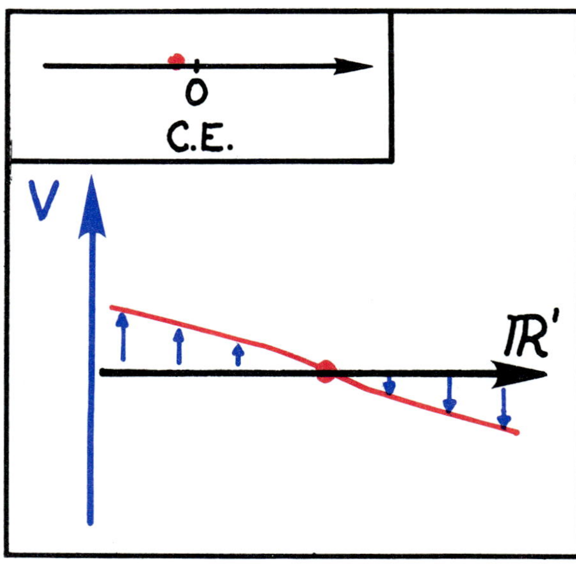

6.1.3. Here is a *weakly attractive* limit point in one dimension. The characteristic exponent (or CE, shown in the inset) is a *small negative number.*

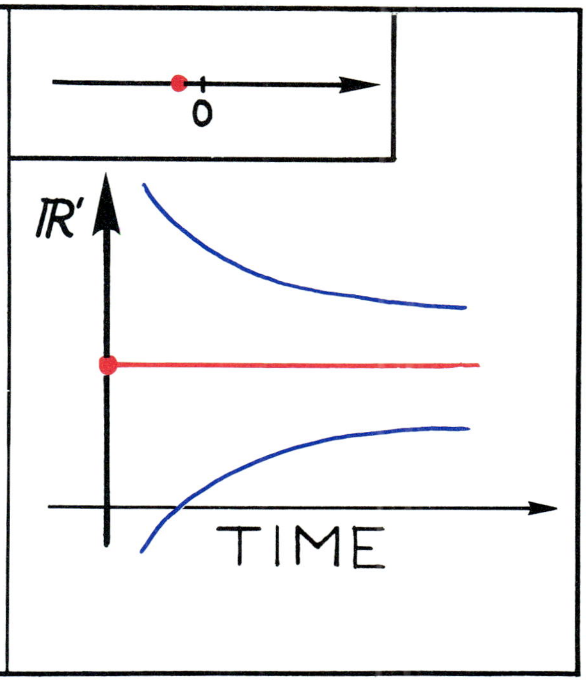

6.1.4. The time series of typical trajectories (see Part One) in this system show a *gradual approach to the weakly attracting point.*

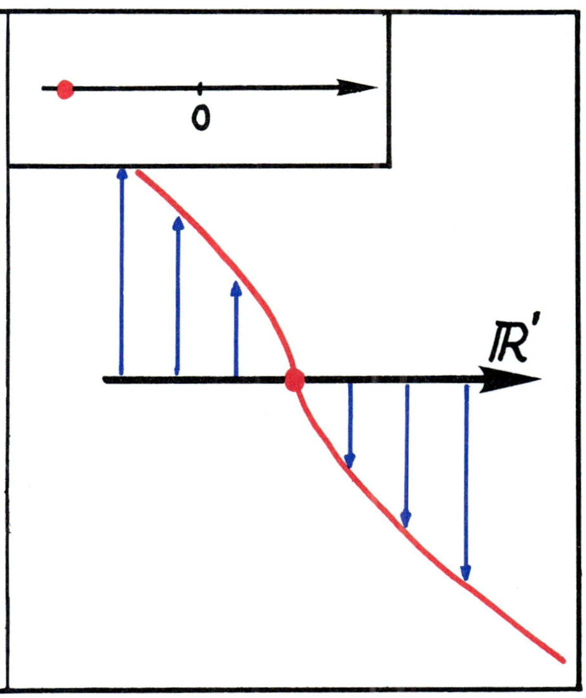

6.1.5. Here, in contrast, is a *strongly attracting* limit point. The characteristic exponent (see inset) is a *large negative number.*

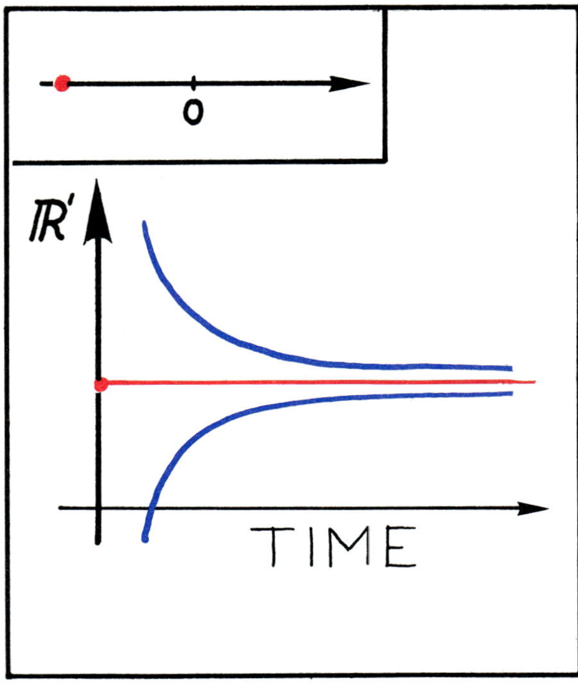

6.1.6. The time series of typical trajectories of this system reveal a *rapid approach to the strongly attracting point.*

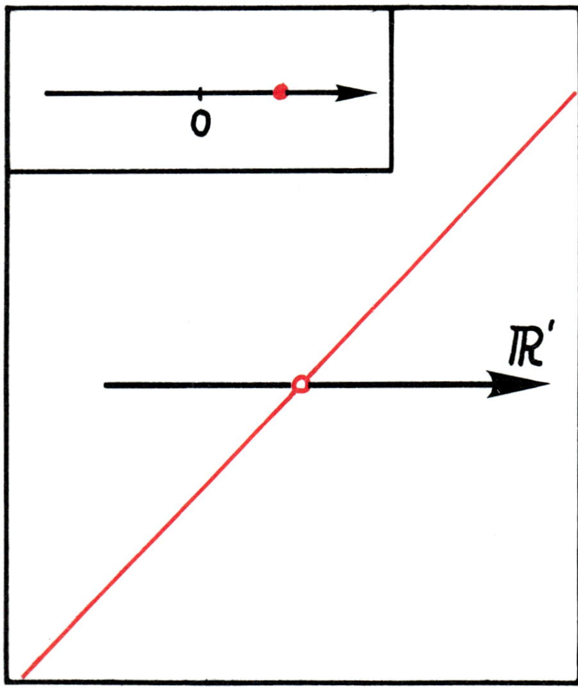

6.1.7. Another possibility is a positive slope, or characteristic exponent, at the critical point. This can occur only for a point repellor, indicated here by a hollow dot. The vectors point away from the critical point.

A limit point in one dimension is called *hyperbolic* if its CE is *not zero.* Some hyperbolic critical points in one dimension are shown below.

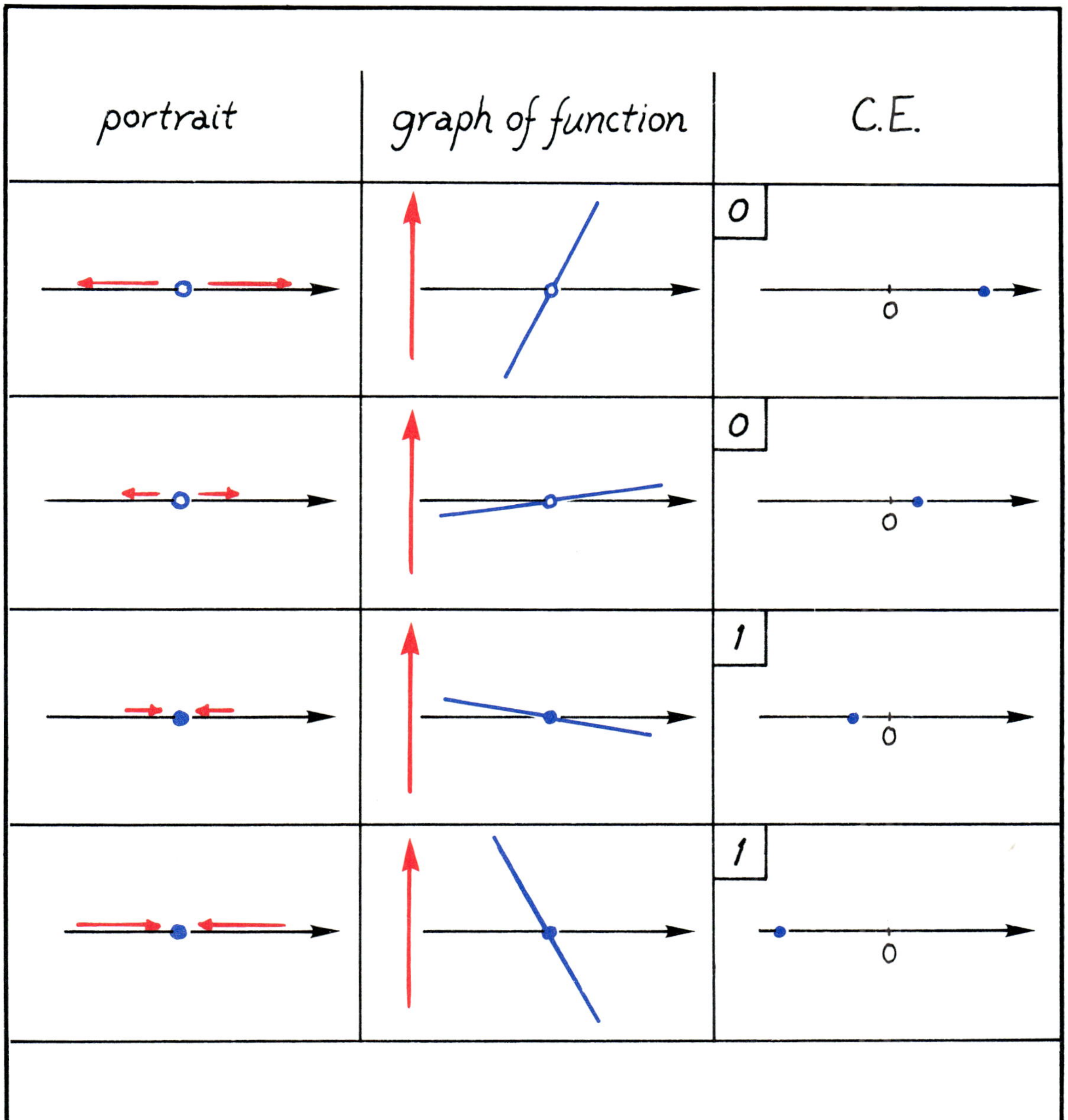

6.1.8. Hyperbolic limit points in one dimension.

What about zero as a CE? Zero Slope implies tangency.

A limit point with zero as its CE is called *nonhyperbolic* or *degenerate.*

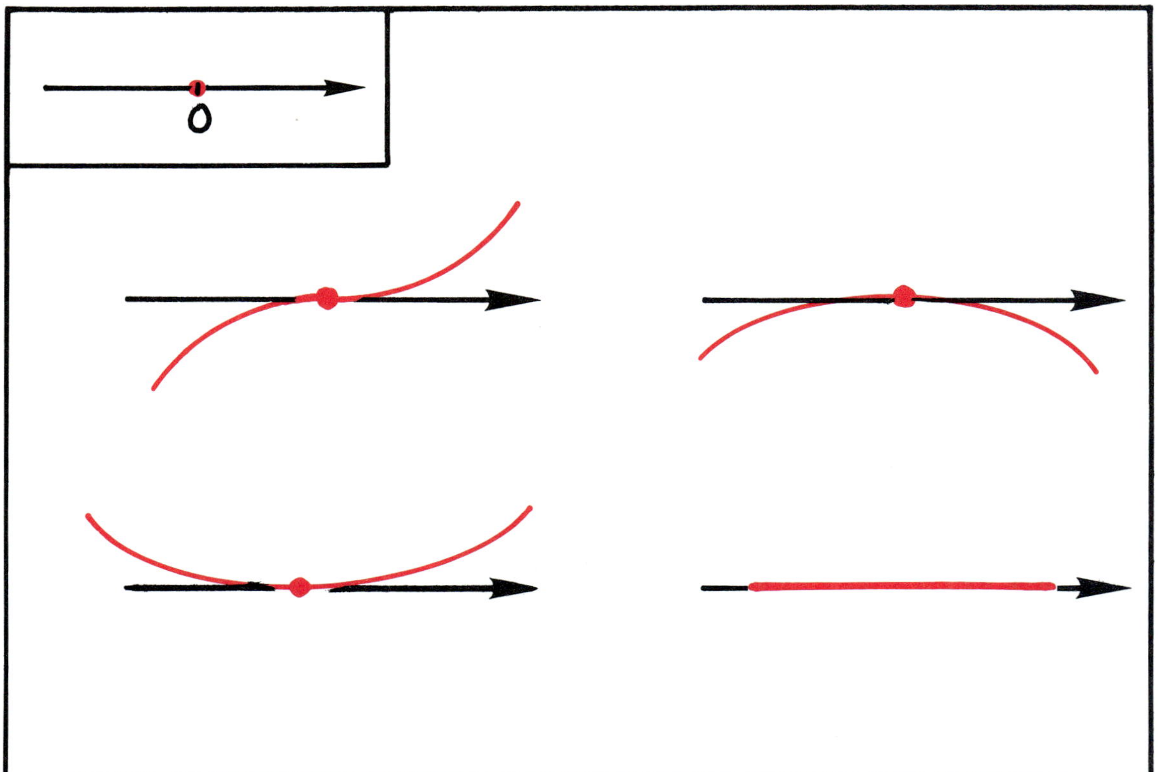

6.1.9. These are the four degenerate critical points on a line. The graph is tangent to the horizontal axis at the critical point.

In Part Three we will explain how these degenerate cases are *exceptional.* Garden-variety dynamical systems do not have them. But they become important in the theory of *bifurcations,* treated in Part Four.

6.2. Saddle Points In Two Dimensions

In Part One, we described limit points in the plane as nodal or spiral. Here we will characterize them in terms of their CEs.

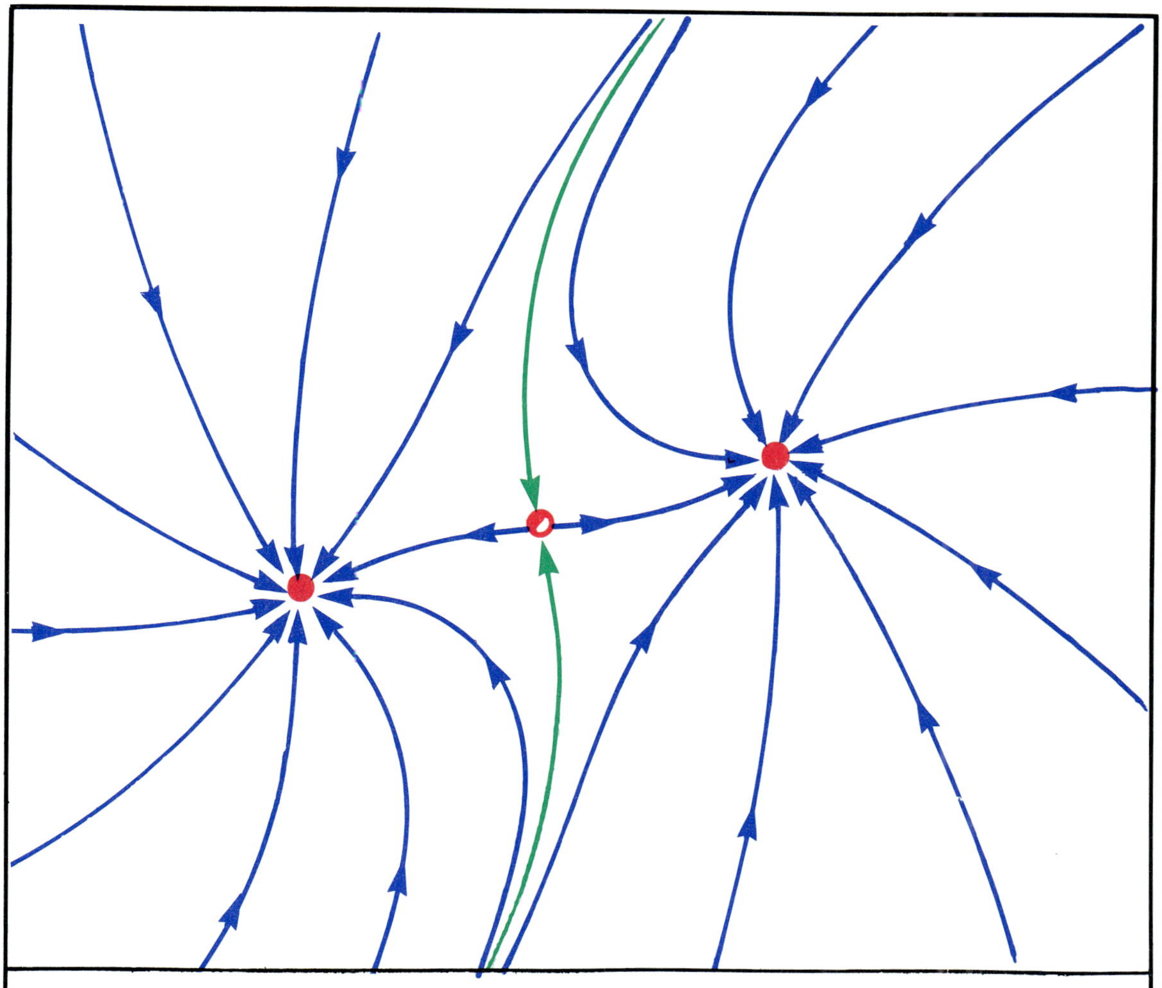

6.2.1. Remember the phase portrait of the gradient system in Part One (1.6.10)? It has three limit points: two point attractors and a saddle point.

Let's start with the saddle point. What are its CEs?

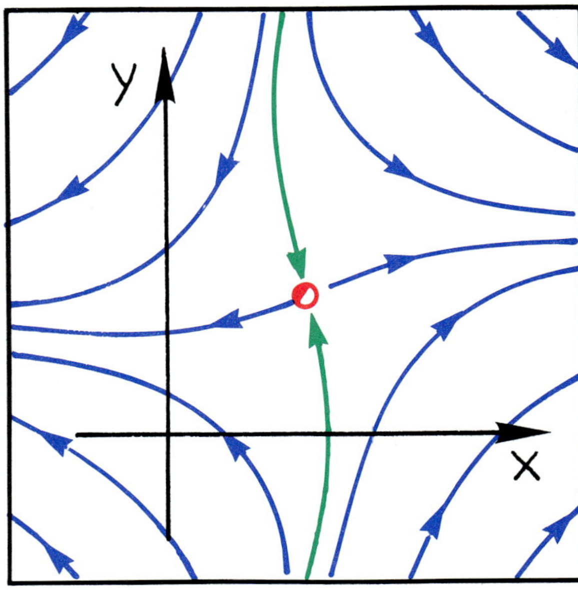

6.2.2. This typical saddle point has an *inset* and an *outset,* both of dimension one. The inset is green, denoting the separatrix.

The *index* of this saddle is one.

To identify the two CEs of this saddle point, let's introduce a new coordinate system, by translation and rotation, so that the new coordinates start at the saddle, and are oriented long its inset and outset.

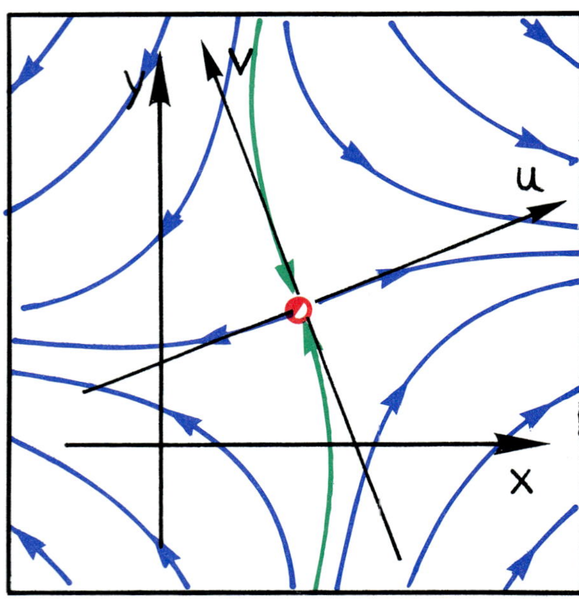

6.2.3. Here is the preceding portrait, with new coordinate axes included.

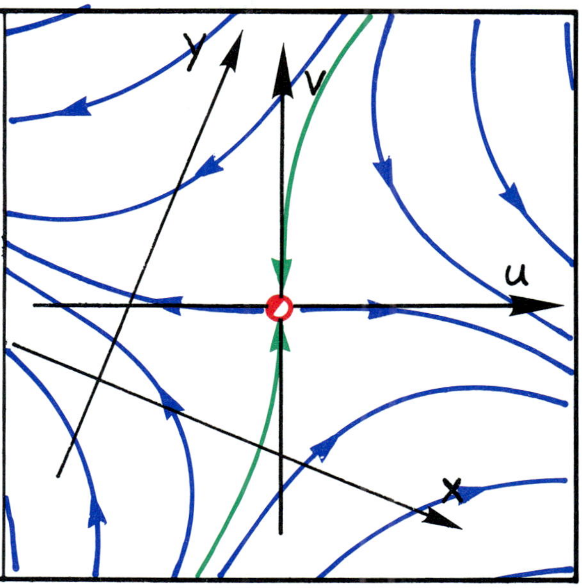

6.2.4. Here is the same portrait, rotated and translated, with the new coordinate axes in standard position.

Now we suppose that we may treat each of the new coordinates separately. This seems to be assuming a lot, but is actually justifiable.

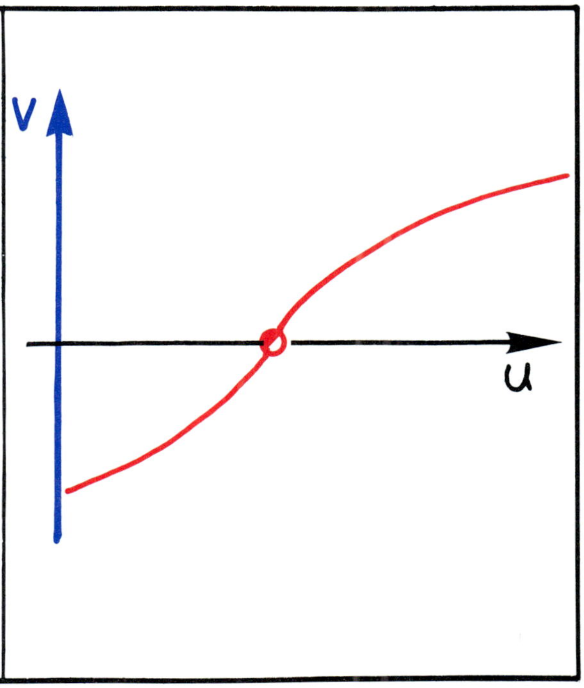

6.2.5. Graphing the horizontal (u) component of the vector field (in the new coordinates) along the horizontal (u) axis, as in the preceding section, we may measure the slope, or *strength of repulsion,* in this direction. *This positive slope is one of the CEs of the saddle point.*

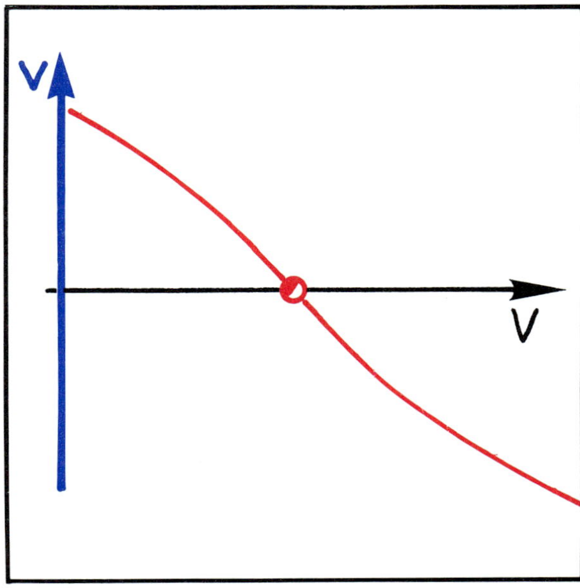

6.2.6. Likewise, graphing the vertical (v) component of the vectorfield, along the vertical (v) axis, we may measure the slope, or *strength of attraction,* in this direction. *This negative slope is the other CE of the saddle point.*

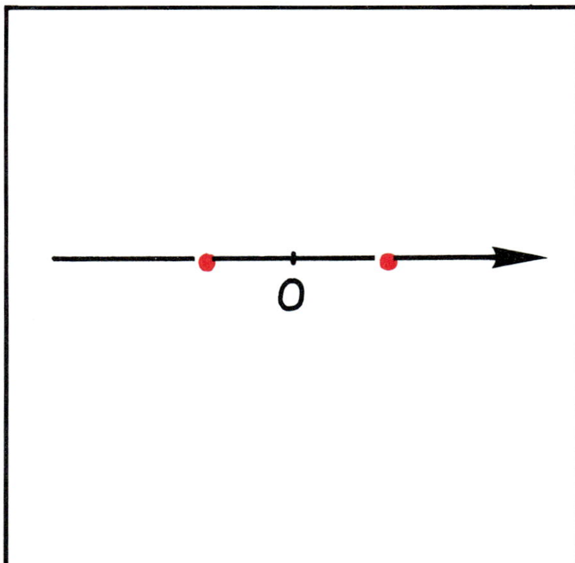

6.2.7. Here are the two CEs of the saddle point, pictured on the line of real numbers. One is negative, the other positive. As in the one-dimensional case, the CE to the left of the origin characterizes attraction, the one to the right repulsion.

Recall that the *index* of a critical point is the dimension of its outset. In this case, *the index is one.* And this is also the number of CEs to the right of zero.

6.3. Nodal Points in Two Dimensions

Next we consider attractive critical points for dynamical systems in two dimensions. There are two types, nodal and spiral, as described in Part One. In this section, we look at the nodal case. What are the CEs in this context? Let's consider a typical nodal attractor, as in the gradient system, Figure 6.2.1.

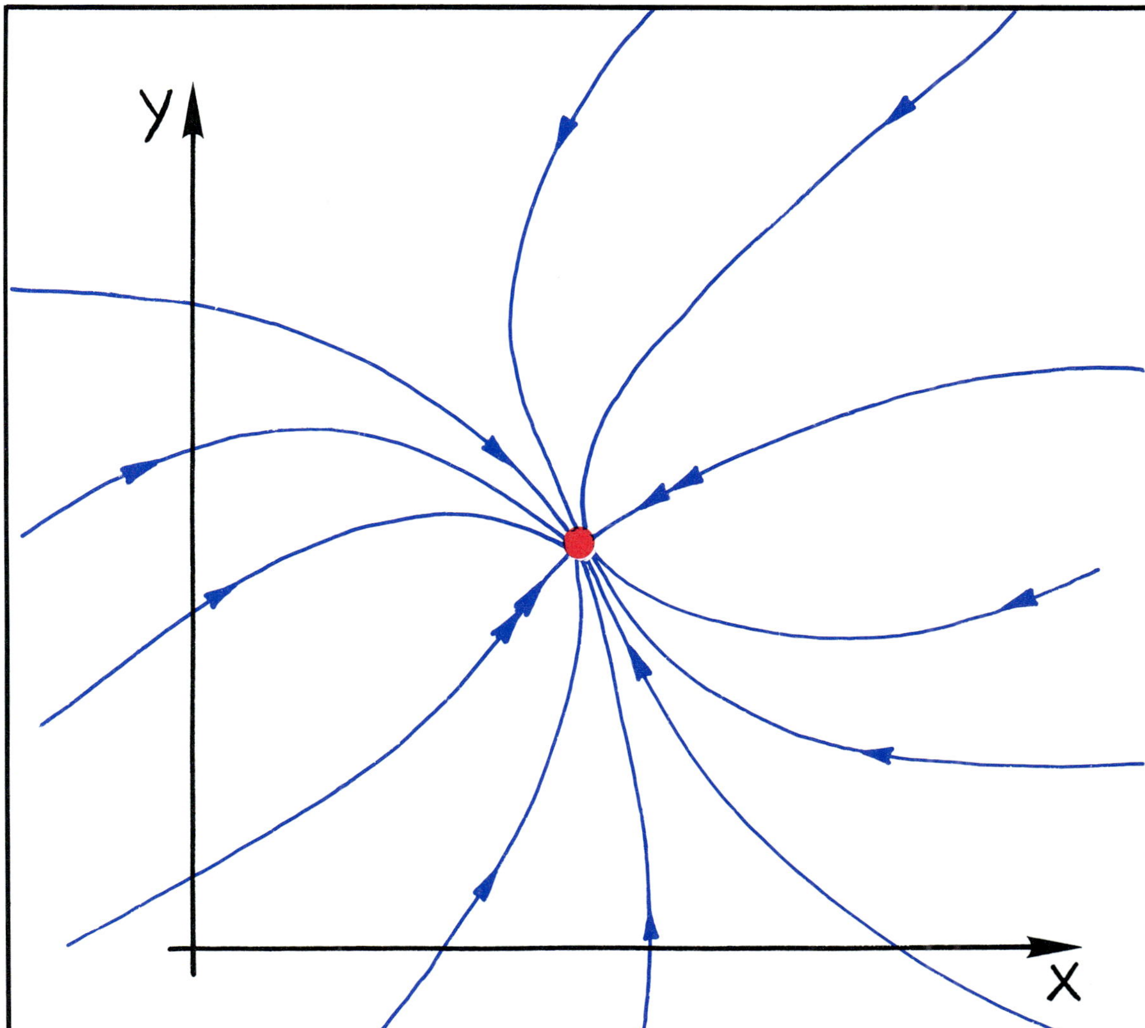

6.3.1. Like the saddle point, this has characteristic directions (red). In one direction indicated by double arrows the attraction is stronger than in the other. We call this the *fast direction,* the other the *slow direction.* The identification of these directions, for a nodal point, is not as easy as in the preceding case of a saddle point. Algorithms of linear algebra are normally used.

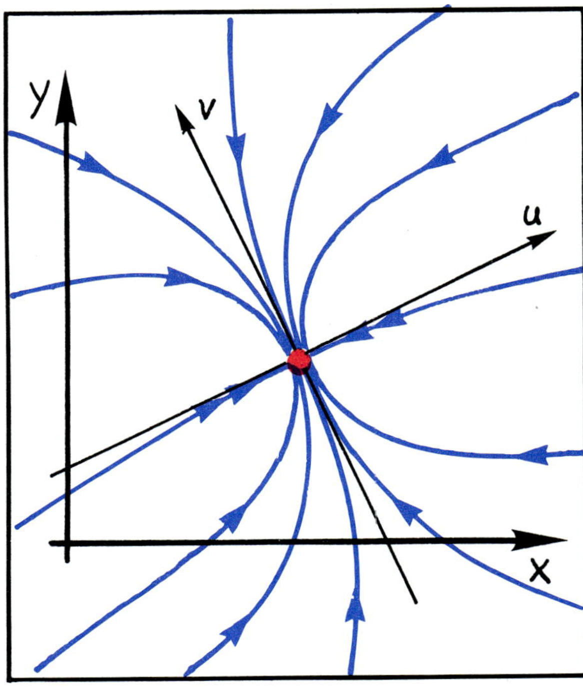

6.3.2. As with the saddle point, we introduce new coordinates, centered at the attractive node and oriented along the fast and slow directions.

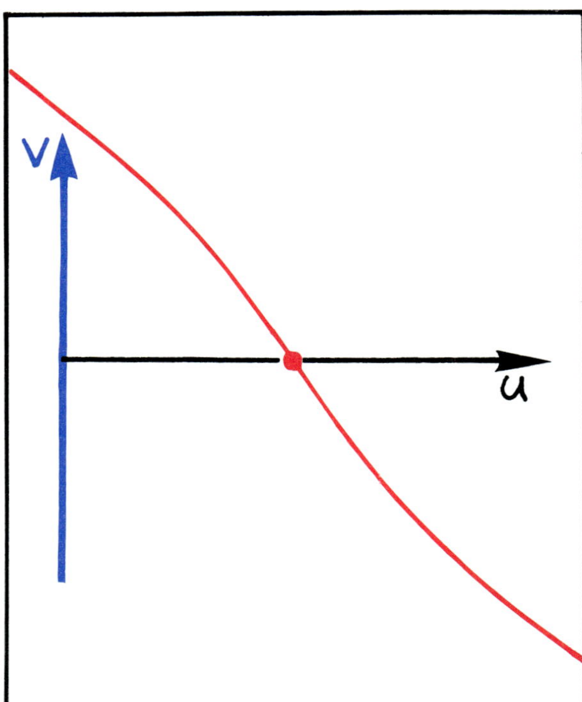

6.3.3. Graphing the components of the vectorfield along the new coordinate axis in the fast direction (u), we measure the slope of the graph at the critical point to obtain the fast CE. It is large and negative, characterizing fast (strong) attraction.

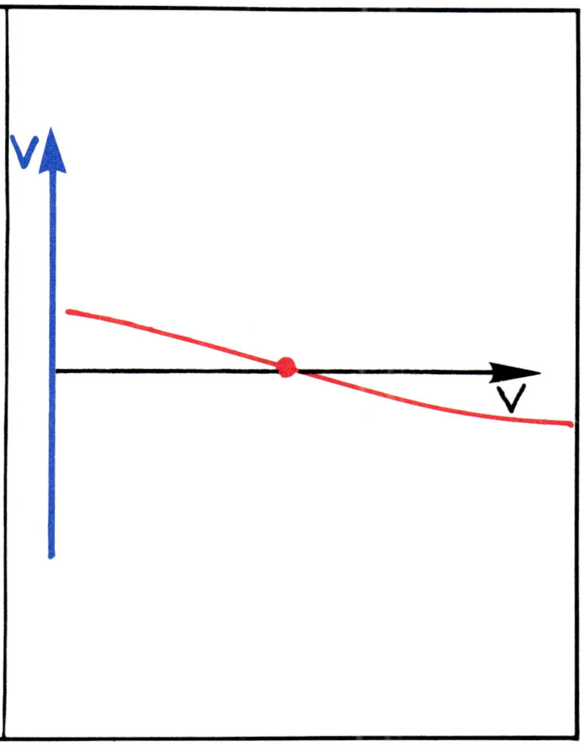

6.3.4. Graphing the component for the vectorfield along the slow (v) axis, the slope at the origin provides the slow CE. It is small and negative, characterizing slow (weak) attraction.

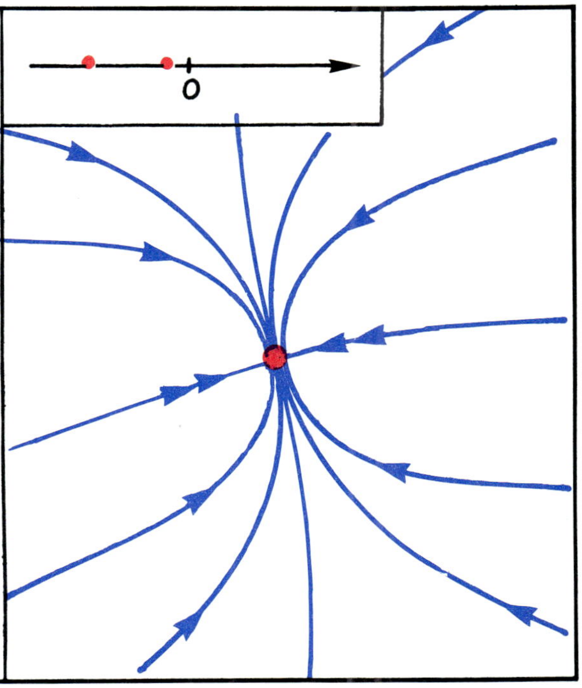

6.3.5. The two CEs are shown in the inset. The one to the left characterizes the stronger attraction of the fast direction. Both are to the left of zero, as the node is an attractor. To the right of zero there are no CEs, and the index of the nodal attractor is zero.

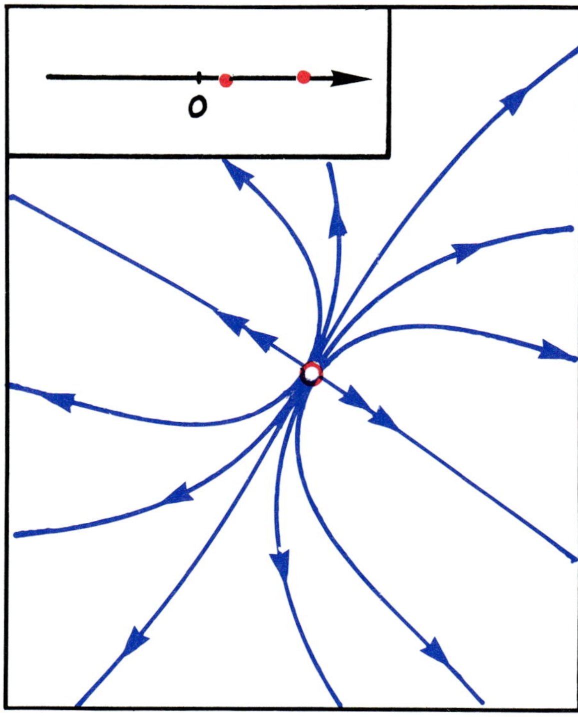

6.3.6. In contrast, the same procedure applied to a nodal repellor provides two CEs to the right of zero. The one to the right characterizes the stronger repulsion of the fast direction (double arrows). The index of this node is two, as the outset is two-dimensional.

In all hyperbolic cases, the index is equal to the number of CEs to the right of zero. This is zero for attractors, one for saddles, and two for repellors.

6.4. Spiral Points In Two Dimensions

In two dimensions, there are two types of limit points, nodes and spirals. The spiral differs from a node in that there are no characteristic (fast and slow) directions.

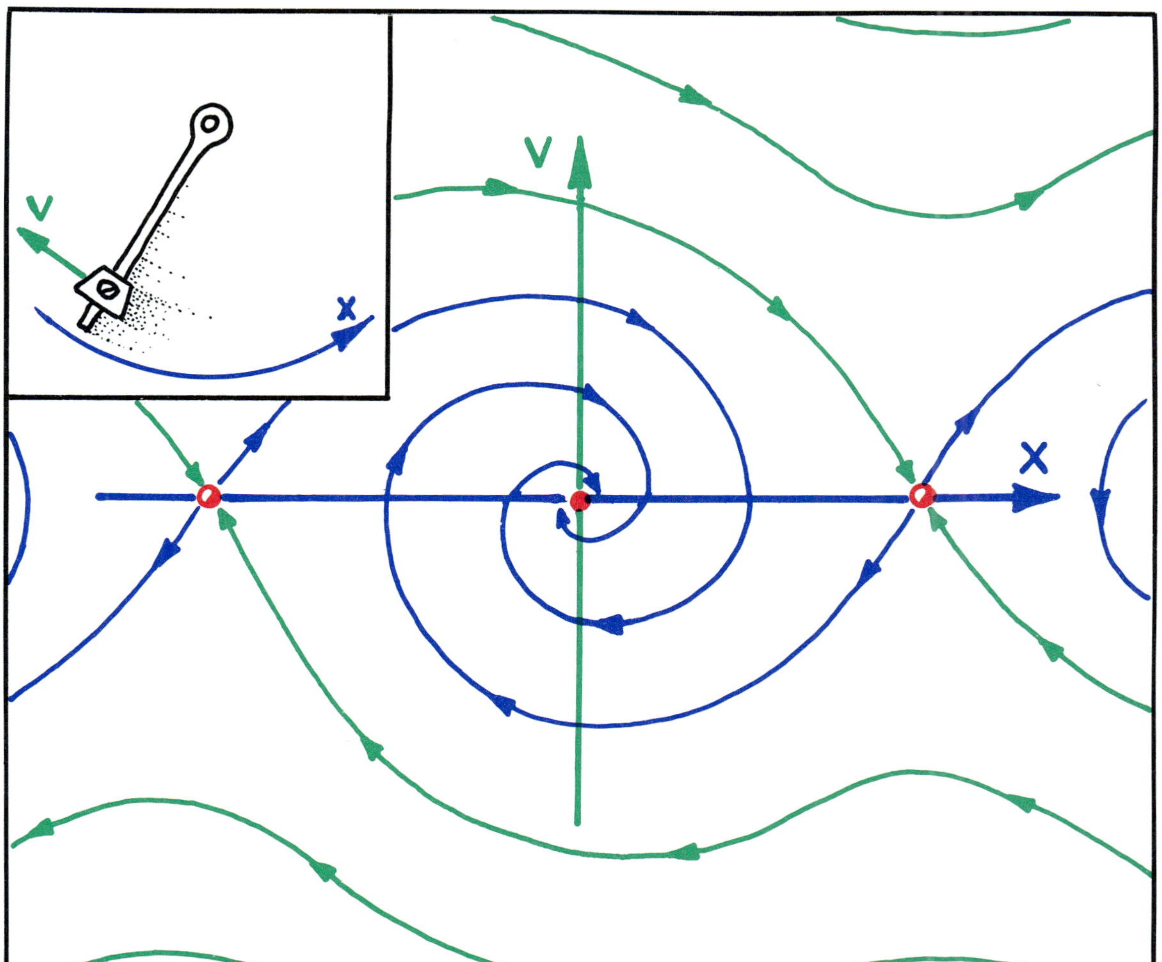

6.4.1. Recall from Part One that the damped simple pendulum has a point attractor at the bottom, corresponding to an oscillation that dies away.

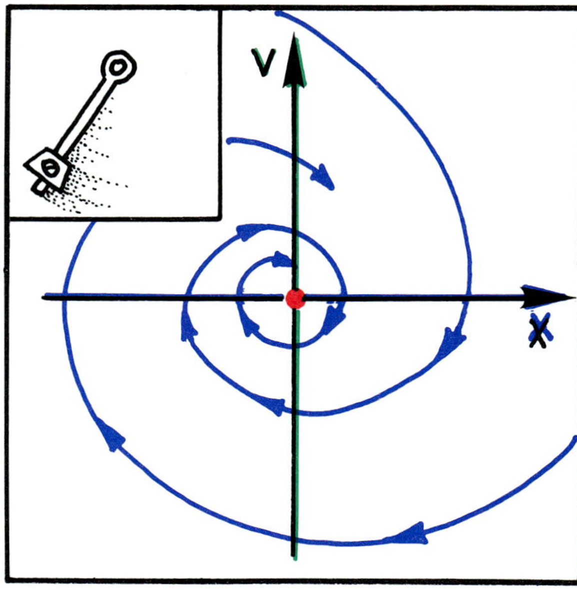

6.4.2. Note that the trajectories asymptotically approach this point, while spiraling around it.

The point attractor of spiral type, in the plane, also has two CEs. But they are not real numbers, and cannot be explained by reducing to the one-dimensional case. Instead, the vector field must be approximated, near the critical point, by a linear vectorfield. Linear algebra, applied to this approximation, provides two eigenvalues. These conjugate complex numbers are the CEs in this case. (For details, see any text on linear algebra.) This is also the actual procedures, using linear algebra, needed to calculate the CEs in the nodal case of the preceding section.

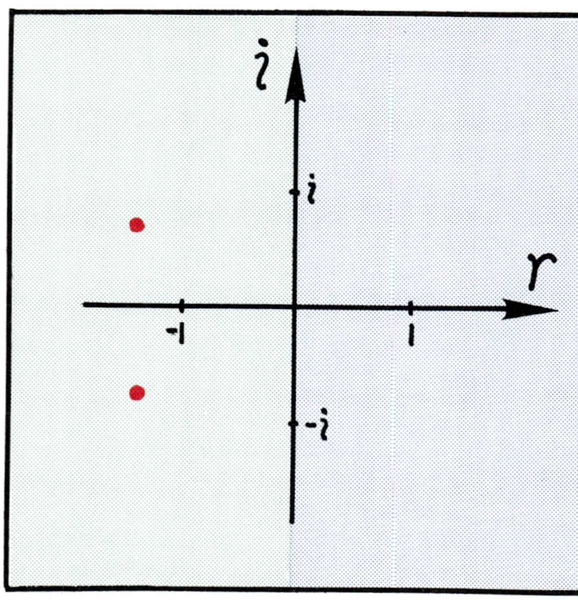

6.4.3. The CEs of a point attractor of spiral type, in the plane, are a pair of conjugate complex numbers. They are in the *left half* (green) of the complex plane because the point is attracting: *the farther to the left, the stronger the attraction.* Meanwhile, their *vertical separation measures the rate of rotation,* or angular velocity of spiraling around the attractor. *Conjugate* means each is the mirror reflection of the other in the horizontal (real) axis, as shown here.

The set of CEs, pictured in the complex plane, is called the *spectrum* of the critical point.

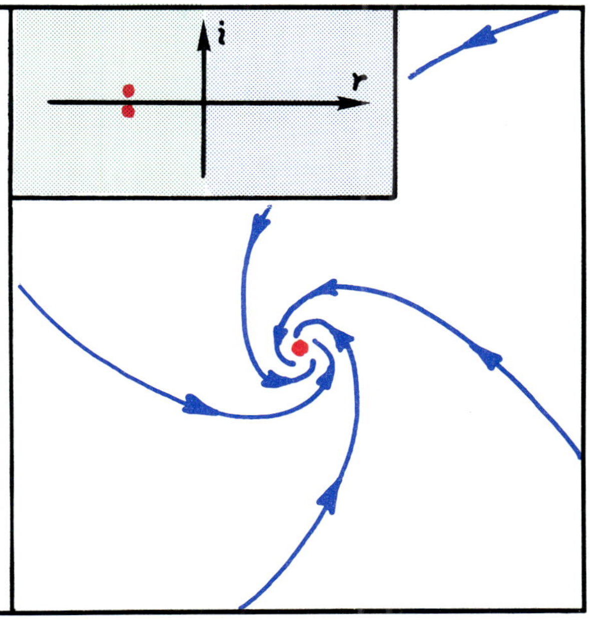

6.4.4. In this example, trajectories zoom toward the point attractor while spiraling slowly. The CE spectrum is shown in the inset.

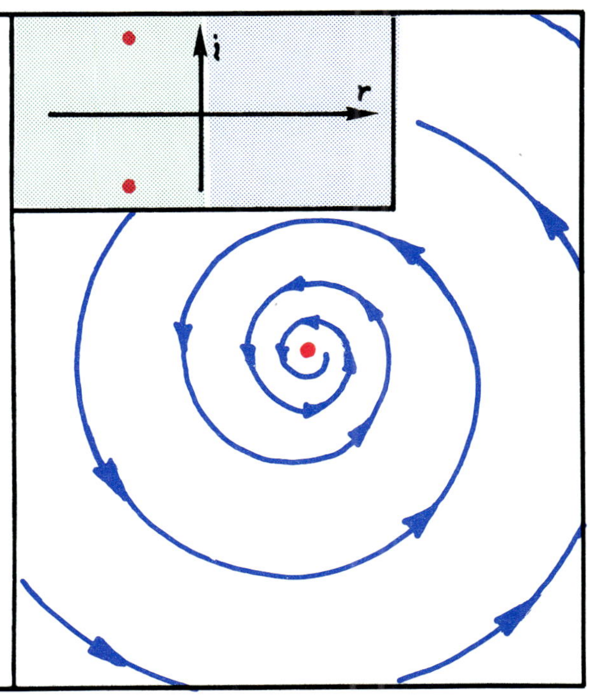

6.4.5. In this case, they rotate rapidly while moving gradually toward the point.

Remember: all attractors have index zero.

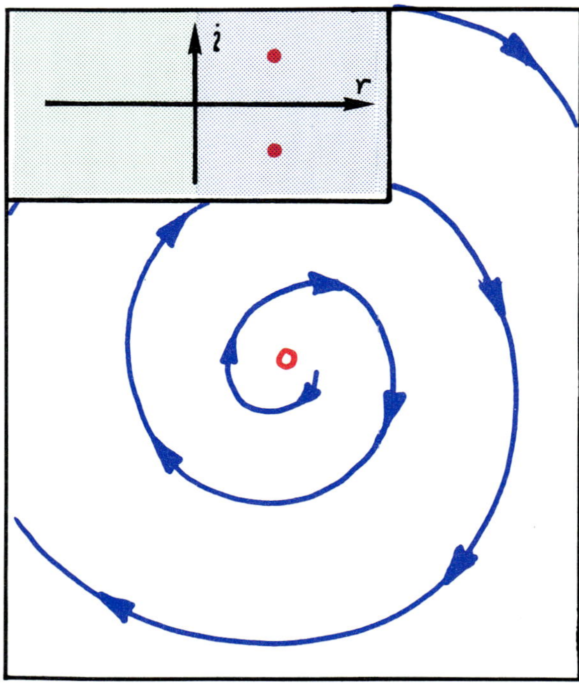

6.4.6. For a repelling spiral point, the conjugate CEs are in the right half of the complex plane. The index of a repellor in the plane is two.

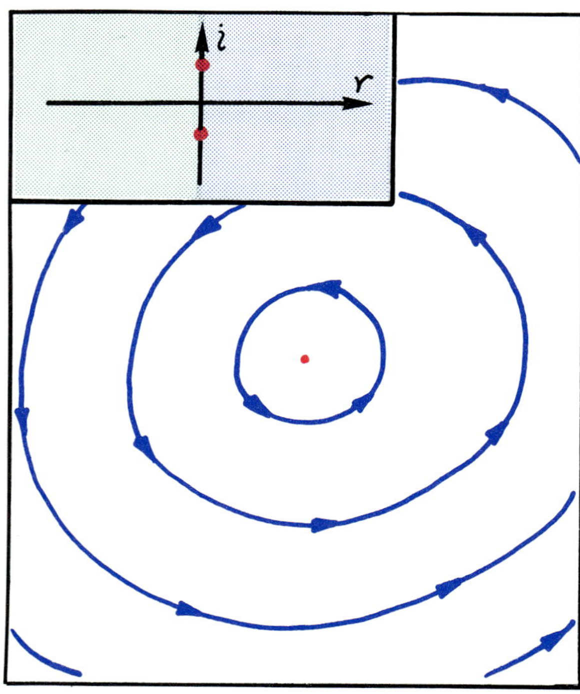

6.4.7. A degenerative spiral or *center,* an intermediate case between attraction and repulsion, is a critical point surrounded by concentric periodic trajectories. This is a *nonhyperbolic* critical point.

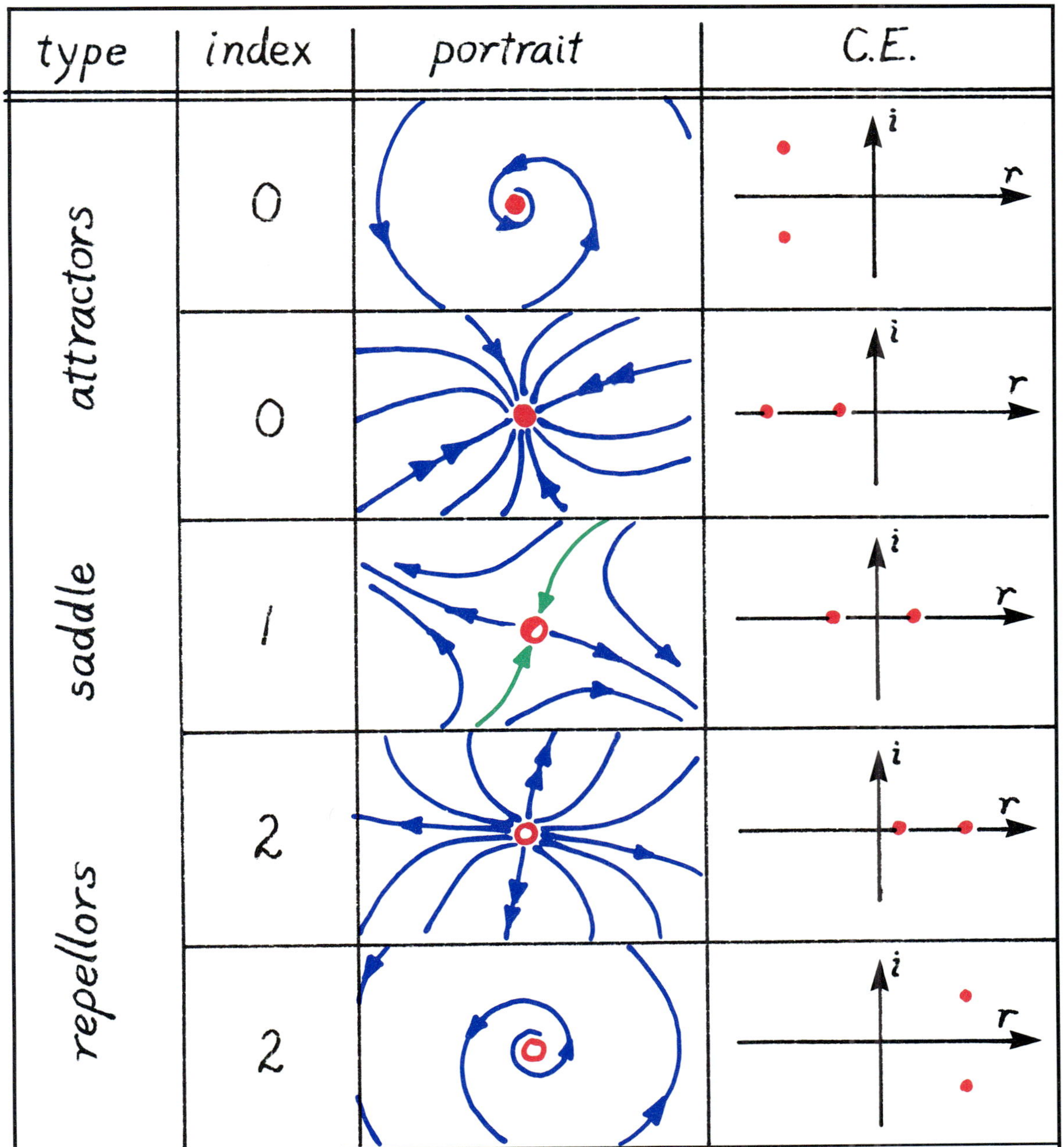

6.4.8. The typical hyperbolic limit points, their CEs, and their indices are summarized in this table.

Actually, two cases are omitted from the table. These are the hyperbolic attractor and repellor with equal (real) CEs. They are classed among the degenerate cases, even though they are hyperbolic, because they are transitional phenomena between the nodal and spiral types. The cases shown are all the *elementary* ones, meaning hyperbolic, with distinct CEs.

6.5. Critical Points In Three Dimensions

As in lower dimensions, critical points in three dimensions are mostly hyperbolic, with nonhyperbolic (degenerate) cases occurring exceptionally. We begin with the typical cases, which are hyperbolic. These may all be constructed by combining the hyperbolic linear and planar critical points of the preceding sections in a single, three-dimensional portrait.

We begin with a three-dimensional saddle point. There are three CEs for a critical point in three dimensions.

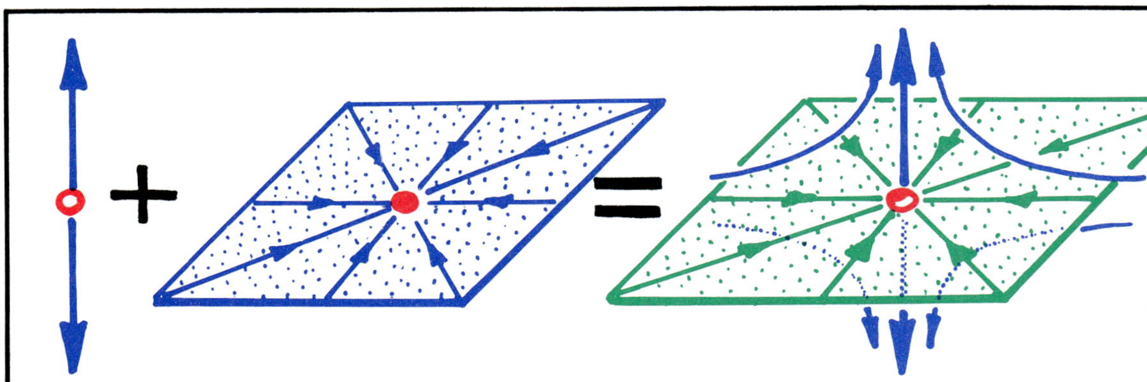

6.5.1. Here, a one-dimensional repellor is combined with a nodal attractor in a plane to form a nodal saddle in three dimensions. In this case the linear component comprises the outset, which is one-dimensional. The planar component comprises the inset, which is two-dimensional. The sum of these dimensions is three, the total dimension of the state space, and the index is one. The trajectories which are neither in the inset nor in the outset fly by hyperbolically. The inset (green) is repelling.

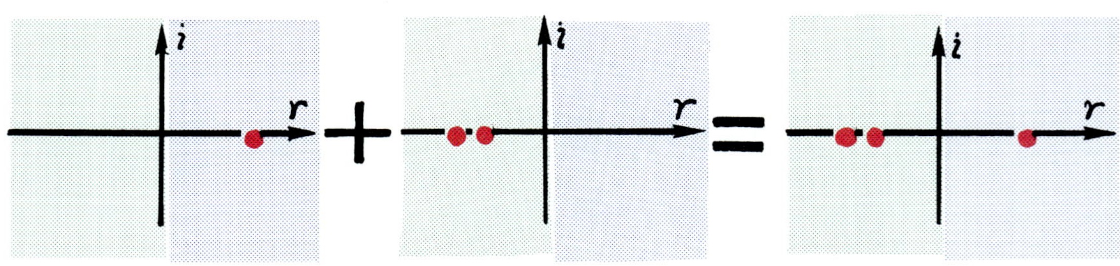

6.5.2. There are three CEs for a critical point in three dimensions. Here are all three, for the nodal saddle point, pictured in the complex plane. All are real numbers in this case. The two in the left half-plane are the CEs of the attractive planar node, comprising the inset. The other is the CE of the linear repellor. Thus, the CEs for the combined critical point are just the CEs of the components.

This is another example of this construction.

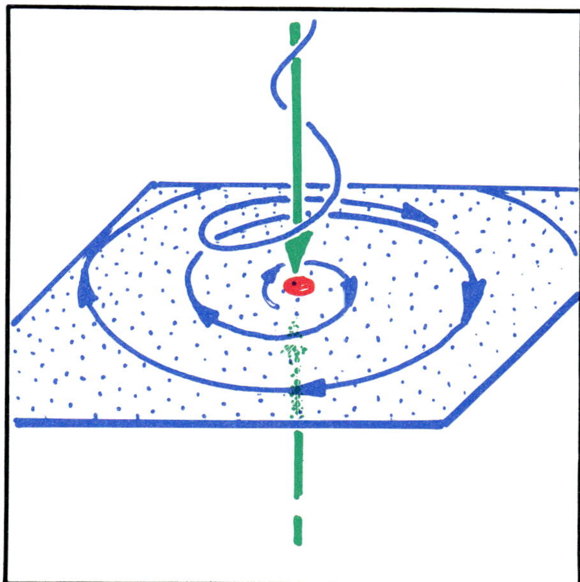

6.5.3. Here a planar repellor of spiral type is combined with a linear attractor, to form a spiraling saddle point in three dimensions. This time, the planar component comprises the outset, which is two-dimensional, so the index is two. The linear component comprises the inset. Trajectories neither in the outset nor the inset fly by, spiraling as they go.

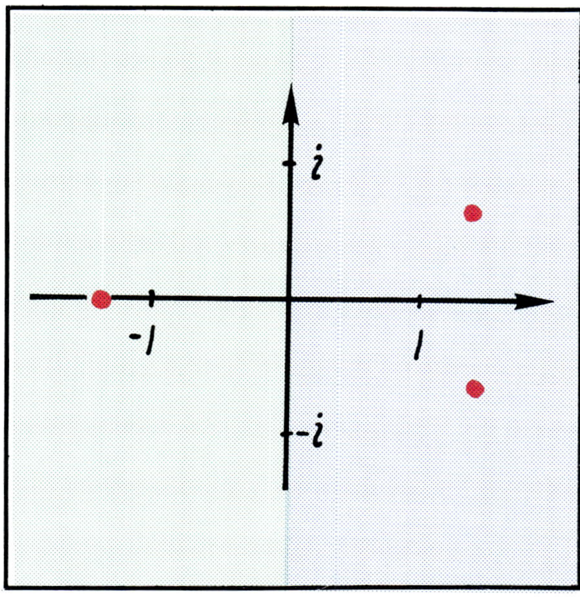

6.5.4. Again, the spectrum of the combined saddle point is the union of the spectra of the components. The conjugate pair in the right half-plane is contributed by the spiral repellor in the outset. The other CE, on the negative real axis in the left half-plane, belongs to the one-dimensional attractor in the inset. Again, the number of CEs on the right equals the index of the combined critical point.

Excluding the degenerate cases with coincident CEs, six more typial hyperbolic critical points may be constructed in this way, for a total of eight. Again, a critical point is called *elementary* if it is hyperbolic, with distinct CEs.

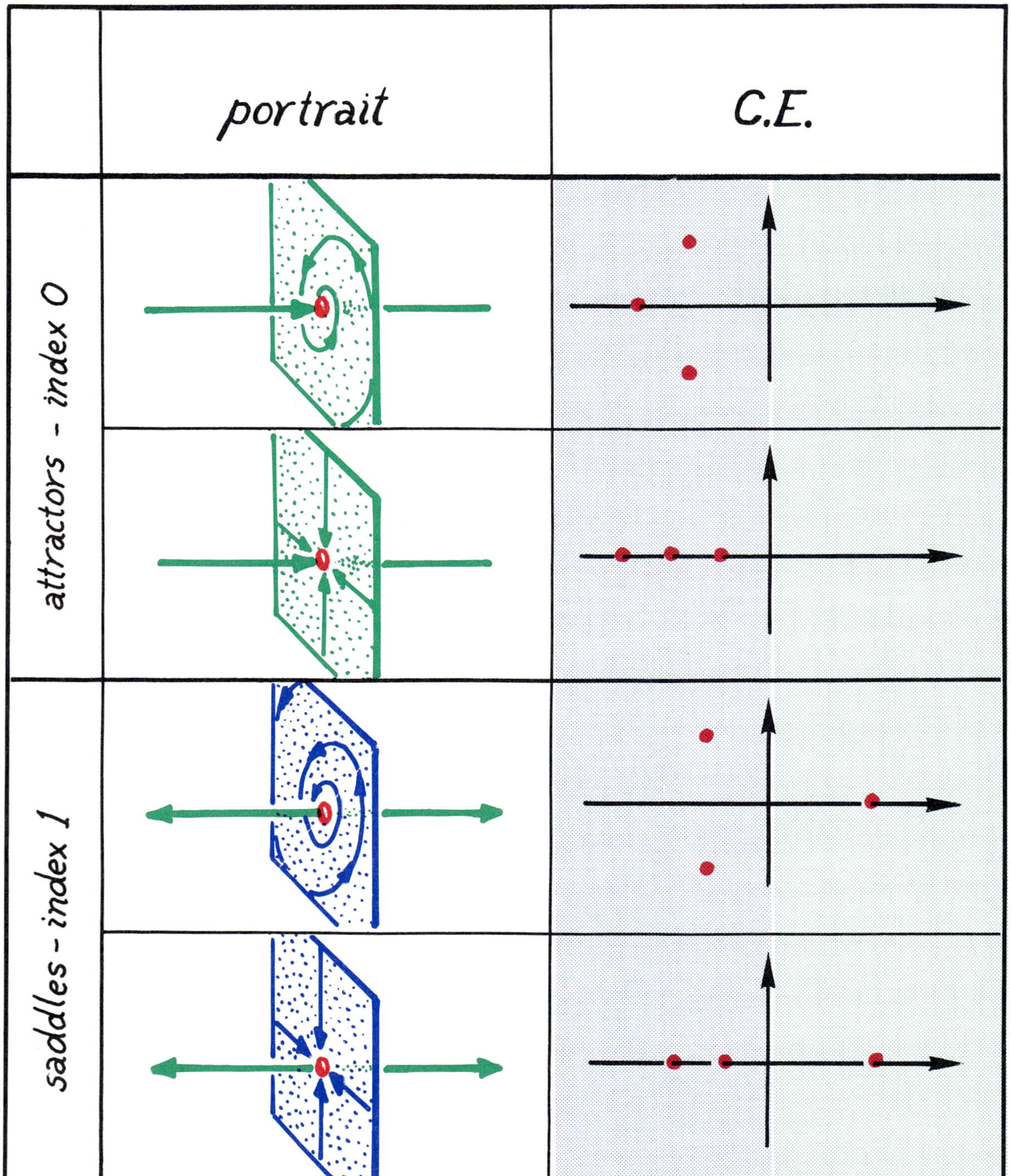

6.5.5. Four elementary critical points are illustrated in this table.

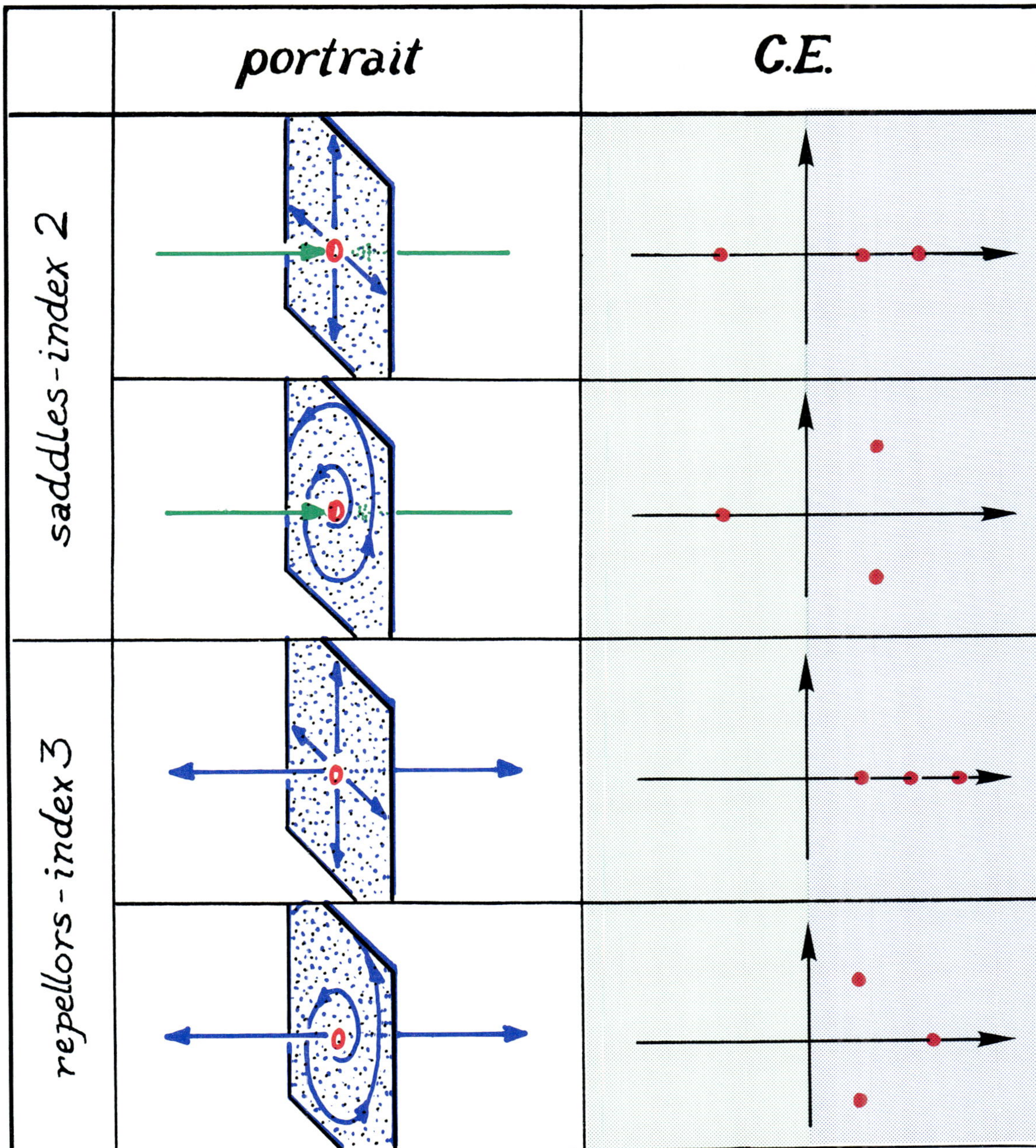

6.5.6. These are the remaining four, in three dimensions.

There are many more degenerate cases in three dimensions than in two. They include the hyperbolic cases with coincident CEs on the real axis. We have not illustrated any of these degenerate critical points, as they are exceptional. That is, they are rarely encountered in applications, except in the context of *bifurcations,* described in Part Four.

7

Periodic Limit Sets and Characteristics Multipliers

We have encountered periodic limit sets, or *limit cycles,* in most of the examples in Part One. In this chapter we will review the typical ones that occur in dimensions two and three. Also, within this review, we will add an important concept, the *Poincaré characteristic multipliers,* or CM's, of these limit cycles. Like the CE's of critical points, these characterize the geometry of the insets and outsets of typical limit cycles.

7.1. Limit Cycles in the Plane

In Part One, we described eight exemplary dynamical systems, five of them with planar state spaces. We now recall the two kinds of limit cycles occurring among them, and describe their CM's. In this context, each limit cycle has a single CM. In general, the number of CM's is one less than the dimension of the state space.

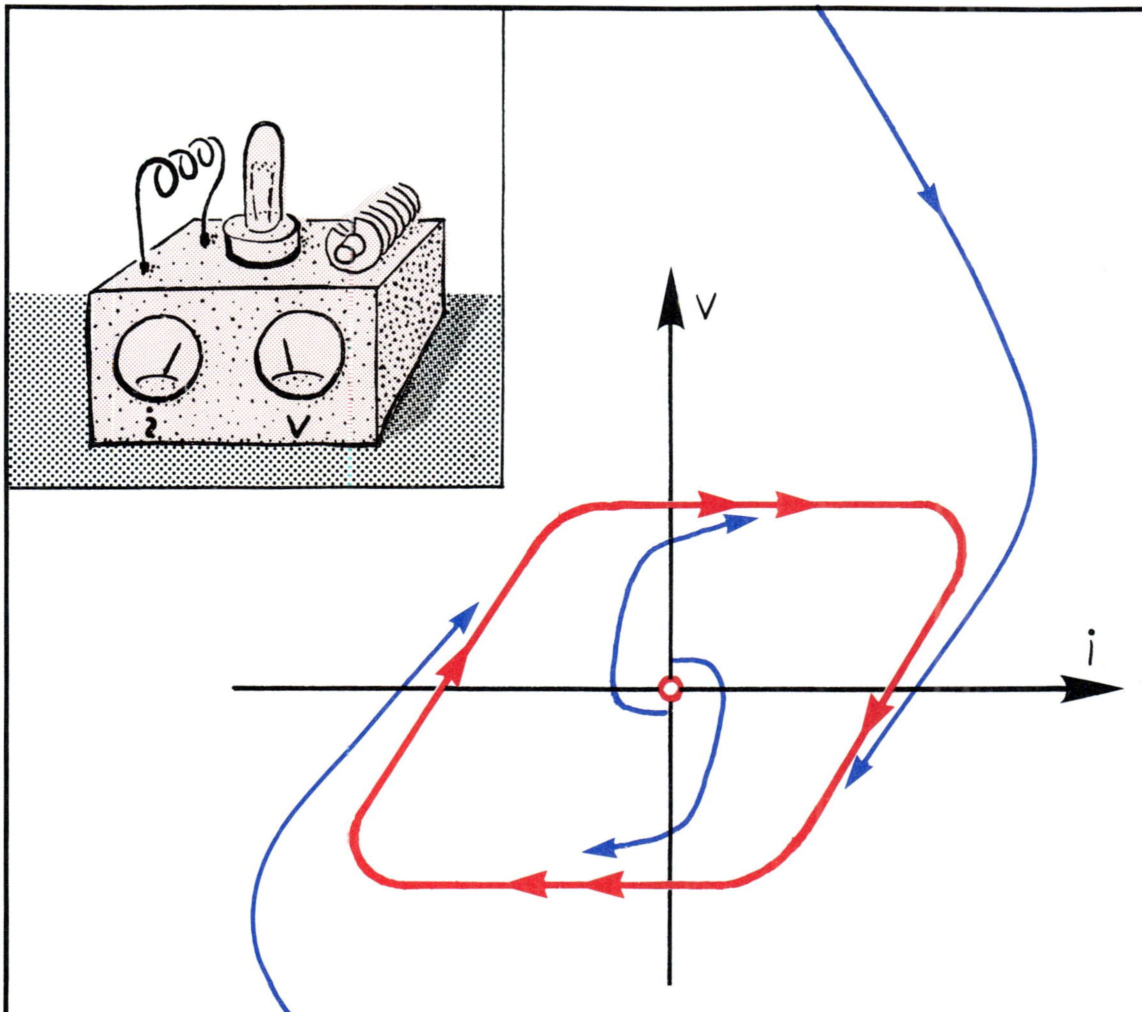

7.1.1. Van der Pol's dynamical model for the radio transmitter contains one limit cycle, which is attractive. It is a *periodic attractor.* All other trajectories (other than the critical point at the origin) tend asymptotically to this limit cycle.

There is no obvious way to define CE's for this limit set. We will describe a subtle way, due to Poincaré.

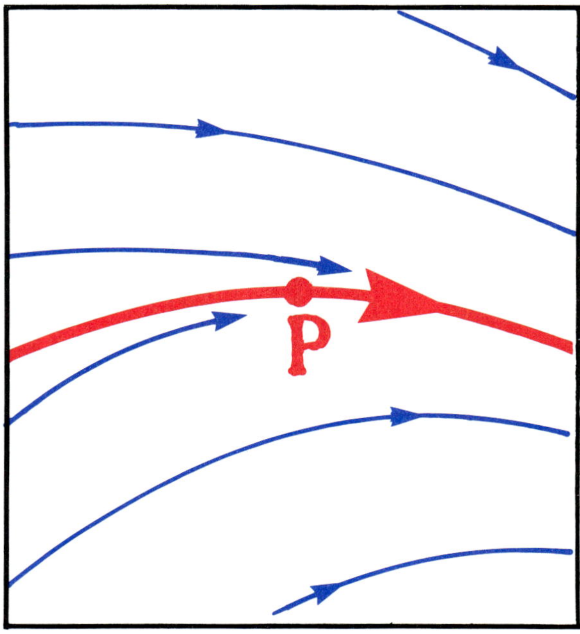

7.1.2. Choose a point on the limit cycle, say P, and enlarge a neighborhood of it. Nearby trajectories are attracted to the limit cycle.

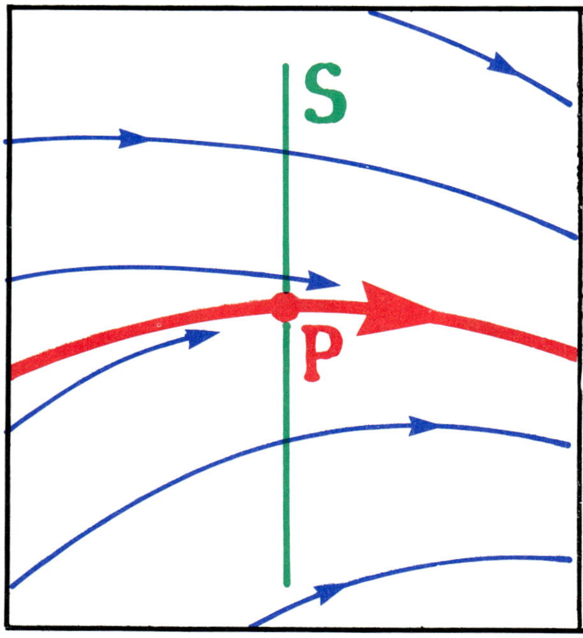

7.1.3. Through P, draw a small perpendicular line segment, S. This is called a *Poincaré section,* as in a "cross-section."

7.1.4. Start at a point in S above P, say x, and follow its trajectory, determined by the dynamical system of Van der Pol. This trajectory follows around near the limit cycle. Eventually, it passes through the section again. Let R(x) denote the first return of the trajectory of x to S. This point, R(x), is above P in S, but is closer to P than x is, as the limit cycle is attractive. Starting at a point y of S below P, the first return of the trajectory of y to S, R(y), is below P, but closer than is y. Likewise, every point of S has a first return to S. This defines a function from S to itself called the *first return map,* R. Note that R(P) = P.

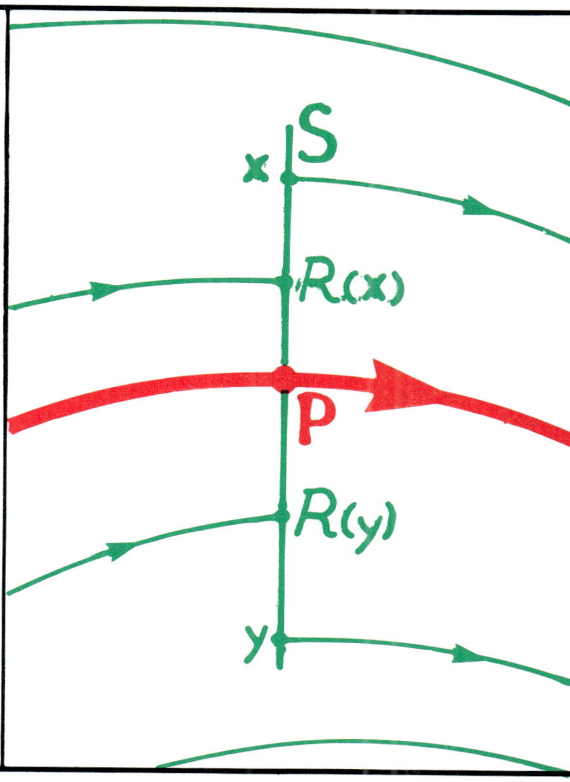

7.1.5. Make a square with the line segment, S, for each side. The graph of R is a curve in this square. It passes through (P,P) because R(P) = P.

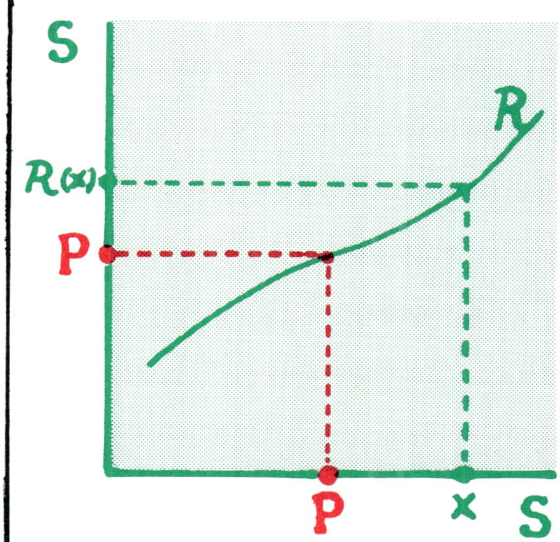

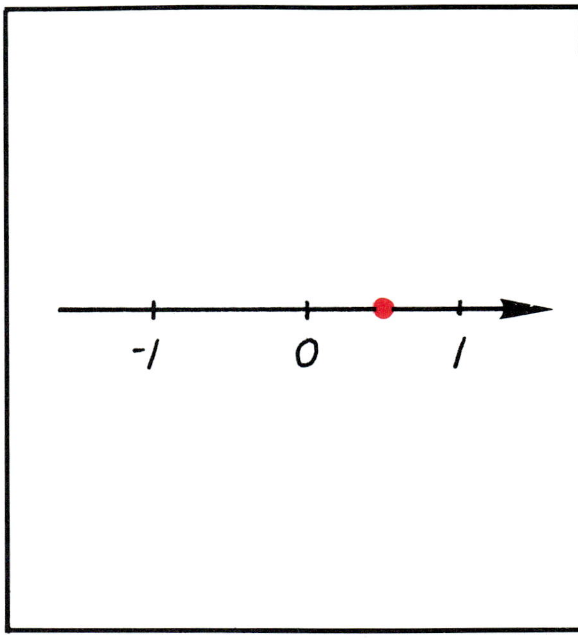

7.1.6. The CM is defined as the slope of this curve over the point P of S. This is necessarily a positive real number, in this context. For example if it's about ½, as shown in this example (Figure 7.1.4.), then R(x) will be about half as far from P as is x. The limit cycle must be an attractor.

For the limit cycle of the Van der Pol system, the CM is between zero and one because the limit cycle is attractive.

Let's consider another system in the plane, with a periodic repellor. The CM of a periodic repellor in the plane is a real number greater than one.

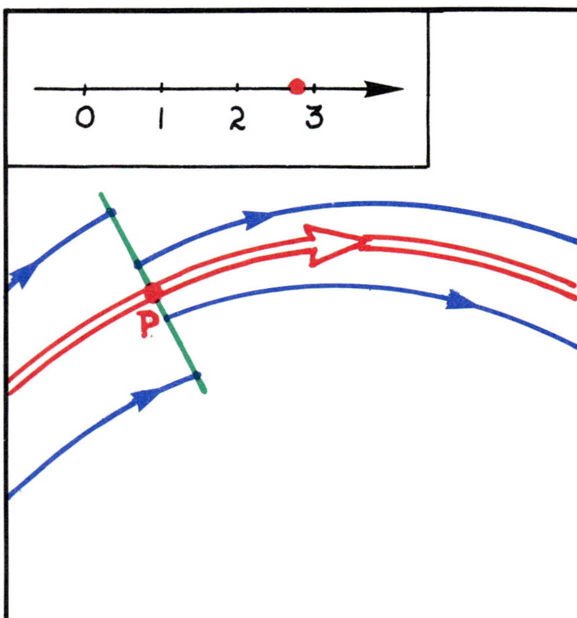

7.1.7. For example, if the CM is about three, as shown here, then R(x) will be about three times as far from P as is x. Thus, the limit cycle must be a repellor.

A limit cycle in the plane is called *hyperbolic* if its CM is not equal to one. There are only two cases. A periodic attractor has its CM between zero and one, while a periodic repellor has its CM greater than one. A limit cycle with CM equal to one may be *neither* an attractor nor a repellor.

7.1.8. A limit cycle surrounded by other limit cycles has a very special return map: the identity function, R(z) = z. The slope of this function is one, so the CM is one in this case. This is a *nonhyperbolic* limit cycle.

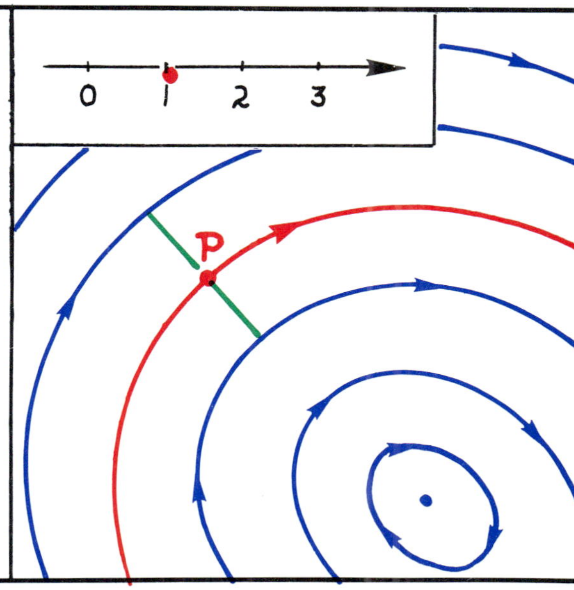

Recall the frictionless pendulum, Figure 2.1.18 in Chapter 2. This is an example of a nonhyperbolic limit cycle.

7.2. Limit Cycles in a Möbius Band

In the preceding section, we found that the CM of a limit cycle in the plane was inevitably a positive real number. Why do we never come across a negative CM? The answer is: we do! But only in a non-orientable surface, as we shall see in this section.

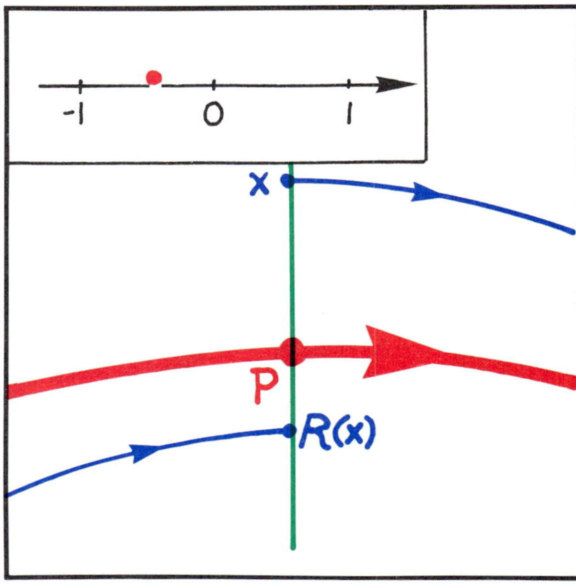

7.2.1. Suppose we have zoomed in on a Poincaré section of a limit cycle in a two-dimensional context, graphed the reverse map, and found a *negative* slope, or CM. Then necessarily, a point in the section above the cycle, like x here, has a first return on the *opposite side* of the cycle, like R(x) here.

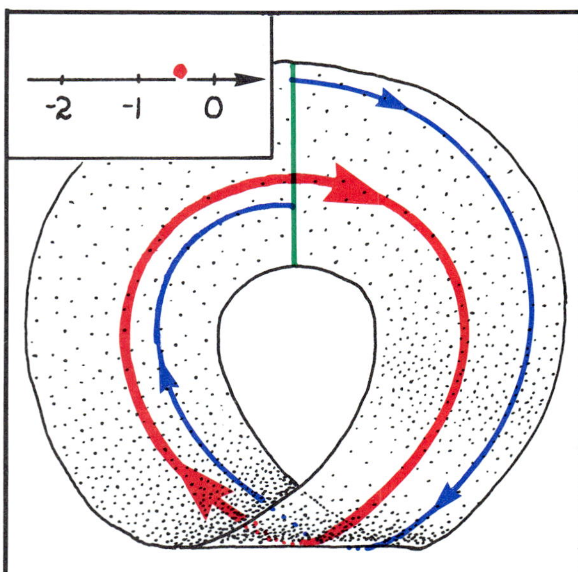

7.2.2. Now let's zoom back out, so we can see the entire limit cycle, and the entire trajectory from x to R(x). There has to be a *twist* in the surface, in order for the return map to *reverse orientation* of the Poincaré section. The Möbius band, shown here, is just one of the many possibilities.

The CM of a periodic attractor around a Möbius band is a negative real number, between −1 and 0.

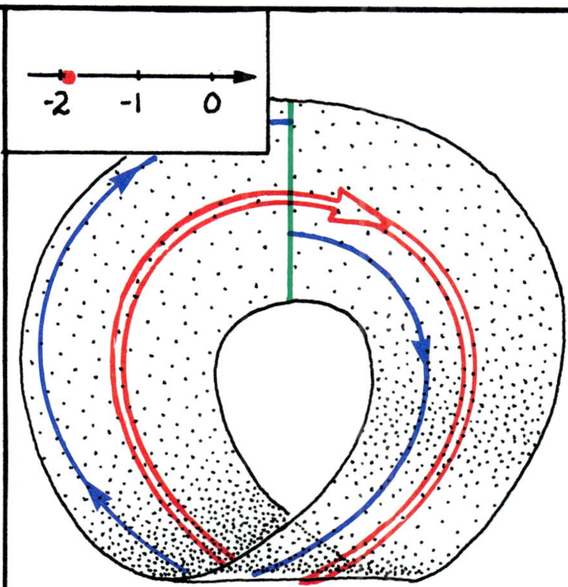

7.2.3. A periodic repellor around a Möbius band has a CM less than −1.

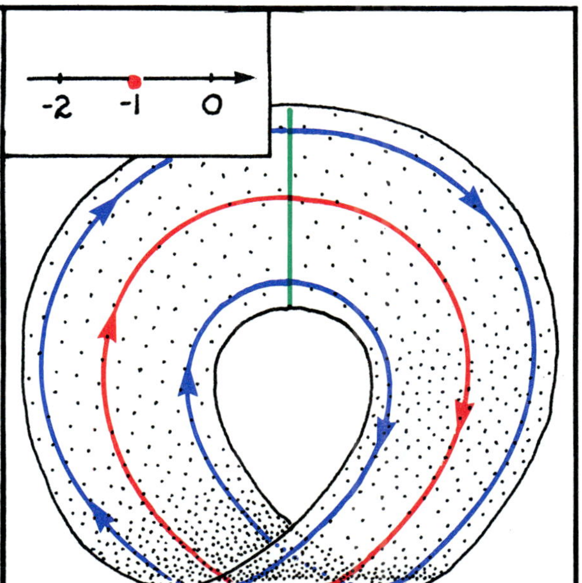

7.2.4. A limit cycle around a Möbius band with a CM equal to −1 is non-hyperbolic. In this case, the limit cycle belongs to a family of "parallel" periodic trajectories.

Of course, not all limit cycles in a Möbius band have negative CM's.

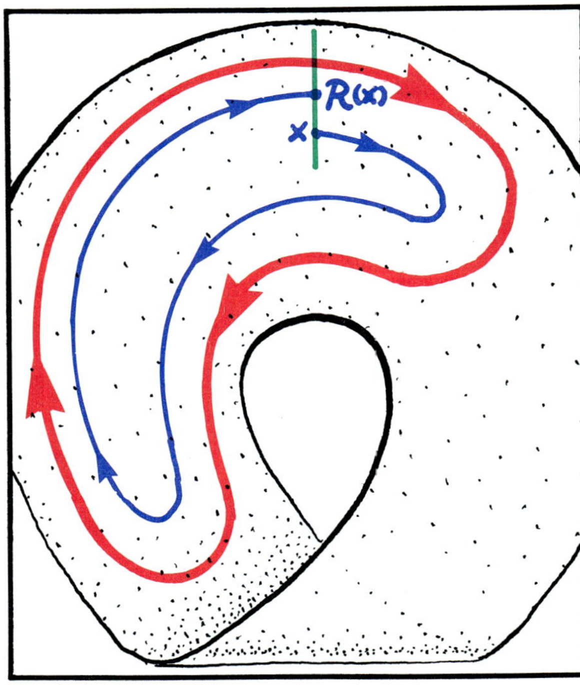

7.2.5. For example, a limit cycle that does not go all the way around . . .

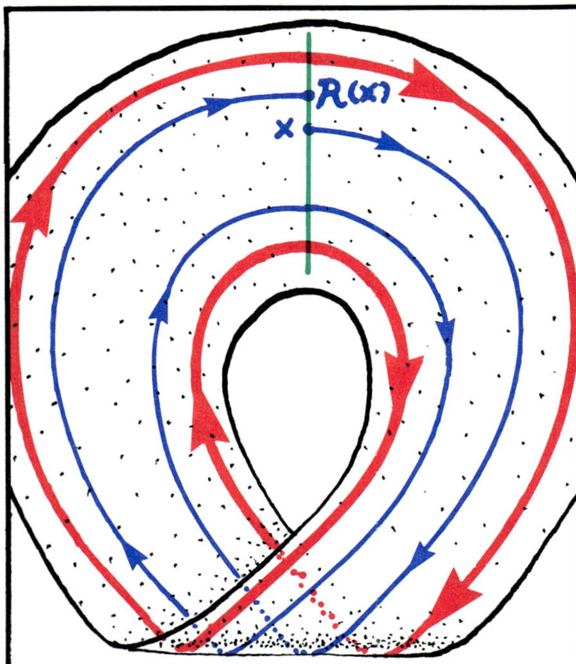

7.2.6. . . . or one that goes twice around.

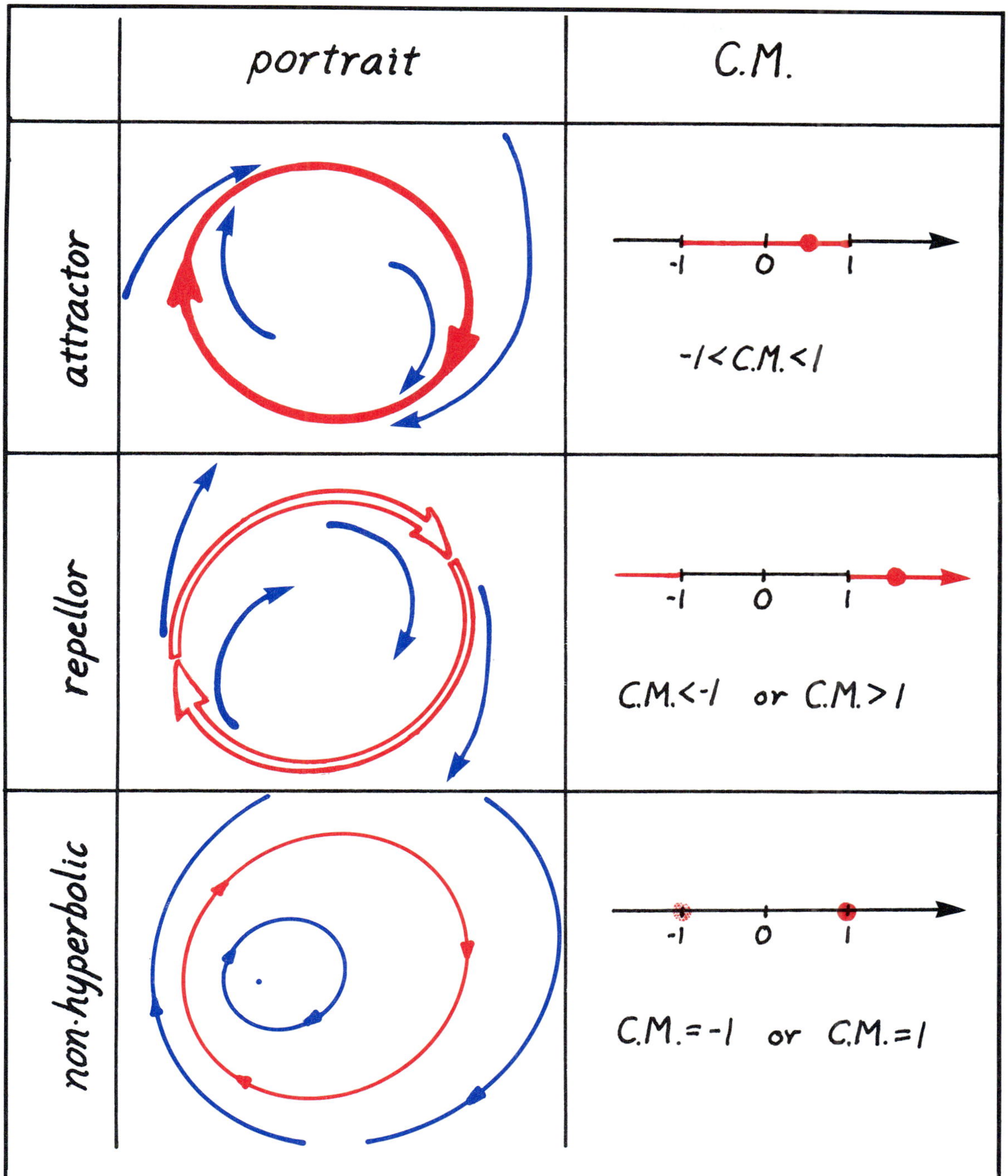

7.2.7. The limit cycles in two-dimensional state spaces are classified in this table, which includes both positive and negative CM contexts. The CM is never zero.

7.3. Saddle Cycles In Three Dimensions

In two-dimensional state spaces, typical limit cycles are either attractors or repellors. In three-dimensional contexts, there are several possibilities.

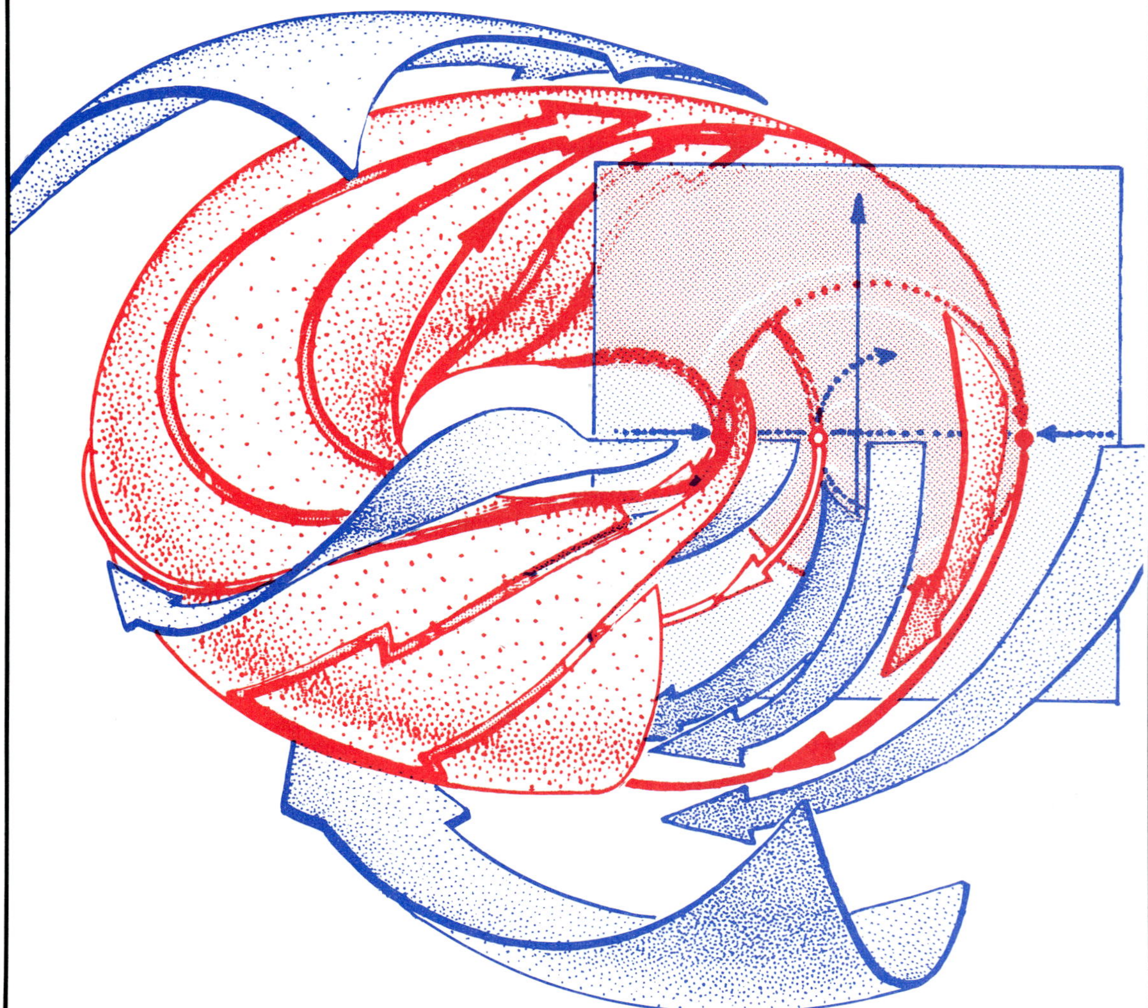

7.3.1. Recall Duffing's model for the forced pendulum, from Part One, Figure 5.4.12. There are three prominent limit cycles in a three-dimensional context: an attractor, a saddle, and a repellor.

Now we are going to characterize these limit cycles in terms of the CM's of Poincaré. His subtle construction of the CM's of a limit cycle works in contexts of any dimension.

We will consider the saddle cycle first.

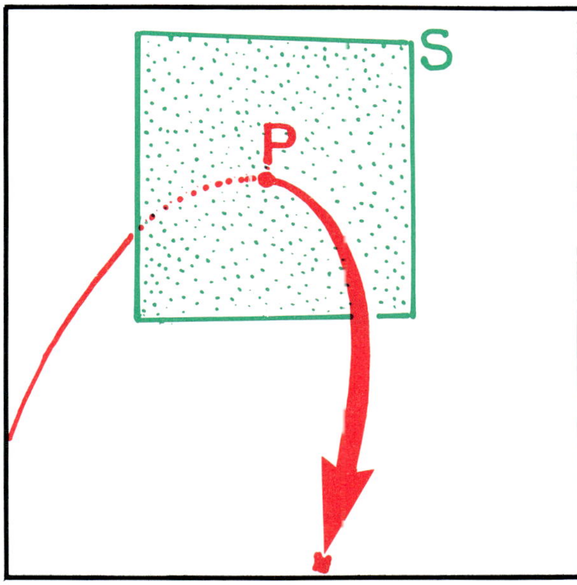

7.3.2. At a point, P, in the limit cycle, construct a small cross-section, S. This time, the Poincaré section is *two-dimensional,* like the *strobe planes* of Chapter 4.

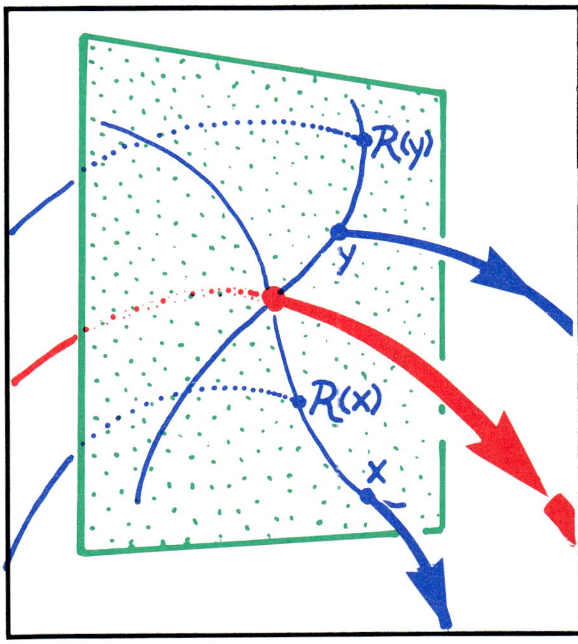

7.3.3. Each point of S (if it's not too close to the edge, at least), when prolonged forward along its trajectory, will have a first return to the section. Here, the point x returns at R(x), while another point y returns at R(y). As x is chosen here in the *inset,* R(x) is closer to the limit cycle. On the other hand, y is in the *outset,* so R(y) is farther from the limit cycle.

This construction defines a function from (most of) S to itself, the *first return map*, R. The graph of R is a surface in four-space, so we cannot draw it pictorially. But according to vector calculus, R may be approximated very well, in a neighborhood of P, by a linear transformation. The theory of these, linear algebra, provides an algorithm for obtaining two complex numbers, the *eigenvalues* of this linear approximation. These characterize the linear transformation exactly and the first return map approximately. These two complex numbers are, in general, either real or a complex conjugate pair. *These are the CM's of the limit cycle.*

7.3.4. In the case of the saddle cycle in Duffing's system, the CM's are positive real numbers. One is *larger* than one, characterizing the *outset*. The other is *smaller* than one, characterizing the *inset*.

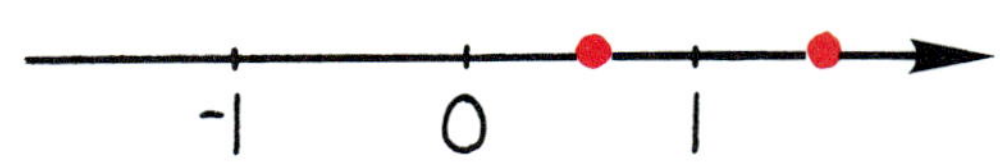

7.3.5. The curve in S corresponding to the smaller CM (shown in the box) is the cross-section of the *inset* of the limit cycle. This surface is *invariant:* trajectories that start on it stay on it. Within this invariant surface, the inset, the limit cycle acts as an attractor. The smaller CM determines t he rate of this attraction. The smaller the CM, the faster the asymptotic approach to the limit cycle.

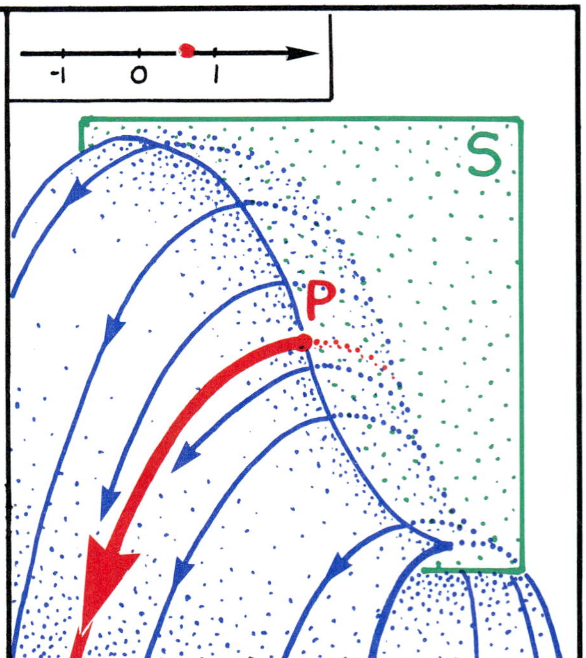

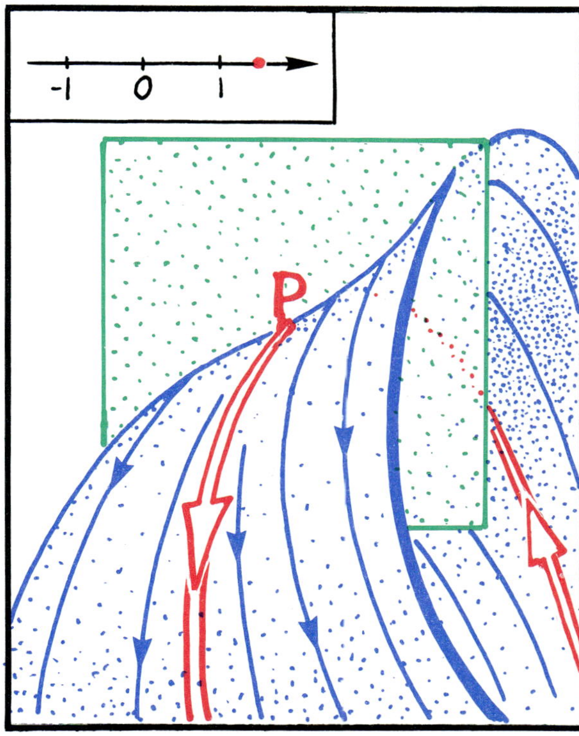

7.3.6. The other curve in S, corresponding to the larger CM (see below), is the cross-section of the *outset* of the limit cycle. Within this invariant surface, the limit cycle acts as a repellor. The larger the CM, the stronger this repulsion.

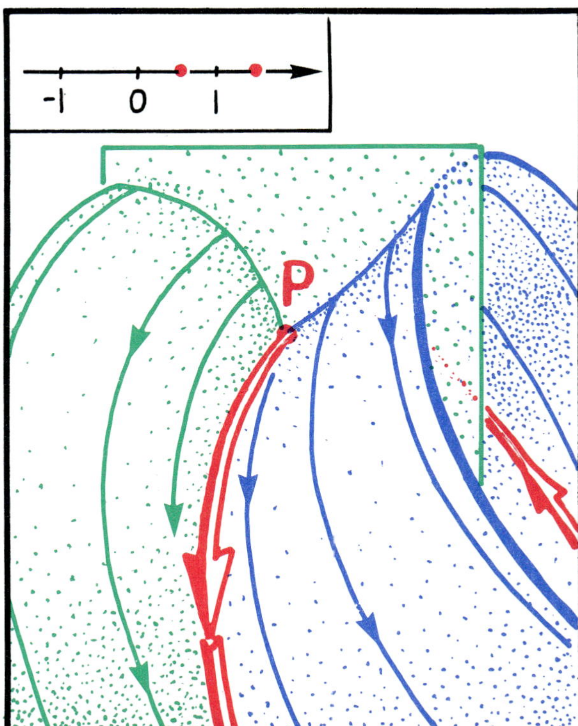

7.3.7 Here are the two invariant surfaces, inset and outset. They intersect in a curve, the limit cycle. Only small portions of them are shown here. They both may extend far out of the picture. Both CM's are shown in the box.

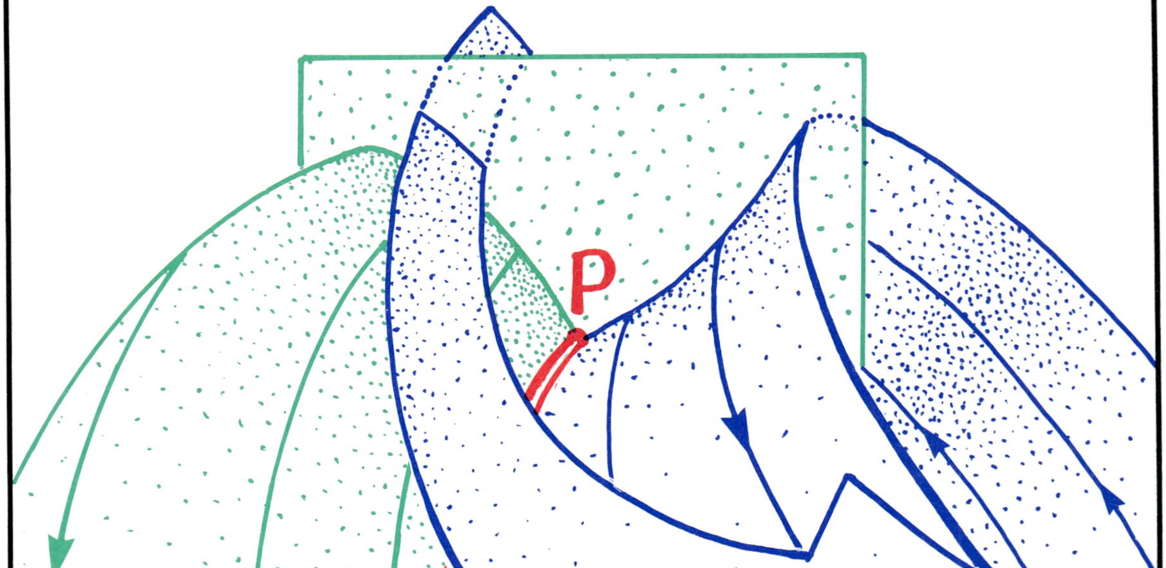

7.3.8. The trajectories of the inset spiral toward the limit cycle, while those of the outset spiral away. Other nearby trajectories just fly on by, spiraling closer for a while, then spiraling away.

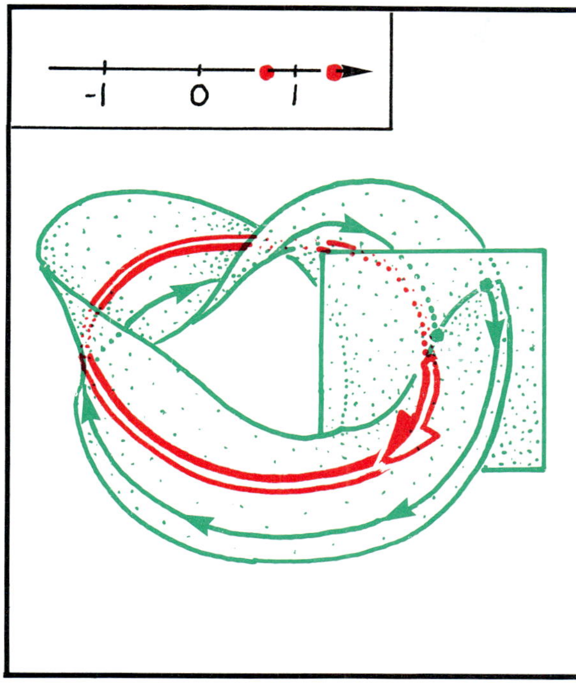

7.3.9. Even though the CM's are positive, the inset and outset cylinders may be twisted. But they must twist an *even* number of twists, if the CM's are *positive*. Here, for better visibility, we have shown only the inset.

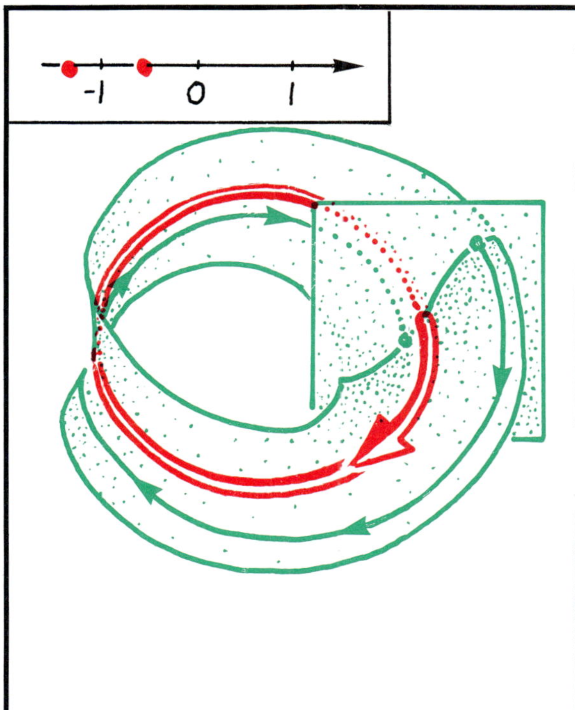

7.3.10. As in the two-dimensional case of the preceding section, *negative* CM's arise when both the inset and the outset are Möbius bands, each with an *odd* number of twists while going once around. Again, the outset band has been deleted for better visibility.

7.4. Nodal Cycles In Three Dimensions

Here we consider variations on the saddle cycle of the preceding section. If the CM's are real and both small (between −1 and 1), we have an attracting or repelling cycle that looks rather like a saddle. These are called *nodal cycles.*

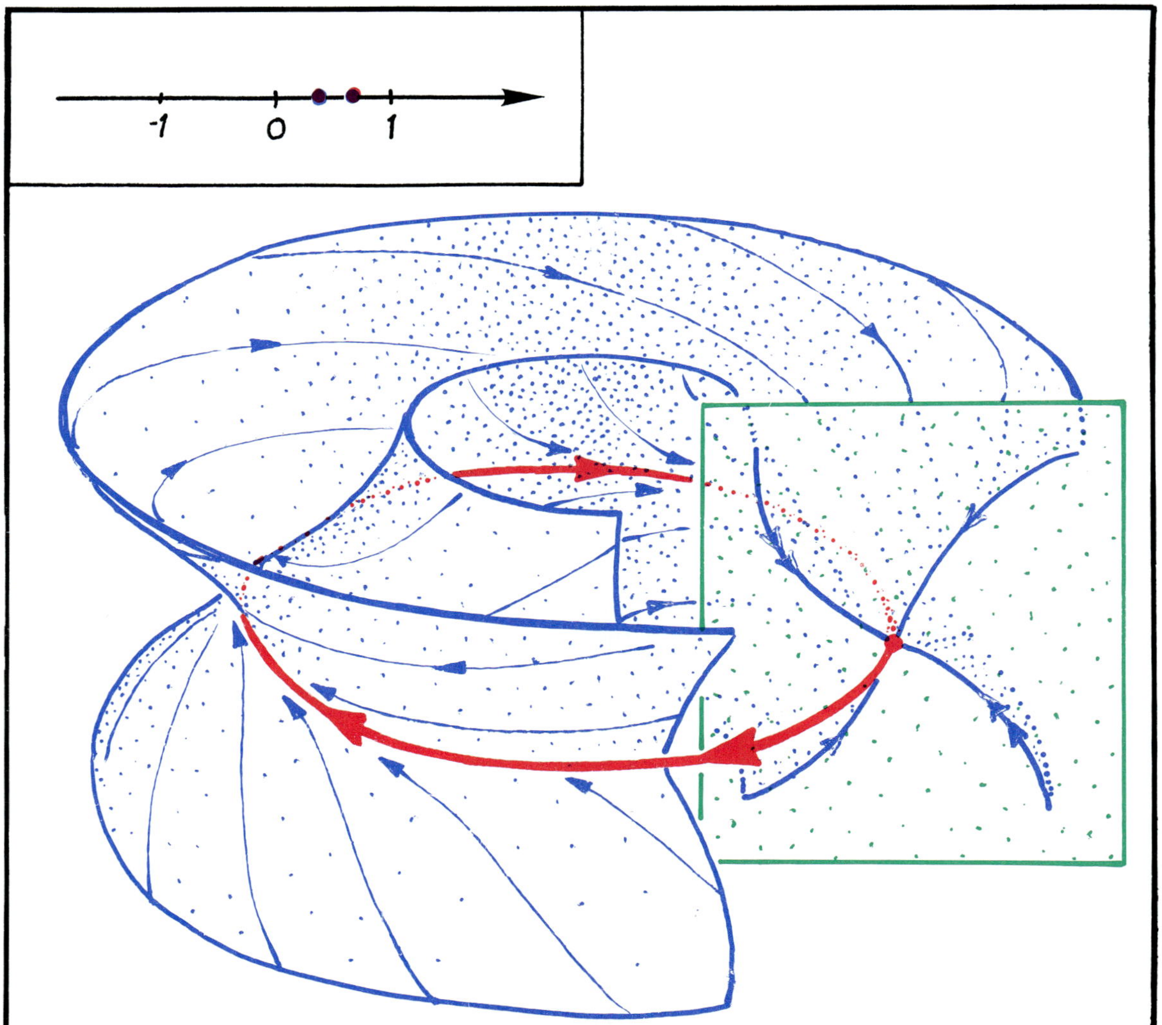

7.4.1. These two small, real CM's correspond to two invariant surfaces, as in the saddle cycle. But the limit cycle acts as an attractor within each. One, with the smaller CM, attracts more strongly than the other. They are *sub-insets,* called the *fast-inset* and the *slow-inset* respectively.

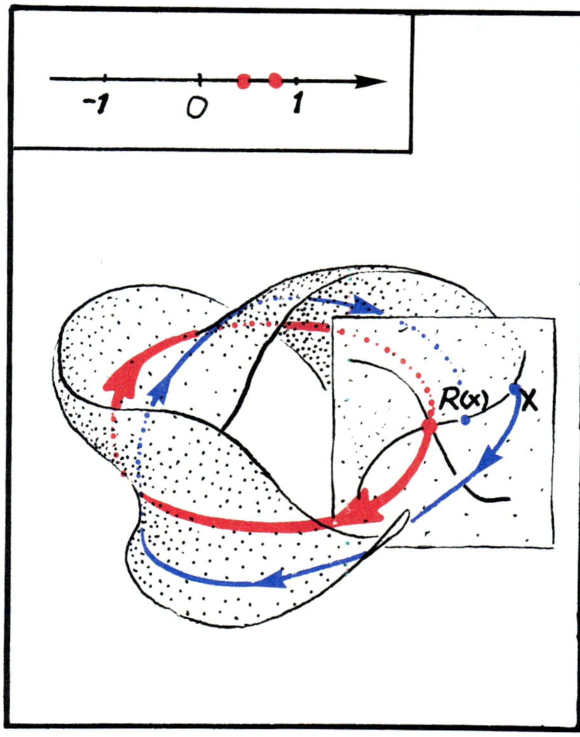

7.4.2. As with saddle cycles, these *positive* CM's could correspond to surfaces with any *even* number of twists. Here, the first return, R(x), is on the *same* side of the fast-inset as the initial point, x.

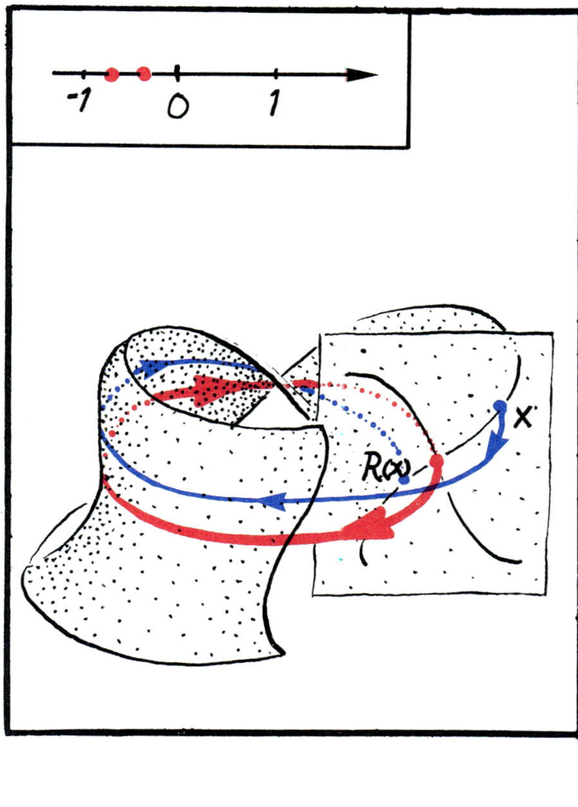

7.4.3. And as with saddle cycles, the invariant surfaces may both be twisted an *odd* number of times. In this case, the CM's are *negative,* and the first return, R(x), is on the *opposite* side from the initial point, x.

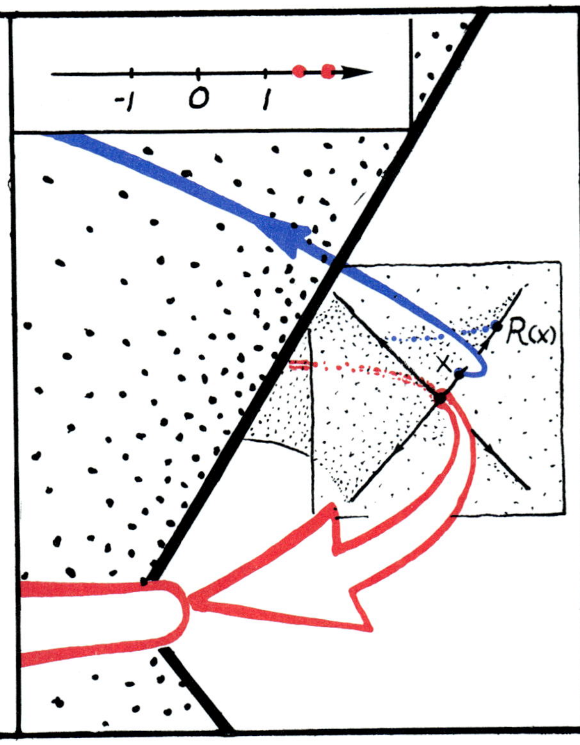

7.4.4. Repelling cycles can be nodal, too. Here, the two large (greater than one), real CM's correspond to two invariant surfaces, the fast- and slow-outsets. But the limit cycle acts as a repellor within each.

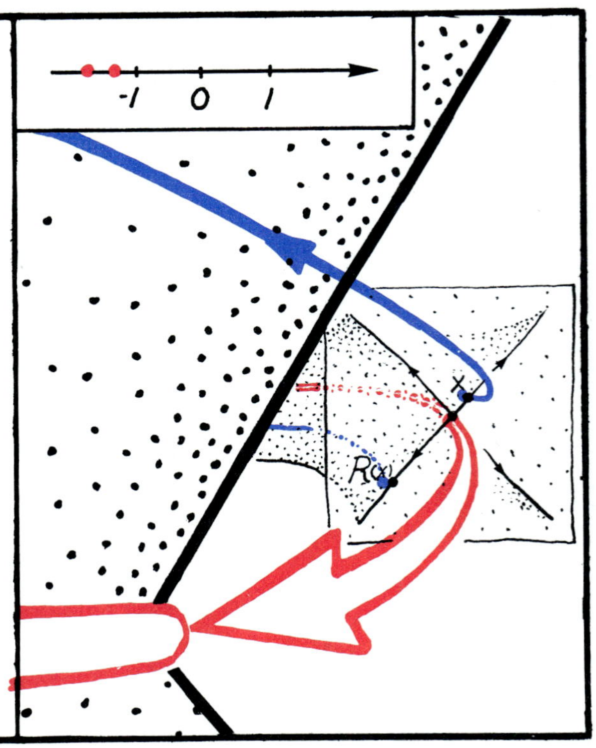

7.4.5. Repelling nodal cycles may also have twisted fast- and slow-outsets, with negative CM's.

7.5. Spiral Cycles In Three Dimensions

As the two CM's of a limit cycle in three-dimensional space are the eigenvalues of a linear transformation, it is possible for them to be a pair of conjugate complex numbers, rather than a pair of real numbers. This is the case with limit cycles of spiral type.

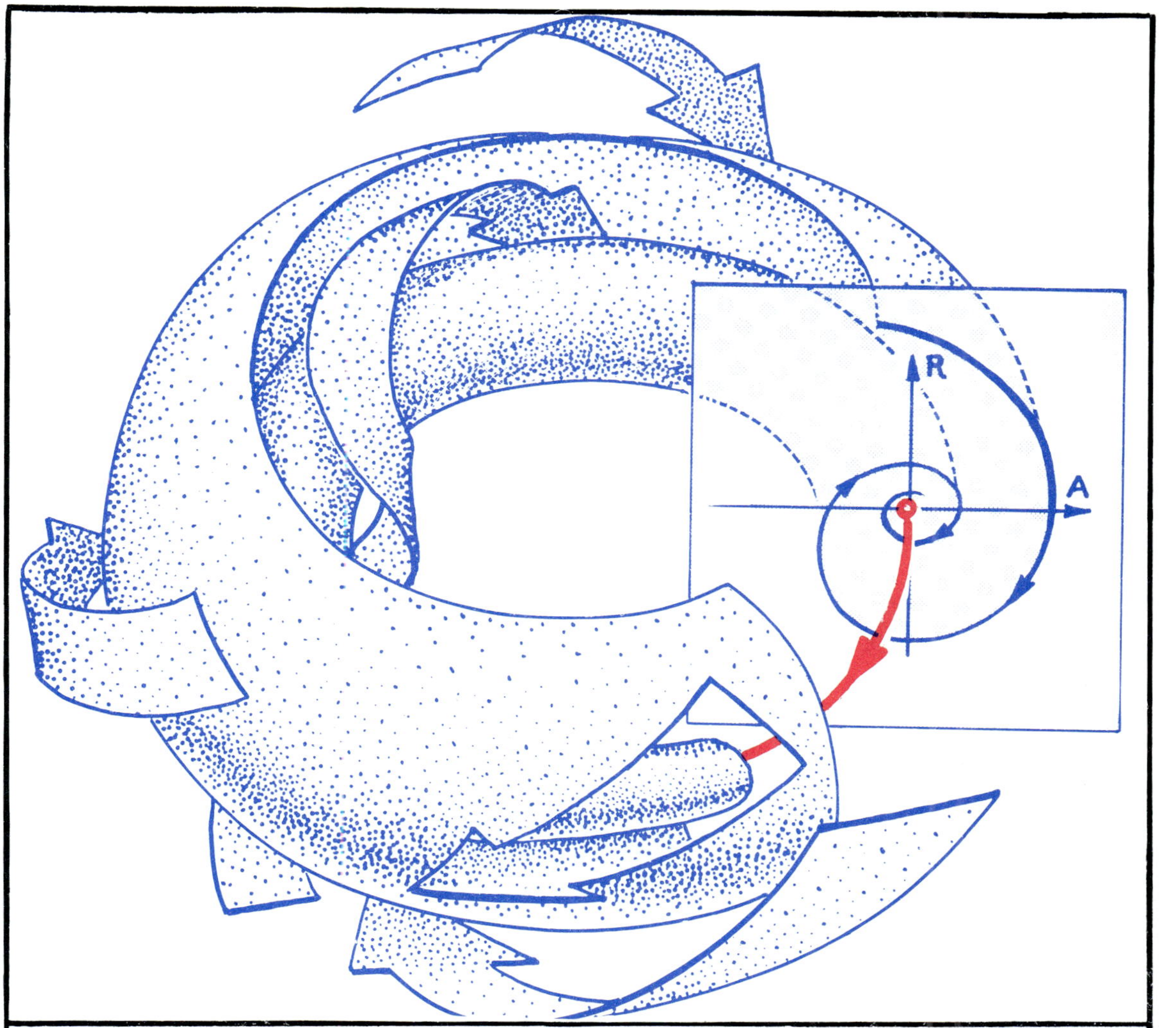

7.5.1. Recall that in the ring model for the damped pendulum (Figure 4.1.12), the periodic attractor around the center was surrounded by spiraling trajectories.

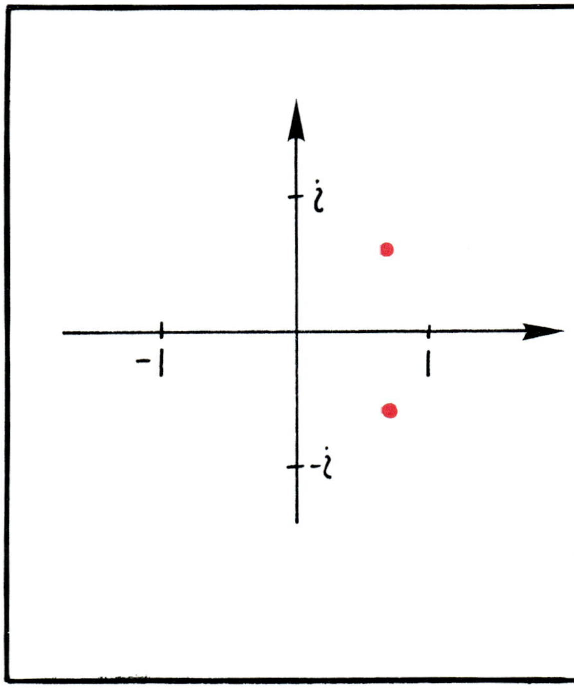

7.5.2. Since there are no invariant surfaces as in the case of a nodal cycle, the CM's cannot be real. They are a conjugate pair of complex numbers. The *magnitude* (distance of either from the origin) characterizes the rate of asymptotic approach to the limit cycle. The *angle* (in the sense of polar coordinates) characterizes the rate of spiraling.

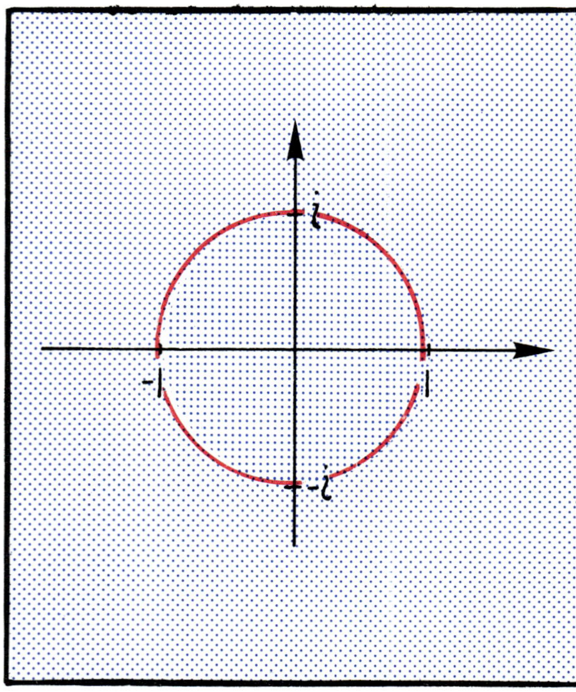

7.5.3. The unit circle in the complex plane discriminates attractors from repellors.

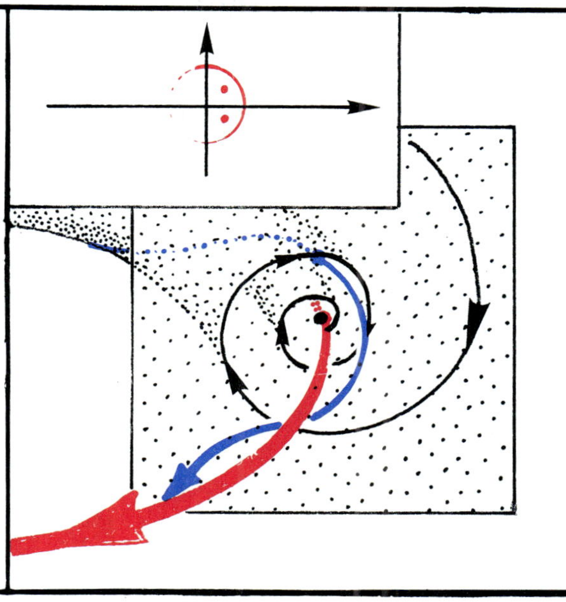

7.5.4. In the case of a spiral *attractor*, the complex conjugate. CM's are *inside* the unit circle.

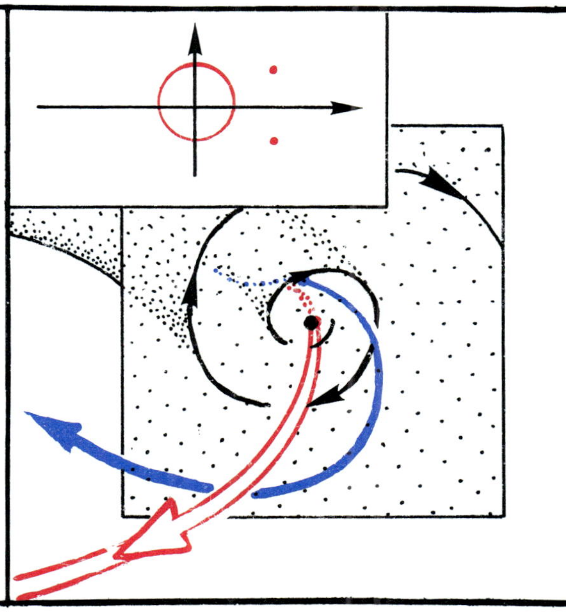

7.5.5. In the case of a spiral *repellor*, the complex conjugate CM's are *outside* the unit circle.

What about CM's *on* the unit circle?

A limit cycle is *hyperbolic* if none of its CM's are on the unit circle. An important example of a nonhyperbolic limit cycle in three-space is a *center*.

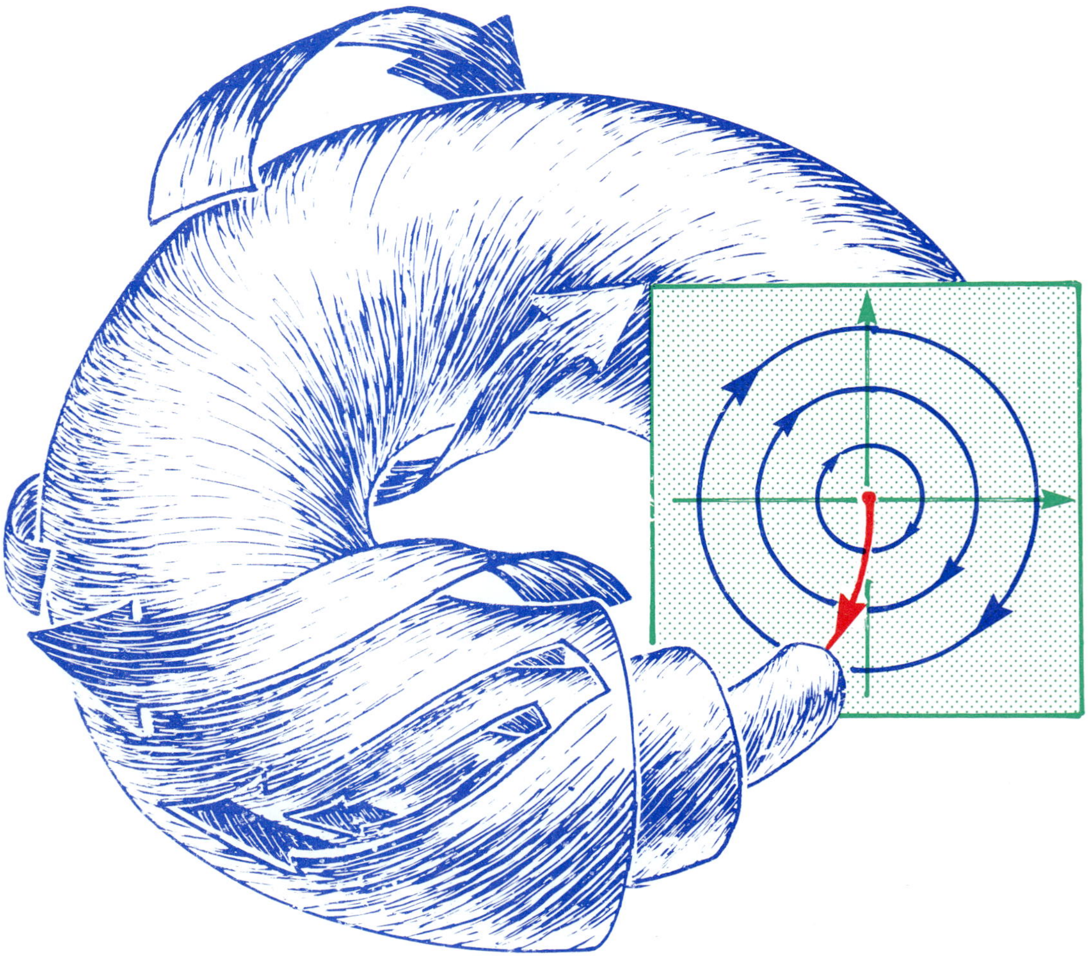

7.5.6. A center is surrounded by concentric tori, which are invariant surfaces. Nearby trajectories stay on these tori, neither attracted nor repelled by the central limit cycle.

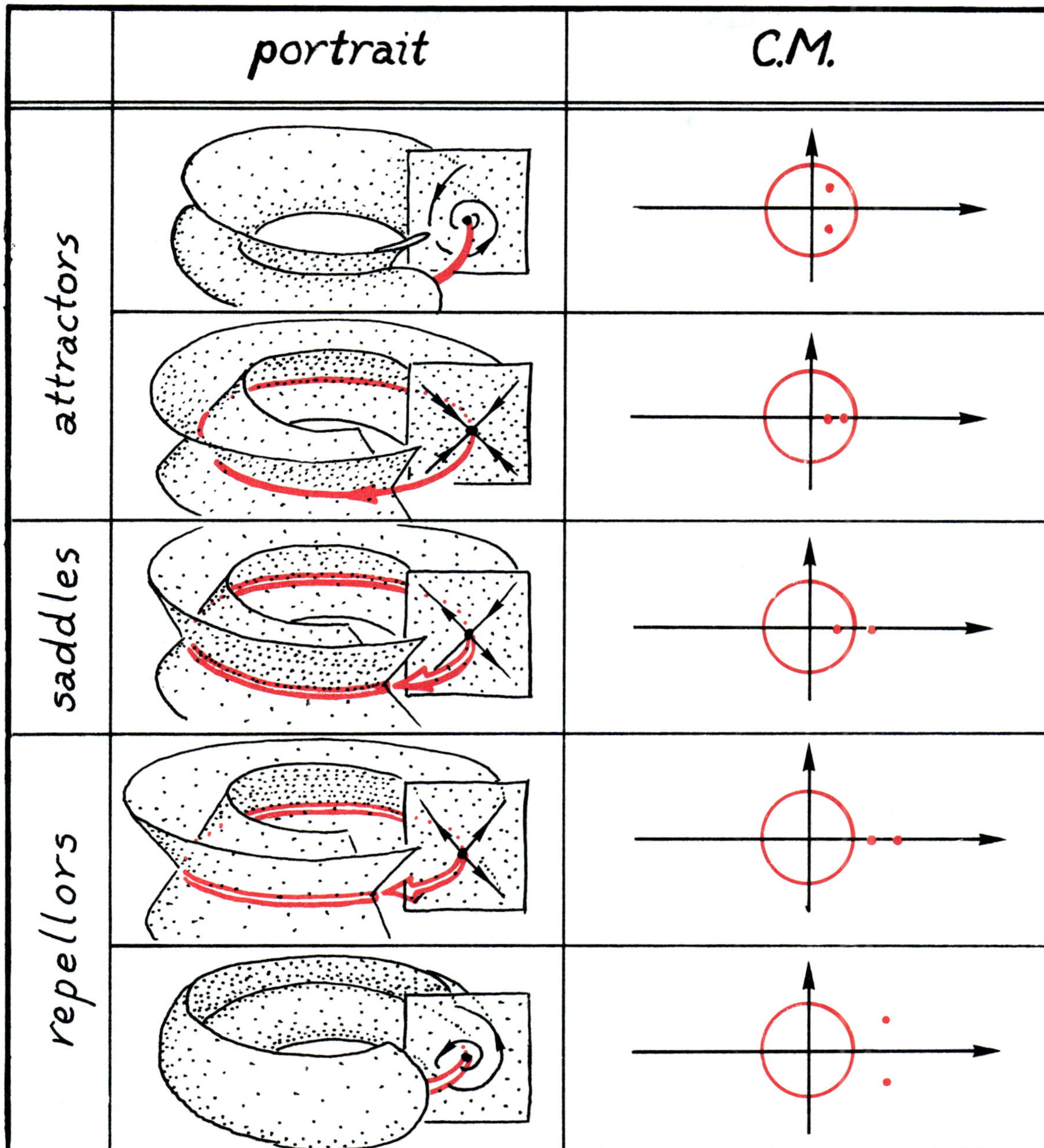

7.5.7. A limit cycle is called *elementary* if it is hyperbolic and its CM's are distinct (no two equal). All the elementary limit cycles in three-space are summarized in this table.

7.6. Characteristic Exponents

In the case of a limit *point,* the local asymptotic behavior is described by the CE's. The number of CE's is equal to the dimension of the state space. In the case of a limit *cycle,* the local asymptotic behavior is described by the CM's. The number of CM's is equal to the dimension of the Poincaré section, one less than the dimension of the state space.

In this section, we develop the relationship between CE's, and CM's.

7.6.1. Recall from the preceding chapter that with CE's the ***horizontal coordinate*** (real) indicates the strength of repulsion (right half) or attraction (left half). The *vertical coordinate* (imaginary) indicates the rate of spiraling. The CE's on the left are *attractive.* Those on the right are *repulsive.* The imaginary axis is excluded, for hyperbolic limit points. We will call this diagram the *CE plane.*

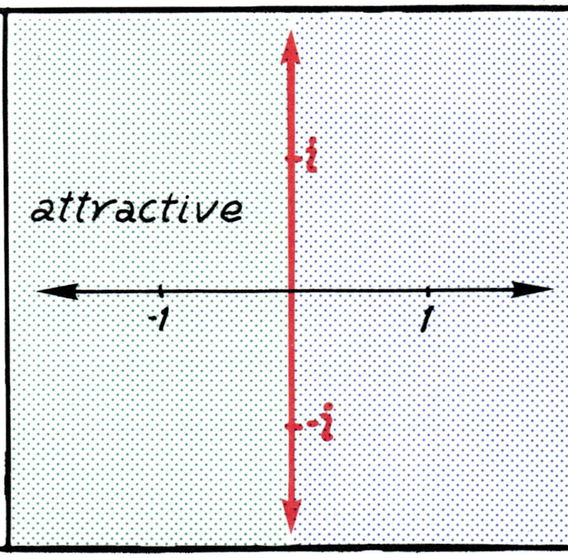

7.6.2. On the other hand, in the preceding section we have seen that, with CM's the *magnitude* indicates strength of repulsion, while the *angle* characterizes spiraling. The CM's are *attractive* if inside the unit circle, and *repulsive* when outside. The unit circle is excluded for hyperbolic limit cycles. We will call this diagram the *CM plane.*

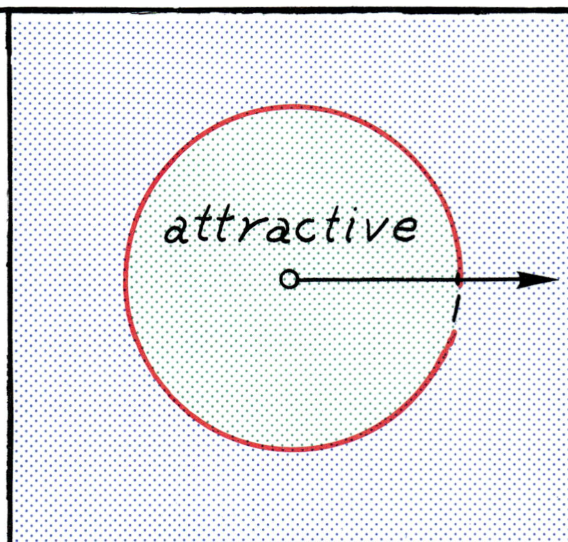

We now describe a function from the CE plane to the CM plane, called the *exponential map,* or *polar coordinates.*

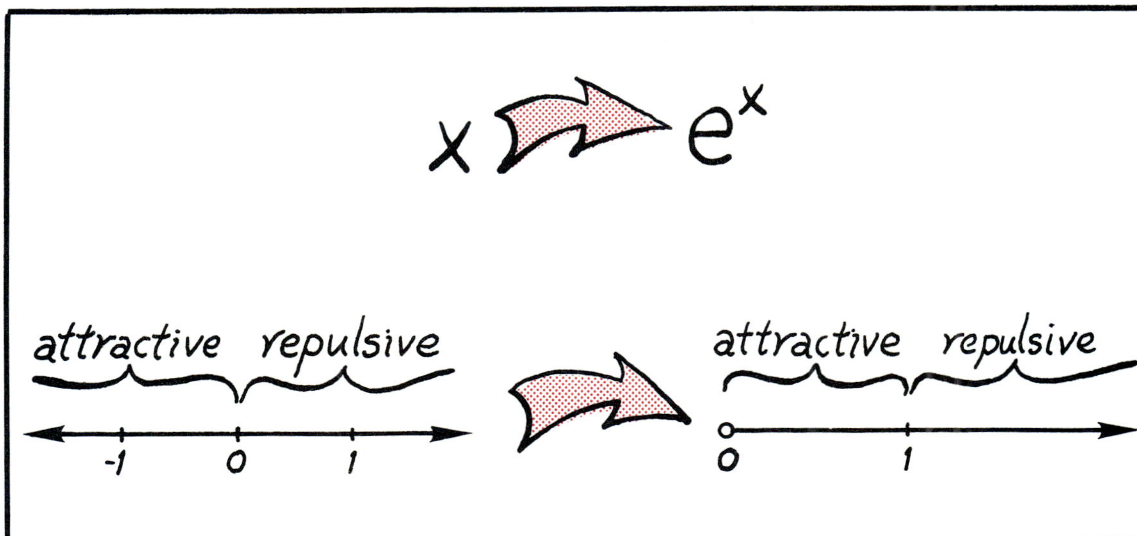

7.6.3. *Step 1.* Map the horizontal axis of the CE plane onto the positive half of the horizontal axis of the CM plane, using the natural exponential function. Note that *zero is mapped to one.*

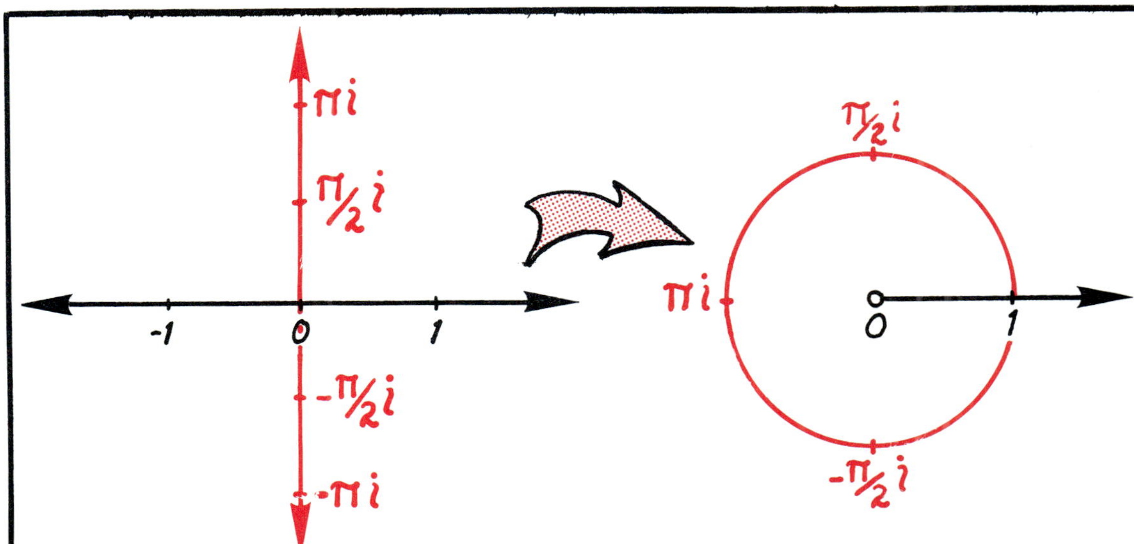

7.6.4. *Step 2.* Wrap the vertical axis of the CE plane around the unit circle of the CM plane. *Linear measure* on the real axis of the CE plane *angular measure* on the unit circle of the CM plane.

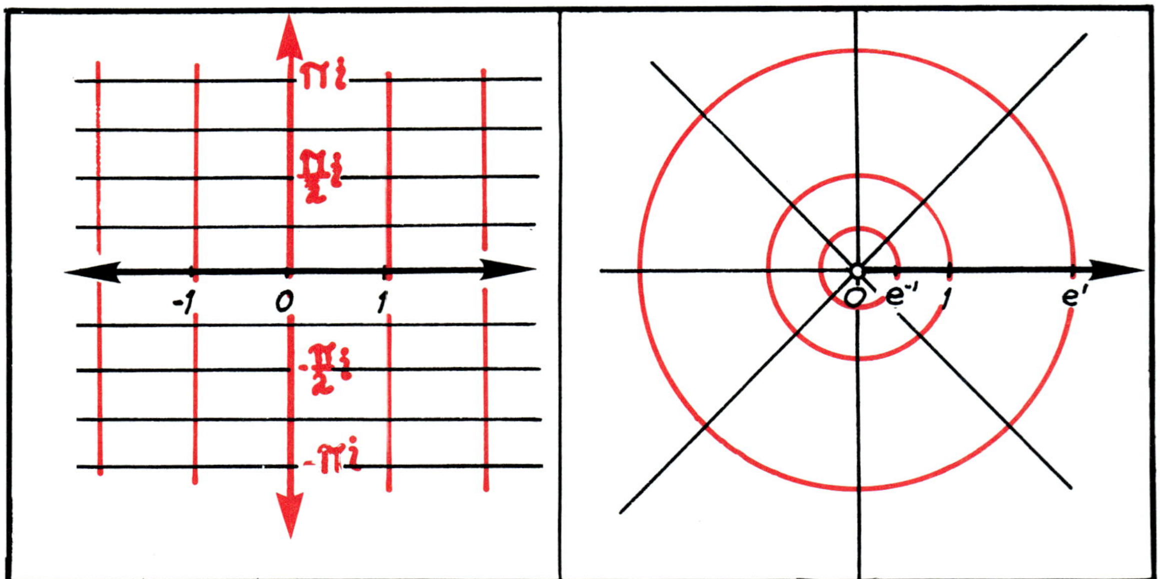

7.6.5. *Step 3. Horizontal lines* in the CE plane are mapped into *rays* in the CM plane, radiating from the origin. *Vertical lines* in the CE plane are wrapped around *concentric circles* in the CM plane. No point in the CE plane goes to the origin of the CM plane.

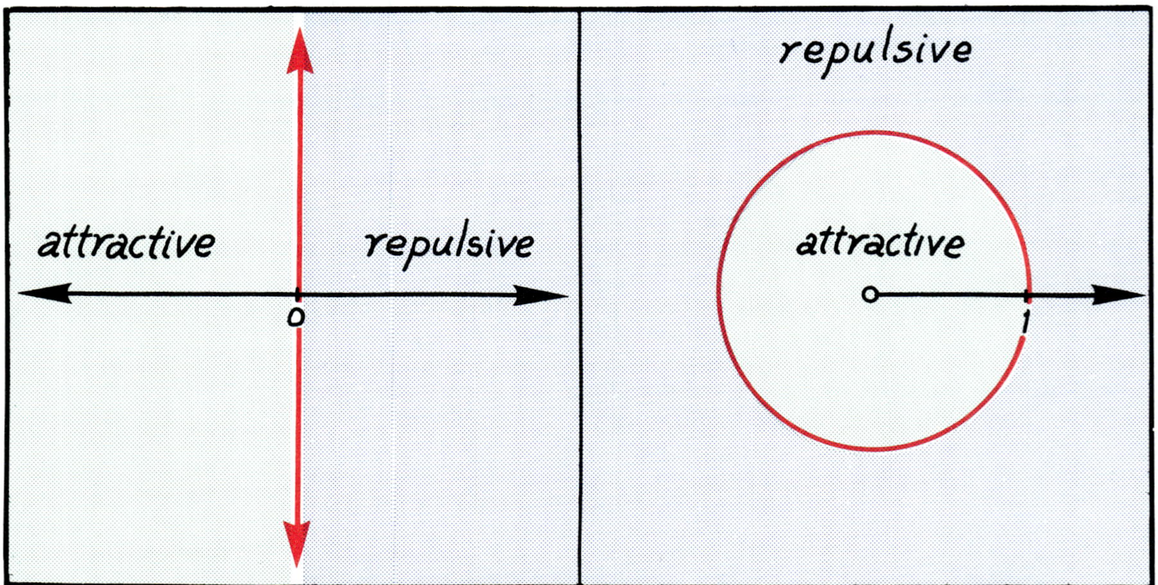

7.6.6. Note that the attractive and repulsive regions of the CE plane are mapped onto the corresponding regions of the CM plane.

We could use this mapping to define CE's for limit cycles, but this will not be necessary.

7.7. Discrete Power Spectra

Before ending this review and expansion of Part One, we pause to emphasize the periodic attributes of attractive limit cycles. This explains why we call them *periodic attractors,* which is not because they come and go!

7.7.1. In Figures 2.3.5 to 2.3.7, we discussed the pitch and volume of *pure tones.* These are attributes of *sinusoidal time series,* or time records of one coordinate of a *regular* (circular) periodic attractor.

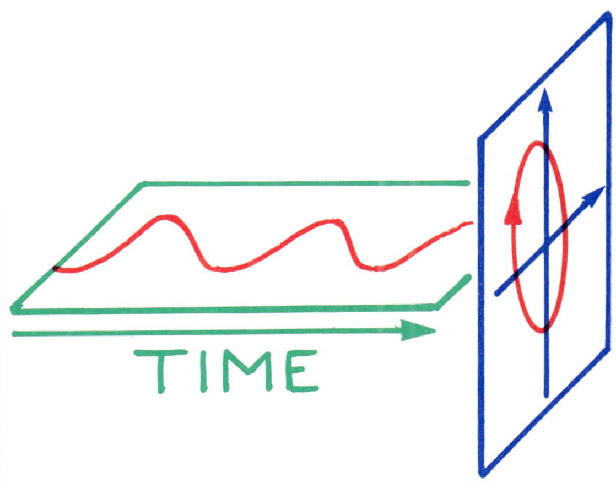

7.7.2. A time series from another, very *irregular* periodic attractor might look like this. It may be *bumpy,* but it is *periodic* in the sense that the same irregular pattern is repeated over and over. Each repetition takes the same period of time—hence *periodic.* The reciprocal of the period of time is the *frequency* of the periodic time series. This one would sound like a complex tone, with a colorful timbre.

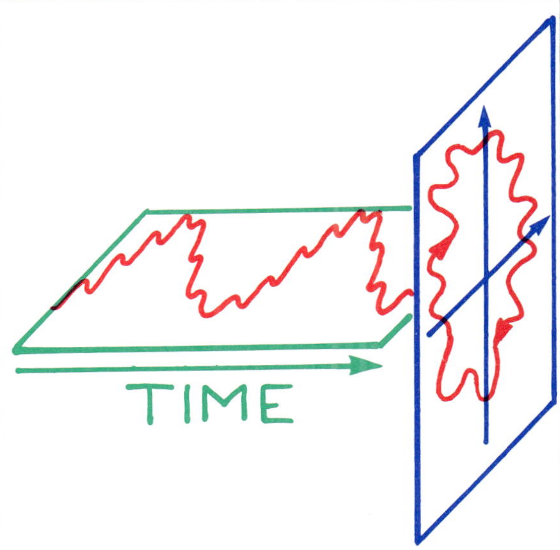

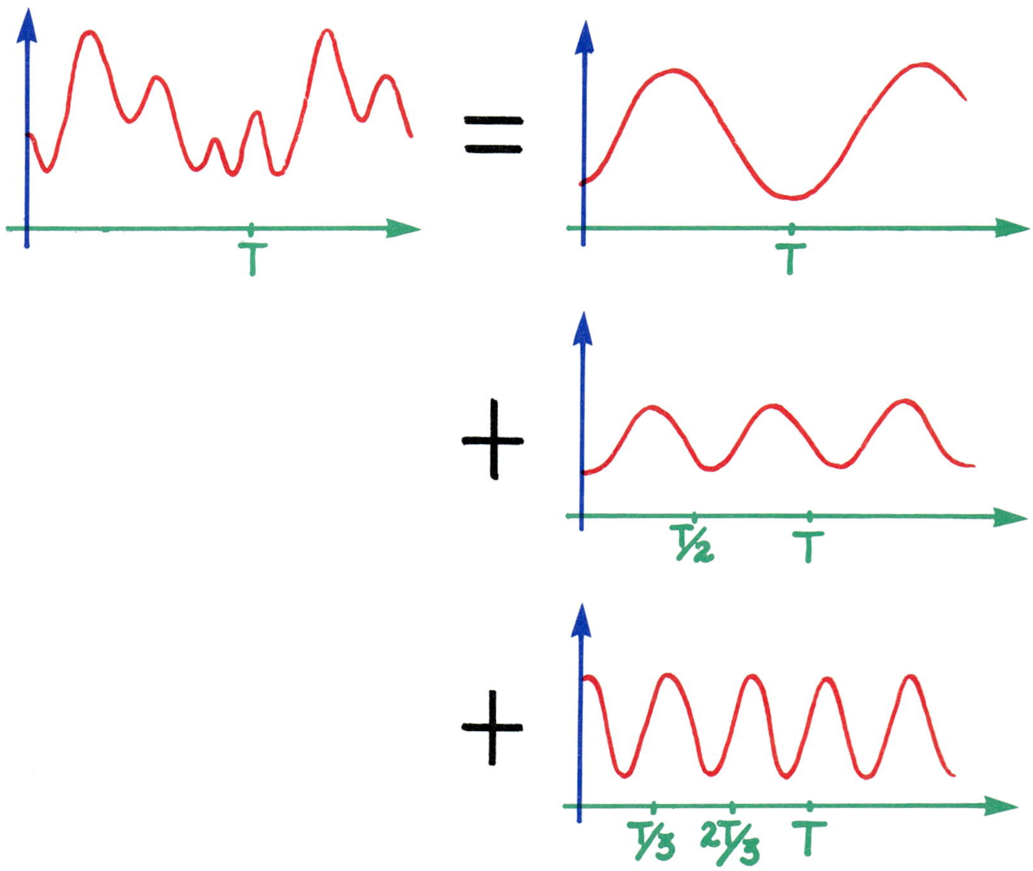

7.7.3. Following the ideas of Fourier analysis, we may represent the irregular periodic time series (or tone) as a sum of regular (sinusoidal) pure tones of various frequencies, amplitudes, and relative phases, all sounded together.

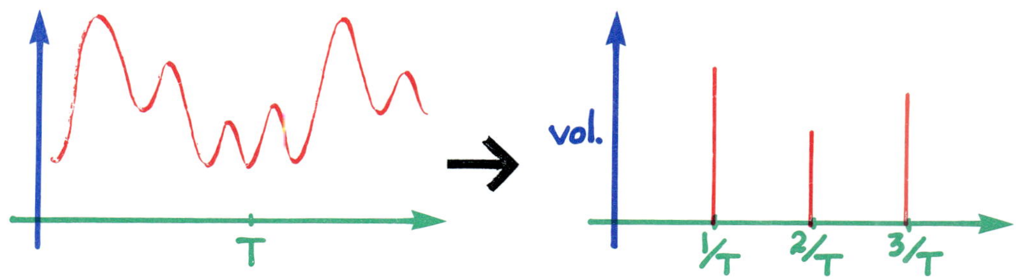

7.7.4. Recording the square of the amplitude, and the frequency, of each of the component pure tones in a diagram such as this is called the *power spectrum* of the periodic time series (complex tone). In this case, the power spectrum is *discrete,* as it consists of discrete vertical line segments. This is characteristic of periodic attractors.

Likewise, continuous power spectra are uncharacteristic of periodic attractors. And yet, they abound in music, and throughout nature, in *noise.* The dynamical model for a noisy time series is the chaotic attractor, to which we turn at last.

8

Chaotic Limit Sets

Points and cycles are not the only limit sets found in dynamical systems. The torus is a limit set proved to be rare in theory but frequently seen in experiments. Reasons are given in Part Three, "Global Behavior." In this chapter, we introduce some of the curious limit sets discovered by experimentalists. Among all kinds of limit sets, only the attractors are directly observable in simulations by analog and digital computers. So this chapter will emphasize chaotic attractors, primarily. The reasons these are called *chaotic* will be described in Chapter 9.

8.1. Poincaré's Solenoid

Poincaré's discovery of *homoclinic tangles,* and their subsequent analysis by Birkhoff and Smale, are described in detail in Part Three. Poincaré predicted, in his original description of homoclinic tangles, that they might be too complicated ever to be understood. The full picture, recently emerged, contains a chaotic limit set of saddle type, which we call *Poincaré's solenoid.* In this section, we describe the superficial appearance of this theoretical object.

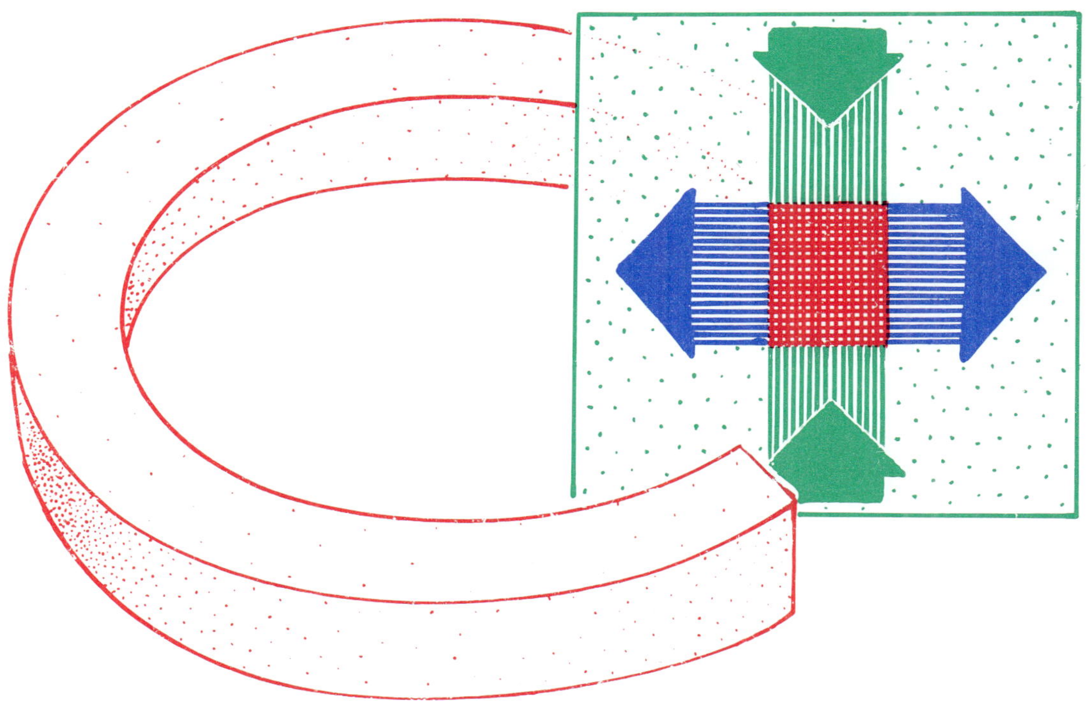

8.1.1. Here is the solenoid. Like an infinite coil of wire, shown here in red with a piece cut away for visibility, it might be regarded as thickening of a periodic trajectory of saddle type. But it has an infinite number of pieces, as we shall see. In this representation, we emphasize its Poincaré cross-section. The inset cross-section is a thickened curve, with an infinite number of pieces. The outset cross-section is also a thickened curve.

Let's build up this picture, piece by piece.

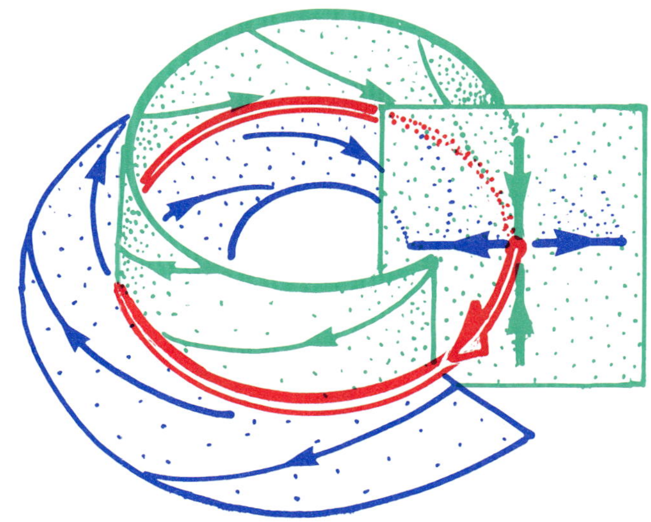

8.1.2. Recall, from Section 7.3, this typical periodic saddle in three-space.

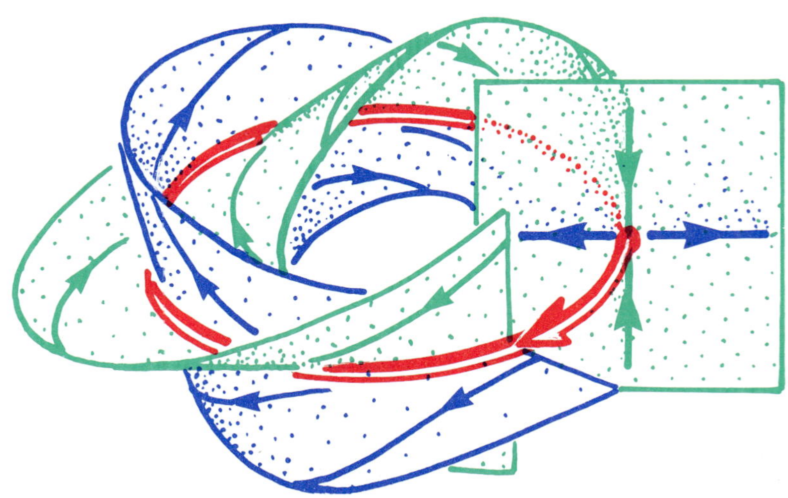

8.1.3. This twisted version has Möbius bands for its inset and outset. Each has a single twist.

Next, consider a twisted periodic saddle going *twice around* before closing. Each half almost closes, but not quite. The inset of each half is twisted, like a Möbius band. These half-insets may thus be parallel, without crossing. The outset of each half is likewise twisted, so these may be parallel as well. We may visualize this as follows.

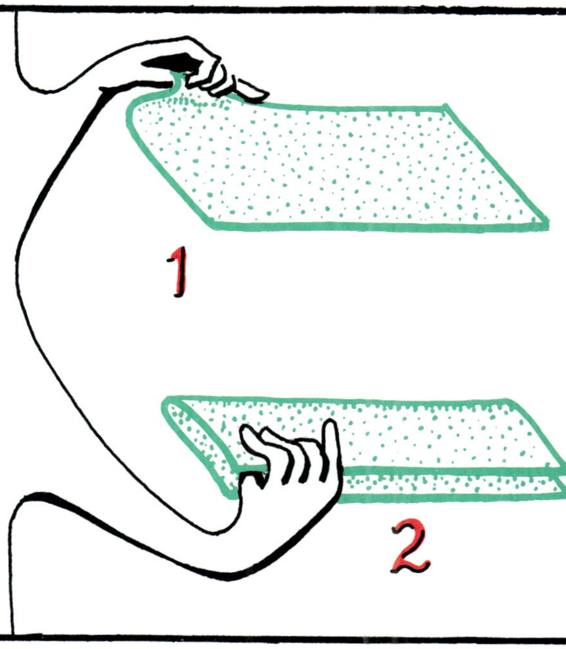

8.1.4. Take a narrow strip of paper (1) and fold it the long way (2).

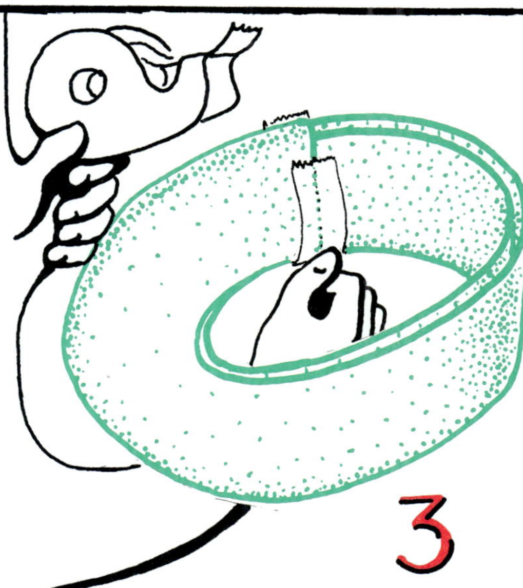

8.1.5. Use this folded strip to make a Möbius band, twisting and taping carefully (3). Note that the fold does not match up at the taped part. This is no problem, because we . . .

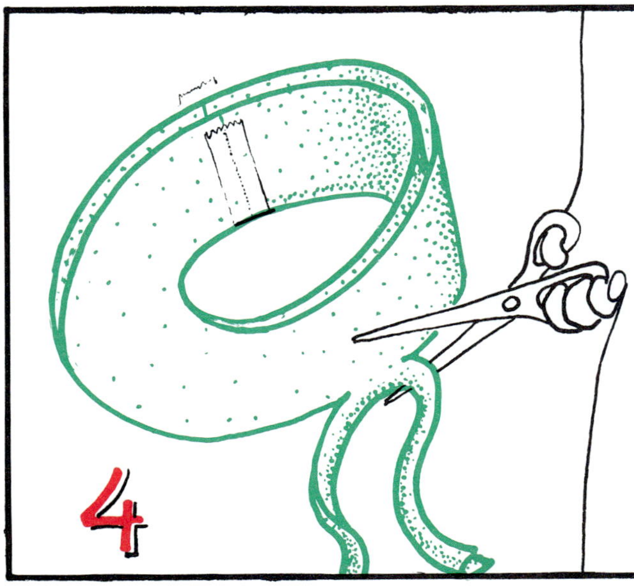

8.1.6. . . . trim the fold off. What remains is a double strip. This is the *inset-model.*

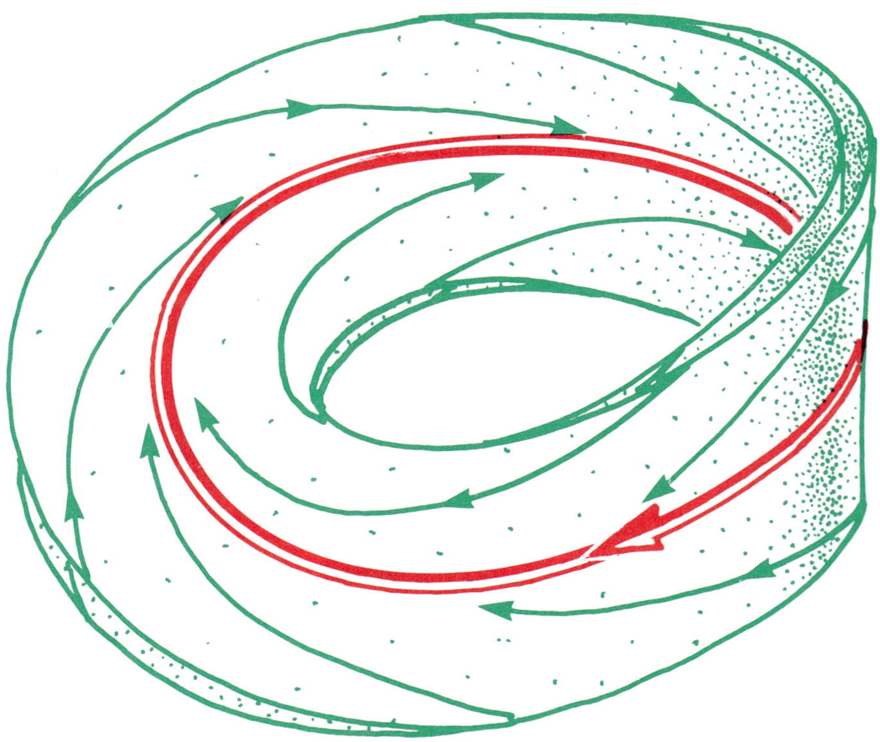

8.1.7. Here is the inset model, with the limit cycle and attracted trajectories drawn on it.

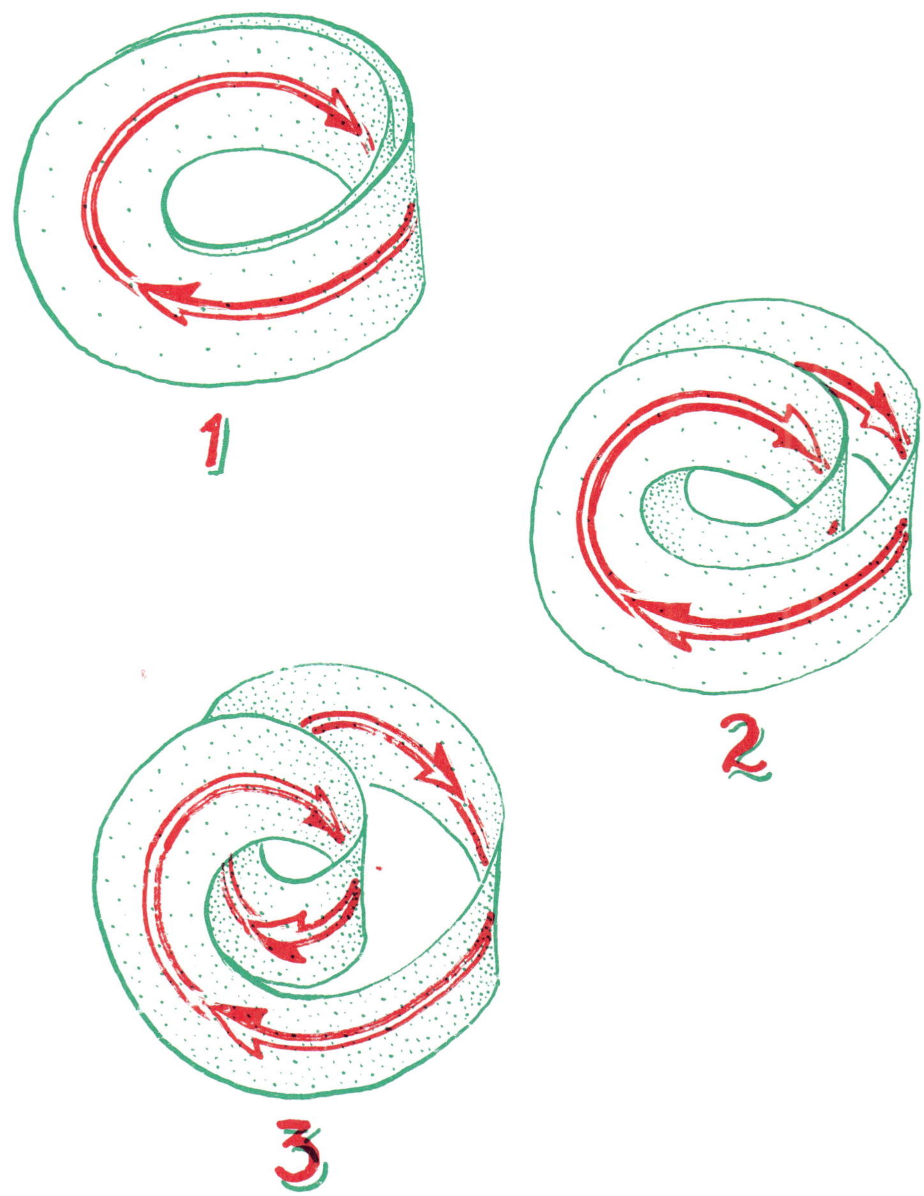

8.1.8. Trying to uncoil the inset model, we find that we have a long band that is *double twisted.*

Now take another narrow strip of paper. Fold, twist, and tape, as above, but with this complementary orientation.

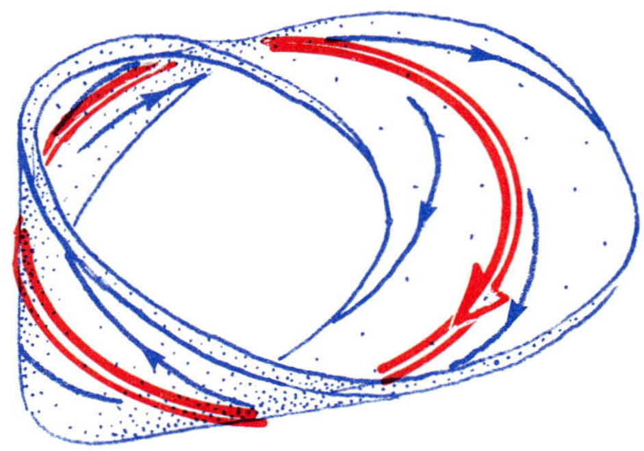

8.1.9. This model is horizontal where the previous one was vertical, and so on. Cutting off the fold, we obtain this model for the outset of our double periodic trajectory. This is the *outset-model.*

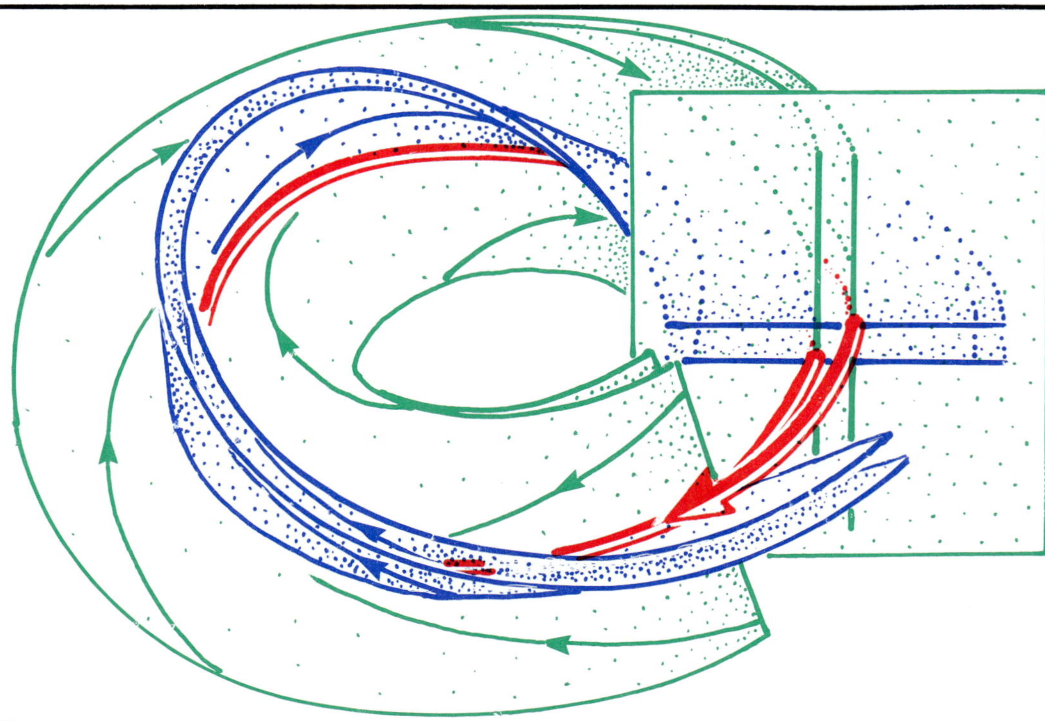

8.1.10. Putting together the two models, we obtain this picture of the double periodic trajectory. The trajectory completes two revolutions before closing, likewise the inset and outset bands.

Note that the inset and outset *must intersect* in this picture. The intersection is another saddle cycle that completes two revolutions before closing. The outset of the new cycle coincides with the inset of the original one, and *vice versa.* This is the complication Poincaré had noticed and despaired. The fantastic consequence of this intersection, obtained by a small perturbation, is known as a *homoclinic tangle.* This is detailed in Part Three, "Global Behavior." For the present, we will ignore this consequence.

8.1.11. We could equally well *fold the strip twice,* like this.

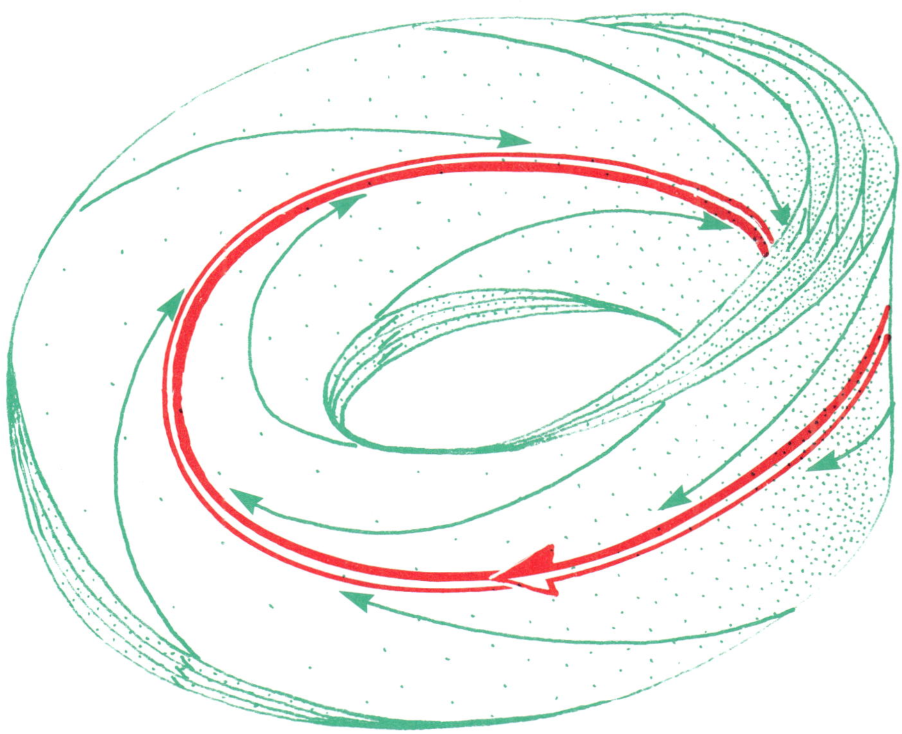

8.1.12. Twisting, taping, and cutting, we obtain an inset-model for a quadruple periodic trajectory.

Iterating this procedure, we may make models for very long periodic trajectories, all coiled so as to fit in the same ring. All these may have their insets parallel, likewise their outsets. Adding trajectories that coil around the ring endlessly *without closing,* we obtain a complicated limit set of saddle type, Poincaré's solenoid, as shown in Figure 8.1.1. The occurrence of this limit set in the forced Van der Pol system (Chapter 5) was suggested by Cartwright and Littlewood, and fully analyzed by Smale, as explained in the Hall of Fame. It is not observed in experiments, because it is *not an attractor.*

We turn now to the chaotic limit sets that are actually attractive.

8.2. Birkhoff's Bagel

Birkhoff picked up dynamical systems theory, in 1912, where it was left by Poincaré's death. Within four years, he had discovered the chaotic attractor, as a theoretical object in discrete dynamical systems (1932). This was his ***remarkable curve,*** found in the dynamics of a map of the plane to itself. This was later studied by Charpentier (1935) and is still an active research topic. This so-called curve is actually thicker than a normal one-dimensional curve, so we call it a ***thick curve.*** Thickness, in this sense, relates to ***fractal dimension,*** explained in the next chapter.

In 1944, Levinson raised the question of the occurrence of this attractor in the context of forced oscillation. Here, the plane would be the strobe plane, a Poincaré section, and the map of the plane to itself would be the first return map. Then the remarkable (thick) curve of Birkhoff would be the strobe section of a thick torus, which we call here a ***bagel.*** He answered this question in the affirmative for a particular system, similar to the forced Van der Pol system (1948, 1949).

Meanwhile, Cartwright and Littlewood announced a similar result (1945) for the forced Van der Pol system, and later published their proof (1949). They pointed out experimental evidence for this strange attractor, which they had found in the earliest electronic simulations of the Van der Pol system, done by Van der Pol himself in collaboration with Van der Mark (1927).

Recent experiments by Holmes (1977) on the Duffing system, and by Robert Shaw (1980) on the forced Van der Pol system (standard and variations) with an analog computer have detailed the folded structure of Birkhoff's attractor, which we present in this section. Our presentation relies heavily on Part One.

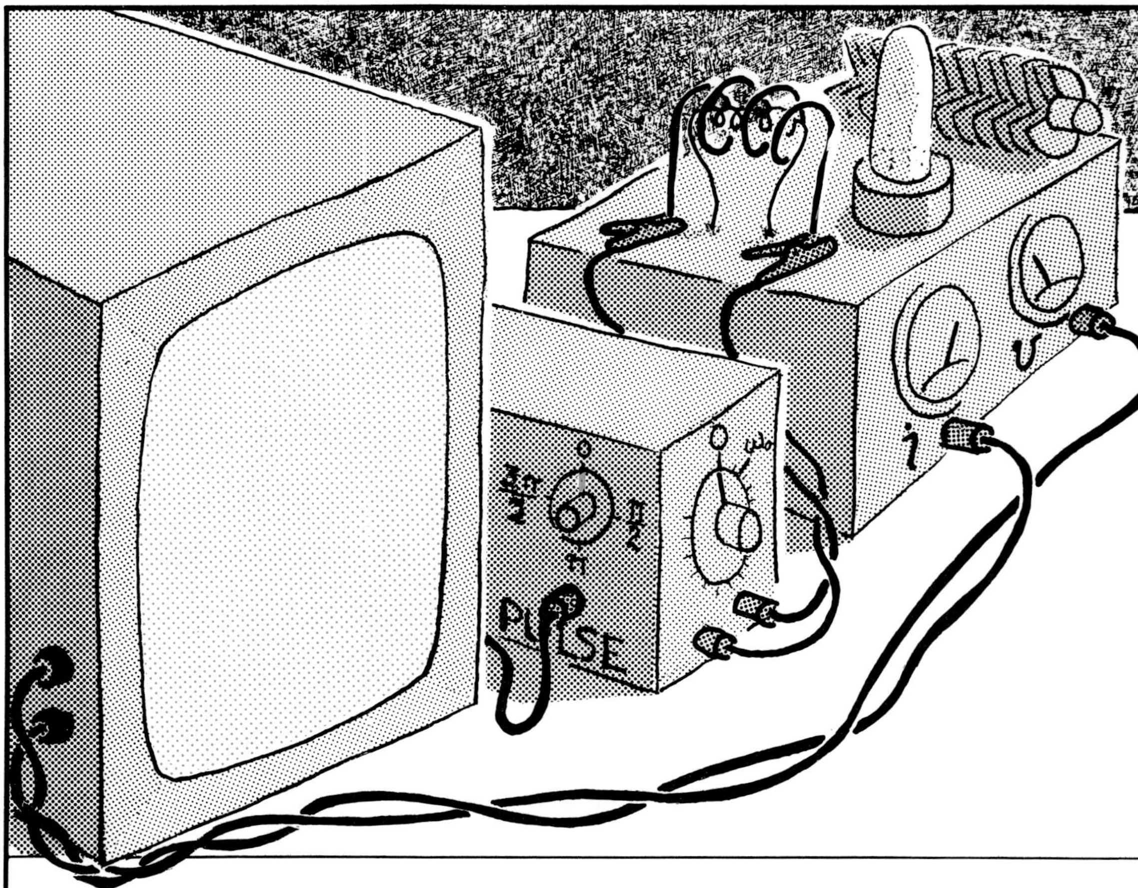

8.2.1. Recall that Van der Pol studied forced oscillators in the context of radio transmitters and receivers. Here is the analog device, as explained in Chapter 5. The transmitter (right) is forced by an isolated oscillator (center). Its plate voltage and current are monitored by a storage scope (left) only during a short pulse sent by the forcing oscillator. This strobe pulse is in phase with the forcing oscillation. Here we have added a control knob to the forcing oscillator (shown high on its left side, with a green marker) for changing the phase of the forcing oscillation at which the strobe pulse is sent.

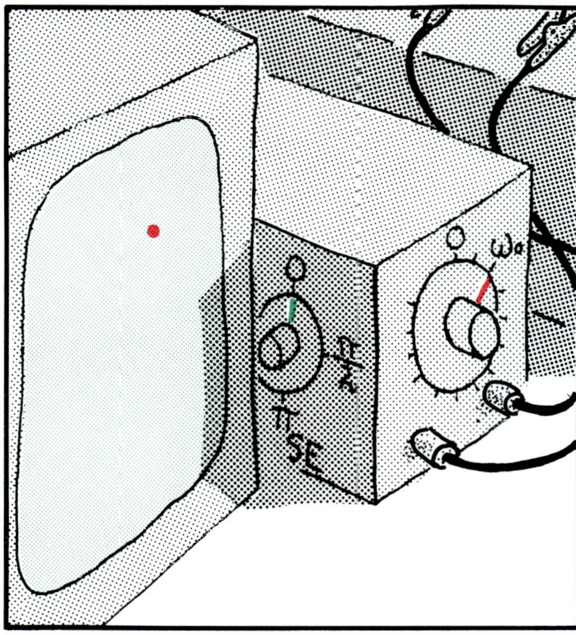

8.2.2. With the frequency of the forcing oscillator close to the natural frequency of the unforced radio oscillator, the combined system may oscillate in an *isochronous harmonic*. In this case, we observe a single red dot on the green storage scope.

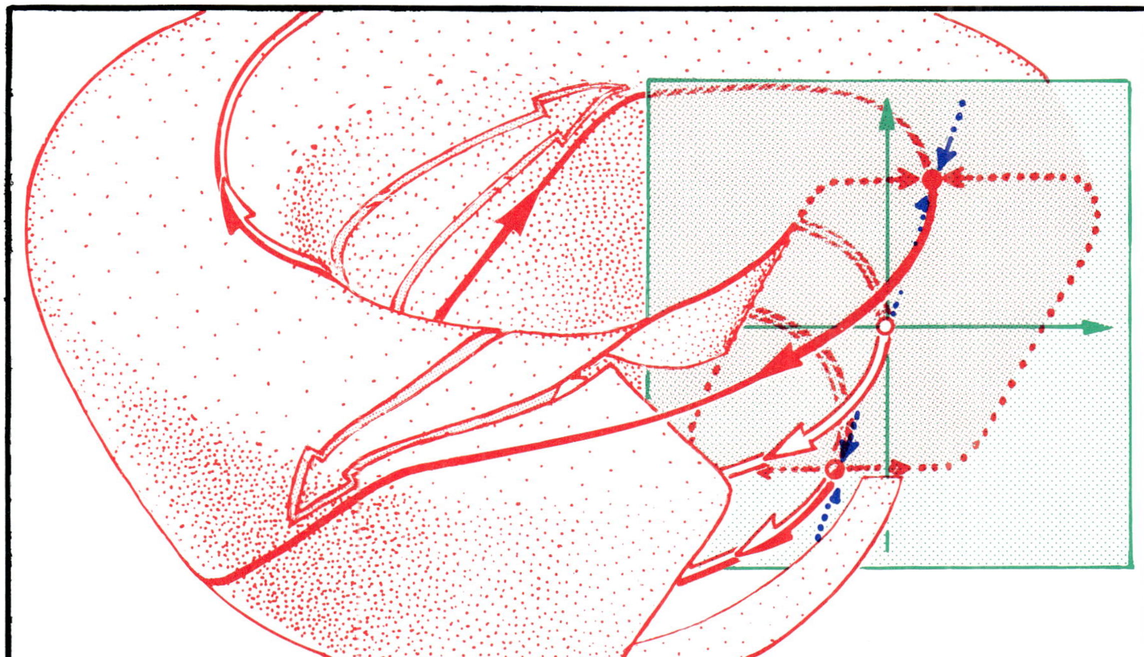

8.2.3. Advancing the phase control from phase zero forward to phase 2π, we see the dot advance around a closed loop. Assembling these planar sections in the large three-dimensional state space of the ring model (see Chapter 5), we see the isolated chronous harmonic as a periodic trajectory on an attractive invariant torus.

By increasing the frequency of the driving oscillator, we may observe many different braids, as described in Chapter 5.

8.2.4. For example, an ultraharmonic would appear, in the strobe plane or storage scope screen, as a discrete point set on the section of the invariant torus. With each pulse from the forcing oscillator, one of the dots is refreshed. In other words, the trajectory hits the strobe plane repeatedly. The successive hits cycle through the discrete point set, repeatedly. The order of successive hits is not simply clockwise or counterclockwise. However, the same order is repeated each cycle. Thus, if we watch for a while, we learn to predict the next hit exactly.

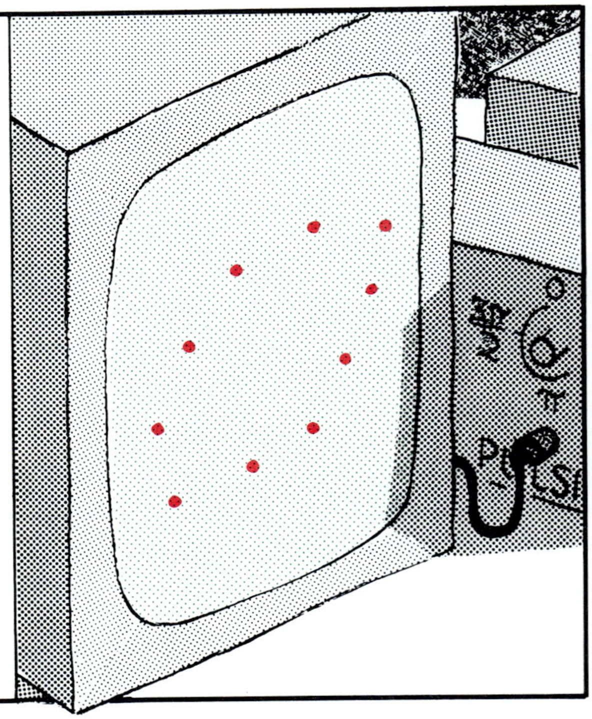

8.2.5. For some higher frequency and carefully chosen amplitude, we find what appears to be a continuous closed curve instead of a discrete point set. This is the *remarkable curve of Birkhoff,* exactly as predicted by Levinson. As far as we know, this was not actually observed until very recently (Shaw, 1980). The trajectory hits the strobe plane in a point that appears to rove randomly around this curve. Unlike the preceding (periodic) situation, we find ourselves unable to predict the next hit of the trajectory on the strange curve, no matter how long we observe it.

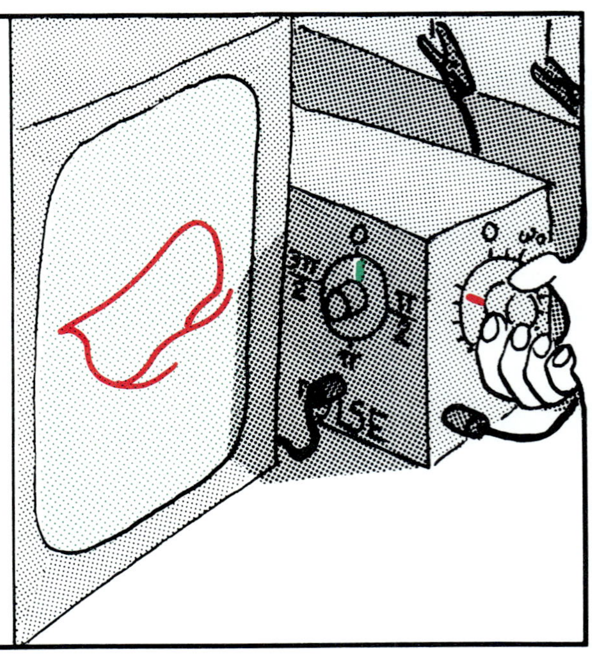

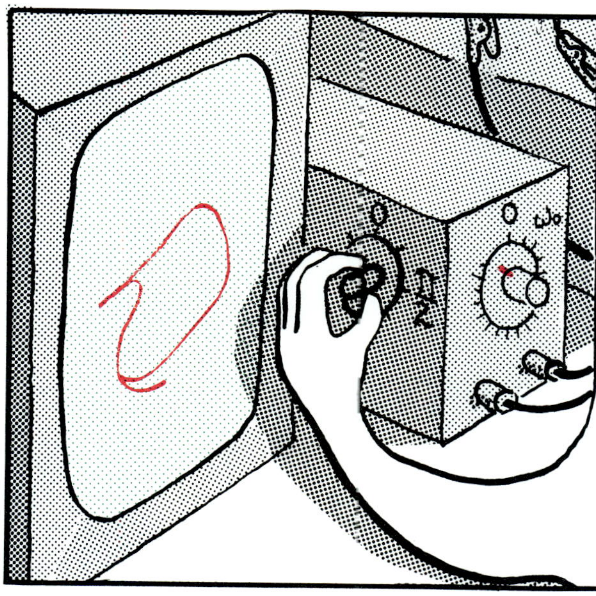

8.2.6. Adjusting the phase control reveals a rotating distortion of the closed curve. The spike slowly crawls to the right along the top, while the lower spurs march briskly to the left along the bottom. One by one, they catch up with the upper spike and merge with it.

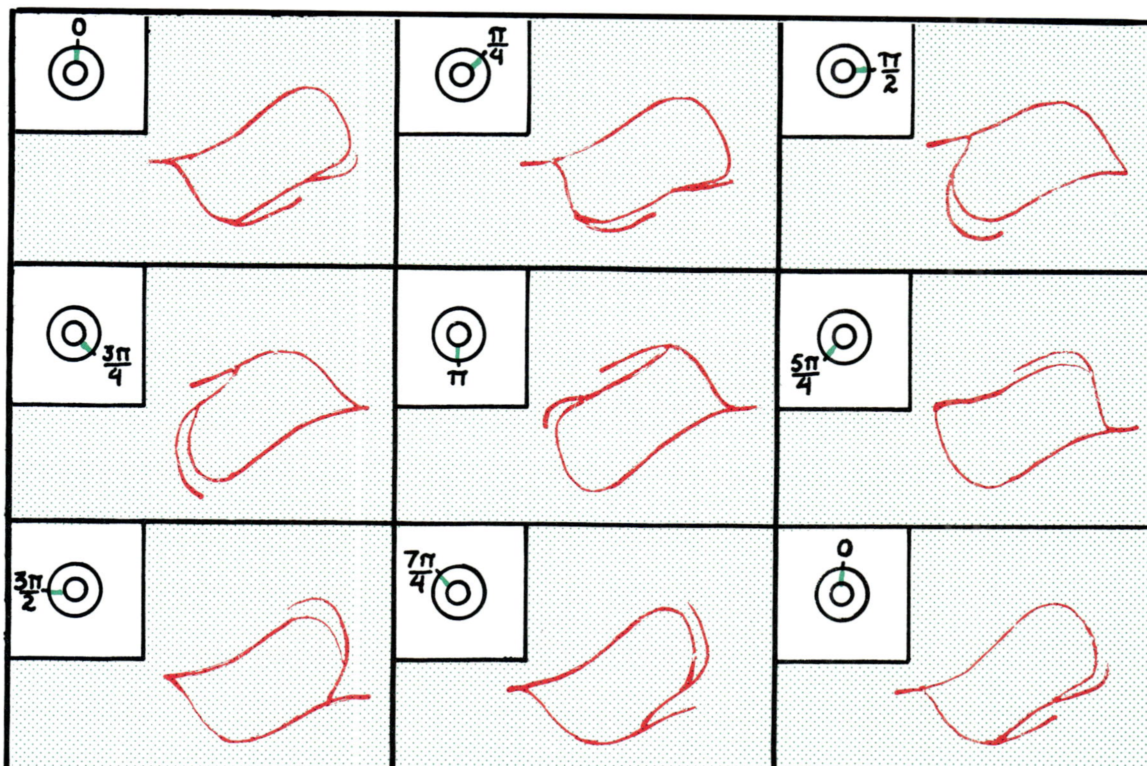

8.2.7. Here is a sequence of eight successive phases, a full cycle around the ring model, exactly as discovered by Robert Shaw (1980) in analog simulation of Van der Pol's system. His equations are reproduced in the Appendix.

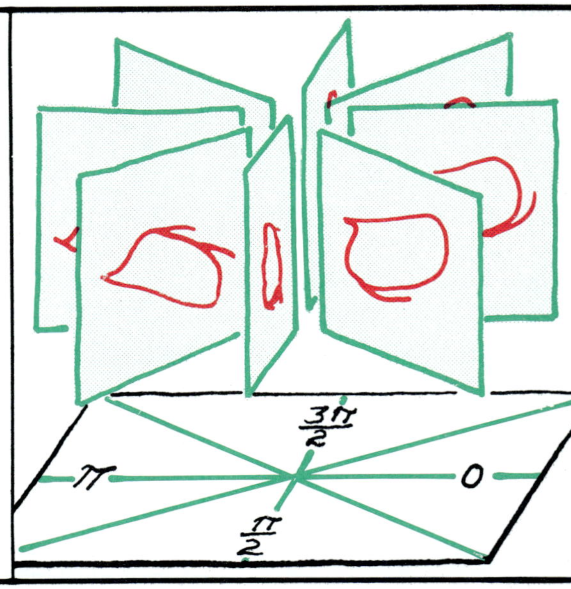

8.2.8. Assembling Robert Shaw's sections in the ring model for the forced Van der Pol system, the bagel emerges.

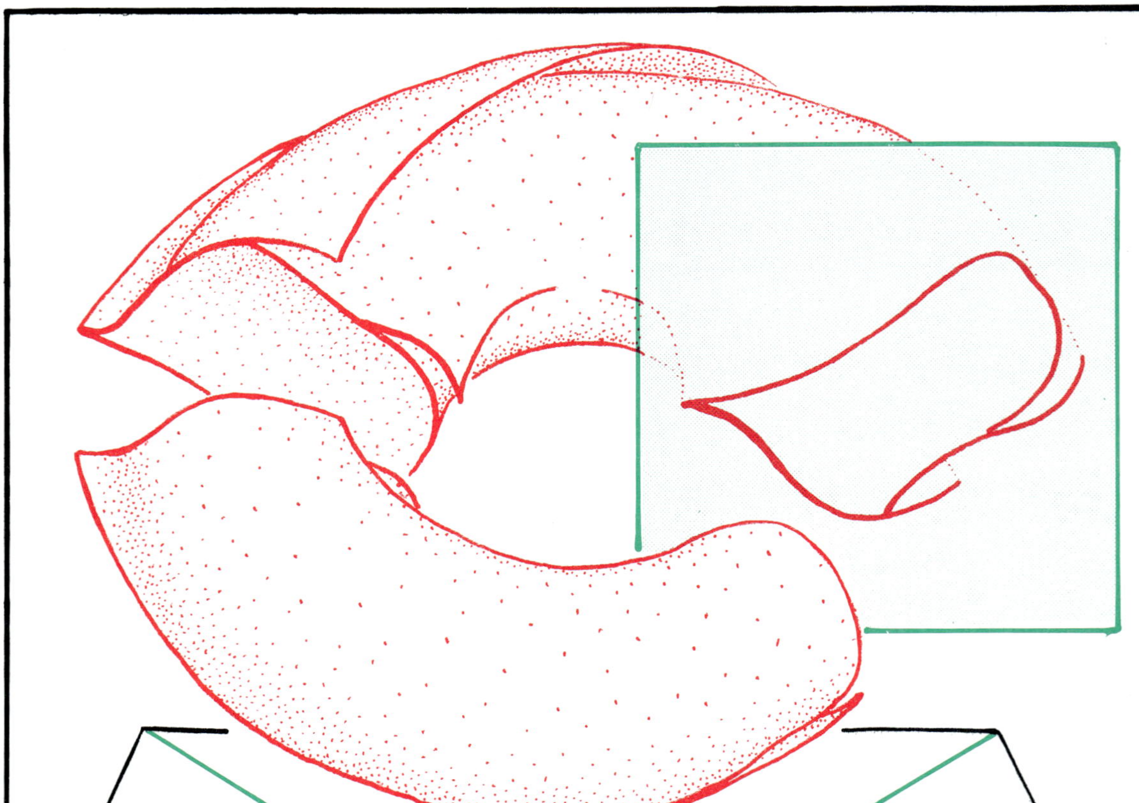

8.2.9. Interpolating more sections and smoothing between them, we obtain this folded, thickened surface. But this is essentially the trace of a *single trajectory. It is not a periodic trajectory.* The missing sections have been cut away for better visibility.

This single trajectory represents a most irregular behavior of the forced oscillator, *chaotic behavior*.

Let's color the bagel red. Now we isolate a small piece on Birkhoff's remarkable curve, color it blue, and follow it once around the bagel.

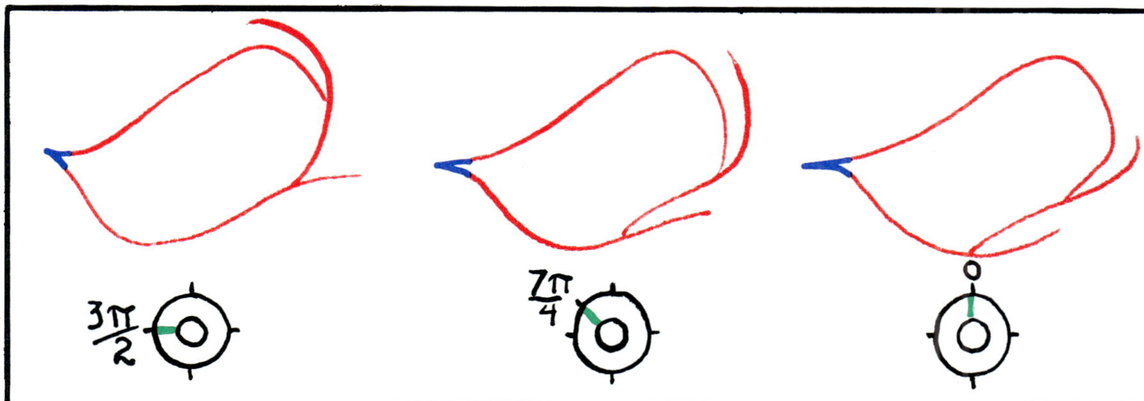

8.2.10. Starting at phase $3\pi/2$, this black piece is pinched to a beak and pulled out to the left, away from the main body of the bagel. It is not yet moving. But the two lower spurs (red) are moving down and to the left.

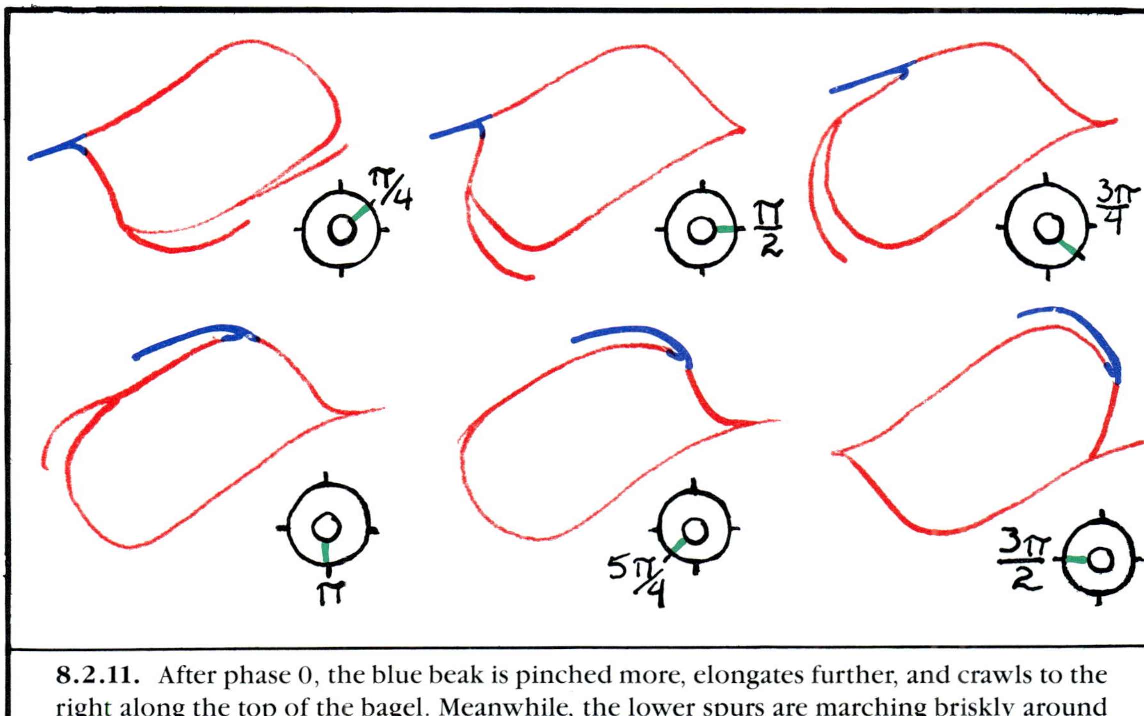

8.2.11. After phase 0, the blue beak is pinched more, elongates further, and crawls to the right along the top of the bagel. Meanwhile, the lower spurs are marching briskly around after the beak. At $\pi/2$, one of them has caught up with the beak and merged with it. A new beak is forming on the right. At $5\pi/4$, the second spur has also merged. The old beak is now a spur. Finally the new beak has fully formed and become a spur, while yet a new beak has been born on the left.

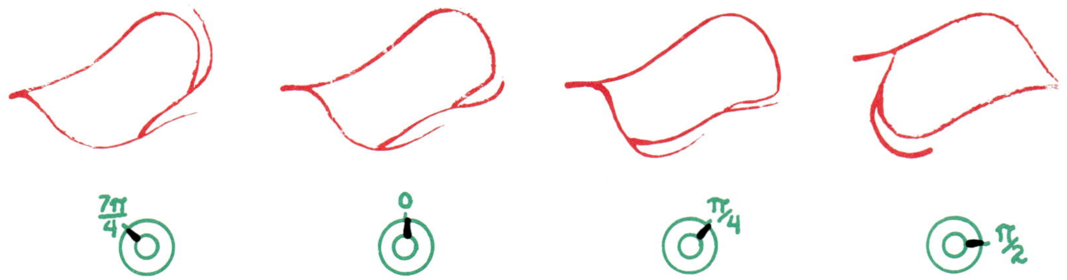

8.2.12. Going on around a bit, the red spur is pressed flat into the side of the bagel, like a pleat. This whole process, from fresh beak to ironed pleat, took one and one-half cycles of the forcing oscillation.

The bagel consists of an infinite number of pleats, almost all of which have been pressed flat against the thickened toroid. This fractal microstructure, characteristic of all the known chaotic attractors, will be clarified in Chapter 9. It is actually responsible for their unpredictable behavior.

8.3. Lorenz's Mask

The advent of digital computers made a great impact on dynamical systems theory. One of the early results was the surprising discovery, by Lorenz (1963), who had been a student of Birkhoff, of an unsuspected type of chaotic attractor. This discovery ocurred during simulation of global weather patterns, and eventually provided science with its first deterministic model of *turbulence.*

8.3.1. The earth, warmed by the sun, heats the atmosphere from below, while outer space, always cold, absorbs heat from the outer shell of the atmosphere. The lower layer wants to rise, while the upper air wants to drop. This causes a traffic problem.

To visualize the air currents in the atmosphere, we construct a cross-section. The plane section, passing through the center of the earth, cuts the atmosphere in a ring. Let us look now at a small piece of that ring.

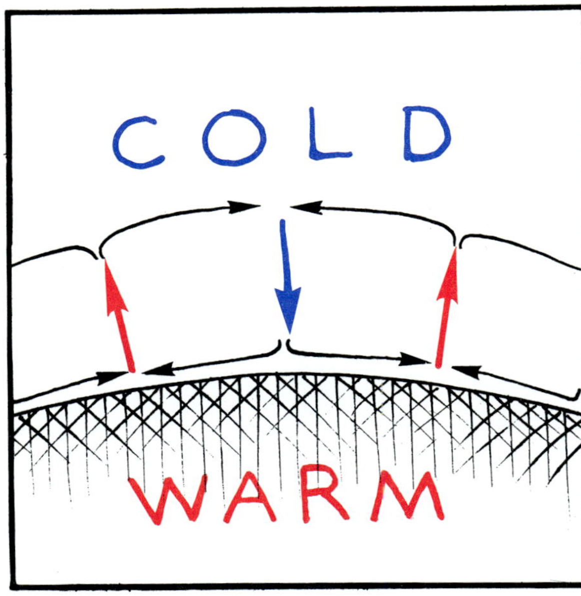

8.3.2. The traffic problem of the competing warm and cold air masses is solved by circulation vortices, called *Bénard cells.*

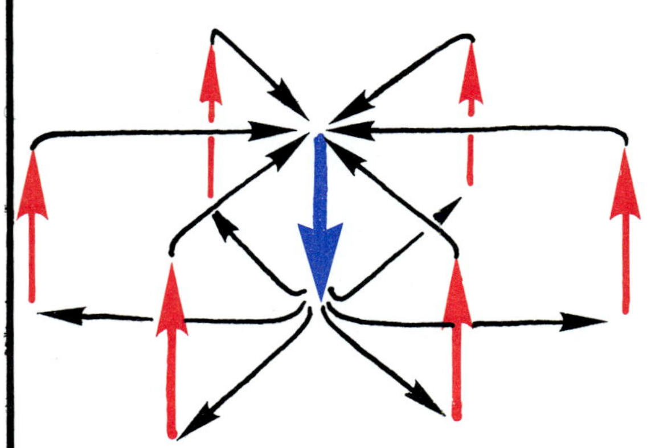

8.3.3. Returning to three dimensions, a typical vortex may have warm air rising in a ring and cool air descending in the center like this.

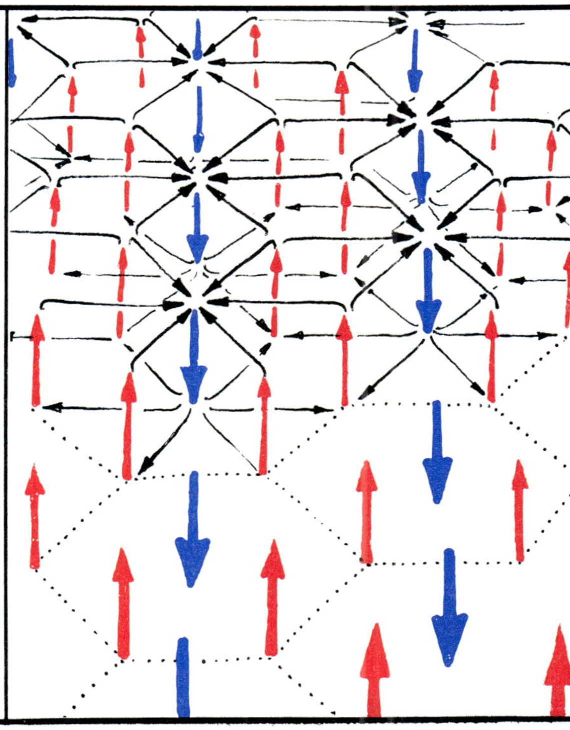

8.3.4. The atmosphere, or at least a portion of its spherical shell perhaps as large as the Sahara Desert, might be seething in a sea of Bénard cells, closely packed as a hexagonal lattice.

8.3.5. In fact, this aerial photo of the Sahara Desert shows, in the pattern of sand dunes, the sculptured footprint of such a sea of atmospheric vortices. Snowfields and icebergs reveal similar patterns, apparently sculptured by invisible Bénard cells.

Lorenz set out to model this atmospheric phenomenon, using a dynamical system derived from the equations of fluid dynamics.

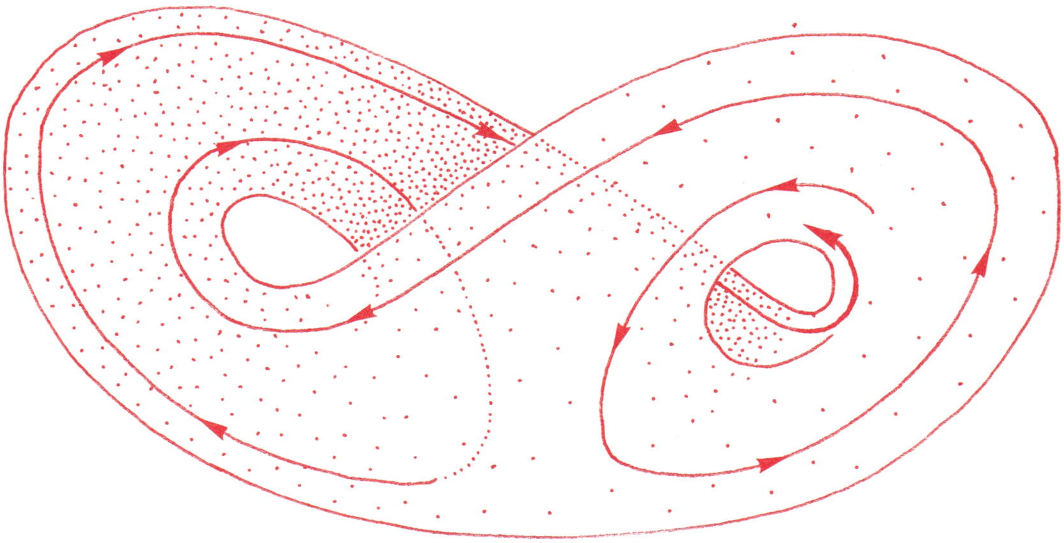

8.3.6. Simulating the model with computer graphics in the early 1960's, he saw this on the screen. Here, as with the bagel in the preceding section, the attracting object is rigidly determined. But the future of a trajectory within it is unpredictable by the observer. In the next chapter, we will explain this paradox.

Like Birkhoff's bagel, this is an attractor that is neither a point nor a cycle.

The behavior of a trajectory following this attractor, as observed by Lorenz, is very erratic. It orbits one of the holes for a while, then jumps to the other for a while, and so on. This is why its behavior is called *chaotic*, as we shall see in more detail in the next chapter. It is so erratic that Lorenz despaired of predicting the weather by simulation of this dynamical model. The chaotic attractor, translated back into the original context of air currents in the atmosphere, provides a model for *atmospheric turbulence.*

The actual microstructure of this object, which we call *Lorenz's mask,* is described further in the next chapter and still further in Part Three.

8.4. Rössler's Band

Inspired by Lorenz's discovery, and aided by an analog computer with stereo 3D display, Rössler set out to find the simplest dynamical systems with chaotic attractors. Among others, he discovered the simple folded band described in this section. Later, Crutchfield et al. (1980) found the same phenomenon in the Duffing model for the forced pendulum.

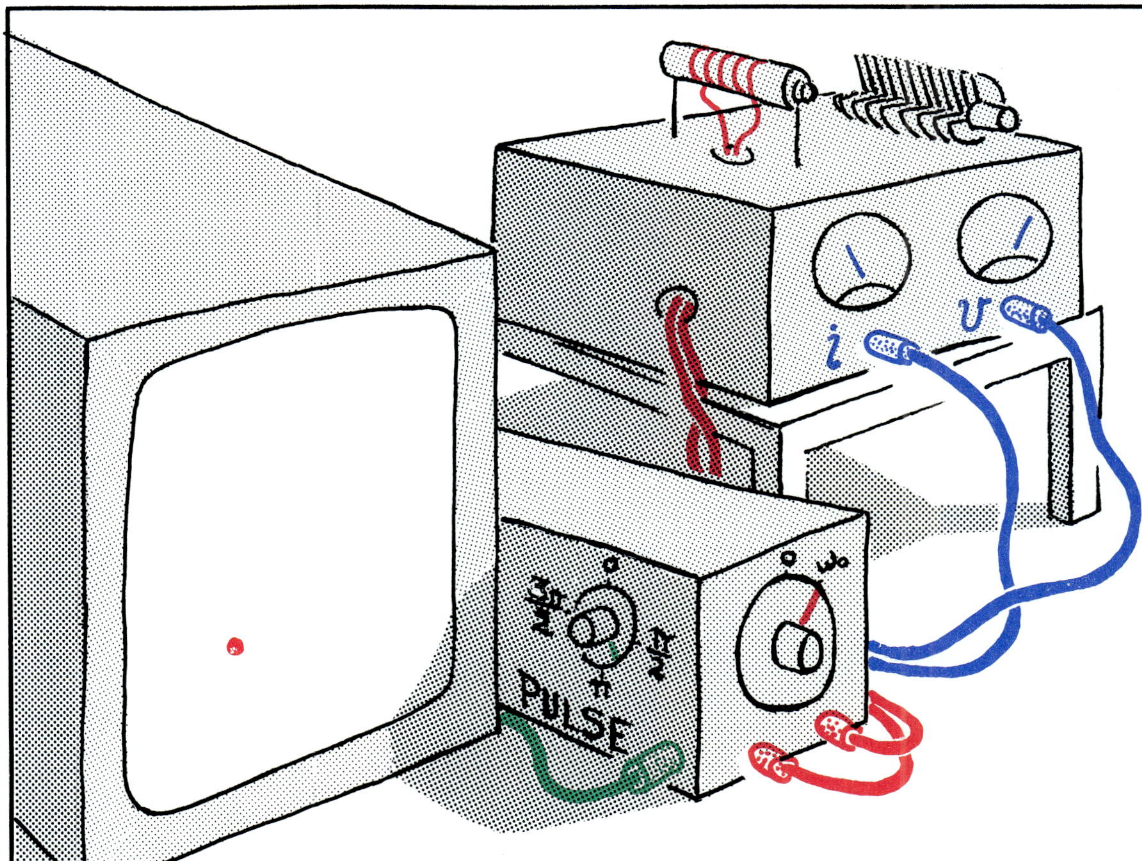

8.4.1. Here is an electronic analog of Duffing's forced pendulum. In the black box, upper right, is an electronic analog of a damped pendulum. The meters read the instantaneous current through, and voltage across, the load resistor on top of the box. The forcing oscillation is generated by the isolated oscillator, center, which also sends a strobe pulse to the storage scope, left. A dot written on this screen will persist.

This is much like Robert Shaw's experimental setup for the Birkhoff bagel, in the forced Van der Pol device described in Section 8.2.

8.4.2. Recall, from Part One, these mechanical devices used by Lord Rayleigh to model the violin string (later studied by Duffing) and the clarinet reed (later studied by Van der Pol). Without being forced, the oscillation of one dies away, while the other's is self-sustained. When forced, both will oscillate.

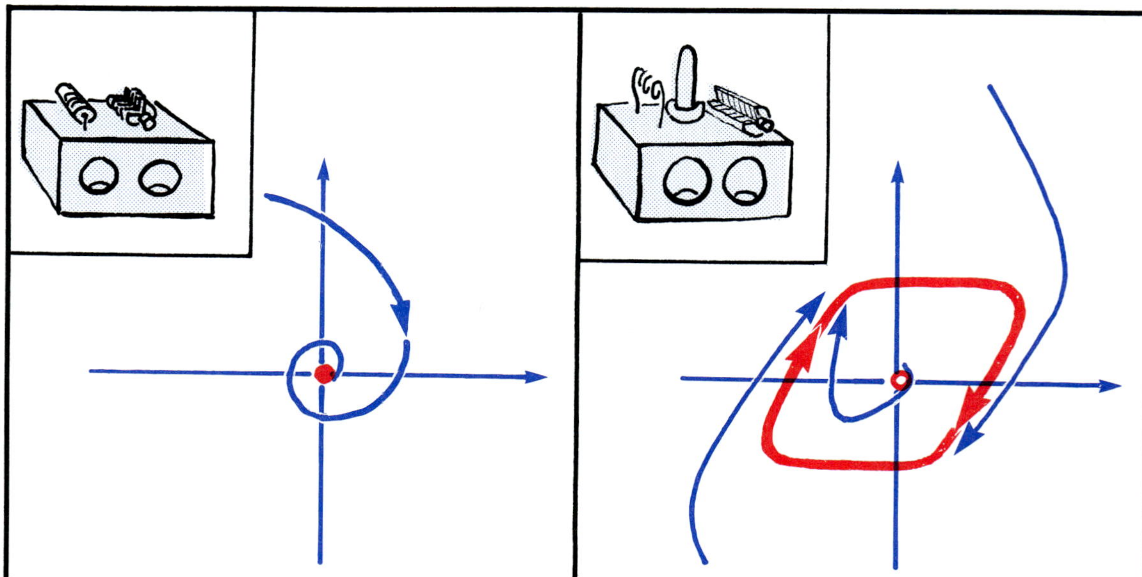

8.4.3. In the context of electronic analog devices, we have these two devices: without forcing, one wants to be still, the other to oscillate, as shown in these phase portraits. When forced, both will oscillate, as described in Part One.

From these two devices, extensively described in Part One, much of dynamics has evolved. We return now to the electronic analog of the Duffing device, starting with a brief review.

With the forcing oscillator set close to the natural frequency of the damped electronic pendulum, there are *two periodic attractors* in the ring model for the three-dimensional state space of the combined system (Figure 4.3.8). Thus, our strobed storage scope will show a dotted curve asymptotically approaching a point. There are two such attractive points in the strobe plane. ***Which attractor*** our strobed (dotted) trajectory ends up at depends on ***which basin*** the initial point is in. (The initial point may be chosen by the experimentalist before turning the device on.) The two basins are separated (in the three-dimensional ring model) by a scrolled surface, the inset of a periodic saddle (Figure 4.3.13). In the strobe plane, they are separated by a teardrop-shaped curve (Figure 4.3.12).

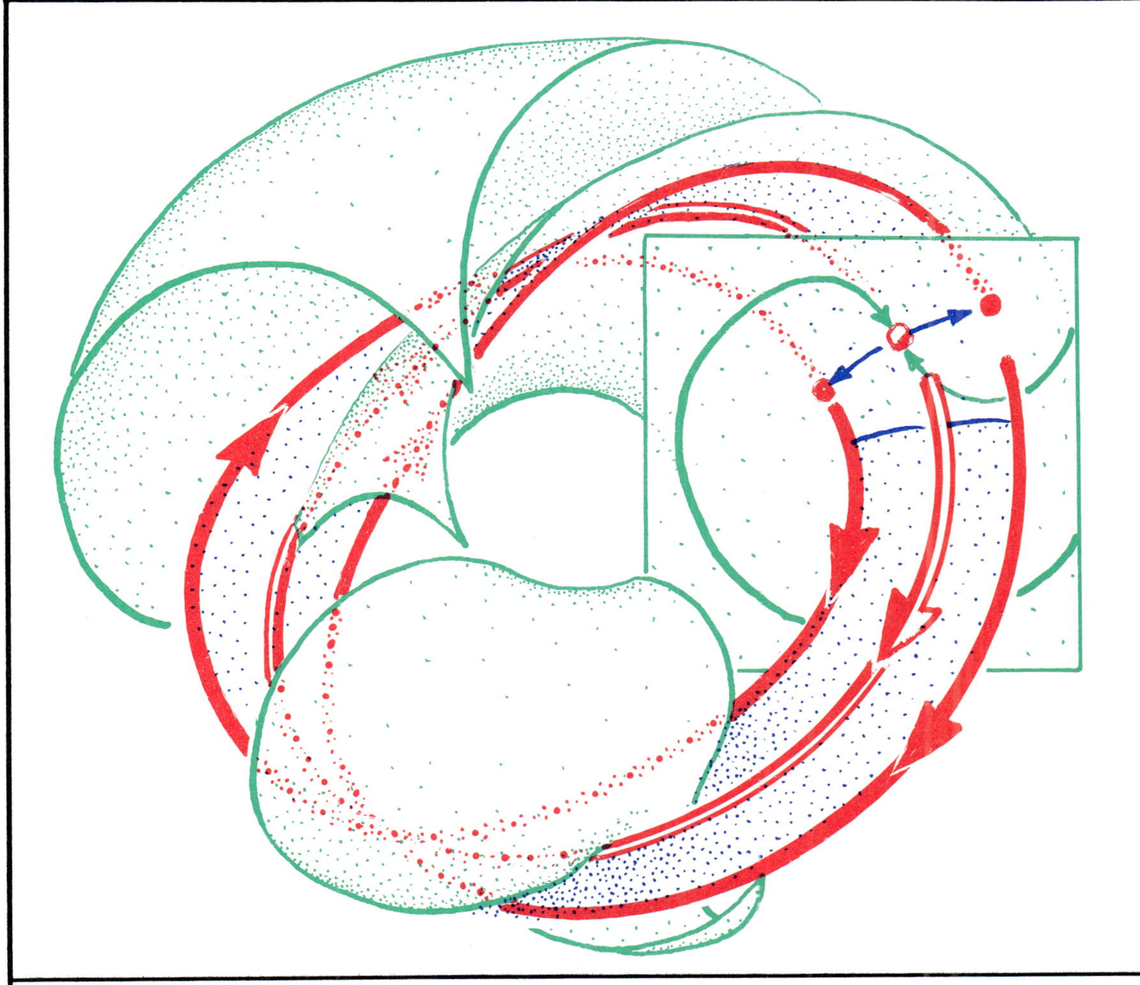

8.4.4. Here is the *separatrix* of the isochronous oscillations of the electronic Duffing device. As the *inset* of a periodic saddle, it is not a limit set, but it is *repelling.* In the strobe plane, it appears as a *teardrop-shaped curve.* Meanwhile, the *outset* of the periodic saddle, although not a limit set either, is *attractive.* It is a *twisted band* (shaded blue in this drawing) bounded by the two periodic attractors. This outset band meets the strobe plane in a short curve segment, shown here in heavy black. Just as the inset (separatrix, red) is repelling, the outset band (black) is attracting.

Now let's watch this attractive outset band, as we increase the frequency of the driving oscillator.

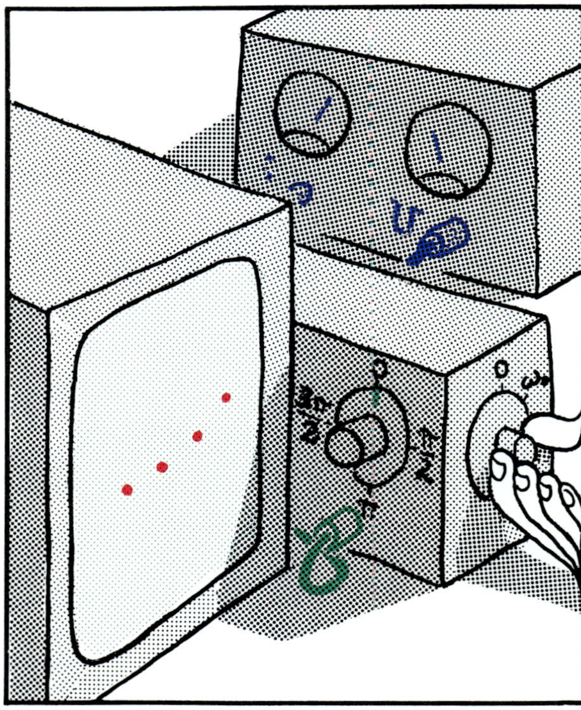

8.4.5. For a while, as the driving frequency increases, we find an increasing number of periodic attractors. These are the ***harmonics,*** described in some detail in Chapter 4 (for example, Figure 4.4.17). In the strobe plane observed by the storage scope, they appear as a growing set of isolated points.

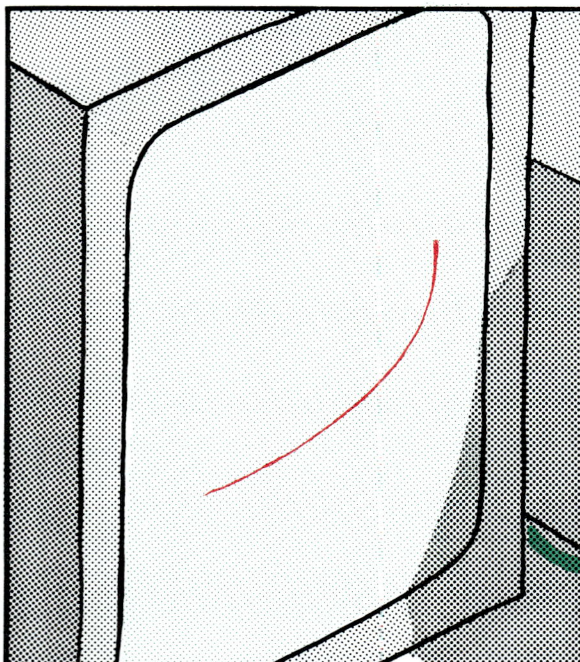

8.4.6. But after we increase the forcing frequency sufficiently, this entire arc appears as an attractor. It is the strobed view of the attractive outset band of the periodic saddle. The attracted trajectory hits it repeatedly, randomly walking throughout its length.

8.4.7. If we leave the *frequency* fixed but change the *phase* of the strobe pulse, we find the arc rotates, flexes, stretches, and folds.

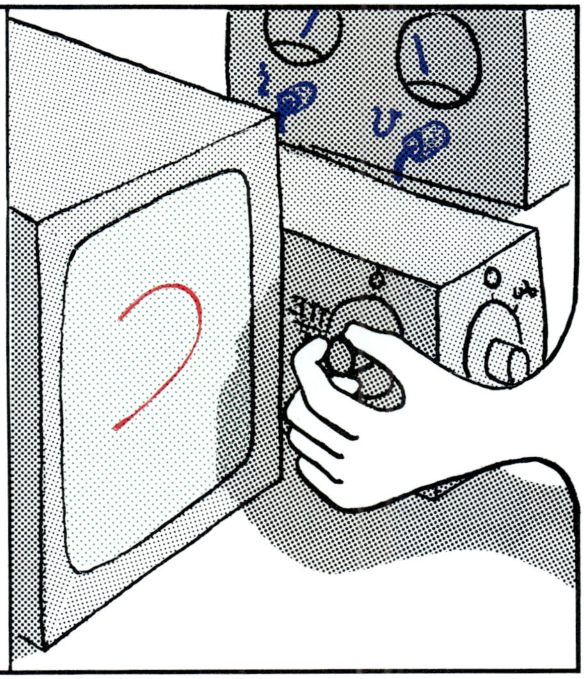

8.4.8. Putting these strobe views together around the ring model, we obtain a three-dimensional picture of this unusual attractor, first observed by Rössler in another dynamical system.

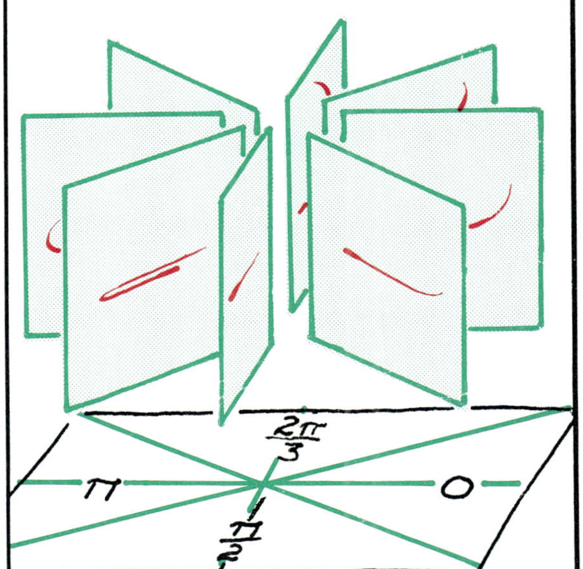

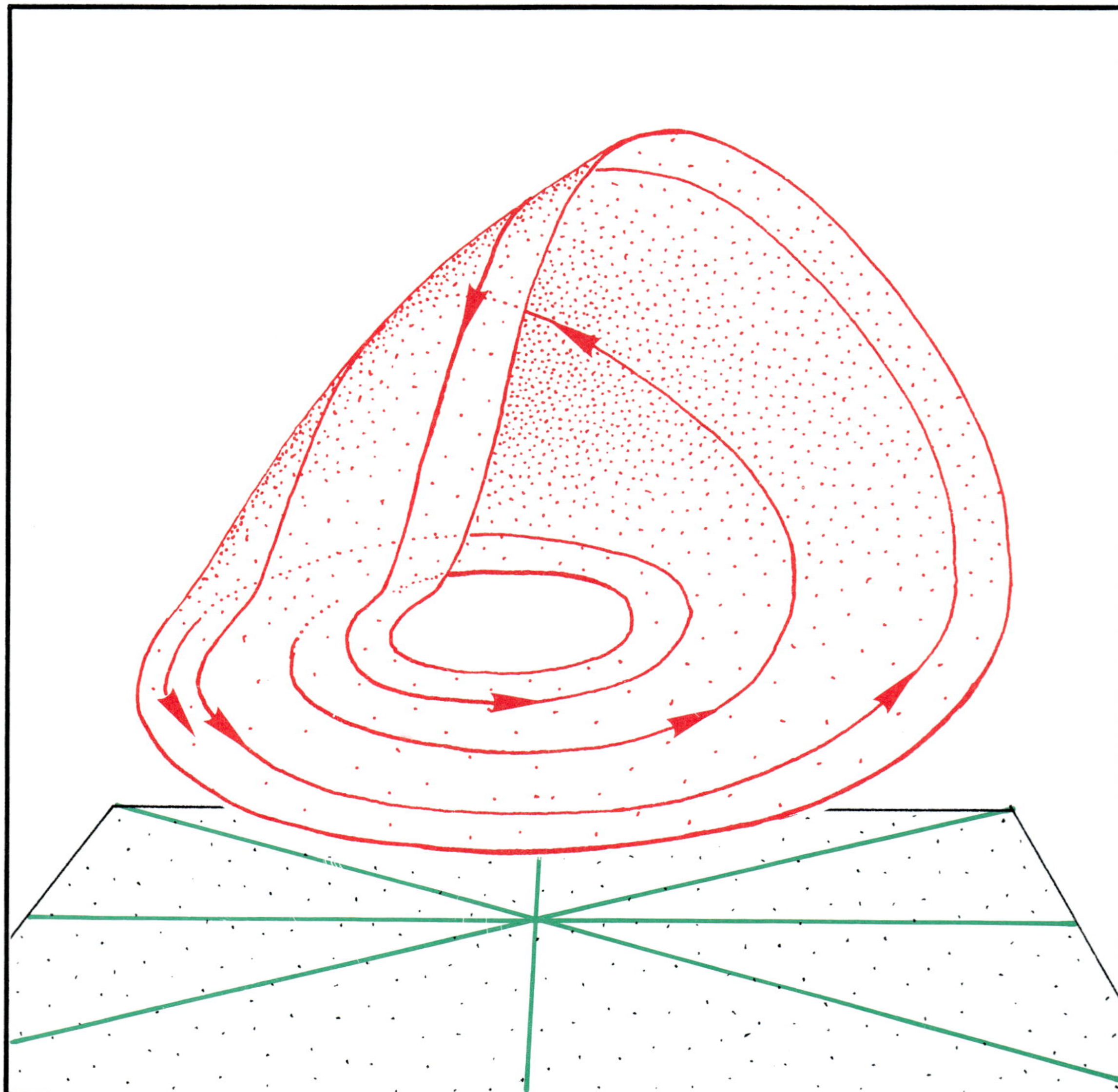

8.4.9. Here is the full Rössler band, in the Duffing system. Like the Birkhoff bagel and the Lorenz mask, it appears to be a slightly thickened surface. A more detailed picture of its rich microstructure is presented in the next chapter. Keep in mind, meanwhile, that these attractors are not static objects. They determine the dynamic behavior of an attracted trajectory.

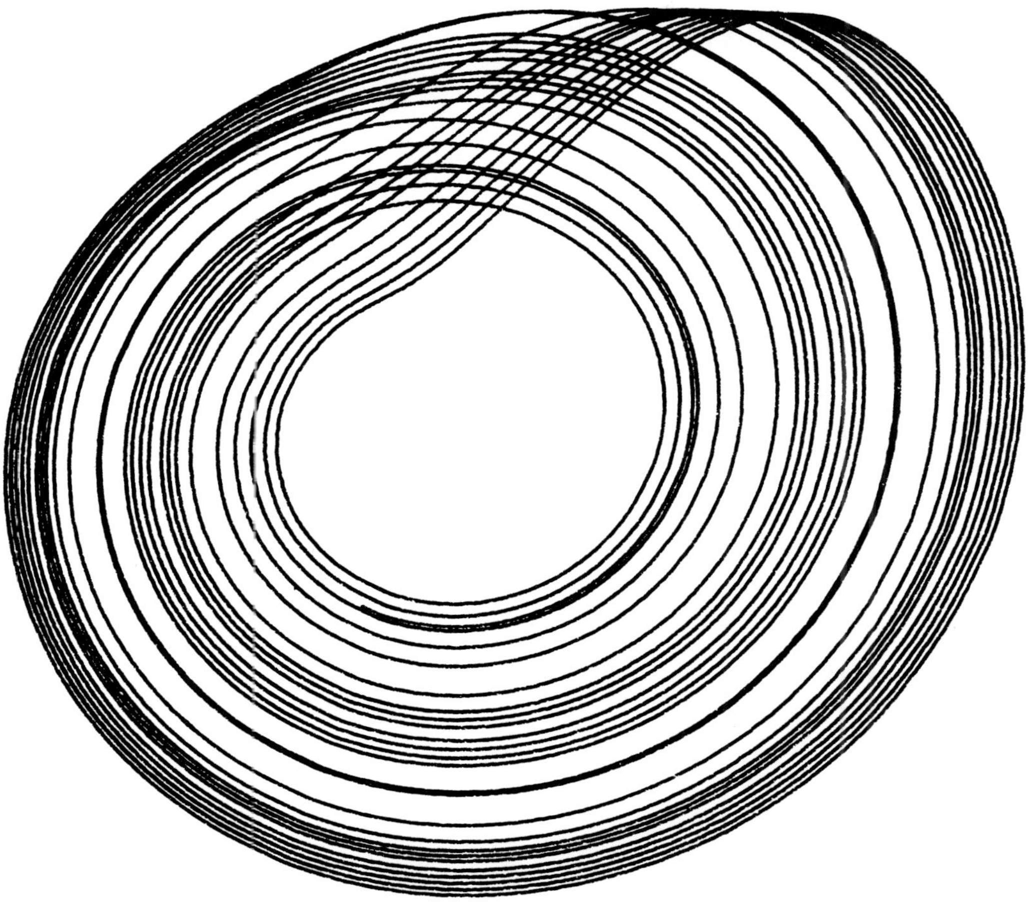

8.4.10. Here is the full Rössler band, as seen by Rössler in his original system. (See the Appendix for the equations; this computer drawing courtesy of Robert Shaw.) This view is straight down from the top, so the height of the fold (shown in the preceding panel) is foreshortened. The trajectories do not actually cross in 3D.

There are many other unusual attractors waiting to be discovered by pioneers with analog and digital computers. This is one of the last frontiers of the *local theory* of dynamical systems.

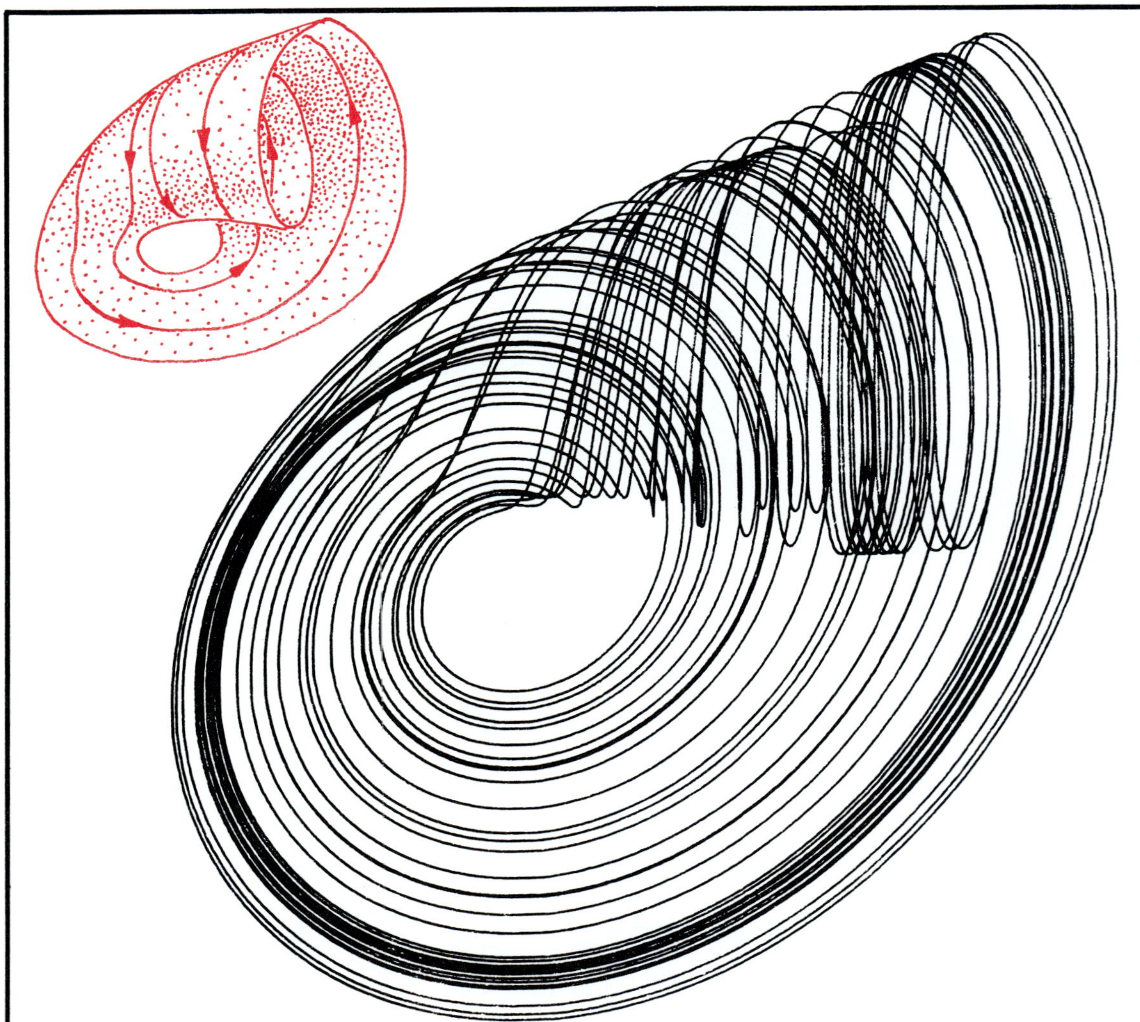

8.4.11. Here is a related experimental object, also discovered by Rössler, which we call the *funnel.* (See the Appendix for the equations; this computer drawing courtesy of Robert Shaw.)

The trajectories attracted to these attractors asymptotically approach almost every location in the thickened surface. Of course, the motion of such a trajectory is precisely determined by the mathematical model. Yet because the smallest uncertainty in the determination of the actual position of the trajectory at a given moment implies an enormous uncertainty later on, the future of the trajectory along the attractor is *apparently chaotic.* Hence, these experimental objects are called *chaotic attractors.* They may not be either attractors or chaotic in any rigorous mathematical sense, however. All this will be clarified in the next chapter.

9

Attributes of Chaos

There are many reasons for calling the unusual limit sets, found experimentally and described in the preceding chapter, *chaotic*. In this chapter we will describe some of these reasons, which have received considerable attention in the literature.

9.1. Unpredictability

Although the large-scale attractors are aspects of a dynamical system that are fully deterministic in the formal mathematical sense, the behavior of a trajectory attracted to such an attractor is totally unpredictable in the long run. This is the reason these extended limit sets have earned the name *chaotic attractors*. In this section we explain this paradox, originally emphasized by Lorenz (1962, 1963).

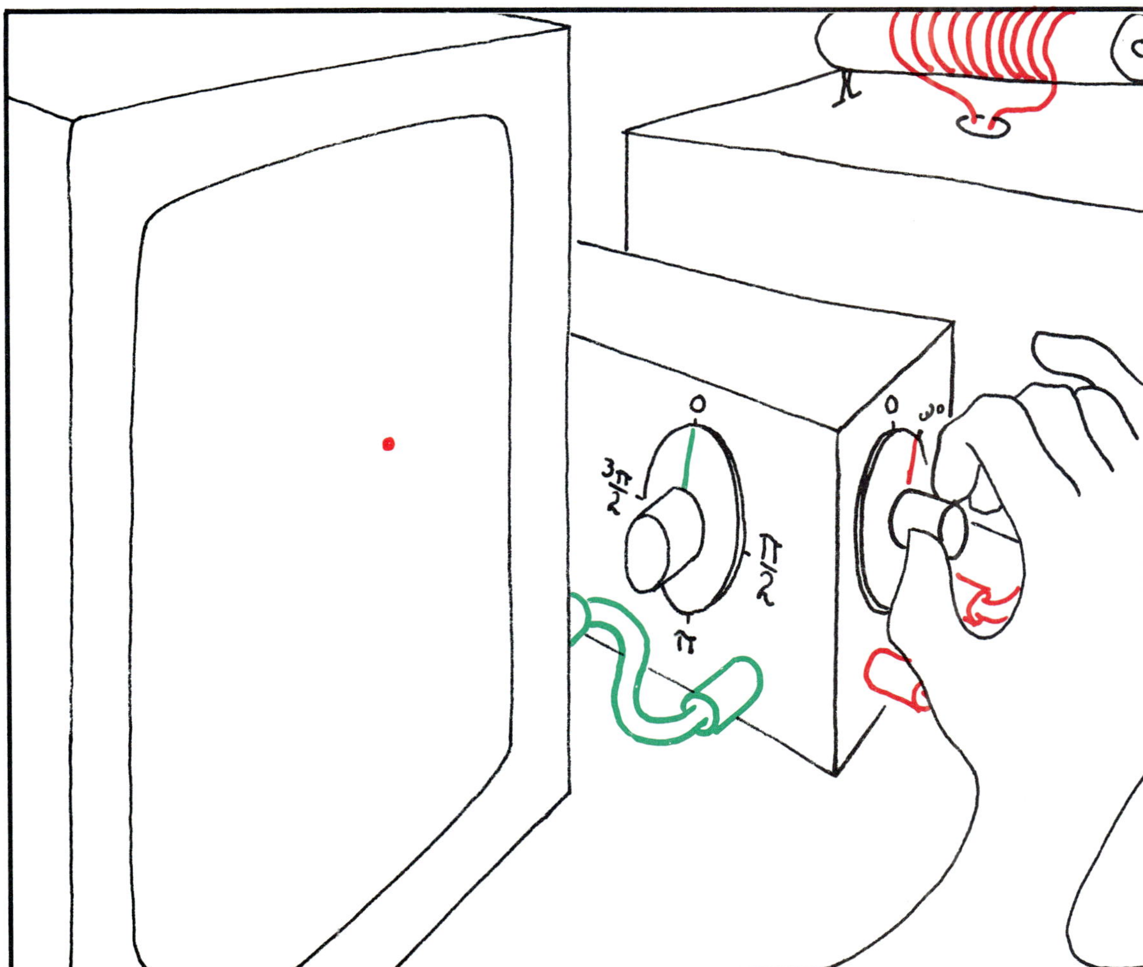

9.1.1. In the preceding section, we observed this situation, in the electronic Duffing device, before the appearance of Rössler's band. The attracted trajectory returns repeatedly, at exact intervals, to the same point in the strobe plane. This is the epitome of *predictable behavior.*

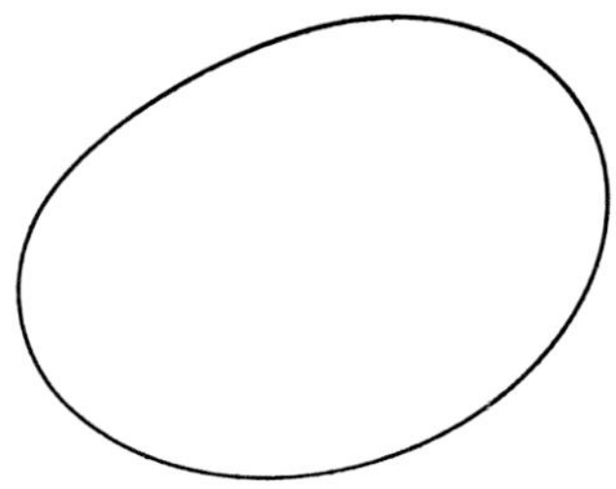

9.1.2. Here is a computer drawing of the periodic attractor of the Duffing system, seen from above. Not only the point of periodic passage through the strobe plane, but all of the states on this periodic attractor, are exactly predictable.

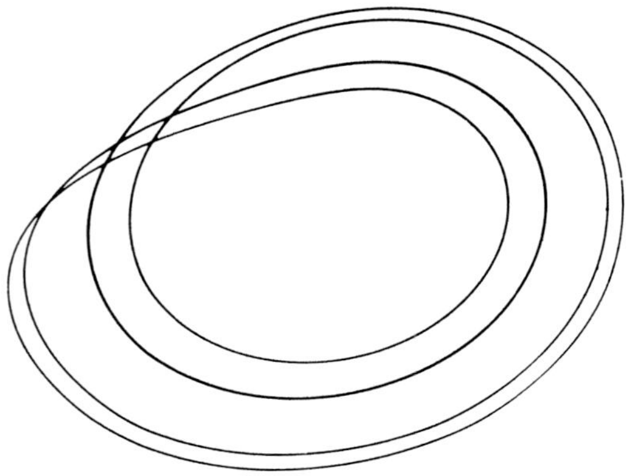

9.1.3. Increasing the forcing frequency, we observe higher harmonics as periodic attractors. Here, in another computer drawing also seen from above, is a periodic trajectory cycling four times around the ring before closing. Its intersection with the strobe plane is a set of four dots, each visited in turn by the orbiting trajectory.

This motion is completely predictable. After observing a few cycles, we can predict, based on one strike of the strobe plane, exactly where the next few will strike.

9.1.4. Increasing the forcing frequency further, we obtain an extended attractor. Also seen from above in this computer drawing, there are several bands within a figure like Rössler's band. The intersection with the strobe plane is a set of thick arcs. (The thickness is explained in Section 4.4.) Although the orbiting trajectory passes through these thick arcs periodically, we are unable to make long-range predictions based on one strike.

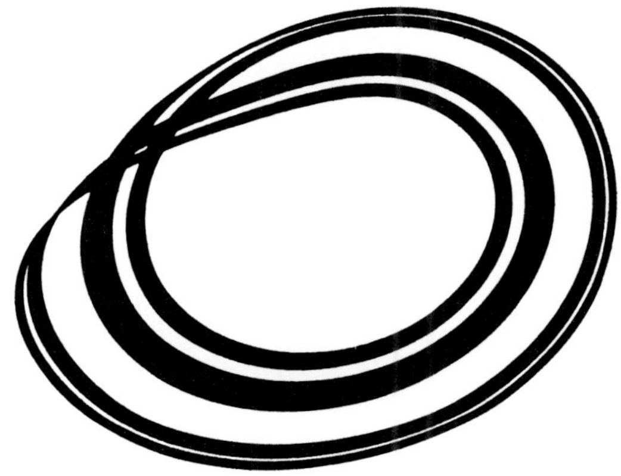

Although the motion is fully deterministic in the mathematical sense, it is rather unpredictable in the experimental sense. This is one reason this motion is called *chaotic.*

The successive passages of the trajectory, orbiting the Rössler-like band, are experimentally rather unpredictable, because (1) we do not know the exact position of the trajectory at a given moment, (2) a small difference in this current position leads to an enormous difference in position later on, and (3) the trajectory will eventually come arbitrarily close to any point on the thick bands.

Property 2, characteristic of all chaotic sets, is called *sensitive dependence on initial conditions.* Property 3, also characteristic of chaotic sets, is called *topological transitivity.* Related to the *ergodic hypothesis* of statistical physics, this just means that a single trajectory pierces every little region in the limit set. These conditions are described further in the next section.

We now explain why these three conditions lead to unpredictability.

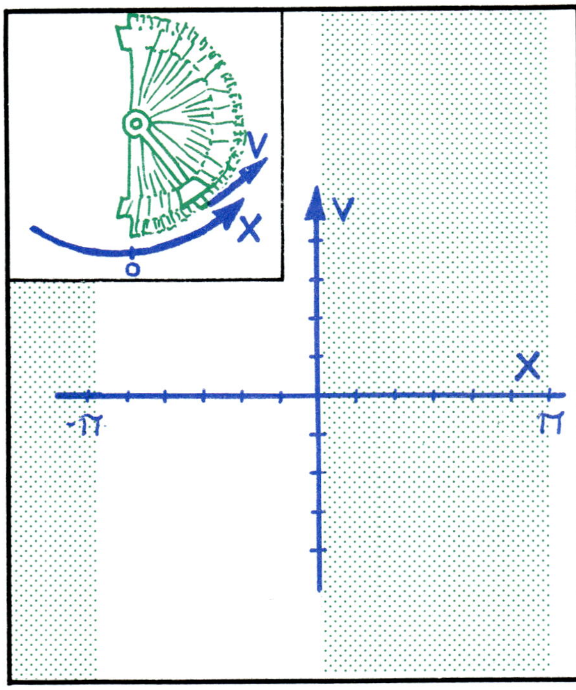

9.1.5. The *precision of our measurement* of the pendulum's state (displacement and velocity), at a given time, may be regarded as an *amount of information*. Here, for example, suppose we know only that the state is in the green shaded area of the state space. Our information, then, is only that the pendulum is to the right of bottom. We know nothing of the speed of its motion, not even its direction. This is a *small amount of information* about the instantaneous state of the system.

The more precise the measurement, the smaller is the region known to be occupied by the state of the system, and the more information we have about it.

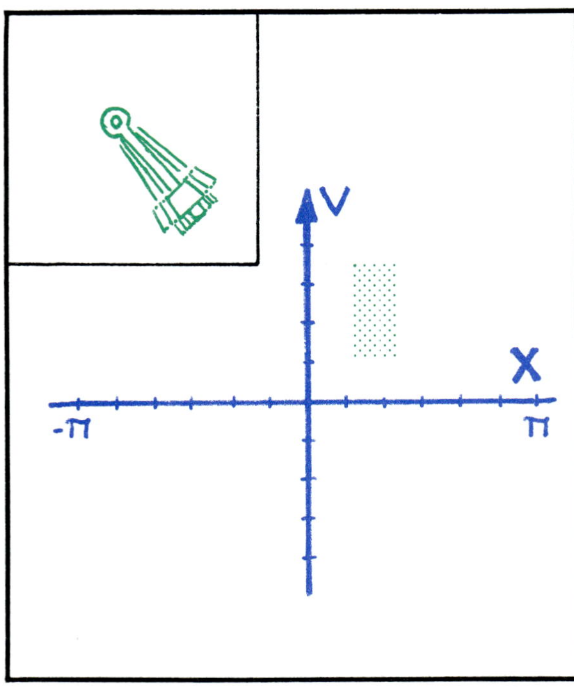

9.1.6. In this case, our measurement is more precise. The state is known to be in the small shaded box. Thus the displacement and velocity are known, within small intervals of error. We have a *considerable amount of information* about the state of the pendulum.

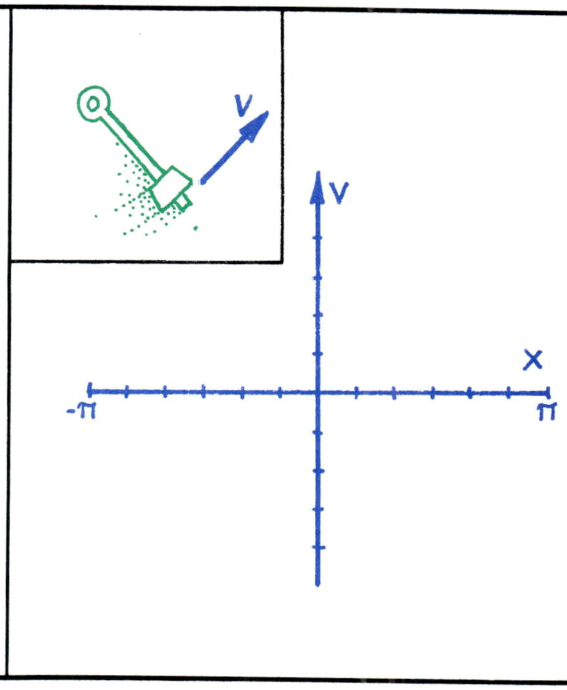

9.1.7. Carried to extremes, if our measurement were infinitely accurate, the region of uncertainty would be a point, and our information would be infinite. This is the assumption in the *mathematical context*.

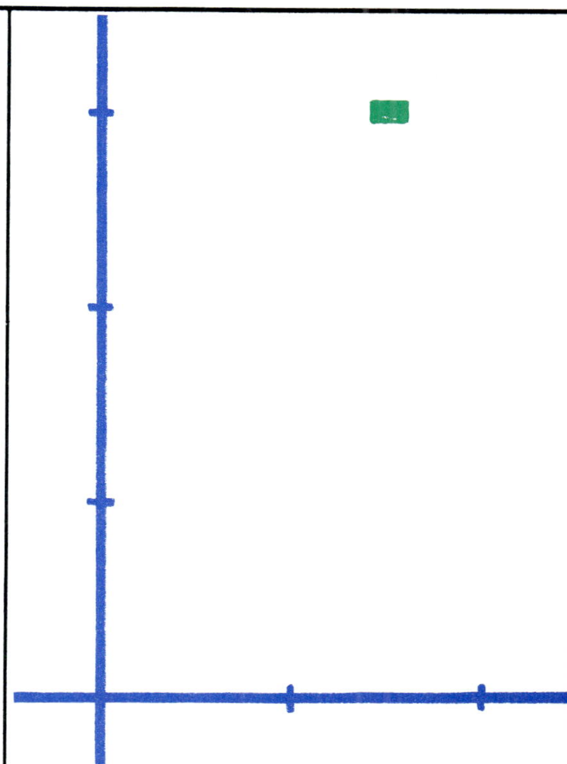

9.1.8. But in the *experimental context*, there is always a *region of uncertainty*, due to the realities of observational instruments, recording media, and the Uncertainty Principle. Here, we see this region by enlarging the preceding drawing.

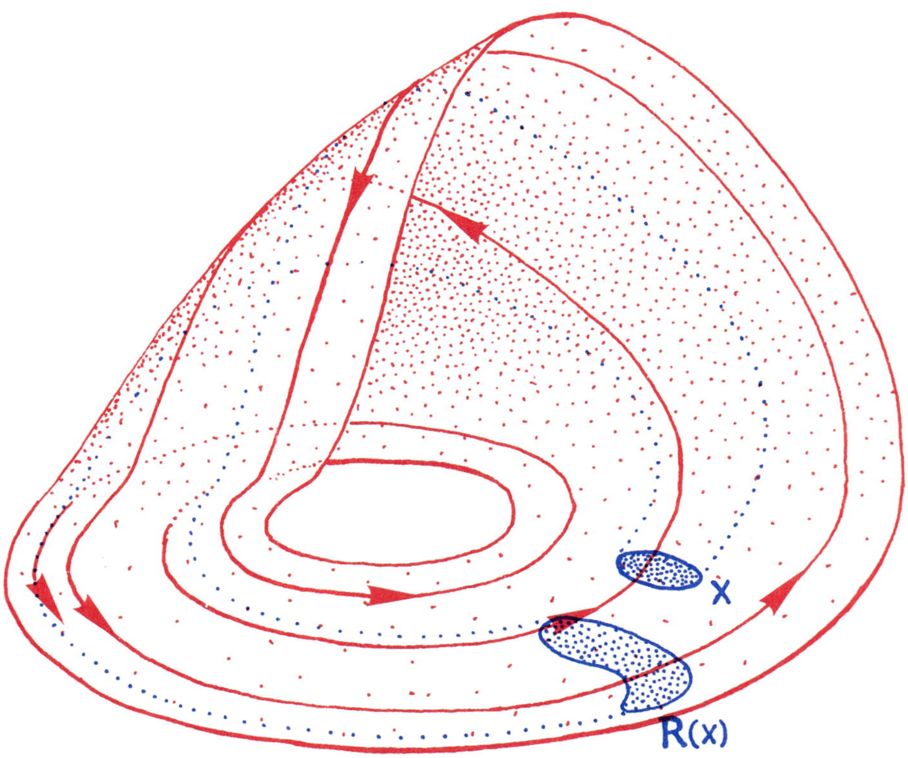

9.1.9. Returning to Rössler's band, let us begin with a region of uncertainty, shown in green here and labeled x, within the strobe plane. After approximately one cycle, this track passes through the strobe plane again in the set shown here in green, labeled R(x). But the track *expands*. So the original region, x, has expanded to the larger set R(x). After this expansion, *we have less information* about the state of the system.

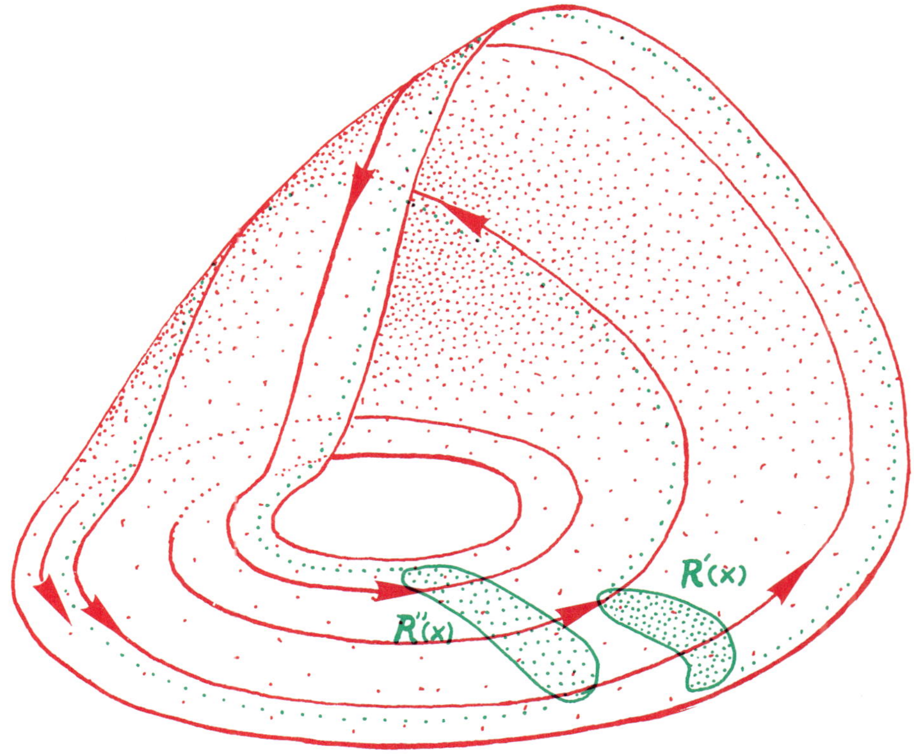

9.1.10. We relabel R(x) by R′(x). The track continues to expand. And after another similar period of time, it has passed through the strobe plane again, in the green set R″(x). The original region x, its first return R′(x), and its second return R″(x), comprise the beginning of an infinite sequence of sets. These are getting wider and wider. And so *we have less and less information* about the state of the system as time goes on.

In fact, no matter how small the initial region, the successive strikes of its trajectories through the strobe plane will spread in this way. The cause of this spreading is explained in the next section.

So this, at last, is the meaning of *unpredictability* in the context of chaotic attractors. *Any small error* in the measurement of the current state (inevitable) eventually *leads to total ignorance* of the position of the trajectory within the chaotic attractor.

9.2. Divergence and Information Gain

The basic dynamical feature of chaotic attractors is *bounded expansion*, or *divergence and folding* together of trajectories within a bounded space. This feature implies *sensitive dependence on initial condition*, originally emphasized by David Ruelle, and *gain of experimental information*, introduced by Robert Shaw. In this section we explain these implications. The connection between divergence and characteristic exponents (CE's) is given in the following section.

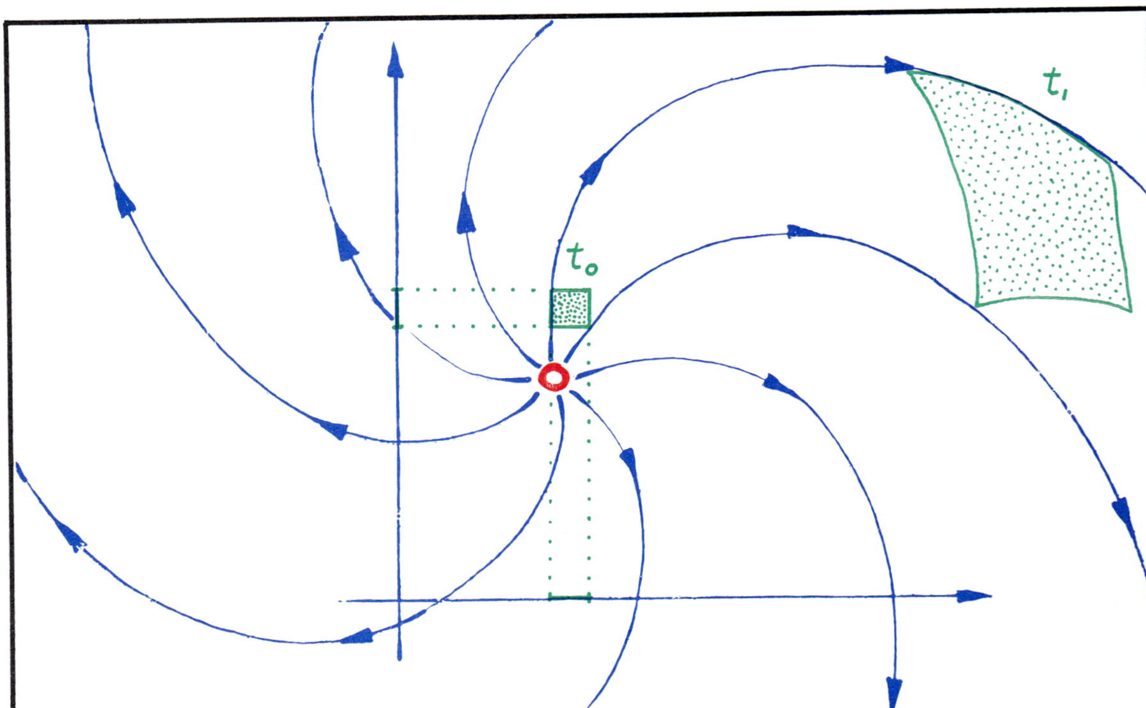

9.2.1. Suppose we have a dynamical system with a planar state space, and a repelling point. After a careful measurement of both variables at time t_0, we have determined that the state of the system is in a *small region near the point repellor.* The track of this region along trajectories of the dynamical system spirals away from the repellor, *expanding as it goes.* Thus, at a later time, t_1, the track of the initial region is a larger region, as shown here. Supposing that *we do not remeasure* the state of the system at this later time. Then the information we originally had, together with our knowledge of the phase portrait of this system, add up to the knowledge that the system is in a state in the larger region, nothing more. But the larger uncertainty means we have *less information at the later time* than we had at the time of the initial measurement. The *divergence of the trajectories* leaving a repellor *implies loss of the information* obtained from an initial measurement.

On the other hand, the divergence of trajectories leaving a repellor, together with new measurements at a later time, implies a *gain of information*. Now let's carefully follow this divergence, and its attendant information gain, in four steps:

- measure the initial state of the system,
- watch this region expand, while initial information is lost,
- remeasure the state of the system, increasing current information,
- extrapolate backwards, gaining information on the initial state.

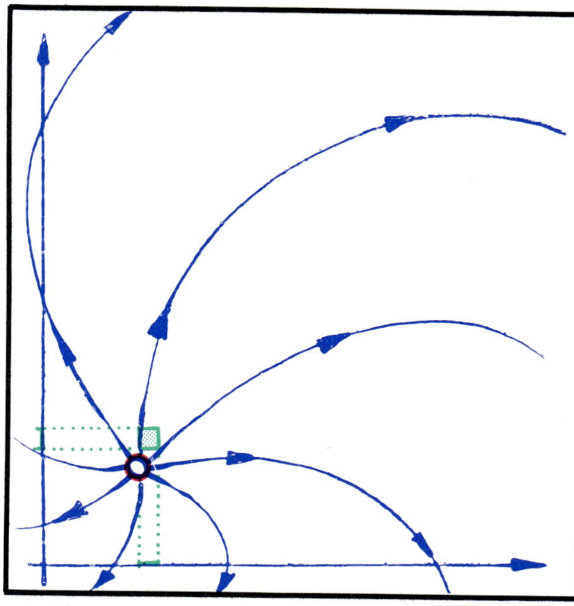

9.2.2. At the initial time, we *measure the state of the system as accurately as possible*, within the limitations of the given laboratory instruments. The position of the initial point in the state space is thus a small region, such as this rectangle. Its sides are the experimental errors in the values of the separate coordinates, such as position and velocity of the pendulum.

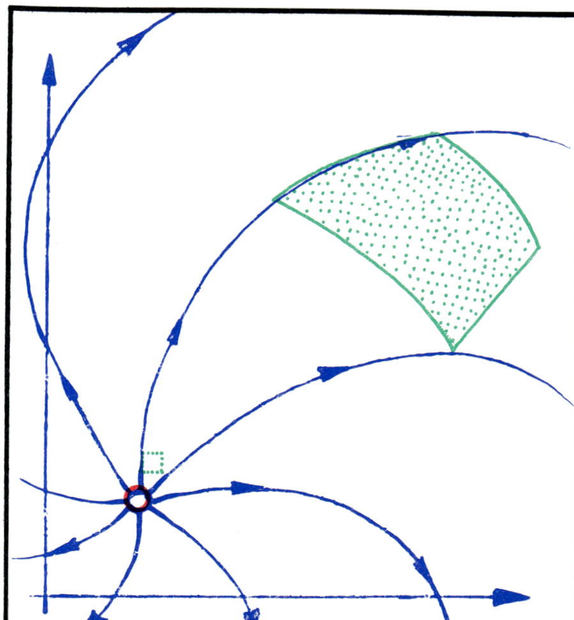

9.2.3. After an interval of time, the initial region flows outward from the repelling point, along the diverging trajectories of the dynamical system, as we see here. The initial region grows into this *larger region*. Information about the current state seems to be decreasing, as we know less and less about the actual state of the system as time goes on.

9.2.4. But at this later time, we *remeasure* the state of the system, again with the best accuracy of the laboratory apparatus. Now we have increased our information about the system at the later time.

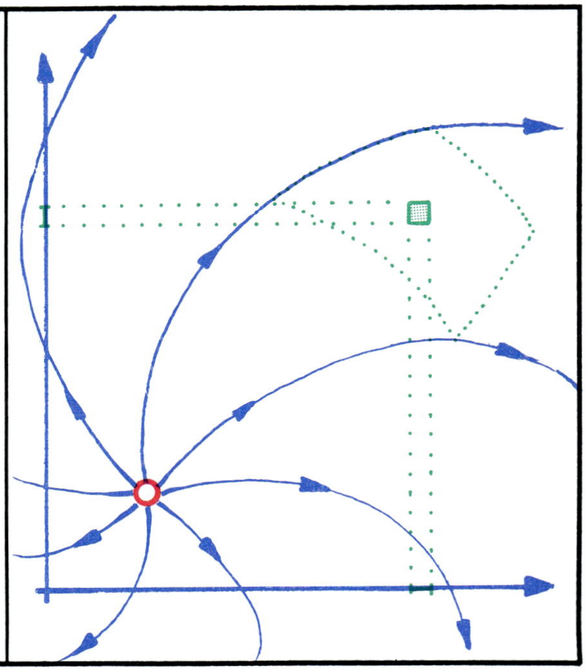

9.2.5. Finally, we flow the larger region *backwards* in time to its original position, as a small rectangle, and along with it we flow the small rectangle it contains. This becomes the immeasurably small dot within the initial rectangle, as shown here. As this is smaller than the initial rectangle, we have *more information* about the initial state, because of the second set of measurements, than we could possibly have known at the beginning.

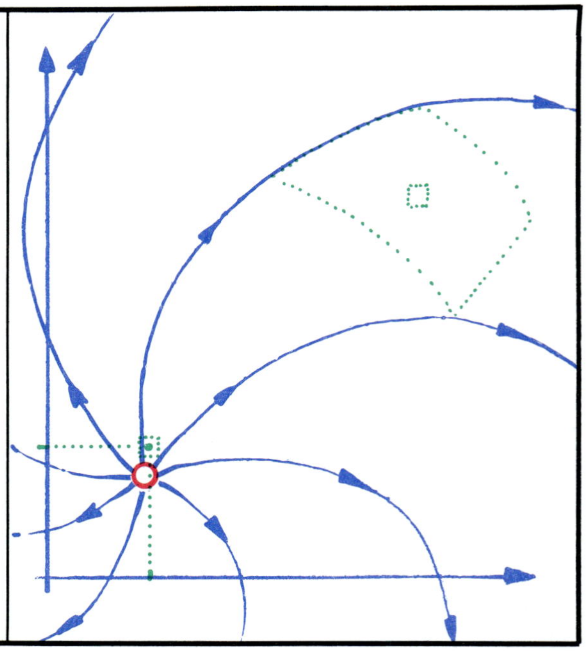

By this sequence—measure, flow, remeasure, reflow backwards—we have obtained *more information* about the state of the system at the initial time than we did with the initial measurement (with the same instruments).

To emphasize the actual gain of information provided by diverging flow, let's repeat this sequence with the state space coarse-grained by cells indicating the limit of precision of the measuring and recording apparatus.

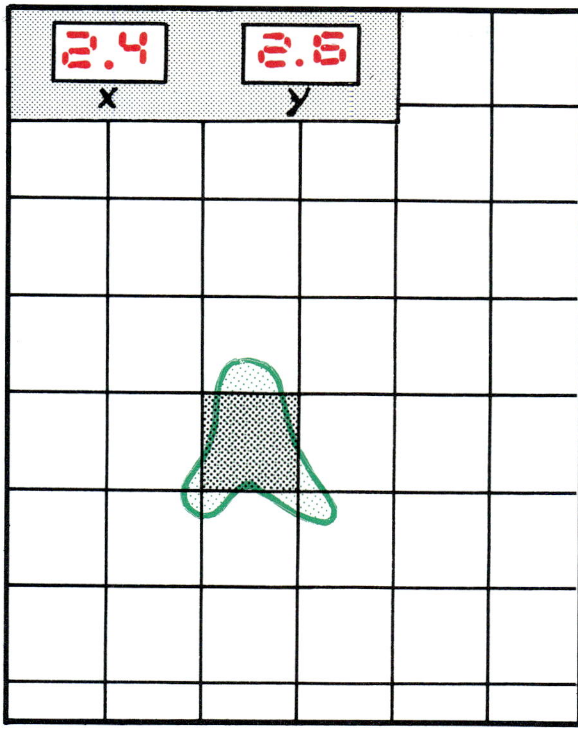

9.2.6. In this armchair experiment, we suppose that the measuring apparatus is digital, with data bins 0.1 unit wide. Thus, all values of x between 2.35 and 2.45 are thrown into the 2.4 bin by the measuring and recording process, and likewise for values of y Here we see the state space, coarse-grained by these bins. The green shading represents a region known mathematically to be occupied by the initial state of the dynamical system. But the black shaded bin represents the region known experimentally, according to the most accurate measurement possible, with the assumed instruments, at the initial time. The system, with almost 100 percent certainty, occupies the cell labeled (2.4,2.6) by the measuring instruments, as shown in the corner box.

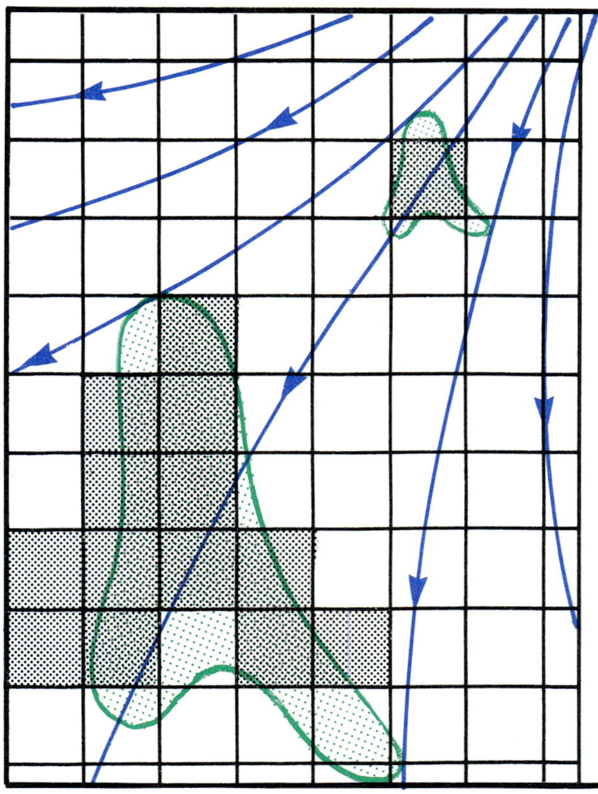

9.2.7. After a period of diverging flow, the initial green cell has expanded so as to cover many experimentally distinguishable cells. Without a new measurement, we have *less information now about the current state* than we initially had about the initial state. We do not know which of the thirteen black shaded bins the system now occupies. However, if we now make a new measurement and extrapolate backwards in time, we will have *more information now about the initial state* than we did originally. Acknowledging our capacity to make measurements any time we wish, we see that *diverging flow provides increasing information about initial states. Information is gained* in diverging flows.

9.2.8. Conversely, near a point attractor, the flow is convergent. An initial region shrinks smaller and smaller. Here, initial points, known through measurement to be distinct, eventually become experimentally indistinguishable. Extrapolation backwards of current measurements may tell us nothing about the initial state of the system. Thus, *convergent flow provides decreasing information about initial states. Information is lost* in converging flows.

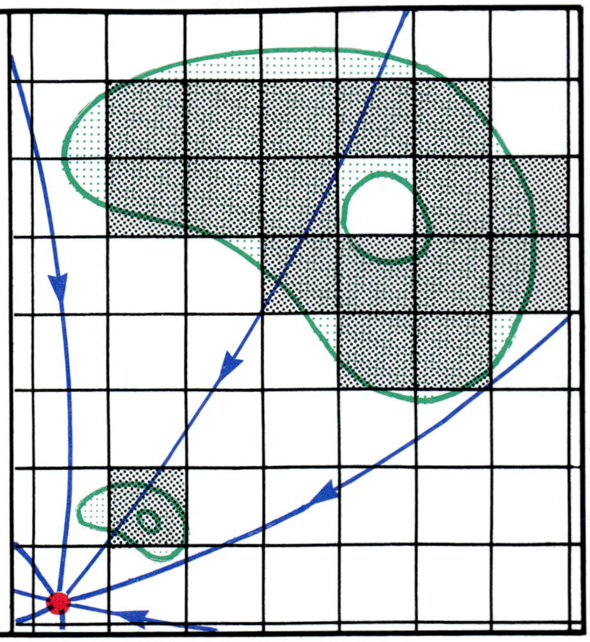

We have illustrated the concepts of *information gain and loss,* or *flow,* in flows near repelling and attracting points. Similar concepts apply to repelling and attracting limit cycles, and, as we shall see in the next section, to chaotic attractors. We turn now to the related concept of *sensitive dependence on initial conditions.*

9.2.9. Suppose we have two adjacent initial conditions, near a repelling point. Because of the divergence of trajectories in this flow, the trajectories from these two initial points diverge. After some time, the two initial states have flowed to final states that are far apart. Thus, in this context, a *small change in initial state results in a large change in final state,* after a short time of evolution of the system. This, called *sensitive dependence on initial conditions,* is a characteristic feature of chaotic attractors. These, like the repelling point illustrated here, have diverging trajectories, as we shall see in the next section.

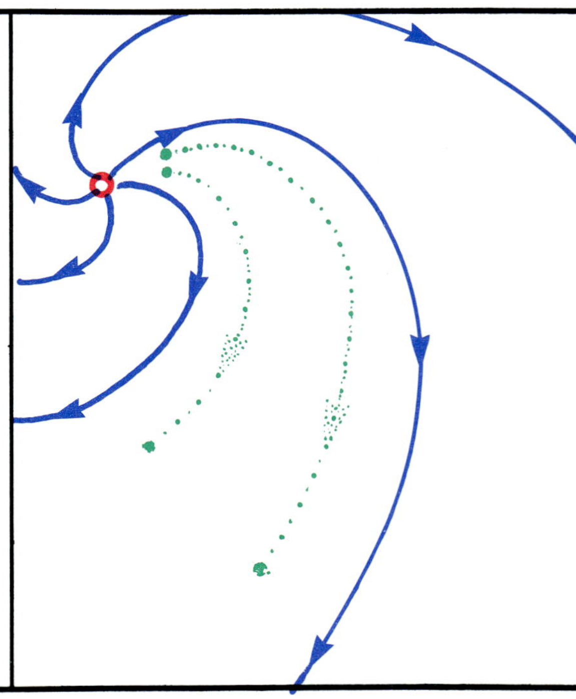

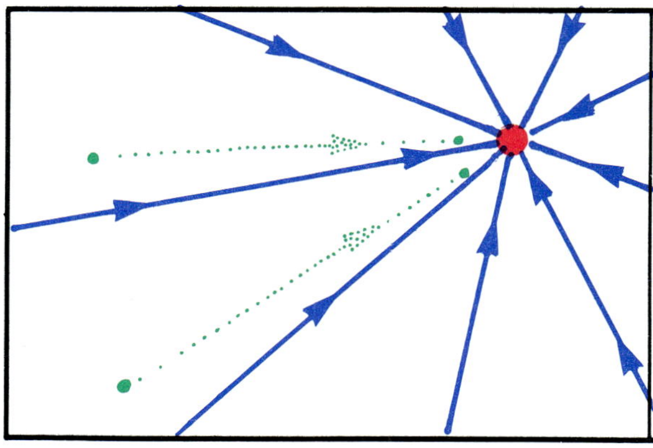

9.2.10. Conversely, in the converging flow near an attracting point, a large change in initial state results in a smaller change in final state, after a short time. We may call this ***insensitive dependence*** on initial conditions.

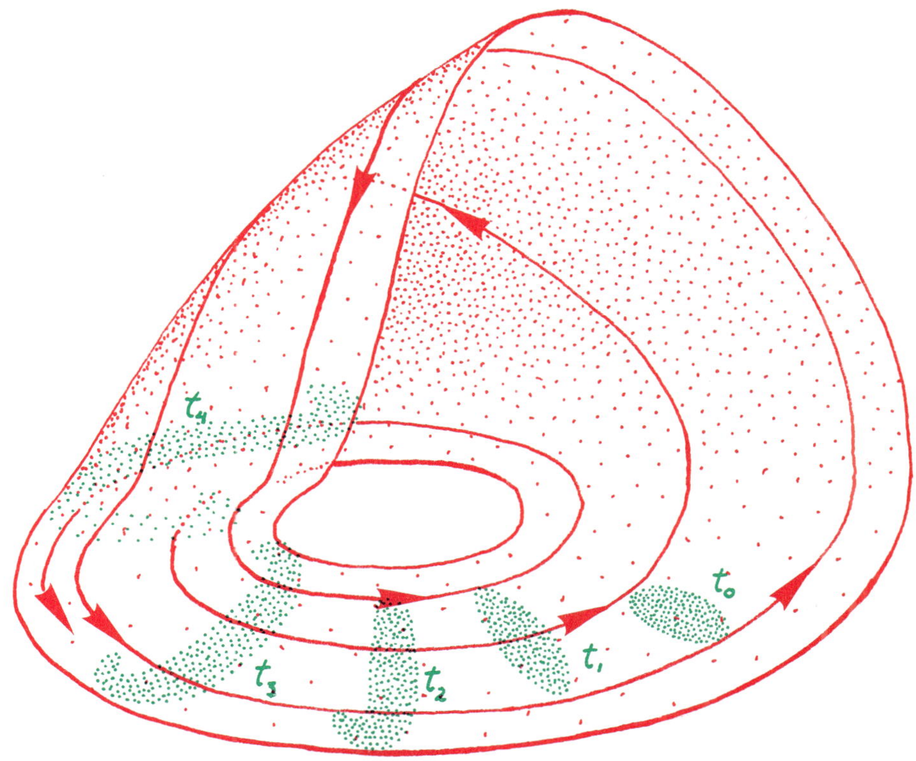

9.2.11. Finally, suppose that a set, such as the Rössler band pictured here, is almost completely filled by a single solenoidal trajectory. Then a small region of initial states at the initial time, t_0, moves along to a more extended region at a later time, t_1. After a long while, this set must be pulled and twisted so as to cover most of the entire band. For the solenoidal trajectory pierces the initial region infinitely often, and these initial points flow along the solenoid and almost fill the band. This situation, called ***topological transitivity***, is not implied by diverging trajectories alone, but is nevertheless an observed feature of all experimentally known attractors. This may be due to the constraints of reality, causing experimental dynamics to study a single trajectory.

9.3. Expansion, Compression, and Characteristic Exponents

Divergence (or equivalently, information gain, or sensitive dependence) plus transitivity imply unpredictability, as described in Section 9.1. In this section, we show how divergence and convergence, or expansion and compression, occur simultaneously near chaotic attractors.

How can compression to the attractor, and expansion along the attractor, happen simultaneously? Here we explain the geometry of this paradox, in the case of Rössler's band.

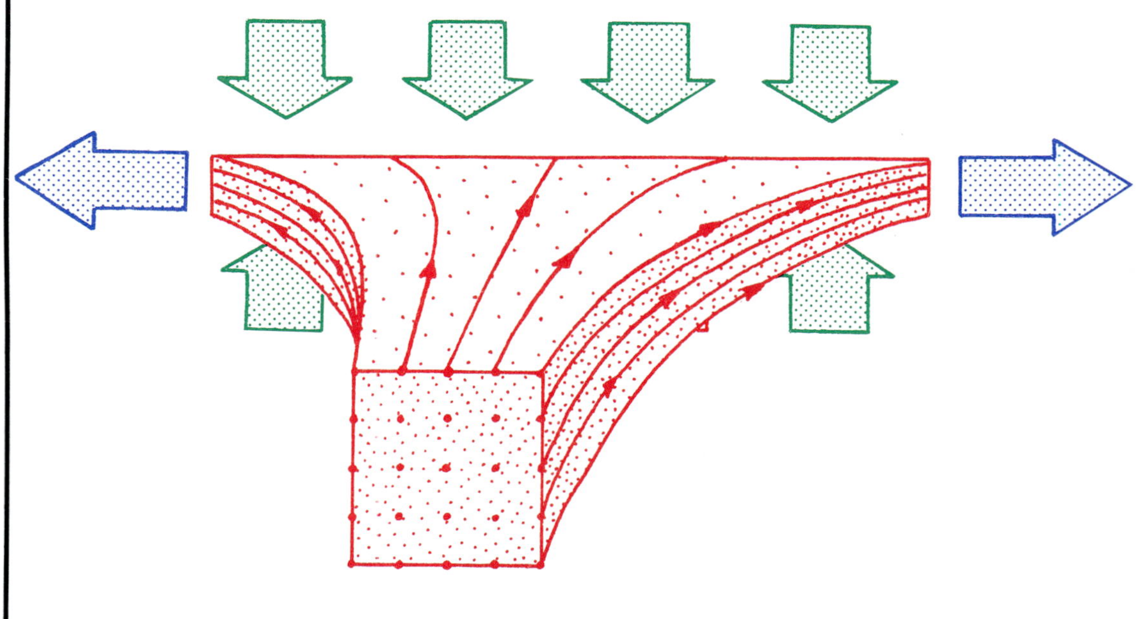

9.3.1. Consider a flow in three dimensions, like the flow near a periodic saddle, which is attractive in one direction and repellent in another. For *periodic saddles*, this is described by two characteristic multipliers: one CM inside the unit circle, the other outside. (See Section 7.3.) Alternatively, we describe the saddle behavior with two characteristic exponents: one CE to the left of the imaginary axis, and one to the right. (See Section 7.6.) For our present purpose, we prefer CE's. But these CE's may be calculated for trajectories that are not closed, as long as they are *recurrent* like the transitive solenoidal trajectory in the Rössler band, illustrated in Section 8.4.

In this flow, horizontal information is gained, while vertical information is lost. The negative CE describes the rate of vertical convergence, while the positive CE describes the rate of horizontal divergence.

All the chaotic attractors illustrated in the preceding chapter are *thick surfaces* in three-dimensional space. (The source of this thickness, an infinite number of surfaces compacted close together, is explained in the next section.) A transitive trajectory in each is saddle-like, as in the preceding panel. But the *attracting direction is perpendicular* to the thick surfaces, which are therefore attractive. Meanwhile, the *repelling direction is tangent* to the thick surfaces, so the flows along these attractors are *diverging*.

At the same time, trajectories *converge* toward the attractor and *diverge* along it, saddle-like. Here is the way the Rössler attractor manages endless divergence within a bounded region, in five steps.

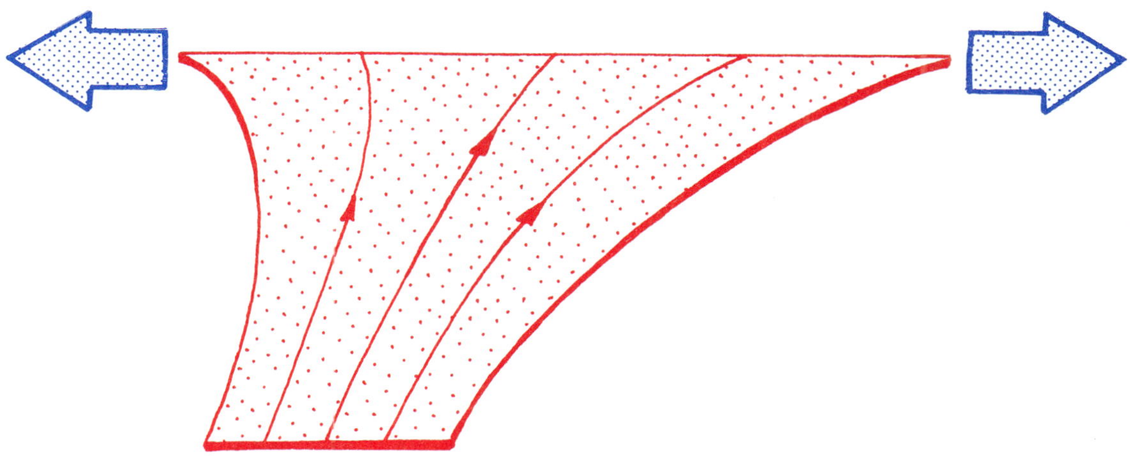

9.3.2. Here we have a horizontal surface, which is a piece of the attractor. The flow is away from us, diverging horizontally for a while, making information, and exhibiting sensitive dependence.

As the flow progresses, this piece of surface grows at the far end. We will track this growth.

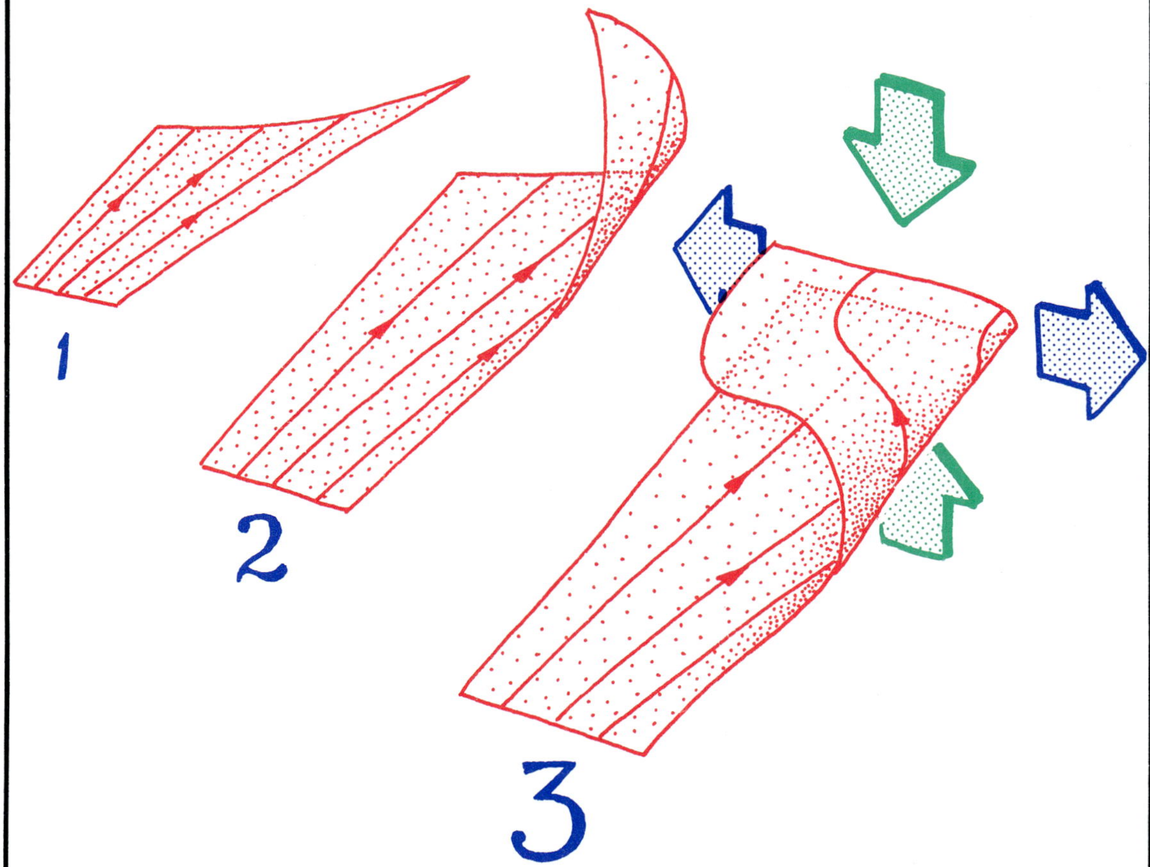

9.3.3. In step 1, the back extends away from us, and widens as it goes because of divergence. The right wing is rising. Step 2, the back extends and widens further, but the widening is all to the right and upwards. The growing right wing is folded up and to the left. Step 3, the back extends further, widens more to the right, and the right wing is folded further upwards and to the left, until it is parallel to the unchanged orientation of the left wing.

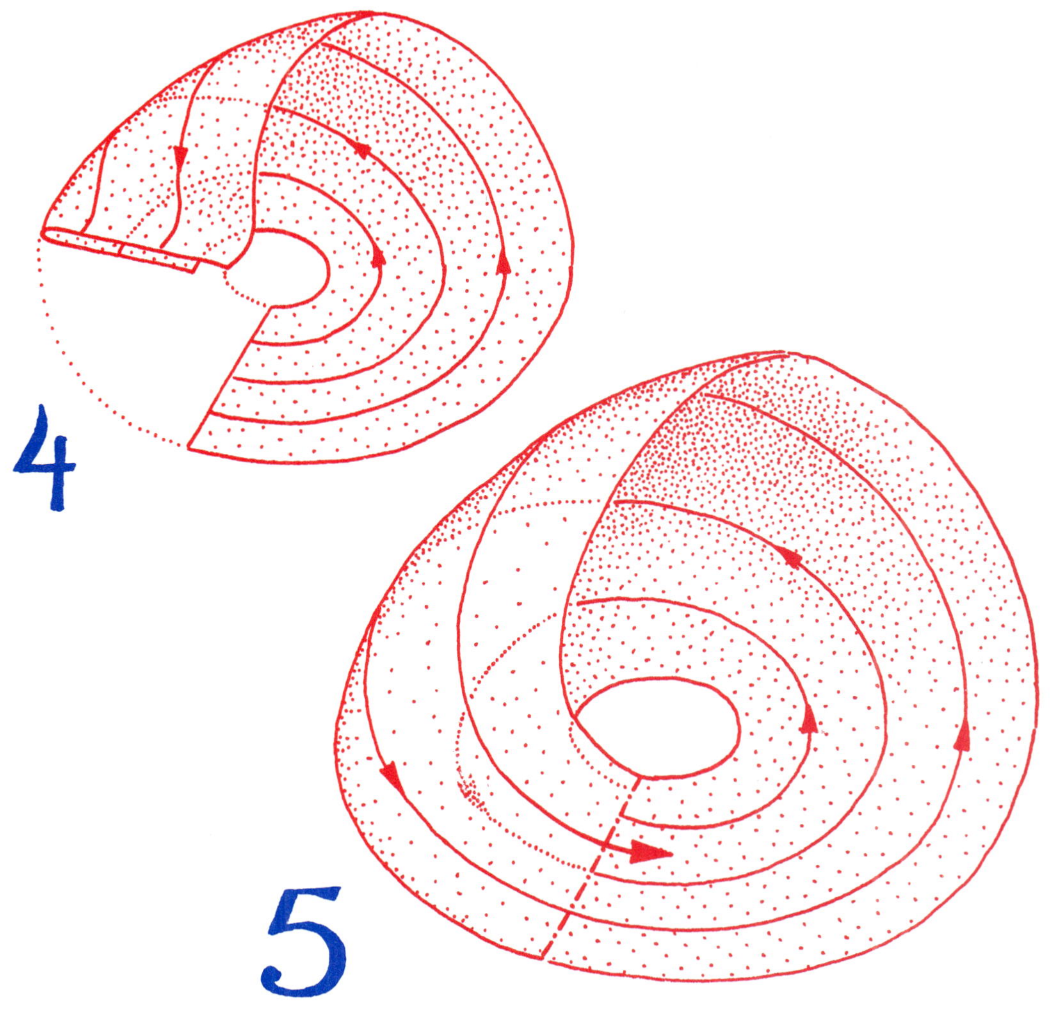

9.3.4. In step four, the process is continued, and we find that the rearward growth of the surface is actually along a circle, bringing it eventually back and around to meet the front. Step five, it closes. In fact, it does not exactly close as a surface. What we have described is a *thick surface,* composed of infinitely many layers, like flaky pastry, filo, or croissant dough. This is explained in some detail in the next section.

Try following two trajectories around the band, and you will see that they endlessly diverge in a bounded space, as transitive solenoids. They keep on diverging from each other in the short term, while succeedinng in getting too far from each other in the long run.

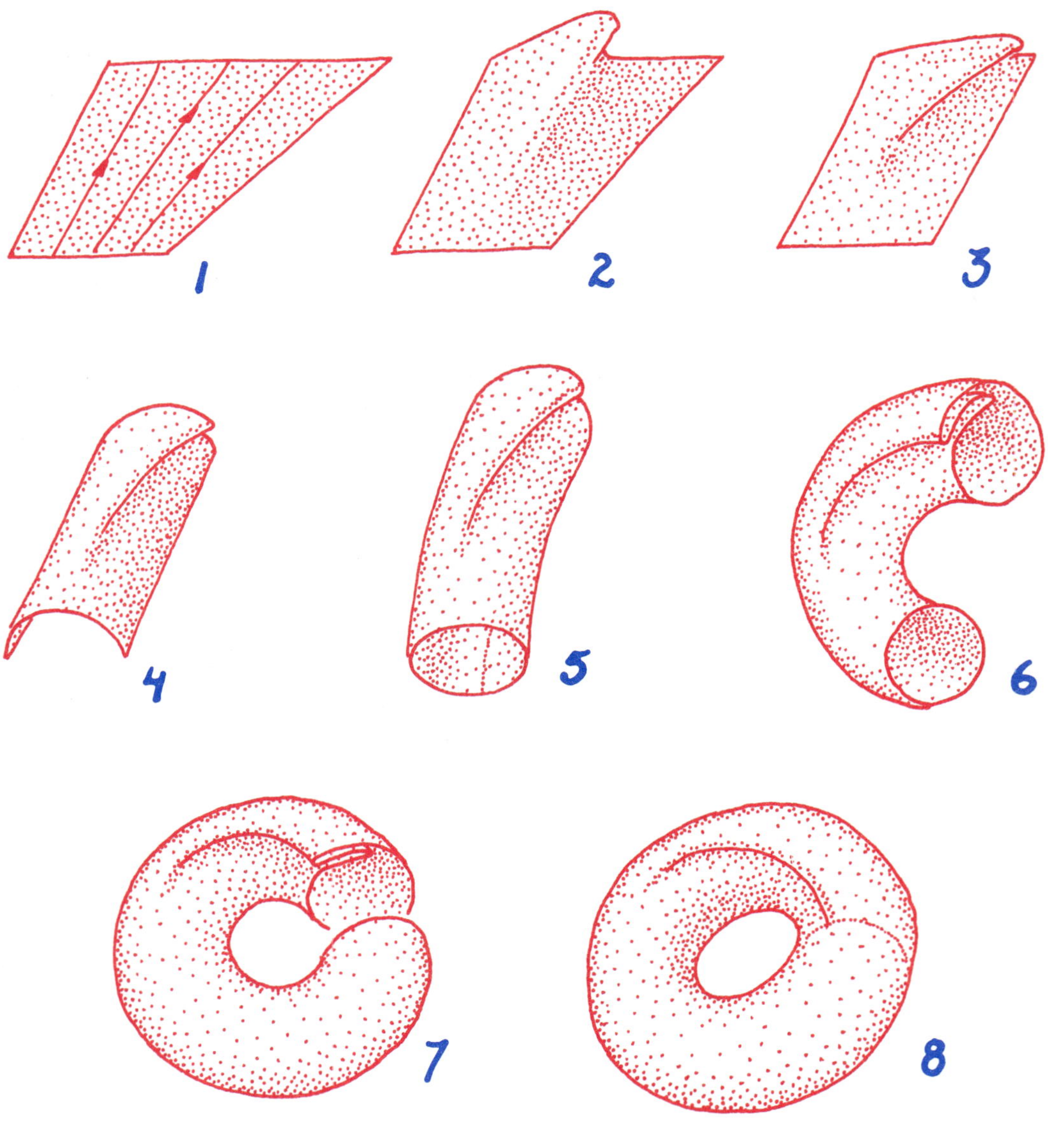

9.3.5. Birkhoff's bagel also expands endlessly in a bounded space, as shown in this eight-step model. In step 1, we start with a diverging flow on a piece of surface, as before. Step 2, the back grows away from us, and a fold begins to form in the middle. In step 3, the back extends further away from us, and the fold is pressed down flat, like a pleat. This drawing has been shrunk a bit to save space. Step 4, the back grows further and the fold is pressed flatter. Taking more of the sides into view, we see they bend downward. The drawing is shrunk again, to fit in the panel. In step 5, more of the object comes into view, and we see that the sides join at the bottom like a pleated cylinder. Step 6, following the growth at the back of the cylinder, we see it is bending around a circle. In step 7, the bent pleated cylinder is joining itself to make a torus. Finally, it closes as a thick surface in step 8.

9.3.6. Lorenz's mask is also endlessly expanding in a bounded surface. We will describe this in seven steps. First, take two copies of our usual piece of surface, with a diverging flow on each. Join them, as shown in step 2.

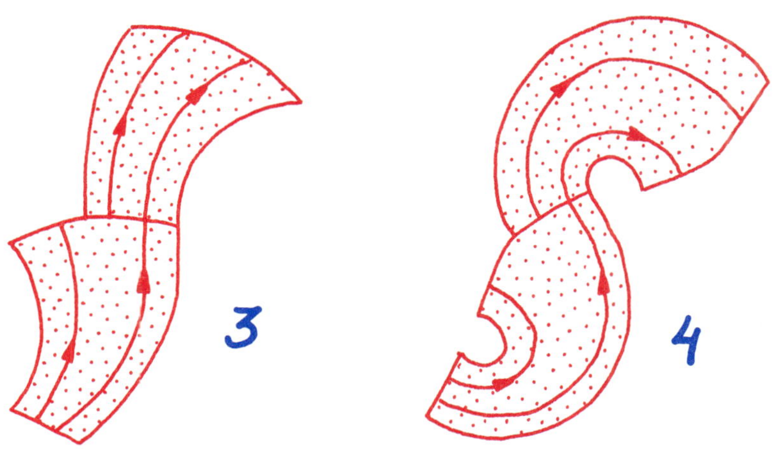

9.3.7. In step 3, taking more of the flow into view, we see the surfaces bending. In step 4, the flow of the back edge away from us begins to curve all the way around. Likewise, the flow into the front edge is seen to come from around a bend, also.

To get ready for the next step, rotate the S-shaped model (step 4) clockwise a half turn.

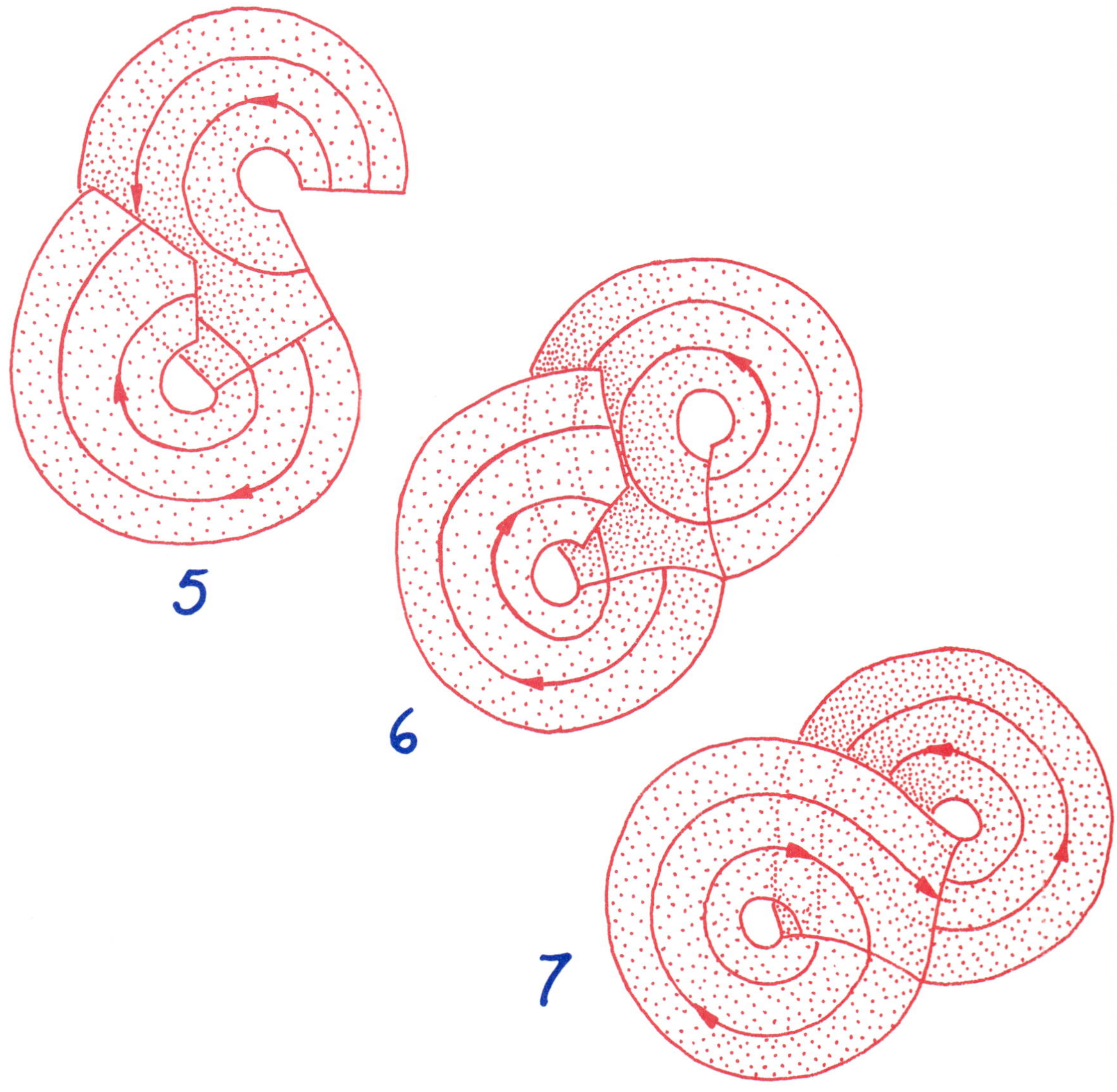

9.3.8. In these last three steps, the front and back edges are extended further along the flow, bending further until they join. These joints are as thick surfaces, as we shall see in more detail in the next section.

These objects have a fractal dimension, somewhere between two and three. They also have CE's, like a periodic saddle.

To complete our models for these three chaotic attractors, we zoom in on the fractal microstructure of thick surfaces in the next section.

9.4. Fractal Microstructure

The folding required to expand endlessly in a bounded region requires an infinite microstructure. Here we explain this feature of chaotic attractors, again using Rössler's band for illustration.

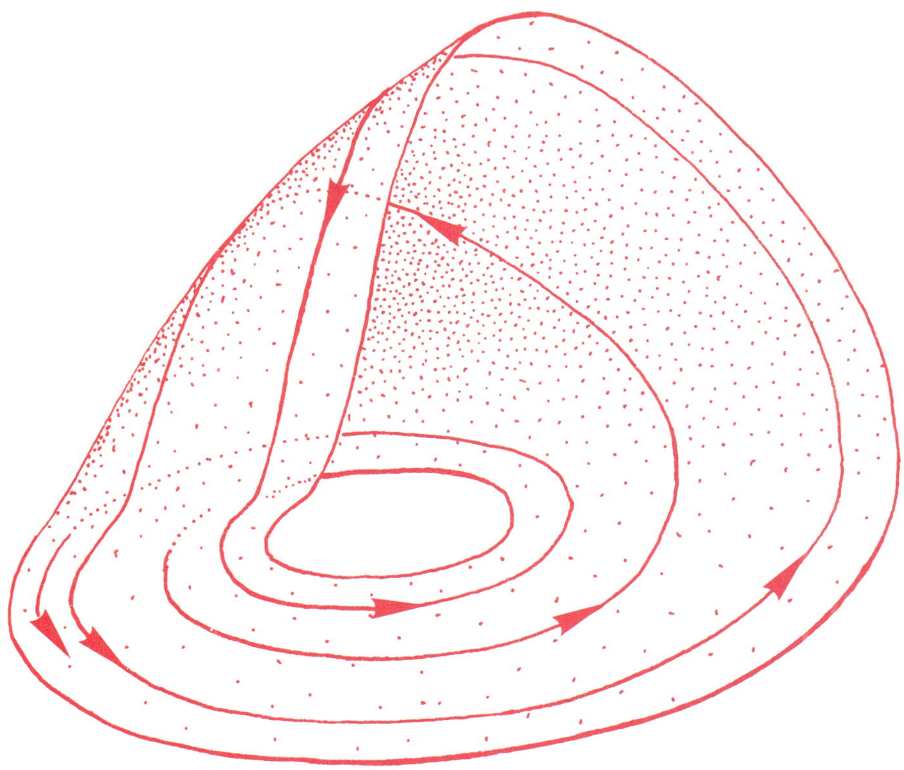

9.4.1. Here is the Rössler band, as the folded thick surface constructed in the preceding section, in five steps. As distinct trajectories of a dynamical system may never join, this object cannot possibly be a simple surface, with two layers glued together.

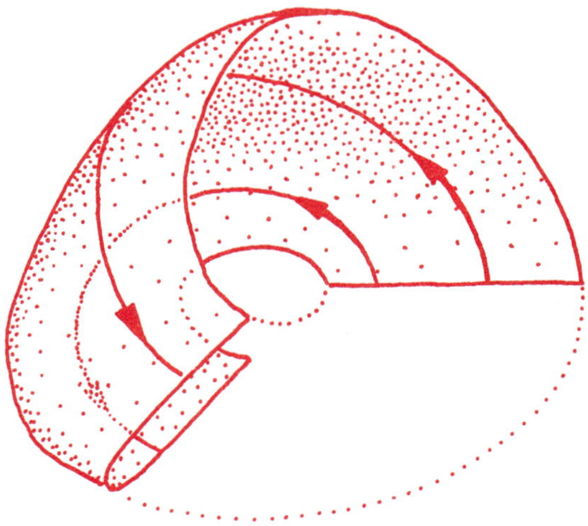

9.4.2. Let's follow the progress, at time marches on, of a *Poincaré* section of the thick surface. This appears to be a line-segment at first. As we shall soon see, it is actually a *thick line-segment*. First it extends backwards away from us, curving to the left and around a circle. At the same time, it widens and folds over. Eventually it reaches double its double width, folding all the way over, and comes back to the Poincaré section as a U-shaped channel.

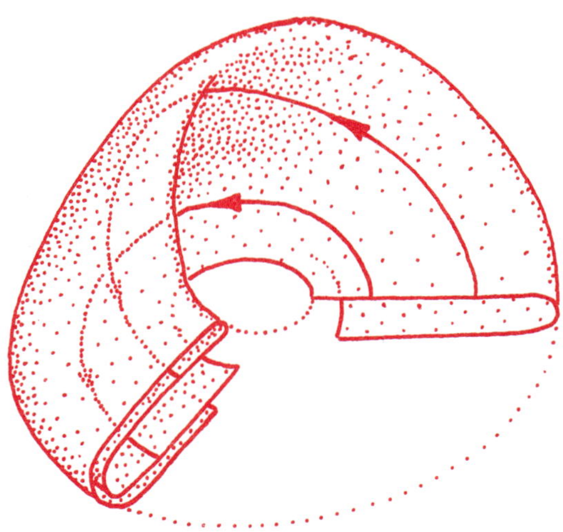

9.4.3. Now we understand that the cross-section of the band is not an interval. It must be a folded interval with two layers. Let's follow the evolution of this folded interval as it goes around, repeating exactly the steps above. Starting another turn as a U-shaped channel, the stretch and fold produces a double-U-shape on return to the Poincaré section.

9.4.4. Now we know that the section is not a line-segment, and it's not a U-channel either. Starting a third turn as a double-U, the folding action of the band creates a quadruple-U on next return to the Poincaré section. And so on!

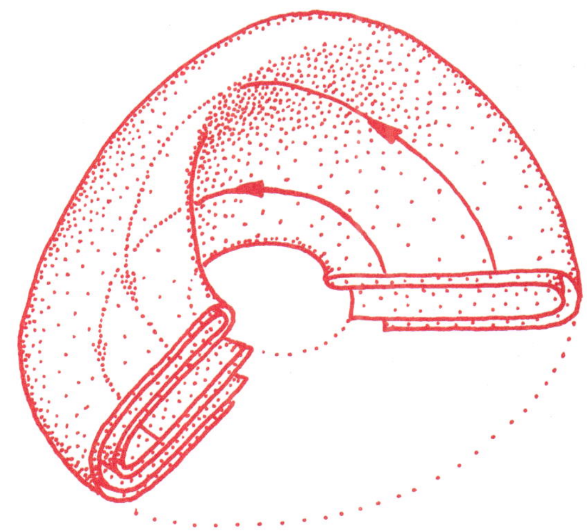

9.4.5. After countless repetitions, the original interval of section has expanded, folded, and returned in countless layers. These accumulate in a certain pattern. To examine this pattern, we cut again in a cross-section. This one, which we call the *Lorenz section*, cuts through the Poincaré section.

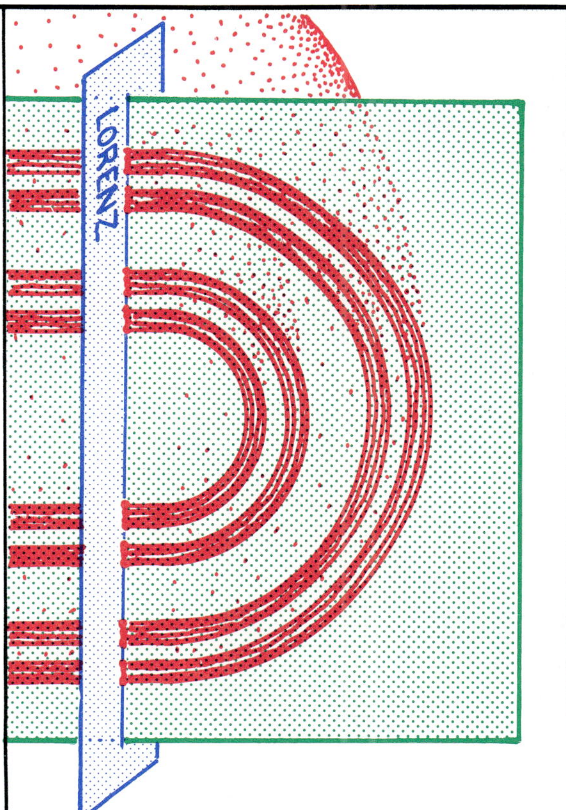

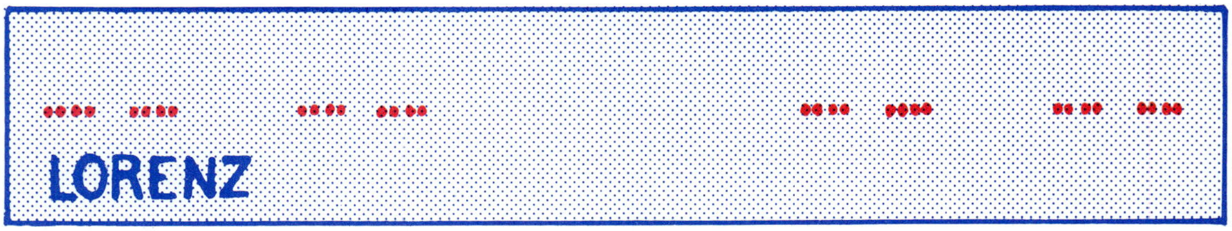

9.4.6. Turning the Lorenz section face front, we see a single dot corresponding to each layer of the filo dough of Rössler's band. The number of dots is infinite. But they have a pattern. There are gaps within gaps within gaps.

9.4.7. The pattern may be reconstructed in a sequence of steps called *Cantor's process.* First take a line segment, at 1. Then remove a smaller segment from somewhere within, making a gap, as in 2. Then repeat this surgery on each of the two remaining segments, obtaining 3, and continue forever. If at each step you take away the middle third of a segment, this is the *middle thirds process.* But there are many other possibilities.

9.4.8. For example, this construction removes two intervals in each surgery. It is the *middle fifths process.* We cannot pin down exactly which process is needed to reconstruct the Lorenz section of the Rössler band. But it is definitely this kind of thing, called in general a *Cantor set.*

9.4.9. So take a Cantor set, rotate to vertical, and attach a horizontal line to each point. This is called a *Cantor one-manifold* by specialists. But we have been calling it a *thick line* all along.

9.4.10. Or instead, attach a piece of surface to each point of the Cantor set. This is a Cantor two-manifold, or *thick surface.*

By *fractal microstructure,* we just mean a thick line or surface, more or less. We also must allow creases, folds, and so on. For more information on fractals and their fractal dimensions, see the excellent books by Mandelbrot and I. Stewart, listed in the Bibiography.

All the experimentally known chaotic attractors are characterized by this kind of fractal microstructure, as well as by the divergence of trajectories, measured by characteristic exponents. Relationships between the fractal dimension and the characteristic exponents are a current research topic.

9.5. Noisy Power Spectra

The normal impulse of an experimentalist, upon sighting a complex time series, is to take its power spectrum. As we have seen in Section 7.7, the power spectrum from a periodic attractor is a discrete, or line, spectrum. This technique has been very successful with the chaotic attractors. But the power spectra of chaotic attractors are continuous, or *noisy*.

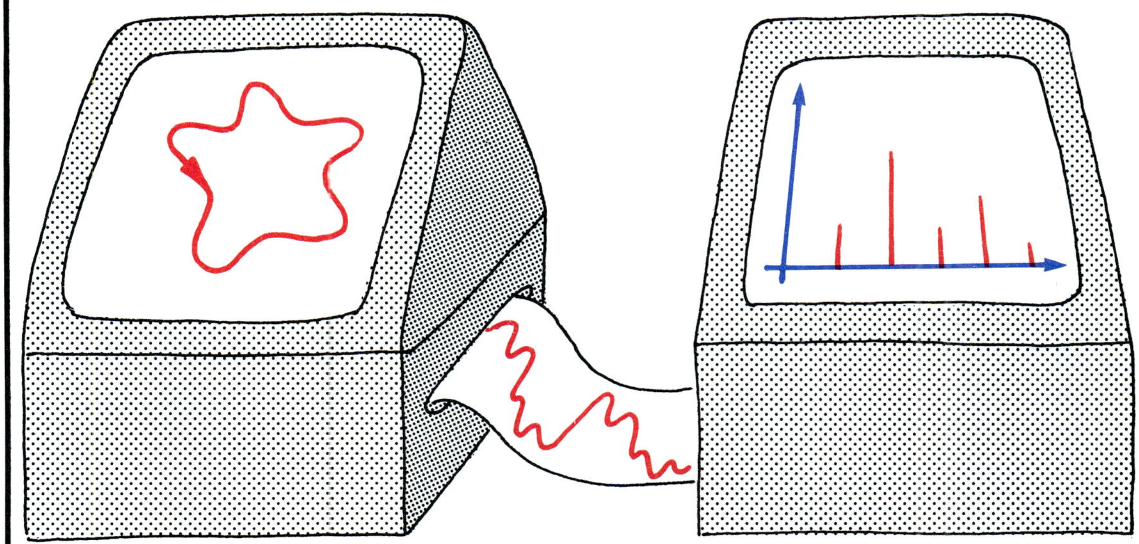

9.5.1. The power spectrum of a time series is computed by an arduous algorithm. The longer the time series input, the more accurate the power spectrum output. The spectrum contains about half enough information to recreate the original time series. It records the frequencies and powers, but not the phases, which are used in the reconstruction, as shown in Section 7.7.

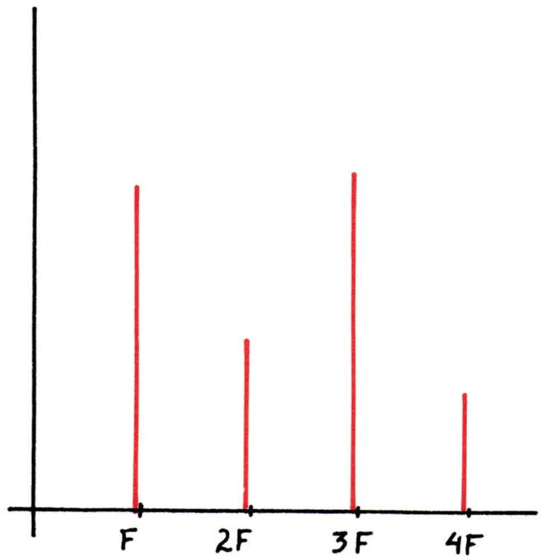

9.5.2. This is the power spectrum of a typical periodic attractor. Each line, over an integral multiple (octave) of the fundamental frequency of the oscillator, indicates by its height the power contributed to the oscillatory motion in that octave. This is called a *discrete spectrum.*

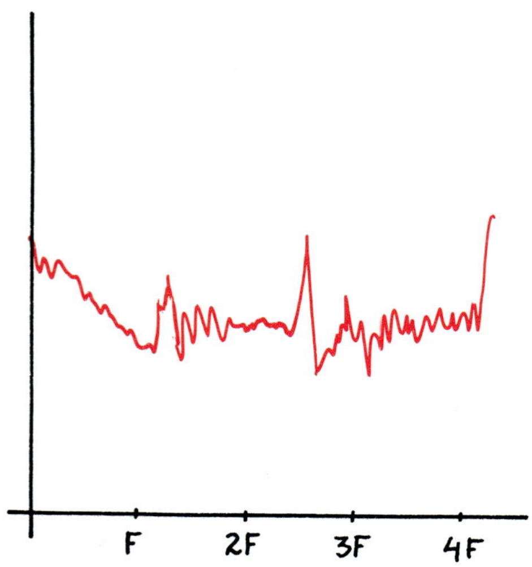

9.5.3. This, in contrast, is the power spectrum of a typical chaotic attractor. There is activity at every frequency. This is called a *continuous spectrum.*

With experience, we might be able to recognize attractors from the power spectra of the time series produced by one coordinate of the state space, if we had enough examples on file. This is like the fingerprint strategy for criminal identification.

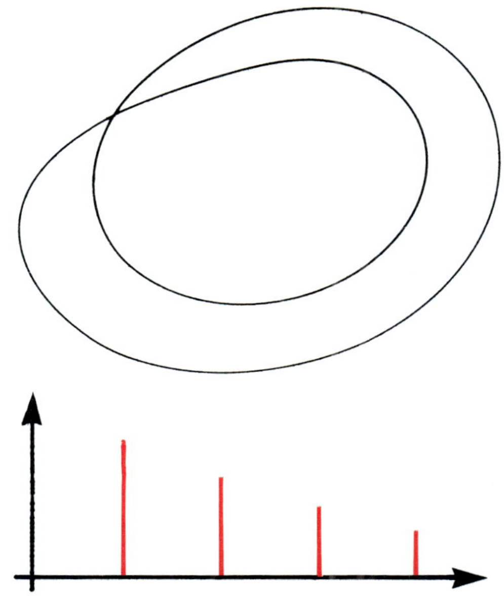

9.5.4. This periodic attractor in three dimensions is fully described by three time series, one for each coordinate. The power spectrum of one of these is shown here.

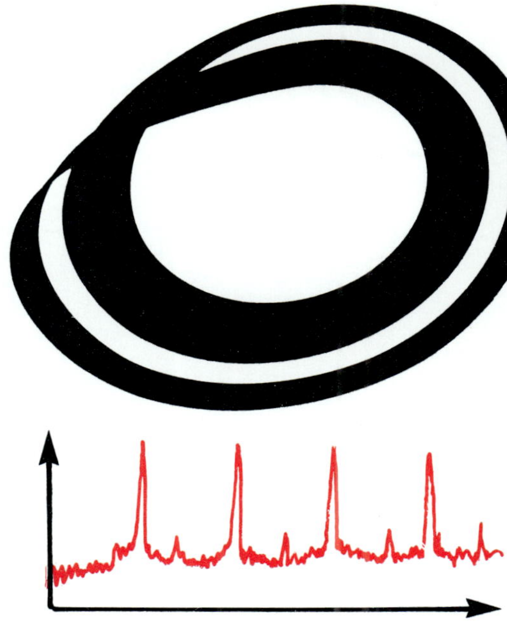

9.5.5. This chaotic attractor in three dimensions is also fully described by three time series, the coordinates of a transitive solenoidal trajectory. The power spectrum of one of these might look like this: a discrete spectrum with a little noise added.

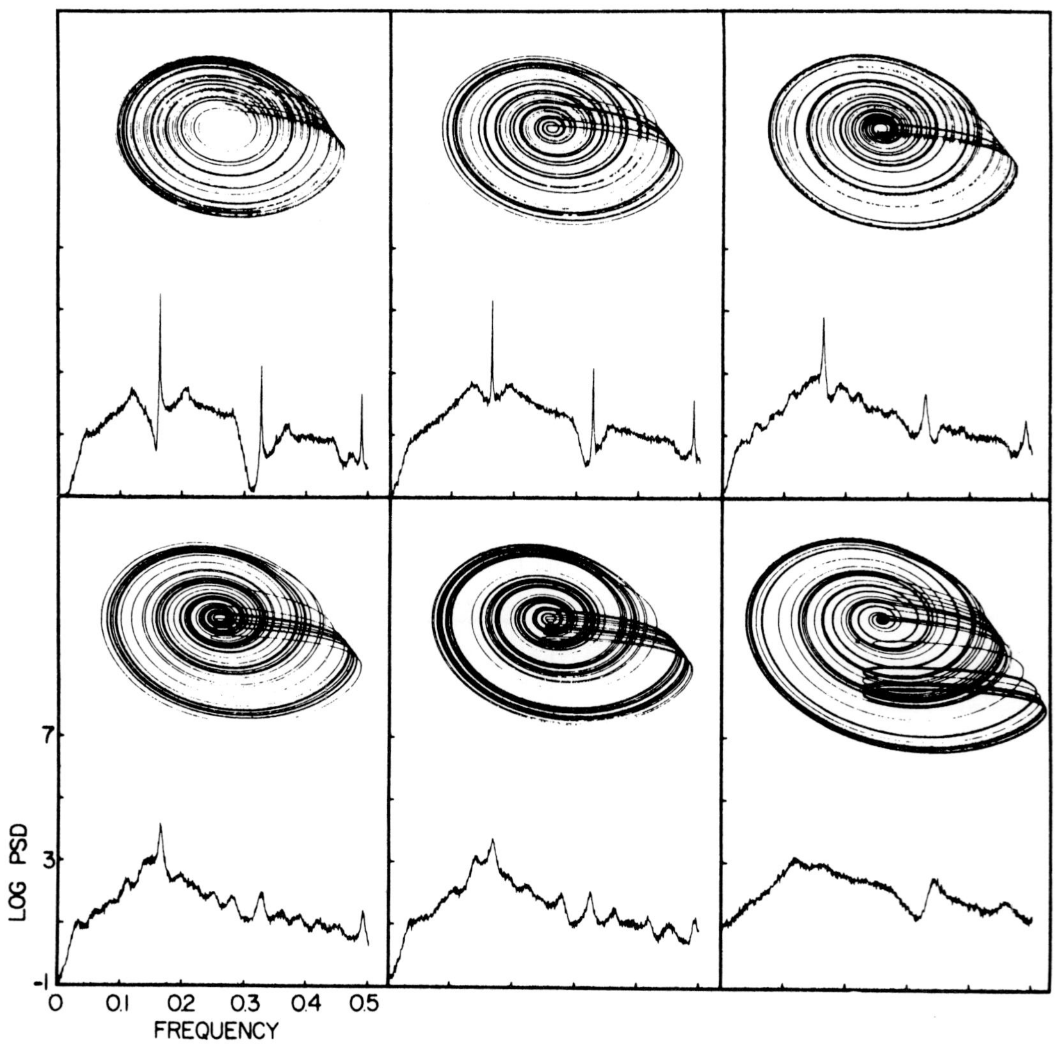

9.5.6. These chaotic attractors (Rössler bands and funnels) are increasingly chaotic, as shown by the increasing levels of the noise (continuous spectra) in their power spectra, shown below in each case. This is actual data from analog simulations (Crutchfield et al., 1980). Note that the spikes, indicating approximately periodic behavior, are lost in the noise in the lower right example.

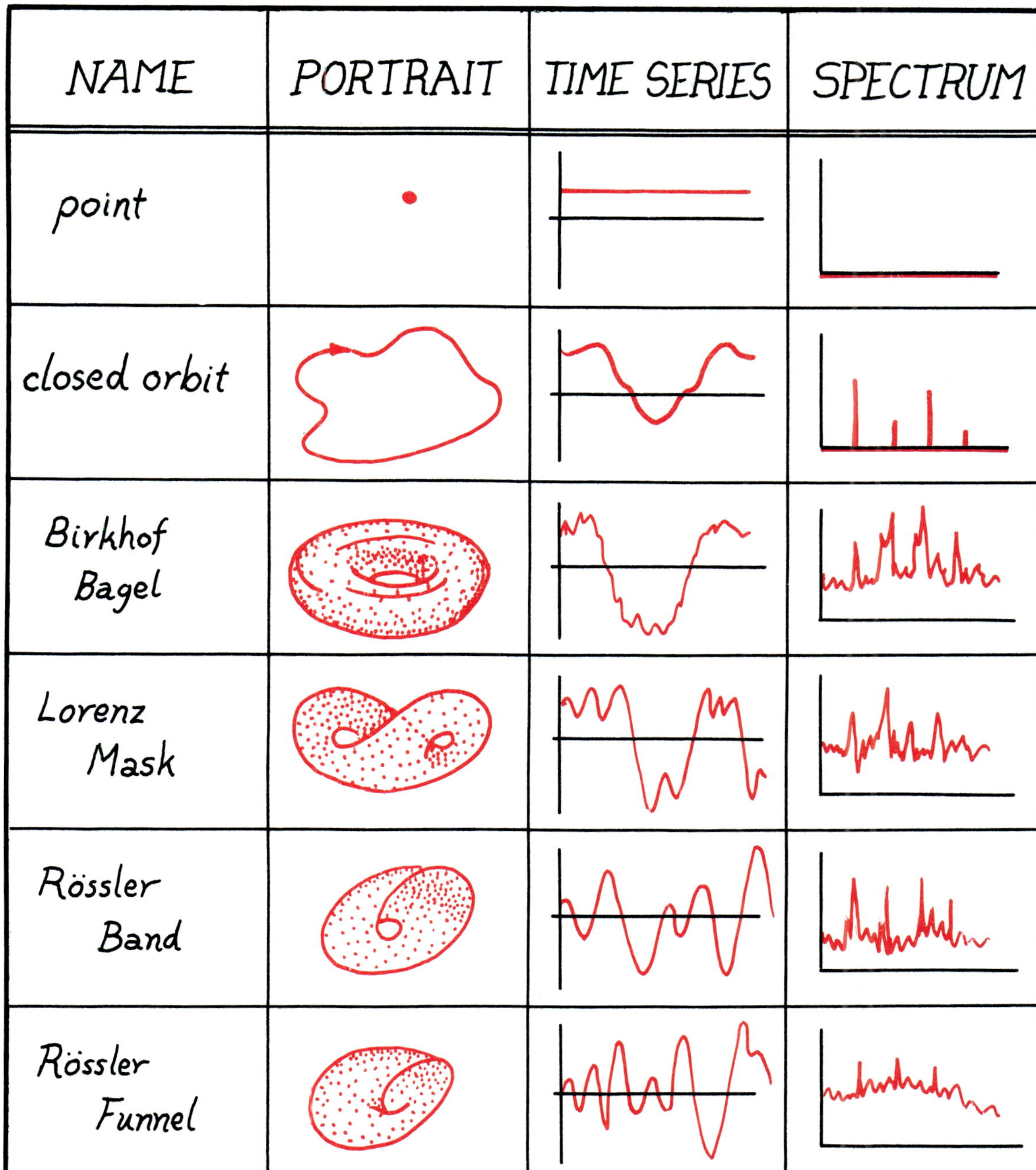

9.5.7. Here is a summary table of the exemplary attractors we have presented, with sketches of their characteristic output. One could wish for an extension of this table, showing all possible attractors likely to arise in experiments and applications. But at this point, that's a big wish.

Conclusion

This completes our visual introduction to the *local behavior* of dynamical systems, begun in Part One. By local behavior, we mean the behavior of trajectories on or near a *single attractor.* It is important to keep in mind that a typical dynamical system has *many attractors* and basins. In applications, the location of these may be more important than the particular local behavior of each attractor (point, cycle, or chaos). We repeat here, in ending this part of our story, that *the identification of experimental attractors with limit sets of mathematical models is not yet fully justified in dynamical theory.*

In Part Three, we will turn to the *global behavior* of dynamical systems.

PART 3

Global Behavior

Dedicated to Mauricio Matos Peixoto

Photograph by Caroline Blakemore

Mathematical Dynamics Hall of Fame

The early days of modern dynamics span half a century, beginning in the 1880's, as described in Part One. At this time in France, Poincaré innovated qualitative methods. More or less simultaneously in Russia, Liapounov pioneered stability methods. These techniques then underwent separate, parallel developments. By the 1930's, important progress had been made in Europe and America, following the lead of Poincaré. Birkhoff, at Harvard, was the outstanding figure. Meanwhile, in Russia, the ideas of Liapounov had grown. Andronov was an important figure in this tradition.

There followed a quiet period. For another quarter century, the tradition of Poincaré dwindled in Europe and America. Developments in Russia were forgotten in the West. During this period, *experimental dynamics* began in Europe, as described in Part Two.

Eventually, through the efforts of emigré mathematicians familiar with the Russian work, such as Lefshetz and Minorksy, the qualitative theory of dynamical systems was revived in America. Beginning in the 1950's, a vigorous mathematical program picked up steam and continues today. The global behavior of dynamical systems is the main theme of this movement, which we may call mathematical dynamics.

TABLE 3.1–THE HISTORY OF GLOBAL THEORY

Date	EUROPE/AMERICA	RUSSIA
1850		
1900	Poincaré Floquet Birkhoff	Liapounov Mandelshtam
1950	Lefshetz, Minorsky, de Baggis, Peixoto, Markus, Thom, Smale, Pugh	Kolmogorov Arnol'd
2000		

Here are some capsule histories:

Henri Poincaré, 1854–1912. Besides pioneering the new methods of dynamics and topology, Poincaré discovered tangles and bifurcations as we know them today.

Aleksandr Mikhailovich Liapounov, 1875–1918. In his Ph.D. thesis of 1892, Liapounov established the Characteristic Exponents of an equilibrium point as the determinants of its asymptotic stability.

George David Birkhoff, 1884–1944. Birkhoff was fascinated by tangles, and wrote several papers about them. In one, he introduced the *signature* of a tangle, making a first step in the historic struggle to untangle them. In another, he showed that homoclinic tangles are always surrounded by myriad periodic trajectories.

Aleksandr Aleksandrovich Andronov, 1901–1952. With co-workers Leontovich and L.S. Pontrjagin, Andronov pioneered the phase portrait point of view. Andronov and Pontrjagin published a five-page paper in 1937 which revolutionized global dynamics. Its main contribution was the definition of *structural stability.* In the same year, Andronov published an influential book, written with C.E. Chaikin, on nonlinear oscillations.

Gaston Floquet. He established the Characteristic Multipliers of a limit cycle as the determinants of stability, parallel to the CE's of Liapounov, in 1879.

Solomon Lefshetz, 1884–1972. In the World War II years, this great innovator of algebraic topology turned his attention to qualitative dynamics. A text on the local theory in 1946 was followed by a global treatment in 1957, in which structural stability was discussed in two-dimensional systems. A native of Russia, he reinjected the tradition of Liapounov into the mainstream of Western mathematics.

Nicolai Minorsky, b. 1883. Like Lefshetz, Minorsky emigrated to the United States in the prewar years. His knowledge of the Russian school of dynamical systems theory, presented in his book of 1952, gave great impetus to the resumption of mathematical dynamics in the United States.

Henry de Baggis, b. 1916. A student of Lefshetz, in 1947 he proved the conjecture of Andronov and Pontrjagin on structural stability in the plane.

Mauricio M. Peixoto. Also a student of Lefshetz, he improved enormously on de Baggis's result in 1959. In doing so, he forged the connection between dynamics and topology which has been so fruitful in recent years.

René Thom, b. 1923. Thom used dynamics in his work in topology, for which he was awarded the Field Medal. In 1960 or so, he began advocating the importance of the concept of structural stability in applications, and his very global view of bifurcations. His program was presented in full in his epochal book, *Structural Stability and Morphogenesis,* in 1966.

Lawrence Markus, b. 1922. Another pioneer in the merger of topology and dynamics, he clarified the meaning of *generic property* in global dynamics, in 1960. This work is described in Section 11.1.

Stephen Smale, b. 1930. Like Thom, Smale used dynamics in his work in topology, which earned a Fields Medal in 1960. He then went on to study dynamics itself, and produced a series of papers in the 1960's which have been very influential ever since. In one of these, he improved substantially on Birkhoff's results on homoclinic tangles, as we explain in Section 14.4.

10

Global Phase Portraits

In Part One, we introduced limit points and cycles in dimensions one, two, and three. The decomposition of the state space into basins of attraction, by the separatrices, was emphasized. In Part Two, the inset structure of the separatrices was developed. The geometry of the exceptional limit sets, determined by their Liapounov characteristic exponents, was described. In this chapter, we review all this and assemble it into a global overview of the phase portrait of a typical system.

10.1. Multiple Attractors

For pedagogic reasons, our discussion has often centered on an attractor. However, generic systems commonly have several attractors. So we begin this review chapter with an explicit acknowledgment of this fundamental feature: multiple attractors.

Let's begin with the simplest case, in which the state space is one-dimensional: a curve.

In this context, limit sets are points. Generically, point attractors and point repellors alternate along the curve. The repelling points separate the basins of the attracting points. An initial state, chosen from one of the basins, tends toward the unique attractor in its basin. The different attractors represent the equilibrium states that may be observed in this system.

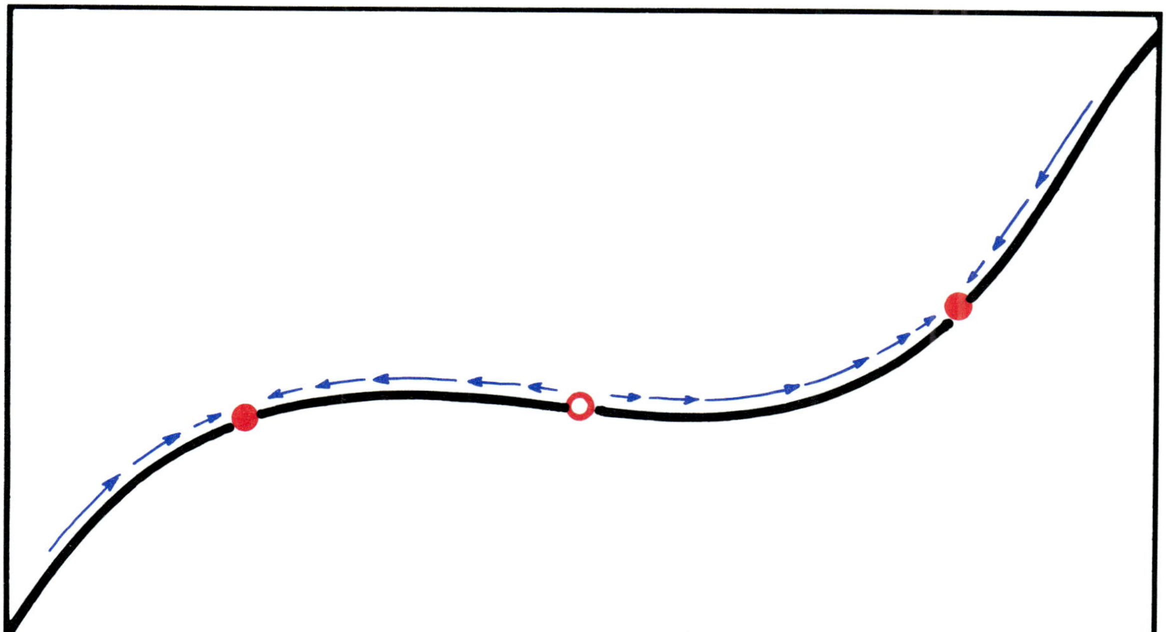

10.1.1. In this example, there are *two* attractive points, each in its own basin. The system is *bistable*, in that two distinct stable equilibria are possible.

In general, a one-dimensional system is *multi-stable*, in that more than one stable equilibrium point is possible.

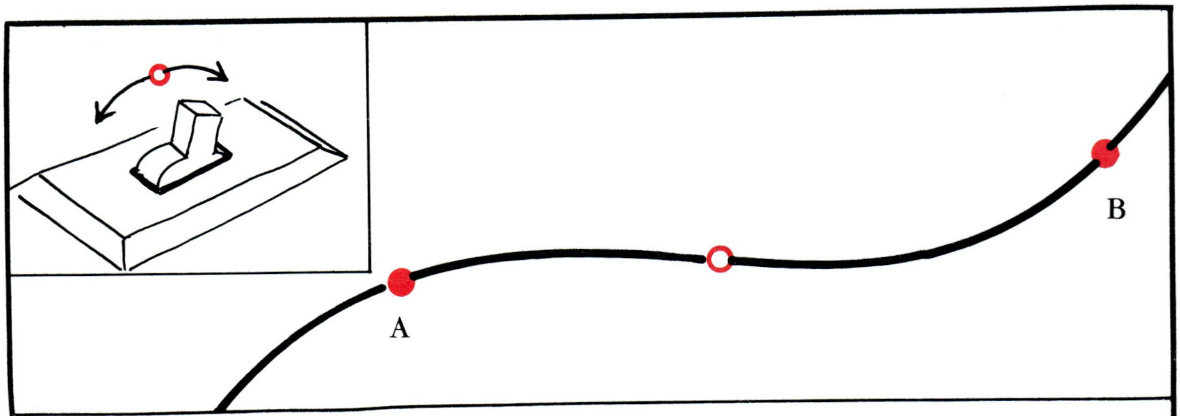

10.1.2. Notice that in this example, the two basins are separated by the point repellor. Initial points slightly to the left of the repellor tend to attractor A, while those slightly to the right tend to attractor B. This behavior is roughly like a mechanical toggle switch.

Now let's move on to two dimensions.

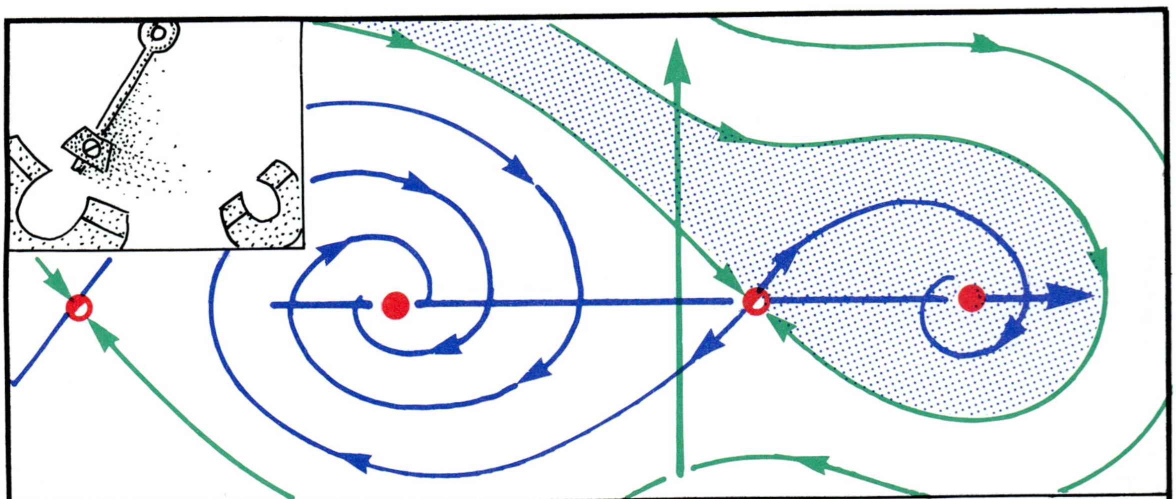

10.1.3. Remember the pendulum? Here is the magnetic bob from Figure 2.1.22. This is also a bistable system. But the two basins are two-dimensional, so the separatrix between them is a curve. This curve is *repelling,* yet not a *repellor.* In fact, it consists of the *inset* of the saddle point between the point attractors. This saddle point represents an *unstable equilibrium* of the bob, balanced between the forces of the two magnets. And its inset represents those improbable initial states which tend to this unstable equilibrium and balance there.

Not every separatrix is the inset of a saddle point.

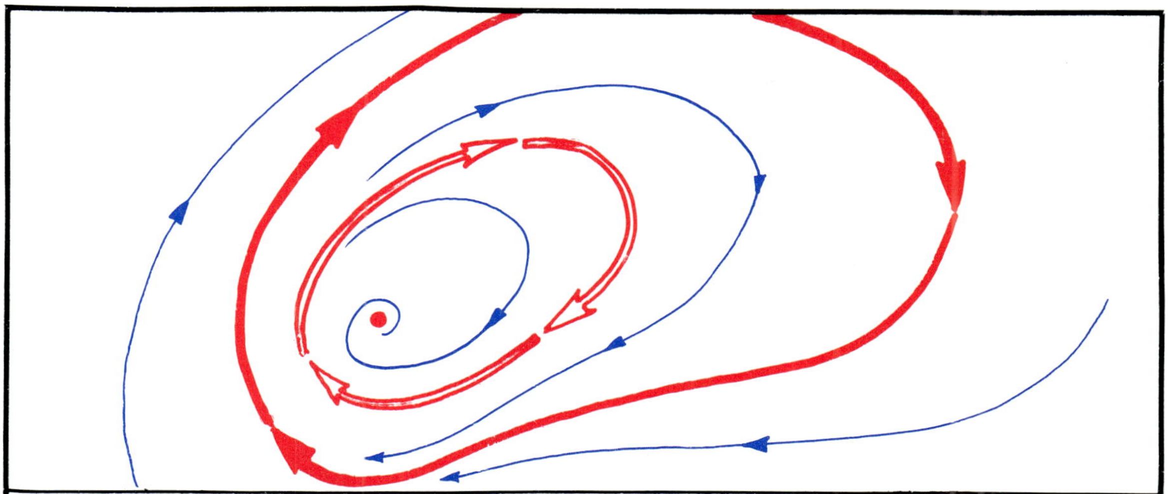

10.1.4. Recall this portrait, from Figure 1.5.8. Here, the periodic repellor bounds the two-dimensional basin of an attractive point. It is a separatrix.

Two dimensions are rather special. Let's have a look at the three-dimensional case, which is more typical.

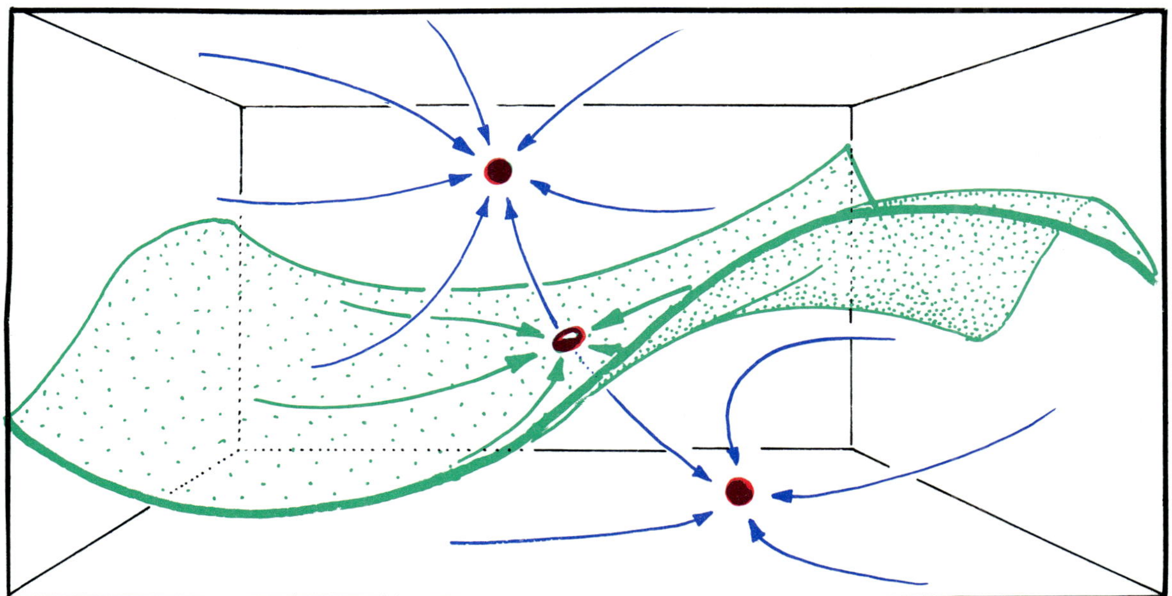

10.1.5. In this portrait of a simple bistable system in 3D, there are again two attractors. Both are rest points. Their basins are three-dimensional, and are bounded by a surface. This surface, the separatrix in this example, is the inset of a saddle point of index 1.

Recall that the *index* of a saddle point is the dimension of its outset.

Separatrices need not be insets of a saddle point. They are, usually (but not always), insets of a nonattractive limit set: point, cycle, or chaos.

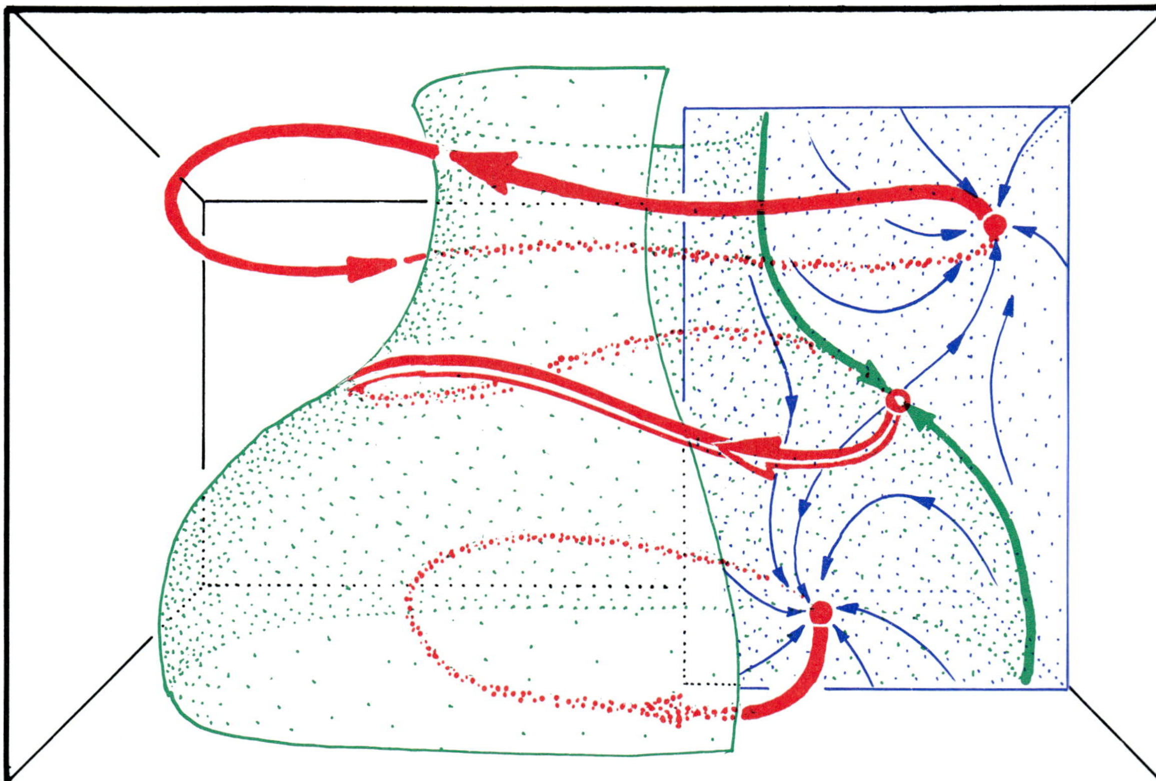

10.1.6. Here, for example, is a bistable system with two periodic attractors. Their basins are bounded by a cylindrical surface, the separatrix. It is the inset of a periodic saddle.

Remember that limit sets can be aperiodic, that is, chaotic. Thus, there may be both chaotic attractors and chaotic separatrices in a typical multistable system. Details are given in Part Two.

10.1.7. This is one of the most famous chaotic attractors.

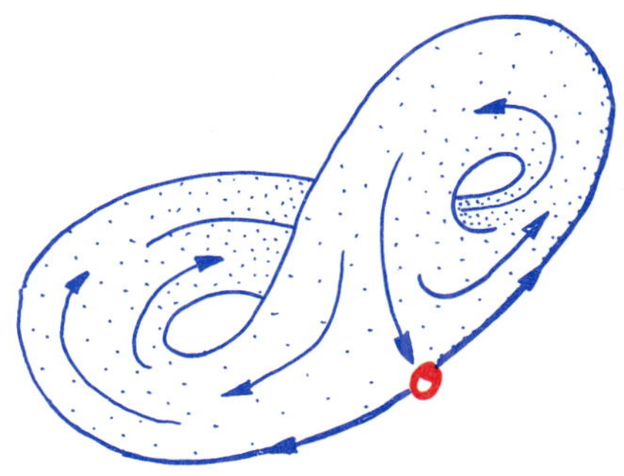

10.2. Actual and Virtual Separatrices

In Section 1.5, we defined the separatrix of a dynamical system as the complement of the basins of attraction. That is, an initial state belongs to the separatrix if its future (omega) limit set is not an attractor. According to this agreement, the separatrix consists of the insets of the non-attractive (or exceptional) limit sets. (See Section 1.5.) But do they, in fact, actually separate basins? If so, they are called *actual separatrices*. But, as we shall see, it may happen that they do not separate basins. In this case, they are called *virtual separatrices*.

Here are some examples, beginning with 1D.

10.2.1. As we have seen in the preceding section, point repellors may separate basins in one-dimensional state spaces.

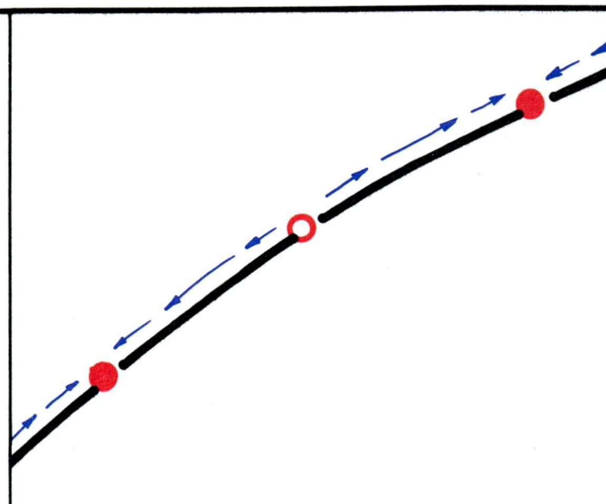

10.2.2. But if we connect the ends of the curve, we have a unistable system! There is only one basin. The separatrix (a single point repellor) bounds it, but does not separate anything. It is a *virtual separatrix*.

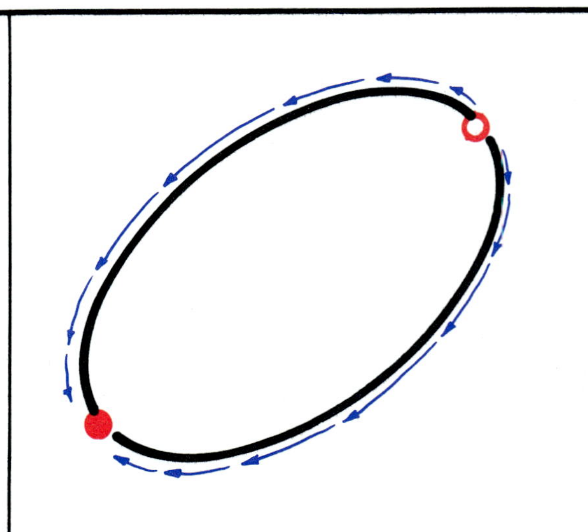

Likewise, in 2D, the separatrix consists of curves that are either insets of saddle points or periodic repellors. Examples of both sorts have been shown in the preceding section. But now look at these.

Recall that in the bistable magnetic pendulum, there is a saddle point near the bottom, as shown in Figure 10.1.3. But like the simple pendulum of Section 2.1, there is also a saddle point at the top of the swing.

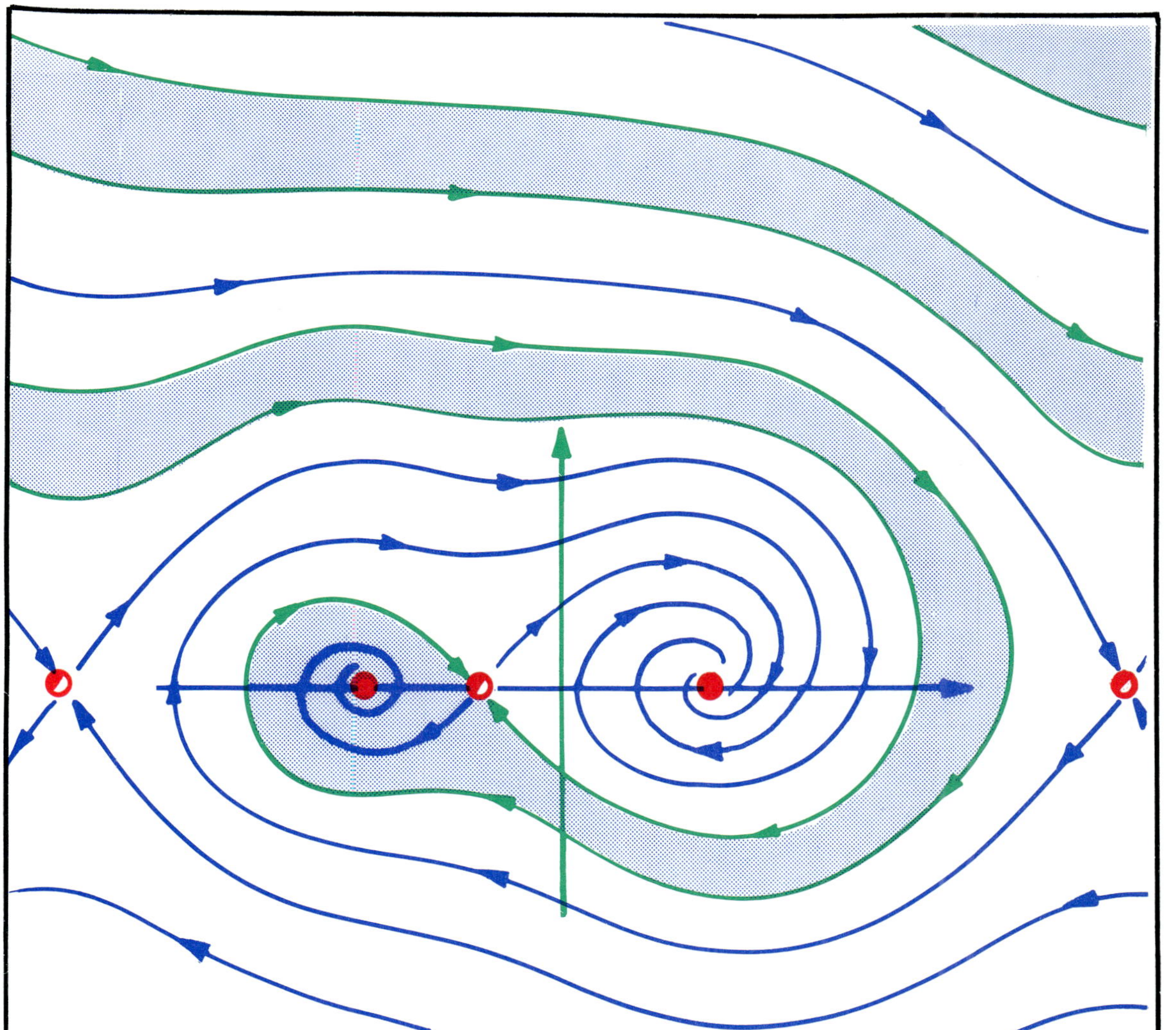

10.2.3. The saddle point at the top of the swing represents the watershed between falling to the right and falling to the left. Its inset consists of those improbable initial states that tend to balance at the top of the swing. As shown here, the initial states close to this inset, to either side, belong to the same (unshaded) basin. Thus, this inset curve is a *virtual separatrix*.

This inset failed to actually separate basins because the state space is a cylinder. Another way an inset may fail to divide basins occurs on the Möbius band.

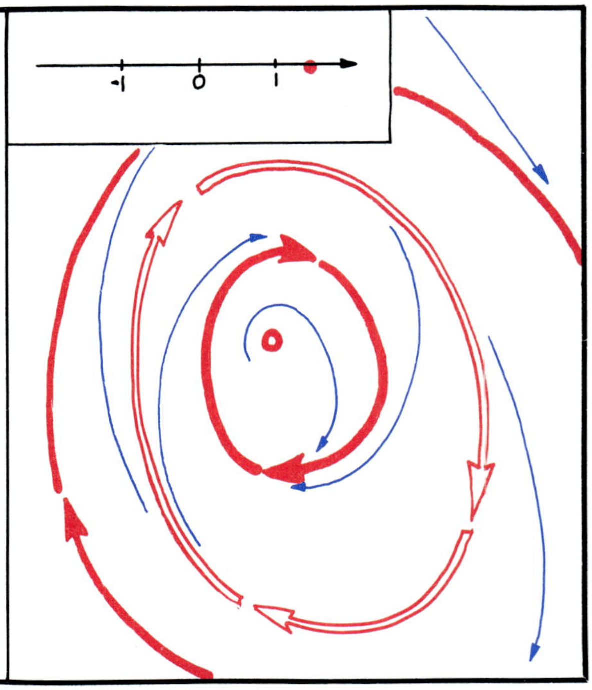

10.2.4. Consider first this bistable system. Two basins of periodic attractors are separated by a periodic repellor. The periodic repellor is an actual separatrix. There is also a point repellor in the center. It is a virtual separatrix.

Now remove the point repellor at the center, cut through the remaining strip, give one end a half-twist, and carefully paste the ends together again.

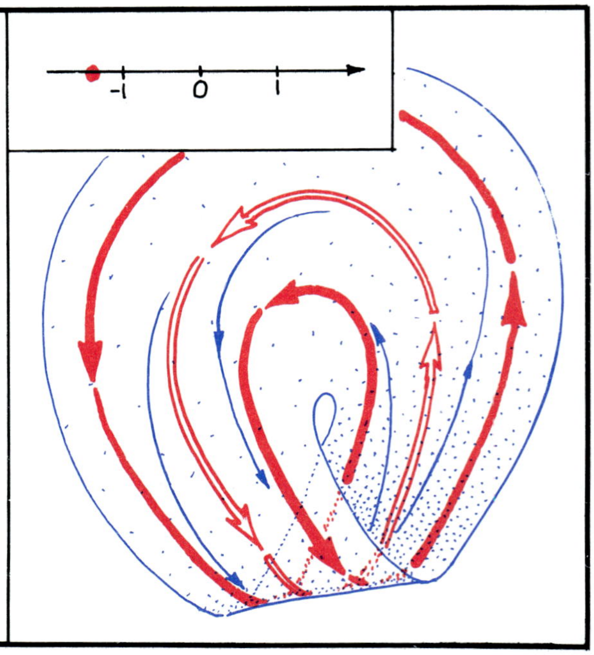

10.2.5. Now we have a monostable dynamical system. There is only one periodic attractor, which goes around twice. The periodic repellor remains, and still goes around only once before closing. It no longer separates two basins. It is a virtual separatrix.
And now, some examples in 3D.

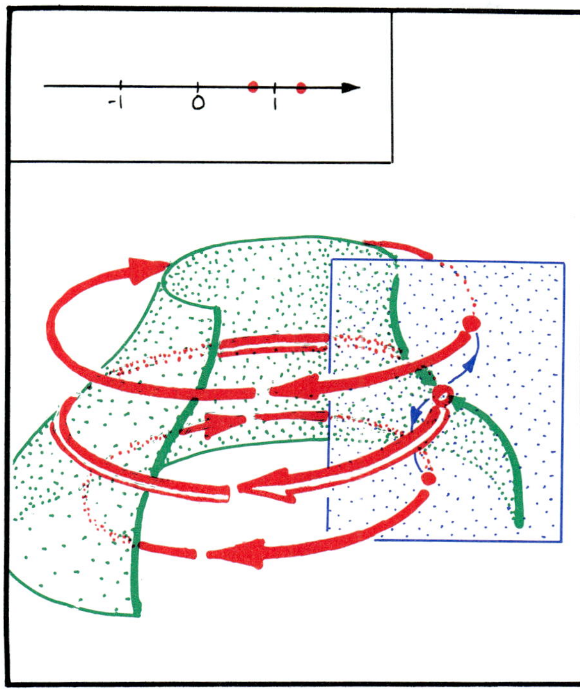

10.2.6. Recall this portrait from Figure 3.9. The inset of a periodic saddle in three-dimensional Euclidean space is twisted an even number of times. It still bounds two basins, and is an actual separatrix.

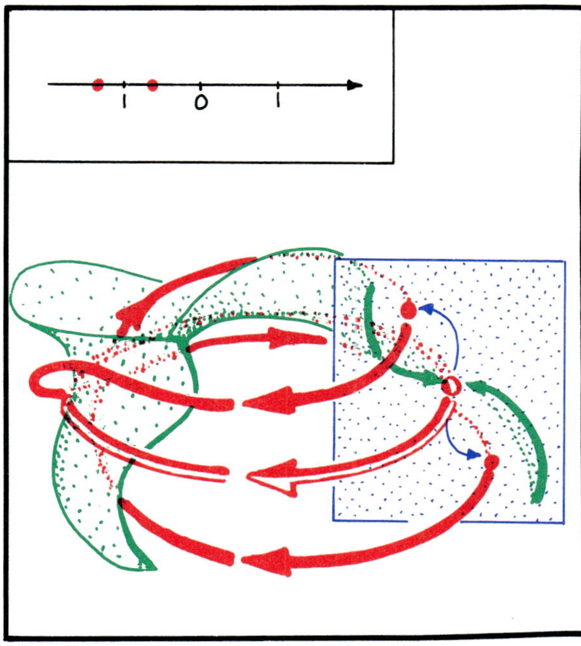

10.2.7. On the other hand, as shown in Figure 7.3.10, it may twist an odd number of times. Then it bounds only one basin, and is a virtual separatrix.

Finally, recall that insets may be thick, or chaotic. Our favorite example, Poincaré's solenoid, was constructed step by step in Section 8.1.

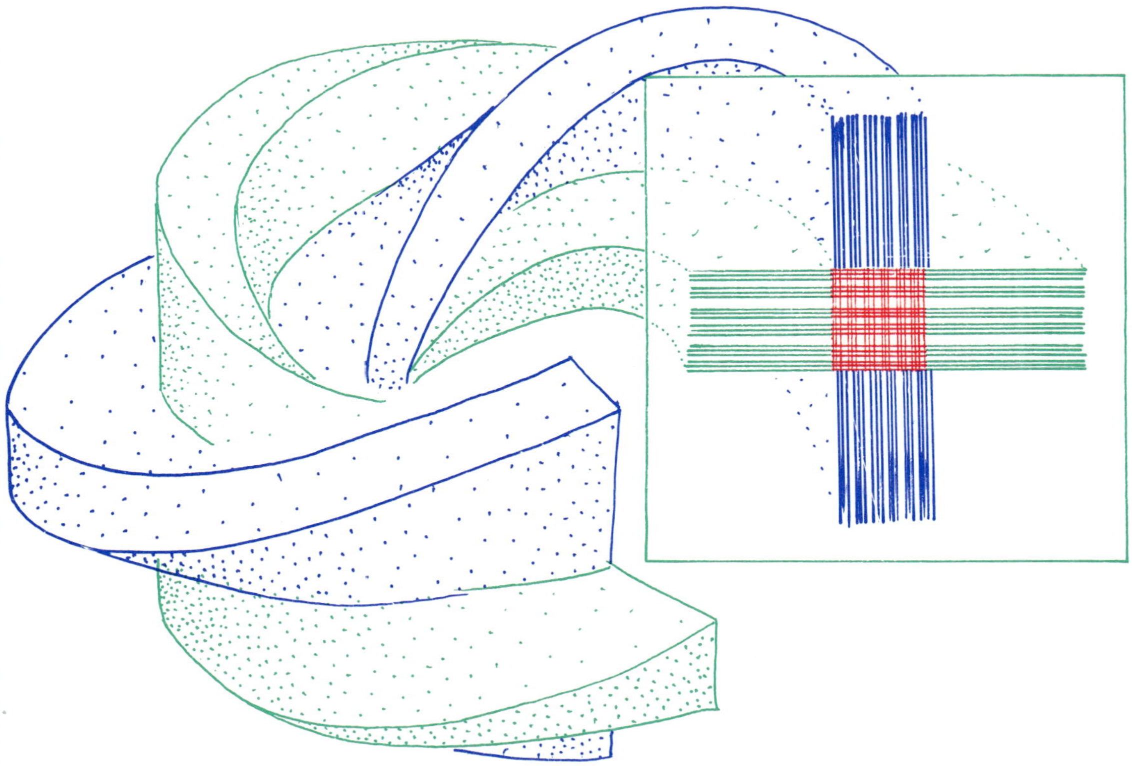

10.2.8. The fractal inset of this periodic saddle of homoclinic type is twisted once, as shown in Figure 8.1.7. It is a virtual separatrix.

11

Generic Properties

We always try to convey the features of typical, garden-variety, dynamical systems. The exceptional cases are more complicated and numerous, and they interrupt the discussion. Moreover, we feel that they should not arise very often in applications, because they are exceptional. This prejudice, shared by all dynamicists, has become a main theme in dynamical systems theory.

The properties characterizing these typical systems are called *generic properties.* Although this name was established early in the program, it turned out that it might have been better to call them *weakly generic properties.* For it has become commonplace to observe exceptional behavior (violating a so-called generic property) very frequently. An explanation for this paradox will be given in Part Four, "Bifurcation Behavior." Meanwhile, with this warning, we will continue to call these properties generic!

A considerable portion of the history of mathematical dynamics has been dominated by the search for generic properties. These define a class of phase portraits that are far simpler than arbitrary ones. The goal of the search is to narrow down the complexity of the portraits enough to allow a complete classification. This was achieved for dynamical systems in the plane by Peixoto around 1959. This gave the whole program a tremendous boost, but the higher dimensional generic systems are still hopelessly complex. This chapter presents the fundamentals of this program, initiated by Andronov and Leontovich in 1934.

The prototypical results, due to Peixoto, apply to orientable (untwisted) surfaces. An early global result for other state spaces was found by Lawrence Markus around 1960. Definitive results were obtained by Ivan Kupka and Stephen Smale in 1964. Now we will describe the essence of this main theme in the theory.

We begin with the definition of the most important global properties of dynamical systems, or vectorfields: G1, G2, and G3. Then, in a final section, we describe the official meaning of *generic property* and state the *Kupka-Smale Theorem: Properties G1, G2, and G3 are generic.*

11.1. Property G1 for Critical Points

To begin, let's recall the distinction between hyperbolic and nonhyperbolic critical points.

In Chapter 6 we created an atlas of limit points. Using their CE's, we carefully distinguished the hyperbolic and nonhyperbolic cases. We brushed aside the nonhyperbolic cases, claiming they are nondegenerate, exceptional, or *nongeneric.* The global formulation of this assertion is the part of the *Kupka-Smale Theorem* asserting the genericity of property G1, for critical points. In this section. we describe this property of critical points (that is, limit points).

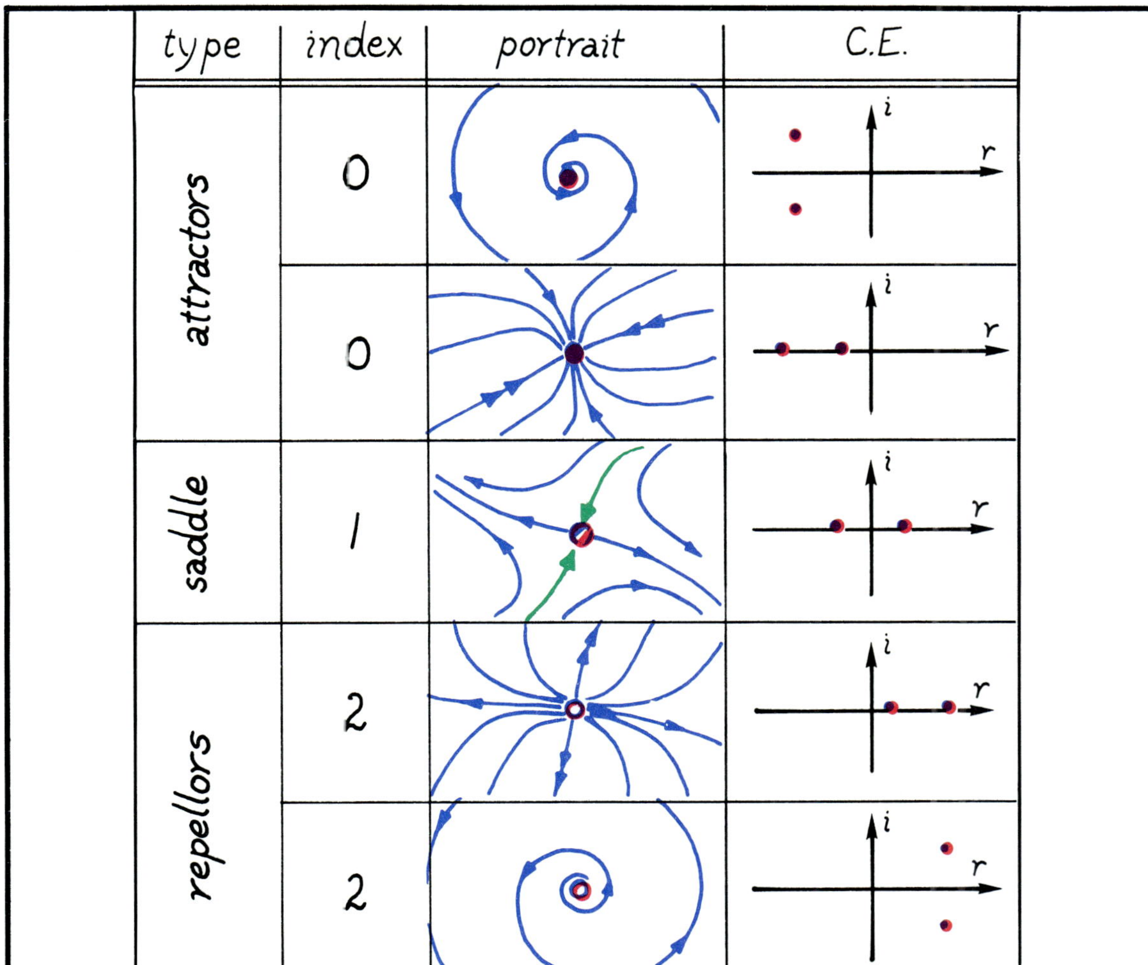

11.1.1. This is Figure 6.4.8, showing the five elementary critical points in 2D. There are seven hyperbolic critical points, namely, these five together with the radial attrator and the radial repellor.

Here, *radial* means that the CE's are real and equal. The radial type is intermediate between the spiral and nodal types.

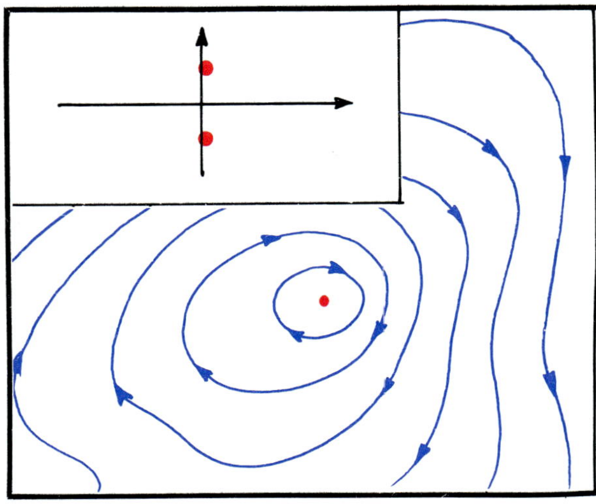

11.1.2. This is a nonhyperbolic critical point called a *center.* The CE's are shown in the inset window.

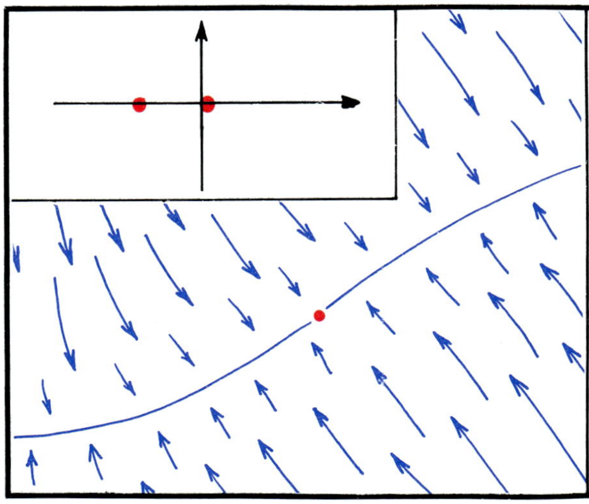

11.1.3. This is another type of nonhyperbolicity.

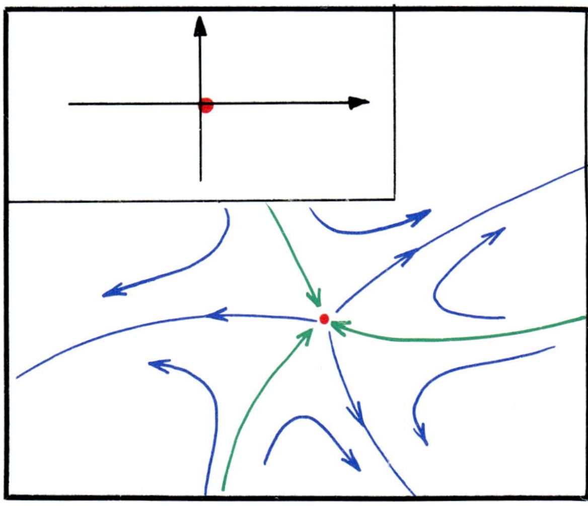

11.1.4. This is the worst case of nonhyperbolicity. Many more different portraits are possible with both CE's zero than in the two cases above.

Now we are ready for property G1.

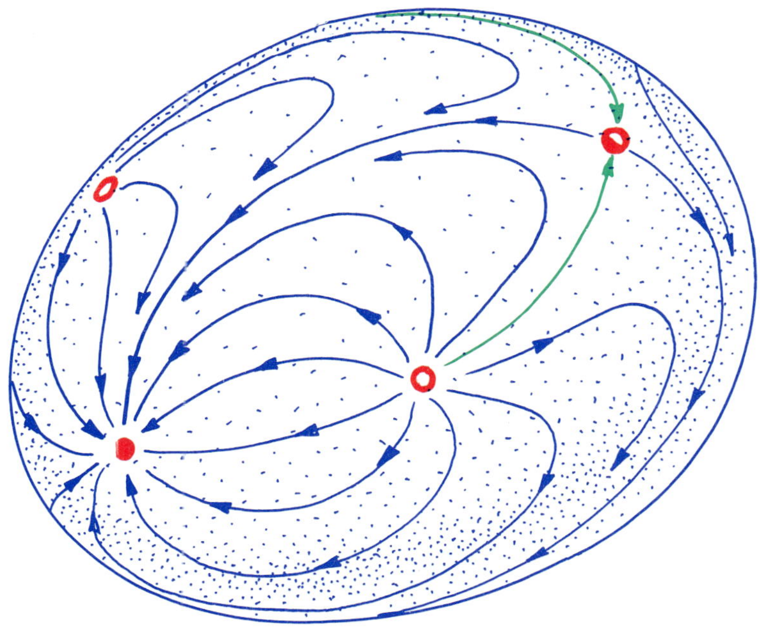

11.1.5. A dynamical system has property G1 if all of its critical points are elementary. In this example, each and every critical point is elementary.

In the literature of dynamical systems theory, this definition usually has *hyperbolic* in place of *elementary.* But this version probably results in a more satisfactory theory, from the point of view of the experimentalist, or in the context of applications.

For the eight elementary critical points that occur in 3D, see Figures 6.5.5 and 6.5.6.

11.2. Property G2 for Closed Orbits

In Chapter 7, we created an atlas of limit cycles. Using their CM's, we carefully distinguished the hyperbolic and nonhyperbolic cases. As in the case of limit points, we neglected the nonhyperbolic cases. The global justification of this neglect is the part of the *Kupka-Smale Theorem* asserting the genericity of property G2, for limit cycles. In this section, we describe this property of limit cycles.

To begin, let's recall the distinction between hyperbolic and nonhyperbolic limit cycles. For 2D, these were shown in Figure 7.2.7.

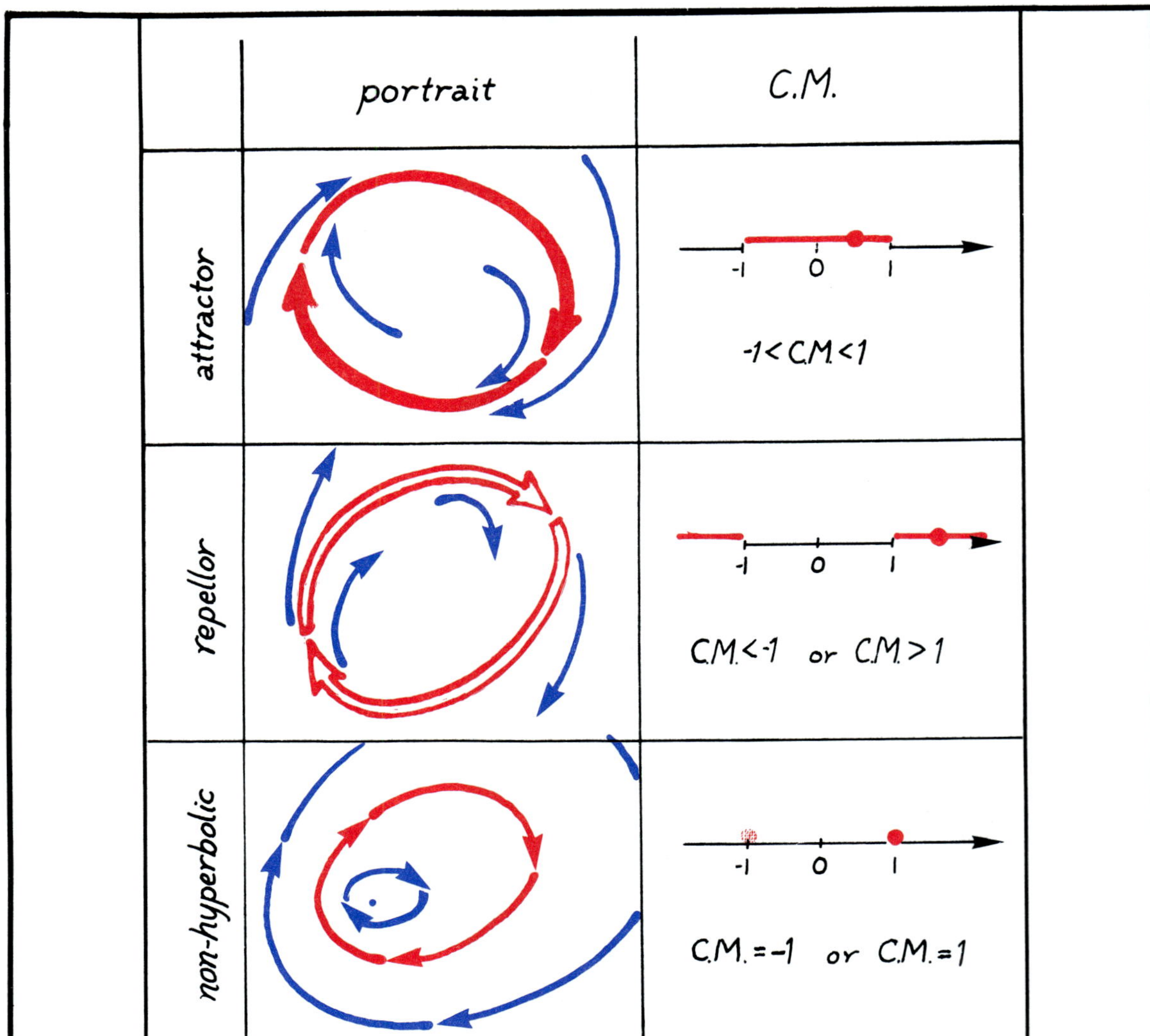

11.2.1. In 2D, a limit cycle has only one characteristic multiplier (CM), which is *real*. These are the only hyperbolic limit cycles in 2D. The absolute value of the CM is smaller than 1 (periodic attractor) or greater than 1 (periodic repellor).

In the nonhyperbolic case, the CM is equal to plus or minus 1, and the limit cycle may be an attractor, a repellor, or neither. Here are two examples, with the CM equal to plus 1.

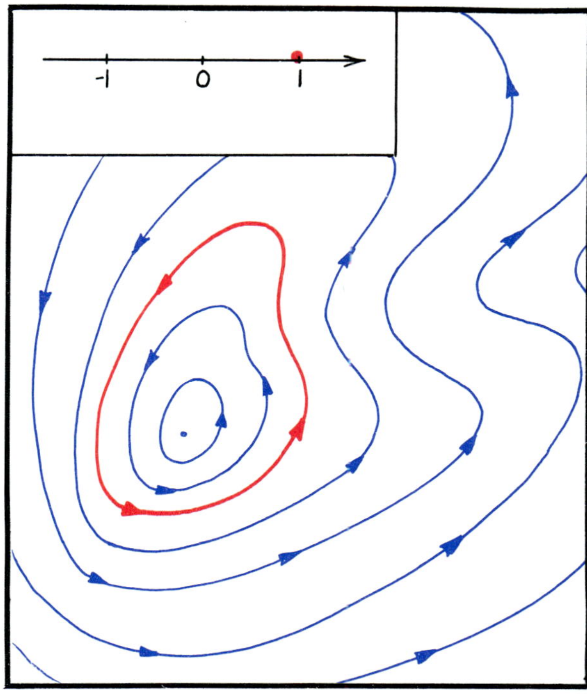

11.2.2. This portrait, called a *center*, has more or less concentric limit cycles. Each of them is nonhyperbolic.

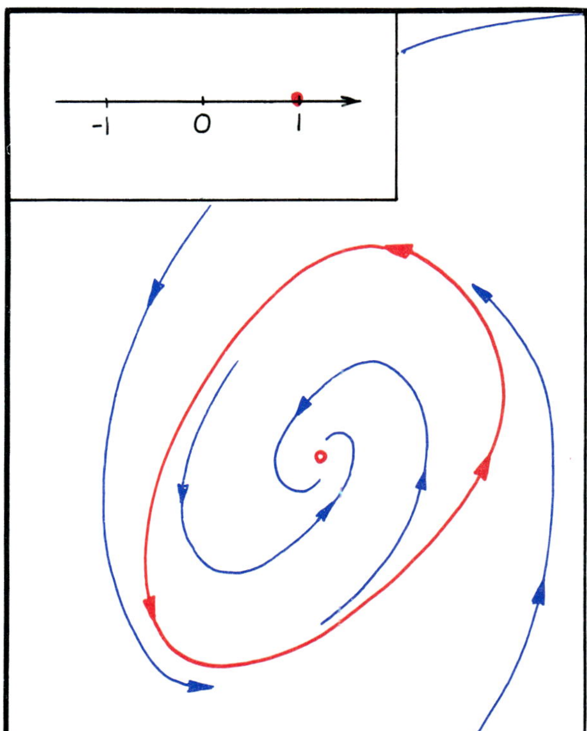

11.2.3. This portrait has a single limit cycle. It attracts on one side, and repels on the other. Its CM of plus 1 is not enough information to predict its attracting/repelling behavior.

This completes our partial survey of limit cycles in 2D. In 3D, each limit cycles has two CM's. They may be conjugate complex, or both real. If they are both real, they may be distinct or identical. This brings up the distinction between *hyperbolic* and the similar idea, *elementary*. The actual definition of *hyperbolic limit cycle* in any one dimension is: there are no CM's of absolute value 1. Elementary is a little stronger. An *elementary limit cycle* is one which is hyperbolic, plus all its CM's are distinct.

All the elementary limit cycles in 3D are shown in Figure 7.5.7.

Here is the definition of G2.

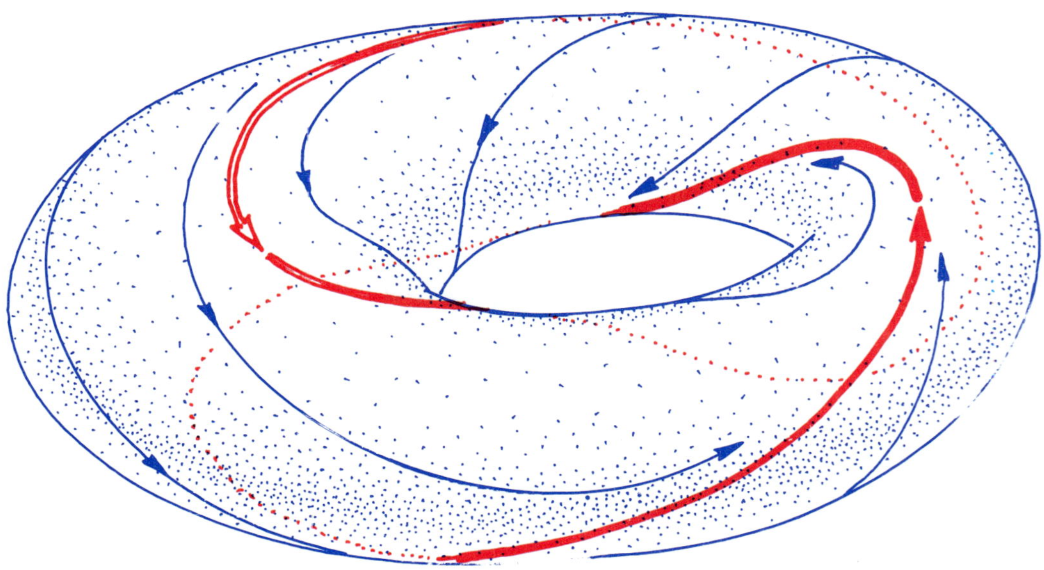

11.2.4. A dynamical system satisfies property G2 if each and every one of its limit cycles is elementary. In this example on the two-dimensional torus, there are several limit cycles in a braid, and each is elementary.

11.3. Property G3 for Saddle Connections in 2D

Every trajectory of a dynamical system comes from somewhere and goes somewhere. That is, it has an alpha limit set and an omega limit set. Every trajectory is in the outset of its alpha limit set, and at the same time in the inset of its omega limit set. Thus, outsets and insets normally intersect each other.

However, most of the time, a trajectory comes from a repellor and goes to an attractor. Exceptionally, one comes from a repellor and goes to a saddle, or comes from a saddle and goes to an attractor. Very exceptionally, a trajectory comes from a saddle and also goes to a saddle. Such a trajectory is called a *saddle connection*, or a *heteroclinic trajectory*. It is even possible for a trajectory to connect a saddle to itself! This is called a *homoclinic trajectory.* Poincaré realized that these trajectories were particularly important in the qualitative behavior of dynamical systems.

Note that a heteroclinic trajectory always belongs to the outset of a saddle (the *donor*), and to the inset of a saddle (the *receptor*) as well. Therefore, the donor outset and the receptor inset must intersect, and their intersection contains the entire heteroclinic trajectory. Generally, the intersection of a saddle outset and a saddle inset contains not just one, but an entire family of heteroclinic trajectories. Property G3 concerns the quality of the intersection of insets and outsets of limit sets of saddle type, especially saddle points and periodic saddles. It requires that these intersections all be *transverse* (that is, cleanly crossing).

In state spaces of one dimension, there are no saddles. In two dimensions, hyperbolic saddle points have invariant curves as inset and outset. There are no periodic saddles. In this section, we briefly explain property G3 *in dimension two only.* The full story is told in detail in Chapters 13 and 14.

In two dimensions, a dynamical system satisfies property G3 if it has no saddle connections at all.

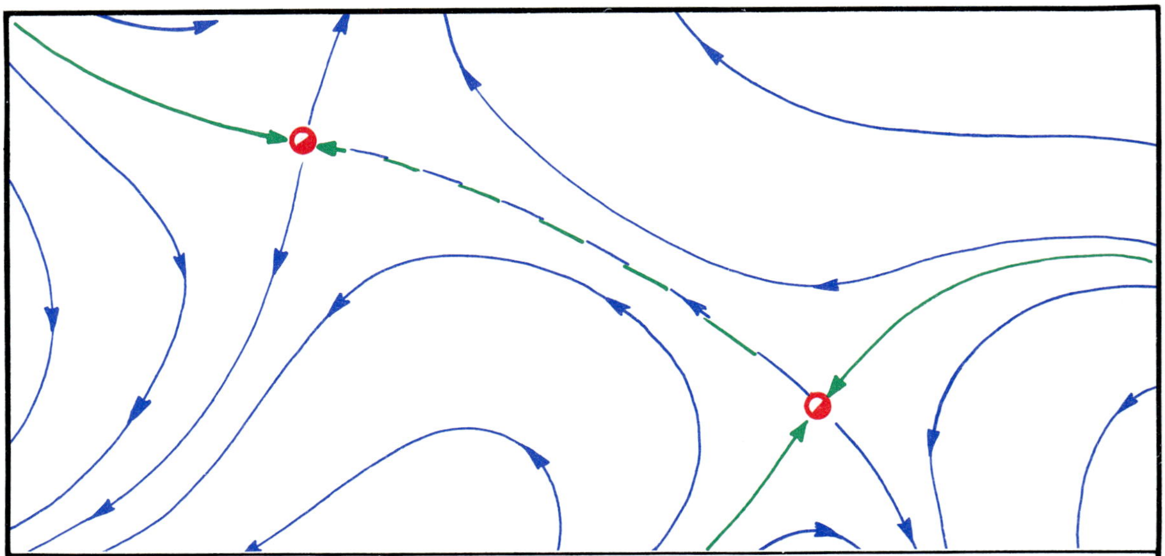

11.3.1. This is a saddle connection in 2D. The dashed trajectory comprises half of the outset of the hyperbolic saddle point on the left, its donor. Simultaneously, it is half of the inset of the hyperbolic saddle point on the right, its receptor. As this system contains a saddle connection, it does not satisfy G3.

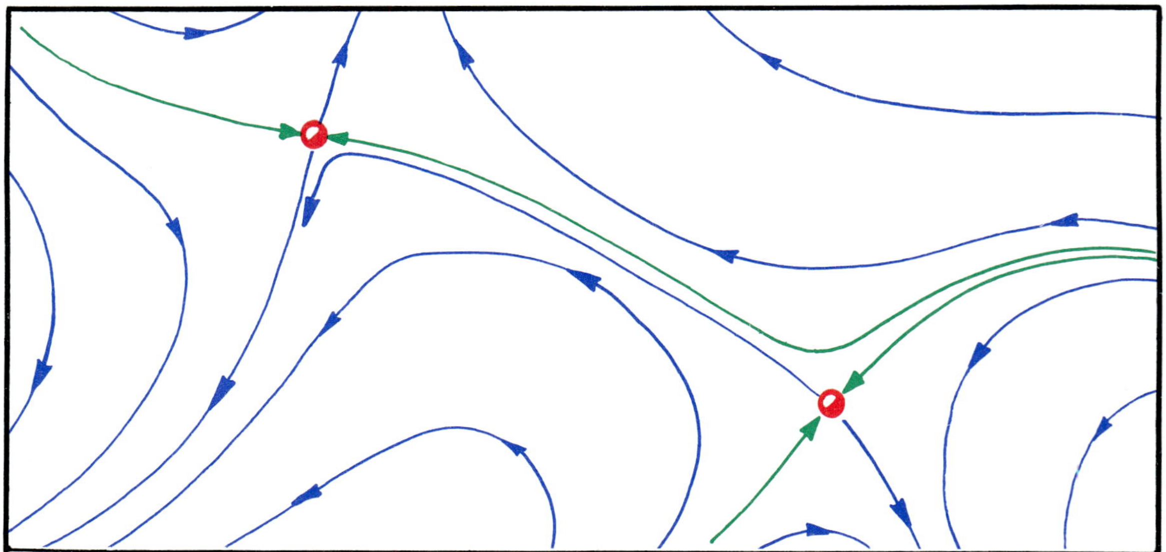

11.3.2. This system has no saddle connection. The outset of the saddle points on the left consists of two trajectories, which go to attractors (not shown). The inset of the saddle point on the right consists of two trajectories, which come from repellors (not shown). One of the trajectories leaving the left saddle narrowly misses one of the trajectories approaching the saddle on the right. This portrait is obtained from the preceding one by a slight perturbation.

Property G3 is a global property. It requires, in two dimensions, that each saddle outset avoid coinciding with any saddle inset.

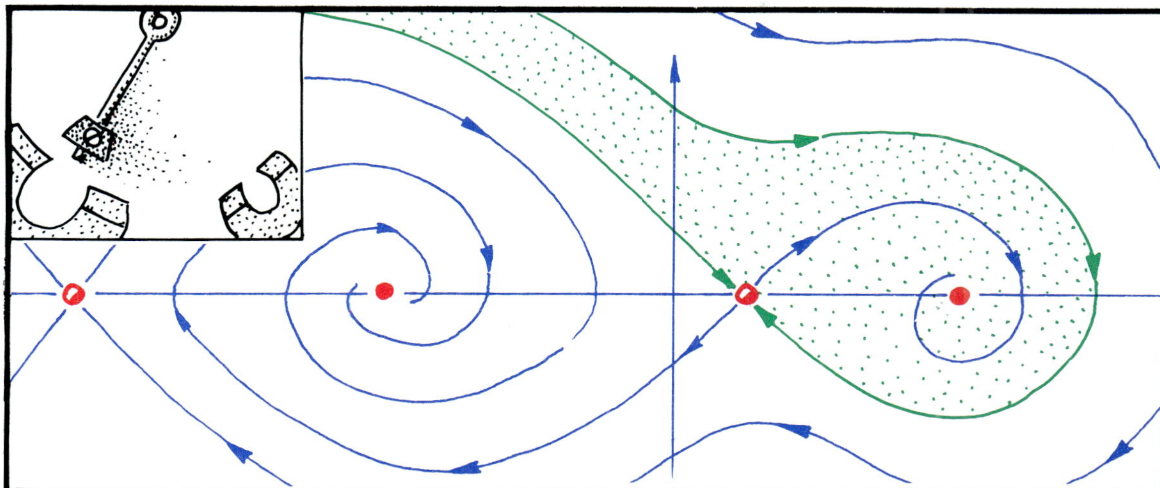

11.3.3. The magnetic pendulum is a global system satisfying property G3. All four saddle outset trajectories successfully avoid all four saddle inset trajectories. (See Figure 2.1.22.)

11.4. Properties G4 and F

Another generic property, G4, will be described in Chapter 15. It was originally formulated by Peixoto, in its oriented, two-dimensional version: The system has *no nontrivial recurrence.* Here is the main example of nontrivial recurrence.

11.4.1. Recall this solenoid, from Figures 1.4.11, 4.4.21, and 4.4.22. All trajectories on this torus are *recurrent* in the sense that their omega (and alpha) limit sets are the entire torus. Thus, if we choose any little disk in the torus, each trajectory recurs, or passes through that disk again and again in its future (and past). We call such a system a *limit torus.*

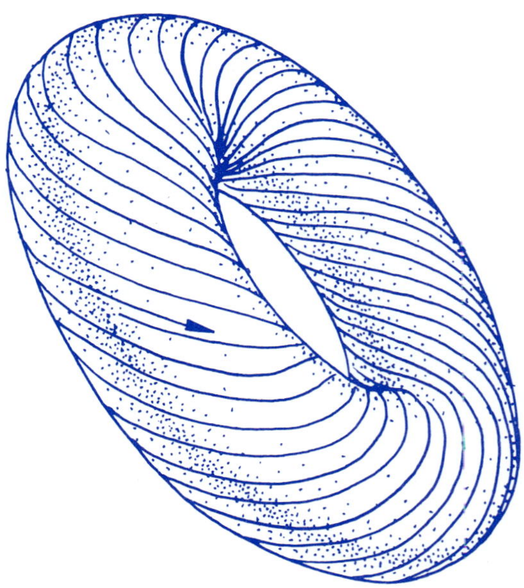

In other words, a limit torus is *topologically transitive,* as described in Figure 9.2.11. It shares this property of all the known chaotic attractors and limit sets. But, it occurs in two-dimensional systems, while chaotic sets do not. So in 2D, the toroidal solenoid is the main example of nontrivial recurrence, while in 3D the situation is much more complicated.

Finally, there is one more generic property we must describe, one which is special to the 2D case.

A dynamical system has *property F* if it has only a finite number of limit sets. In the 2D context, limit sets must be limit points, limit cycles or limit tori. This is a classical result of two-dimensional dynamic systems theory, known as the *Poincaré-Bendixson theorem.* Thus a 2D system satisfying G4 (no limit tori) will also satisfy property F if it has only a finite number of limit points and only a finite number of limit cycles.

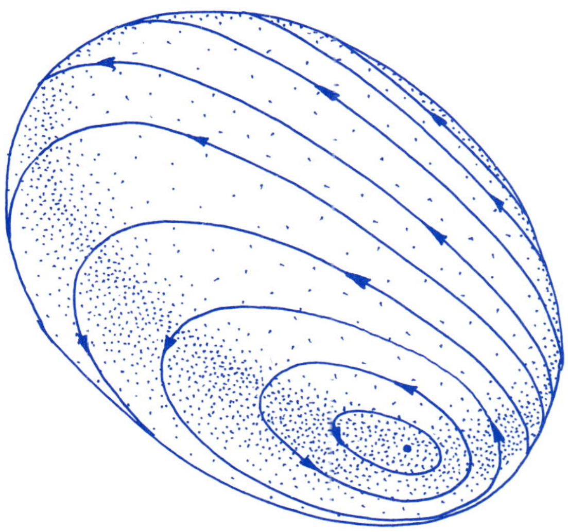

11.4.2. Here is a 2D system violating property F. It has a center: an infinite number of limit cycles, arranged as concentric cycles around a limit point. See Figures 2.1.18, 2.2.3, and 2.2.5 for examples.

These properties G1, G2, G3, G4, and F, were all introduced by Andronov, de Baggis, and Peixoto in their historical works on structural stability in 2D. We now turn to that subject.

12

Structural Stability

In the applications of dynamics in various fields, the dynamics—that is, the actual vectorfield—can never be specified exactly. In fact, outside of a few cases in theoretical physics, one basically makes a rough guess. The mathematical theory of dynamical systems might be useful anyway, if it can describe features of the phase portrait that persist when the vectorfield is allowed to move around. This idea, now called *structural stability,* emerged early in the history of dynamics.

12.1. Stability Concepts

The idea of structural stability seems to have appeared first in the 1930's, in the writings of Andronov and collaborators, in Russia. It was introduced to North America by Lefshetz, the great topologist, and has played a central role in the development of the subject here ever since.

12.1.1. The criteria for structural stability rely upon two supplementary notions: perturbation and topological equivalence. A *perturbation* of a vectorfield means the addition to it of a relatively small vectorfield, frequently unspecified. Here we show the effect of a perturbation, at a single point in the state space.

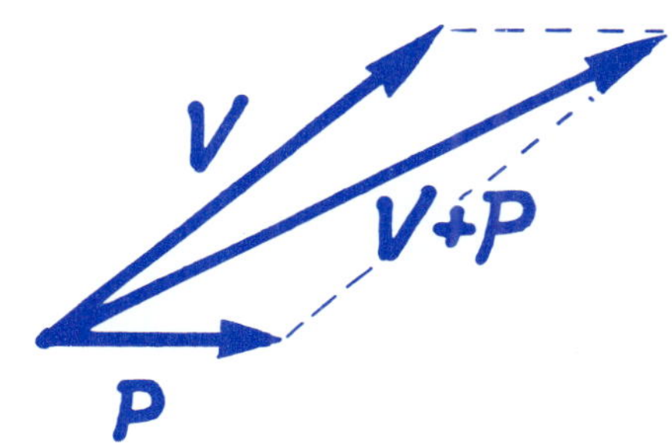

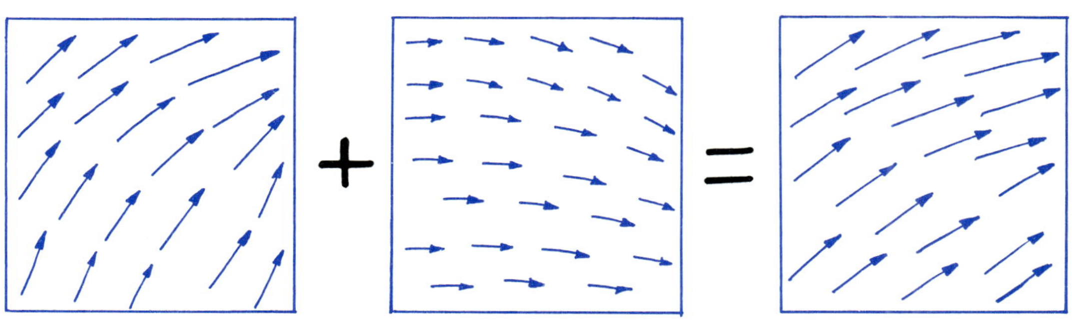

12.1.2. Here we show the effect of a global perturbation. The perturbation is itself a vectorfield, as shown here. The effect of adding this perturbing vectorfield to the original one (on the left) is to modify it at every point in the state space.

Topological equivalence, or synonymously, *topological conjugacy*, of two phase portraits, means there is a *homeomorphism* of the state space, or continuous "rubber sheet" deformation, which maps one of the portraits to the other, preserving the arrow of time on each trajectory.

Here are some topologically equivalent portraits in two dimensions.

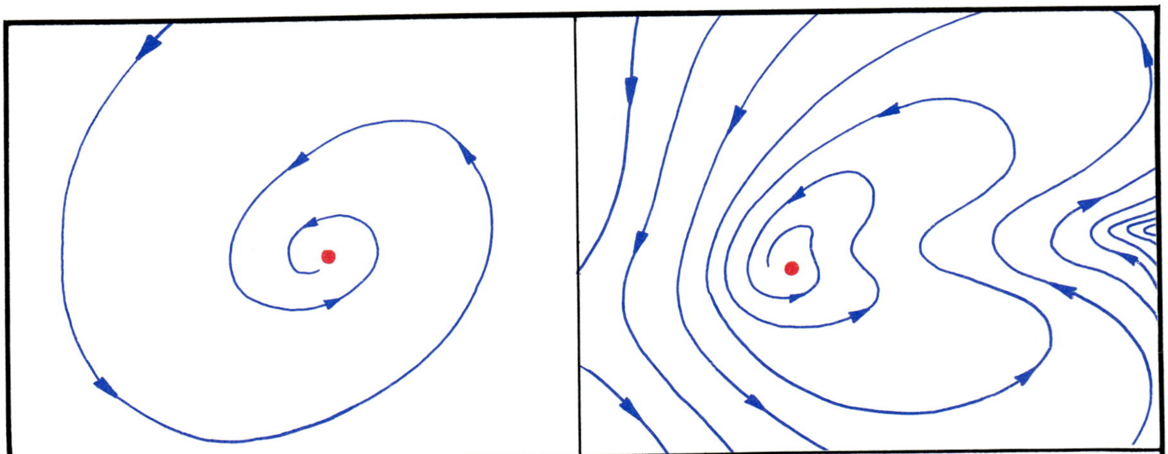

12.1.3. These two point attractors are *topologically equivalent*. A homeomorphism can deform one into the other, preserving the integral curves.

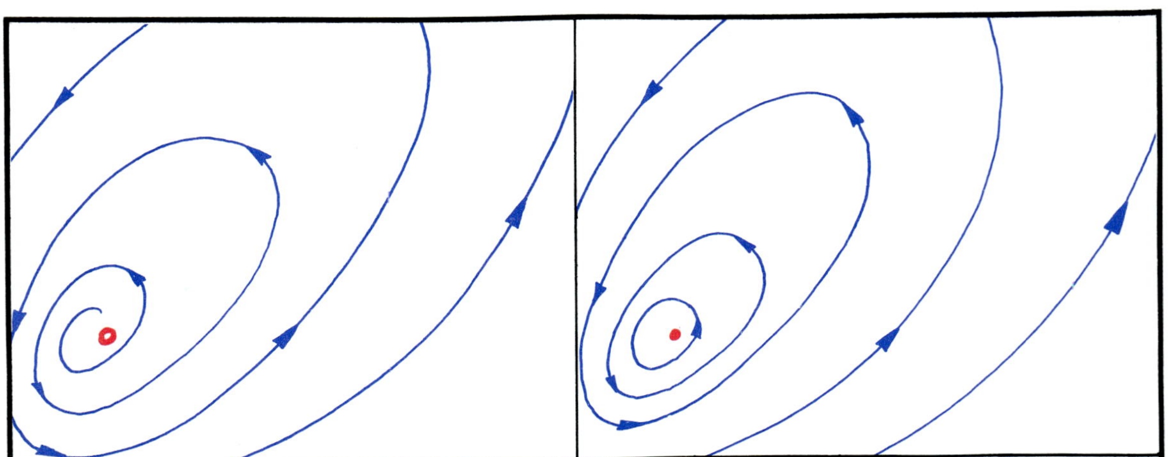

12.1.4. But the point repellor on the left is *not topologically equivalent* to the center on the right. A homeomorphism cannot map a spiral onto a circle.

To be faithful to the theory in higher dimensions, we will need also the concept of *epsilon equivalence*. This is a topological equivalence of dynamical systems, in which the deforming homeomorphism only stretches or slides the state space a small amount (measured by epsilon). Likewise, in the spirit of classical mathematics, we will call a perturbation a *delta perturbation*, if it is small (measured by delta).

Now we use both of these stability concepts, delta perturbation and epsilon equivalence, to introduce the idea of structural stability.

A vectorfield has the property of *structural stability* if (choosing epsilon) all delta perturbations of it (sufficiently small) have epsilon equivalent phase portraits.

Here is a simple example.

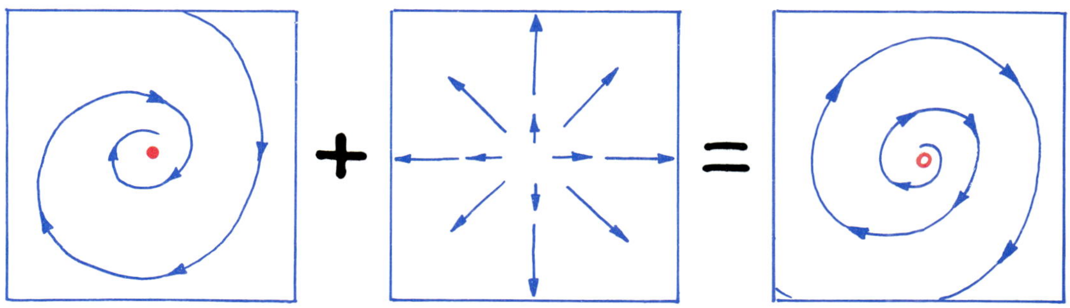

12.1.5. Imagine a system with a spiral attractor which attracts *very weakly*. By adding a medium-sized perturbation pointing outward, we might be able to change it into a spiral repellor.

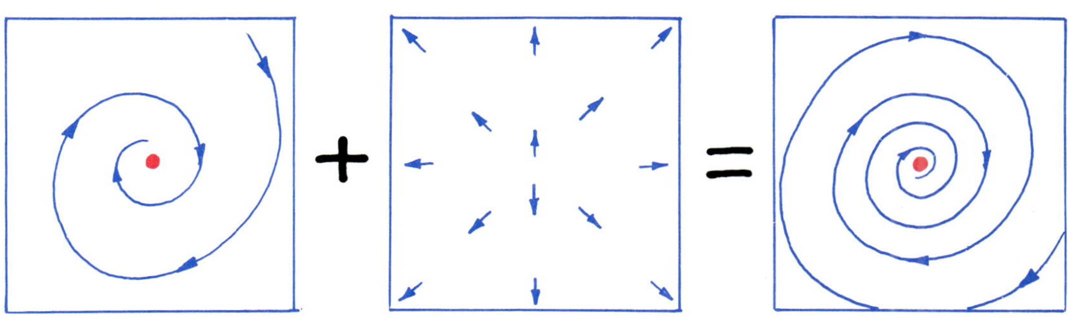

12.1.6. But adding a delta perturbation pointing outward (sufficiently feeble) may make our attractor weaker, but it still attracts. It is topologically (in fact, epsilon) equivalent to the original system. This is an example of a *structurally stable system*.

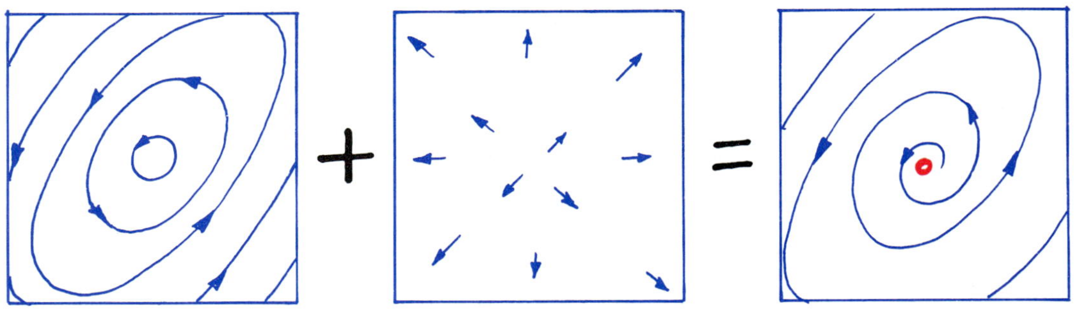

12.1.7. Now consider this dynamical system, a center. The addition of a delta perturbation pointing outward (no matter how weak) results in a point repellor, which is not topologically equivalent to the center. This is a primary example of a ***structurally unstable system***.

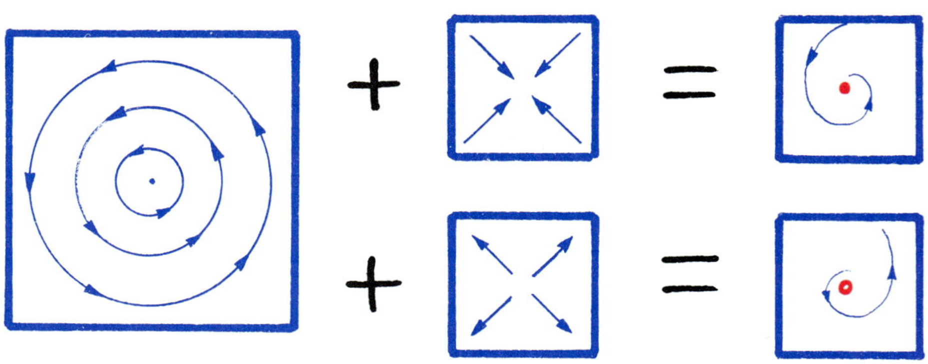

12.1.8. In fact, the center may be perturbed into either a point repellor or a point attractor, depending on the inclination of the perturbation.

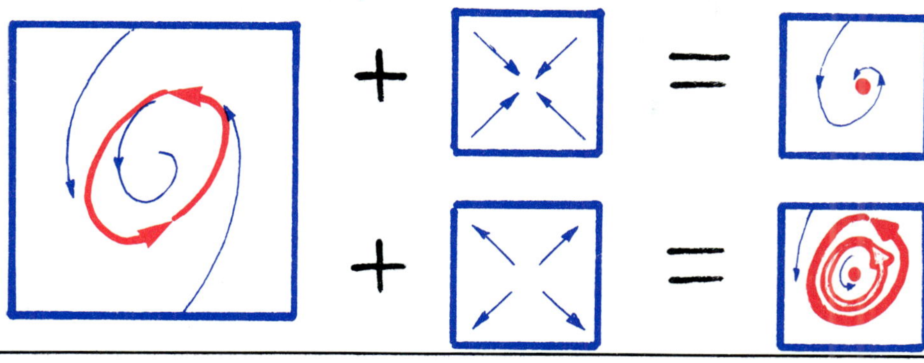

12.1.9. On the other hand, this portrait is structurally stable. The inclination of the perturbation may make the periodic attractor smaller or larger, but the perturbed portraits are all topologically equivalent.

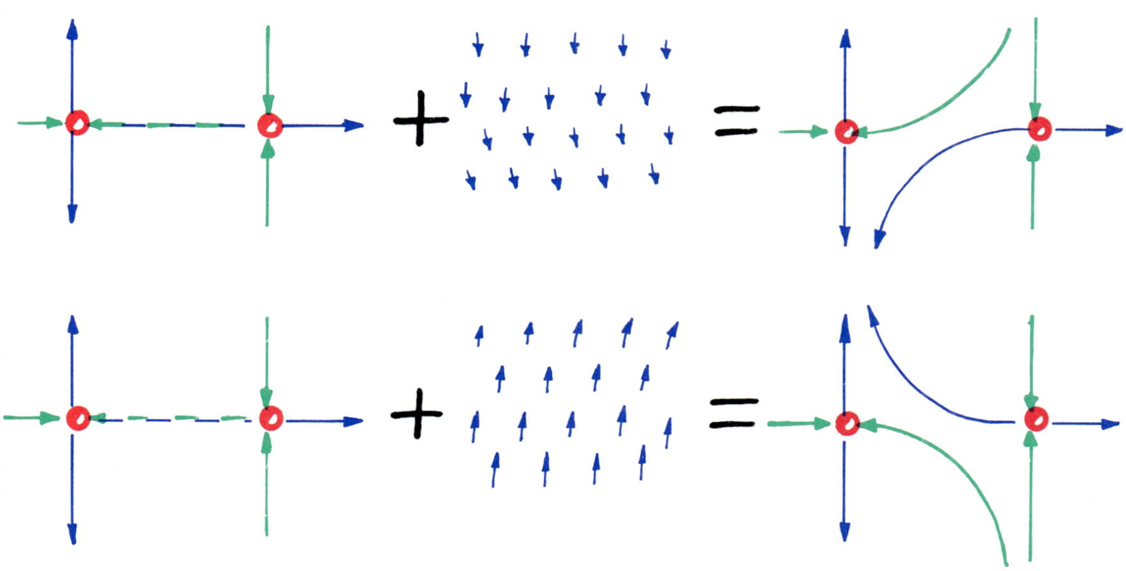

12.1.10. Here is another important example. Consider a system with a saddle connection, as in Figure 11.3.1. Adding a delta perturbation pointing downward (or upward), we destroy the saddle connection. The resulting phase portrait is not topologically equivalent. These two examples illustrate all basic types of structural instability in 2D.

12.2. Peixoto's Theorem

Now we go on to Peixoto's historic theorem, relating the generic properties of the preceding chapter to structural stability in 2D

A watershed in the history of dynamics, Peixoto's work brought together different topology and classical dynamics, ushering in a new age of mathematical dynamics. The attempts to extend his 2D results to 3D and beyond characterized the early days of this new approach, in the 1960's.

Peixoto's result applies to a very restricted class of state spaces, called *compact, orientable surfaces.* We start with these.

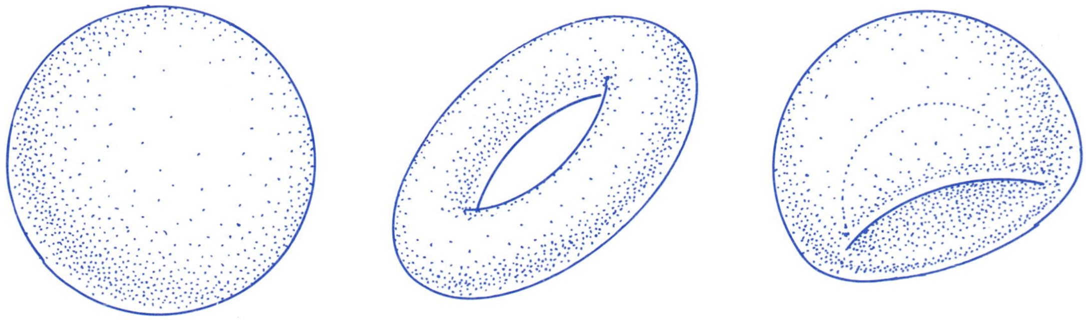

12.2.1. A state space is called *compact* if it can be described as a surface (of whatever dimension) in a finite region of Euclidean space (of a higher dimension) which is a closed set. Here, *closed* means no holes or loose ends. A surface is *orientable* if it has two sides inside and outside. The surfaces shown here are all compact and orientable. *All state spaces in this section will be assumed to be compact orientable 2D surfaces.*

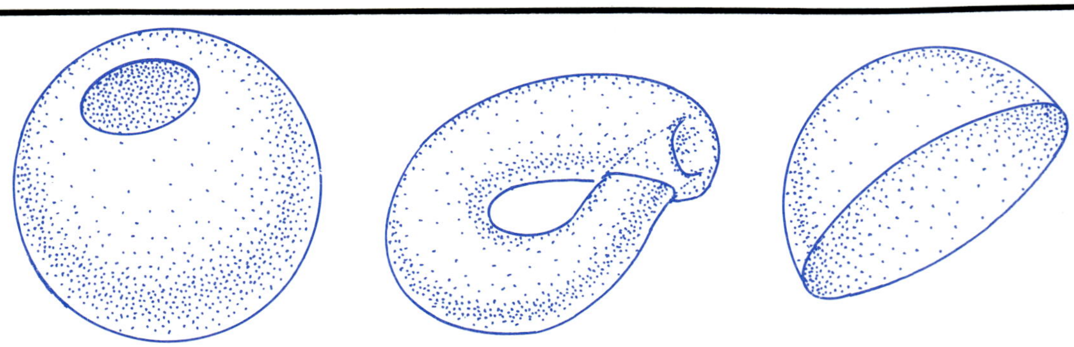

12.2.2. This excludes a sphere with a hole, the Klein bottle, the upper hemisphere, and so on. Nevertheless, the theory described here has been extended to many of these spaces as well.[1]

Now we are ready to state Peixoto's theorem. We will use *property S* as a synonym for structural stability.

Peixoto's theorem: among all smooth dynamical systems on a compact, orientable surface,

A. properties G1, G2, G3, G4, and F are generic,

B. property S is equivalent to these properties (A), and

C. property S is generic.

Clearly C follows from A and B, but this is the most exciting aspect of the theorem. For it says that in applications, this strong kind of stability is to be expected as the typical case, while structural instability is pathological.

Here, *generic* is a technical term, which we translate as *typical* sometimes. However, the atypical cases (especially those in which property G4 is violated) are so frequently observed in experiments that we should use *weakly generic* as the technical term, and understand *typical* as meaning *slightly more probable* than the exceptional cases. The reason for this paradox is that the Kronecker (solenoidal) flows on the torus (See Part One) occur for a *fat fractal* or *thick Cantor* set of leaves in Thom's big picture.[2] This will be explained in more detail in Part Four.

Part A was generalized promptly to higher dimensions, except for the genericity of F, which failed, along with C. Part B also was generalized, by Smale and Palis. More about this in later chapters.

Peixoto's proof is outlined in the next section. Here, we give some examples.

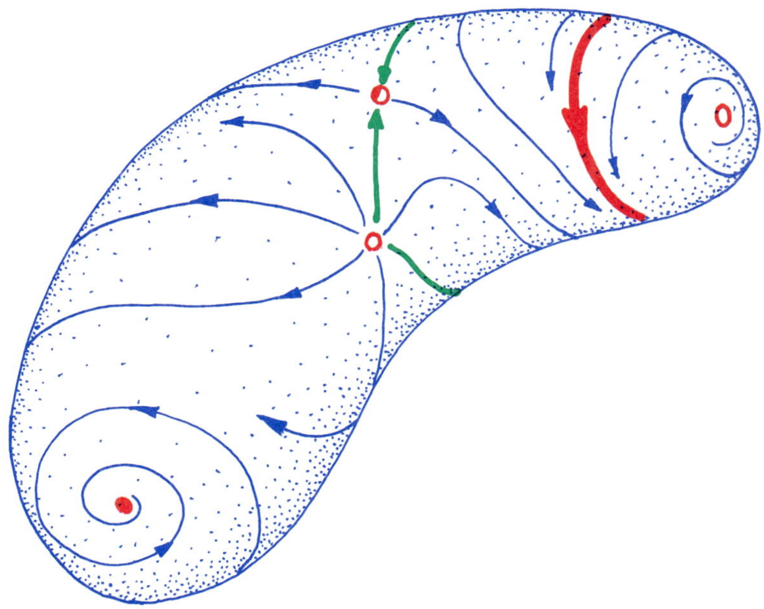

12.2.3. Here is a system exhibiting G1–G4 plus F, and thus S.

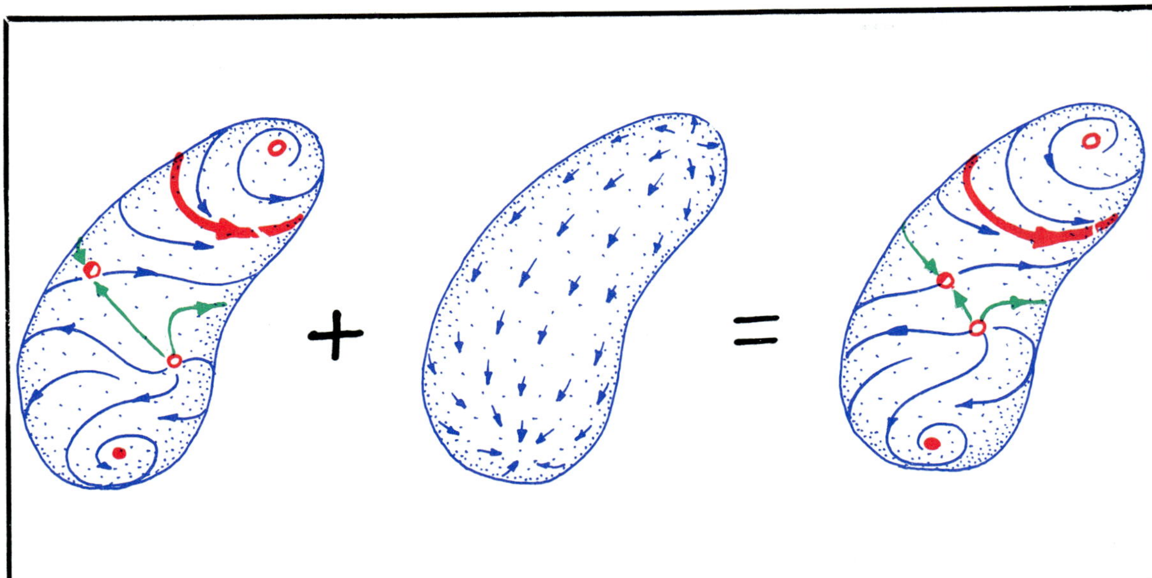

12.2.4. A delta perturbation yields an epsilon equivalent portrait.

But Peixoto's theorem says more: saddle connections are structurally unstable, as we saw in Section 11.3.

Peixoto's theorem says still more: nontrivial recurrence (solenoidal flow on a torus) can be perturbed (in Thom's big picture) into a structurally stable system.

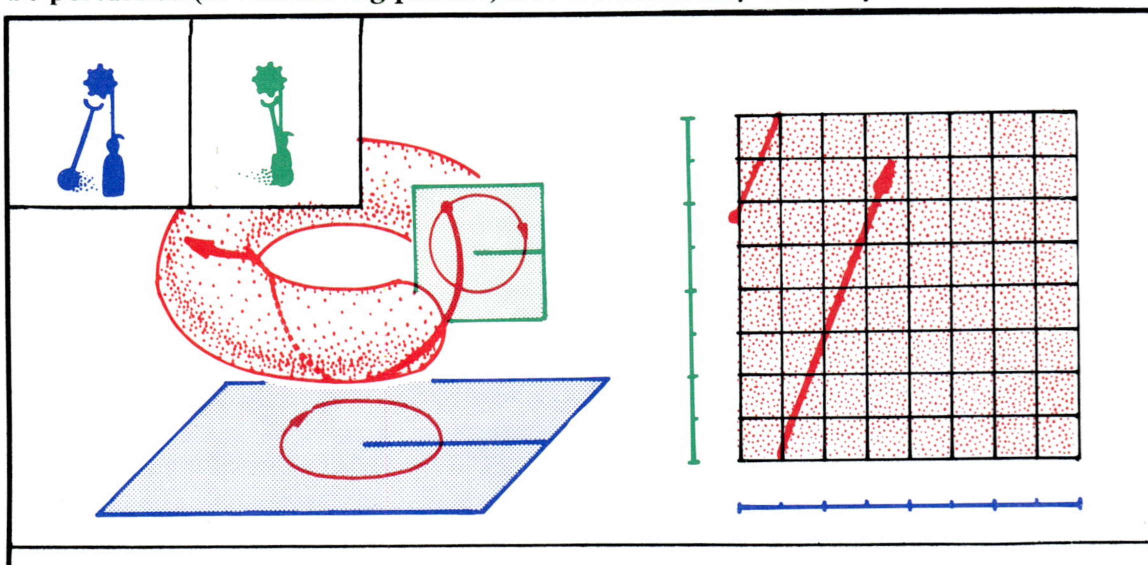

12.2.5. Here is a torus with a solenoidal flow. It violates property G4, so by part B of Peixoto's theorem, it is not structurally stable. By part C, it can be changed to an S system by a delta perturbation. *Warning:* This delta perturbation may be rare, or hard to find, since it belongs to the complement of a thick Cantor set, as explained in Part Four.

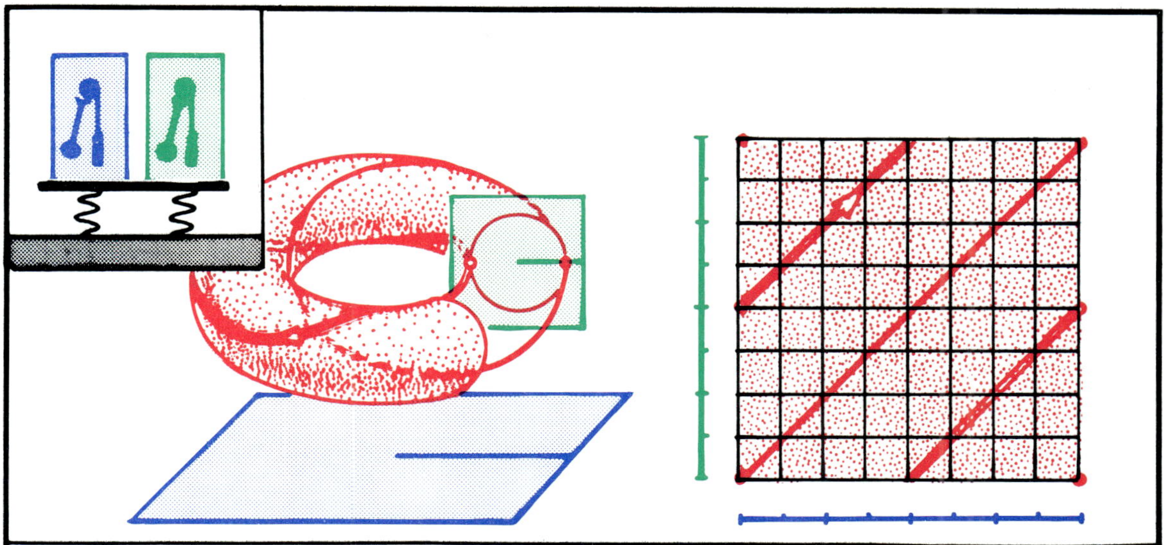

12.2.6. This S system will not have any limit points or limit tori, but it must have limit sets. So, there are some limit cycles, braided around the torus. They occur in pairs, alternately attracting and repelling. The implications for frequency entrainment of coupled oscillators are discussed in detail in Chapter 5.

12.3. Peixoto's Proof

We break the proof into five steps:

1. G1 implies FP (finite number of limit points).
2. G2 implies FC (finite number of limit cycles).
3. G4 implies no limit tori.

Therefore, G1, G2, and G4 imply F.

4. G1, G2, G3, and G4 (and hence F) imply S.
5. S implies G1, G2, G3, and G4.

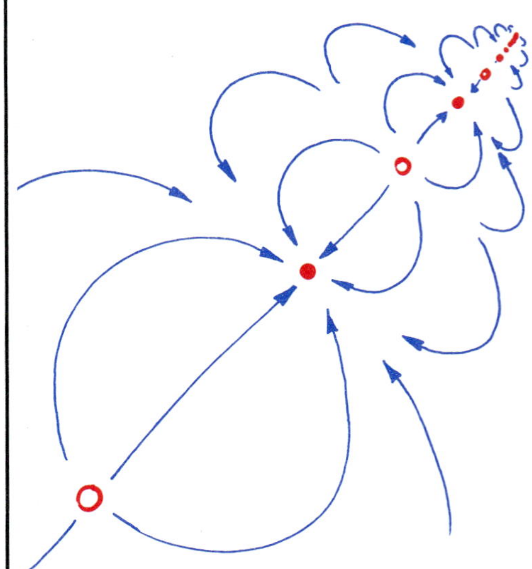

12.3.1. *Step 1: Generic property G1 implies there are only a finite number of limit points.* For in the compact state space, an infinite number of critical points would have to contain a convergent sequence as shown here. And the critical point at the end of the sequence will have to violate G1.

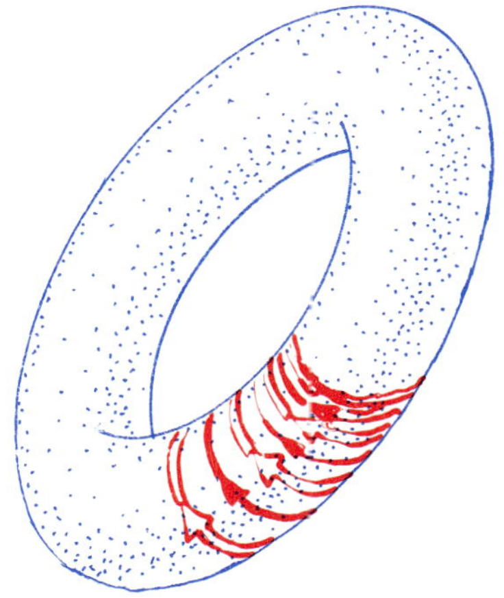

12.3.2. Similarly, generic property G2 implies there are only a finite number of limit cycles. This is special to two dimensions, where an infinite number of limit cycles would be forced to "pile up." That is, either they must converge to a limit cycle, as shown here (violating the generic condition G2—hyperbolic limit points).

The proof of this step used topology and calculus, and is not terribly difficult.

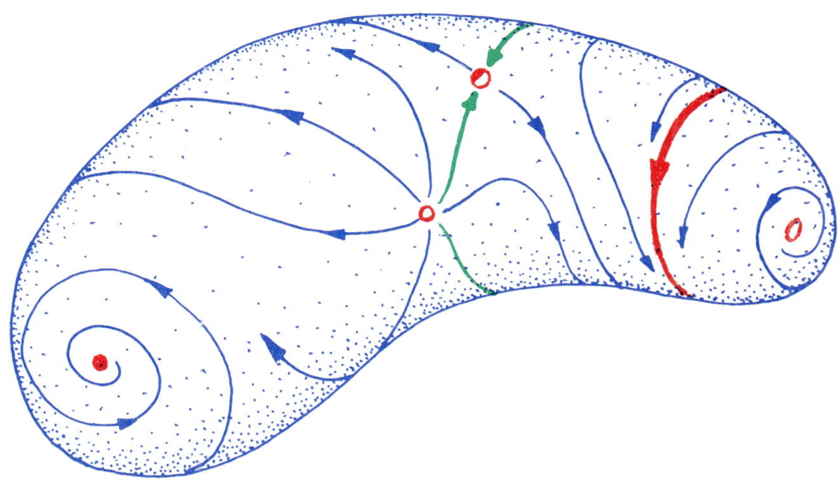

12.3.3. *Step 2: If the system is generic (G1, G2, G3, and G4), then it has only a finite number of limit points, a finite number of limit cycles, and no other limit sets.* This is called *property F.* Further, they are all hyperbolic, and there are all hyperbolic, and there are no saddle connections. Here is a typical portrait of this type.

The proof of Step 2 requires the infamous *Closing Lemma*. This is used to eliminate the possibility of a toroidal limit set. First proved in the present context by Peixoto, it has been wonderfully generalized by Pugh and Robinson.[3]

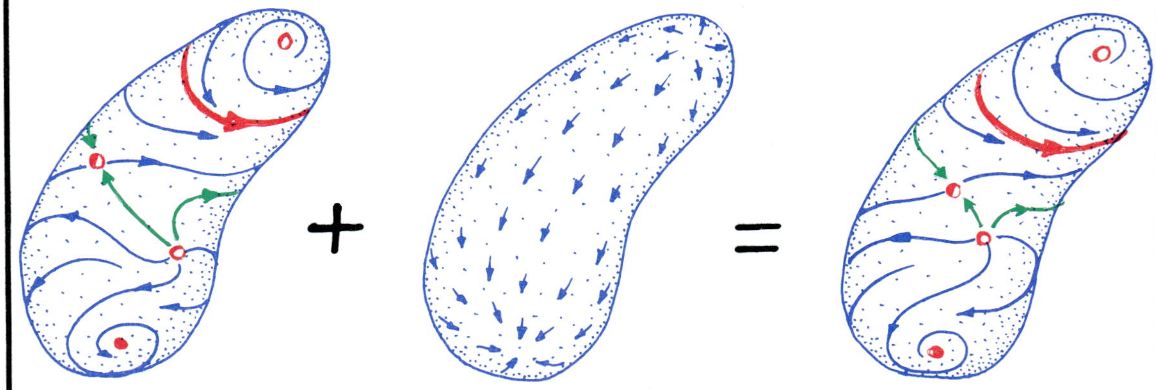

12.3.4. *Step 3: These generic properties (G1, G2, G3, G4, and perforce F) ensure structural stability.* An arbitrary small perturbation of the portrait shown in the preceding panel produces an equivalent portrait.

The proof of this step requires the actual construction of a topological deformation from the original portrait to the perturbed one, but is not too difficult.

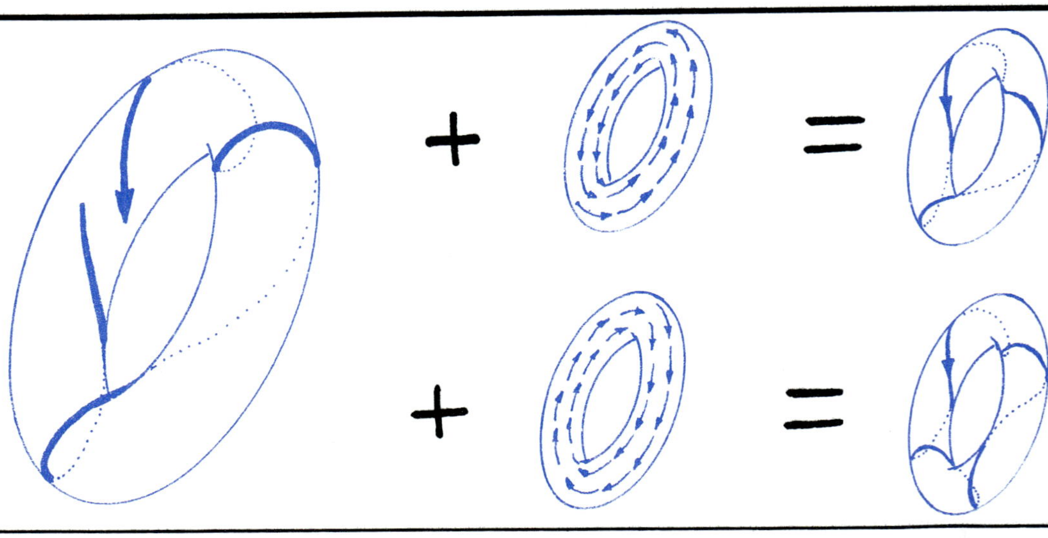

12.3.5. *Step 4: Structural stability ensures the generic properties (G1, G2, G3, G4, and necessarily, F).* The preceding section gives examples showing how structural stability ensures the first three properties. Here is an example showing how G4 is ensured. The center portrait has a toroidal limit set with no limit cycles or limit points. The only limit set is the entire state space, a torus. Small perturbation can produce the two portraits shown below, which are not topologically equivalent. The difficult Closing Lemma is used in this step also. *Warning:* Again, the perturbations producing these structurally stable (braided) flows from the solenoidal flow can be rare, or hard to find, because of belonging to the complement of a thick Cantor set.

13

Heteroclinic Tangles

Limit points and cycles of saddle type may be distributed throughout the state space. Each has insets and outsets, which wander around near each other. Intersections are not unlikely. These, called *saddle connections*, consist of trajectories of the dynamical system that lead from one saddle (called the *donor*) to another (the *receptor*). This connecting curve is called a *heteroclinic trajectory* if the donor and receptor saddles are different, or a *homoclinic trajectory* if they are the same. This chapter is devoted to saddle connections by heteroclinic trajectories which satisfy the generic property *G3*, or *transversality*. The *homoclinic case* (a trajectory connects a saddle to itself) is described in the next chapter.

In state spaces of one dimension, there are no saddles. In two dimensions, there are generic saddle points with one-dimensional insets and outsets. In the three-dimensional cases, there are generic saddle points and cycles, of which the insets or outsets may be surfaces. In this chapter, we will describe all of the transverse heteroclinic saddle connections in two and three dimensions: limit point to limit point, limit point to limit cycle, and cycle to cycle.

13.1. Point to Point

First, consider phase portraits in the plane, with two hyperbolic limit points of saddle type. The insets of each are curves, likewise their outsets. These curves are trajectories of the dynamical system.

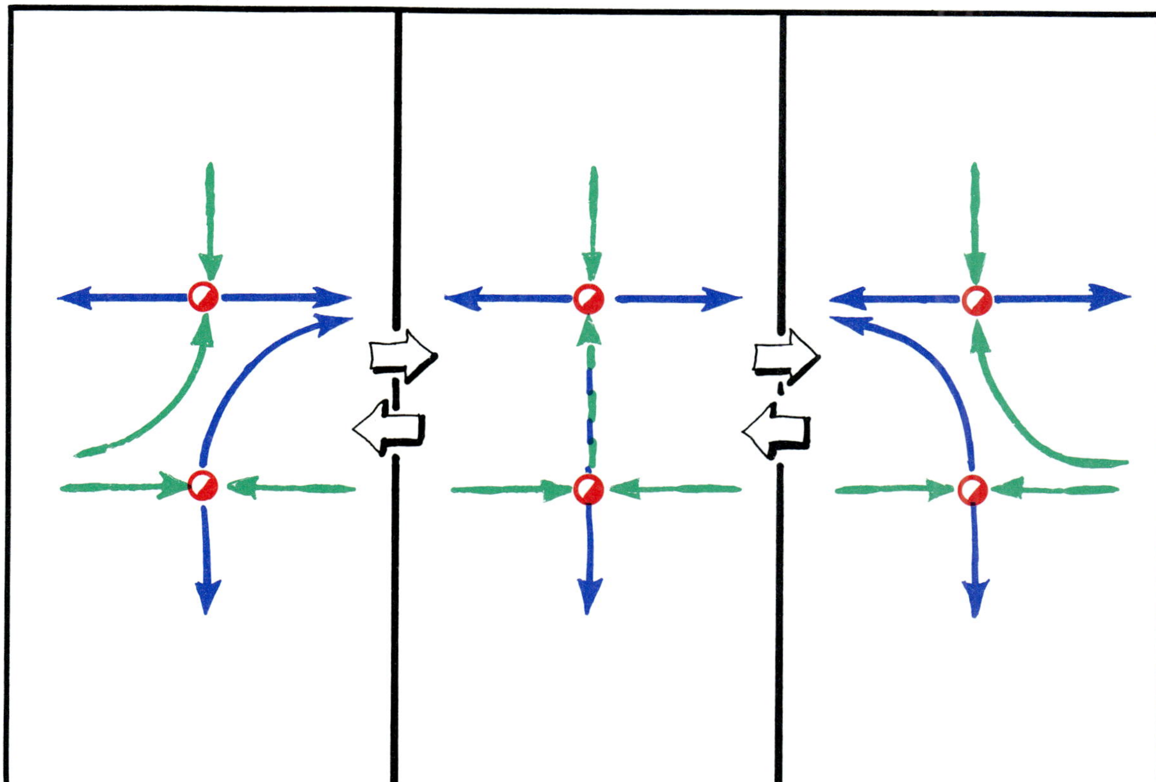

13.1.1. These three phase portraits each have two hyperbolic limit points of saddle type. The end ones have no saddle connection, while the one in the center has a single heteroclinic trajectory. The sequence has occurred previously in Part One, uunder the name *saddle switching*. It represents the actual coincidence of the outset from the left saddle and the inset to the one on the right. The transverse intersection of two curves in the plane must be in isolated points. Therefore, this intersection is not transverse. It is a *nongeneric* saddle connection. There are not transverse saddle connections in the two-dimensional case.

And now, on to two hyperbolic saddle points in 3D.

In the three-dimensional case, there are several possibilities. There are two types of topologically distinct hyperbolic saddle points: index 1 (inset two-dimensional, outset one-dimensional) and index 2 (inset one-dimensional, outset two-dimensional). Each can be a donor or receptor of a saddle connection. But *transverse* saddle connections, in 3D, only occur between two-dimensional outsets and two-dimensional insets. Such an intersection consists of a single curve, a trajectory.

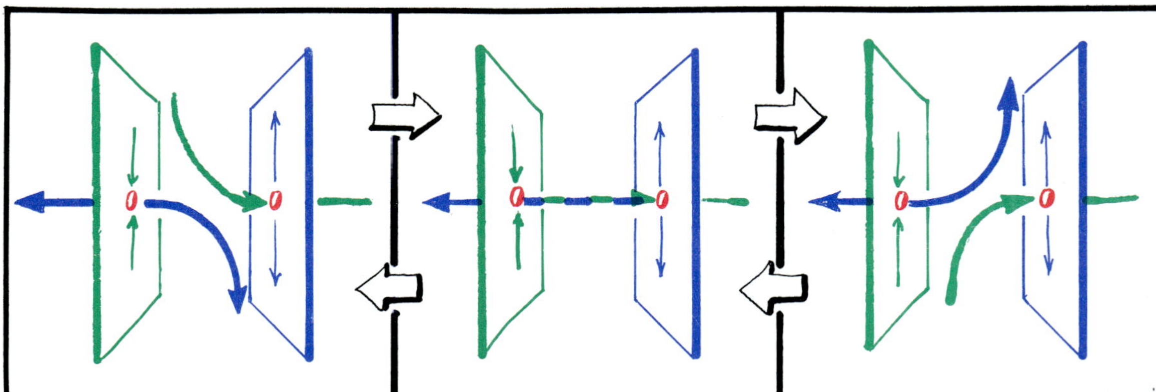

13.1.2. A saddle point of index 1 cannot have a transverse connection to a saddle point of index 2, in three dimensions. Three closely related portraits are shown here, in analogy to saddle switching in the two-dimensional case. The one in the center has a nontransverse heteroclinic trajectory connecting the two saddle points.

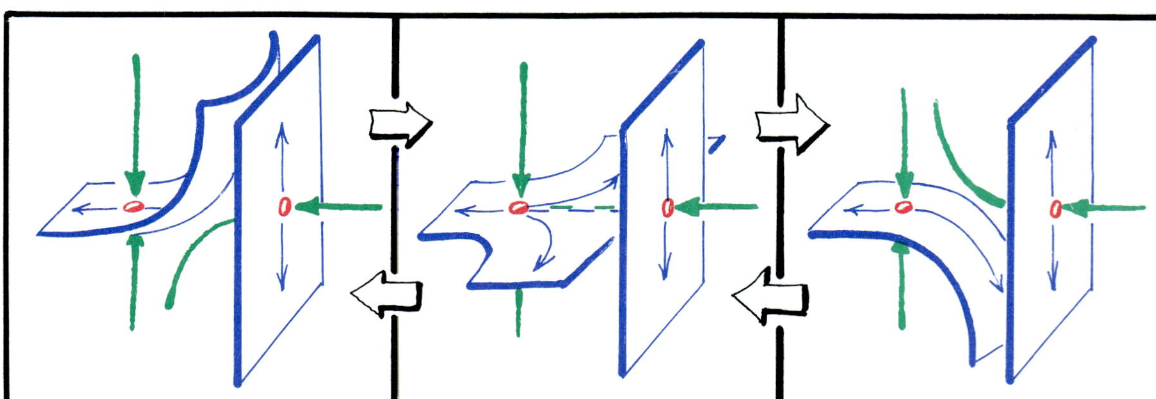

13.1.3. The next donor, a saddle point of index 2, cannot have a transverse connection to a saddle point of index 2 (same receptor as above), in three dimensions. Here again, three similar portraits are shown. The one in the center is an example of a nontransverse heteroclinic trajectory.

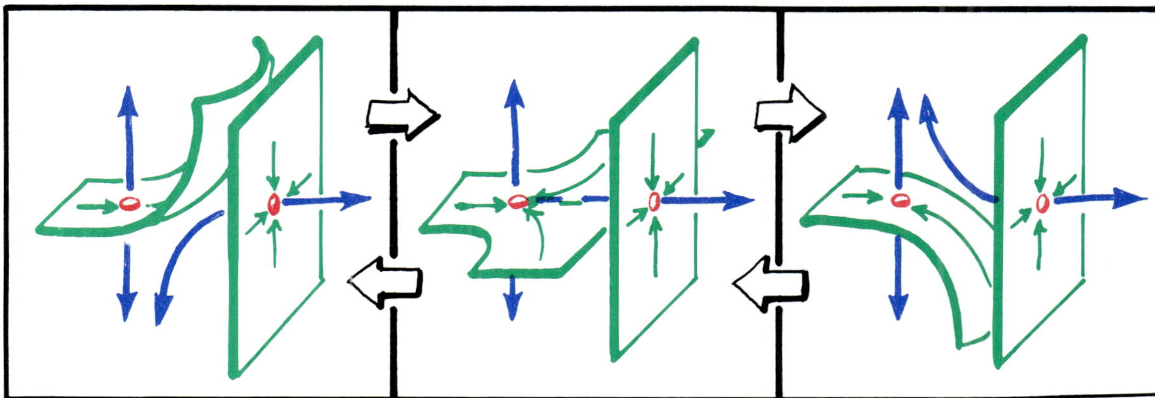

13.1.4. Transverse connection from a saddle point of index 1 to a saddle point of index 1 (like the case of index 1 to index 2, and index 2 to index 2, described above) cannot occur in three dimensions.

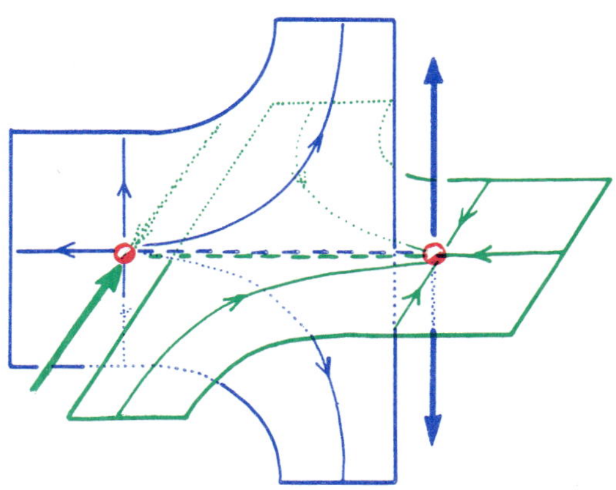

13.1.5. In this fourth case, a heteroclinic trajectory leads from a saddle point of index 2 to one of index 1. The outset of the donor and the inset of the receptor are both two-dimensional. Thus, a transverse intersection of them in a one-dimensional curve (necessarily a trajectory of the dynamical system) is possible. A nontransverse intersection along a heterclinic trajectory is also possible—for example, the two surfaces could be tangent to each other, along their intersection. Here, the transverse case is illustrated. This is the only generic (transverse) connection between saddle points in three dimensions.

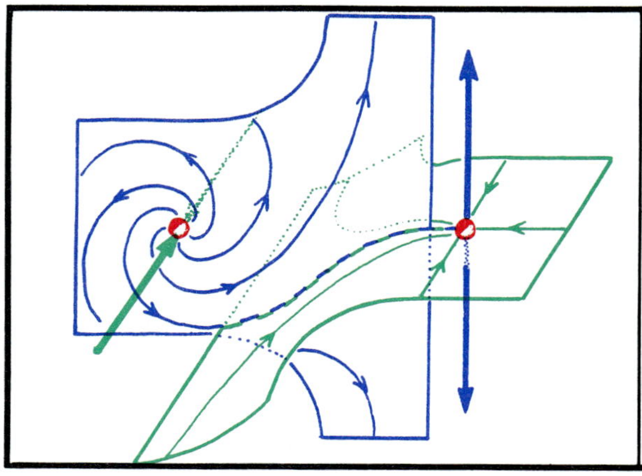

13.1.6. The preceding illustration shows the transversely connected saddle points, assuming both are of the radial (nonspiral) type. Here, the donor has been replaced by a spiral type. This is topologically equivalent to the preceding portrait.

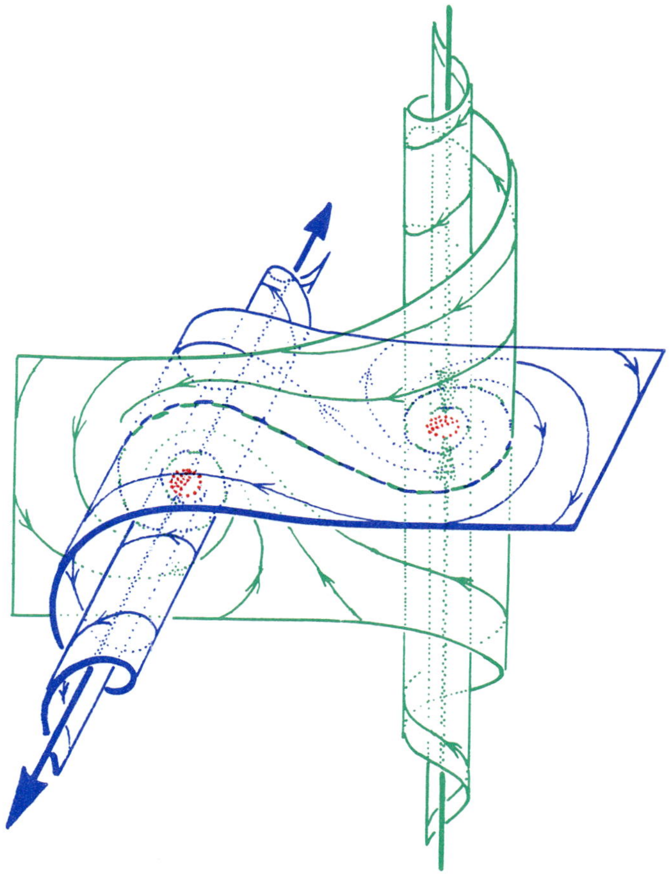

13.1.7. In this example, both the donor and the receptor are of the spiral type. Again, this is topologically equivalent to the preceding portraits.

13.2. Outsets of the Lorenz Mask

Recall the Lorenz mask, from Part Two. This was the first chaotic attractor to be firmly established in experimental dynamics. It is actually made of tangled outsets. Here, developed in stages, is the complex of point-to-point tangles found in the Lorenz system.[1] There is a radial saddle point of index 1 (the *receptor*) situated between two spiral saddle points of index 2 (the *donors*). The outset surfaces of the two donors are heteroclinically incident to the inset surface of the receptor.

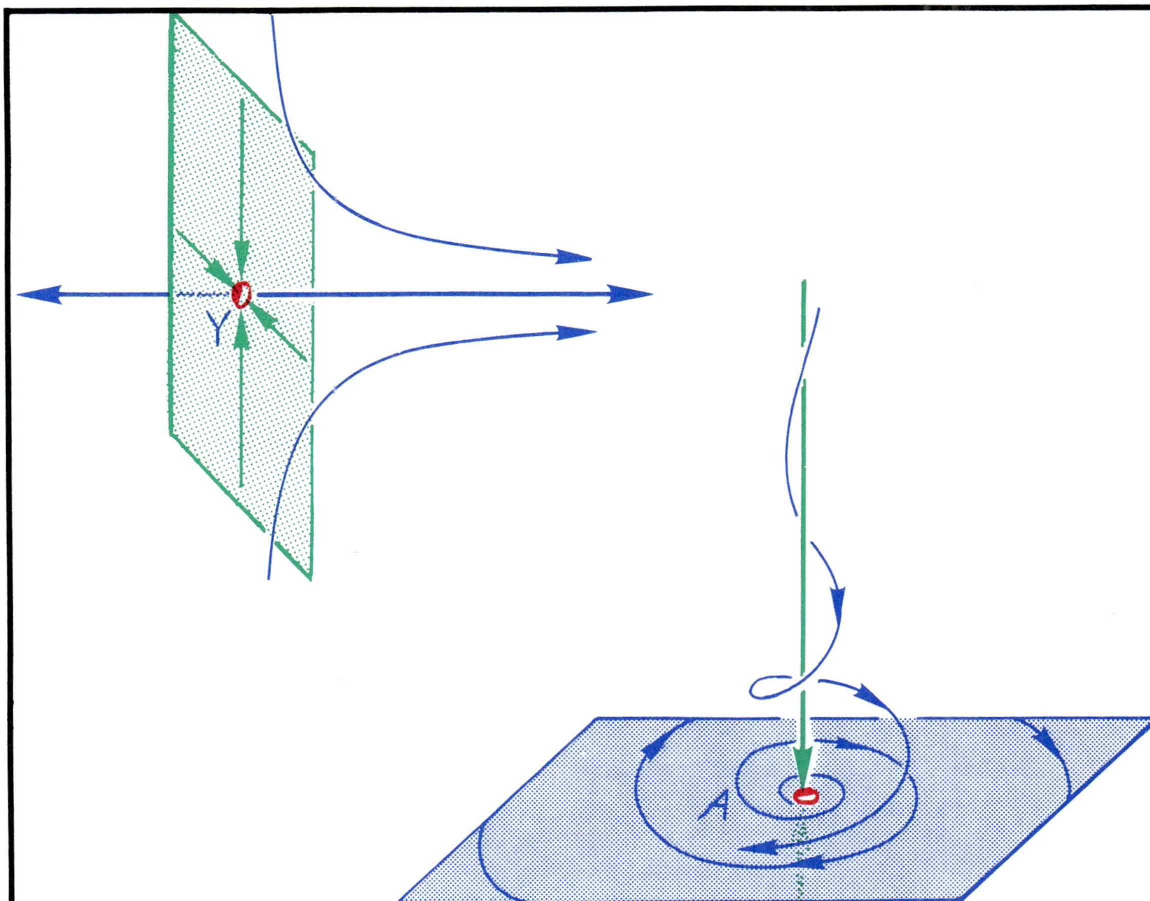

13.2.1. Here are two saddle points, A and Y. They are hyperbolic, in three dimensions. One, A, has index 2, with spiral dynamics on its planar outset (shaded), Out(A). The other, Y, has index 1, with nodal dynamics on its planar inset (dotted), In(Y). The two outsets are attractive, as shown by the neighboring trajectories. As Out(A) and In(Y) are both two-dimensional, they could intersect transversely in three space. If they did, the transversal intersection would have to be a trajectory, called a *heteroclinic trajectory*.

Next, we will build up this complex, step by step.

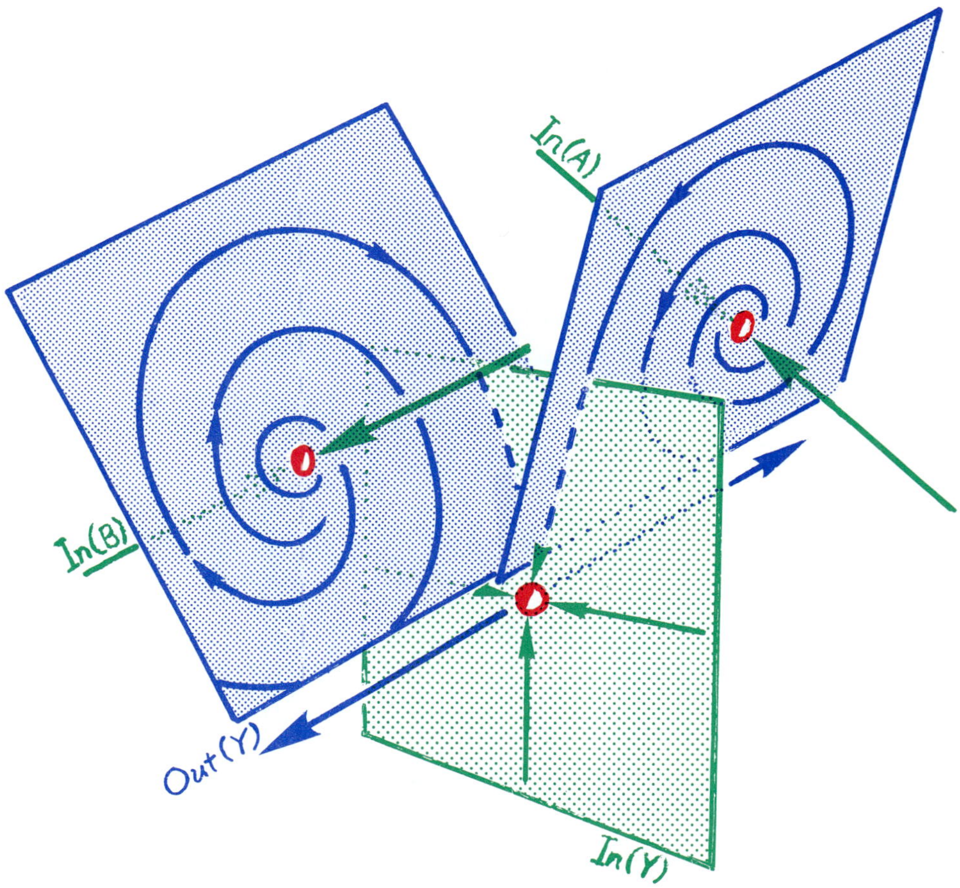

13.2.2. Adding another saddle point, B, essentially identical to A, we make a yoke like this. Both A and B are heteroclinic to Y. They are *transversely heteroclinic*, as the two planar outsets (shaded) intersect the planar inset (dotted) transversely. There are *two heteroclinic trajectories* in this yoke. Note that the arriving outsets are incident upon the departing outset, at Y. We call this a *neat yoke*. Next, we will see where these outsets end up.

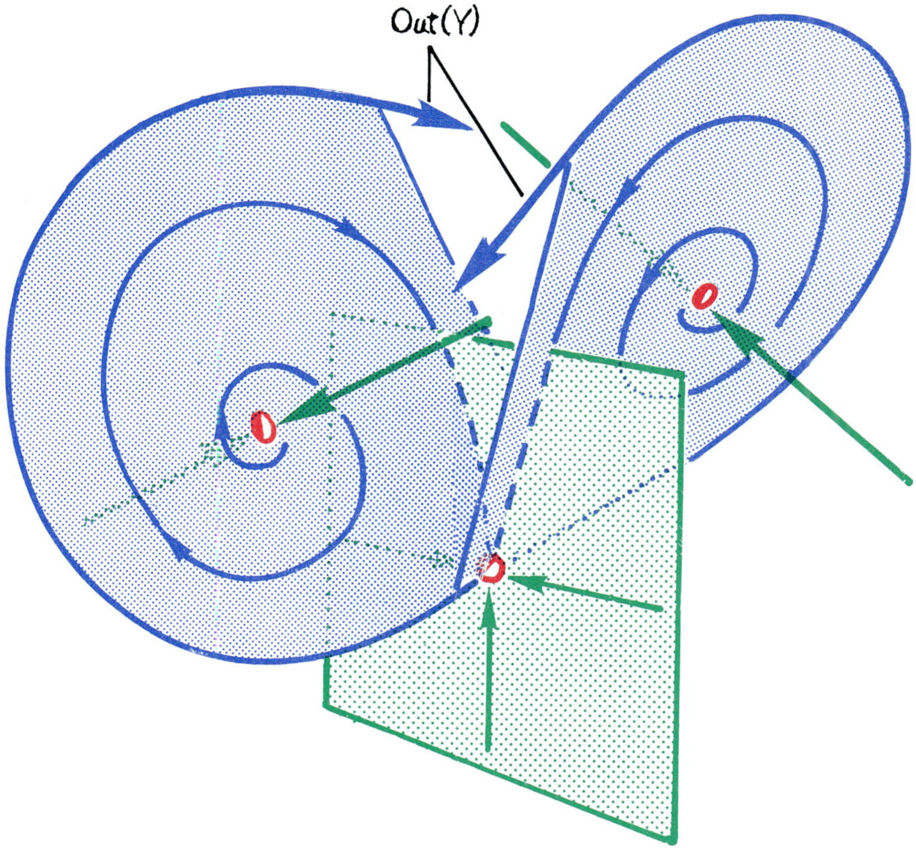

13.2.3. As the arriving outsets, Out(A) and Out(B), both have spiral dynamics, the departing outset that bounds them, Out(Y), swirls around and reinserts, as shown here. It cannot go off to infinity, as the Lorenz system has a repellor at infinity.

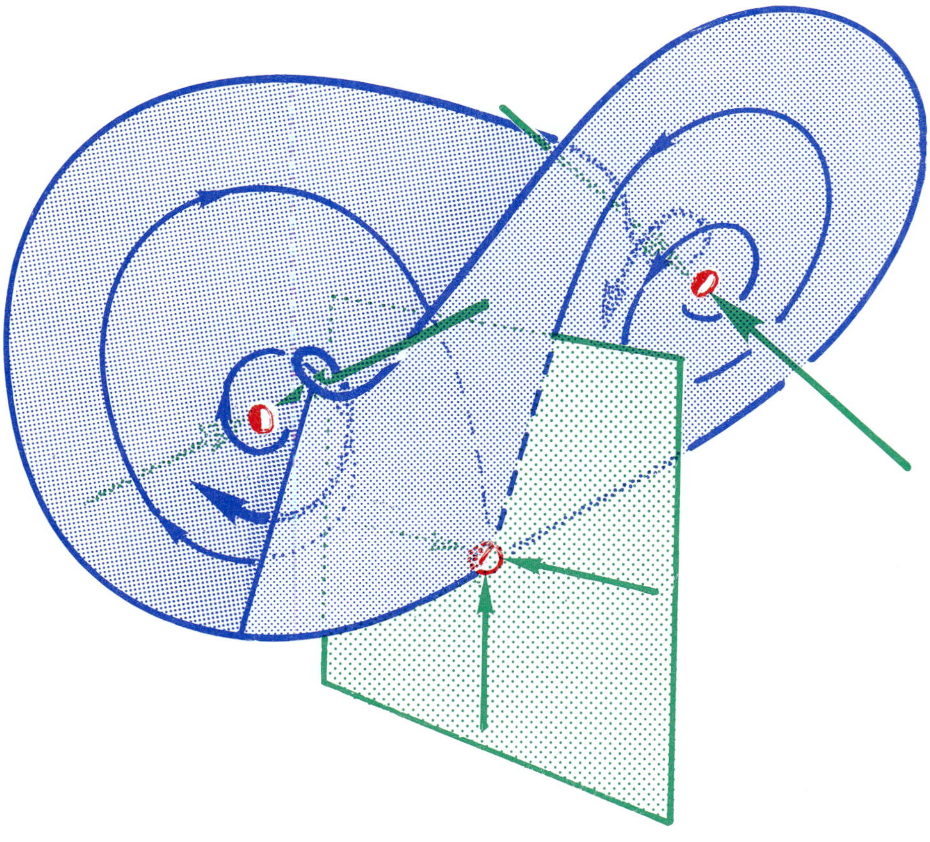

13.2.4. The result of reinserting is this: as each branch of Out(Y) swirls around one of the shaded outsets, it approaches near the other shaded outset. It gets attracted, as outsets are attractive. Thus, the omega limit set of Out(Y) is within the closure of the union of the three yoked outsets.

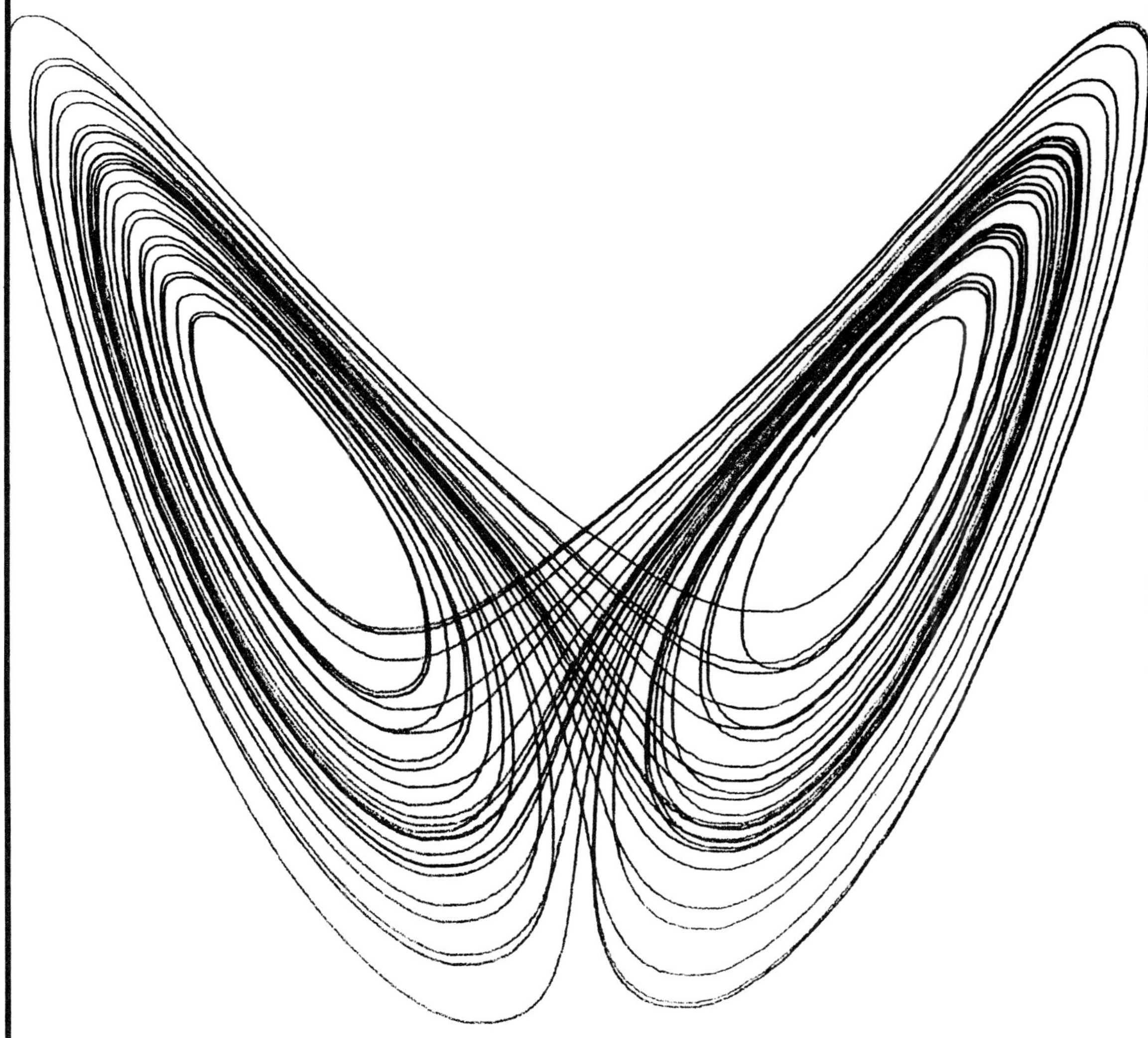

13.2.5. And here, for comparison, is a computer drawing by Robert Shaw of the Lorenz attractor. Inspection of the equations reveals the three distinguished saddle points, right where we want them. But the planar inset of the saddle point in the lower center is qualitatively invisible. It is a kind of a separatrix. Now we will add it to the picture, with its full extension.

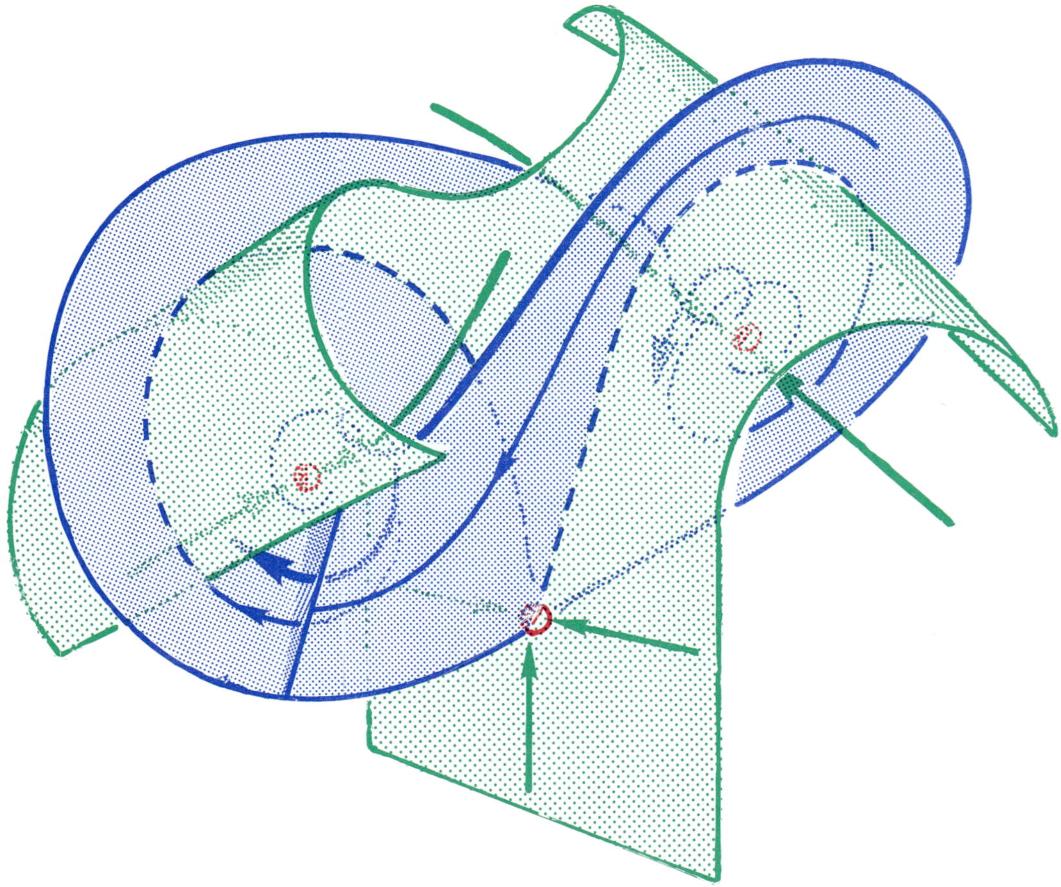

13.2.6. Referring to Figure 13.2.4, we run the flow backwards in time, to extend the planar (dotted) inset outward from Y. It follows the heteroclinic trajectories (dashed) back to the yoked saddles, A and B, scrolling as it goes.

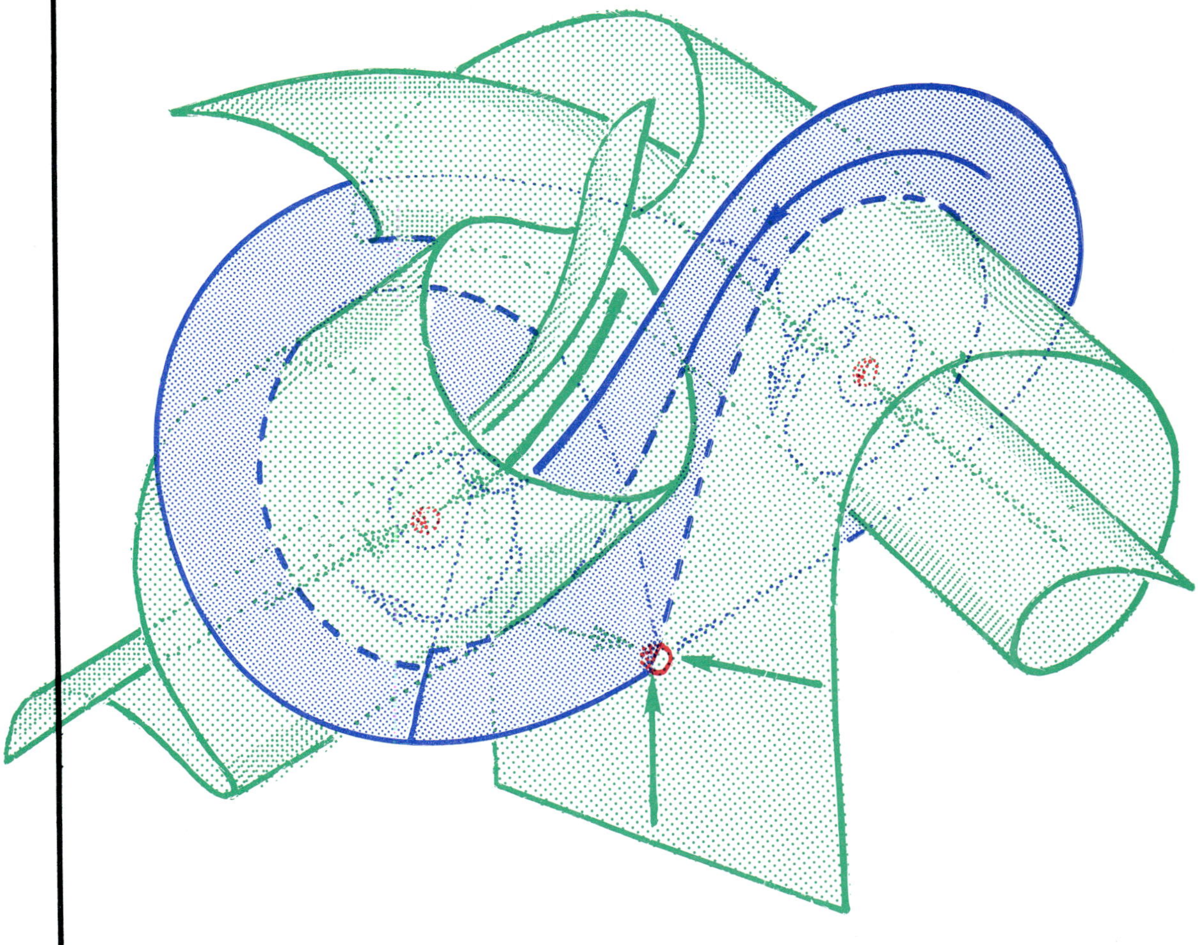

13.2.7. Extending the dotted inset farther backwards in time, it scrolls up tightly around the *one-dimensional insets* of A and B, In(A) and In(B).

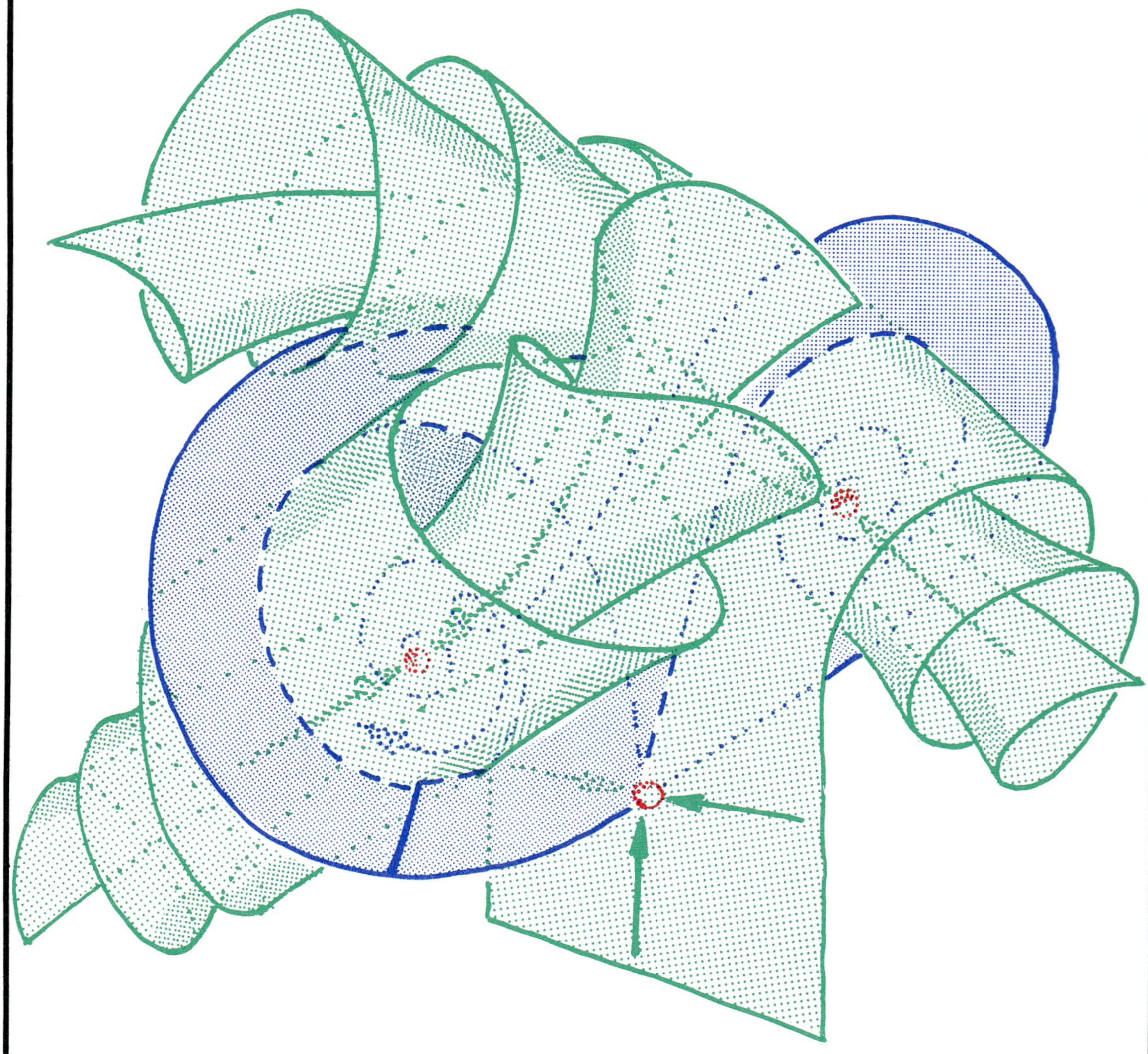

13.2.8. Extending the dotted inset farther backwards still, the four ends of the scrolls are pulled out along the curves, In(A) and In(B), toward their source at infinity.

The chaotic Lorenz attractor is composed of a yoke of tangles, folded into itself. Perhaps all of the familiar chaotic attractors have such an *outset structure*. But even in nonchaotic systems, the tangles are very important features.

We resume now our excursion into tangles.

13.3. Point to Cycle

There is only one kind of hyperbolic saddle cycle in 3D: index 1 (two-dimensional inset and outset). The two-dimensional outset of a hyperbolic limit point of index 2 can have a transverse intersection with the two-dimensional inset of such a limit cycle.

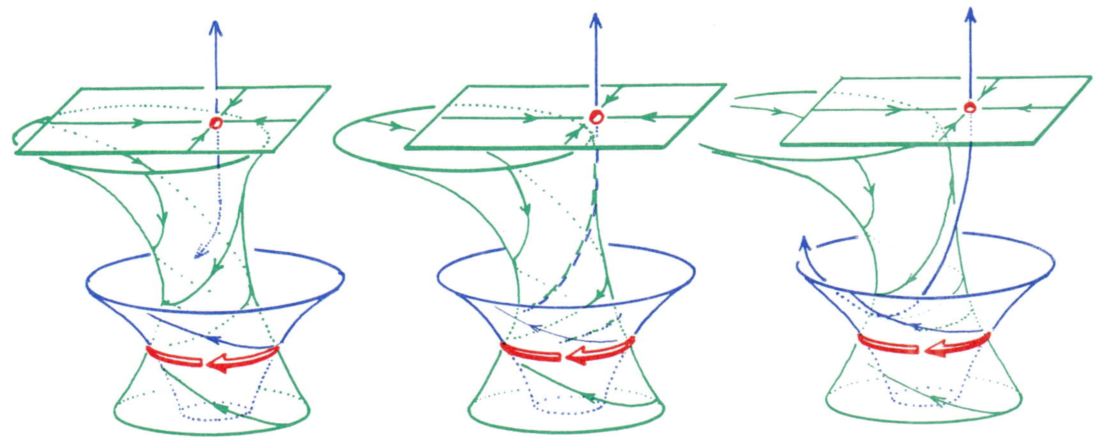

13.3.1. A heteroclinic trajectory from a saddle point of index 1 to a saddle cycle can never be transverse in three dimensions. Here is a nongeneric portrait, in the center, flanked by two nearby generic ones.

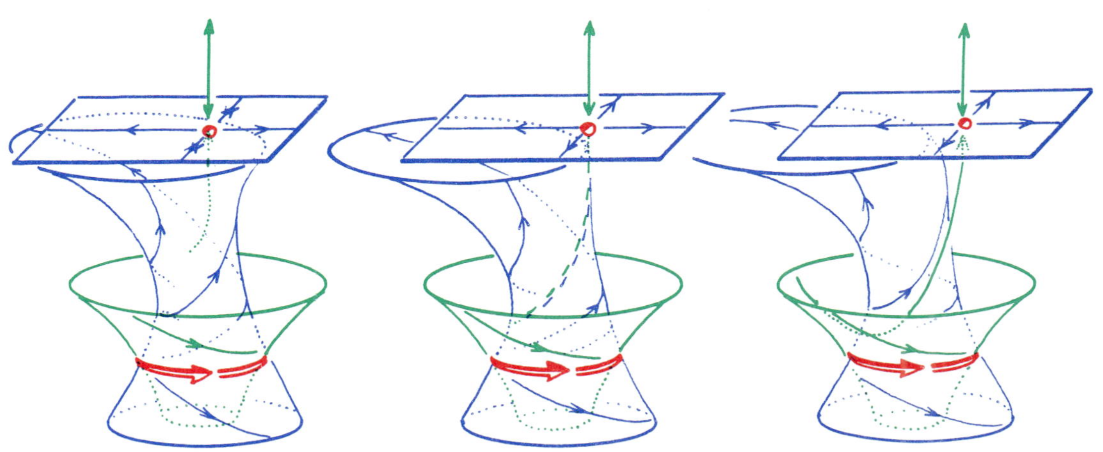

13.3.2. Similarly, a heteroclinic trajectory from a saddle cycle to a saddle point of index 2 is nongeneric.

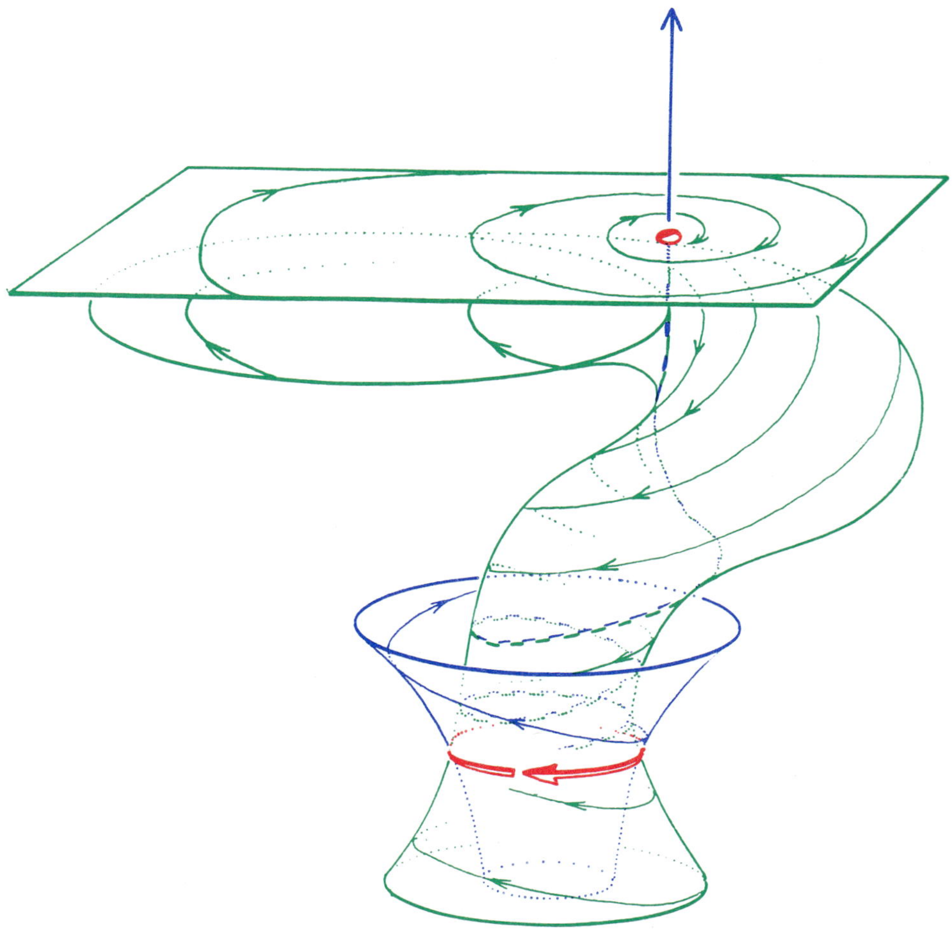

13.3.3. The two preceding panels illustrate nongeneric connections between a saddle cycle and a saddle point of the radial type. Here is an analog, with the radical point replaced by a spiral.

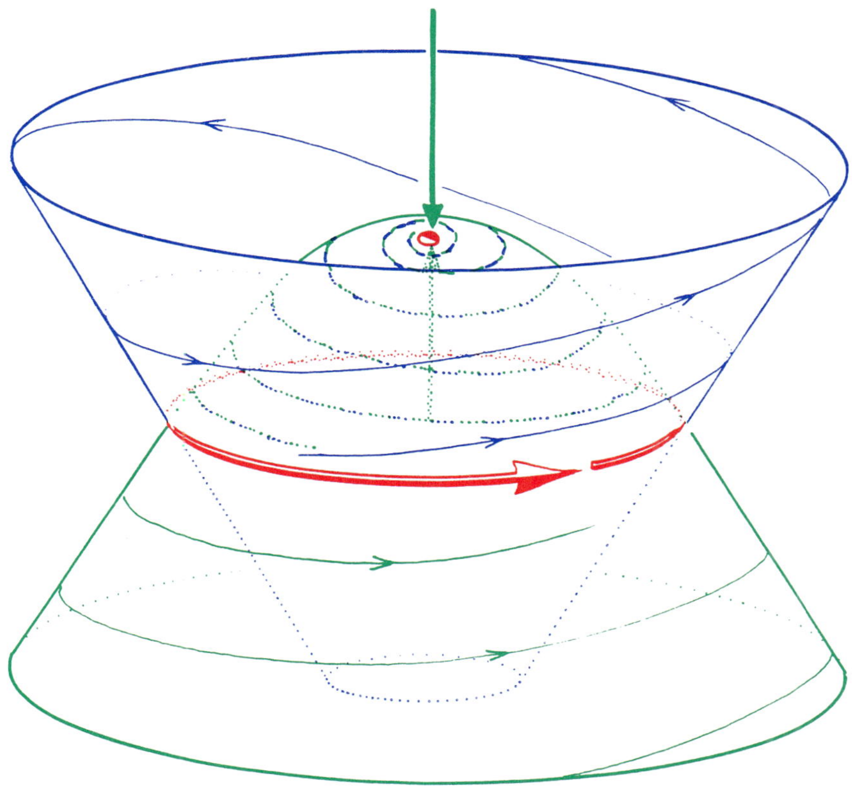

13.3.4. In this example, the outset of a saddle point of index 2 actually coincides with the inset of a saddle cycle. These nongeneric examples illustrate *a degeneracy of order 1:* only one condition of genericity has been violated.

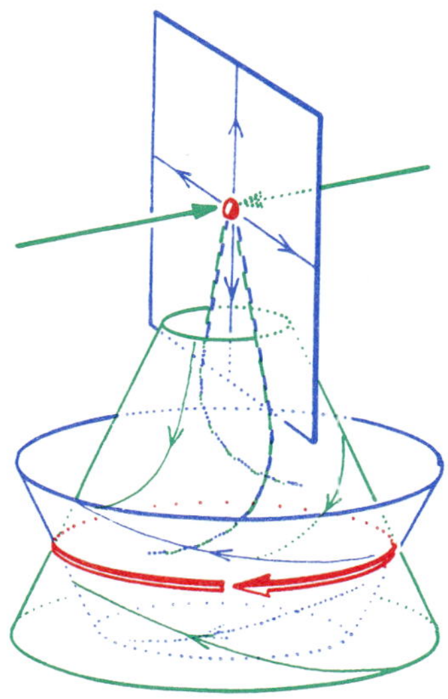

13.3.5. Nevertheless, heteroclinic connection from a saddle point of index 2 to a saddle cycle can occur generically in three dimensions. Here is the first step in the visualization of this configuration.

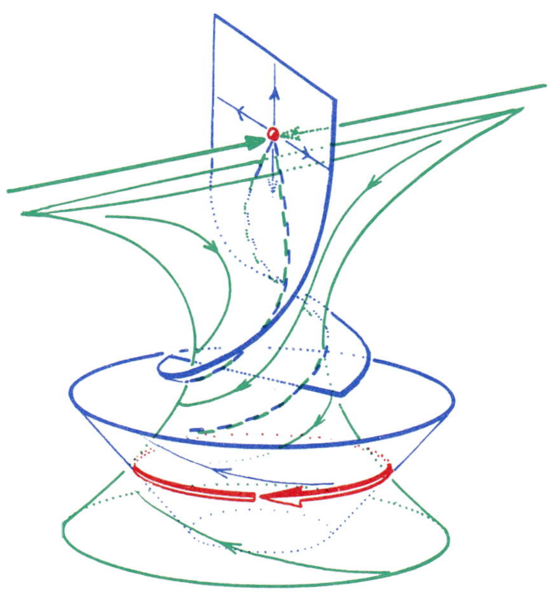

13.3.6. To generate more of the picture, the inset of the limit cycle (upper cone above) must be extended further into the past, to see how the trajectories spiraling into the limit cycle must have come from near the inset trajectories of the limit point.

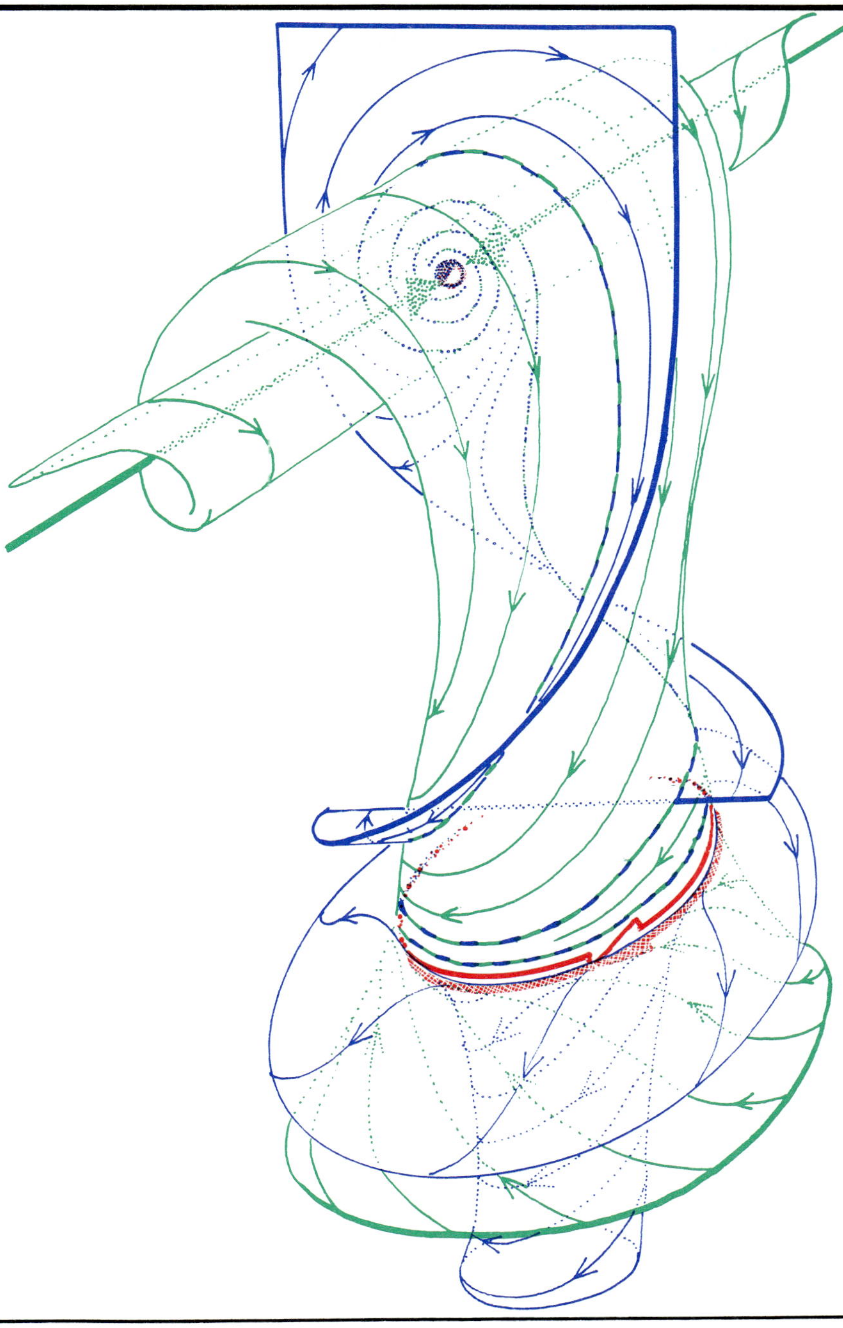

13.3.7. Before, the saddle point of radial type was shown. Here, it has been replaced by a spiraling one. These two distinctive types of heteroclinic behavior are topologically equivalent, however.

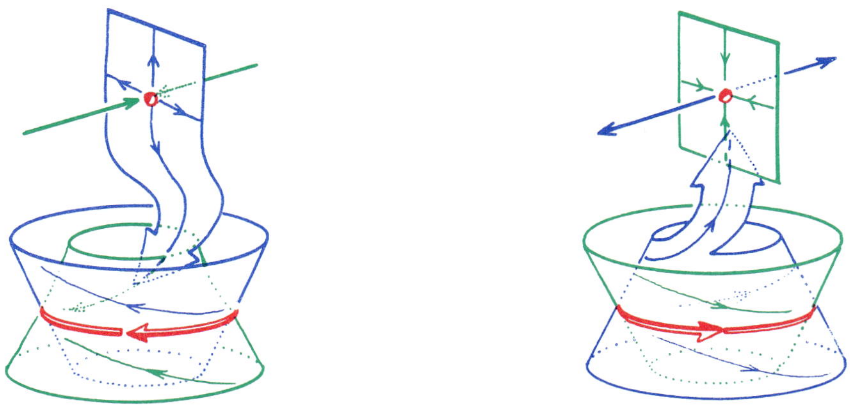

13.3.8. The heteroclinic portraits just described can be transformed into two other generic portraits by reversing the direction of time. Thus, the prior connection, on the left, suggests a new sort, on the right, in which the heteroclinic trajectory goes from a saddle cycle to a saddle point of index 2.

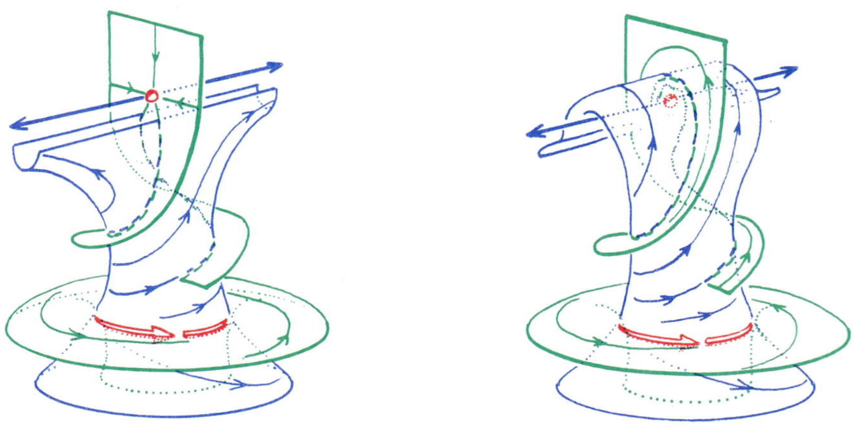

13.3.9. These two forms, radial and spiral, of the generic saddle connection result. As above, they are topologically equivalent.

All of the forms of this section could be reversed, by changing the direction of time, to provide examples of heteroclinic tangles from a limit cycle to a limit point: *cycle to point*.

13.4. Cycle to Cycle

Thus far, three generic and topologically distinct saddle connections have been described:

—saddle point index 2 to saddle point index 1,

—saddle point index 2 to saddle cycle,

—saddle cycle to saddle point index 1.

In three dimensions, there is just one more.

13.4.1. The outset of a saddle cycle (two-dimensional) can intersect the inset of another saddle cycle (also two-dimensional) transversely, in a (one-dimensional) curve of intersection, necessarily a spiraling trajectory. This fourth type of generic heteroclinic behavior is decidedly complicated.

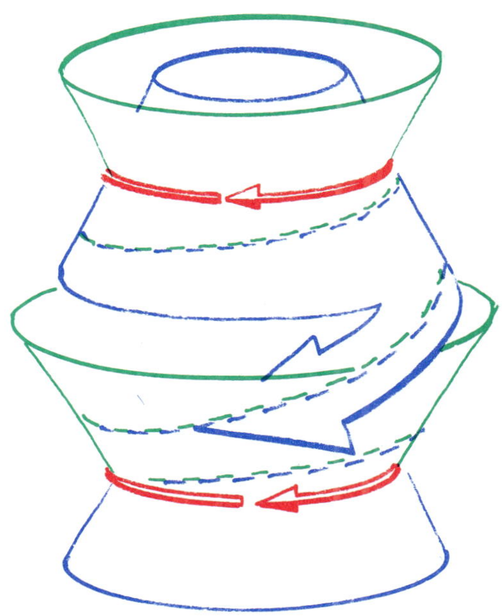

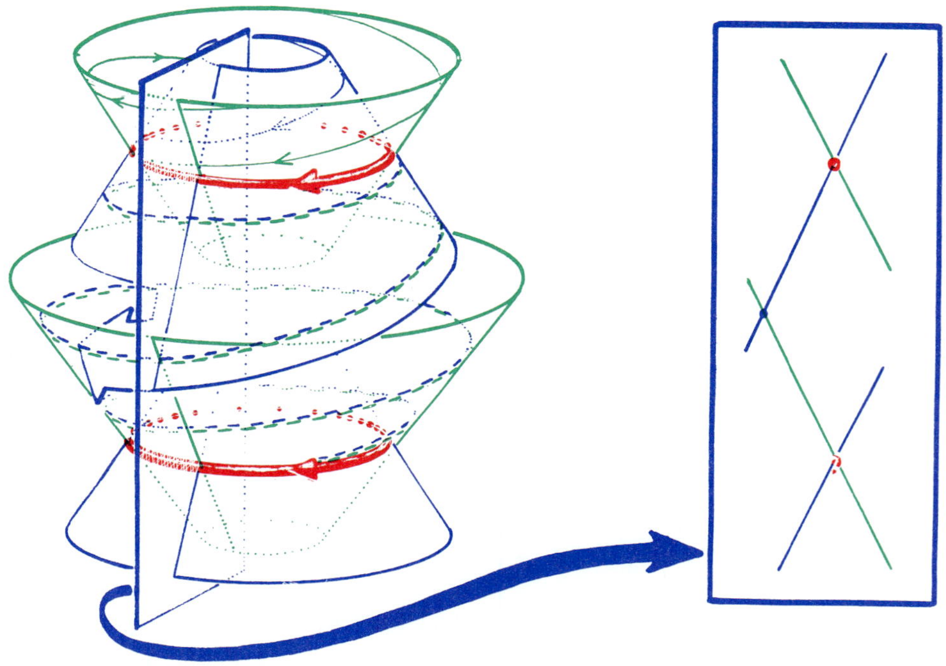

13.4.2. To dissect the complicated structure of such a connection between limit cycles, Poincaré introduced the *transverse section*, and the *first return map*. Within the cross-section (the *Poincaré section*) the two limit cycles are represented by points, and their insets and outsets by curves. The intersection of the outset of the donor cycle (above) and the inset of the receptor cycle (below) is a heteroclinic trajectory, represented in the Poincaré section by the point designated *H*.

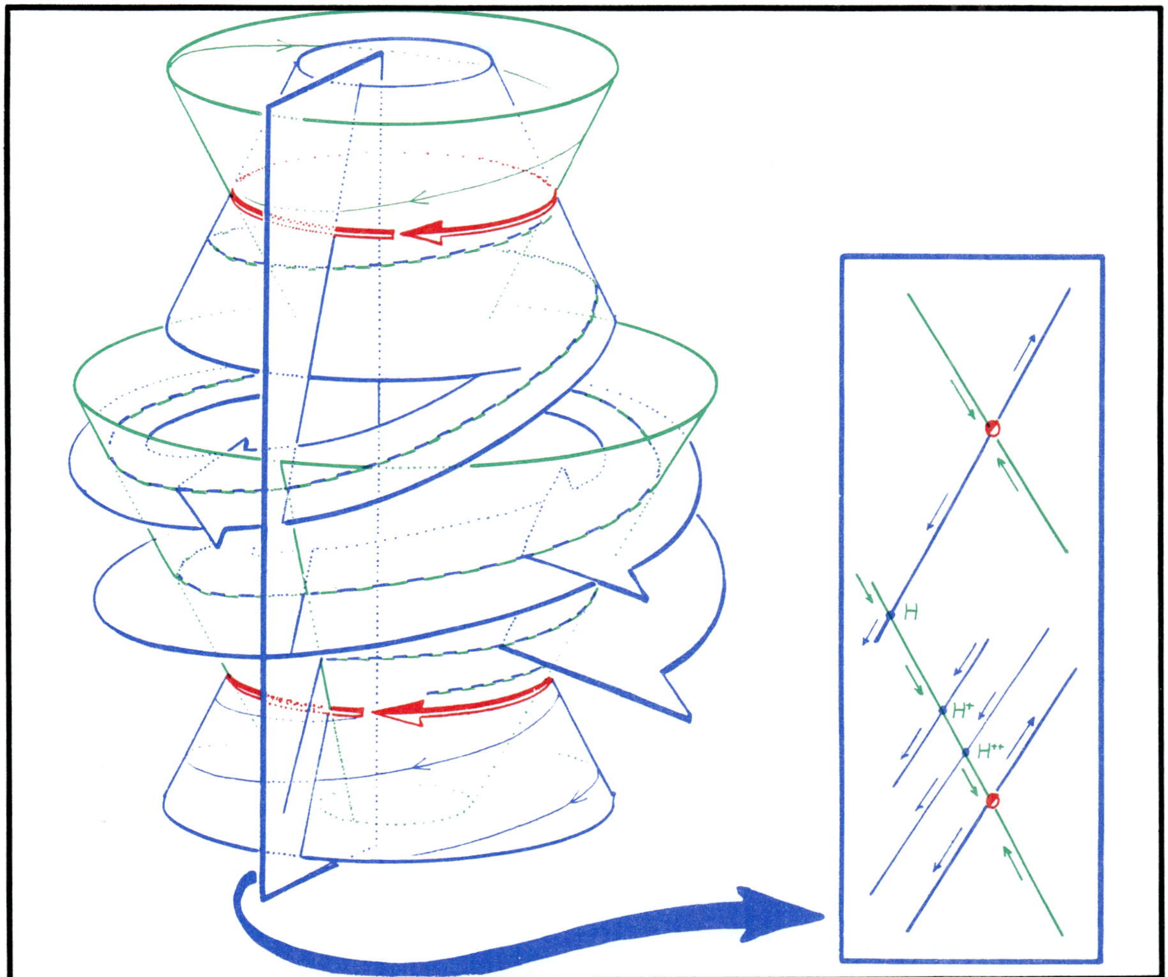

13.4.3. This picture, understood by Poincaré and fully analyzed by Birkhoff and Smith,[3] involves a doubly infinite sequence of intersections of the curves representing the inset and outset. For the marked point, *H*, representing the heteroclinic trajectory, is mapped by the Poincaré first return map into another point, H+, which is also in both curves. This point, H+, is actually on the same heteroclinic trajectory as H, at a later time. Further, the image of H+ is another point, H++, through which both curves must cross.

The completion of this drawing, showing the full tangle of curves within the Poincaré section, was carried out brilliantly by Birkhoff. His topological analysis of this picture reveals that between the points of intersection, *H* and *H*+, there must be, assuming G3, an odd number of others.

This construction of Birkhoff is carried out in the next section.

13.5. Birkhoff's Signature

The successive intersections of the inset and outset, curves within the Poincaré section, shown above, are all points belonging to a single heteroclinic trajectory. However, there may be (in fact, must be) other intersections, belonging to other heteroclinic trajectories. Our task now is to chart all of these, and the course of the inset and the outset curves between intersection points.

13.5.1. Here is a close-up view of two successive intersections, H and $H+$, belonging to a single heteroclinic trajectory. They are shown here on a piece of the inset curve of the saddle point on the right, representing the receptor saddle cycle. Through $H+$ passes a short piece of the outset curve of the saddle point on the left, representing the donor saddle cycle. How can we fill in the entire donor outset curve, connecting these short segments?

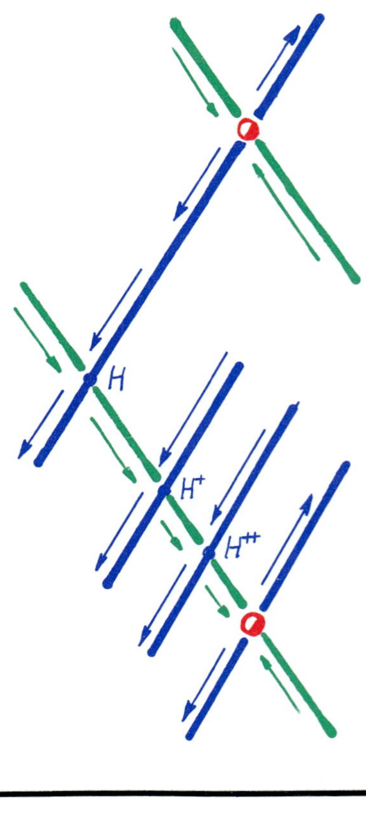

Notice the arrows on the outset segments, indicating the out-directions on the outset curve, away from the donor.

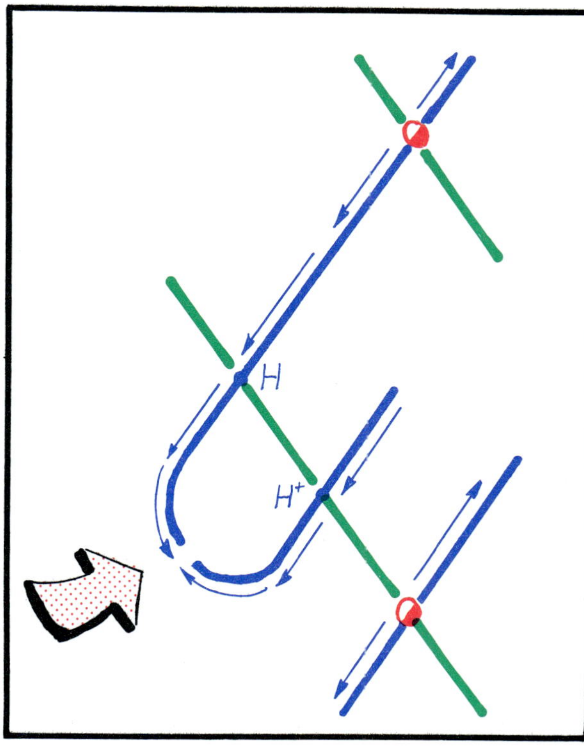

13.5.2. The simplest solution might be just to connect up the loose ends, as shown here. Unfortunately, this does not work. The out-directions must connect properly, without conflict.

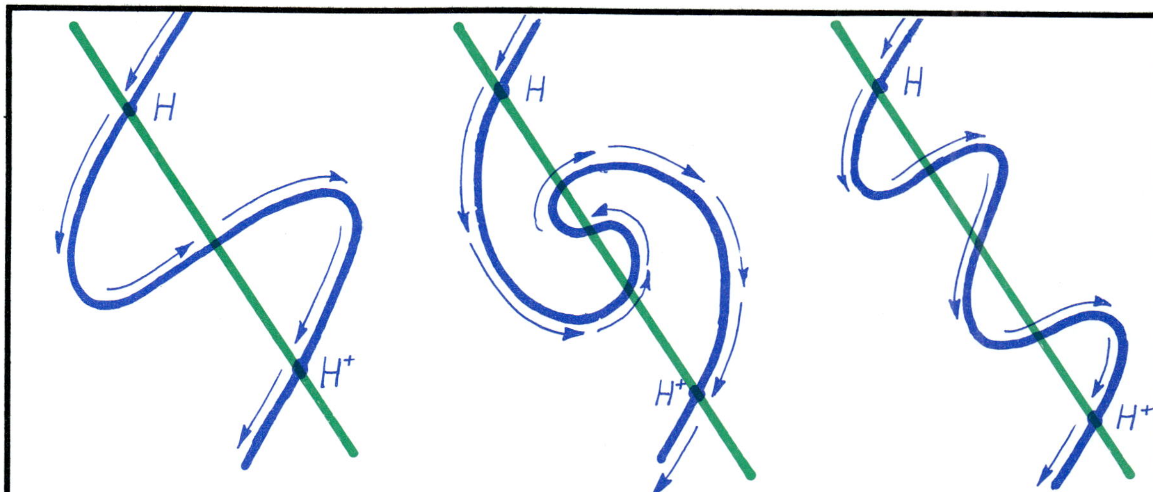

13.5.3. This drawing shows three possible connections for the outset curves, joining the short segments without conflict of the out-directions. The complete outset segment, joining two successive points corresponding to the same heteroclinic trajectory, H and $H+$, cuts through the inset segment joining the same two points in an odd number of points, all heteroclinic, but belonging to different heterclinic trajectories. The two complete segments, joining H and $H+$, comprise the figure Birkhoff called the *signature* of the saddle connection.

Some more complicated examples are given in the next chapter.

13.5.4. This shows the simplest possible Birkhoff signature. The odd number of interpolated heteroclinic points is only 1. This point, *I*, represents another heteroclinic trajectory, sharing the same donor and receptor, and possessing its own signature (not shown).

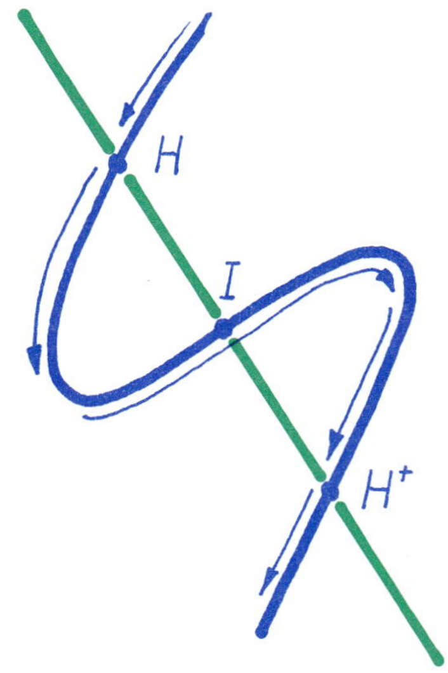

13.5.5. Reinserting this Birkhoff signature into the starting picture of this section, together with two of its forward images under the first return map, we have a roughly complete idea of the donor outset. There are many possibilities for the future of the outset, but here we have used only the simplest signature, as shown in the preceding panel. In this case, there is an infinite sequence of points of intersection, *H, H+, H++, . . .*, all belonging to a single heteroclinic trajectory.

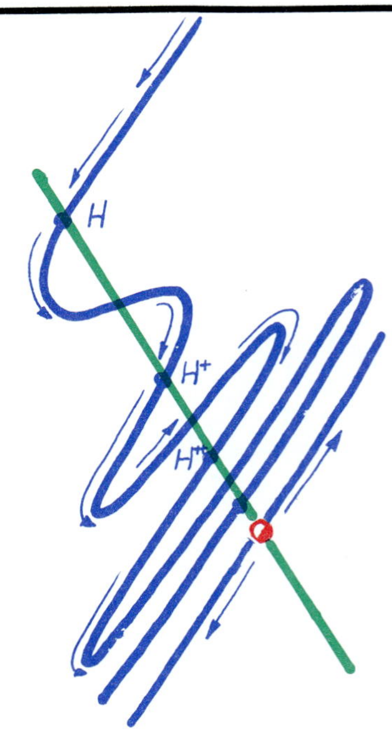

Meanwhile, the inset curve of the receptor is still only half-drawn. Where is its past?

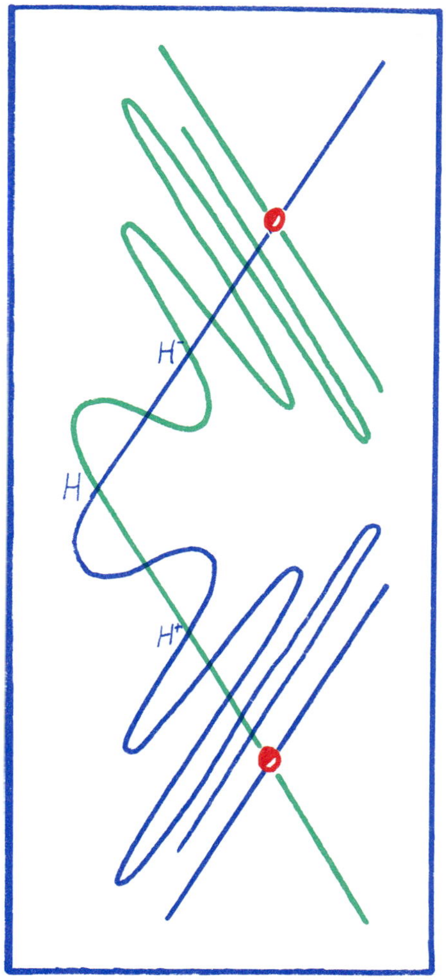

13.5.6. Extending the receptor's inset backwards in time, we obtain the predecessor of *H*, *H*−, its predecessor, *H*− −, and so on. This completes a doubly infinite sequence, *corresponding to one full heteroclinic* trajectory. Likewise, the interspersed heteroclinic trajectory contributes a complementary doubly infinite sequence *as shown here, in the Poincaré section.*

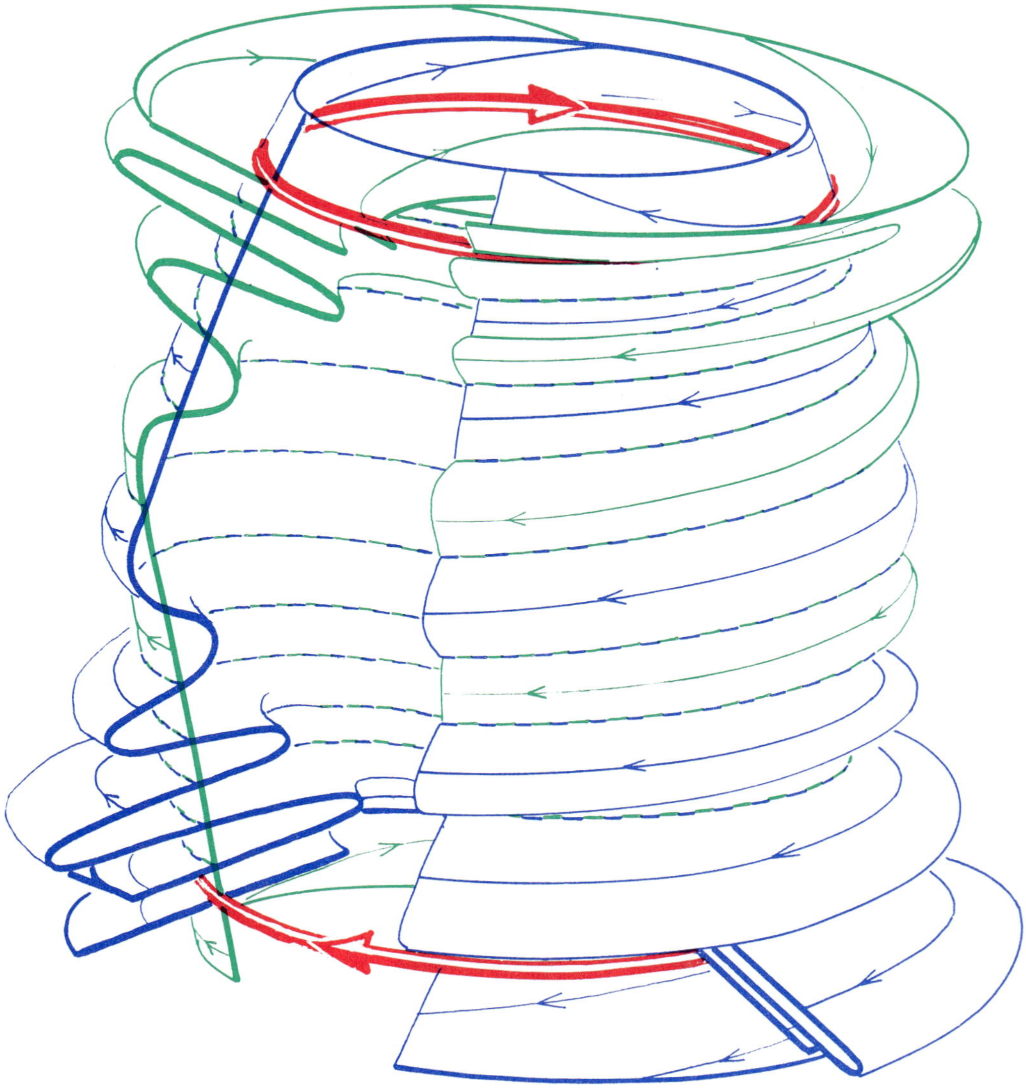

13.5.7. The doubly infinite sequences each correspond to a heteroclinic trajectory of intersection of the donor's outset and the receptor's inset, in the original three-dimensional context. Here, the generic connection of saddle cycles in three dimensions is shown, with all its complex structure. A section has been removed here, for improved visibility.

If this object were set down upon a rotating phonograph turntable, it would look rather like a bolt being screwed down.

The behavior of the trajectory passing by a cycle-to-cycle heteroclinic tangle is a spiraling asymptotic approach along the inset of the donor, followed by a period of entrapment, spiraling along the screw thread of the heteroclinic tangle, and finally an asymptotic escape, along the outset of the receptor. Thus, the heteroclinic tangle provides a model for ***transient oscillation.***

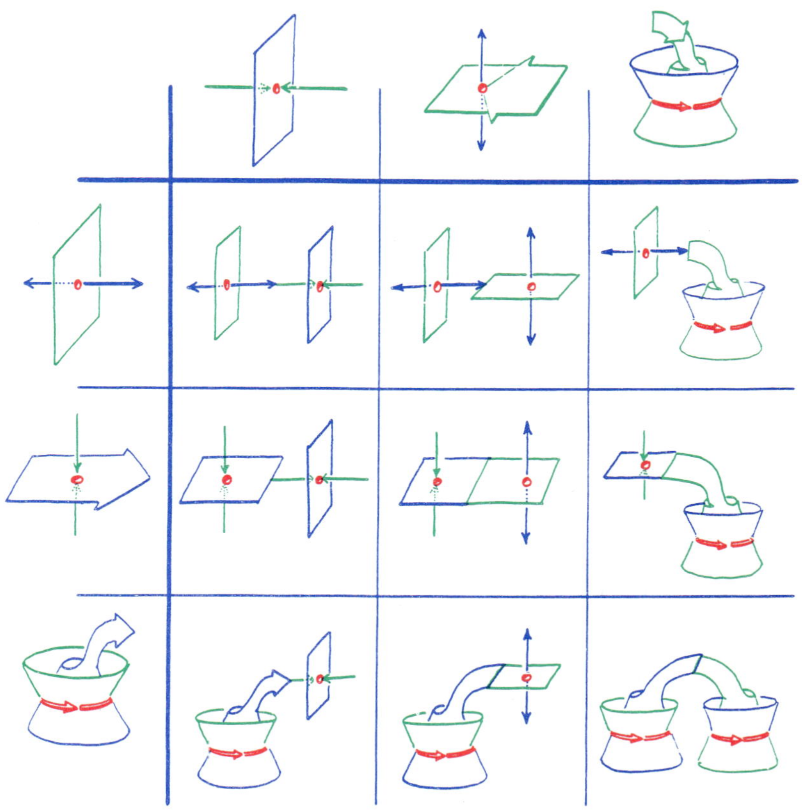

13.5.8. In the three-dimensional case, there are several possibilities, summarized in this table. The two types of topologically distinct hyperbolic saddle points (of index 1 and 2) and the unique hyperbolic saddle cycle are each possible donors, or receptors, of a saddle connection. The nine possibilities are pictured here, with the donors down the left, and the receptors along the top. Note the order and orientation of the donors is not the same as those of the receptors.

In summary, there are no generic saddle connections in two-dimensional dynamical systems. In three dimensions, there are four topologically distinct types. In higher dimensions, the situation is even more complicated.

The generic property G3 for dynamical systems is this: *all inset and outset intersections are transverse.* The genericity of this property, like the properties G1 and G2, is established by the theorem of Kupka and Smale.

14

Homoclinic Tangles

In addition to the four kinds of transverse saddle connections described in the preceding chapter, there is one more that can occur in three dimensions. This is the connection from a saddle cycle to itself, called a ***homoclinic connection.*** Homoclinic connections are much more important than heteroclinic ones, as ***they occur as exceptional limit sets*** within separatrices. Further, as shown by Birkhoff and Smith,[1] they are full of limit cycles. The study of this complicated case, initiated by Poincaré, is still in progress. An advance was made by Smale[2] in 1963. Many topologically different forms are possible. This chapter describes the main ideas of the three-dimensional context, including some constructions not previously published.

14.1. Homoclinic Cycles

By definition, a homoclinic trajectory must belong to the inset and outset of the same limit set. In the generic context of *properties G1, G2,* and *G3,* this limit set may not be a point. The simplest generic case is a limit cycle of saddle type, in three dimensions. In this section, we dissect this case.

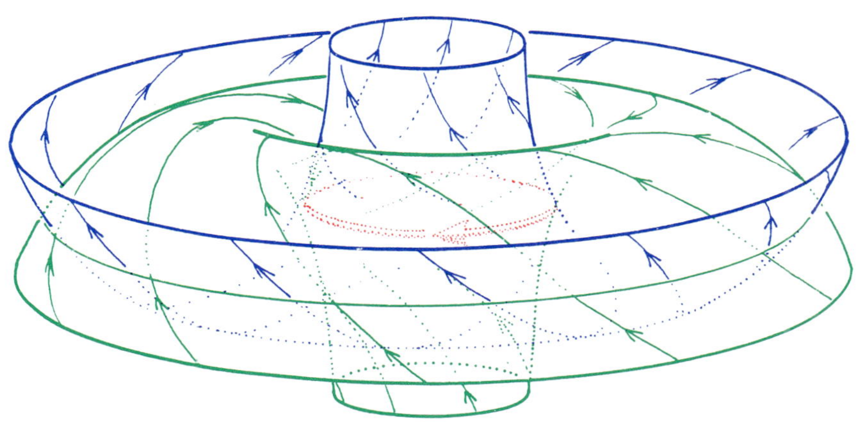

14.1.1. Here the outset of the limit cycle, at the top, is pulled down like a sleeve turned inside out. The inset, below, is likewise pulled up. Then, they are pushed through each other, to produce the beginning of an extensive intersection.

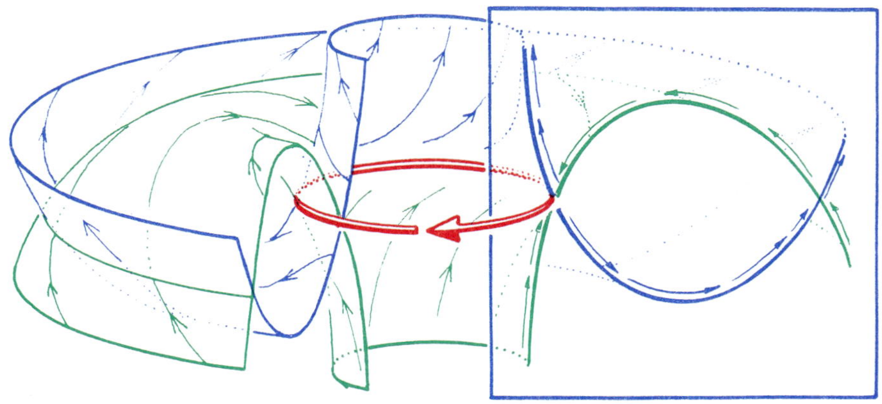

14.1.2. To visualize the intersection, we cut through it with a Poincaré section. The procedure is the same as the heteroclinic case, described in the preceding chapter (see 13.4.2.).

The key to the analysis is the first return map, which maps the Poincaré section into itself, corresponding to one revolution around the limit cycle.

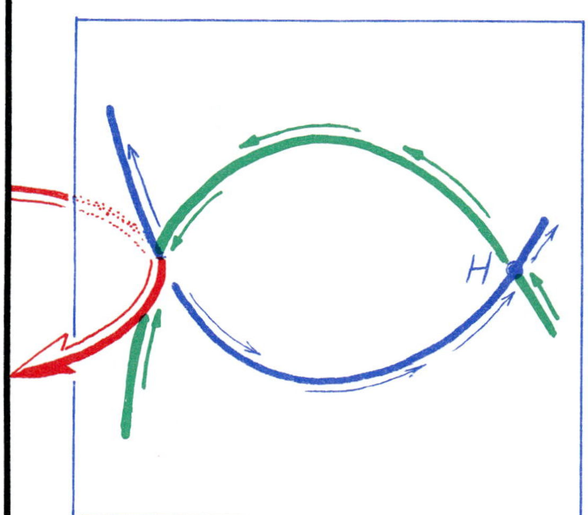

14.1.3. As in the preceding chapter (see 13.5.1.), the outset surface of the donor limit cycle and the inset surface of the receptor limit cycle (in this case, they are the same cycle) intersect the Poincaré section in two curves, the outset and inset curves. These curves intersect once at the point cut by the limit cycle (shown as a curved arrow here), and again at a point cut by the homoclinic trajectory, such as the homoclinic point H, shown here.

What happens to the homoclinic point after another revolution around the limit cycle?

14.1.4. As in the heterclinic case (again, see 13.5.1), this point is mapped to another, $H+$, closer to the limit point. This image point is on the inset curve, as this curve is mapped into itself by the first return map. Further, this curve consists of all the incoming points. However, the image point must also be on the outset curve, which is also mapped into itself by the first return map, and which consists of al outgoing points. The homoclinic points, H and $H+$, are both outgoing and incoming, by assumption. Thus through the image point, $H+$, there must also pass a piece of the outset curve, shown here with its out-direction indicated by an arrow.

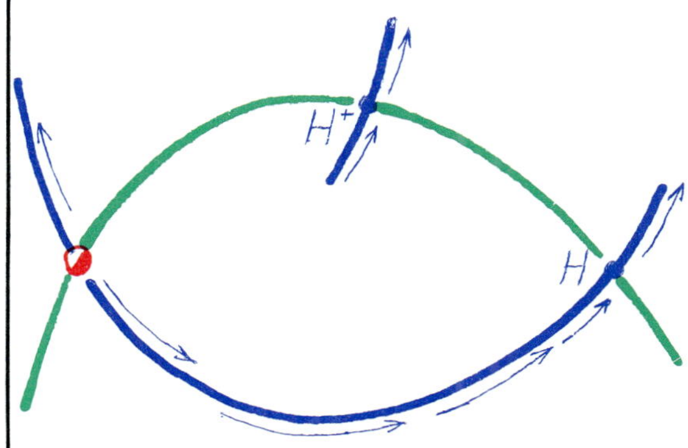

How may these outset segments be connected, so as to obtain the entire outset?

14.1.5. As in the heteroclinic case (see 13.5.2.), direct connection leads to a conflict of out-directions. Thus . . .

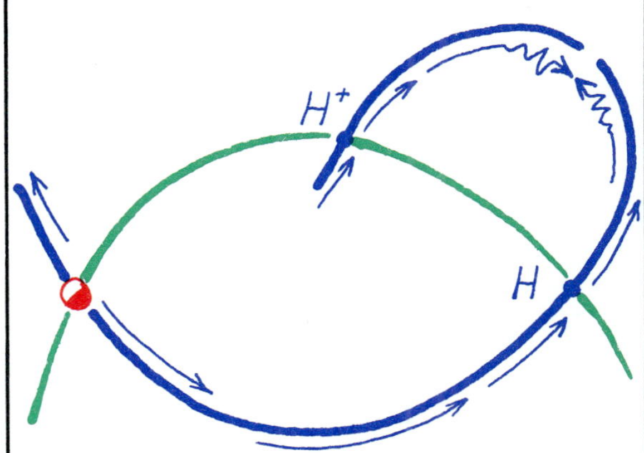

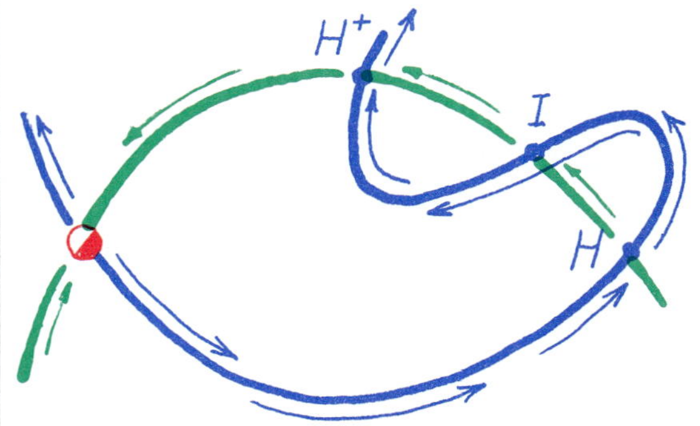

14.1.6. . . . as in the heteroclinic case (see 13.5.3.), the outset segment from H to $H+$ must cross the inset segment (between the same two points) an odd number of times. This is the simplest legal construction, illustrating the *Birkhoff signature* in the homoclinic case.

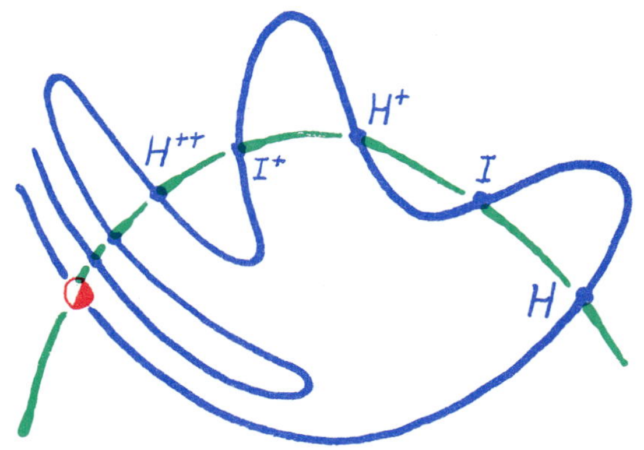

14.1.7. Reiterating the first return map again and again, the outset segments push up against the inset curve, near the limit point.

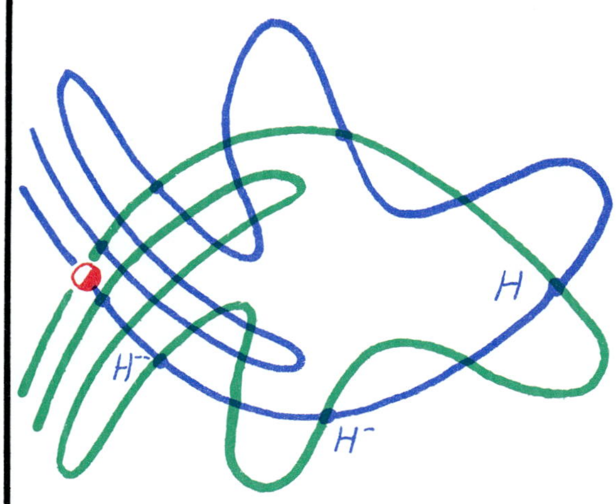

14.1.8. Repeating the construction for negative times (iterating the *prior return map*), the inset segments pile up against the outset curve, again near the limit point. Thus, we obtain a full picture of the entire *homoclinic tangle,* as shown in this drawing of a tangle studied by Hayashi,[3] the greatest master of experimental tangle art.

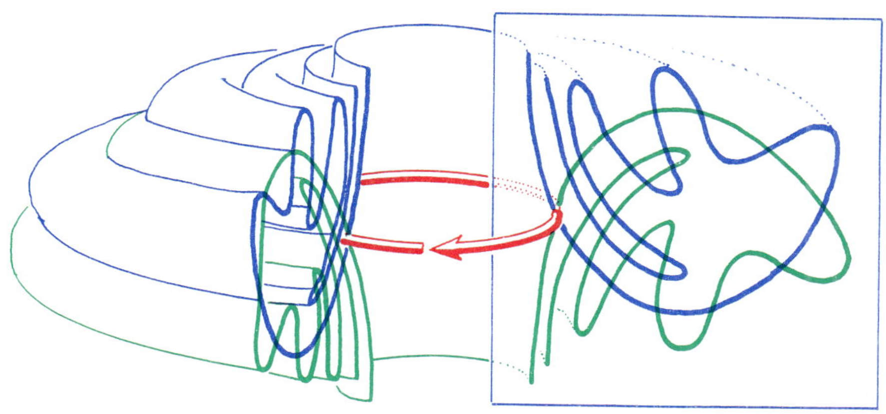

14.1.9. Here is the tangle within the Poincaré section, replaced within the original 3D context (compare with 13.5.7). The behavior of a nearby trajectory is a spiraling asymptotic approach, along the non-tangled half of the inset surface, followed by a period of chaotic motion, entrapped within the tangle, and finally a spiraling asymptotic escape, along the non-tangled half of the outset surface. Thus, the homoclinic tangle provides a model for *transient chaos.*

This tangle, based on the simplest Birkhoff signature, reveals additional intersections of inset and outset loops. This deeper structure is not determined by the Birkhoff signature. Thus, to fully describe the structure of the tangle, additional signatures must be specified.

In the next section, we introduce a sequence of signatures, published here for the first time, for the full description of a homoclinic tangle in 3D.

14.2. Signature Sequence

During the preparation of a preliminary edition of this work in 1980, we tried to deform the Hayashi tangle (shown in the preceding panel, 14.1.8.) into the Smale horseshoe (described in the next section; see also Figure 8.1.10). Although the two homoclinic tangles have the same Birkhoff signature, we were unable to deform the Hayashi tangle into the horseshoe.

In trying to understand the difference between these two exemplary tangles, we developed an infinite sequence of signatures. The first of these is the Birkhoff signature, which is the same for the two examples. The second, however, is different. Thus, they could not be deformed, one into the other. This led to our *signature conjecture: if two tangles have the same signature sequence they are topologically equivalent.*

In this section, we construct the signature sequence, step by step, for the Hayashi tangle. In the next section, we will apply it to the Smale horseshoe tangle.

14.2.1. Here, again, is the Hayashi tangle. It is not a mathematician's pipe dream, but was laboriously drawn by Hayashi, from extensive simulations of the Duffing system (for the forced pendulum, see Part One) with an electronic analog computer. How can we give a full characterization of this tangle? Let's single out a homoclinic point, such as *H,* and its image *H+*.

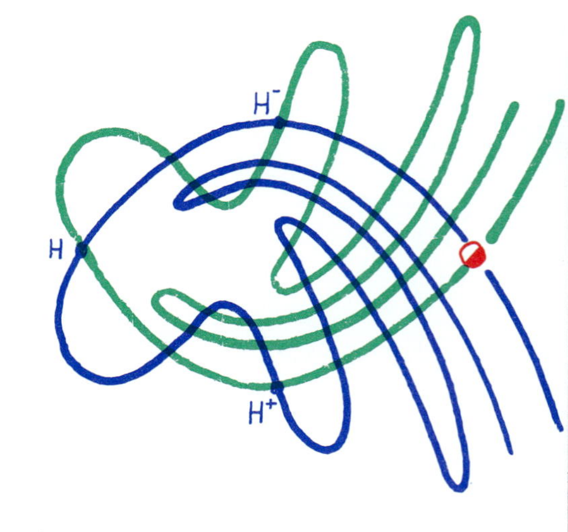

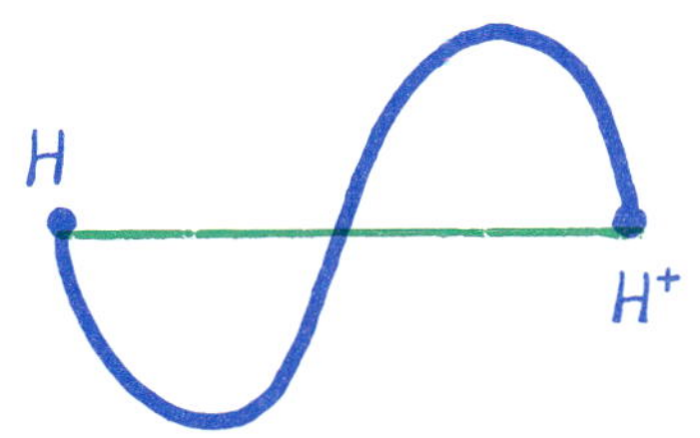

14.2.2. Extracting the inset and outset curve segments bounded by these two points, we obtain the Birkhoff signature. Again it is the simplest possible one (see Figure 13.5.3 for three alternatives). To some extent, it characterizes the chief feature of the tangle.

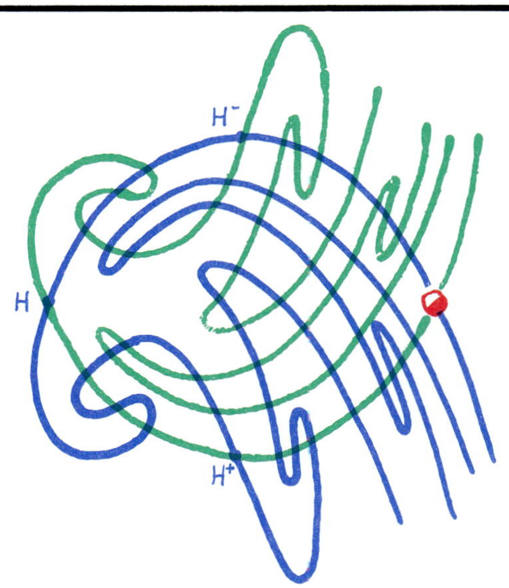

14.2.3. For example, here is another tangle. At first glance, it appears significantly different from the preceding one.

14.2.4. Extracting a Birkhoff signature, we see that it is indeed different. And this does seem to capture the chief feature of this new tangle.

14.2.5. Now let's return to the old tangle. Notice how inset loops may cross through several outset loops. We want to capture a signature of this larger-scale behavior of the figure, corresponding to its minor features. We will proceed in steps.

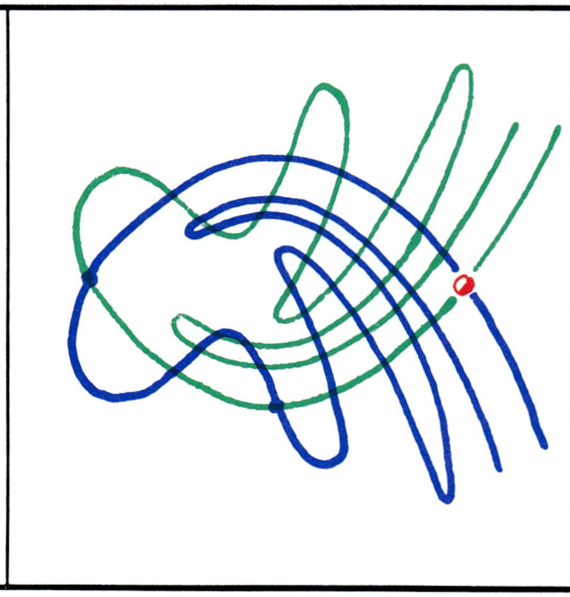

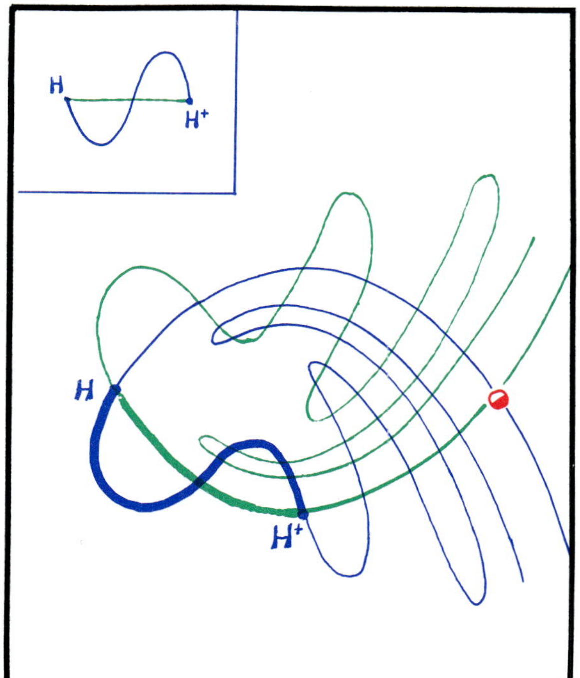

14.2.6. *Step 1.* Single out a homoclinic point and its image, and draw the Birkhoff signature they determine. Draw it again, straightened out, as shown in the inset.

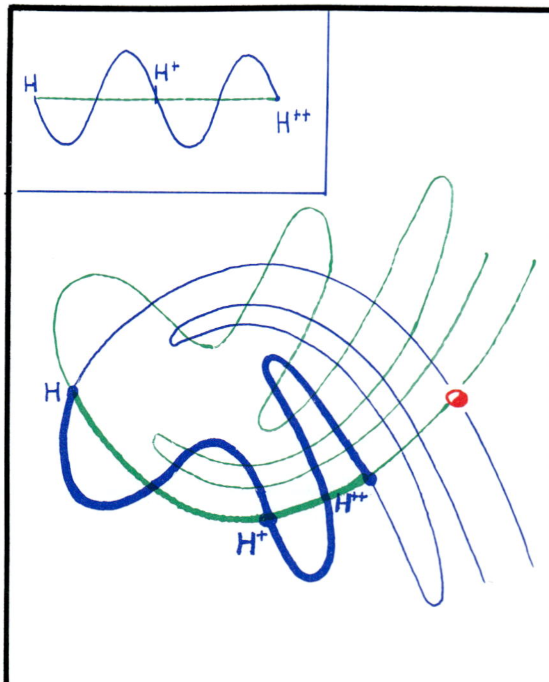

14.2.7. *Step 2.* Extend the signature another iteration, by applying the first return map. Here we see both the Birkhoff signature and its entire image. Straighten out this figure also, as shown in the inset.

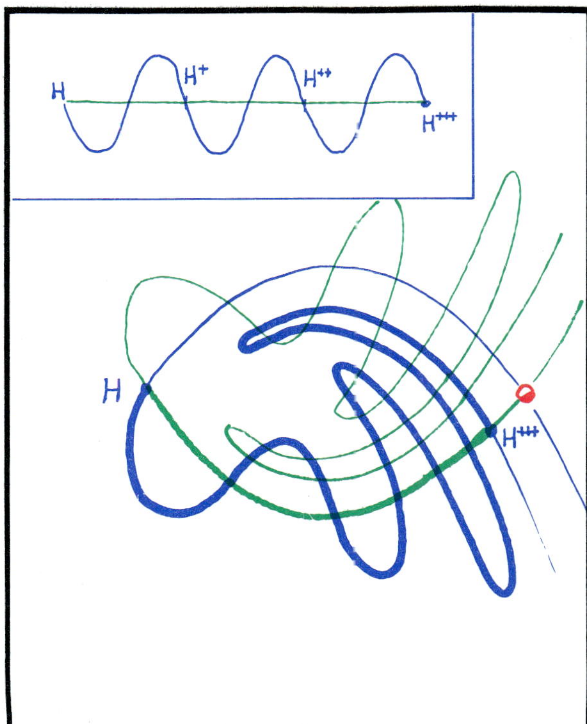

14.2.8. *Step 3.* Repeat the preceding step again, and in general, as many times as you can, as long as the elongating outset loops never come back to cross a part of the original signature, or its forward iterations.

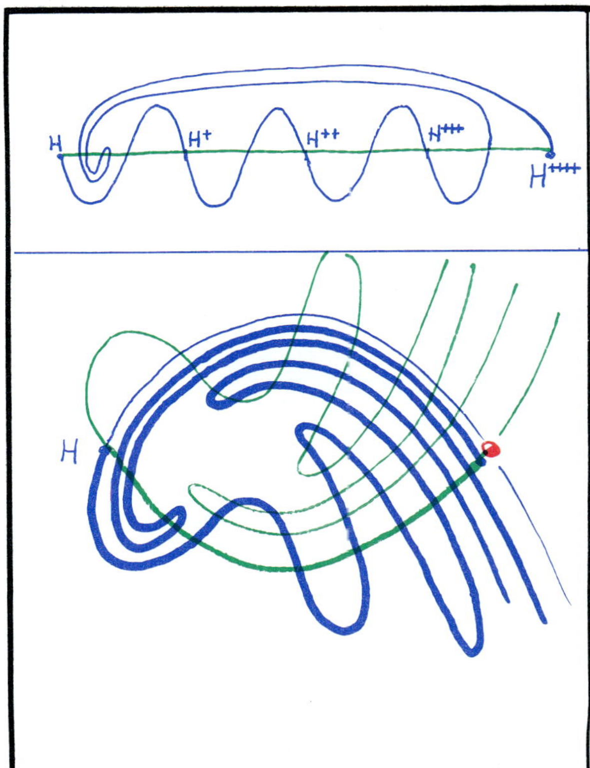

14.2.9. *Step 4.* Repeat the single iteration step once more. This time, one of the elongating upper loops will make a new intersection with a segment of the inset curve belonging to the original Birkhoff signature, or one of its forward iterations. In this example, four new crossings have all appeared at once.

The Birkhoff signature is the first of our sequence. The figure in the inset above is the second. Let's try out these two on another example.

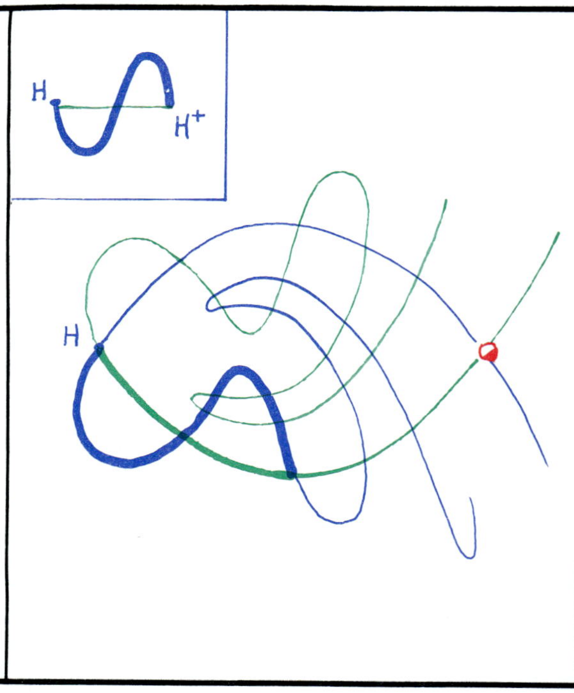

14.2.10. This is yet another tangle. It looks like Hayashi's, but is not. The Birkhoff signature (shown in the inset) is the same. Will our second signature reveal the difference?

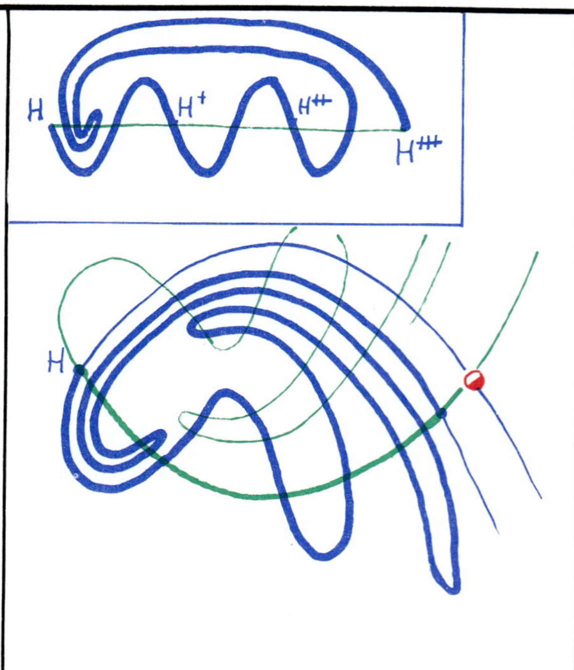

14.2.11. The first iteration of the fundamental (Birkhoff) signature comes close to the fundamental, but does not cross it. The second image crosses the fundamental. The figure in the inset is the second signature of this tangle.

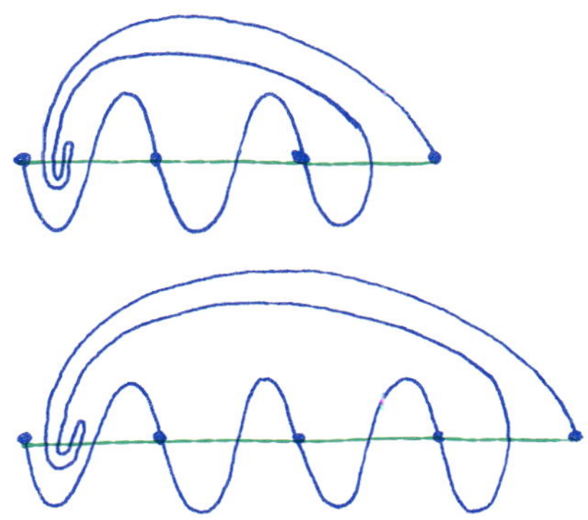

14.2.12. Here, for comparison, is the second signature of the two tangles. The new tangle has two humps, while the Hayashi tangle has three, under the arching inset loop. They are topologically inequivalent tangles.

These two signatures are the first of an infinite sequence. See if you can draw the third signature in the examples above.

14.3. Horsehoes

In this section, we will tackle another tangle, called Smale's horseshoe. This third example originated as a geometric construction, but was subsequently observed in the forced Van der Pol system,[4] and many others. Along the way, we will give an idea of the third signature of a tangle.

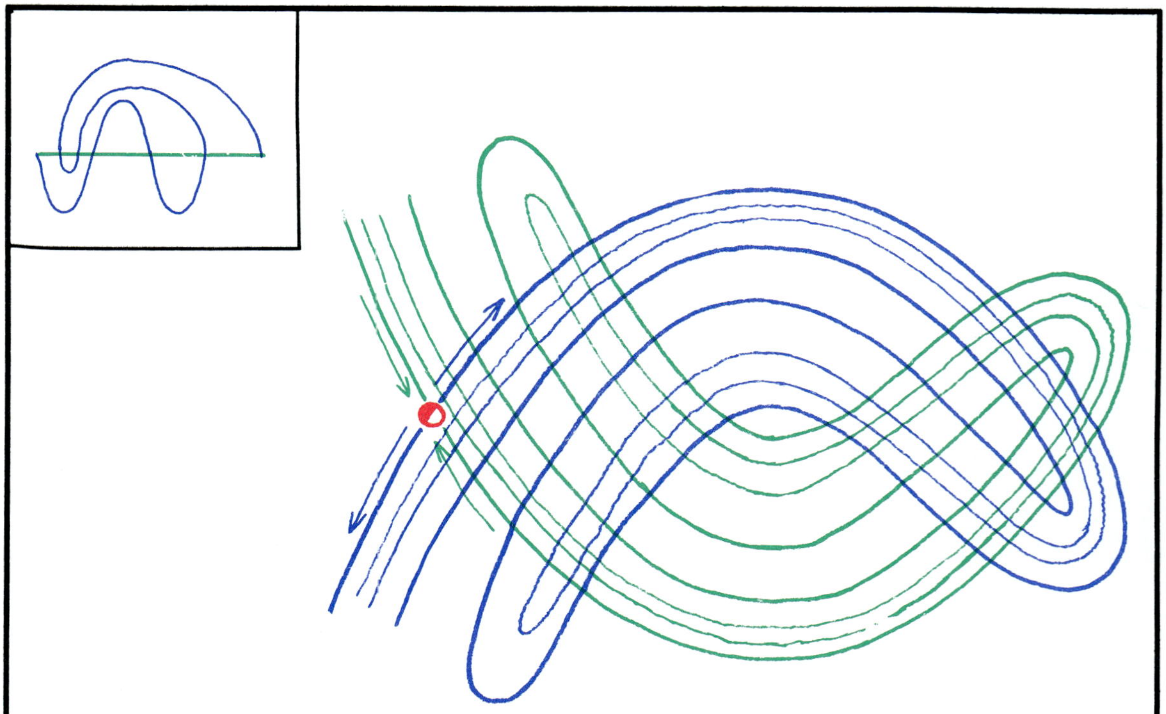

14.3.1. Here is yet another homoclinic tangle, the famous *horseshoe of Smale.* Note that the first signature is the familiar simplest one. But in the second signature, shown in the inset, the hump has been twisted back, creating two new intersections. To further characterize this tangle, we must draw the *third signature.*

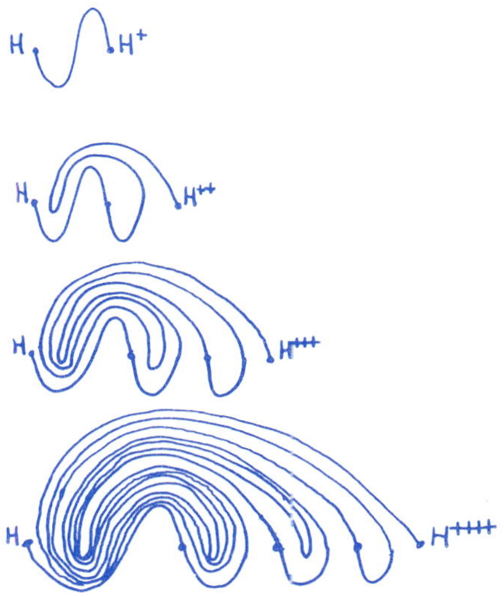

14.3.2. Here are the first four signatures of our signature sequence, for Smale's horseshoe. The third signature is not identical to the third signature of the preceding example (try it and see).

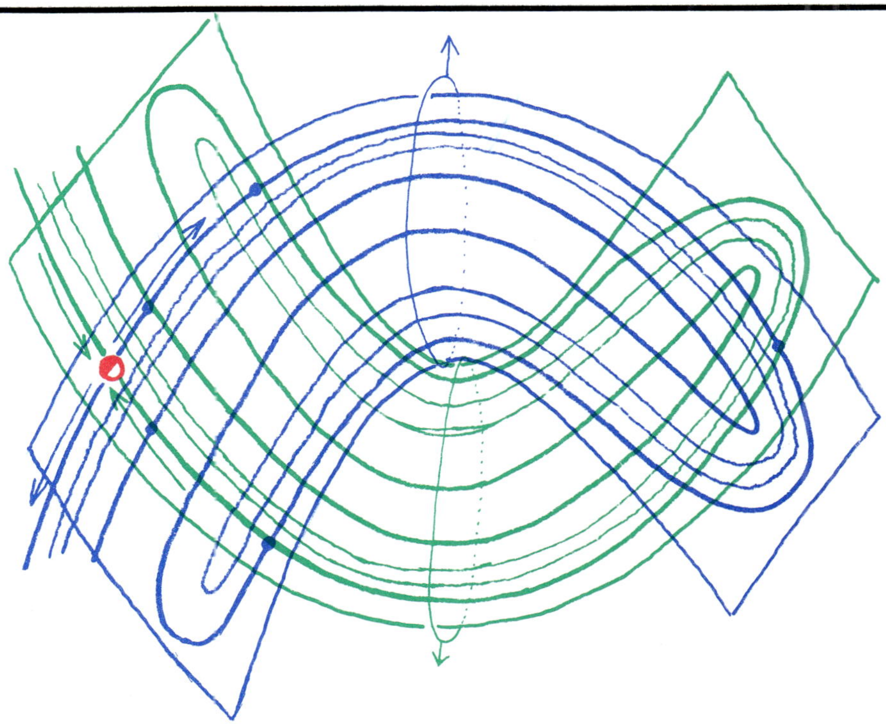

14.3.3. The horseshoe has been untangled by Smale[5] in a most ingenious way. Choosing a curved rectangular patch in the Poincaré section with some care, and applying the first return map yields another rectangular patch crossing the original patch at each end. Now, deform the whole picture by lassoing the two patches around the waist and pulling gently.

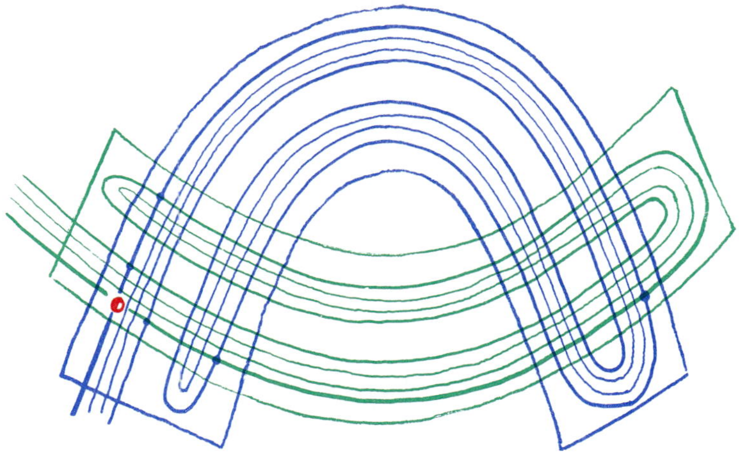

14.3.4. Continue to pull the upper patch upwards by the waist, while pushing down on the ends. The idea is to straighten out the lower patch.

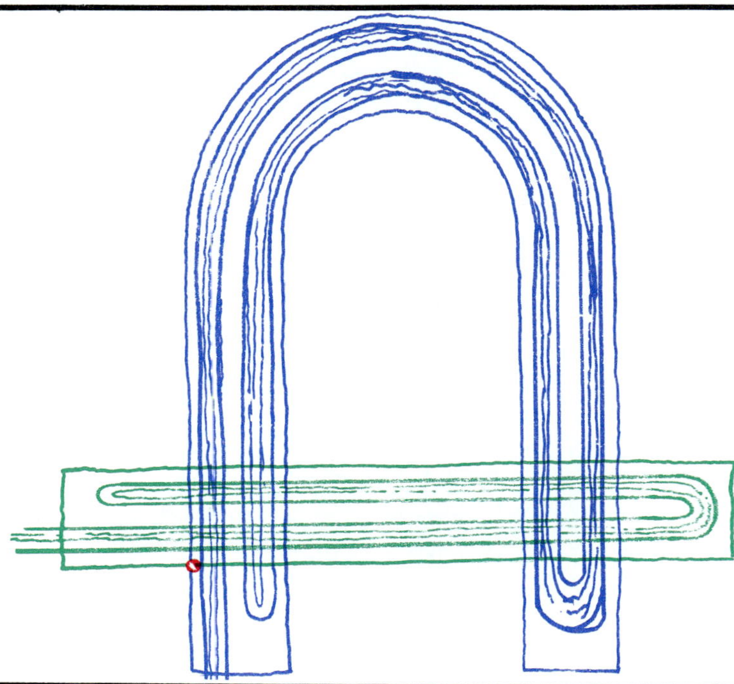

14.3.5. There is the fully untangled tangle, the horseshoe of Smale. It is topologically equivalent to the messy original tangle, yet it admits a full analysis, as shown by Smale.

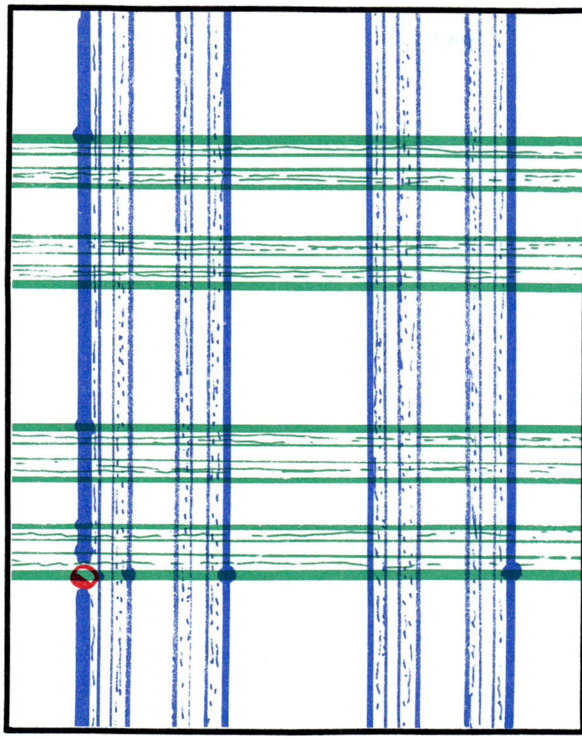

14.3.6. The analysis is based upon a clever scheme for labeling all the points of intersection of the insets and outsets within the Poincaré section.

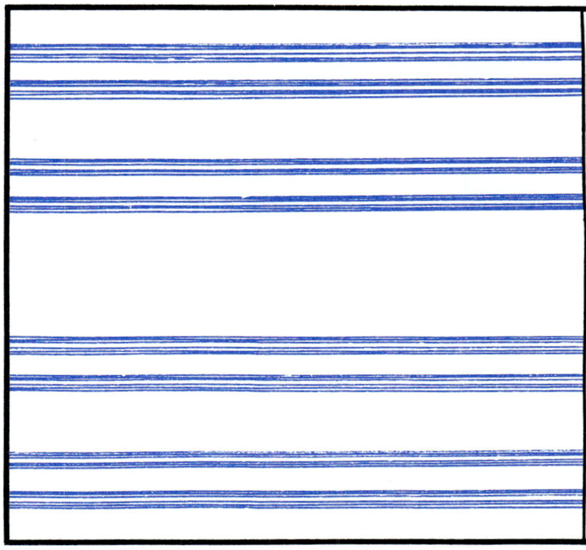

14.3.7. Looking at a portion of the outset through a microscope, we see an infinite set of horizontal lines. Their intersection with a vertical line (such as the left edge of the box here) is much like Cantor's middle thirds set (see Figure 9.4.7).

Smale's analysis of this particular tangle, based on combing it out and applying symbolic dynamics, might be applied to other homoclinic tangles, through careful use of the signature sequence.

The theory of homoclinic tanges is very important, and yet little known. Even in three dimensions, the lowest in which they occur generically, there are outstanding problems. In higher dimensions, little is known. Poincaré expressed the fear that they might defy analysis forever, but the theory of horseshoes, and the work of Zeeman, Newhouse, and others[6] on more general shoes, gives hope.

14.4. Hypercycles

An even more complicted situation occurs generically in dimension three or more. The insets and outsets of these may have transverse intersections, tangles, and heteroclinic trajectories in a daisy chain, called a *hypercycle*, or *heteroclinic cycle*.

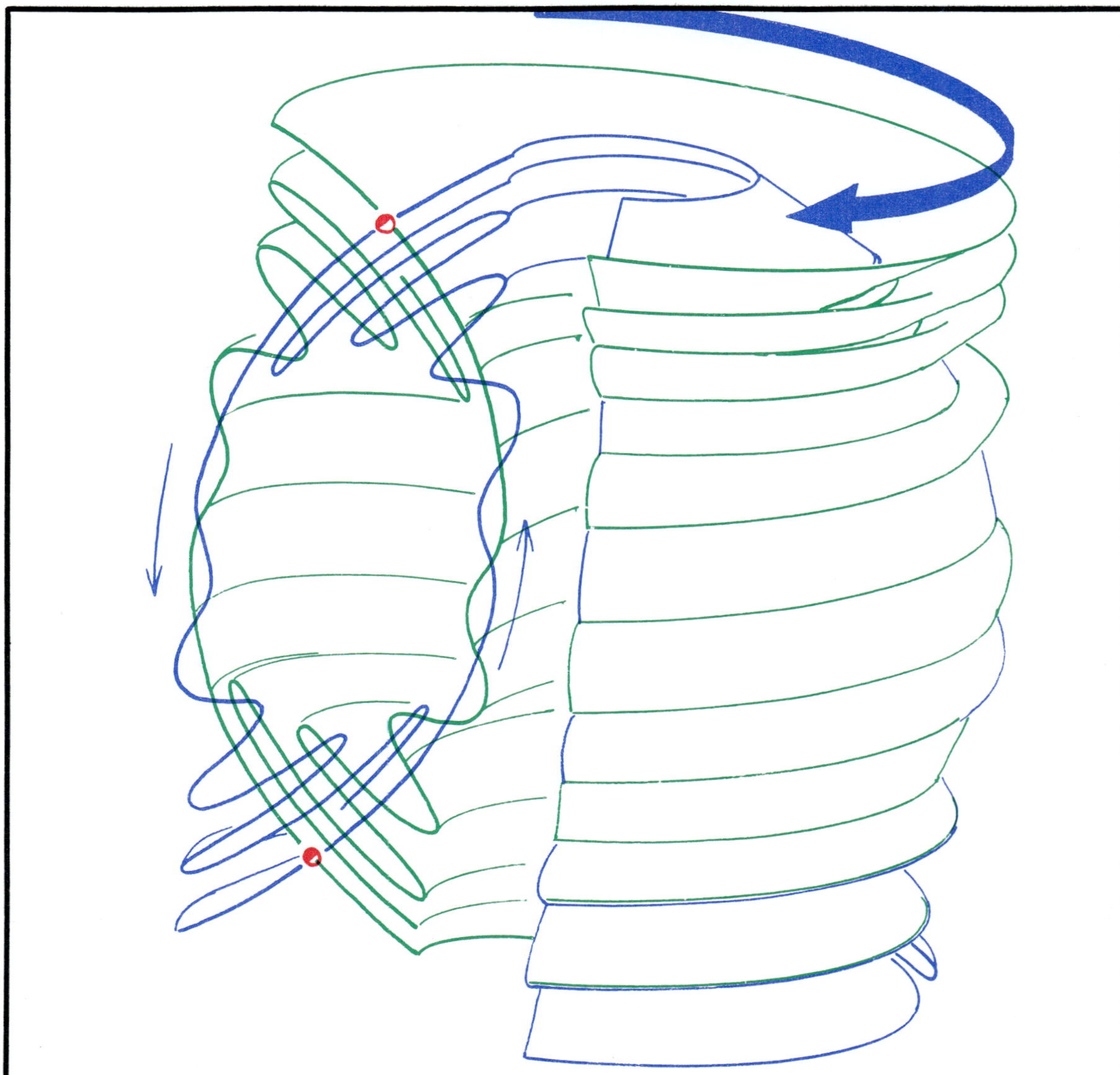

14.4.1. Here is the simplest example of a cycle. In a three-dimensional state space, two closed orbits of saddle type (index 1) have heteroclinic trajectories, each to the other.

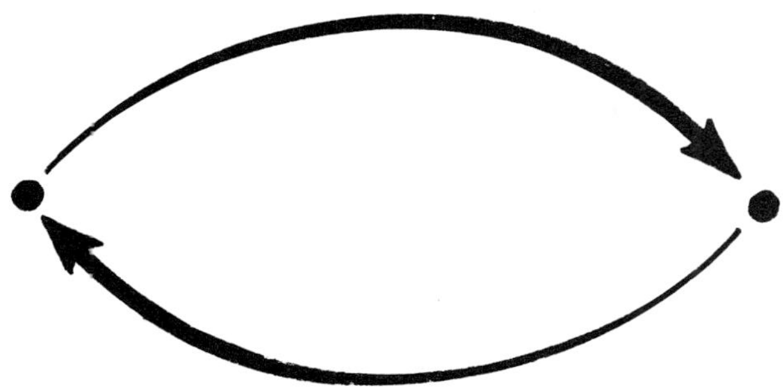

14.4.2. This situation may be described by this diagram, called a *directed graph*, or *quiver.* This has a vertex for each of the limit cycles, and was introduced by Peixoto[7] to describe generic systems in two-dimensional state spaces.

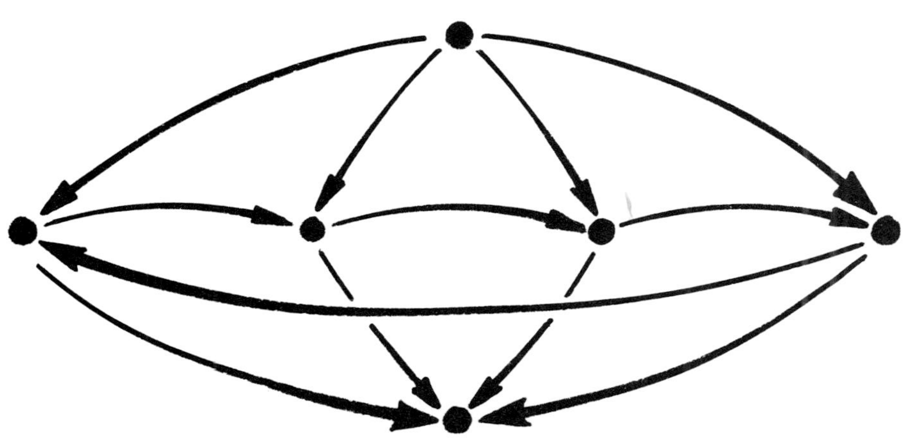

14.4.3. More complicated cycles may involve larger sets of critical points, closed orbits, and even more complicated limit sets, in a daisy chain of saddle connections.

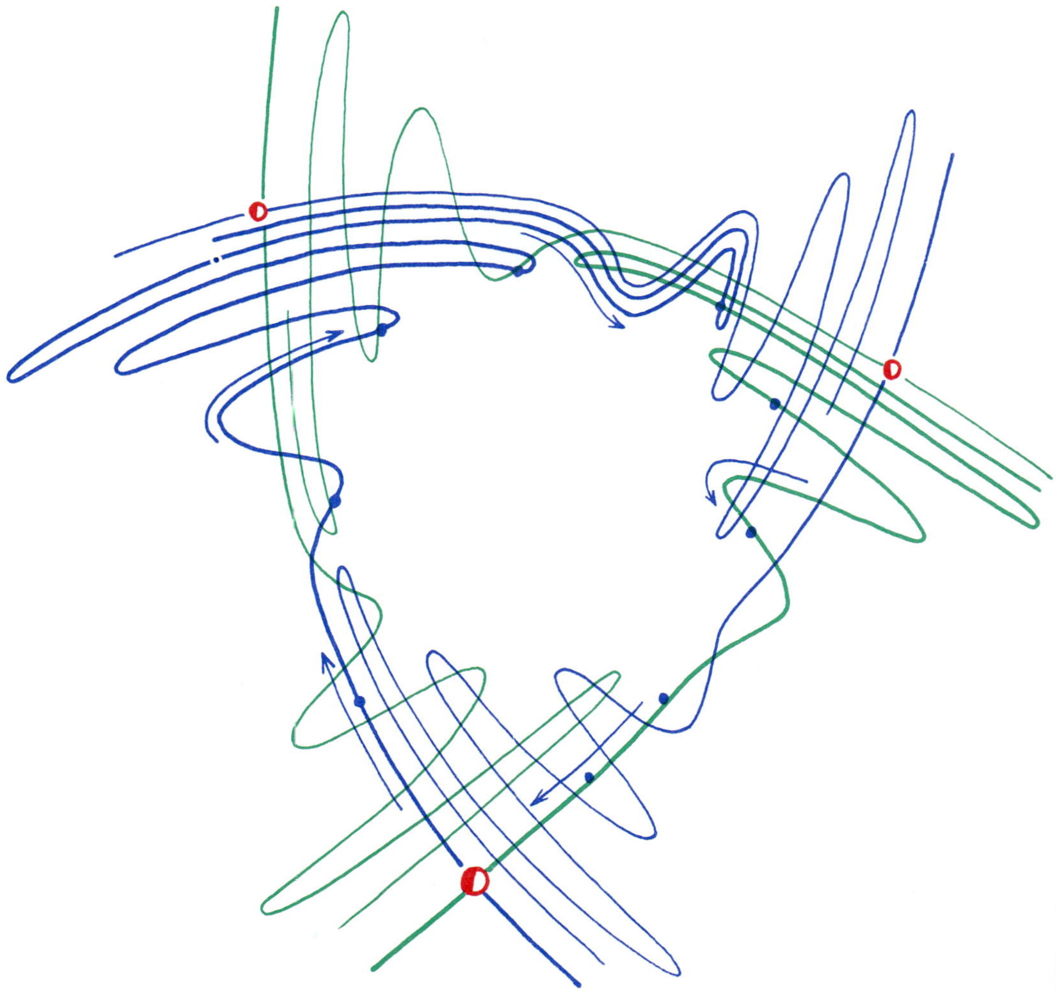

14.4.4. Here is a hypercycle involving three limit cycles of saddle type. Each is heteroclinic to each of the others. In all of these situations, it can be proved, by topological analysis, that each of the limit sets involved is actually homoclinic. That is, membership in a heteroclinic cycle implies homoclinicity.

Cycles of heteroclinically related critical points are endemic in real dynamical systems, and are vitally involved in chaotic motions.

15

Nontrivial Recurrence

In the history of dynamics, as in philosophy, the concept of *recurrence* frequently recurs. A periodic trajectory has the recurrence property: every one of its states will recur again and again. This is called *trivial recurrence.* The recurrence property also applies to more complicated (aperiodic) trajectories. This is called *nontrivial recurrence.* This concept already surfaced in the generic property G4, described in Section 11.3, and in the chaotic attractors of Part Two. In this chapter, more versions of this important phenomenon will be described.

15.1. Nearly Periodic Orbits

Recall that generic property G4 limits the types of almost-periodic motions. Discovered by Peixoto in two dimensions, its genericity was established by Pugh in higher dimensions.[1] Suppose that we take a sequence of points in the state space, converging (approaching asymptotically closer and closer) to a point, and that each of the points belongs to a closed orbit (limit cycle, or periodic trajectory). Topological consequences of the generic condition G3 (transversality) force the periods of these periodic trajectories to get longer and longer. Thus, the oscillations they represent have frequencies that get lower and lower. The limit point of the original sequence lies on a trajectory that need not be periodic. But it is *nearly periodic*, in that observations cannot distinguish it from a low-frequency oscillation. We will denote the set of all nearly periodic points of the dynamical system by *NP*.

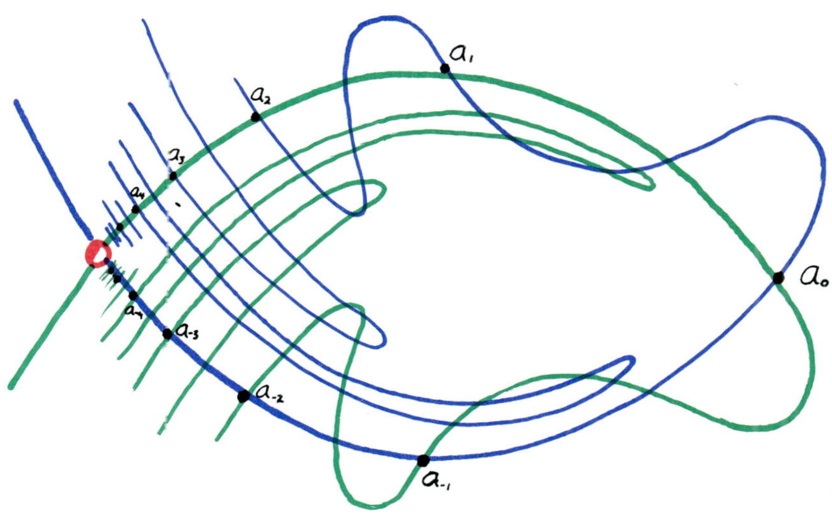

15.1.1. A homoclinic limit cycle provides good examples of nearly periodic points. Here is a Poincaré section of a homoclinic tangle. Look carefully at the trajectory of the point a_0.

Expansion of the tangle shows how the periodic orbits fit into this picture, from the cover of Hayashi's collected works.[2]

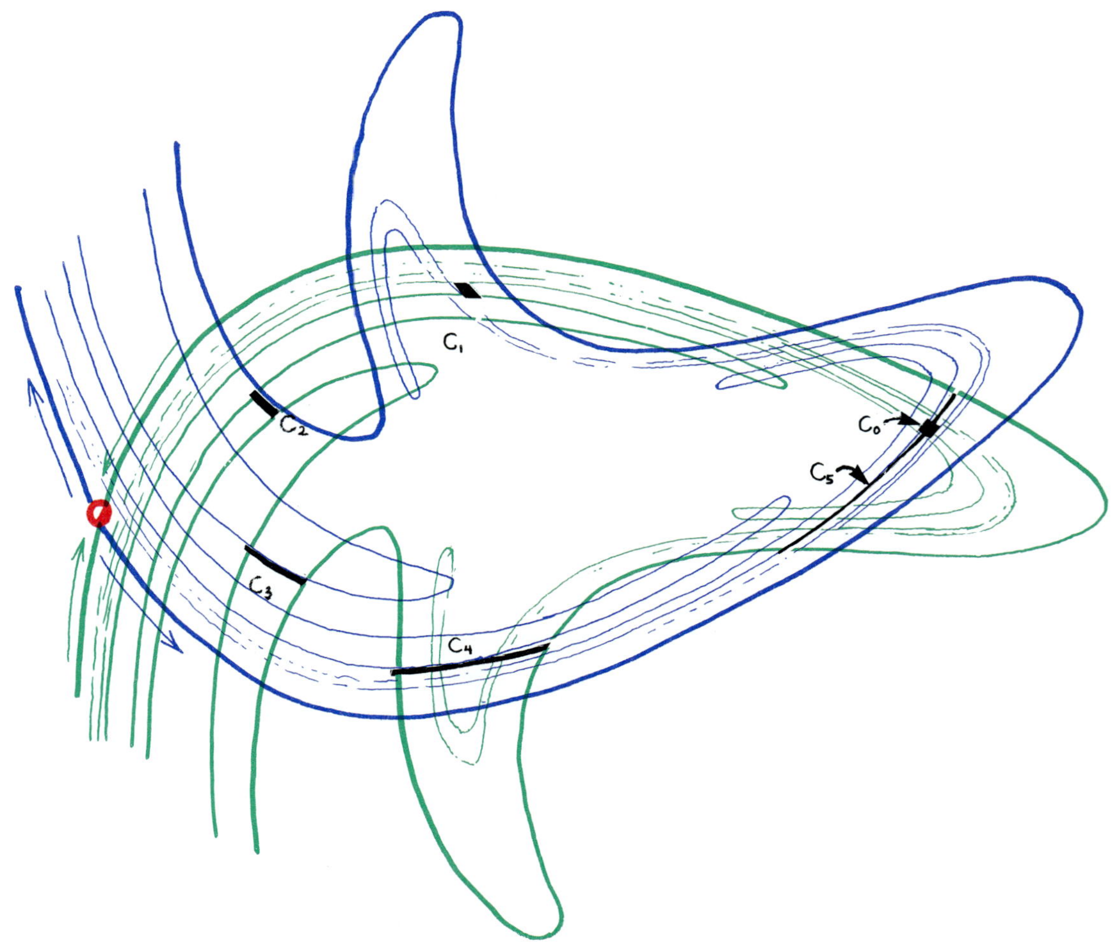

15.1.2. Inside this tangle, there must be a periodic orbit.[3] Let's follow the small red rectangle, marked c_0. Its sides are segments of insets and outsets. After one revolution around the ring, its first return to the Poincaré section is again a small rectangle c_1. Note that it is stretched in one direction and compressed in the other. Now follow its next five revolutions, noting that inset segments are stretched to longer inset segments, and outset segments are compressed to shorter outset segments. Note that c_5 intersects c_0.

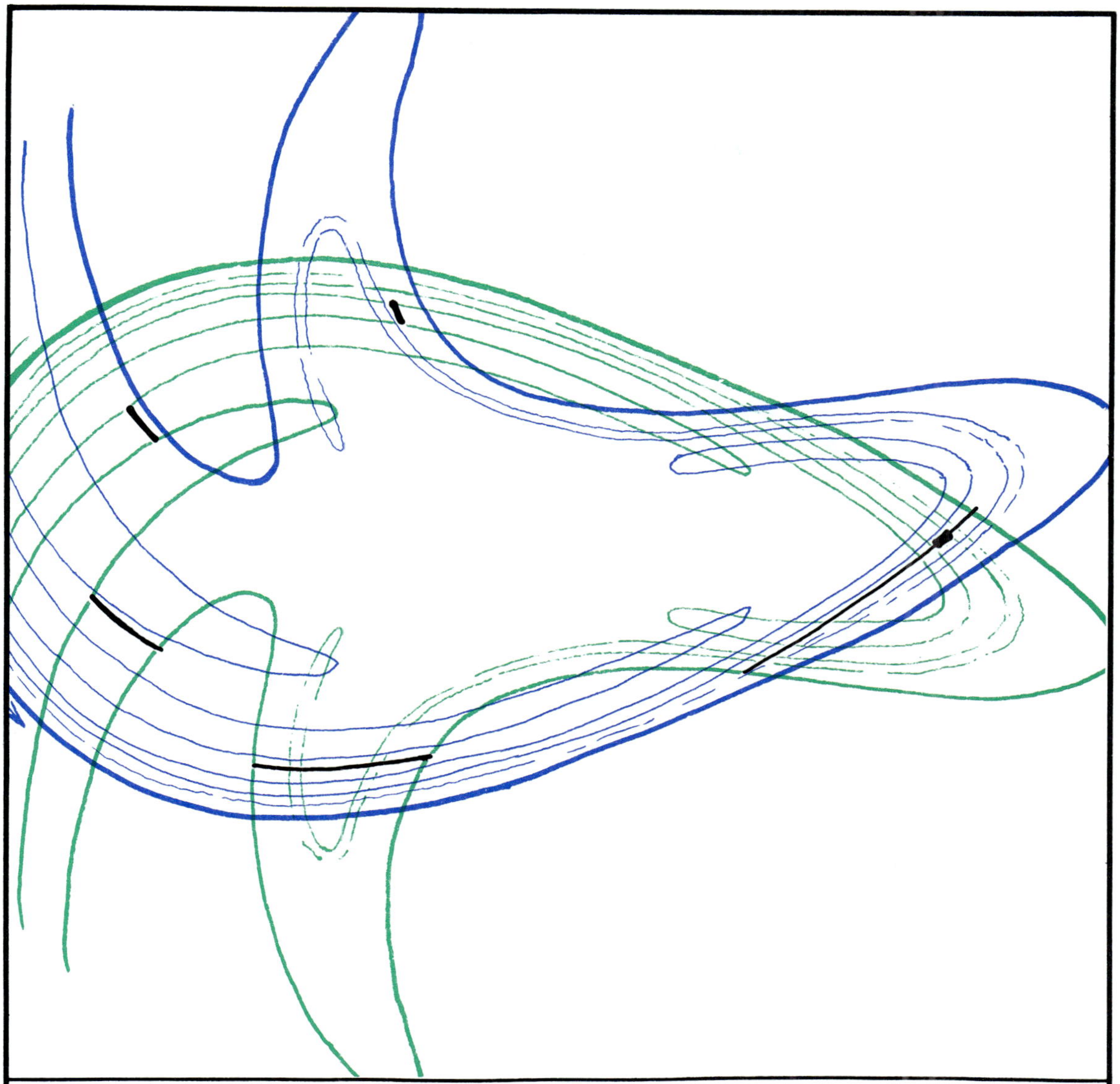

15.1.3. Now take the little piece of c_0 intersected by c_5 and follow it around five times. It will again pass through the initial rectangle. Continuing in this way, we obtain a sequence of nested boxes, which converge to a periodic point of period five, as predicted by the theorem of Birkhoff and Smith.[4]

We may use the expansion of the tangle as a magnifier, to zoom into the microstructure of the tangle.

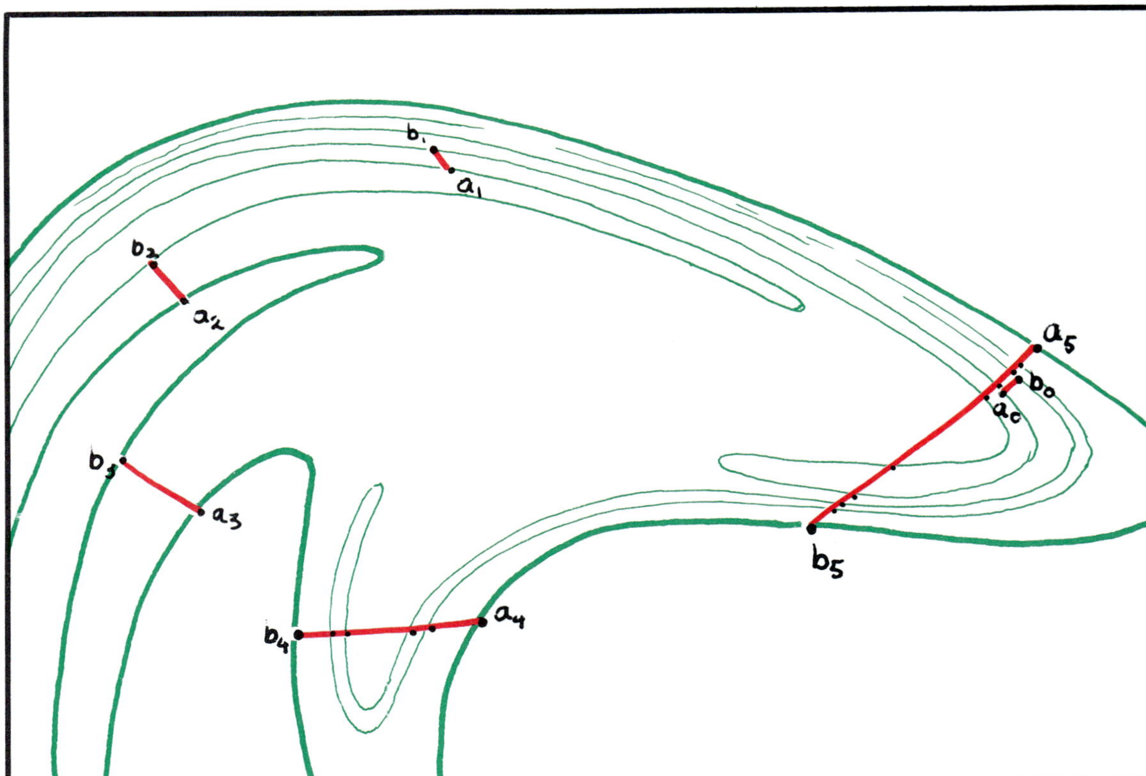

15.1.4. Now let's select two points a_0 and b_0, and follow their fates. The line segment a_0b_0 becomes, after five revolutions, the segment a_0b_0.

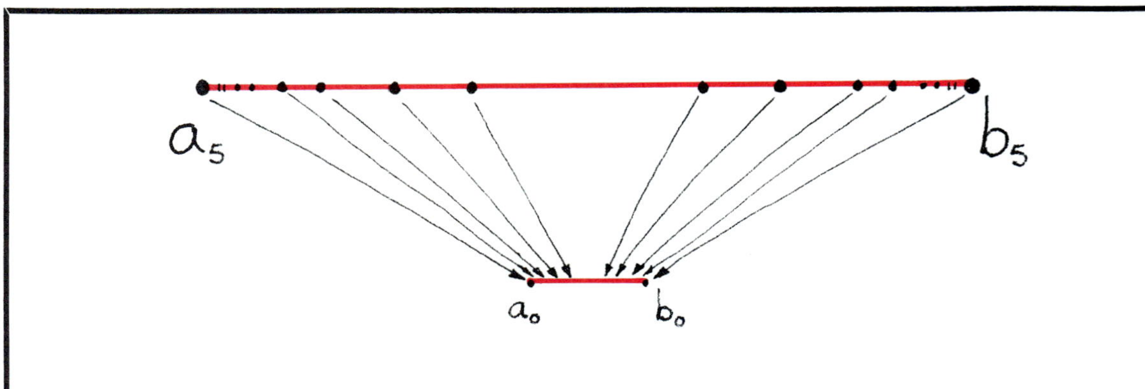

15.1.5. All the intersections of the inset within this stretched segment ab, must also be found in the shorter segment ab, but they are *five generations smaller.*

We may continue to zoom into this microscopic structure of the tangle.

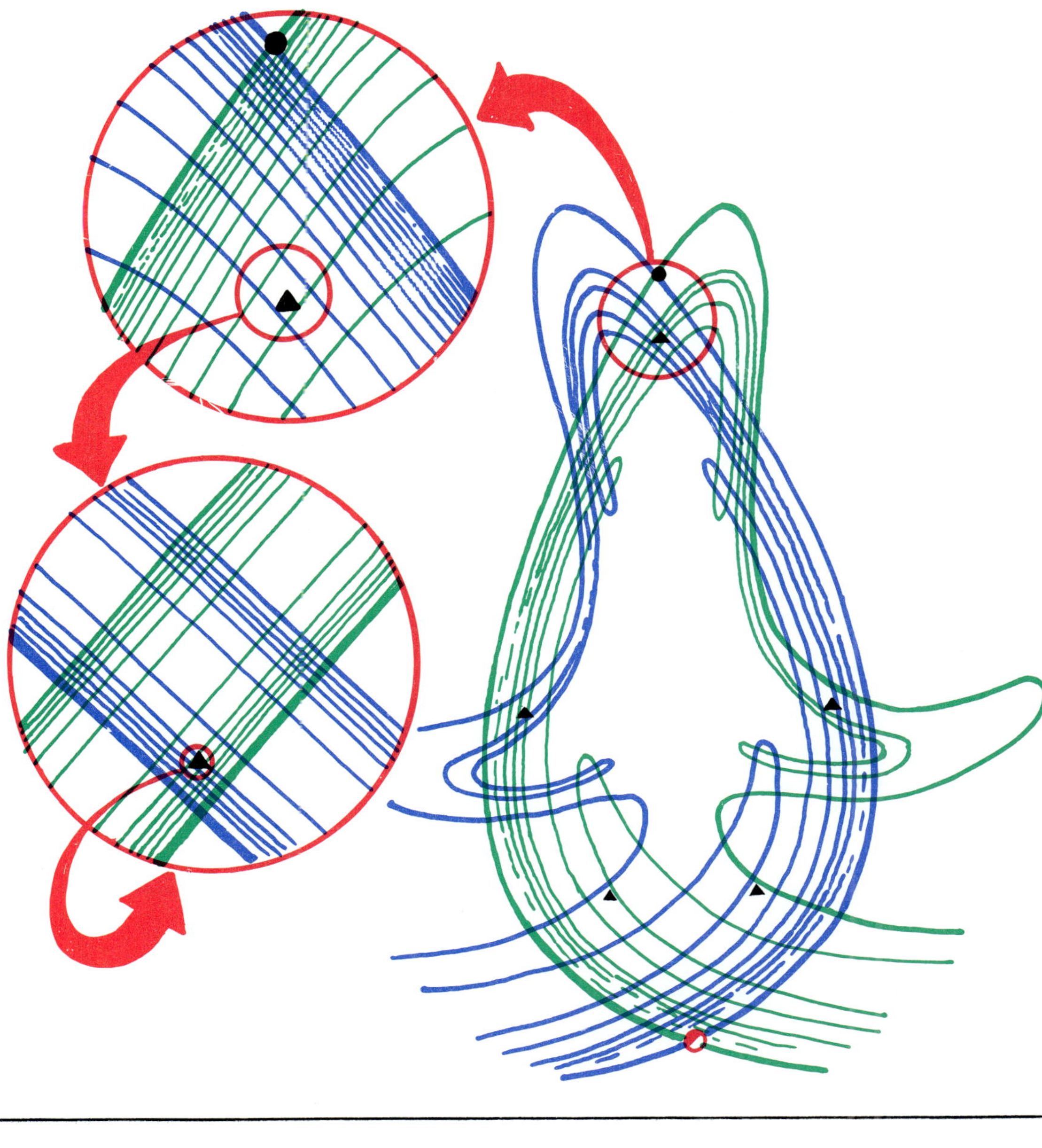

15.1.6. A few repetitions of the magnification method suffice to locate the periodic point as accurately as needed. It is within the small tangle.

By starting with other small rectangles and making judicious use of the zoom method, additional periodic points may be found.

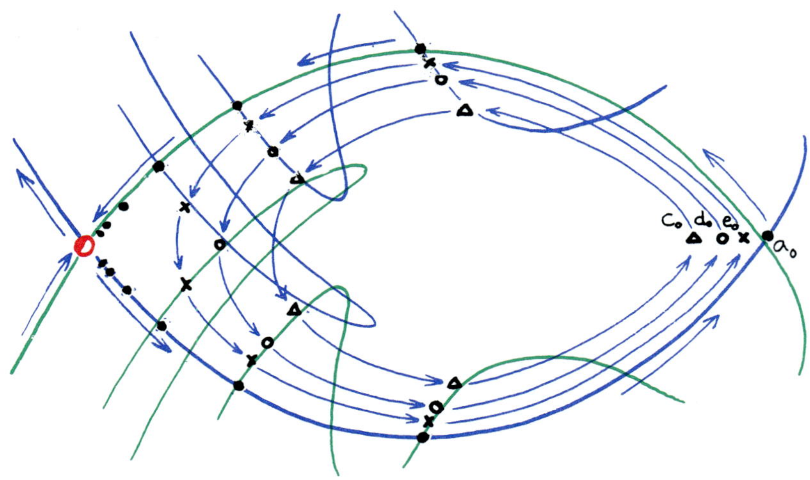

15.1.7. Closer to the homoclinic point a_0 there must be another periodic point with a higher period, such as c_0 shown here. And even closer, another with an even higher period, such as d_0. These may be located as accurately as needed by the zoom method described above.

Thus the original homoclinic point is the limit of a sequence of periodic points in the Poincaré section. In the three-dimensional state space, a sequence of closed orbits (periodic trajectories) asumptotially approach the homoclinic trajectory. Thus, every point on the homoclinic trajectory is *nearly periodic,* yet not periodic.

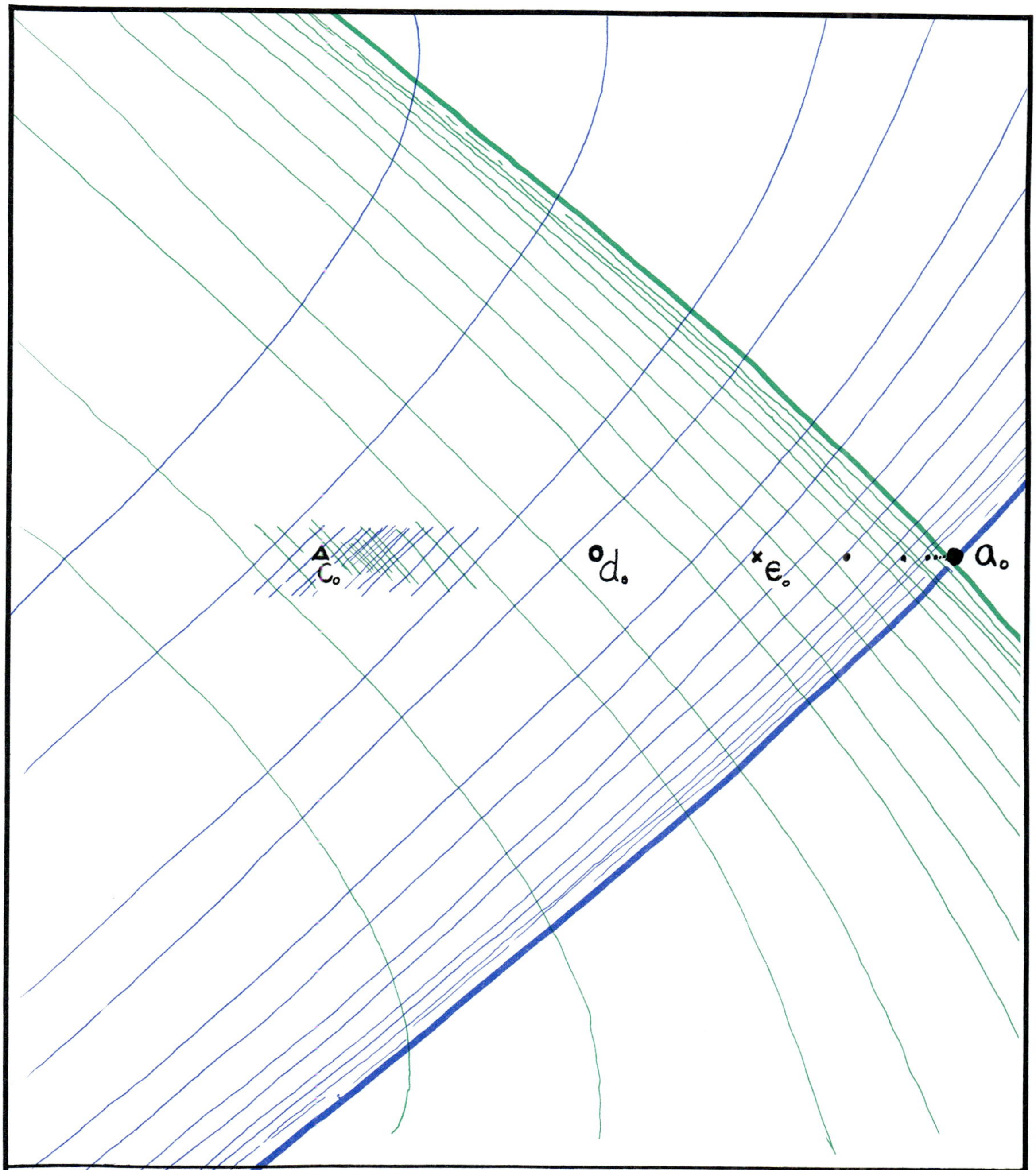

15.1.8. Here, highly magnified, is a sequence of periodic points approaching closer and closer to a homoclinic point, which is nearly periodic, yet not actually periodic.

15.2. Why Peixoto's Theorem Failed In 3D

As described in Section 12.2, Peixoto's theory of structurally stable systems is restricted to the two-dimensional case. In the case of state spaces of three dimensions or more, it is still true that structurally stable systems must have the four generic properties: G1, G2, G3, and G4. But these conditions no longer ensure structural stability. In fact, structurally stable systems are rare (that is, hard to find) in higher dimensions. A complete characterization of structural stability in three-dimensional systems (having a global section) has been accomplished recently.[5] This section describes the failure, and the remnants of Peixoto's theory that apply in higher dimensions.

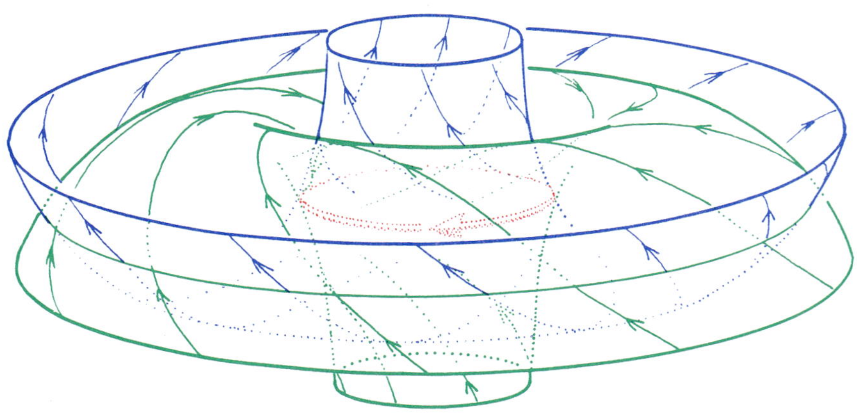

15.2.1. It is Step 1 in Peixoto's proof which is specifically two-dimensional. That step established that there are only a finite number of closed orbits (limit cycles) in the two-dimensional case. Here is an example of a *generic* portrait in three-dimensions. The homoclinic tangle forces the occurrence of an infinite number of limit cycles, as described in the preceding section. This example makes Step 2 wrong as well, as Step 2 is a simple consequence of Step 1.

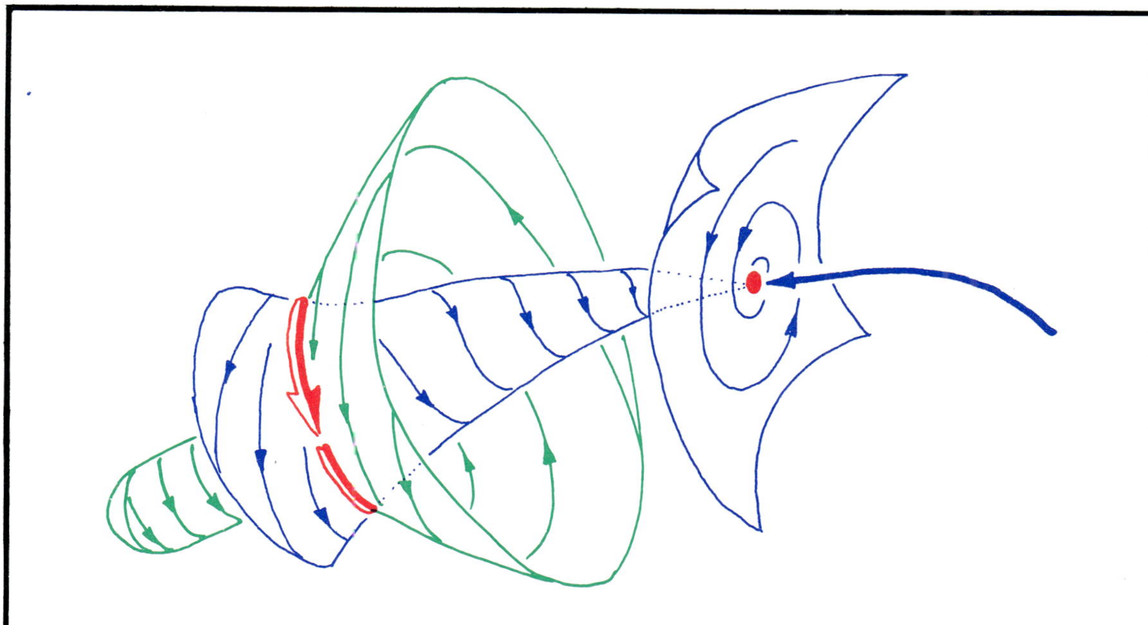

15.2.2. Step 3 remains true in higher dimensions. It assumes *property F:* the limit sets consist of a finite number of limit points and limit cycles only, as well as the four generic conditions. These are sufficient to ensure structural stability. This is a difficult result, due to Palis and Smale.[6] Here is an example of such a portrait, in three dimensions.

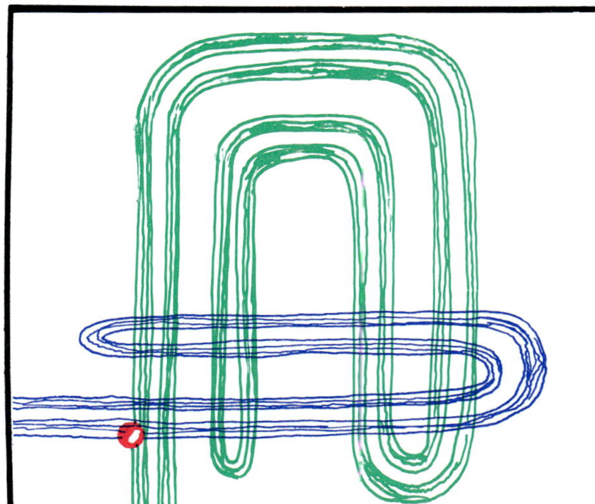

15.2.3. Step 4 fails in higher dimensions. Structural stability does ensure the four generic conditions. This is a relatively easy result, due to Markus and Robinson.[7] But structural stability does not ensure property F, the finiteness of the limit sets. The generic homoclinic tangles can be structurally stable, as Smale has shown for the example shown here.[8]

The progress of dynamical systems theory stalled briefly at this point, until it occurred to Smale to regard a homoclinic tangle as a generalized limit cycle and propose generic properties for it as a unit. He called this a *basic set.* The main example is the horseshoe, dissected in the preceding chapter. This was a prototype for the chaotic attractors, described in Part Two. One of the fundamental properties of a basic set is nonwanderingness, described in the next section.

15.3. Nonwandering Points

One of the most restrictive versions of the recurrence property is near-periodicity, defined above, in Section 15.1. And one of the least restrictive versions is the property of *nonwandering*, defined in this section.

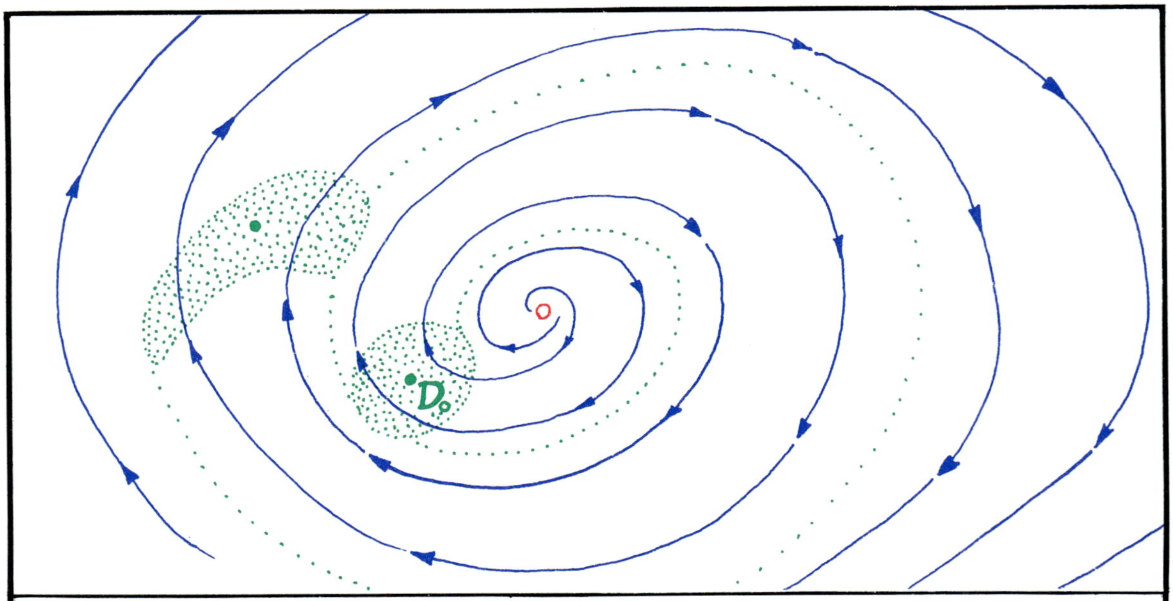

15.3.1. Suppose, having picked out a point in the state space and a little disk centered on it, that we follow the future meandering of the entire disk. If wide enough, it may meet up with itself along its meander.

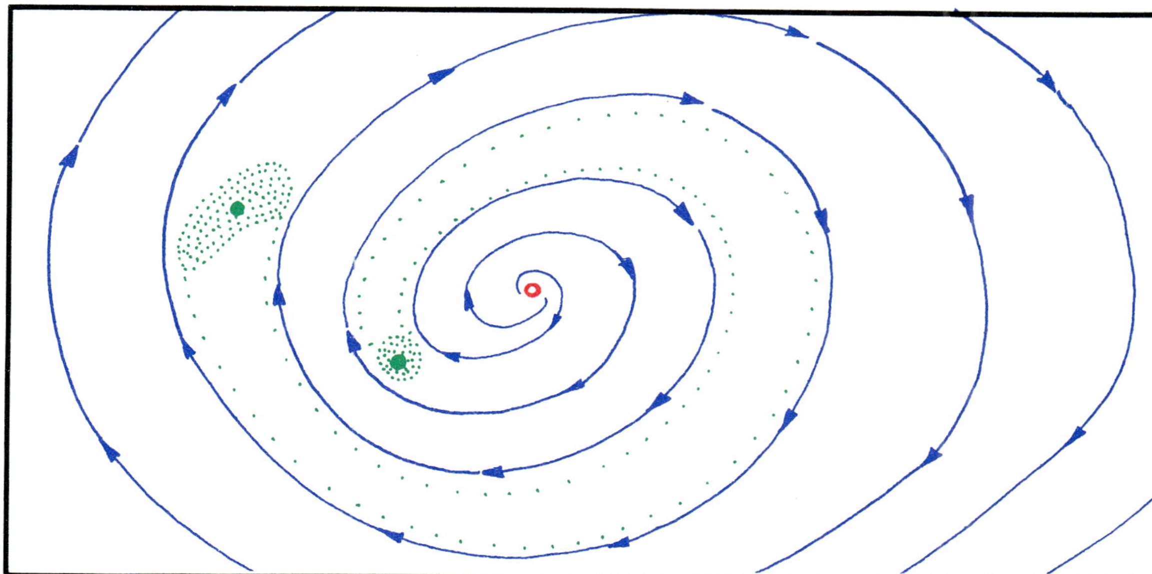

15.3.2. If so, start with a smaller disk, and repeat the construction. If now the meandering disk leaves its original position, wanders away, and never returns to overlap its original position, then the original point at the center of the disk is called a *wandering point.*

On the other hand, it may happen that, no matter how small you draw the original disk, it always comes back to overlap itself. Or, it may never cease overlapping itself, no matter how long you wait. In these cases, the original point is a *nonwandering point.* The set of all nonwandering points of a given dynamical sytstem will be denoted by *NW.*

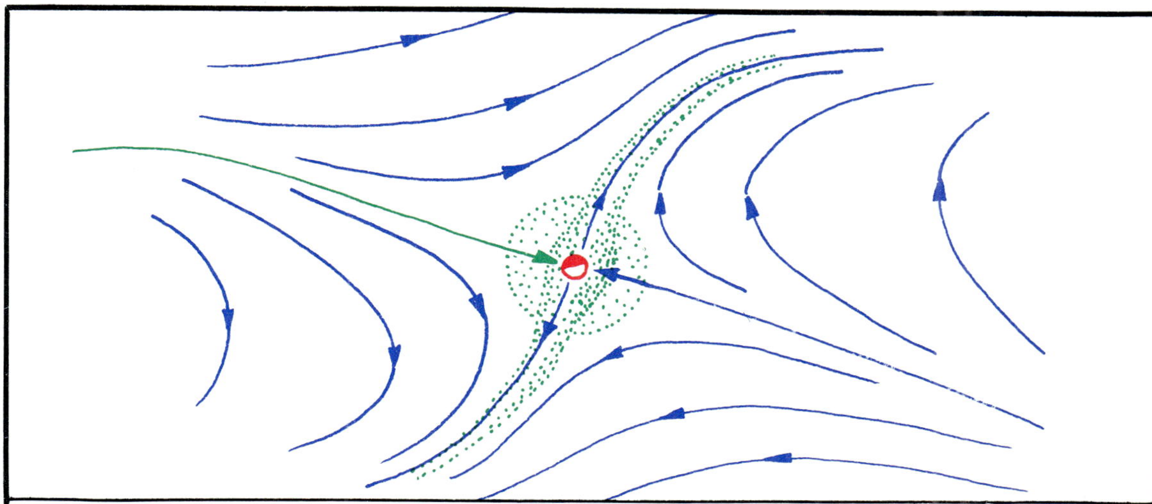

15.3.3. For example, a limit point (equilibrium) is nonwandering. The little disk is tied down at the center.

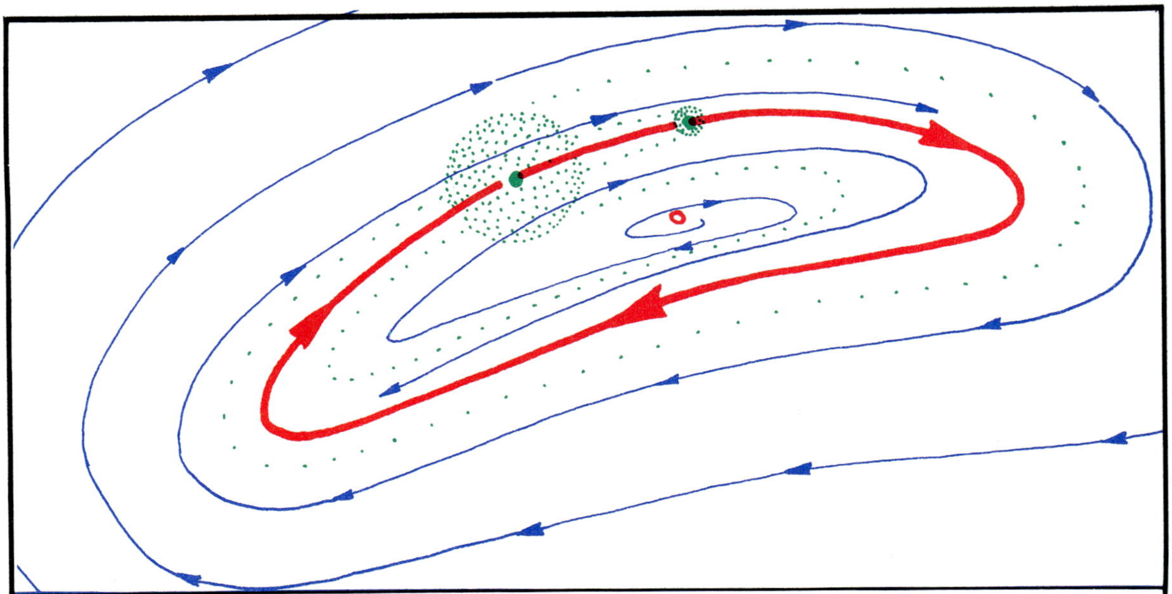

15.3.4. Similarly, a closed orbit (limit cycle) is nonwandering. The center of the little disk keeps passing through the initial point, again and again. In fact, the set of nearly-periodic points, *NP,* is contained in the set of nonwandering points, *NW,* for topological reasons.

15.3.5. Here is an outstanding example of a nonwandering point which is *not nearly periodic.* In this solenoidal flow on the torus, called a *Kronecker irrational flow,* every point is nonwandering, yet no point is periodic, or even nearly periodic.

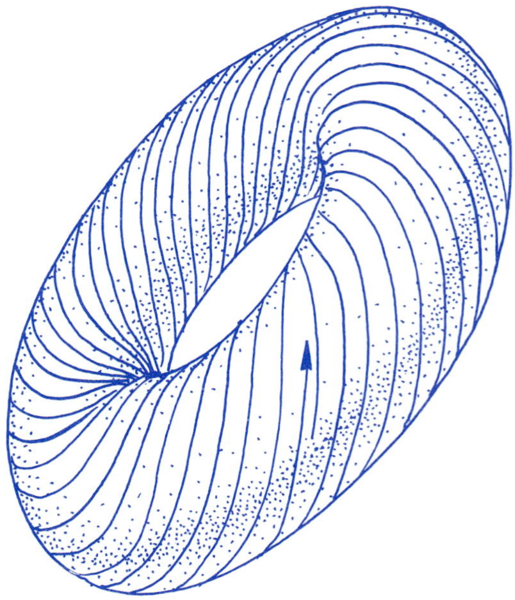

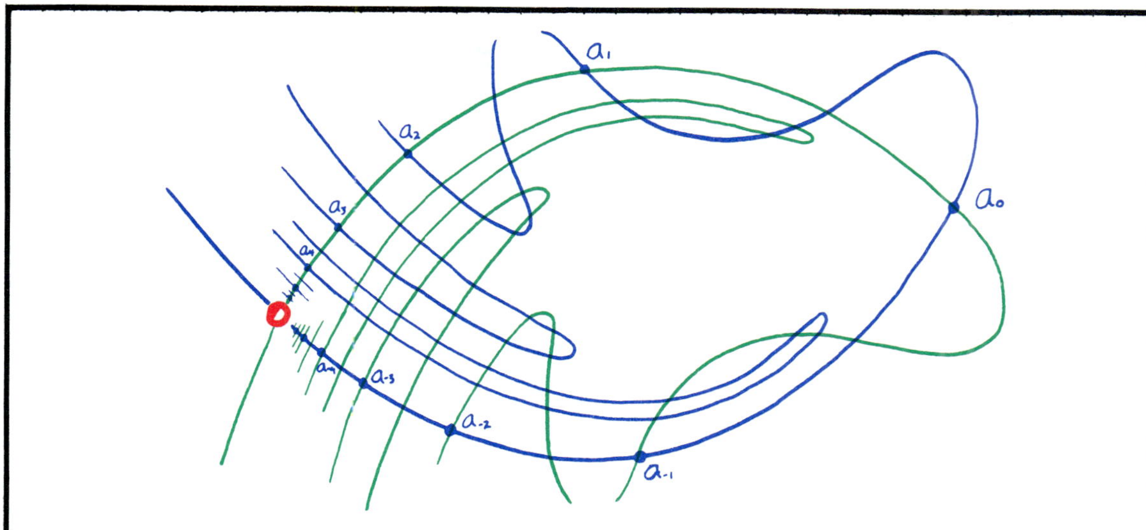

15.3.6. This is an example of a nonwandering point which is not itself recurrent in any sense. The flow has a limit cycle of saddle (index 1) type, which is homoclinic, and satisfies G3 (transversal intersection). The heteroclinic trajectories within this tangle are nonwandering.

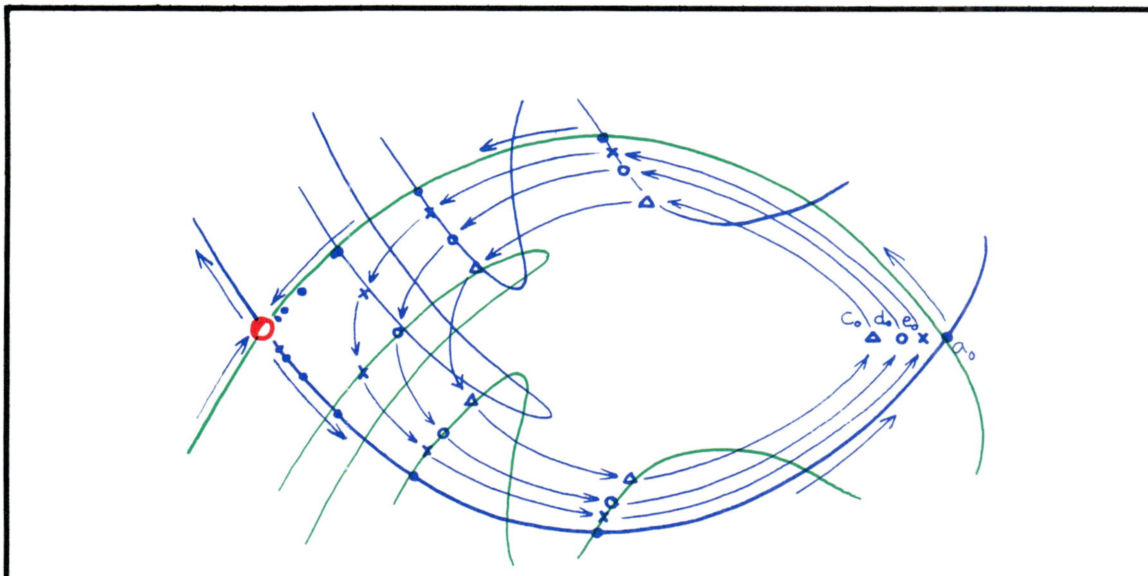

15.3.7. The theorem of Birkhoff and Smith, later generalized to higher dimensions by Smale, shows that these trajectories are nearly periodic. That is, they are approximated by limit cycles of very low frequencies. The heteroclinic trajectories belonging to a heteroclinic cycle of tangles are also nearly periodic.

Generic property G4, discussed previously in Section 11.4, can now be simply stated: *NP* = *NW*. That is, a dynamical system has property G4 if its every nonwandering point can be approximated by periodic points (points belonging to limit cycles). This property is generic, as proved by Peixoto (in 2D) and Pugh (in higher dimensions).

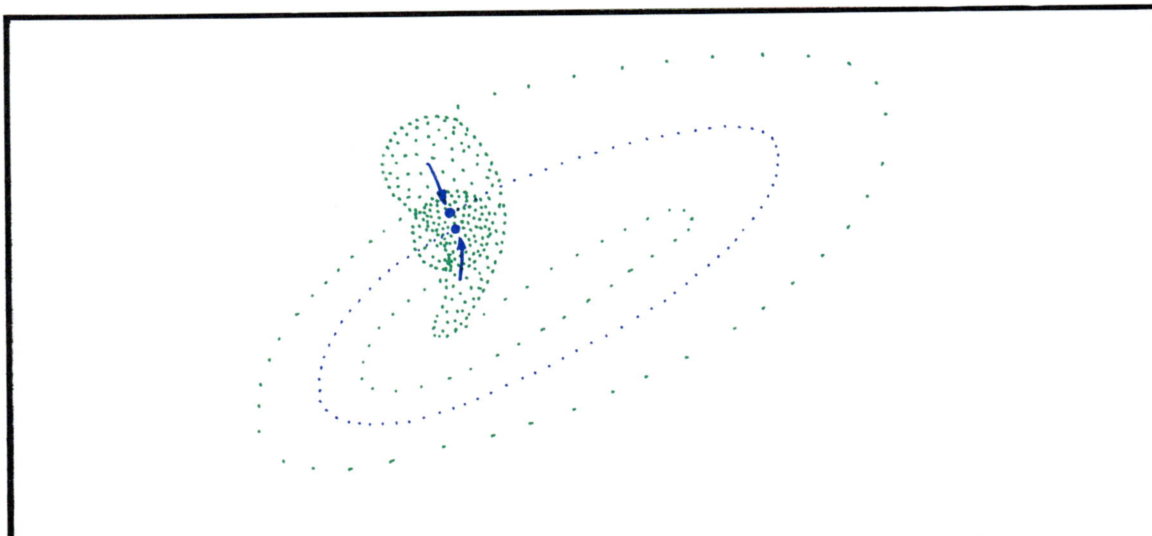

15.3.8. The proof of the genericity of this property is intuitively simple, yet it is one of the most difficult in the whole literature of mathematical dynamics to carry out in detail. The key step, called the *Closing Lemma,* makes small changes in the vectorfield, so that a closed orbit is found in the disk that meets itself.

***Warning:* As described briefly in Chapter 12, this property is generic only in a very weak sense. The reason is that the violation of G4 by persistent solenoidal flows (equivalent to irrational Kronecker flows on invariant tori) occurs with positive expectation. Thus, in the sense of probability, G4 violation is also generic. We may call this the *G4 paradox.* It will be explained further in Part Four.**

PART 4

Bifurcation Behavior

Dedicated to René Thom

Photograph by Ismael Selim Khaznadar

Bifurcation Hall of Fame

Bifurcation concepts emerged early in the history of dynamics. Soon after Newton the first case, the pitchfork, was discovered. Eventually, bifurcation theory bifurcated into two branches, dealing with similar phenomena in the contexts of ordinary differential equations (ODE's) and partial differential equations (PDE's), respectively. ODE's comprise the type of model introduced by Newton for mechanics, and this branch has evolved in this century into dynamical systems theory. PDE's were introduced by d'Alembert in 1749 to model the continuous (that is, spatially extended) mechanics of the vibrating string.

Recently, thanks to global analysis, these two branches have reunited. This reunification was effected by reinterpreting a PDE in a finite-dimensional physical space as an ODE in an infinite-dimensional space of functions. In this section we give capsule biographies of some of the historically important personalities. Further description of their roles in the history of the subject may be found in Chapter 16.

TABLE 4.1.—THE HISTORY OF BIFURCATION THEORY

Date	ODE	PDE
1600		
		Hooke Newton
1700		Clairaut Maclaurin Simpson d'Alembert
1800		Jacobi
		Tchebychev
1900	Poincaré Liapounov Andronov Hopf Thom	Liapounov Poincaré Couette Taylor
2000		

Here are some capsule histories.

Robert Hooke, 1635–1703. In 1683, Hooke guessed that the Earth was flattened at the poles.

Isaac Newton, 1642–1727. In 1687, Newton assumed that the Earth was a spheroid, flattened at the poles. To calculate its eccentricity, he devised his *principle of canals.*

Alexis Claude Clairaut, 1713–1765. Clairaut examined the possibility that Newton's oblate spheroid was a relative equilibrium for a blob of fluid.

Colin Maclaurin, 1698–1746. Using Newton's principle of canals, he established, in 1742, the relative equilibrium of a rotating ellipsoid of homogeneous fluid, subsequently known as a *Maclaurin spheroid.*

Thomas Simpson, 1710–1761. By careful analysis, he actually showed in 1743 that *two distinct Maclaurin spheroids* were relative equilibria, implying a bifurcation in the possible figures of the Earth.

Jean d'Alembert, 1717–1783. He explicitly analyzed, in 1768, the bifurcation implied by Simpson in 1743.

Carl Gustav Jakob Jacobi, 1804–1851. In 1834, he discovered a new equilibrium figure for a rotating fluid blob, the *Jacobi ellipsoid.* Also, he introduced the word *bifurcation* in this context, to describe the relationship between the Maclaurin spheroids and the ellipsoidal figures.

Jules Henri Poincaré, 1854–1912. The question of the *stability* of the figures of the Earth was introduced by Poincaré in 1885, in response to a problem posed in 1882 by Tchebychev on the evolution of the figure in the case of a gradually increasing angular momentum. He also carried over this concept into our current context of dynamical systems.

Aleksandr Mikhailovich Liapounov, 1857–1918. Liapounov considered the problem of Tchebychev also, and created his classical theory of stability in this context.

Aleksandr Aleksandrovich Andronov, 1901–1952. Andronov created a complete theory of bifurcations of dynamical systems in the plane.

Eberhard Hopf, b. 1902. Hopf published, in 1942, a rigorous proof of the first excitation event, after which it became known as the *Hopf bifurcation.*

René Thom, b. 1923. The publication of Thom's revolutionary book, *Structural Stability and Morphogenesis,* in 1972 marked a major turning point in the importance of nonlinear dynamics and bifurcation theory to the sciences: physical, biological, and social. Without doubt he is the most important pioneer in this area since Poincaré and we are all deeply in his debt. In recognition of this, we have dedicated this Part to him.

16

Origins of Bifurcation Concepts

In Part One, ''Periodic Behavior,'' limit points and cycles in dimensions one, two, and three were introduced. The decomposition of the state space into basins of attraction, by the separatrices, was emphasized. In Part Two, ''Chaotic Behavior,'' the inset structure of the separatrices was developed. The geometry of the exceptional limit sets, determined by their Lyapounov characteristic exponents, was described. In Part Three, ''Global Behavior,'' the fundamental idea of structural stability was introduced, along with the related notion of generic property.

All of this material is basic to the theory, experiments, and applications of dynamics. However, the most important of all, from the point of view of applications, are the *bifurcations of* dynamical systems being changed by a control parameter. This is the subject of Part Four, and in the preceding parts we have selected topics so as to create the minimum background needed for this theory.

In this chapter, we trace the history of the bifurcation concept from darkest antiquity.

16.1. The Battle of the Bulge

Our knowledge of the shape of the Earth has grown throughout history, and out of this history emerged the concepts of bifurcation theory. We begin with a capsule version of this story. A splendid 150-page version may be found in Jones and the associated mathematical details in Todhunter.[1]

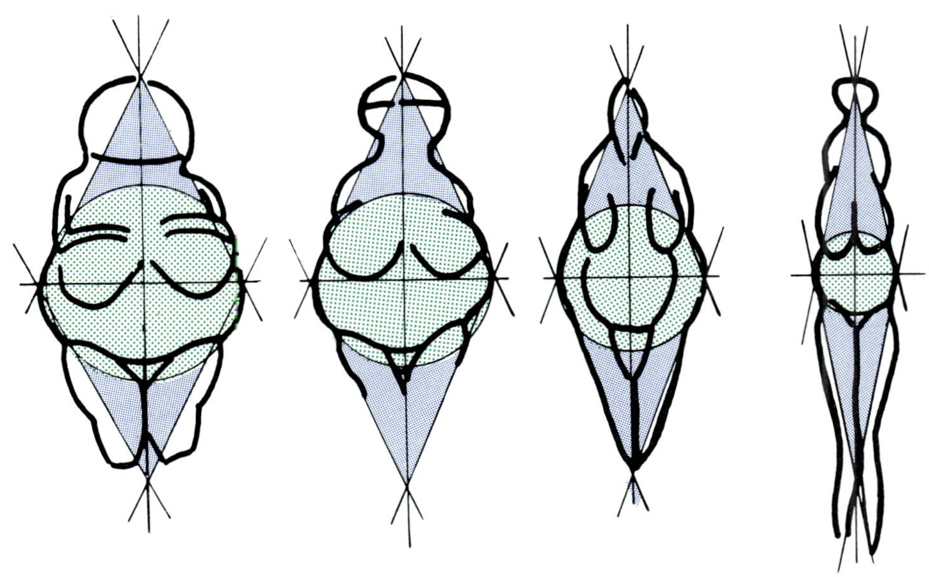

16.1.1. The Earth, Gaia, Goddess: what is her figure? The Venus figurine from the Gravettian culture, found throughout Upper Paleolithic Europe, is generally assumed to be a fertility amulet. (Reproduced from Leroi-Gourhan, 1967.) Was it also a geographical model of the Earth?

We do not know when or how the globular shape of our home planet was first discovered, but we do know that Aristotle knew it by 350 BCE. And by 225 BCE, Eratosthenes (the Alexandrian librarian) knew its circumference within one percent! Thus begins the early history of our subject. Things changed little until the dawn of the Baroque, although confidence in Eratosthenes had waned by the time of Columbus.

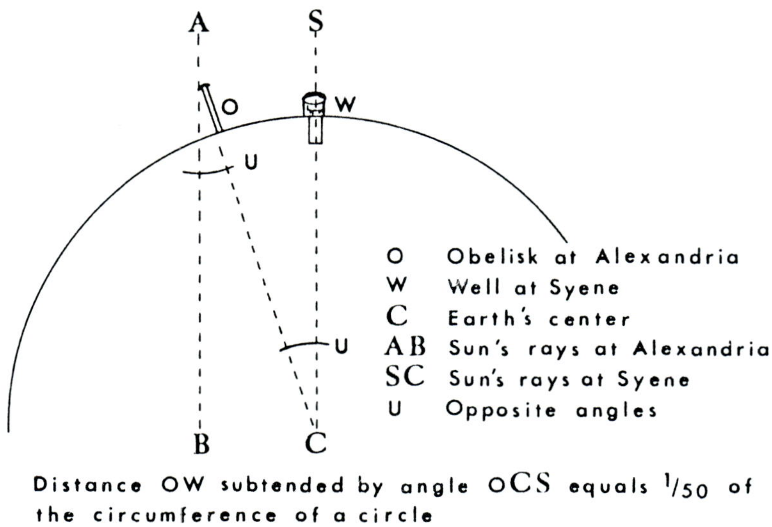

Calculation of the earth's circumference by Eratosthenes.

16.1.2. Throughout this period, the Earth was thought to be roughly spherical. Then paradoxes began to accumulate.

One problem was the discrepancies in the measurements of the circumference, some of which were done with great care. Another difficulty was the accuracy of pendulum clocks, which were found to slow down at the equator. Here is a summary of the events casting doubt on the spherical hypothesis.

1522 Magellan sails all the way around
1525 Fernel calculates circumference
1577 Halley's comet
1610 Galileo
1617 Snell measures one degree latitude
1635 Norwood measures latitude
1650 Riccioli measures latitude
1657 Firenze academy founded
1665 Halley's comet
1666 Paris academy founded
1669 Picard measures latitude accurately
1669 Dom Cassini (I) moves to Paris
1672 Richer goes to Cayenne, pendulum too slow
1677 Halley goes to St. Helena, pendulum slow
1680 Halley comes to Paris to work with Cassini

16.1.3. Early Baroque events leading up to the "Battle of the Bulge."

NEWTON

453
16.1.4. In 1683, Robert Hooke suggested an oblate spheroid figure for the Earth. This *onion model* was immortalized in Newton's *Principia* in 1687.

A *spheroid* is what you get by spinning an ellipse about an axis. An *oblate spheroid* results from spinning around the shorter axis. A spanish onion or a bun has this shape. A *prolate spheriod* results from spinning around the longer (major) axis. A lemon has this shape. An *ellipsoid* is not made by spinning, but has elliptical sections when cut.

CASSINI

16.1.5. Dom Cassini, the first of four generations of outstanding astronomers, countered in 1700 with a prolate spheroid, or *lemon model*, and the Battle of the Bulge was on!

1683 **Robert Hooke makes onion hypothesis**
1686 **Fontenelle publishes *Plurality of Worlds,* popularizing Descartes**
1687 **Isaac Newton publishes onion hypothesis with mechanical arguments, computations**
1690 **Huyghens supports Newton**
1691 **Dom Cassini observes oblateness of Jupiter**
1700 **Dom Cassini publishes the lemon hypothesis**

16.1.6. The events (1680–1700) leading up to the controversy, which nearly resulted in World War in Europe and the Americas.

The Paris Academy of Science decided to resolve the crisis by sending expeditions to the Arctic Circle in Lapland and to the Equator in Peru, to make definitive measurements of meridional arcs of one degree of latitude. These would be north-to-south arcs of about 110.5 kilometers (68.7 miles) length, assuming a spherical figure. Between endpoints determined by observing the angle to the Sun at noon (as in celestial navigation), the measured length of the southern arc should be longer than 110.5 kilometers for a prolate spheroid, and shorter for an oblate one.

1718 **Jacques Cassini publishes measurements supporting the lemon hypothesis**
1732 **Maupertuis and Clairaut support Newton's onion hypothesis**
1733 **La Condamine proposes expedition to Equator (Cayenne)**
1734 **Godin suggests expedition to Equator (Ecuador)**
1735 **Expedition leaves Paris for Ecuador with La Condamine and Bouguer**
Maupertuis proposes expedition to Arctic Circle (Lapland)
1736 **Expedition leaves Paris for Lapland with Maupertuis, Clairaut, and Celsius**

16.1.7. Here, in a nutshell, is the sequence of events during the early years of the conflict, 1718–1736.

When the measurements finally reached Paris in 1744, the onion team had won.

1737	**Lapland measurements reach Paris**
	Algarotti publishes popular account of Newton's optics
	D. Bernoulli analysis fluid cylinder
1738	**Clairaut finds a formula for the equilibrium of a rotating fluid blob**
	Euler writes analysis of the fluid onion
1739	**War of Jenkins' Ear, and pyramids**
1740	**Maclaurin proves the onion for a rotating blob of homogeneous fluid**
	Cassini publishes new measurements supporting the lemon hypothesis
1743	**Expedition leaves Ecuador for home**
1744	**Bouguer arrives in Paris**
	Ecuador measurements completed by La Condamine and Bouguer
	Maupertuis finds the principle of least action
	d'Alembert analyzes the fluid blob
	Celsius dies
	Cassini de Thury capitulates
	La Condamine returns to Paris

16.1.8. The last eight years of battle, during which hydrostatics and bifurcation theory were born.

Degree of Latitude in Meters compared with early measures
(after *Geographical Journal* 98 (1941) p. 292)

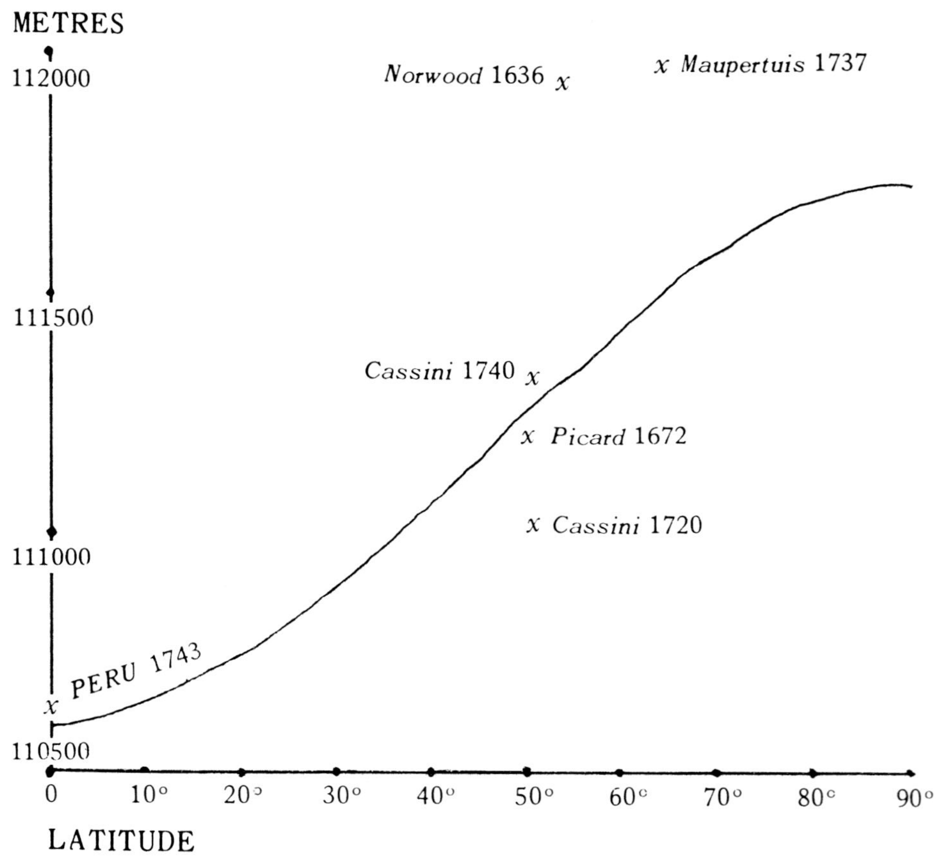

16.1.9. Here is a summary display of the best known measurements of a one-degree arc, taken from the *Geographical Journal*.[2] The measures at all latitudes would fall along a horizontal line, if the world were spherical. The rise toward the pole (on the right) confirms an oblate (onion) figure.

Cassini I	1625	1712
Huyghens	1629	1695
Newton	1642	1727
Cassini II	1677	1756
Bach	1685	1750
Bouguer	1698	1758
Maupertuis	1698	1758
Maclaurin	1698	1759
La Condamine	1701	1774
Clairaut	1713	1765
Euler	1707	1783
Simpson	1710	1761
Clairaut	1713	1765
Cassini II	1714	1784
d'Alembert	1717	1783
Lagrange	1736	1813
Cassini IV	1748	1845
Laplace	1749	1827

16.1.10. The cast of characters in order of their appearance.

We turn now to the emergence, in this strange context, of the bifurcation concepts which are fundamental to the modern theory of bifurcation behavior.

16.2. The Figure of the Earth

From this lively story of scientific conflict and creativity that dominated the activities of the first scientific societies of Europe throughout their early years, we here extract the mathematical events leading up to the development of the bifurcation concepts of modern dynamical systems theory. These considerations deal with a rotating homogeneous fluid mass (or *blob*) and apply equally to the cosmogenic problems of stellar evolution and galaxy formation. For additional details, see Hagihara[4] and Lyttleton.[5]

From 400 BCE until 1683, the Earth was thought to be roughly spherical.

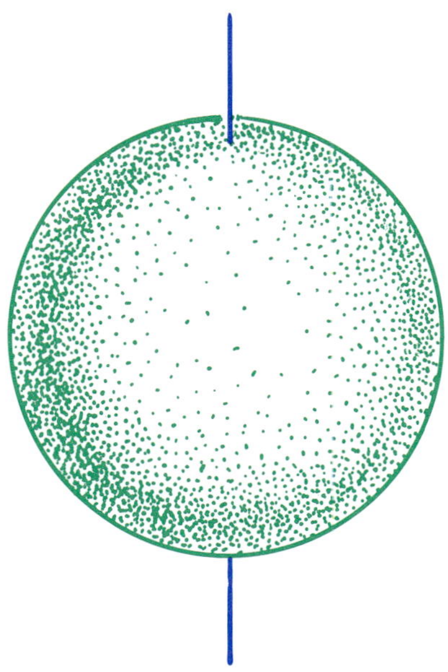

16.2.1. In fact, if the blob is not rotating, then a sphere is its only relative equilibrium, and it is stable, as Liapounov showed in 1884.

But it *is* rotating. Newton introduced the *principle of canals* in his *Principia* of 1687, to analyze the dynamics of a spinning oblate spheroid (onion shape).

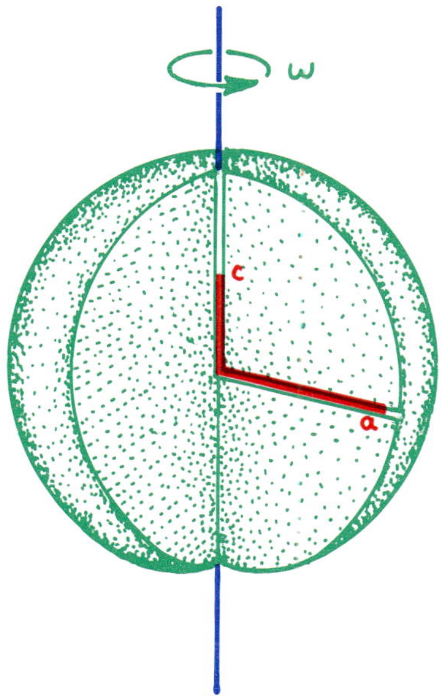

16.2.2. *The principle of canals.* In the spherical Earth, assumed solid, drill two long tunnels that meet in the center: one from the North Pole, and another from some point on the Equator. Fill nearly full with water. Due to the centrifugal force (named by Huyghens but successfully analyzed by Newton as a graduate student) the Equatorial tube of water will be pulled outward, rising to a higher level than the polar tube.

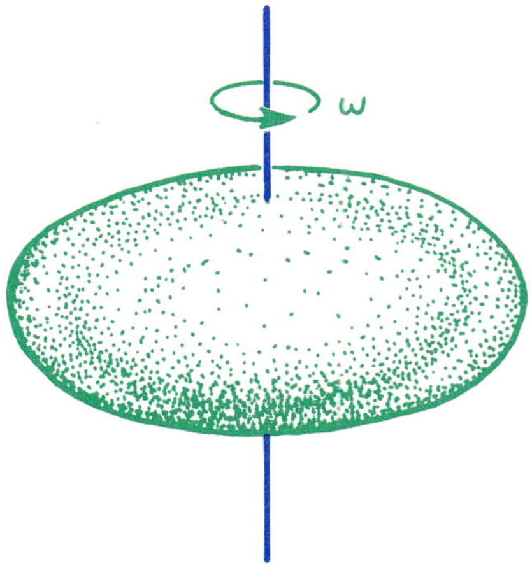

16.2.3. Using the principle of canals, Newton calculated the eccentricity needed by the onion (as a function of its angular momentum) to maintain equilibrium. Maclaurin proved this condition was necessary for hydrostatic equilibrium in 1740, and Clairaut generalized his result to inhomogeneous blobs.

The analysis of the onion by the principle of canals was completed by Simpson[4] in 1743. Let 2a denote the length of the major axis of an ellipse, 2c the length of the minor axis, and 2d the distance between its foci. Then $a^2 = c^2 + d^2$. Recall that the *eccentricity* of the ellipse is the ratio d/a. The ratio is zero for a circle, and is always less than one. The eccentricity increases with the ratio a/c.

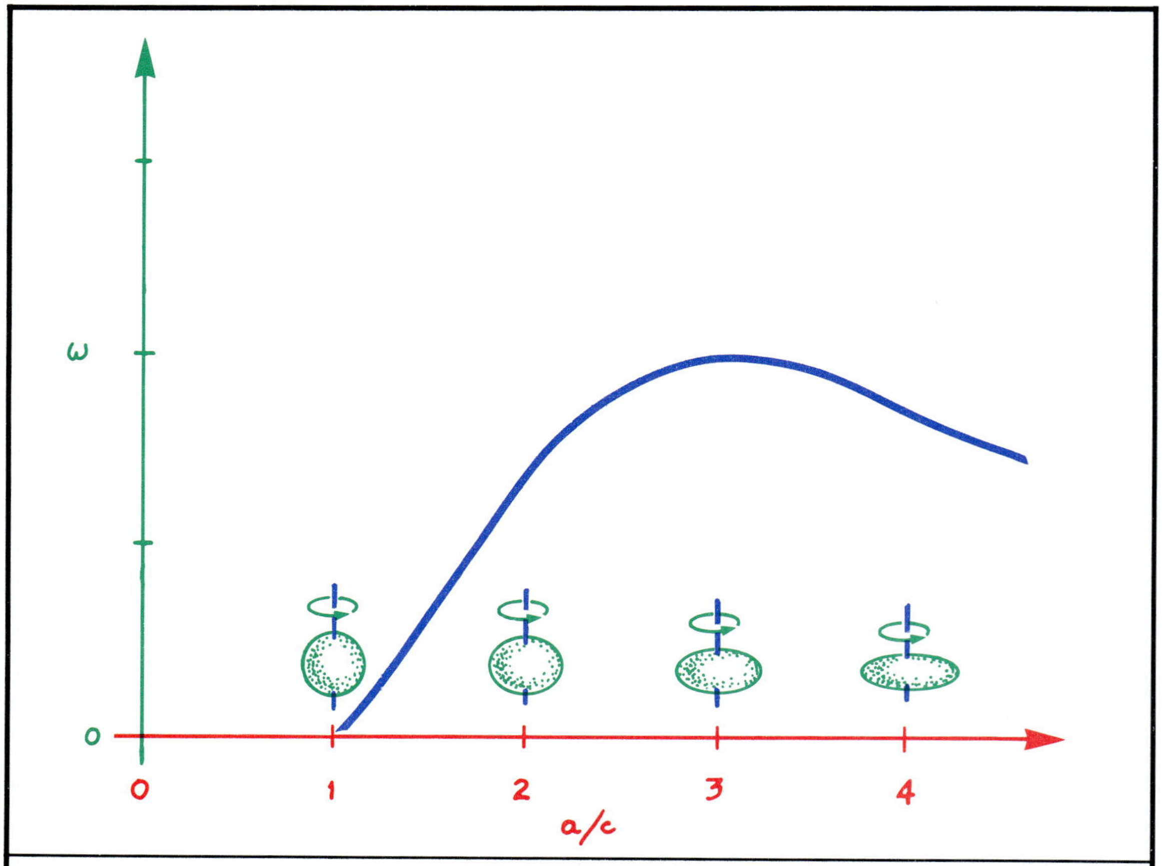

16.2.4. As the eccentricity of the homogeneous spheroidal blob increases, the angular velocity traces out this curve, obtained by Simpson, with a maximum at e = 0.9299, or a/c = 2.7198.[8] Thus for smaller angular velocities there are two equilibrium Maclaurin spheroids having the same rate of rotation, and there is a maximum rate. However, the *angular momentum* goes on increasing along this curve,[6] which describes the *Maclaurin series* of spheroids.

The next major advance in our story was the discovery by Jacobi in 1834 of a new figure, which defies intuition in that it does not have rotational symmetry.

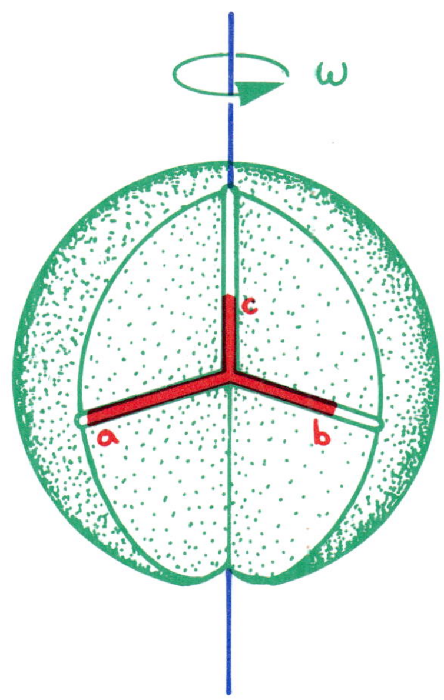

16.2.5. Drilling three canals in a spherical Earth, the two Equatorial canals would be expected to have the same equilibrium water level. Jacobi showed that this is not necessarily true.

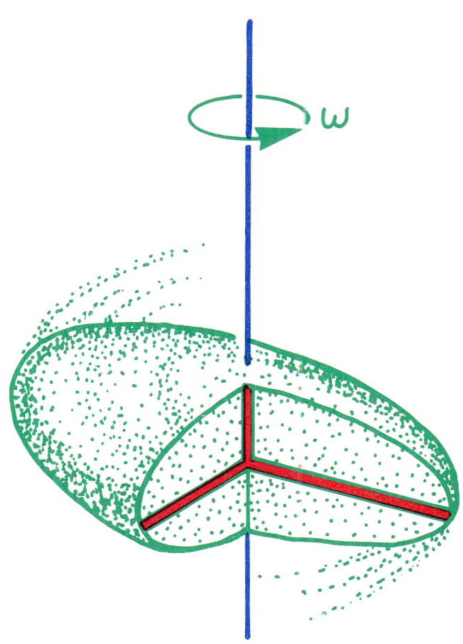

16.2.6. The rotating ellipsoids, with three unequal axes (rotating about the shortest of the three) will be in hydrostatic equilibrium at the correct rate of rotation. These are now called the *Jacobi ellipsoids.* They are found in a curve of increasing ellipticity, called the *Jacobi series* of ellipsoids.

The two series of equilibrium figures actually cross. That is, at one special case of the Maclaurin series, the Jacobi series *branches off* from the Maclaurin series.
It is for this crossing point that Jacobi invented the word bifurcation.

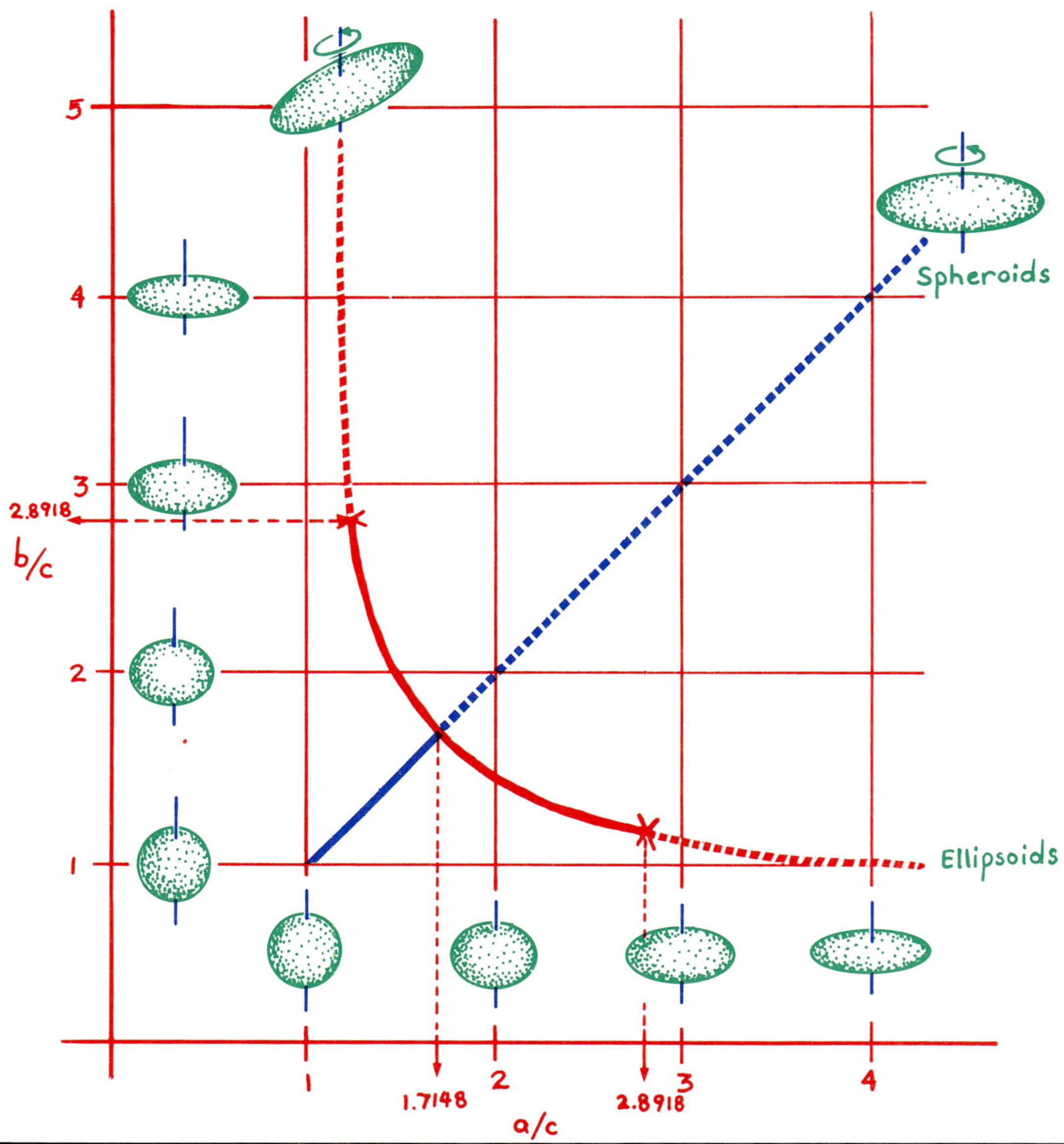

16.2.7. Here are the two series, Maclaurin in blue and Jacobi in red. The coordinates represent the diameters of the equatorial ellipse, if the polar diameter is taken as *one.* Note that the Maclaurin series occupies the diagonal, along which the two equatorial diameters are equal. That is, the spheroids are special cases of the ellipsoids.[7]

In 1882 Tchebychev asked how the equilibrium figure changes as the angular momentum is gradually increased. This is the essential question in most applications of bifurcation theory today, and we may take this moment as the alpha point of the history of bifurcation theory.

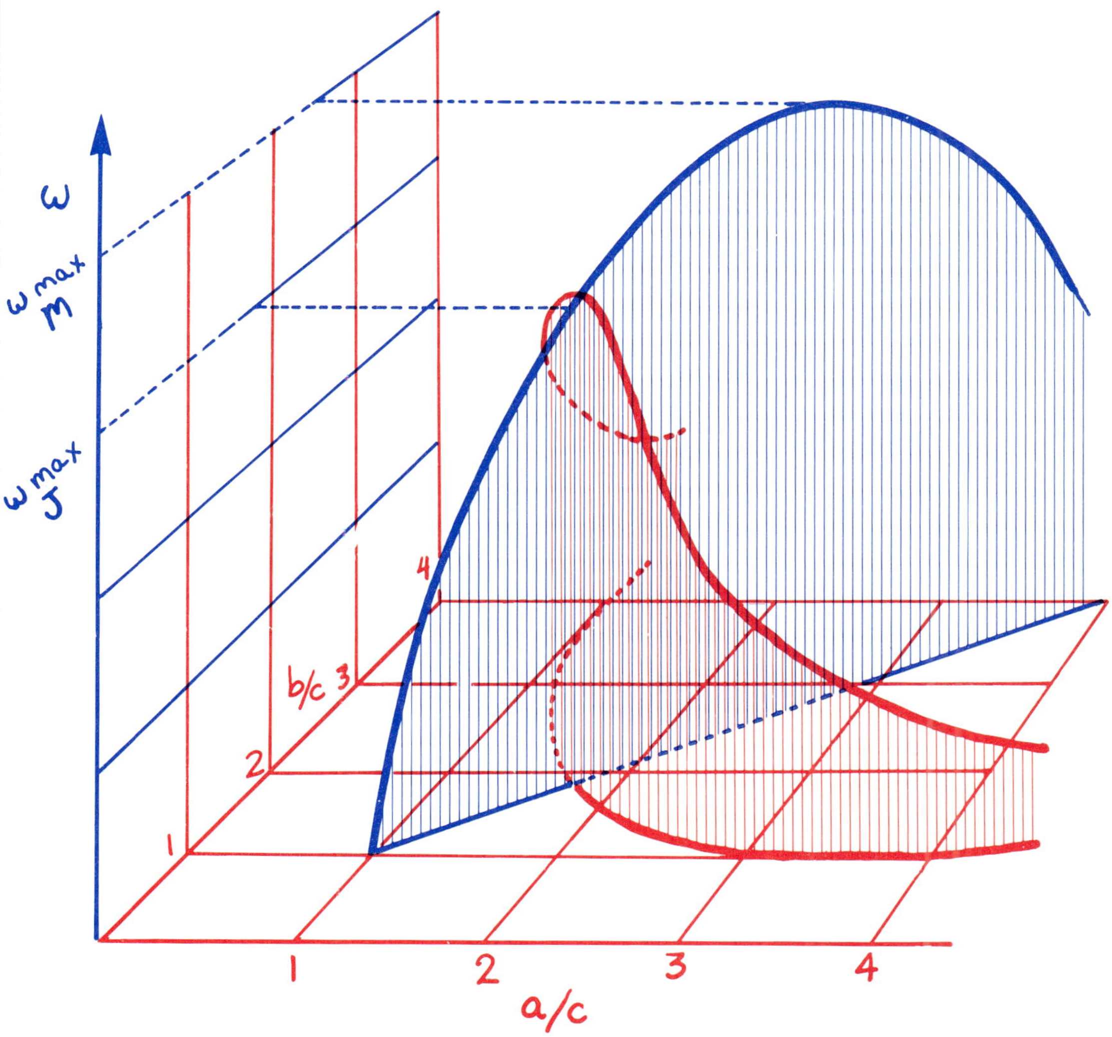

16.2.8. Adding angular velocity as a third (vertical) coordinate, the two crossing series are draped like this. The Maclaurin series (blue curve) has a maximum, as discovered by Simpson, and the Jacobi series also has a maximum velocity, but smaller than the Maclaurin maximum. This was discovered by Sir G. H. Darwin[8] in 1887.

Relative equilibria of a dynamical problem may be stable or unstable. The unstable ones, in general, will not be observed. Thus, the relevance of the mathematical models of Newton to the actual figures of the planets and stars will depend critically on stability.

And yet, the stability problem was never analyzed until Poincaré and Liapounov attacked Tchebychev's problem in 1885. After forty years, the last part of the problem was resolved by Cartan[9] in 1922.

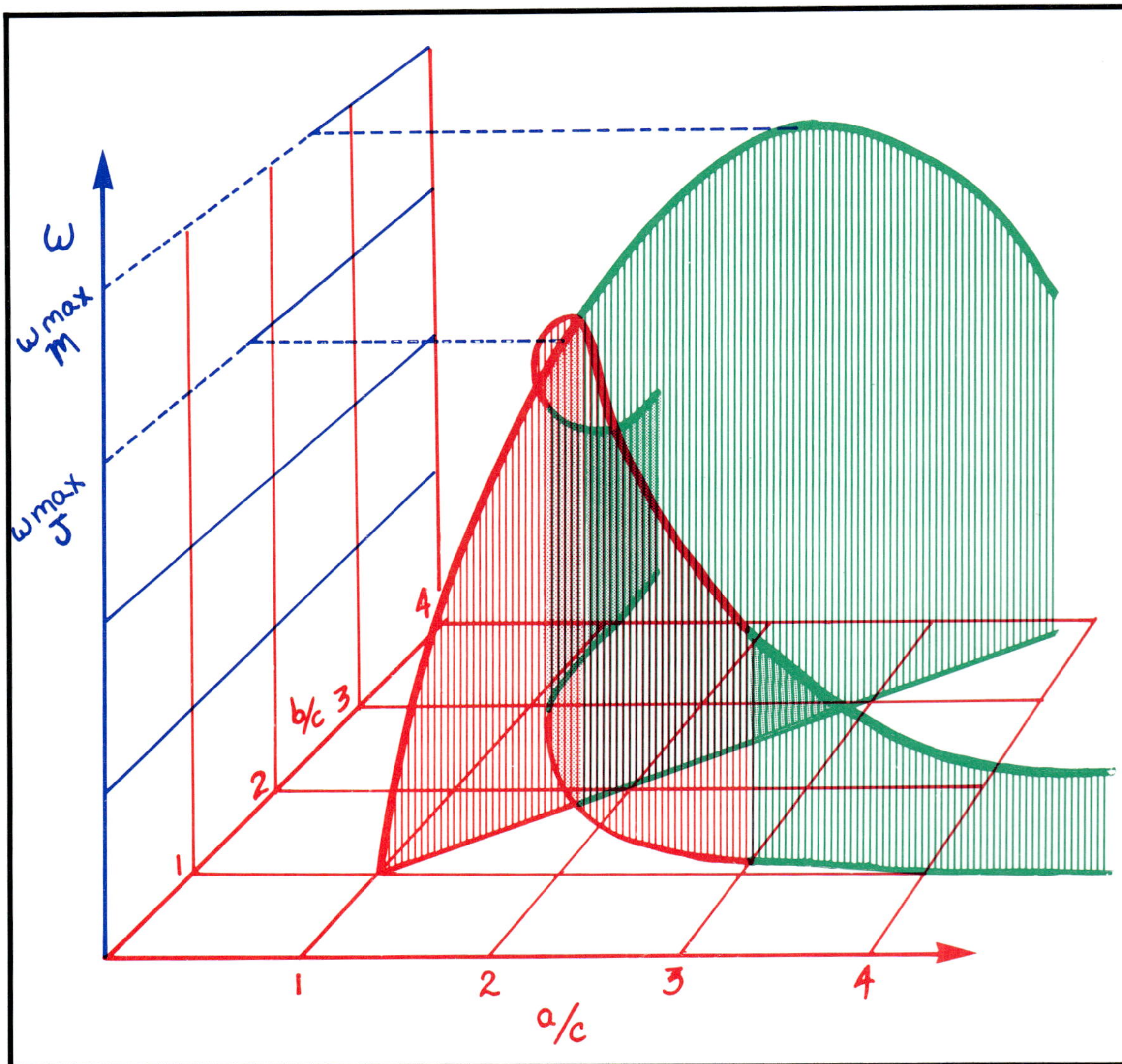

16.2.9. Here again is the plot of the two series, in the space of eccentricity and angular velocity, as before. But here, the stable branches are shown in red, the unstable ones in green.

This is the final stage in the emergence of the first bifurcation diagram in history, the *pitchfork*. But, there is more to our history of the figures of the Earth.

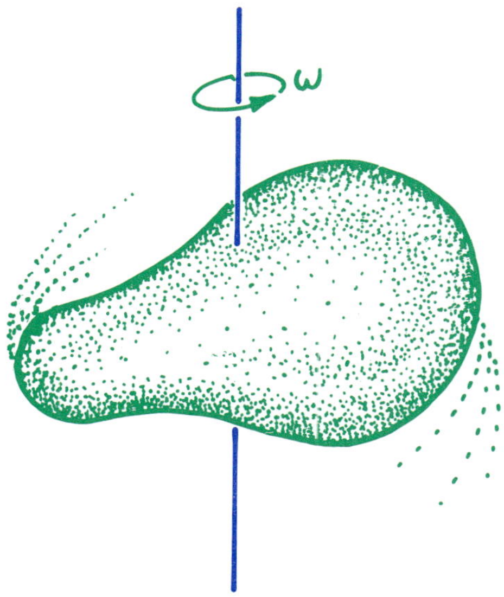

16.2.10. While studying the stability of the Jacobi series of equilibria, Poincaré discovered yet another equilibrium figure, the *pear shape* or piriform figure. These occur in series, called the *Poincaré* series, which branch off from the Jacobi series.

The stability of the Jacobi series ends at its crossing with the Poincaré series. Poincaré believed the pears were stable, and could explain the origin of double stars and planetary satellites by a kind of hydrostatic bifurcation. But Liapounov was convinced they were unstable, and he was right.

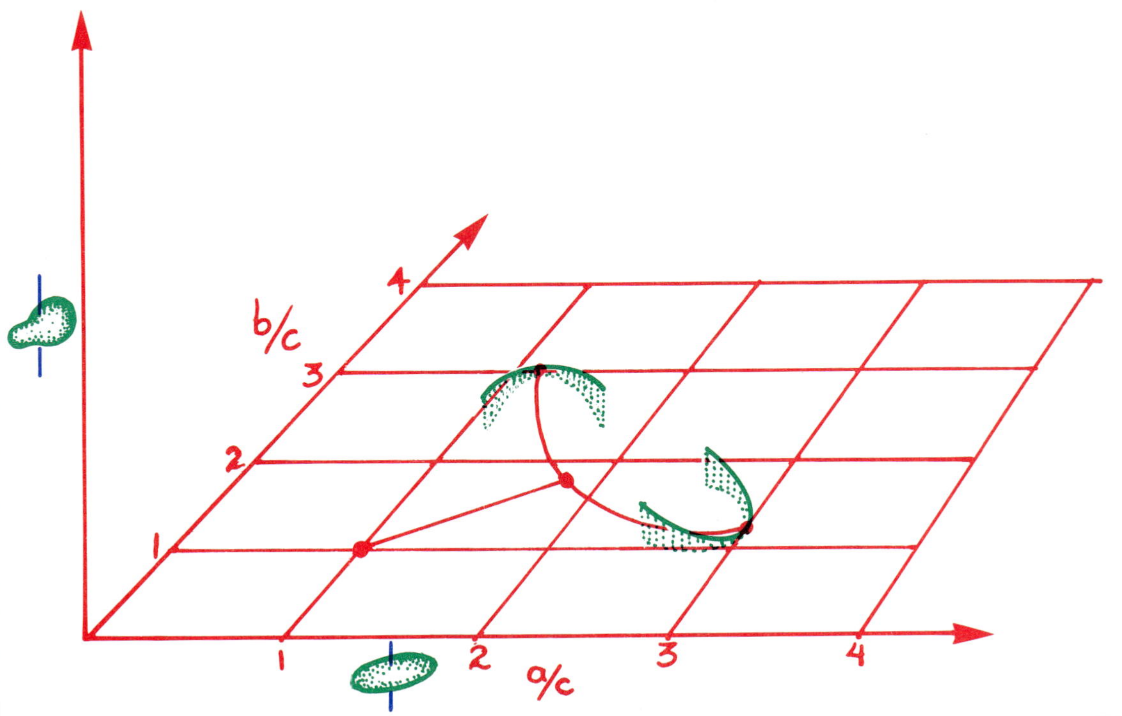

16.2.11. Here is a plot of the three series, showing stability in red. The horizontal axes are the overall ellipticity of the Equatorial section as before, while the vertical axis represents pearness. The arrival of the unstable branches of the Poincaré series at the Jacobi series kills its stability. This is an example of a *catastrophic* bifurcation, and is related to the pitchfork bifurcation at the branching of the Jacobi series from the Maclaurin series.

Both of these historic bifurcations were discovered by Poincaré[10] in 1885.

16.3. The Stirring Machine

As interest in rotating fluids heated up, kitchen experimentalists inevitably began to carefully observe their soup pots, coffee cups, and martini glasses, while vigorously stirring with a spoon or swizzle stick. Eventually, the professionals created a super-sophisticated version, *Couette's stirring machine,* capable of reproducible phenomena[14]. And among these phenomena were found a host of reliable examples of the bifurcation effects discovered by Poincaré among his analytical formulae.

Stellar evolution had come to Earth!

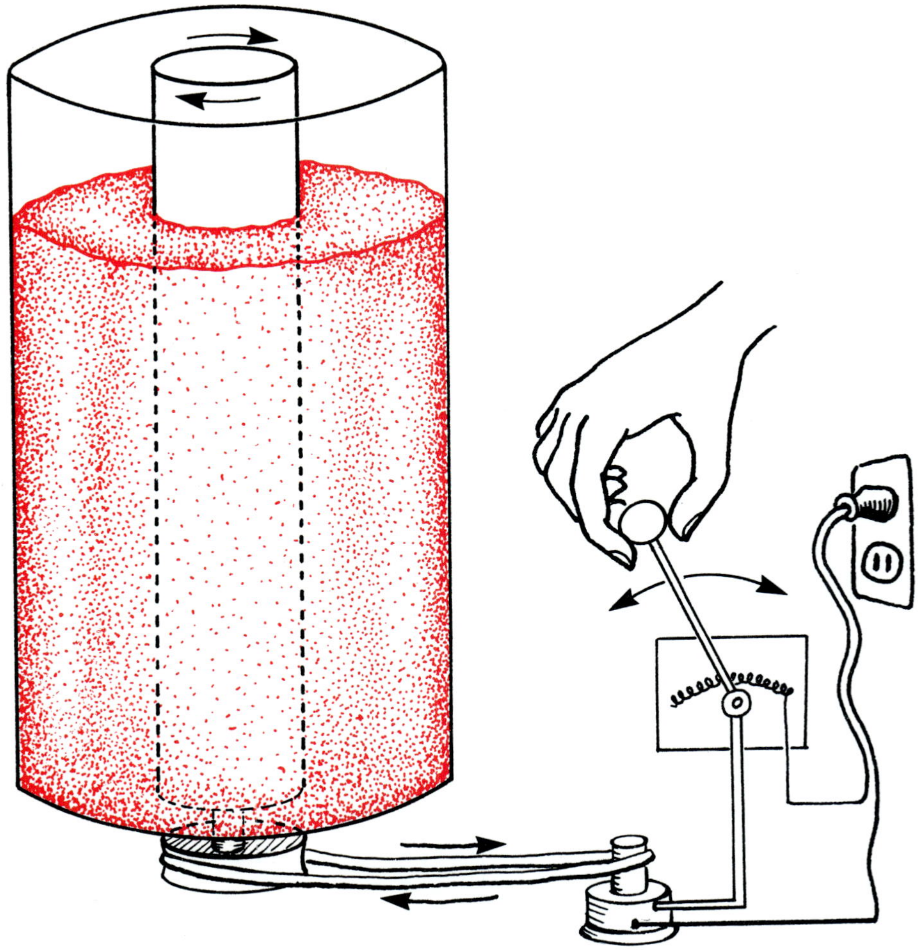

16.3.1. One way to stir water in a glass (not necessarily the best) is with a rotating cylindrical rod. Although inefficient at the cocktail bar, great progress has been made in experimental fluid dynamics this way. Since its earliest days a century ago, this experiment of Mallock and Couette has been repeated over and over, with ever improving rods, cylinders, motors, and observing methods. As the speed of rotation is gradually increased, an experimental comment on the problem of Tchebychev is reliably obtained.

Although the shape of the outer envelope of the fluid is constrained by the glass and the rod, the inner structure is seen to depart very quickly from a homogeneous form (latered rotating cylinders) to a highly structured form of nonuniform motion. We show this inner structure in three representations:

(a) the fluid seen from the outside,

(b) the velocity vectorfield within the fluid, and

(c) as a point in the state space.

First, we begin with a still fluid and no rotation.

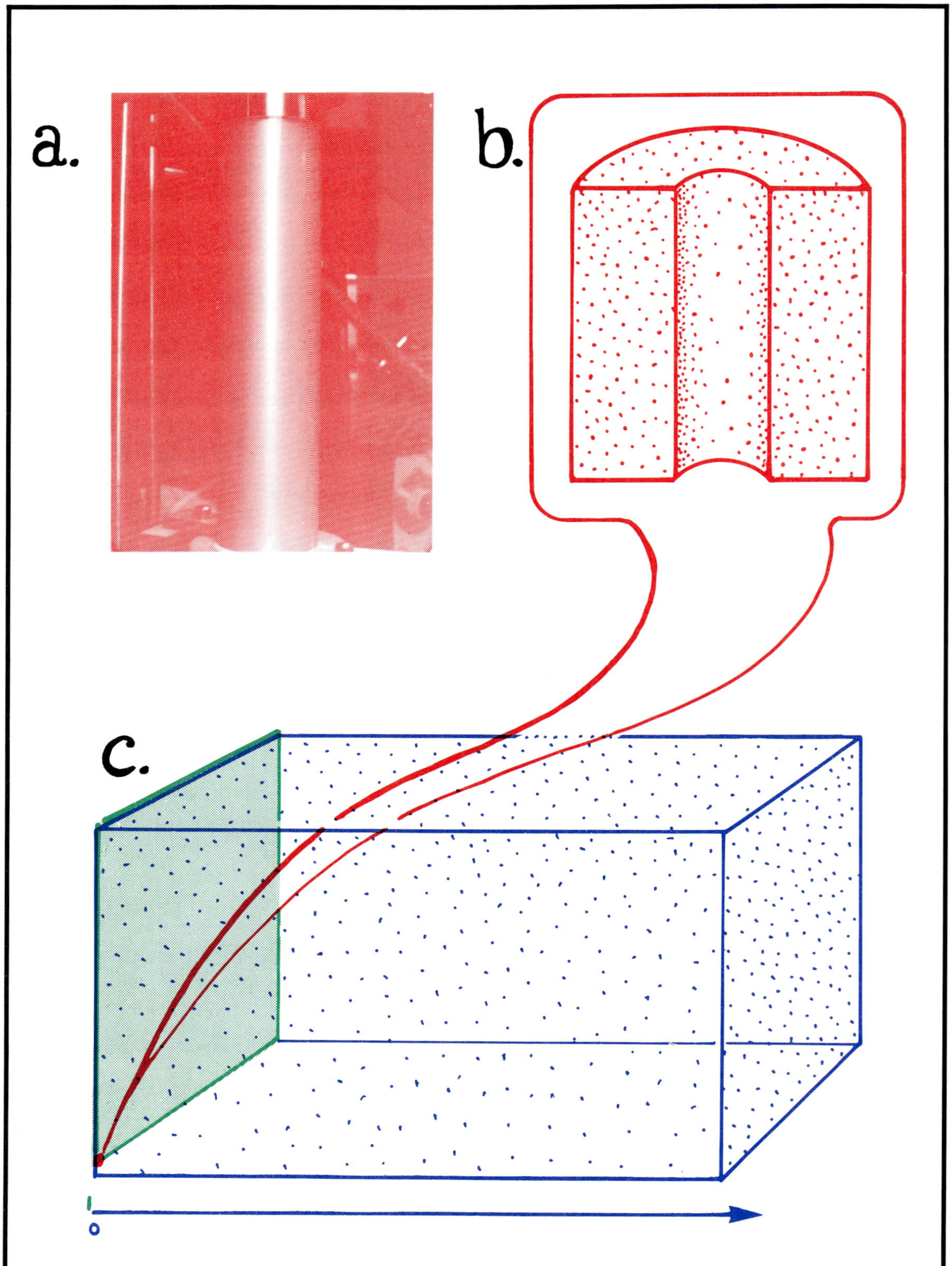
a.
b.
c.
0

16.3.2.a. Through the side of the outer cylinder, we see a homogeneous fluid mass at rest. The fluid fills a three-dimensional space in the shape of a thick tube, which we will call the *fluid domain.* (We are grateful to Rob Shaw and Russ Donnelly for these photos of an actual Couette machine.)

16.3.2.b. One mathematical model for the state of the fluid is the *velocity vectorfield* in the fluid domain (here shaded in red). At each point in this domain is drawn a vector representing the velocity of the particle of the fluid passing through the selected point at the instant of observation. As in this first instance the fluid is at rest, the velocity vectors are all zero. They are shown as red dots in this illustration.

16.3.2. c. The entire velocity vectorfield may be regarded as a mathematical point in a huge (infinite-dimensional) space of all possible velocity vector fields, the *state space.* And since the velocity vector at each point in the cylinder of fluid is zero, this point in the state space of all vector fields is zero as well. Thus, it is located at the *origin* (or zero vector) of the infinite-dimensional state space. Here we show the state space as a vertical plane (in schematic rather than pictorial representation) outlined in green. The observed state of the fluid, represented here by the red dot in the green plane, is actually an *attractor* of a dynamical system (infinite-dimensional vectorfield) on the green state space, which we will call the *superdynamic* (for example, the Navier-Stokes equations) according to fluid dynamical theory[12].

The third dimension, extending to the right, represents the *control parameter:* the speed of rotation, or equivalently, the Reynolds number. This composite picture of the state space (vertical plane) and the control parameter (horizontal line) may be called the *response space,* here shaded blue. Within it, the *response diagram* will be drawn, showing the loci of the attractors as the controls are varied.

In the response space, the zero velocity vectorfield is shown as a point, the red dot, in the state space furthest to the left (outlined in green) corresponding to the zero value of the control parameter (no rotation).

Next, we gradually begin a slow rotation of the rod, and let the system relax into a state of constant stirring.

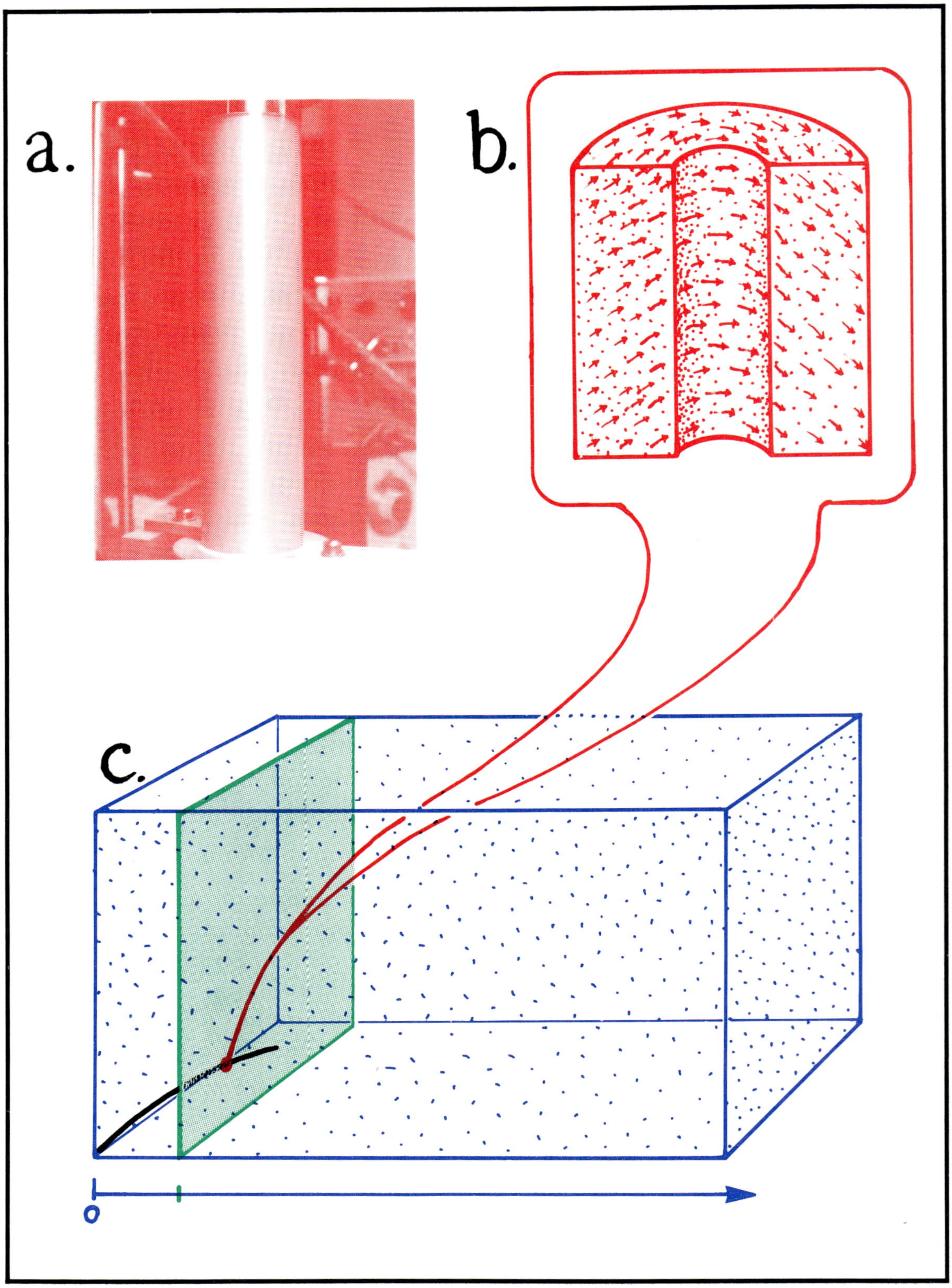
a.
b.
c.
0

16.3.3. a. Through the side of the outer cylinder, we see a homogeneous fluid mass in uniform rotation. Each particle of fluid is moving along a horizontal circle at a constant rate. The rate is zero for the largest circles at the outer glass cylinder, it is fixed by the speed of rotation of the rod at the smallest circles, and in the interior of the fluid, it varies uniformly between these extremes.

16.3.3. b. Here is the velocity vectorfield in the fluid domain, revealed by some trajectories of fluid particles, drawn in red. It shows the circular motion of the fluid, with the smaller velocities outside, and the higher ones inside. At any chosen point in the fluid domain, this velocity vector is at rest, it does not change with time while we are observing.

16.3.3. c. Here we view the velocity vectorfield as a single point in the response space. And since the velocity vector at each point in the fluid domain is at rest, this point in the response space is at rest as well. This red point is located within the vertical plane (state space of all velocity vectorfields) shaded in green, corresponding to a small value of the control parameter. And within that green plane, it is near the origin, indicating a small range of actual fluid velocities within the fluid domain. The black line passing through the red dot represents the track of the red dot as the control parameter is varied, and is called the *locus of attraction.* Generally, the *response diagram* signifies the record of all the loci of attraction known for the given experimental system, drawn within this blue response space. The job of the experimentalist is to create this record.

We will now try to discover a locus of attraction for the stirring machine, by continuing to increase the speed of rotation.

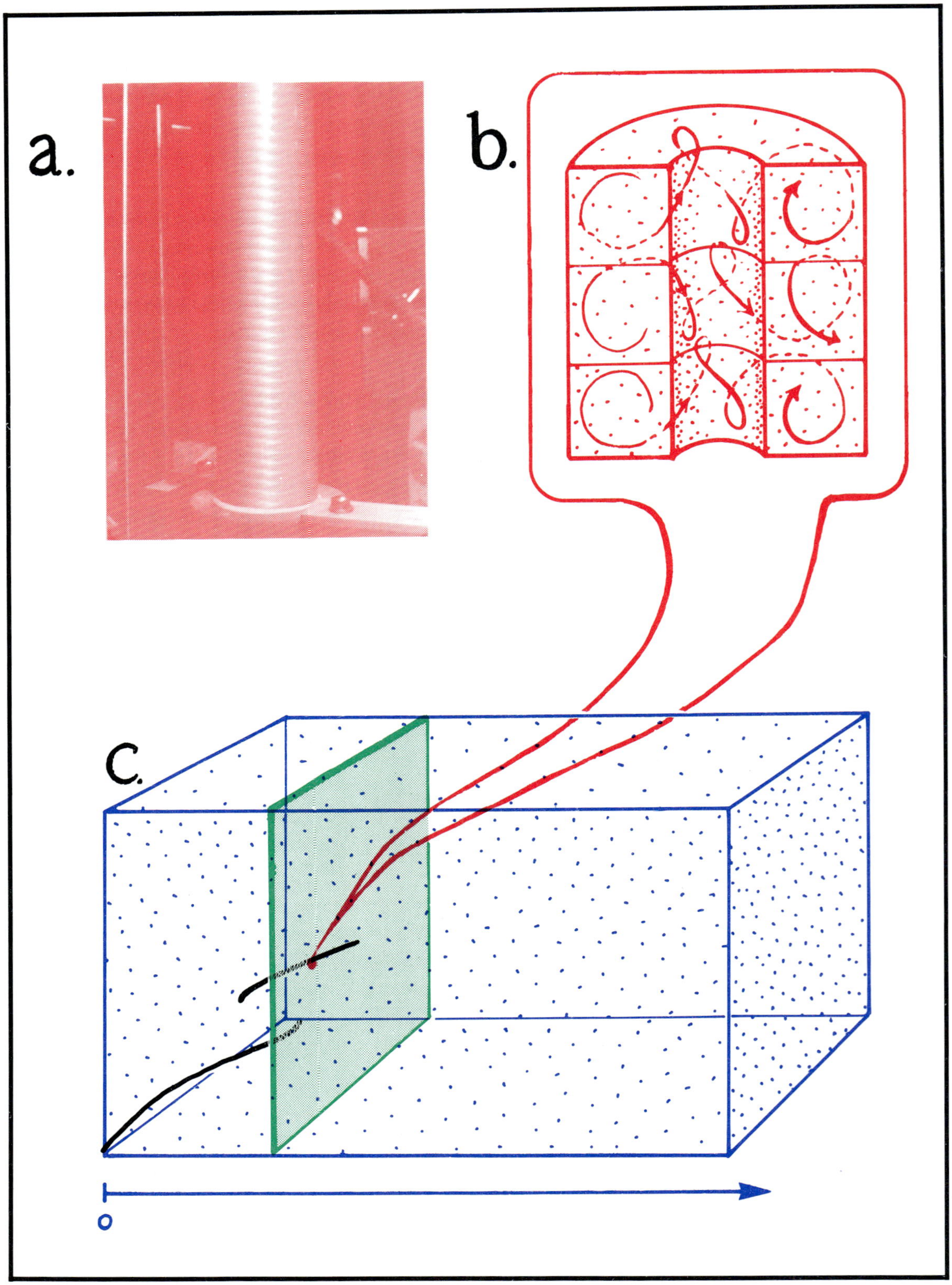
a.
b.
c.
0

16.3.4. a. After a moderate increase in the rate of stirring, this is one mode we are likely to see through the walls of the stirring machine. The fluid motion has separated into a stack of ring-shaped cells, divided by flat boundaries evenly spaced along the axis of rotation. The boundaries are still. These are the *Taylor cells* discovered by Taylor in 1923.

16.3.4. b. Upon closer inspection (sometimes aluminum powder is put in the fluid to show the motion more more clearly) we see that the fluid motion in each cell is *solenoidal.* One cell is a clockwise vortex, the next counter-clockwise, and so on. Here, the velocity vectorfield is indicated by some exemplary red trajectories. In spite of the increasingly complicated fluid motion, the vectorfield is still *stationary.* It is a point attractor of the superdynamic.

16.3.4. c. Here is the current situation, represented in the response diagram. The current state (velocity vectorfield) is shown as a red dot (point attractor) in the appropriate state space (shaded in green) somewhat to the right of the state space outlined in the previous sequence. Note that the locus of attraction (black curve through the red dot) has a gap just to the left of the yoke. This is due to a bifurcation event called the *static fold,* related to the pitchfork bifurcation discovered by Jacobi in 1834 in his study of the figure of the Earth. The observation of this event requires careful experimental work, repeatedly turning the speed control up and down.

Now we make substantial increase in the rate of the stirring rod, to see what turns up.

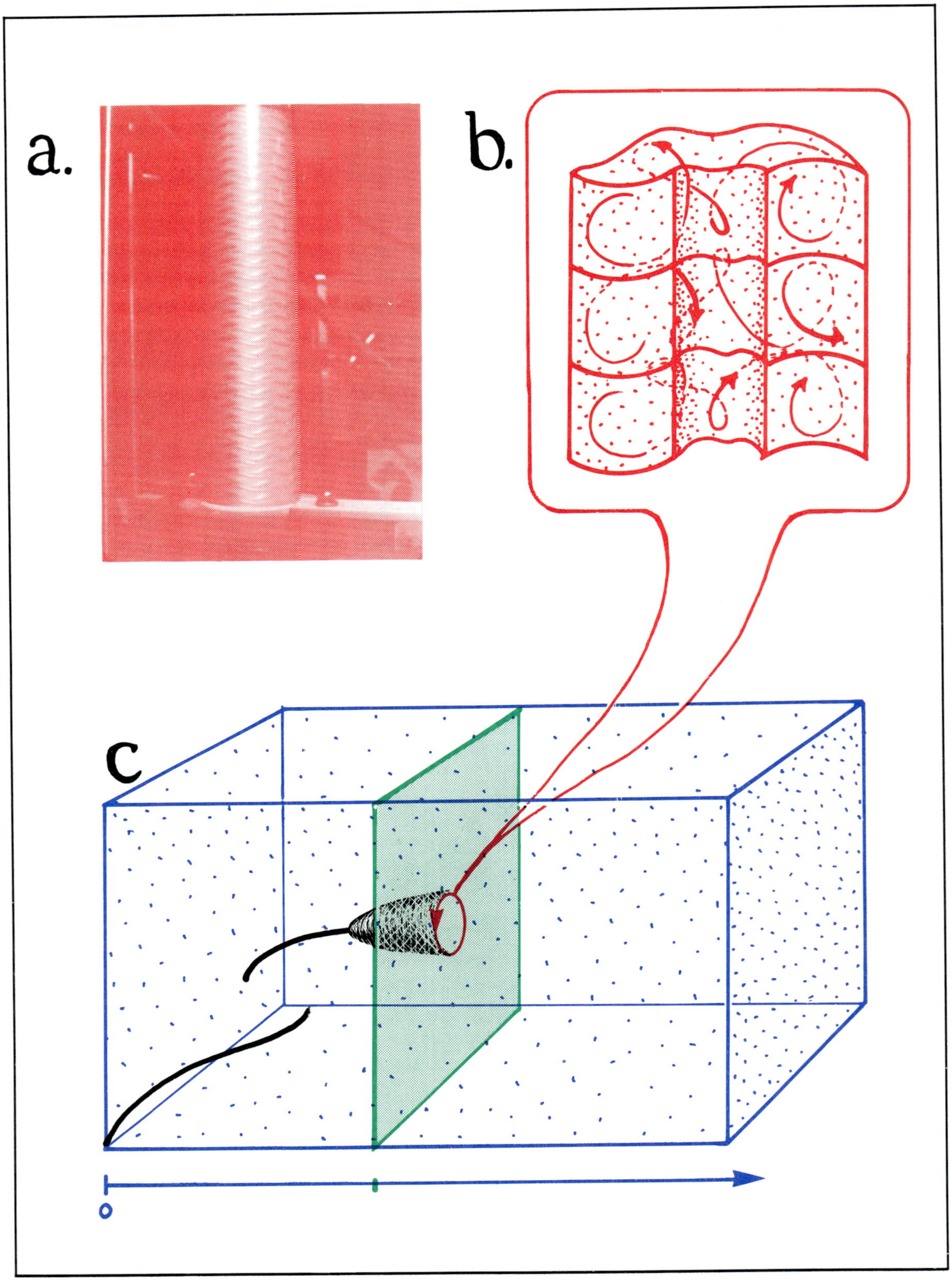
a.
b.
c
0

16.3.5. a. We now see, looking directly at the fluid through the side of the outer cylinder, that the Taylor cell boundaries have developed waves, and the wavy cells are slowly rotating around the central axis[13].

16.3.5. b. This *wavy vortex* phenomenon is represented by a velocity vectorfield in the fluid domain, shown here in stop-motion drawing, which is slowly (and periodically) varying in time. The pattern shown here repeats every few seconds.

16.3.5. c. Representing this stop-motion vectorfield as a red dot in the state space (here outlined in green as usual) and waiting a few seconds, we would see it move around a small cycle. The periodic change in the fluid velocity vectorfield indicates that we are observing a *periodic attractor* of the superdynamic. And the black locus of attraction, to the left of this red cycle, shows a change from a static to a periodic attractor. This exhibits yet another bifurcation, discovered by Poincaré in 1885 and successfully analyzed by Hopf in 1942. This event will be described in further detail later in this volume.

Now we make a further increase in the rate of rotation of the stirring rod, and discover chaos!

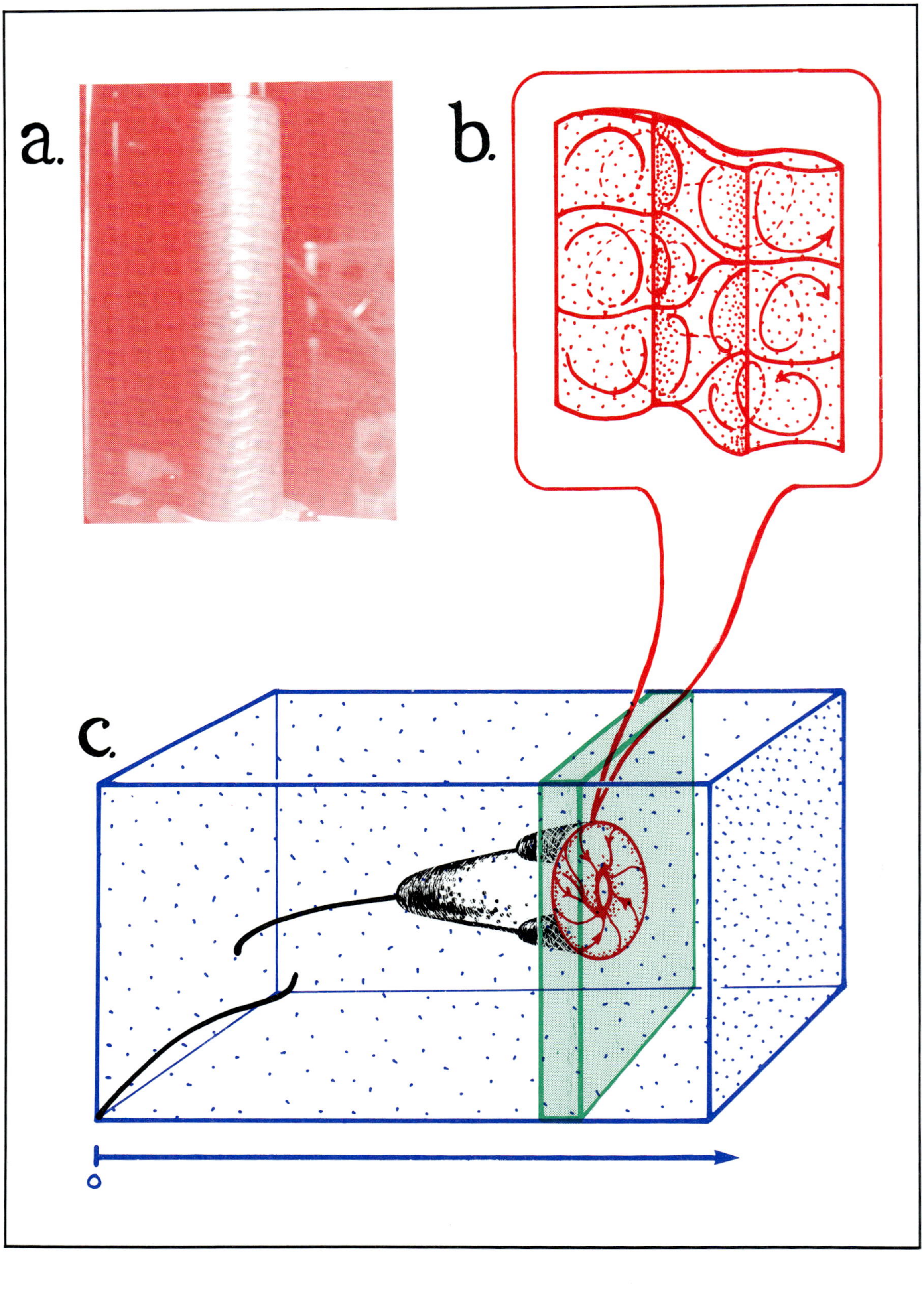
a.
b.
c.
0

16.3.6. a. After things settle down as much as they ever will, we may see tht some rings are pinched off. The complete rings are still wavy vortices, but they wave irregularly. Worse still, the pinched cells may jump around. This is genuine turbulence!

16.3.6. b. The velocity vectorfield, at a single instant of observation, is very complicated, but still understandable. Over time, however, it wanders erratically about, and never returns to an exact copy of an earlier state.

16.3.6. c. The wandering of the red dot within the green state space (shown here as a three-dimensional box, to give us adequate space in which to represent its shape) fills out with a thickened torus, perhaps, or some other *chaotic attractor* (see Part Two) of the superdynamic. The black locus of attraction has suffered some further bifurcation, the *onset of chaos.* Its entire history, from the far left to this point, is called a *chaotic scenario.*

Finally, we thrust the speed control to the maximum.

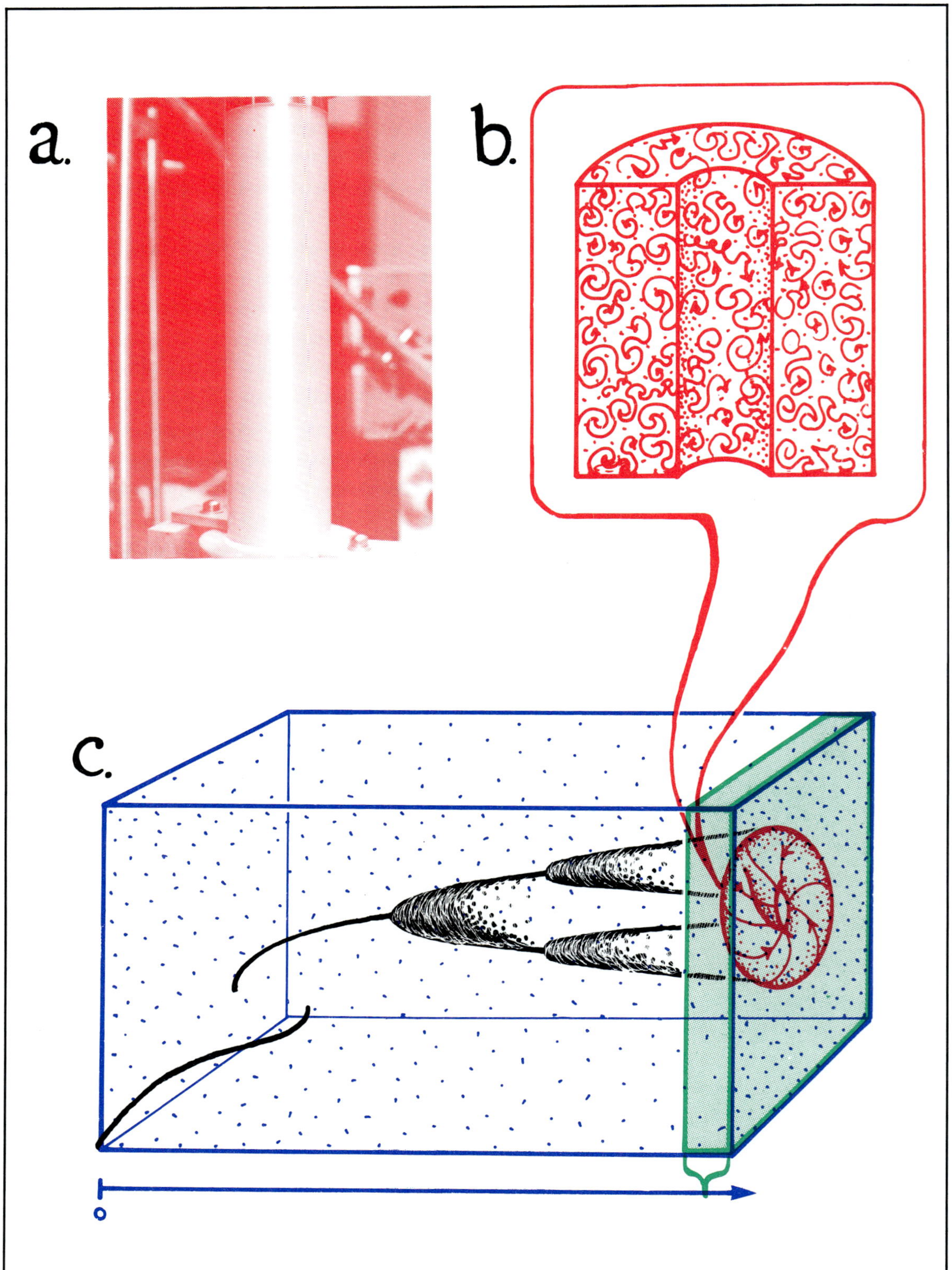
a.
b.
c.
o

16.3.7. a. Hang on to your kayak, there is white water everywhere!

16.3.7. b. The velocity vectorfield, drawn here in stop motion, is beyond understanding.

16.3.7. c. The motion of the red dot within the green state space, again traced over a few seconds, describes a chaotic attractor of the syperdynamic as before. Here the attractor is shown as a bagel (Section 8.2), although in reality it might be much more complicated. This is a fully developed turbulence! The black locus of attraction has suffered further bifurcations, from one type of chaos to another, which are little understood at present.

16.4. The Big Picture

In the first two sections of this historical introduction to the concepts of bifurcation theory, we kept our feet on the ground. We spoke about physical phenomena and their simplest mathematical models. In the third section we introduced a fourfold visual representation:

1. the physical phenomena,
2. their simplest mathematical representation (velocity vectorfield in fluid domain, stop-motion or time-varying),
3. more abstract mathematical representation (red dot moving upon red attractor in green state space),
4. green state space moving within blue response diagram, while red attractor drags along black locus of attraction.

We now wish to take one last leap, to a bigger picture of bifurcation theory, introduced by René Thom in his revolutionary text of 1972[14] on structural stability.

We will again use the Couette machine as an example, but the rotating blob or the game of bob (see Part One) would do as well. What we must keep in mind in these examples is that the fundamental model is not a dynamical system, such as the superdynamic of fluid mechanics. Instead, the central object of bifurcation theory is a *dynamical scheme,* that is, a dynamical system *depending upon control parameters.* For example, in the dynamical model for the Couette machine, the *superdynamic depends on the stirring speed.* Thus also, the phase portrait of the dynamical system depends on the controls, and putting these pictures side by side generates the response diagram.

We are now going to leap to the ultimate abstraction: the superspace of all superdynamics.

In the preceding section we imploded an instantaneous state of a complex physical system and regarded it as simply a single point of a geometrical model, the state space. The dynamical model for the evolution of these instantaneous states in time consists of a dynamical system, which we have been calling the superdynamic in this context. In general, it is just called the dynamic, or the dynamical rule. This representation assumes that the control parameter is fixed at a constant value.

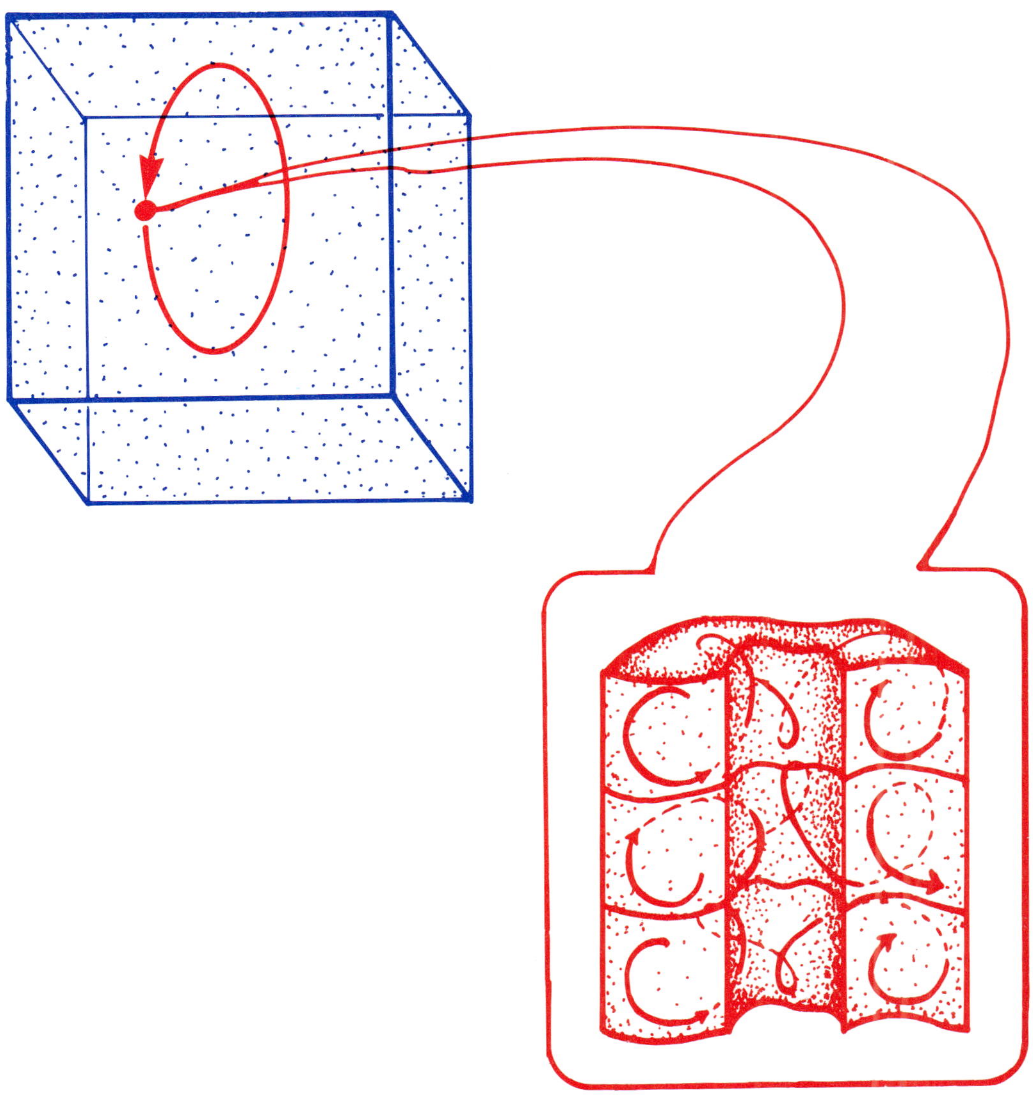

16.4.1. Here we show the instantaneous state scrunched to a single red point of the state space, on which the dynamic is indicated by the blue curves. The red cycle denotes a periodic attractor, such as that observed in Figure 16.3.5.

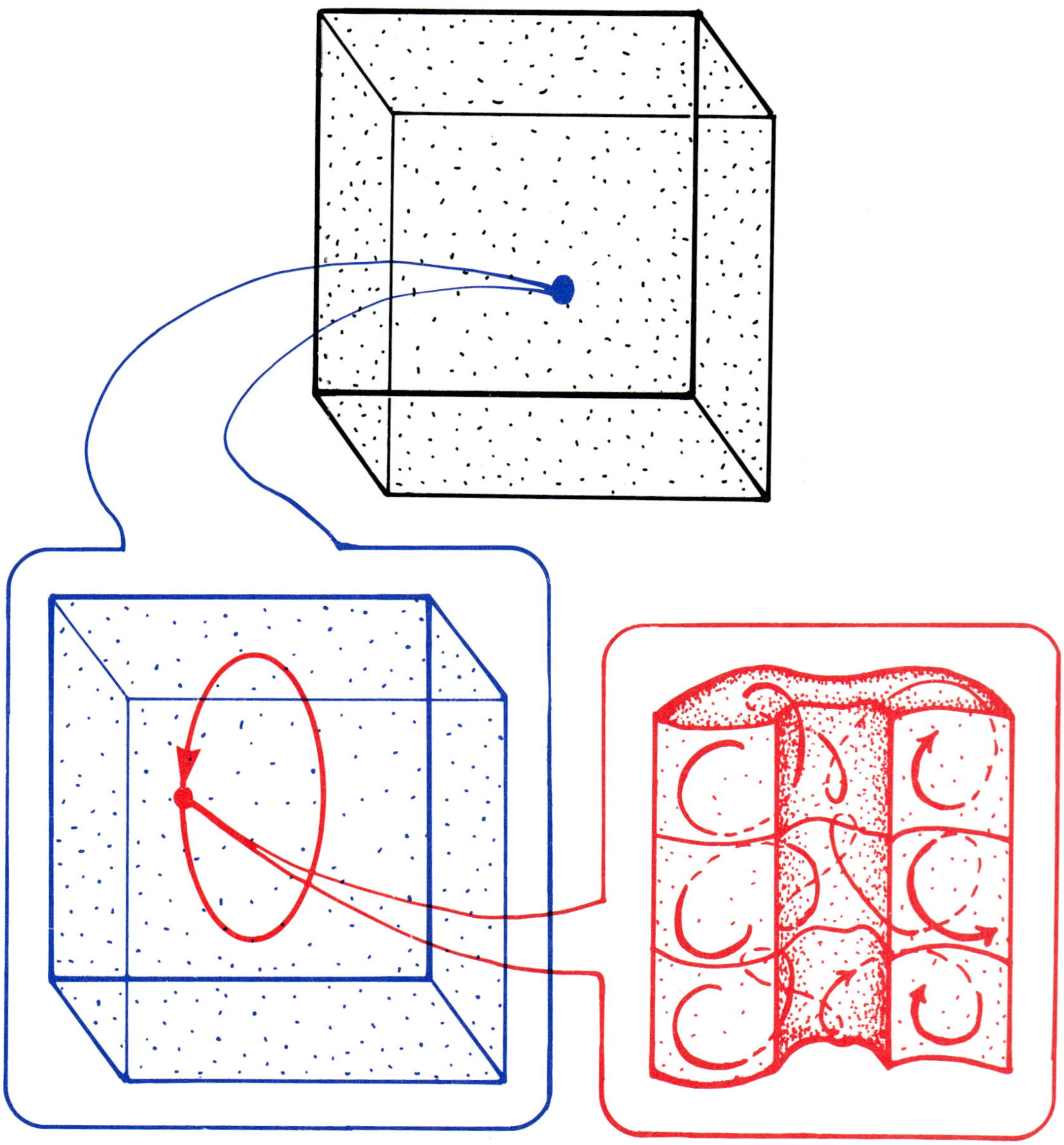

16.4.2. Here we have scrunched once again, so that the superdynamic becomes represented by a single point in *dynamical superspace*, Thom's Big Picture, within which every superdynamic on the given state space is represented by a single point. Every state space has its own personal superspace.

We now use superspace to construct an alternative to the response diagram as a representation of a dynamical scheme, or dynamical system depending on control parameters. We start with a small piece of the response diagram developed in the preceding section, in Fig. 16.3.5.

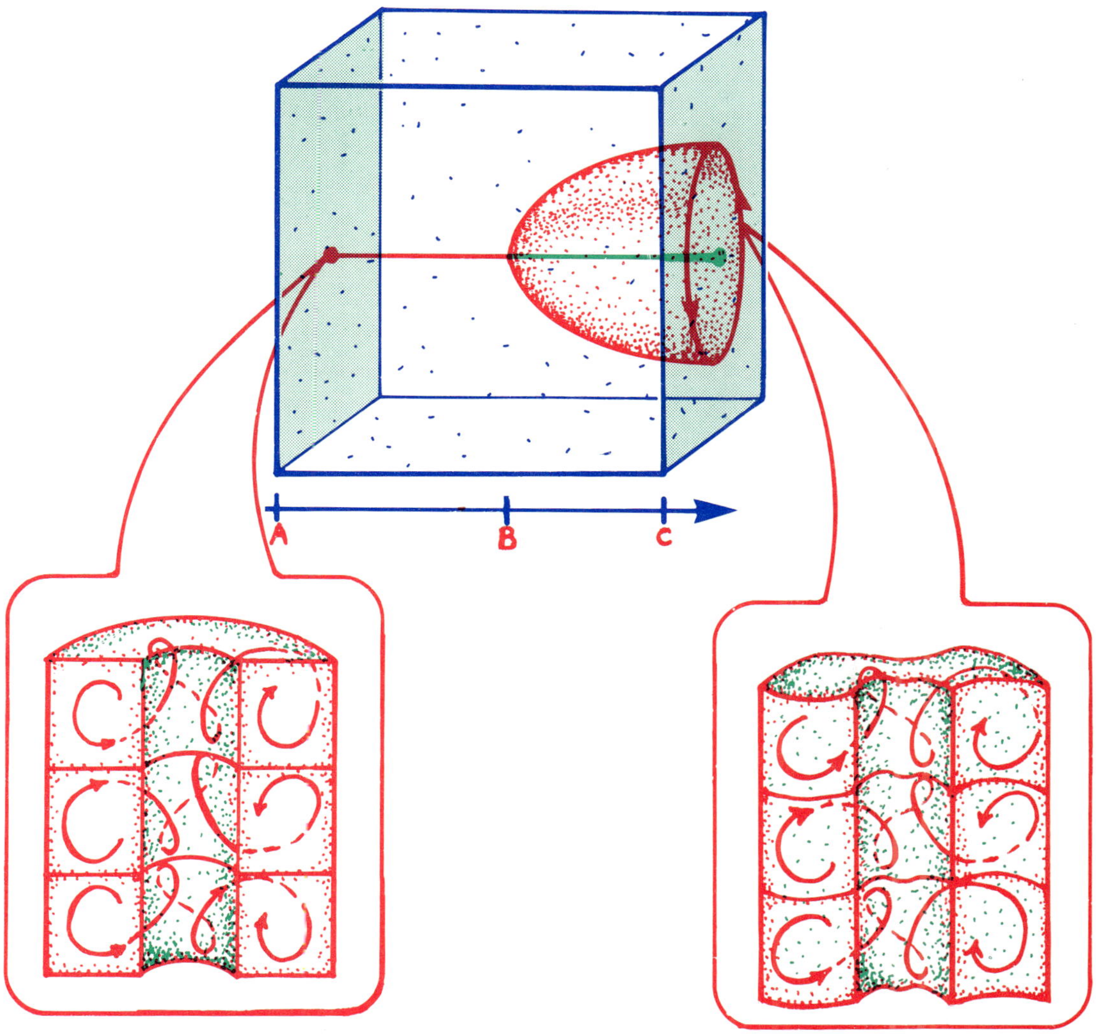

16.4.3. Here the control parameter increases a small amount from A to C, the cylinder of fluid changes from flat vortices (point attractor) to wavy vortices (periodic attractor) as the control passes B, and the locus of attraction has the form of a goblet. This is an example of a bifurcation event called *excitation,* also known as the *Hopf bifurcation.* It is treated in more detail in the next section.

We now transform this picture into superspace. Each and every value of the control parameter specifies a copy of the state space having its own superdynamic. That is, the control parameter changes the dynamical rule but not the state space. In this scrunch, each of these superdynamics becomes a single point in superspace. We may think of this scrunch as a movement of the control space (the blue interval in this panel) into superspace. The result is the blue curve in the following picture of superspace.

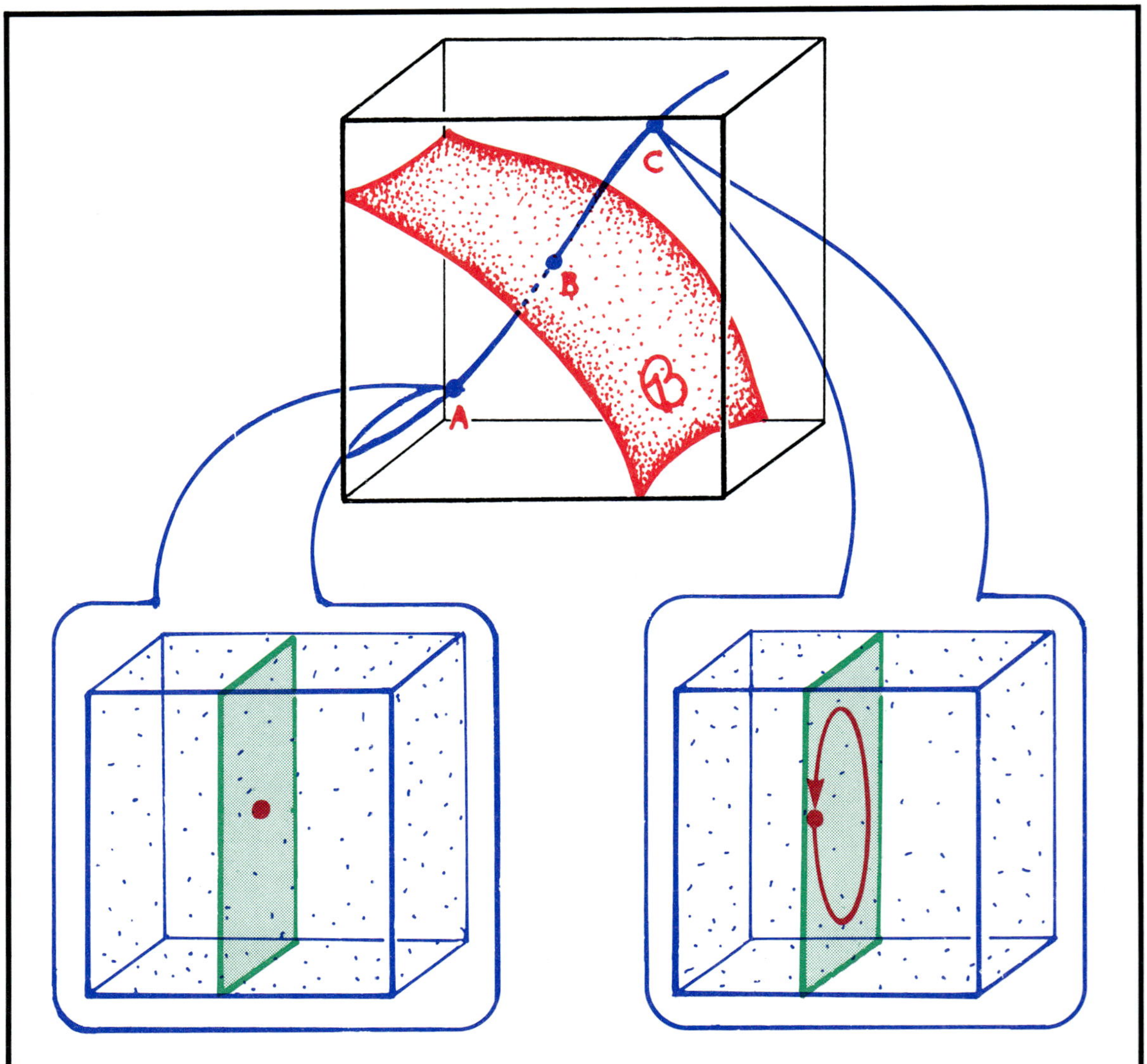

16.4.4. See that the starting superdynamic determined by A corresponds to the lower endpoint of this curve, the final superdynamic determined by C corresponds to the upper endpoint of this curve, and the bifurcation point B corresponds to an intermediate point in the curve, shown here at the intersection of the blue curve and a red surface. This red surface is part of the *Bad Set*.

As Andronov discovered in 1937, the Bad Set consists of dynamical systems which are *not structurally stable* (see Chapter 12 for the definition). One of the numerous contributions of René Thom to dynamical systems theory is a Big Picture of the Bad Set within superspace, a strategy for analyzing its structure, and a recognition of this structure as a kind of universal Platonic model for morphogenetic (evolutionary) processes in nature.

Now let's take a guided tour of superspace, using the stirring machine, to develop a better idea of the structure of the Bad Set in the Big Picture.

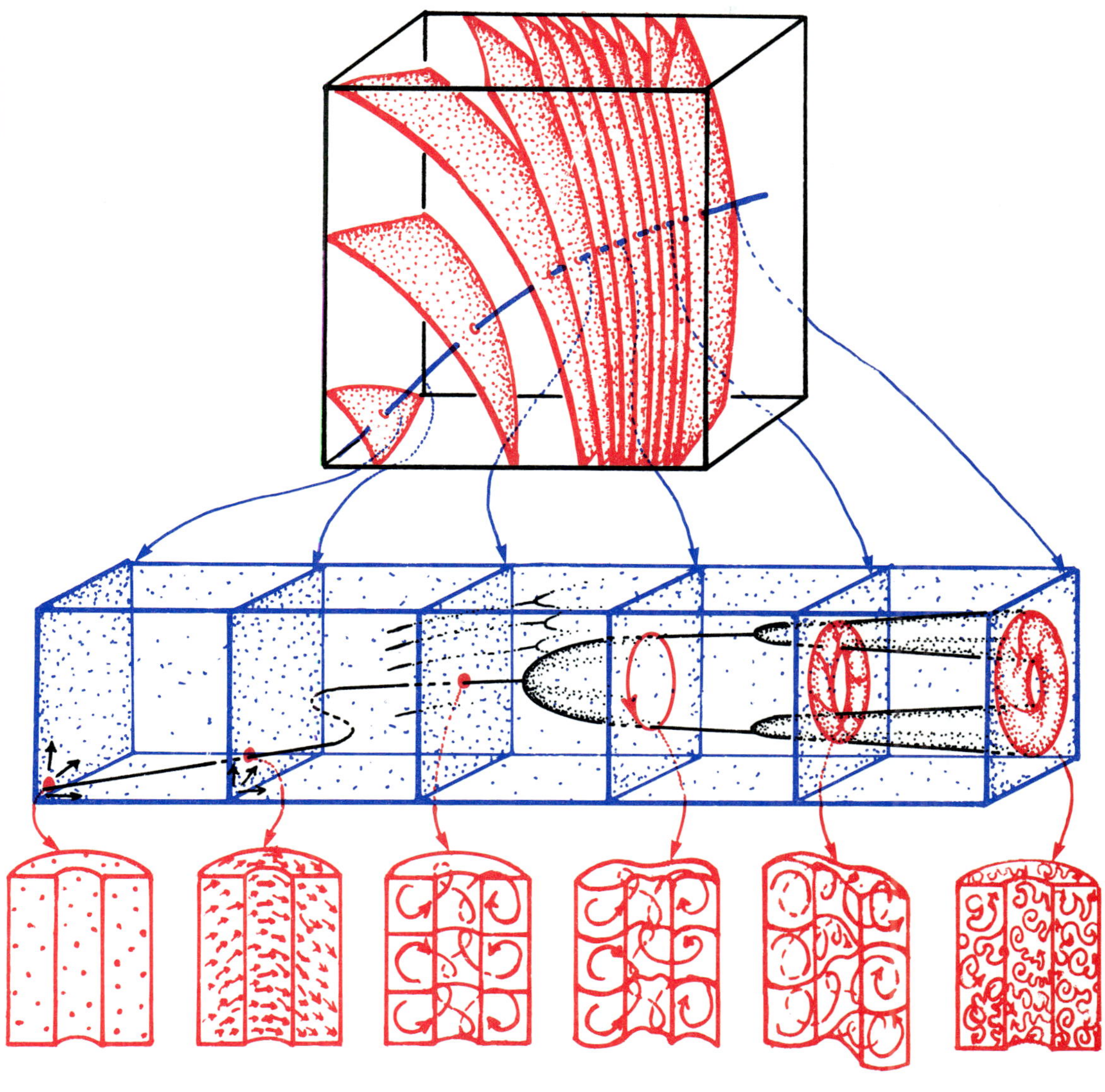

16.4.5. Here is the entire response diagram developed in the preceding section, all scrunched onto a curve in the Big Picture. Each and every bifurcation located in the preceding discussion identifies a sheet of the Bad Set. There are infinitely many of these sheets, which accumulate in fat fractals, as we shall see later in this book.

The balance of this work is a pictorial atlas of bifurcations that are generic for dynamical schemes depending upon a single control parameter. They identify only the largest structures of the Bad Set.

17

Subtle Bifurcations

The early history of the bifurcation idea unrolled in the context of physical systems demanding infinite-dimensional dynamical models, based upon partial differential equations. In the revolutionary work of Poincaré, the connection was made with dynamical systems of finite dimension, based upon ordinary differential equations. The recent history of bifurcation theory has therefore been developed along two parallel tracks. Throughout this work, we concentrate on the finite-dimensional case, *dynamical bifurcation theory (DBT)*. The parallel theory for systems of partial differential equations, which we occasionally call *classical bifurcation theory (CBT)*, may be developed in a similar fashion. Unfortunately, we will not be able to indicate the fantastic importance of this theory in applications. Our view, often expressed in other writings, is that the response diagrams of these atomic bifurcation events are our most important source of models for the dynamical processes of nature. For some support of this view, see the seminal works of Thom and Prigogine. Thus, *Part Four* is the culmination and main motive for the entire book. The bifurcations of DBT may be classified in three types: *subtle bifurcations, catastrophic bifurcations (catastrophes)*, and *explosive bifurcations (explosions)*.

In this Chapter, we present the subtle bifurcations in their simplest versions.

2.1. First Excitation

We have already encountered the first excitation, known to Poincaré in 1885 and analyzed by Hopf in 1942, as the second event in the bifurcation sequence of the stirring machine (see Figure 16.3.5). It is most frequently called the *Hopf bifurcation*. There are at least two complete texts on this phenomenon alone. One of these[2] contains a section on the recent history and numerous applications of the event. The other[3] develops very early applications to the speed governor (for the steam engine invented by Watt in 1782) ascribed to Airy (1840), Maxwell (1868), and Vyshnegradskii (1876). Both have very detailed analyses and extensive bibliographies and are highly recommended.

We will describe simple excitation in the context of radio transmitters developed in Sections 3.3, 5.6, and 8.2.

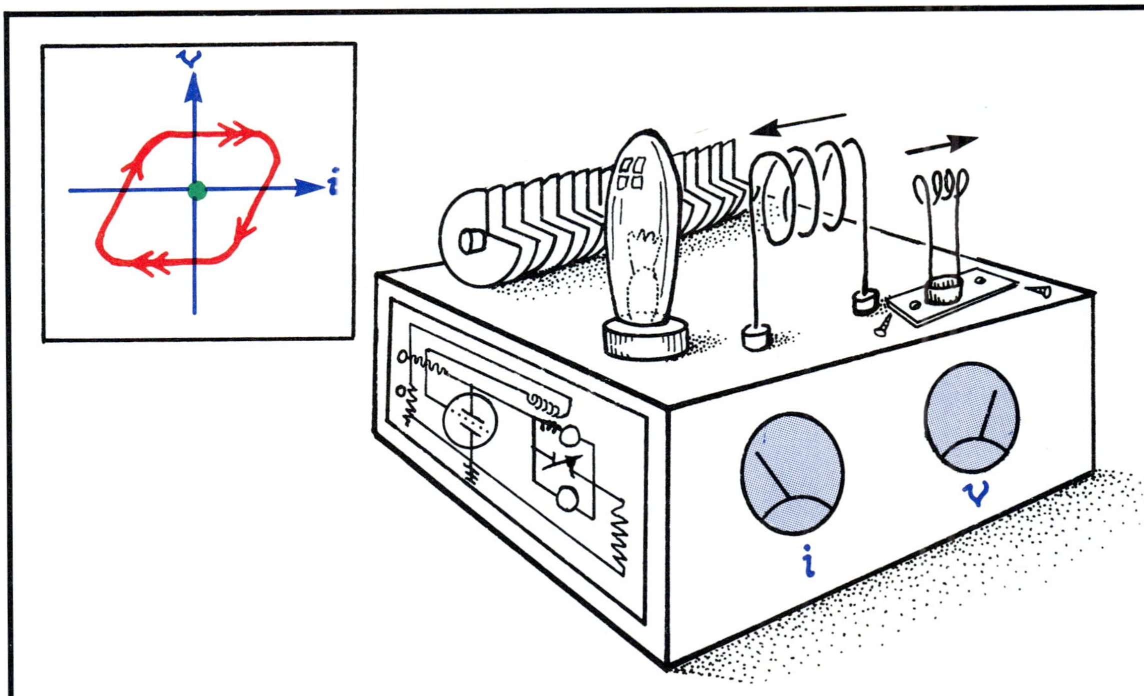

17.1.1. Here is the antique transmitter, as used by Marconi and Van der Pol. We have specially prepared it for this experiment by loosening the screws at the base of the small feedback coil so that it may be easily rotated a half turn or so in either direction. In this figure, the coil is in the normal position, the current and voltage in the larger tank coil are oscillating, and the transmitter is "on the air." A plot of the current and voltage (shown in the window) is tracking a periodic attractor clockwise.

In this exercise, the small coil will be our control knob.

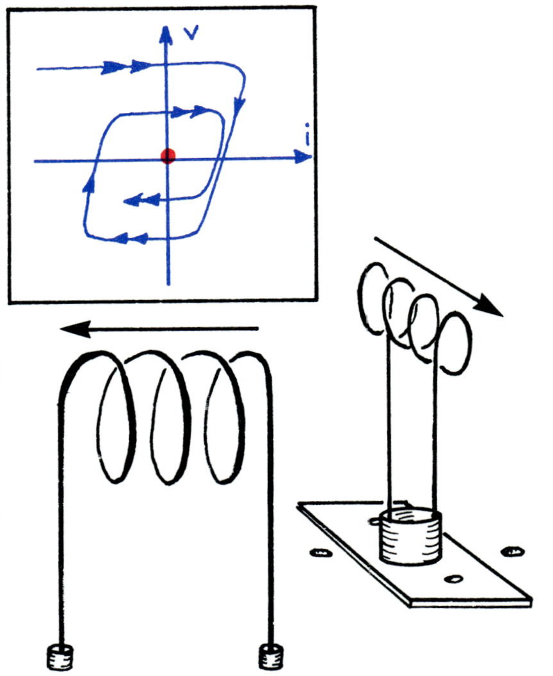

17.1.2. Now we grab the small coil from above and twist it more than a quarter turn to the right. If not electrocuted, we find that the transmitter quits, the oscillations gradually die out, and the plotted measurements follow an exponential spiral down to the point attractor at the origin of the state space. We are off the air!

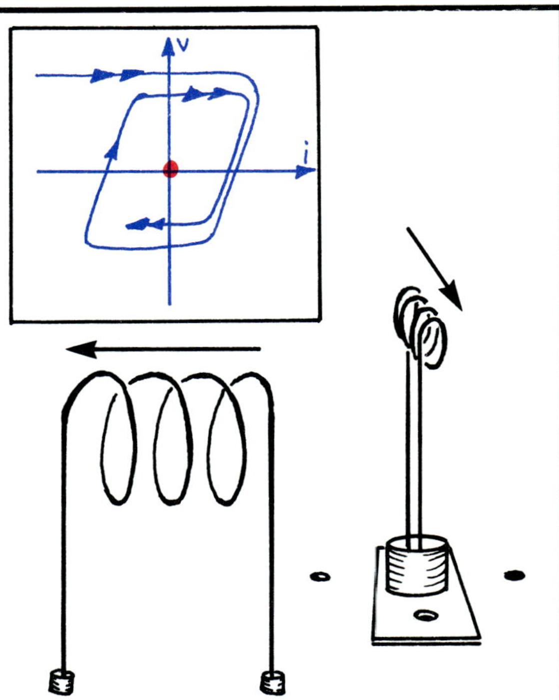

17.1.3. Grabbing the small coil again, we slowly untwist it a few degrees, trying to undo the damage. No change in the situation, so we keep on untwisting.

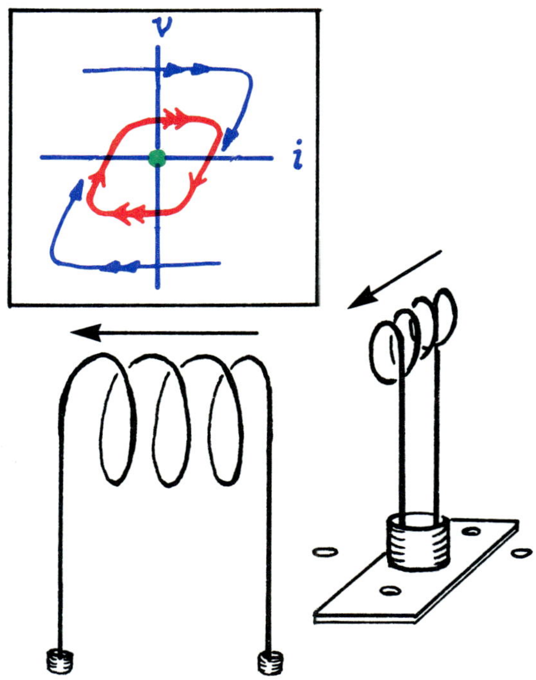

17.1.4. After we pass a quarter turn by a few degrees, the oscillations resume, but very feebly! The phase portrait (window) shows a very small periodic attractor encircling a point repellor at the origin.

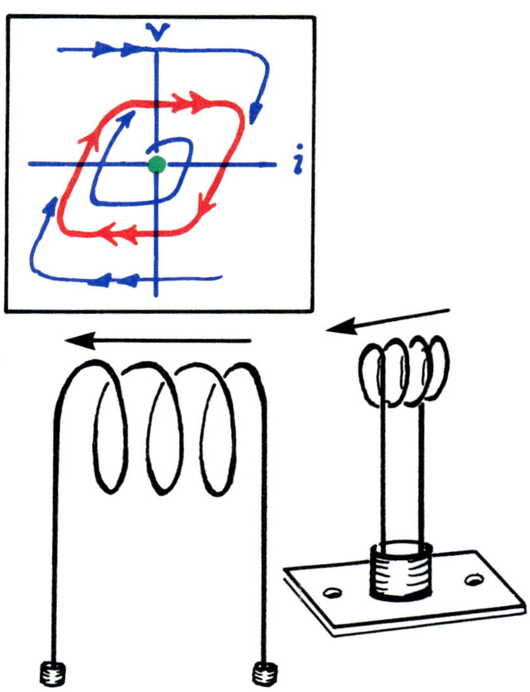

17.1.5. Now slowly untwist the coil some more, gradually returning to the original position. The oscillation grows stronger, and the audience may hear a feeble broadcast, which grows stronger and becomes normal.

This is a *subtle* bifurcation, as the qualitative change in the behavior of the transmitter is almost imperceptible. The small coil orientation is the control parameter, and for each orientation there is a different dynamical system modeling the transmitter, according to Van der Pol. (The parameter is represented by B in Example 4a of the Appendix. The other parameters are fixed.) We now put together, side by side, the phase portraits shown in the window of the last few panels. To relate the phenomenon to the qualitative theory of rest points, we will add to each portrait the CE's of the critical point of the origin (see Section 6.4).

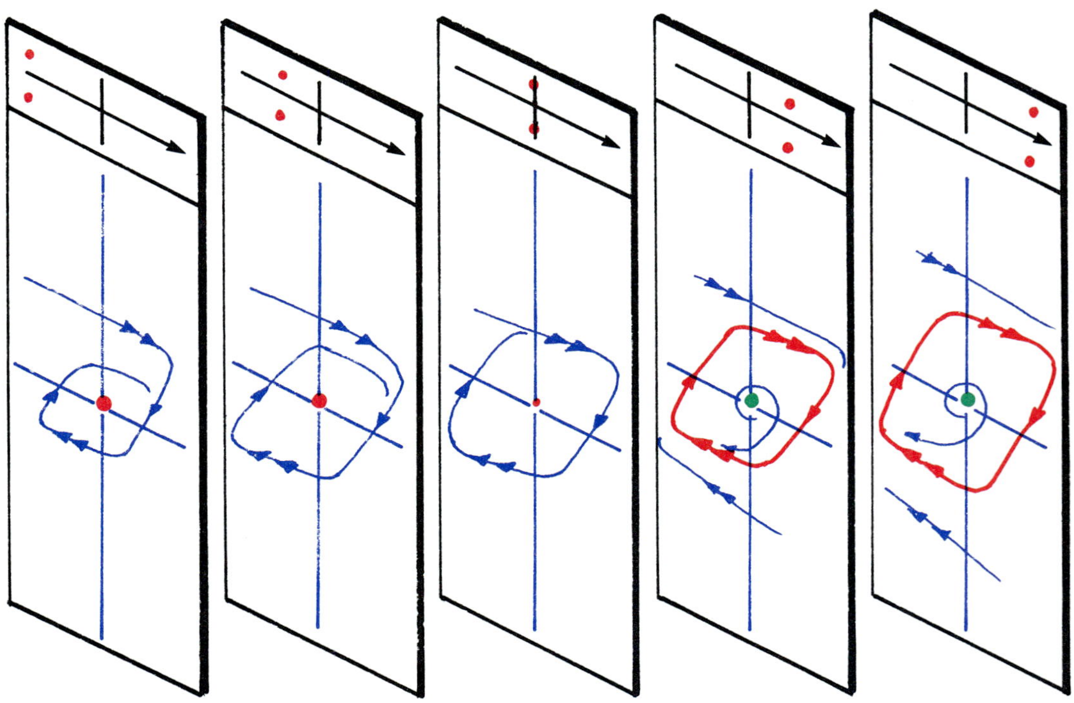

17.1.6. Arranging the phase portraits in order, we obtain this record of behavior, with the CE's shown in the windows. They may be obtained from the equations defining the scheme, by means of algorithms given in texts of linear algebra.[5] Here we have added, at the moment of bifurcation, an additional phase portrait. This portrait, shown as a *center* (concentric periodic trajectories), is generally a very weak spiral in or out. But it will look like a center to the casual observer.

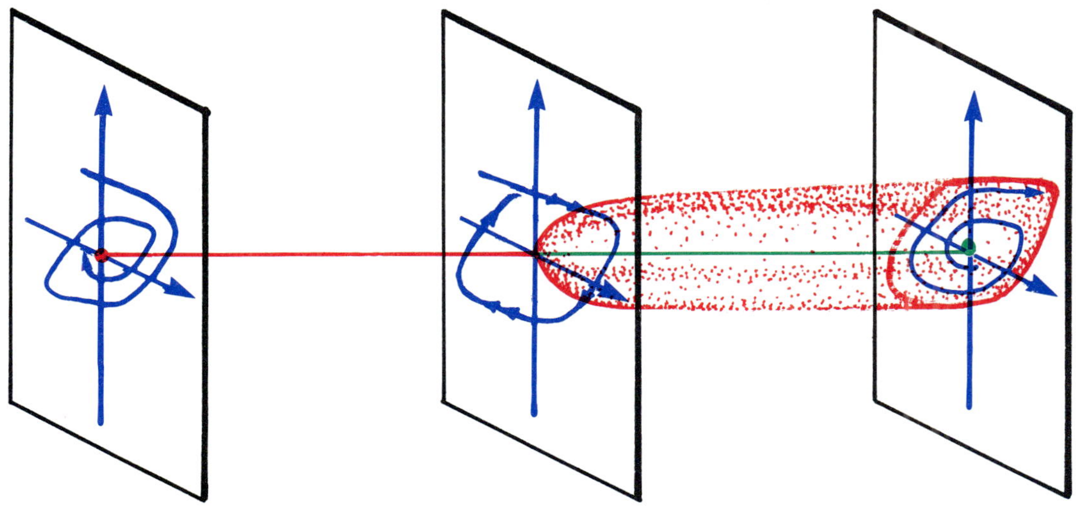

17.1.7. We now erect these portraits and arrange them in their proper places within the response diagram of Van der Pol's dynamical scheme. We call this display a *side-by-side* representation. It is a sort of skeleton of the full response diagram.

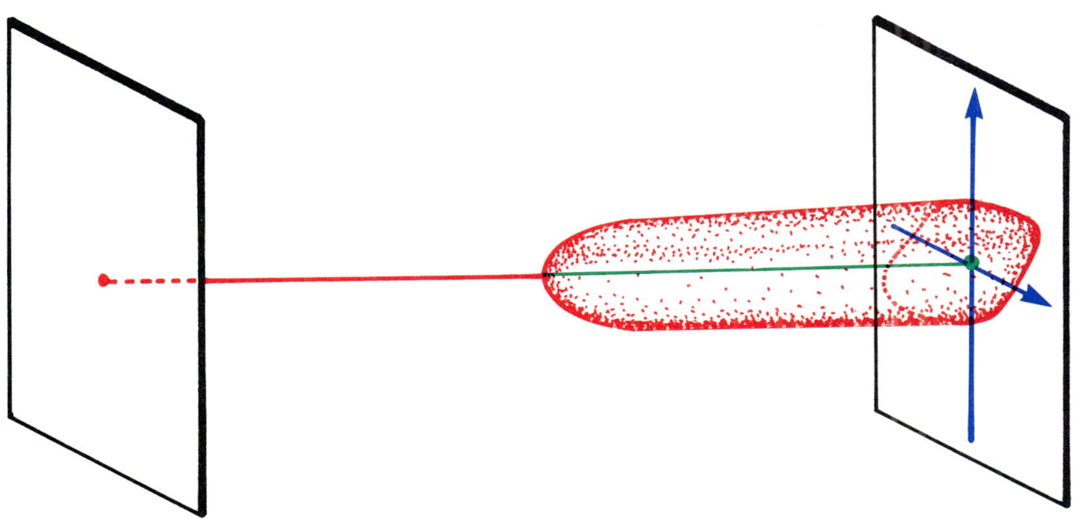

17.1.8. Finally we strip away the technical detail and connect up the skeletons to form this mnemonic version of the response diagram, called the *cutaway* representation, suitable for framing. The joint of the goblet, where the stem joins the cup, is the bifurcation event. Near it, the goblet has a smooth parabolic shape.

SUMMARY. In the first excitation, a point attractor of spiral type gradually weakens (its complex CE's move toward the right, approaching the imaginary axis) and destabilizes (the CE's cross to right half-plane) while emitting a periodic attractor. The period of oscillation of the new attractor is determined by the CE's of the origin, at the moment of its creation. The amplitude of the new attractor grows gradually as the control parameter continues to increase, creating the parabolic goblet shape shown in Figure 17.1.8.

The *bifurcation point* (the critical value of the control parameter at which the CE's cross the imaginary axis) is a quarter turn of the coil in this case. At the bifurcation point, the critical point at the origin is not elementary, it is a center (see Section 11.1). Thus, by Peixoto's theorem, the dynamical system corresponding to this bifurcation point is not structurally stable (see Section 12.2). The arc in the Big Picture described by this dynamical scheme (see Figure 16.3.4 above) pierces the Bad Set at the bifurcation point only.

The locus of this new periodic attractor *branches off* from the locus of the critical point. We may say that *stability is lost* by the critical point at the origin, and passes to the limit cycle. This event is also described as the *excitation of a mode of oscillation,* a mode that (before excitation) is implicit in the point attractor of spiral type. Such an attractor may be viewed as an attractive oscillator of amplitude zero. More details may be found in the literature, which includes several volumes devoted entirely to this event.

17.2. Second Excitation

The kind of excitation we have just described can happen to an oscillator as well as to a static attractor. This event has been known since Poincaré at least, but is usually attributed to Neimark in 1959. Sometimes it is called the *secondary Hopf bifurcation*. It occurs in the bifurcation sequence of the stirring machine, but we omitted it in our description for economy.

We will now need all we have learned about the art of toral arrangement from the preceding chapters.

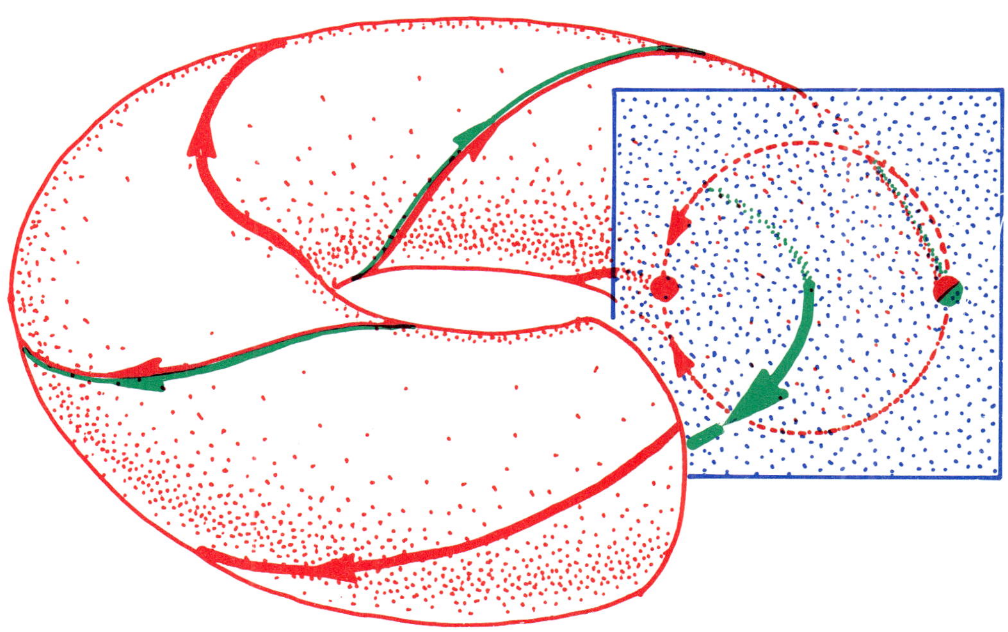

17.2.1. Recall this attractive invariant torus (AIT) from the discussion of forced oscillators in Section 5.3. We will now explode an innocent periodic attractor of spiral type in 3D (see Section 7.5) into one of these gems.

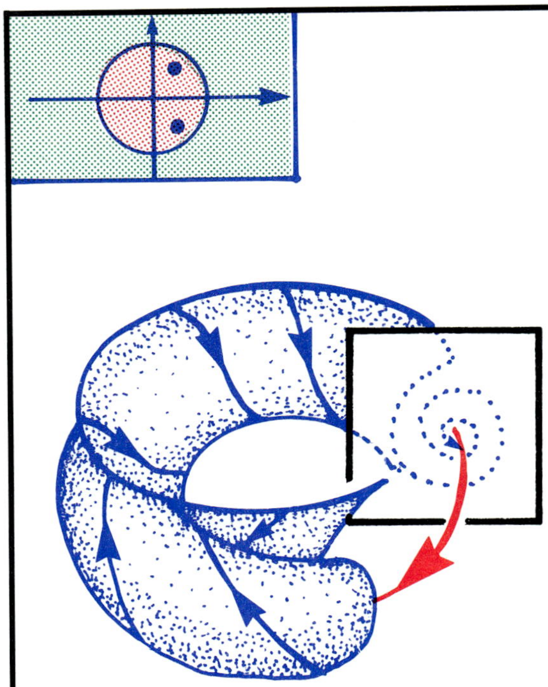

17.2.2. BEFORE: Here is your garden variety spiral cycle in 3D (red for attractors) with a piece of the asymptotically attracted trajectory foliage (blue) cut away to reveal a 2D section (strobe plane, see Section 4.1). And in the window, the CM's within the unit circle, indicating the strength of attraction (Section 7.5).

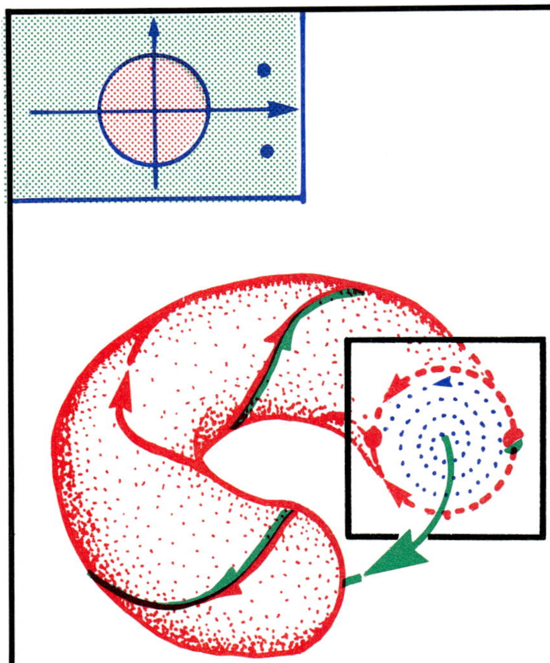

17.2.3. AFTER: The cycle has turned green, indicating repulsion, while an enclosing AIT has appeared, red indicating attraction. It is attractive, yet not necessarily an attractor. (It contains attractors, like the red curve spiraling it.) Again, the window shows the CM's of the central cycle, now outside the unit circle, showing the strength of repulsion.

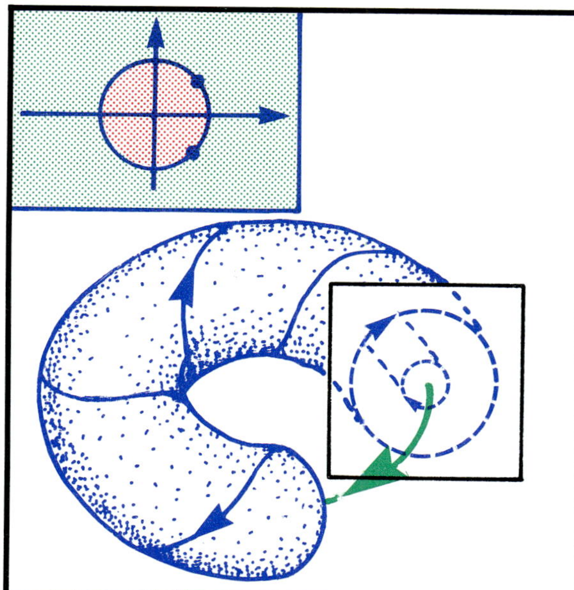

17.2.4. DURING: The bifurcation occurs as the CM's pass through the unit circle. At this moment, the central limit cycle is neither attracting nor repelling. In fact, the 2D strobe plane is filled with spirals (in or out) which are wound so tightly that they may appear to be nested invariant circles. These tight spirals correspond to scrolled insets or outsets which may appear to be nested invariant tori in the 3D flow.

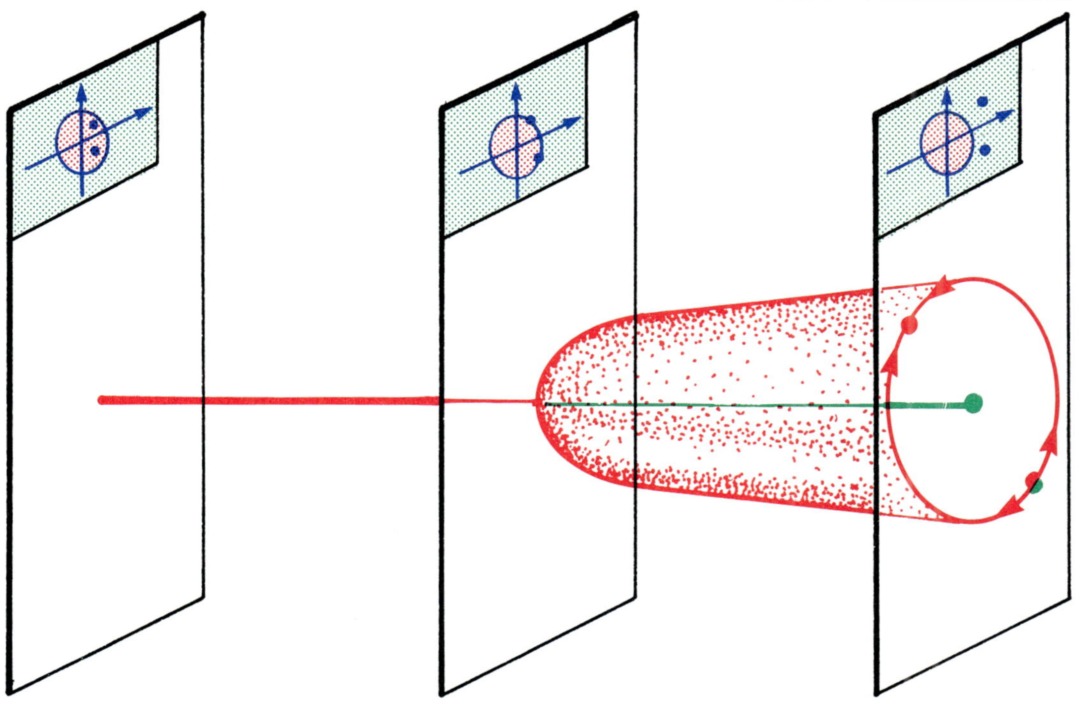

17.2.5. To construct the response diagram of this scheme, we will extract the strobe planes and erect them side by side in our usual fashion. Note that it looks like the response diagram for first excitation at the end of the preceding section. But here we are stacking strobe planes, rather than state spaces.

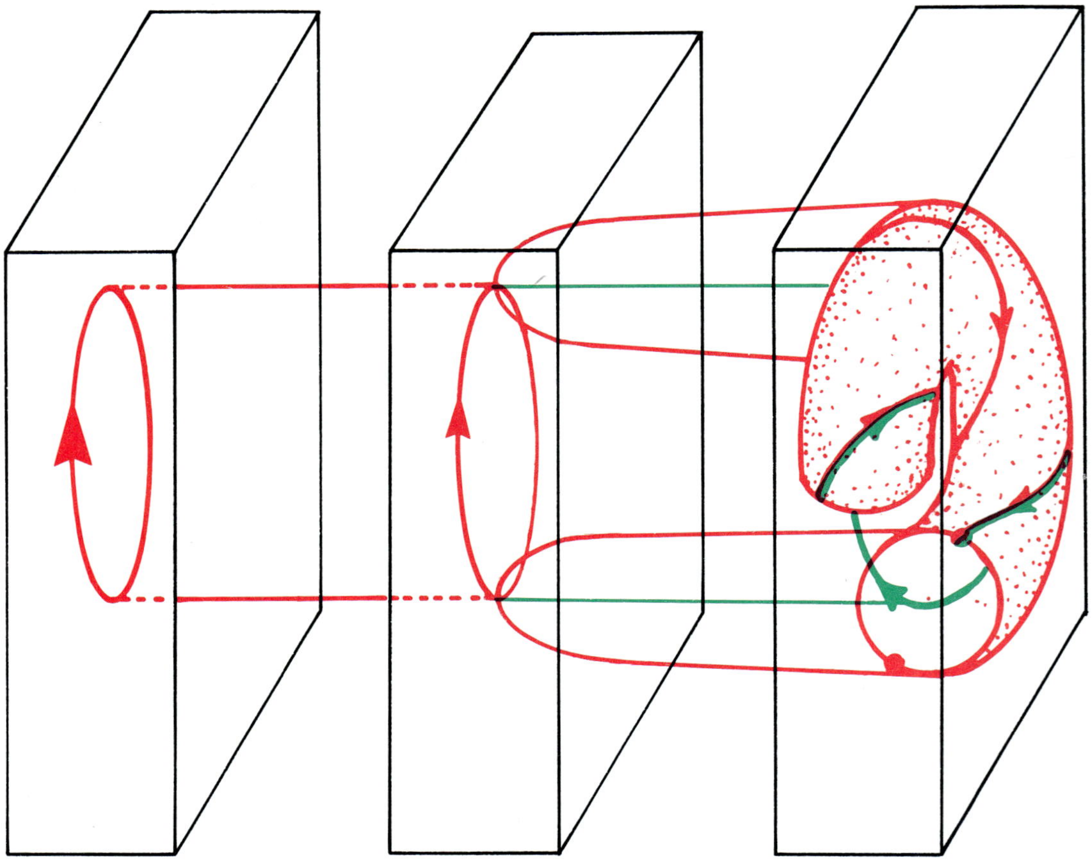

17.2.6. Finally, here is the cutaway view of the response diagram. We have abandoned the strobe planes, and replaced each with a full 3D state space, seen in 2D perspective view.

SUMMARY: In the first excitation, stability is lost by a critical point, and its implicit oscillator emerges from hiding. This event is characterized by the CE's of the critical point. But a critical point in 2D is rather like a limit cycle in 3D, as the Poincaré section (strobe plane) technique shows. And in this way a point attractor of spiral type in the plane (see Section 6.4) corresponds to a periodic attractor in 3D of spiral type (see Section 7.5). And the two complex conjugate CE's of the spiral point in 2D correspond to the two CM's of the spiral cycle in 3D. So by analogy, we may see in such an oscillation in 3D an infinitesimally thin torus, hiding and ready to jump out, should the attraction of the oscillator weaken. When this torus jumps out, it represents a *compound oscillation of two modes*, such as we have studied in the context of two coupled oscillators in Chapter 5. If the original oscillator is considered the first mode, then the new torus may be regarded as the combination of the original mode with a new, second mode. Hence the name, *second excitation.*

17.3. Octave Jump In 2D

The main feature of this event is the replacement of a periodic attractor by another one of twice the period. If this happened while a musical instrument was holding a note and some parameter was being adjusted, you would hear a very soft tone begin an octave below and gradually increase in volume.

Here are some computer plots of actual simulations of an octave jump in a 3D system, provided by Rob Shaw.

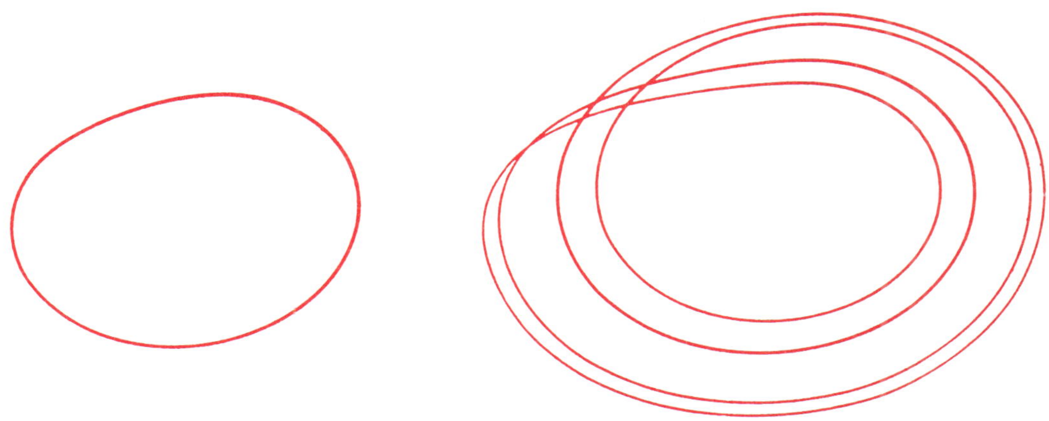

17.3.1. BEFORE: Trajectory tracing around a periodic attractor in the 3D state space, but projected into the 2D screen of the oscilloscope.

AFTER: Trajectory tracing around a different periodic attractor. Note that this one follows closely the track of the previous one. But after one revolution, it does not quite close. After a second circuit, it closes upon itself exactly. Note that the trajectory does not cross itself, as that is highly illegal for dynamical systems. But it appears to, because of the projection onto the 2D viewing screen.

Now we will replot these trajectories in an intrinsically 2D context. A Möbius band is neccesary to accommodate the negative real CM, as explained in Section 7.2.

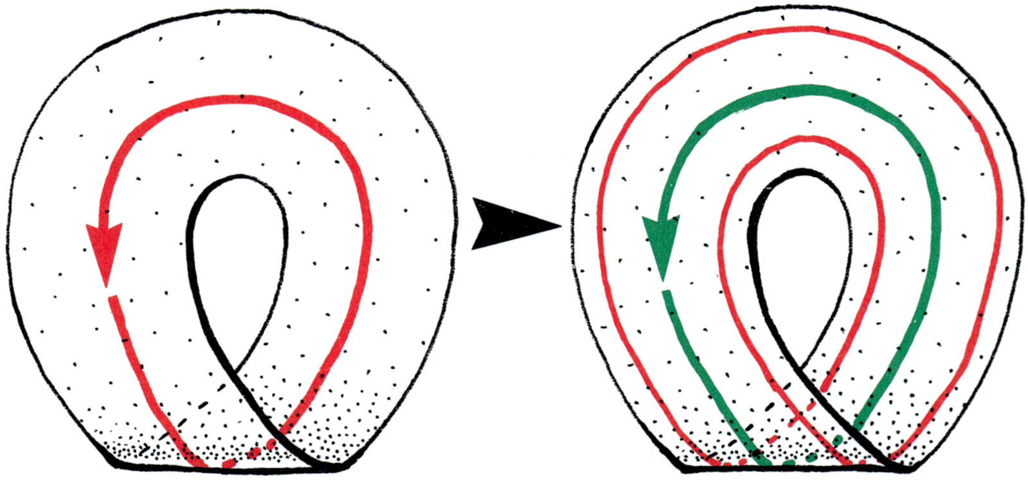

17.3.2. BEFORE: We start with a dynamical system on the band having a single periodic attractor (red) that goes once around.

AFTER: This new system has a single periodic attractor that goes twice around the band, without crossing itself. The former attractor still exists as a repellor (green).

Here is a review of the CM's of the central cycle, from Section 7.2.

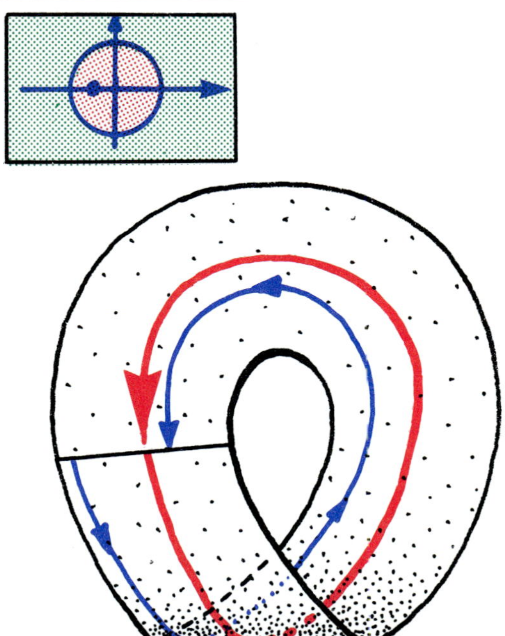

17.3.3. BEFORE: Draw a strobe line (black) across the band. The negative real CM means that a near by trajectory (blue) starting on the strobe line to the left of the red cycle will return to the right (and closer) after one circuit. Thus the strobe line is reversed by the first-return map, and the inset must be twisted.

The CM is shown in the window. While negative, it is within the unit circle (blue), signifying attraction.

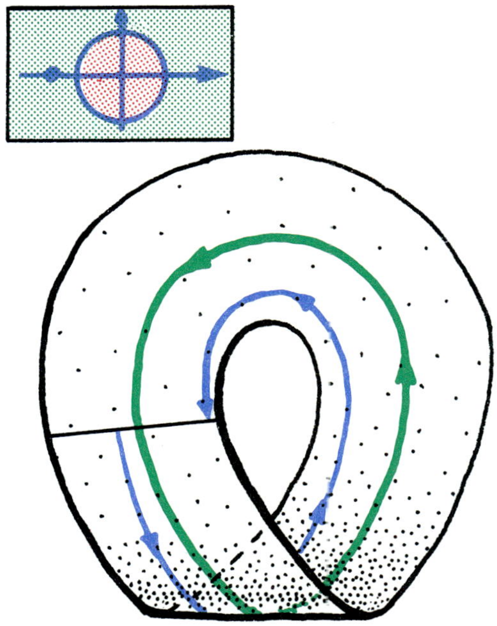

17.3.4. AFTER: Again draw a strobe line across the band. The CM is still negative. A nearby trajectory starting on the strobe line to the left of the red cycle will still return to the right (but further away) after one circuit.

The CM is again shown in the window. While negative, it is outside the unit circle, signifying repulsion. The cycle is drawn in red and green, as it is still attractive to some curves off the band, and repulsive to those on the band.

Between these two portraits there is a bifurcation.

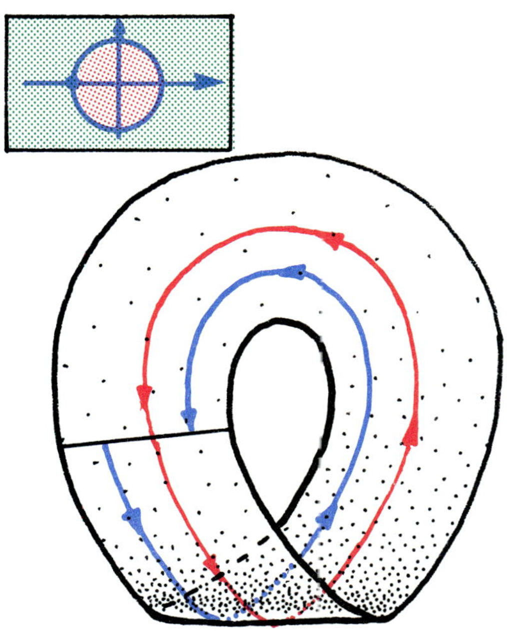

17.3.5. DURING: The CM, still negative, is actually sitting at minus one on the negative real axis. A nearby trajectory starting on the strobe line to the left of the red cycle will still return to the right (but at the same distance from the red cycle) after one circuit.

The CM is again shown in the window. While negative, it is on the unit circle, signifying neither attraction nor repulsion. This situation is structurally unstable, according to Peixoto's theorem (see Section 12.2).

We now put these portraits together, in a side-by-side representation of the response diagram.

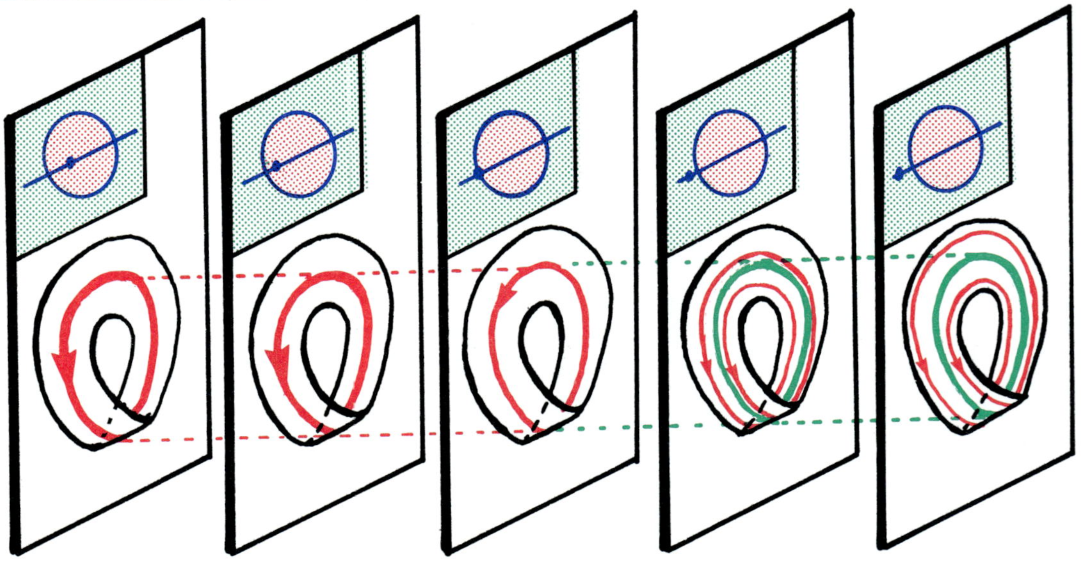

17.3.6. Reading from left to right, the red cycle turns green at the instant of bifurcation. A new red cycle branches off gradually, but it must go twice around to close on itself. Note that the CM of the new periodic attractor is positive and within the unit circle.

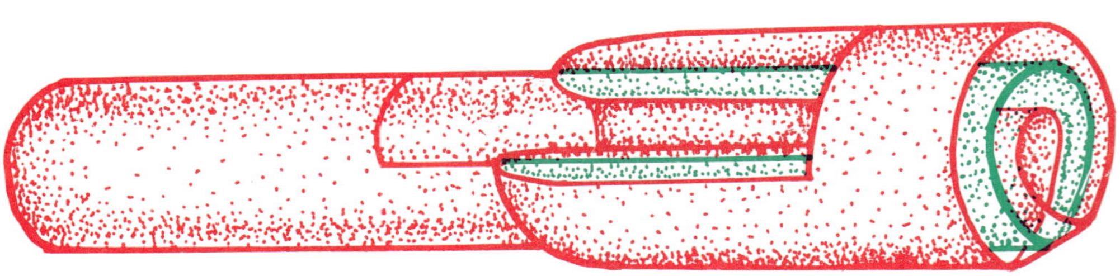

17.3.7. And finally, a cutaway representation, for the collection.

SUMMARY: A periodic attractor on a Möbius band responds to a control parameter by losing stability. Its CM journeys outward, seeking to escape. Upon the CM crossing the unit circle, this limit cycle becomes a repellor and a new periodic attractor is born. This has twice the period, hence half the frequency. Its fundamental is one octave down. However, it traces closely around the same track twice before closing, so its second harmonic (same tone as the recently vanished attractor) is strong. Thus, this bifurcation event is subtle, in that its detection is possible only after the new behavior grows strong. One cannot detect the exact moment of bifurcation by casual observation. This is in marked contrast to the explosive and catastrophic events, as we shall soon see. Some people like to call this event a *period doubling bifurcation*.

17.4. Octave Jump in 3D

Of course the octave jump can happen in 3D, 4D, and so on. In this section we illustrate one of several scenarios in 3D.

We begin with a periodic attractor of nodal type.

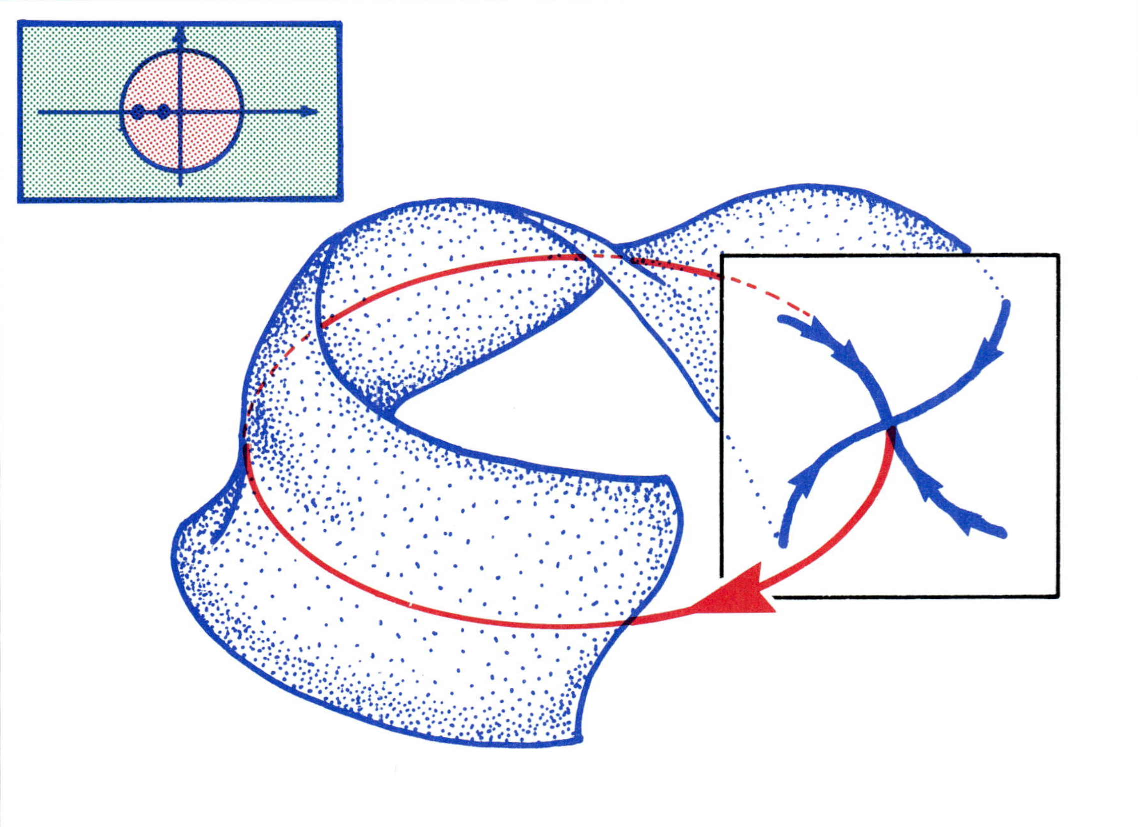

17.4.1. BEFORE: Recall this nodal saddle in 3D from Figure 7.4.3. Note that both of its CM's are negative reals within the unit circle. Its inset, a solid 3D ring, contains two invariant surfaces, both Möbius bands (blue). One of these, the *fast band,* corresponds to the smaller CM (closer to the origin). Trajectories on this band spiral toward the red attractor faster than the others. The other, the *slow band*, corresponds to the other CM (for slow traffic only).

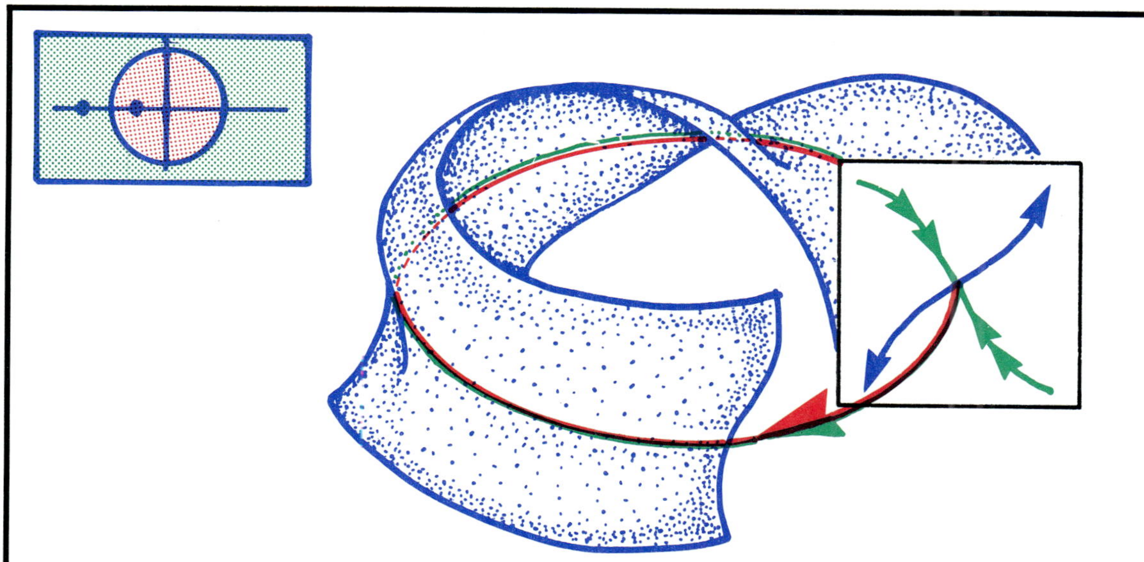

17.4.2. AFTER: Here is another old acquaintance, a saddle cycle in 3D from Figure 7.3.10. Both CM's are negative real, but only one is within the unit circle. It corresponds to a twisted band as before, but now this band is the entire inset. The other CM is outside the unit circle and corresponds to the outset, another twisted band. These two bands are oriented exactly like those in the BEFORE panel, but now one of them is the outset, rather than the slow band.

But this AFTER portrait has an additional feature, which we shall reveal with strobe sections.

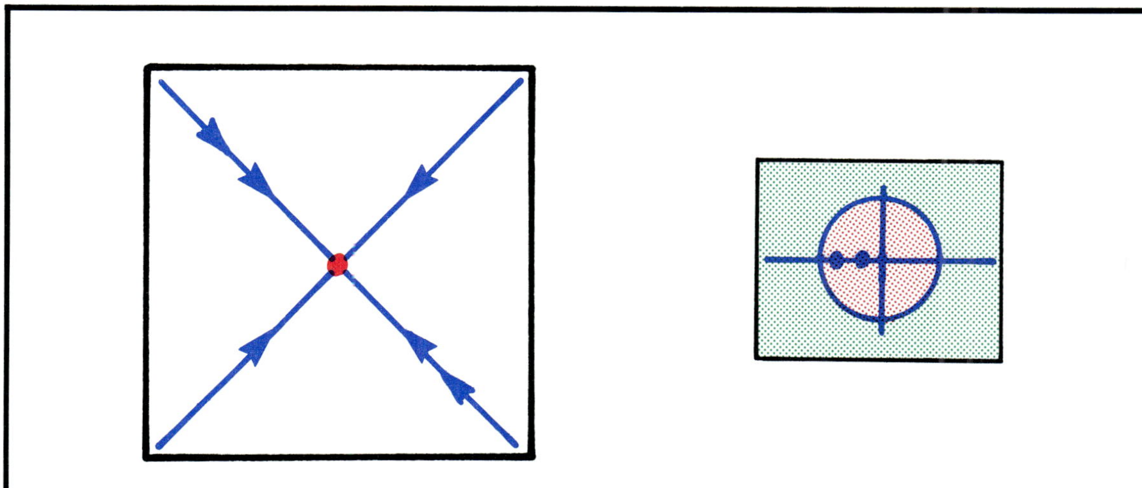

17.4.3. BEFORE: Here is a section of the periodic attractor (red) of nodal type, showing the section curves (green) of the fast and slow bands within its inset, and the CM's.

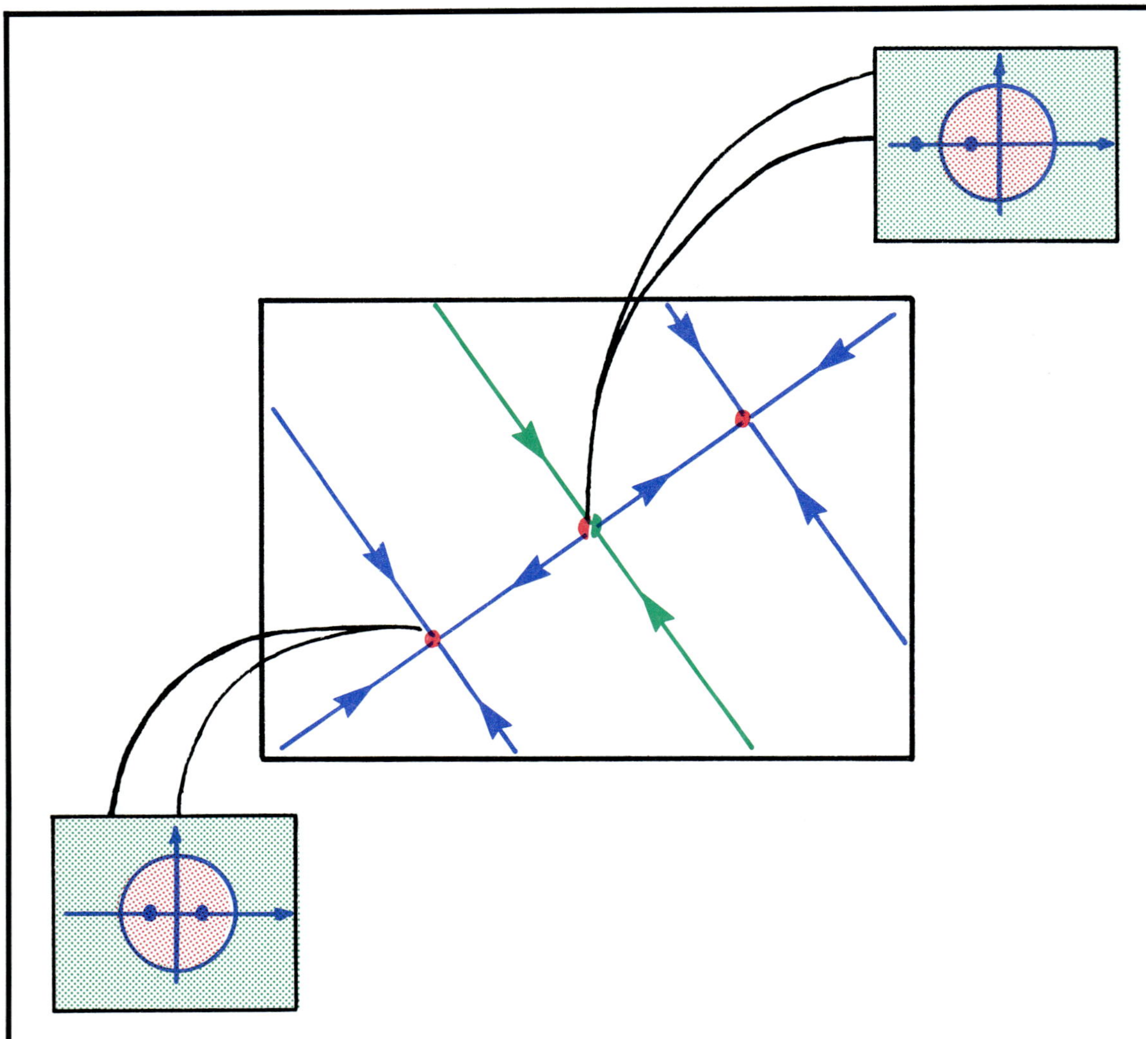

17.4.4. AFTER: And here likewise is a section of the saddle cycle (red and green) with its inset (green) and outset (blue). But here we have added an additional feature to the sectional portrait. Notice the two red dots. These are two successive passages of the same trajectory, a periodic attractor, through the strobe plane. The CM's of the new attractor are shown in the window on the left.

To understand this 3D phase portrait, just take the AFTER portrait of Figure 17.3.2 and imbed it here as the outset of the saddle. This octave jump in 3D is identical to the 2D event on a Möbius band, but taking place entirely on this fixed band through our central cycle. The attraction along the fast inset does not change during this event. This 3D bifurcation is an extension of the preceding 2D bifurcation.

Here is the view of the response diagram made from erecting the strobe planes side by side.

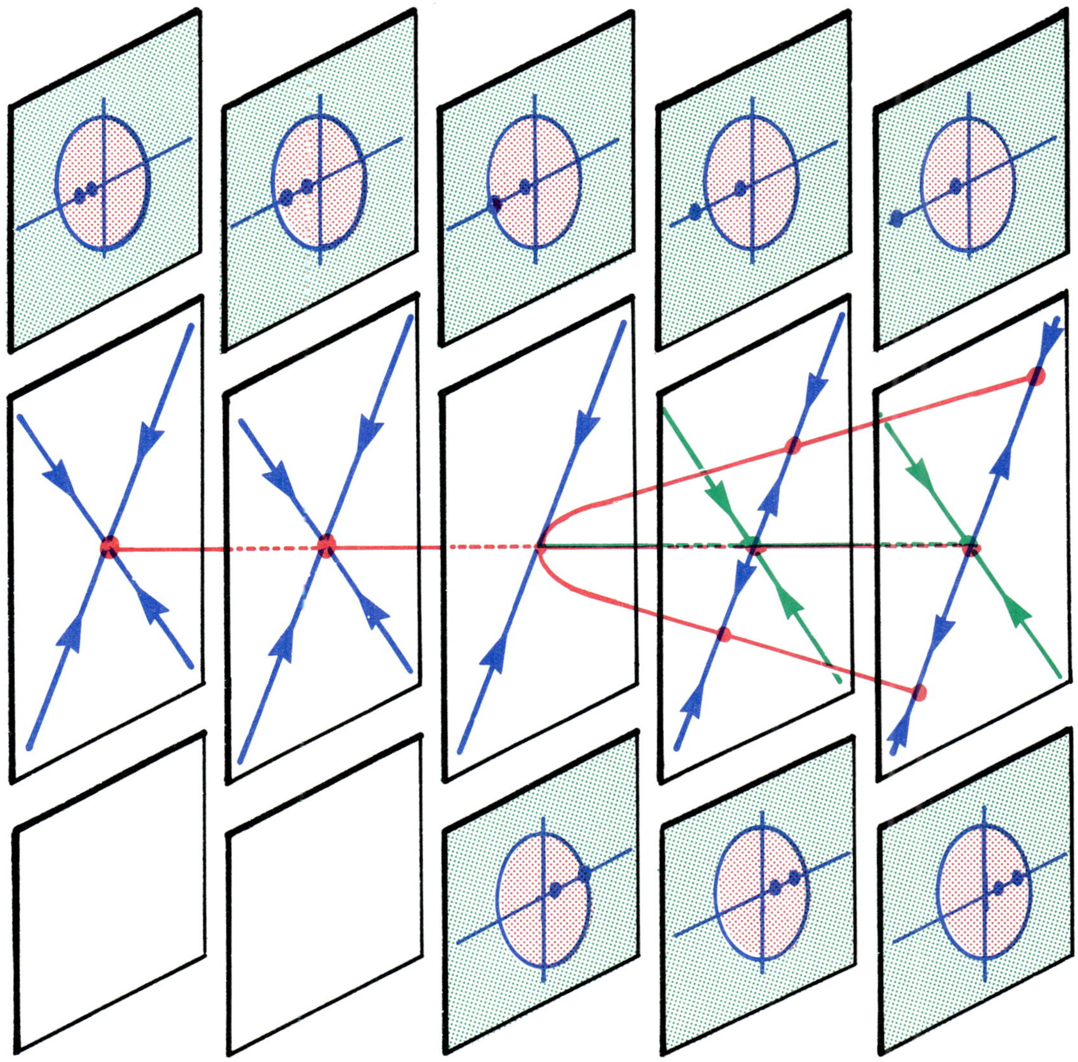

17.4.5. From left to right, the strength of attraction of the slow band of the nodal attractor progressively weakens (outer CM moves further out) and turns to repulsion at the bifurcation point (as the outer CM transits the unit circle). The speed of attraction of the fast band (inner CM) is unaffected by the control parameter. The progress of the CM's of the original cycle is shown in the upper window.

A new attractor is subtly born, which closes only after two circuits of the 3D ring. This is shown, after the event, as two disjoint red points. They spread roughly parabolically as the control parameter continues to increase. The CM's of the new attractor are roughly the square of those of the original cycle. These are shown in the lower window.

SUMMARY: This event is not a new entry in our encyclopedia of generic bifurcation diagrams for single control schemes. It is just the extension to 3D of the preceding entry, the octave jump in 2D, to suggest the variety of possible presentations of a single universal form in the Big Picture.

18

Fold Catastrophes

As explained earlier, there are three kinds of bifurcations with one control in DBT: subtle, catastrophic, and explosive. The previous chapter surveyed the simplest occurrences of the best known subtle bifurcations, and there are not many known bifurcations in this class. But catastrophic bifurcations are very numerous, and this chapter and the following two will be devoted to them. The chief feature of a catastrophe is the disappearance of an attractor, along with its entire basin. This can occur to any type of attractor—whether static, periodic, or chaotic—in a variety of ways. The catastrophic bifurcations of static attractors comprise the subject matter of *elementary catastrophe theory (ECT)*.[1]

In this chapter, we introduce the simplest one: pairwise annihilation, also called the *fold catastrophe*.

18.1. Static Fold in 1D

A favorite way for an attractor to disappear, as the control parameter of a dynamical scheme is varied, is like the moth and the flame. The attractor drifts slowly towards the separatrix at the edge of its basin. When it arrives, three things disappear simultaneously: the attractor, its basin, and its separatrix. In this section we illustrate the simplest case of this type of catastrophe: the 1D (one-dimensional) case.

We will make use of the CE of the critical point of a dynamical system in 1D.

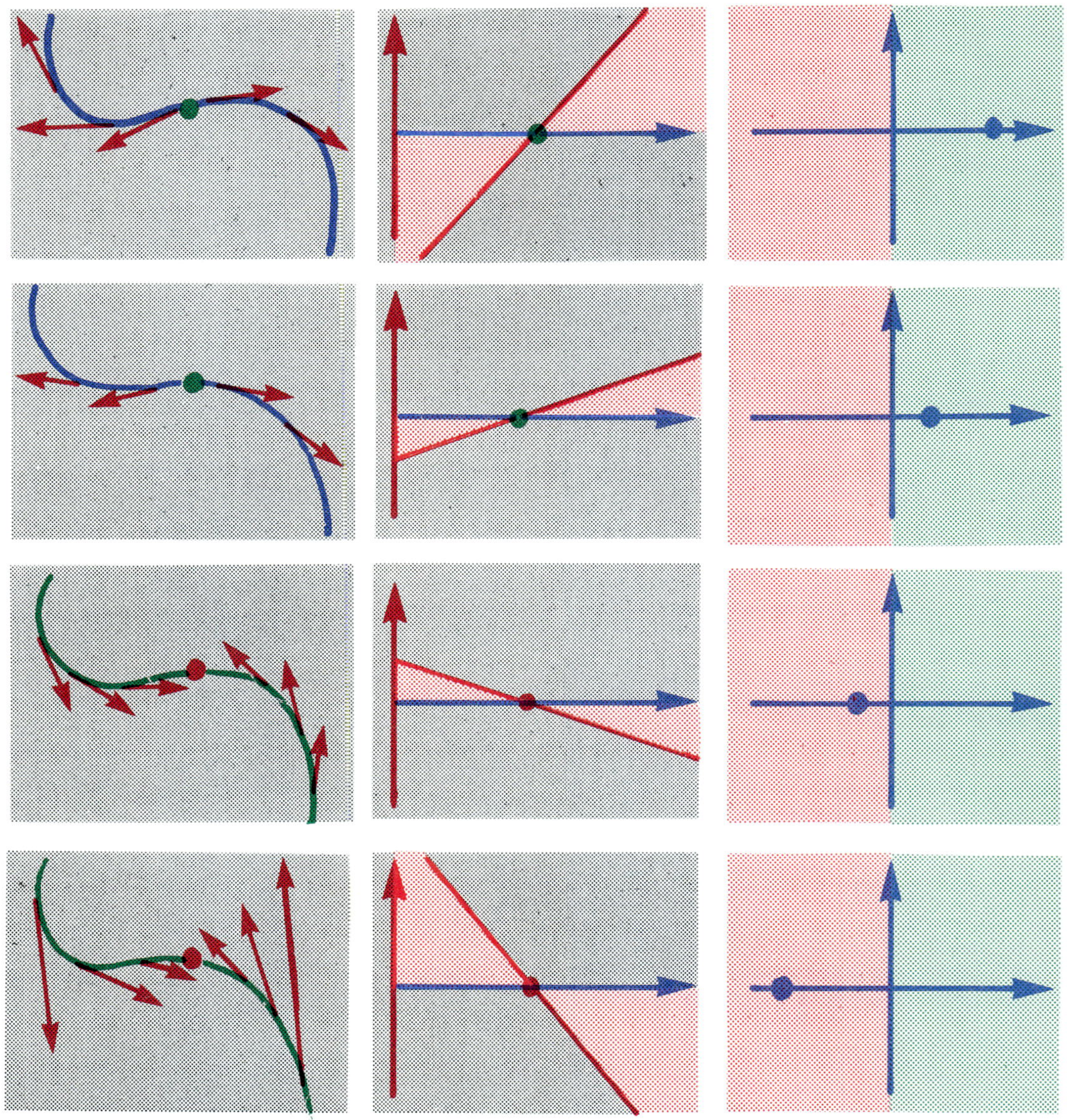

18.1.1. Recall this tabulation of the hyperbolic critical points in 1D from Figure 6.1.8. Here we introduce our predominant color code: attractors are red (all trajectories *stop*) and repellors are *green* (all trajectories *go*). At the same time, insets are green (gr-in) and outsets are blue (bl-out). Meanwhile, the velocity vectors in the first column are red (a temporary expedient) and the inclined red lines in the middle column are the graphs of the vectorfield as a function of position in the (horizontal) state space. The right column indicates the position of the CE (blue) in the CE plane of complex numbers. The CE (blue point) in the green region indicates repulsion. The one in the red region indicates attraction. See Part Two for more explanation of the CE's.

Having recalled this technical background from Part Two, we are now ready for our first fold catastrophe: the static fold.

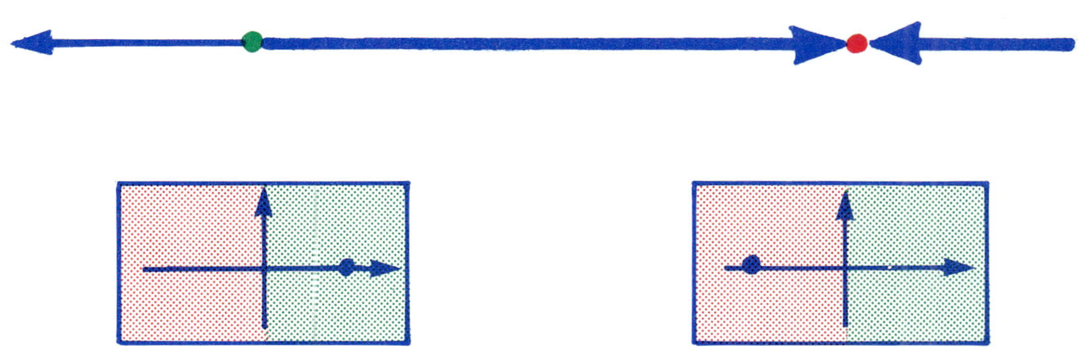

18.1.2. BEFORE: Here is a relatively simple phase portrait of a dynamical system in 1D. As closed orbits are impossible, the only limit sets are critical points. There are only two of them. Their CE's, shown in the windows, are hyperbolic. This system is structurally stable. There is only one attractor, and its basin is shaded green. The repellor is the separatrix, and the upper outset consists of a trajectory going to infinity. We could regard infinity as an attractor in this case and the blue segment as its basin.

18.1.3. AFTER: what can we say? The attractor has disappeared into the blue!

How can we go smoothly from BEFORE to AFTER by simply turning a control knob? The clue, hinted in the first panel of this section, is in the CE's.

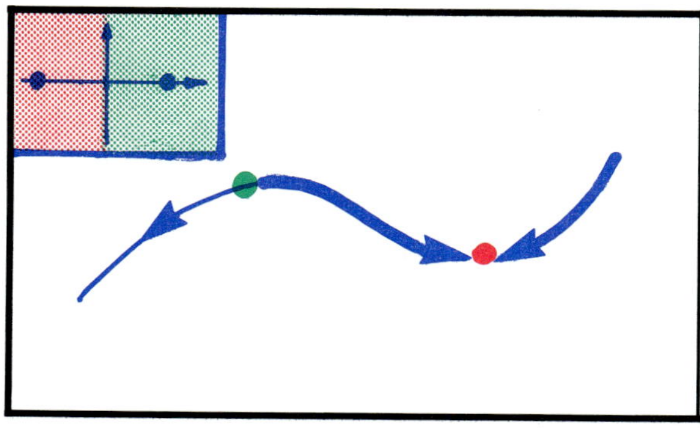

18.1.4. Here is BEFORE again, but we have bent the state space so that the flow of the dynamical system may be understood as raindrops trickling downhill. A puddle collects at the point attractor. The CE's are shown in the windows.

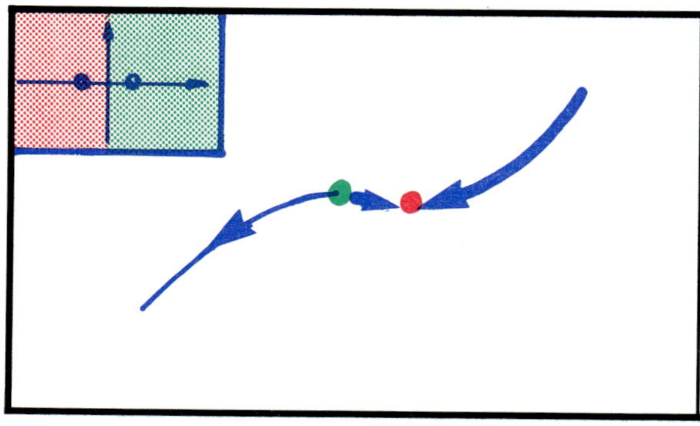

18.1.5. Pulling up the right hand end of the hill a bit brings the two critical points closer together: the red moth approaches the blue flame. The puddle decreases. Note that the CE's are getting intimate as well.

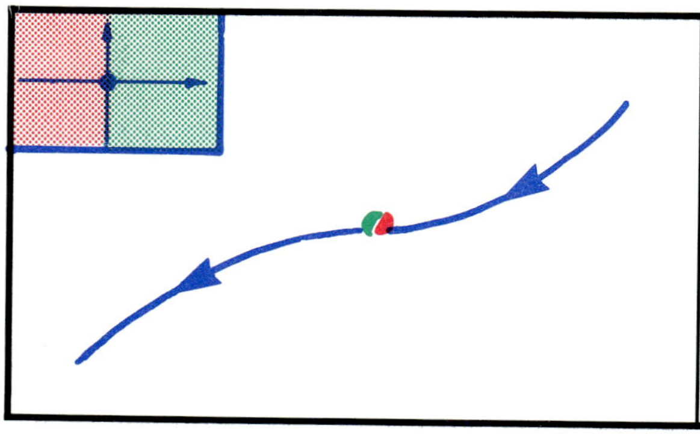

18.1.6. Pulling on the right end some more, the puddle and the two critical points are gone. The catastrophe has occurred!

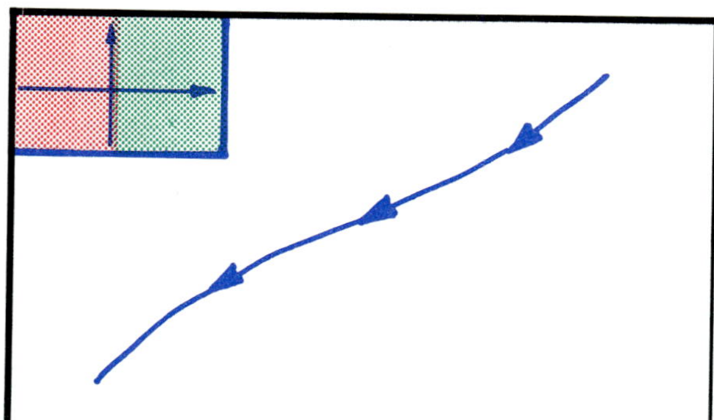

18.1.7. Pulling firmly up on the right end some more increases the slope a bit but makes little difference to the surface water. All rain goes downhill to the left forever.

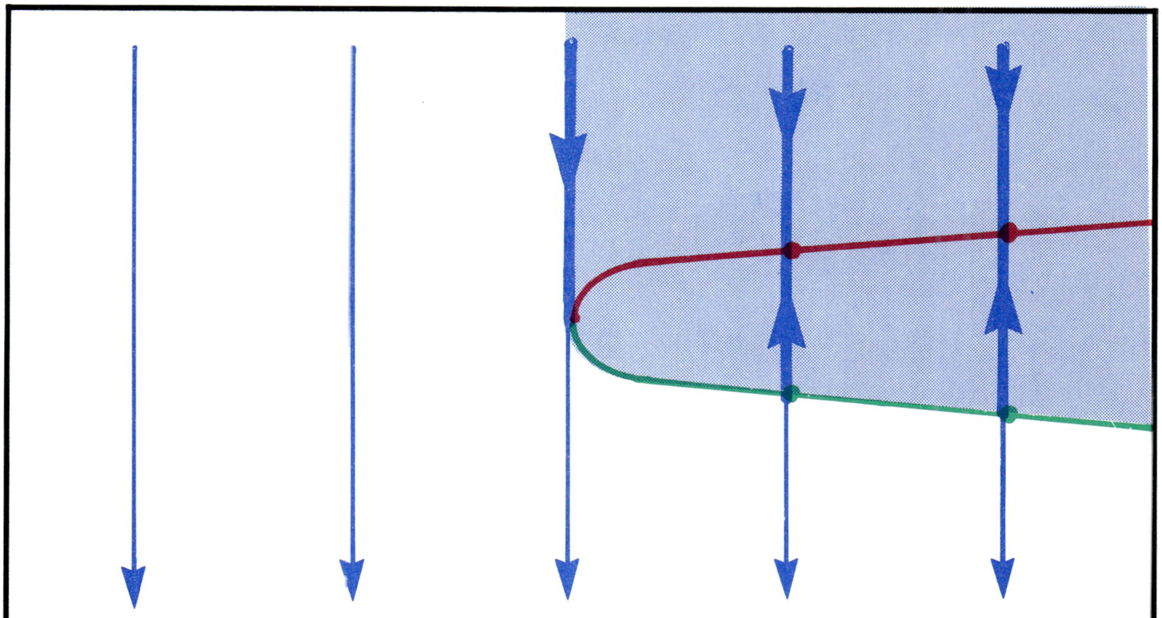

18.1.8. We now turn these phase portraits upright, place them side by side in their proper places in the response diagram of the dynamical scheme we have constructed, and interpolate a few more portraits. At the moment of bifurcation, the red and green critical points meet at an inflection point of the curve (state space).

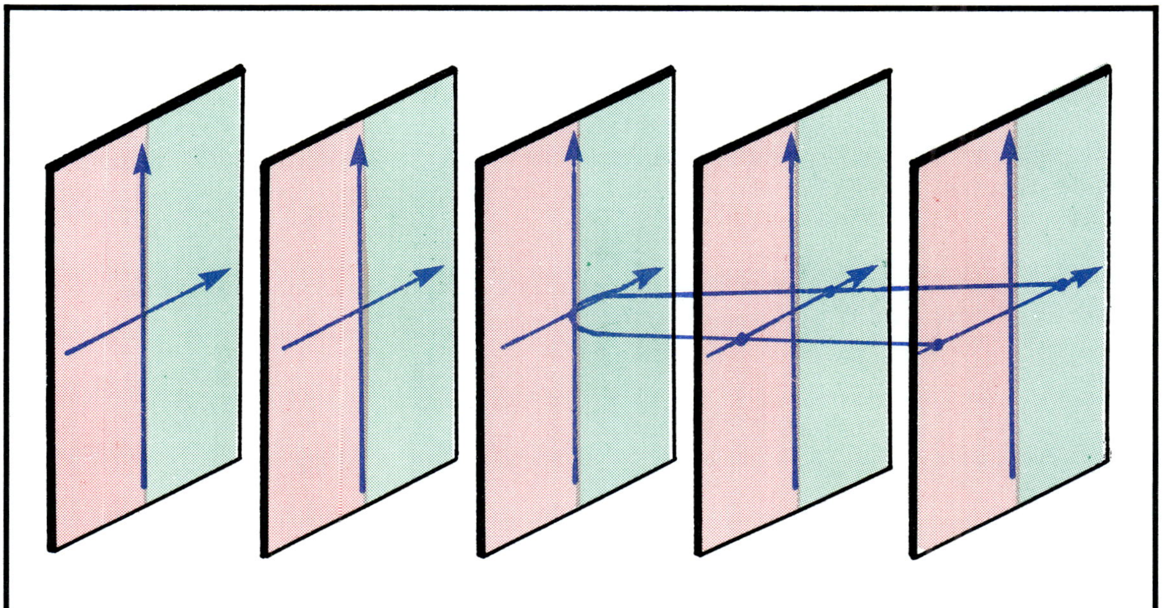

18.1.9. Plotting both of the blue CE's in the same red/green CE plane directly under the corresponding state space, we obtain this curve as a record of their dependence on the control parameter.

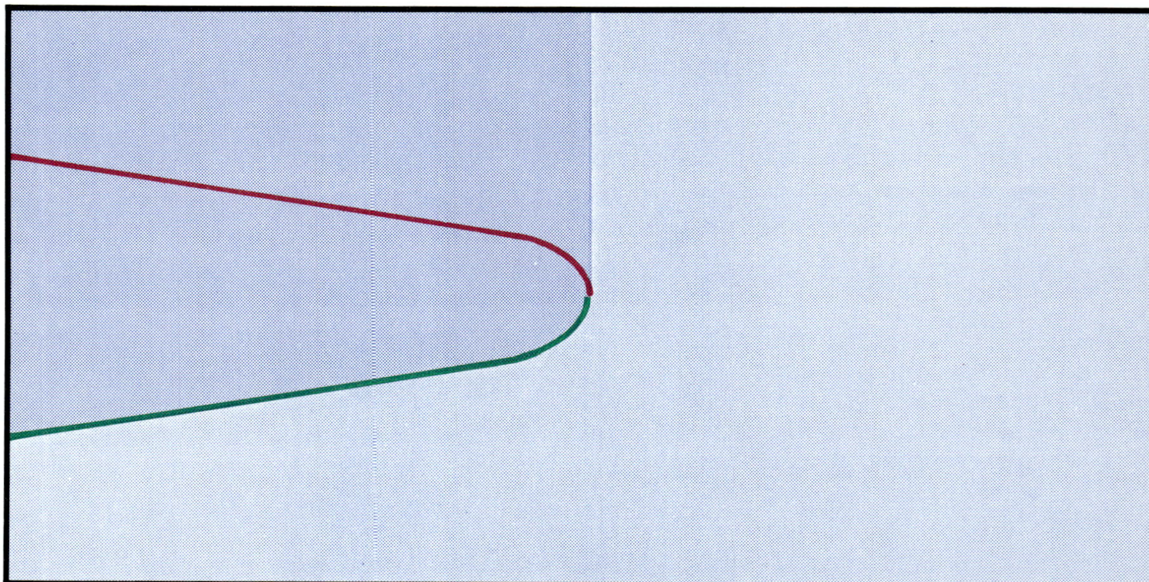

18.1.10. Filling in the continuous locus of each critical point as the control knob is smoothly changed provides this parabolic curve, the full response diagram of the static fold catastrophe in 1D.

Elementary catastrophe theory (ECT) is a beautiful subject, crucially important for the progress of many scientific subjects. Also, it boasts several superb expository texts, which are largely responsible for the development of the basic bifurcation concepts presented in this volume.[2] Study of the early chapters of these texts is strongly suggested for those who wish eventually to understand this atlas of bifurcations. BEWARE: Most of ECT deals with schemes having two or more control parameters, a context well beyond our present purview. But as our agreed context includes more complicated attractors than ECT allows, we will see challenging complications, even with only one control.

There is a growing literature of multiparameter bifurcations, and in due time we may present a pictorial atlas of some of them.

18.2. Static Fold In 2D

Here we present not a new bifurcation, but simply a review of the preceding event in a different context: a 2D state space. We will use the characterization of a critical point in 2D in terms of its CE's, as summarized in Figure 6.4.8. Again we are looking for the drift of a point attractor toward a fatal assignation with its separatrix. In 2D, recall that a separatrix must be either a periodic repeller or the inset of a saddle point. It is the latter case that occurs here.

In this event the point attractor drifts toward the saddle point of its separatrix.

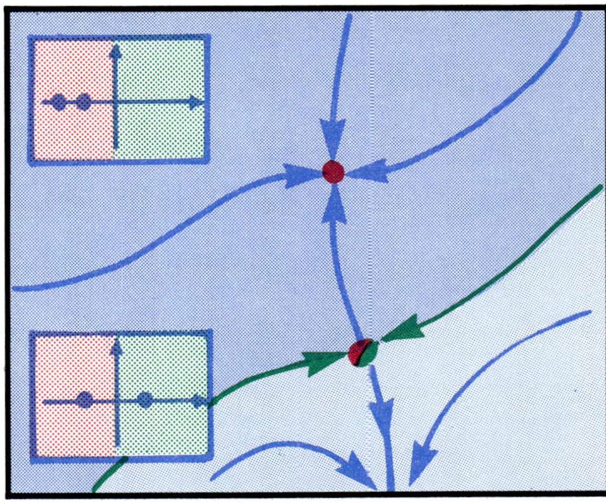

18.2.1. BEFORE: There are two critical points, a saddle (red/green) and an attractor (red). The inset of the saddle (green) is the separatrix of the basin of the attractor. The CE's are shown in the windows.

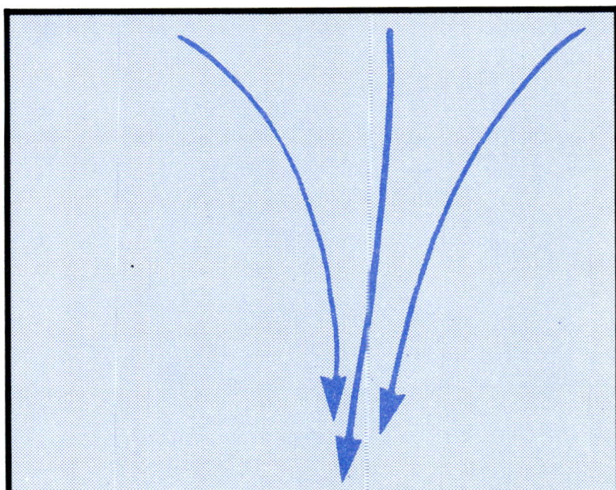

18.2.2. AFTER: There are no attractors, except for an infinite ocean in the South. All trajectories disappear into the blue. Note that a test droplet resting in equilibrium at the attractor before the event now finds itself in the blue basin of infinity, very far from equilibrium. It must now get under way and begin a major journey. This is the reason for the name, *catastrophe,* used in French to describe this behavior.

The mechanism of this event may be intuited from the fold in 1D. Just imagine the two critical points on a north/south train track, which is attractive to trajectories off the track to the east and west.

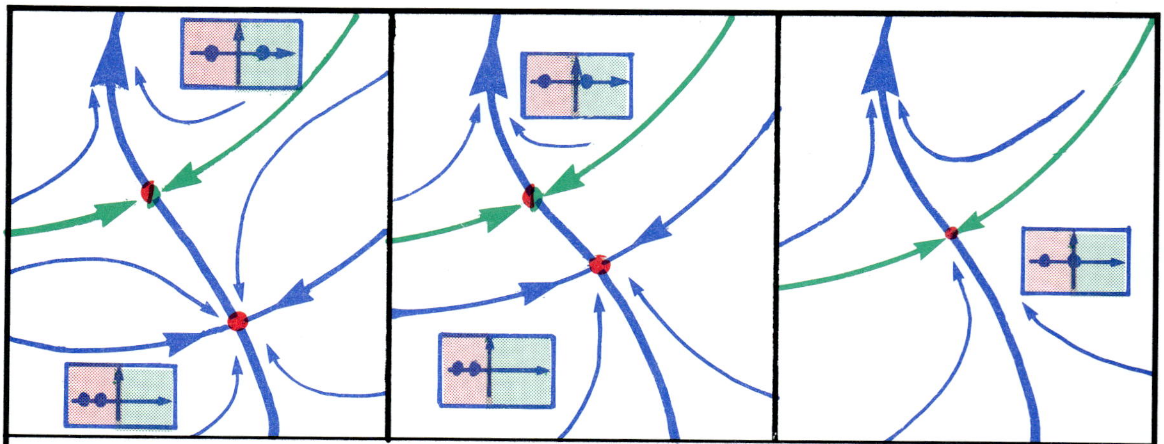

18.2.3. As the control knob is adjusted, let the two critical points fold together along the track, exactly as in the 1D event of the preceding section, culminating at the instant of bifurcation. But in this case there are *two* CE's for each of the two critical points. Only one of the CE's for each of the critical points is affected by the variation of the control. The affected CE corresponds to the strength of attraction or repulsion along the track. The other CE of each critical point indicates the attraction of the track for trajectories off the track, and is unaffected by the control. At the moment of catastrophe, there is only one critical point, and it is nonhyperbolic, as one of its CE's is zero.

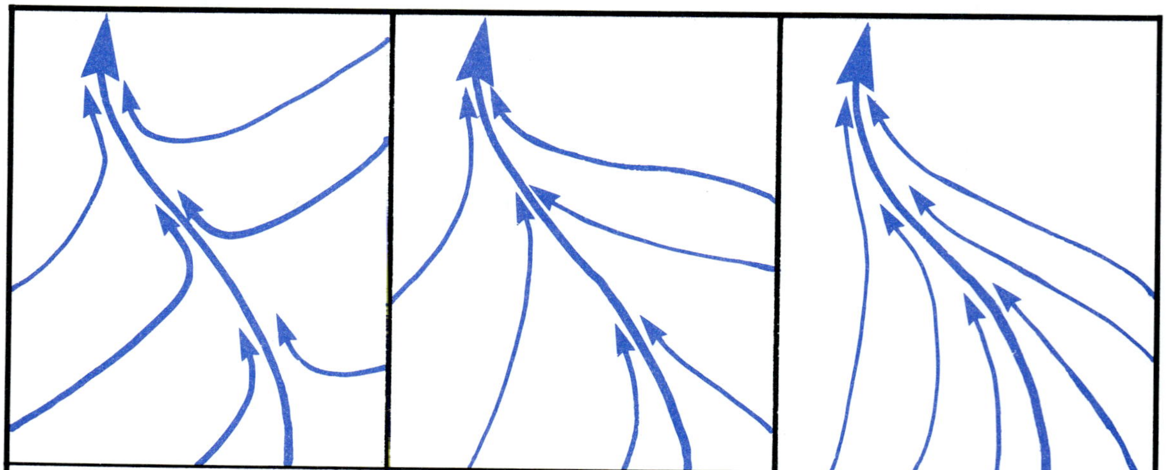

18.2.4. As the control knob continues to turn, the nonhyperbolic critical point vanishes into the blue, and the flow smooths over the shadow of the event.

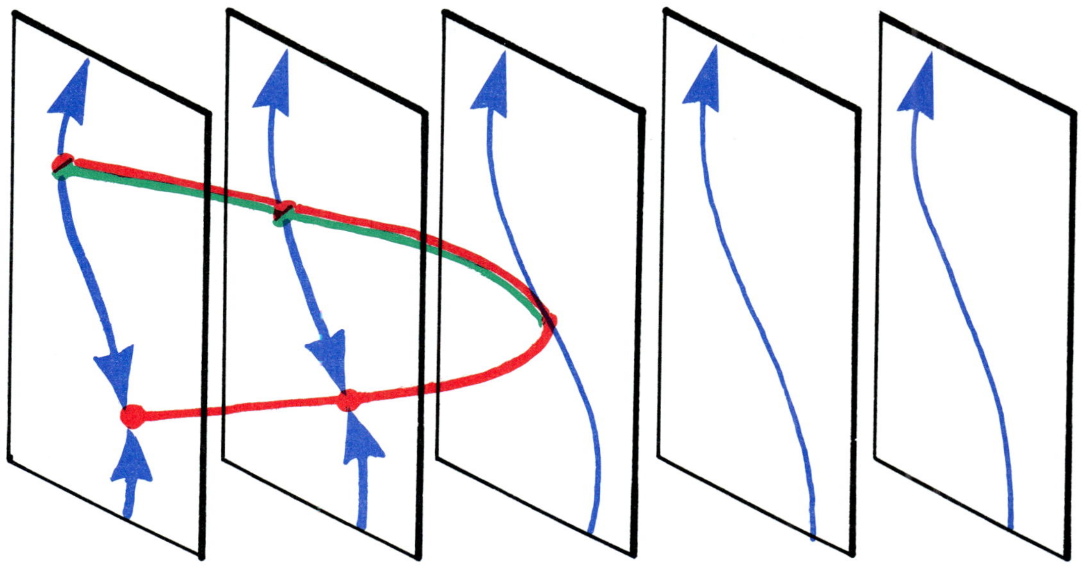

18.2.5. Erecting five phase portraits side by side in their proper positions in the space of the response diagram, we obtain this skeleton of the full response of the scheme.

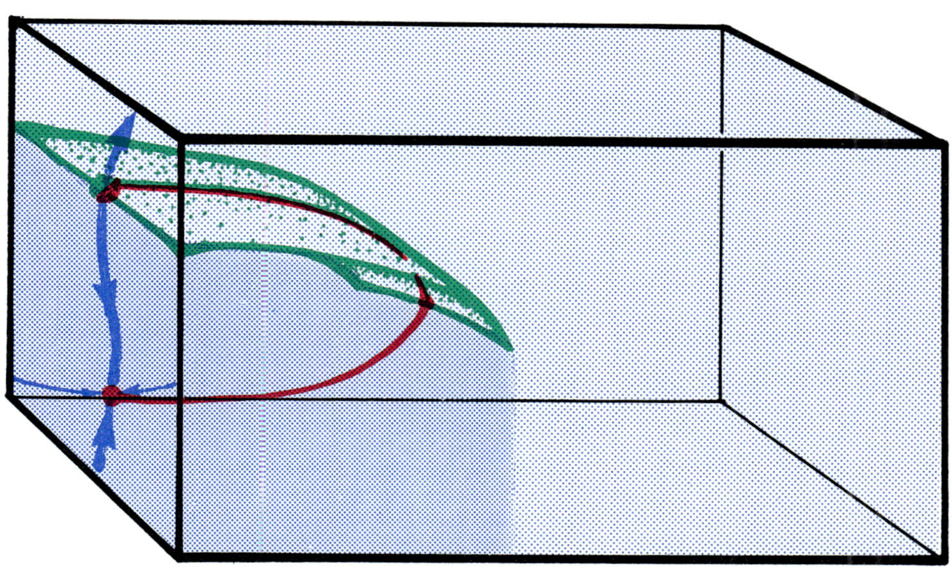

18.2.6. Interpolating the remaining details, we have this cutaway picture of the full response diagram.

Note that if the control is moved to the right, the catastrophe may be described as the drift of an attractor to a saddle point within its separatrix. At contact, there occurs the simultaneous disappearance into the blue of three things: the attractor, its entire separatrix, and its entire basin. But reading from right to left, the catastrophic event consists of an increasingly evident slowing down of the blue flow in a certain region, then the magical emergence (out of the blue slow region) of an attractor, with its full-blown basin and separatrix. Thus, the fold is sometimes called an *annihilation/creation event.*

SUMMARY: This 2D version of the fold is may be called the *extension* of the 1D version presented in the preceding section. This is the same relationship that we have seen at the end of the prececing Chapter, in which the octave jump event was presented first in 2D, then again in 3D.

We want to point out, before going on to another event, that the bifurcation events we are describing one at a time in small boxes of euclidean space are *atomic events.* They are to be expected, in actual dynamical schemes encountered in applications, in *molecular combinations* comprising complex response diagrams, such as that of the stirring machine. A further complication encountered in practice is that the phase portrait of the system, for a fixed value of control, will be a global one with multiple attractors and basins. Generally, at a given bifurcation, only a small part of the global picture will be affected. Our atomic response diagrams may thus be encountered in a small piece of the garden variety response diagram.

Here is a global version of the 2D fold catastrophe. As usual with catastrophic events, the total number of basins is altered by the bifurcation.

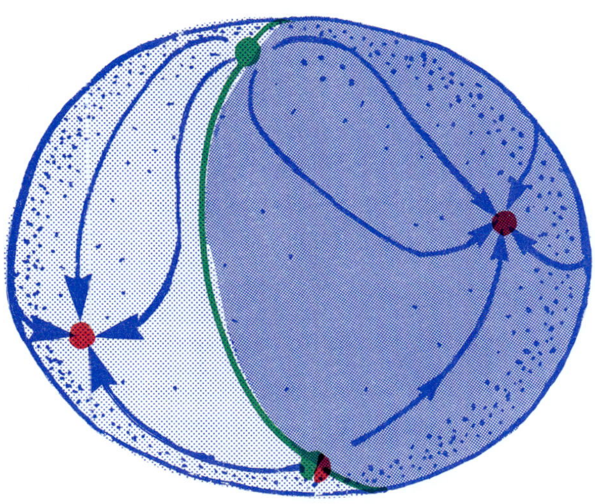

18.2.7. BEFORE: A flow on the two-sphere has two basins (green and blue). Each attractor is static. The separatrix is the inset of a saddle point, completed by a repellor at the North Pole.

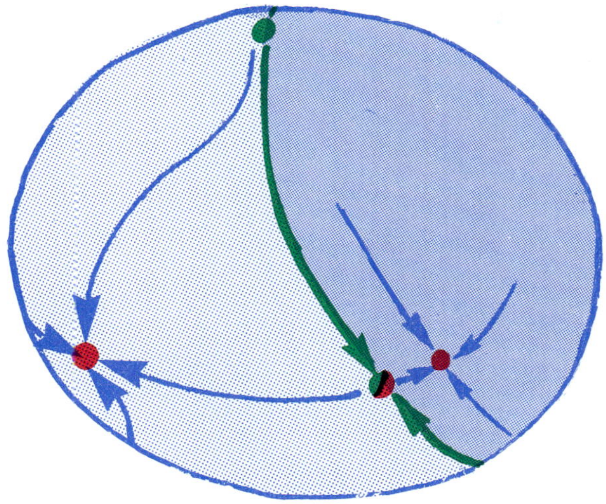

18.2.8. APPROACHING: One of the attractors has drifted close to the flame. The green basin has shrunk about to about half its former greatness. The blue domain has gained, but near its attractor nothing has changed.

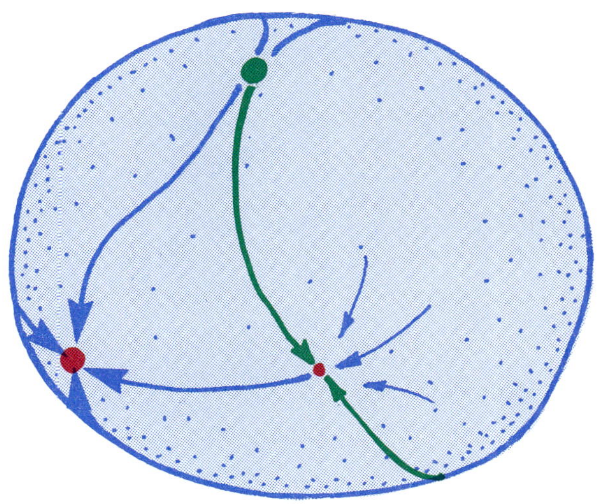

18.2.9. BIFURCATION INSTANT: Briefly, there is a degenerate (nonhyperbolic) critical point on the separatrix, which is now virtual, within the blue basin of the one-and-only attractor. This is the shadow of the blue basin, shrunk completely onto its former separatrix.

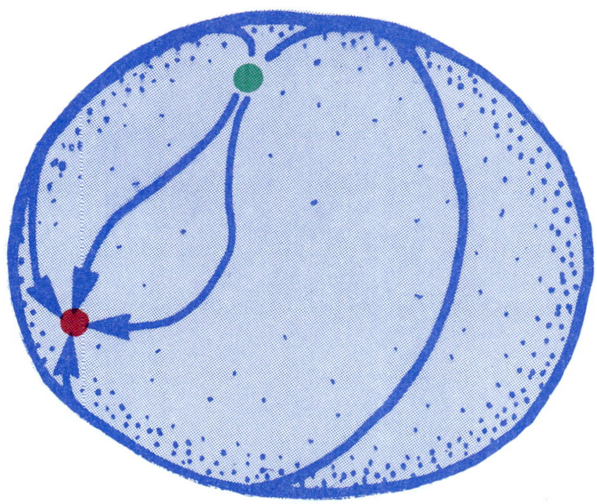

18.2.10. DEPARTING: There is a slow spot in the flow, a feeble memory of the departed basin.

A response diagram for this scheme is easy to imagine, but difficult to draw. Only the study of numerous examples of real systems, such as are found in the literature of experimental and applied dynamics, can give an idea of the enormous variety of response diagrams that may be constructed from the atomic bifurcations we have presented so far.

But furthermore, some of the atomic events presented in the following drawings get pretty complicated by themselves.

18.3. Periodic Fold In 2D

The fold catastrophe we have seen in the two preceding sections for static attractors has an analog for periodic attractors. One way to understand this periodic fold catastophe is by a rotation of the static fold in 1D around a circle.

We now depart the domain of elementary catastrophe theory forever.

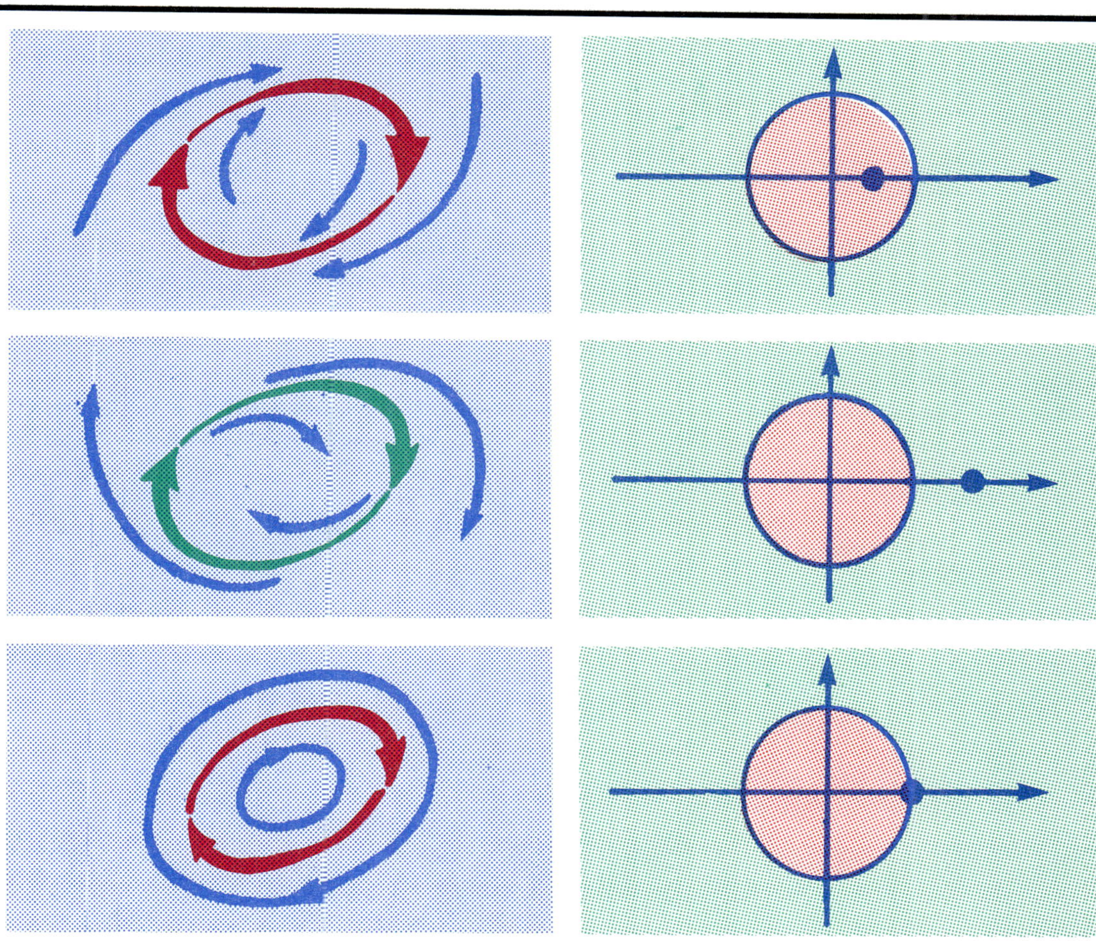

18.3.1. Recall the characterization of a limit cycle in 2D by means of its CM. Here is a table of three cases from Figure 7.2.7: attractor, repellor, and a highly degenerate intermediate case. The CM's on the right correspond to the limit cycles on the left.

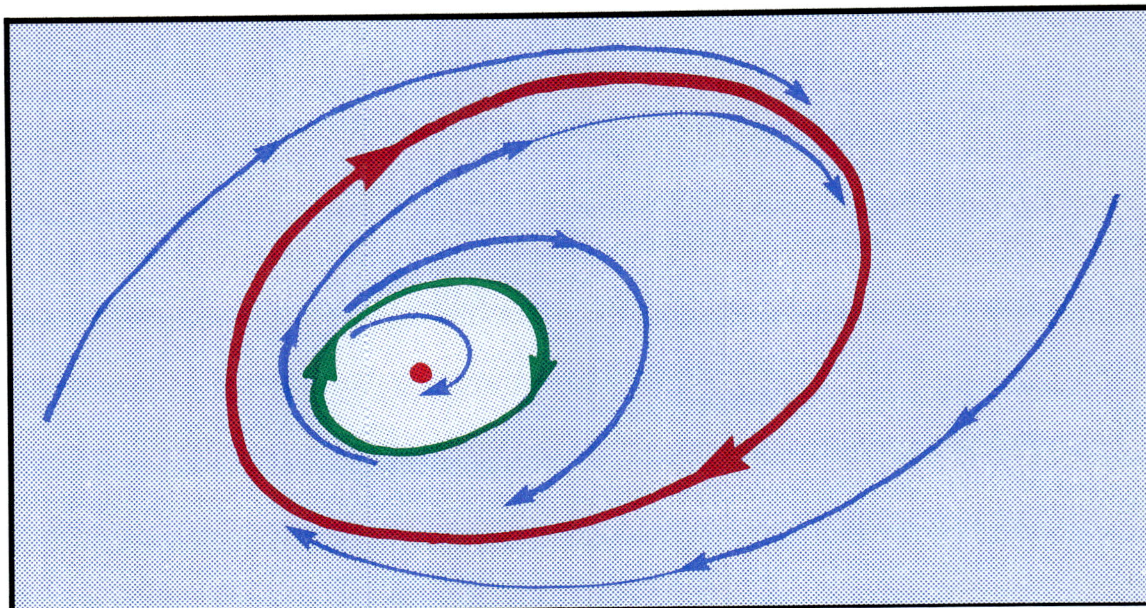

18.3.2. BEFORE: Here is a simple phase portrait in 2D, with two attractors. One is static (light blue basin), the other periodic (dark blue basin). A periodic repellor serves as separatrix, dividing the two basins. If a test droplet is thrown into this dynamic, it will evolve towards a rest state ("off") if it falls initially into the light blue and toward oscillation ("on") if it falls to the dark blue. A toggle switch to turn on a motor might have a model of this *bistable* type.

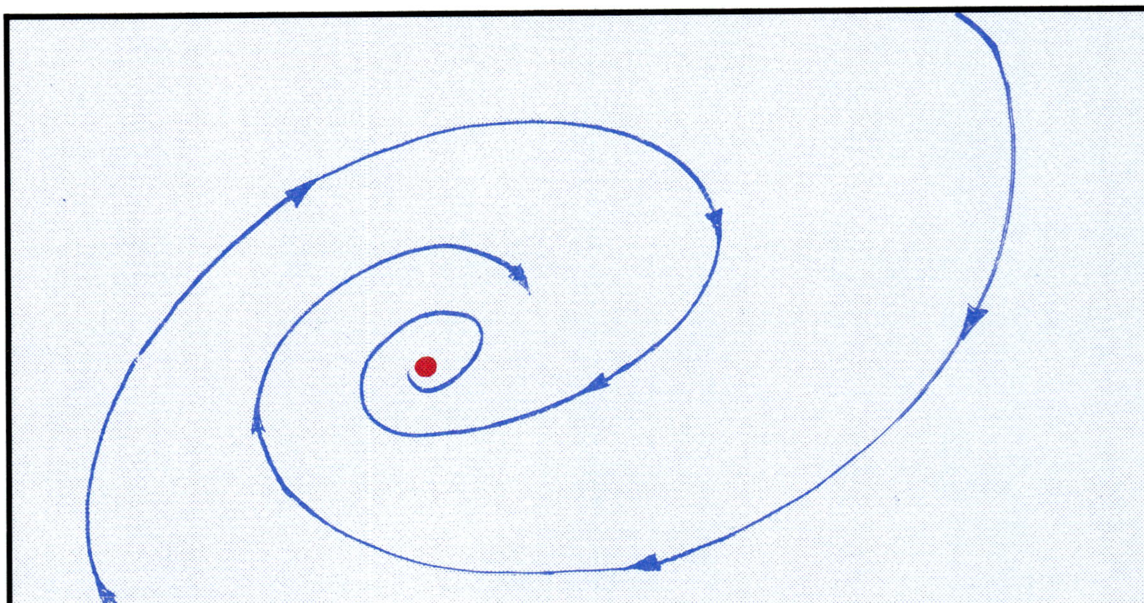

18.3.3. AFTER: This simpler phase portrait is *monostable*. Any initial state will settle to the "off" attractor as its transient dies away.

The annihilation of the periodic attractor takes place in a periodic fold catastrophe.

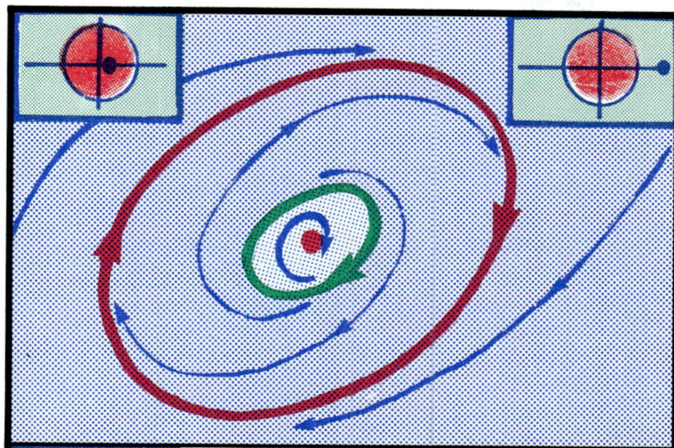

18.3.4. BEFORE: The first step is the bistable phase portrait just described. Here we have shown the CM's of each limit cycle in the windows.

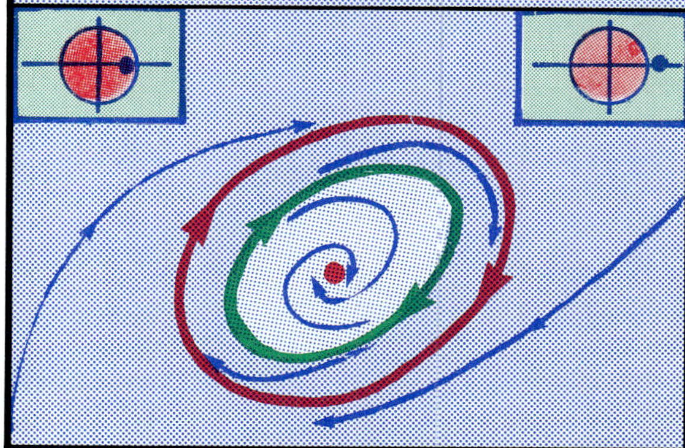

18.3.5 APPROACHING: Next, the periodic attractor and its separatrix (the periodic repellor) drift toward each other. Note the progress of the CM's toward each other as well.

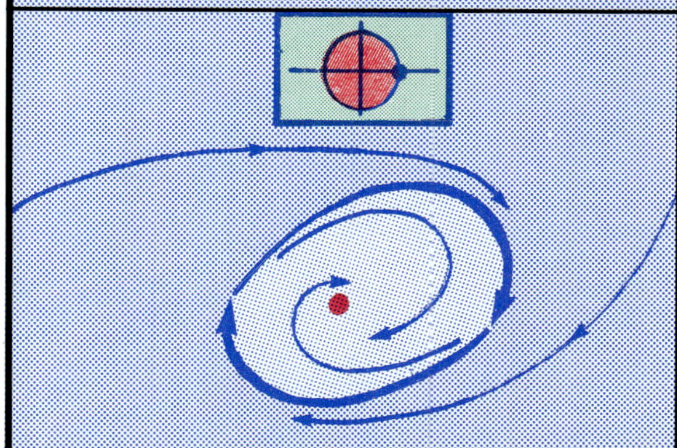

18.3.6. BIFURCATION: Conjunction of the two hyperbolic cycles creates a nonhyperbolic cycle with CM equal to one. It still bounds the basin of the point attractor.

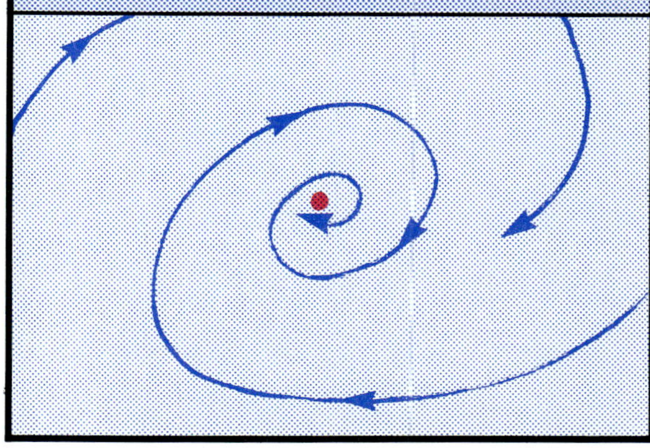

18.3.7. AFTER: No limit cycles. The whole plane is the basin of the one attractor, which has been unaffected by the control parameter.

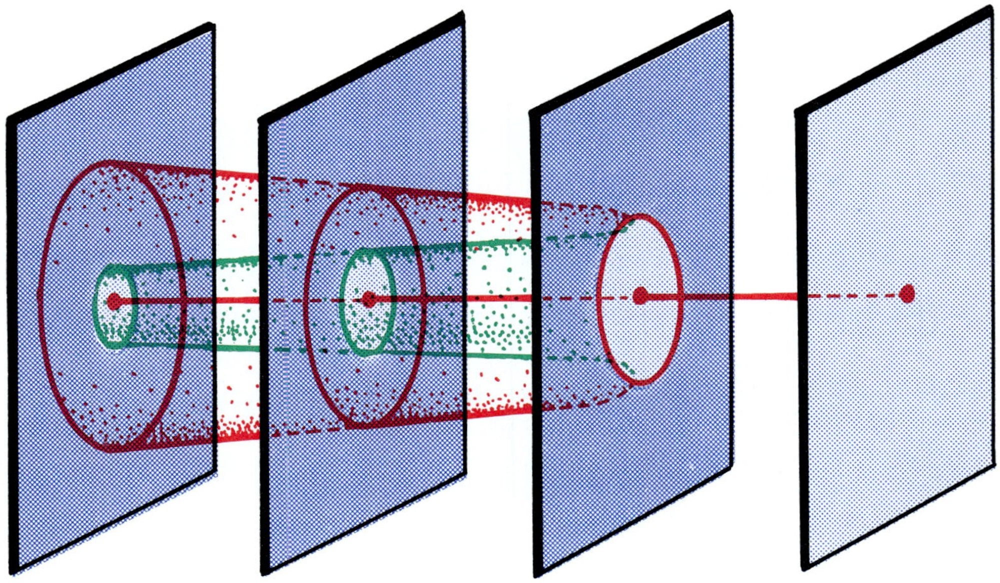

18.3.8. Here we stack the four phase portraits just described side by side in their proper positions in the response diagram of the hypothetical scheme under discussion.

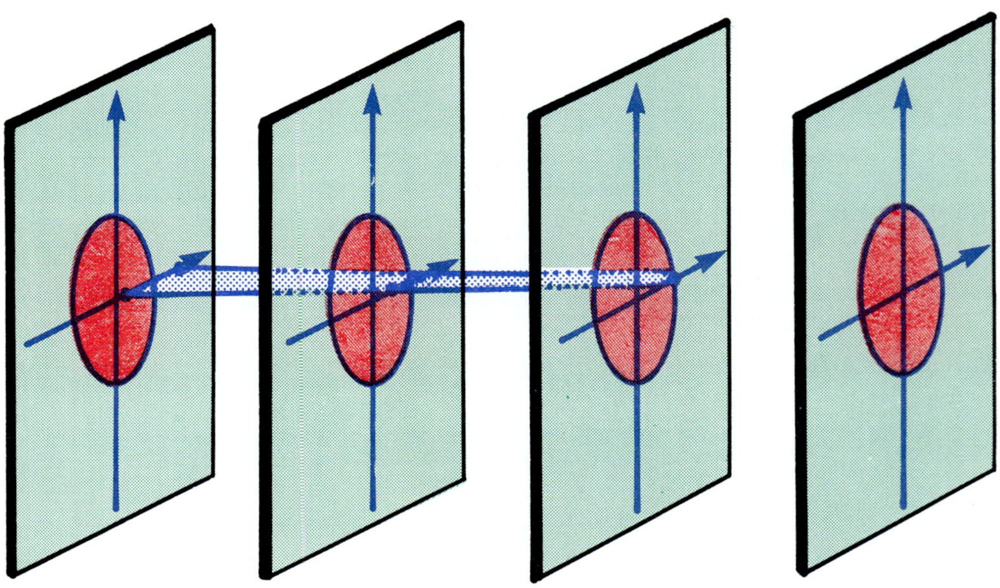

18.3.9. Here, suitably enlarged, we stack the CM's of the two limit cycles, this time in the same CM plane. Note the similarity to the CE movie of the static fold in 1D (Fig. 18.1.9).

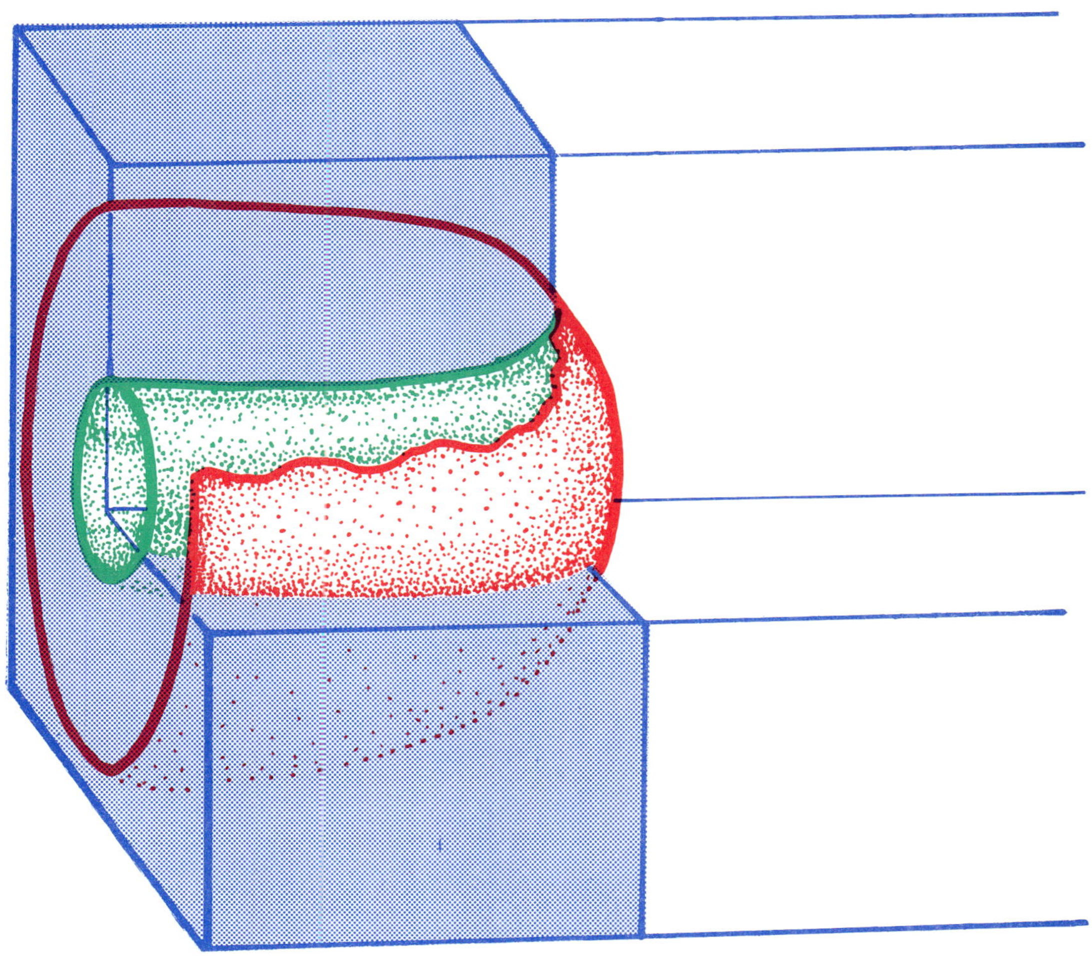

18.3.10. Finally, we interpolate a continuum of phase portraits, for the museum edition of the response diagram of this event. For easy viewing, we have omitted the blue filling denoting the locus of the blue basin. Also, we have omitted the static attractor in the center of the blue basin entirely, as it is not really part of this atomic bifurcation event.

For the response diagrams of bifurcations involving periodic attractors, it is sometimes helpful (or even essential) to extract a strobe section movie of the entire event.

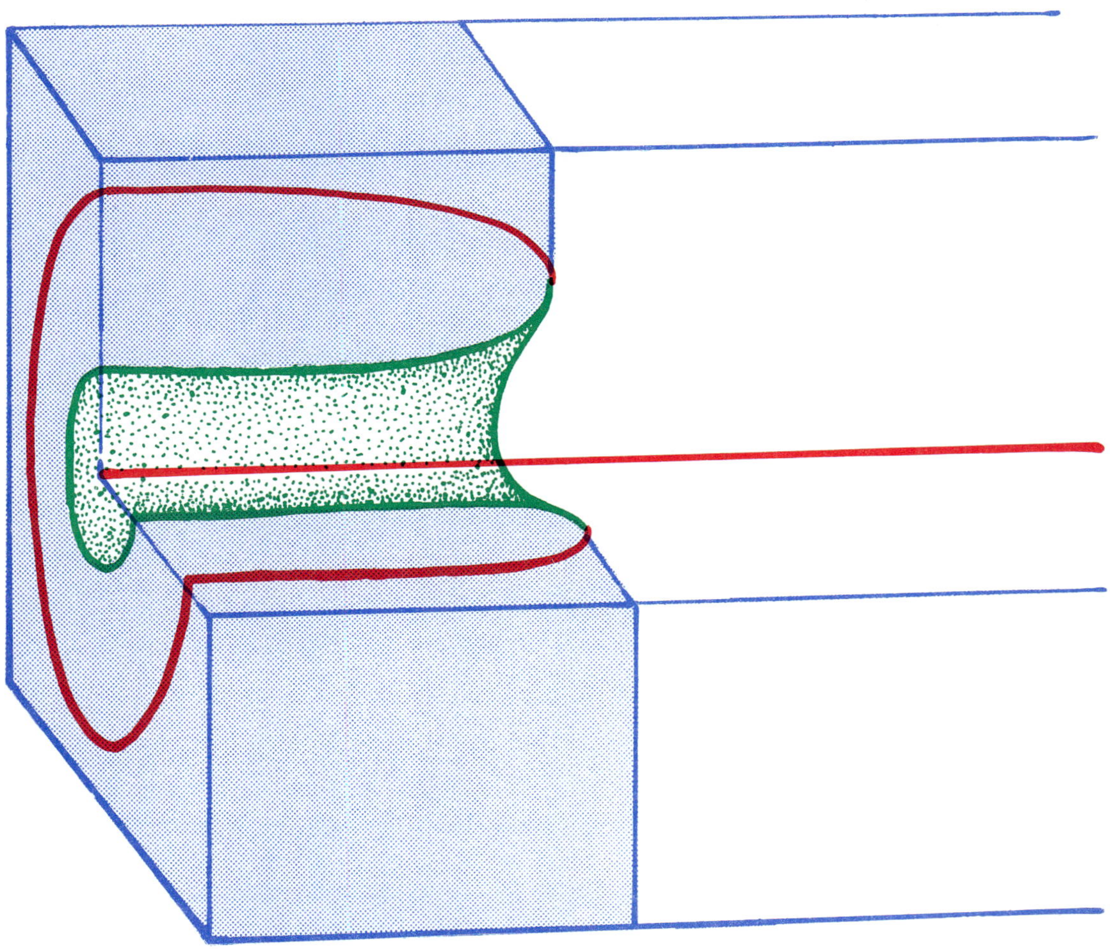

18.3.11. A *strobe section line* in each phase portrait generates a *strobe section plane* of the whole response diagram. Here we show a strobe section plane embedded within the whole response diagram as a vertical wall.

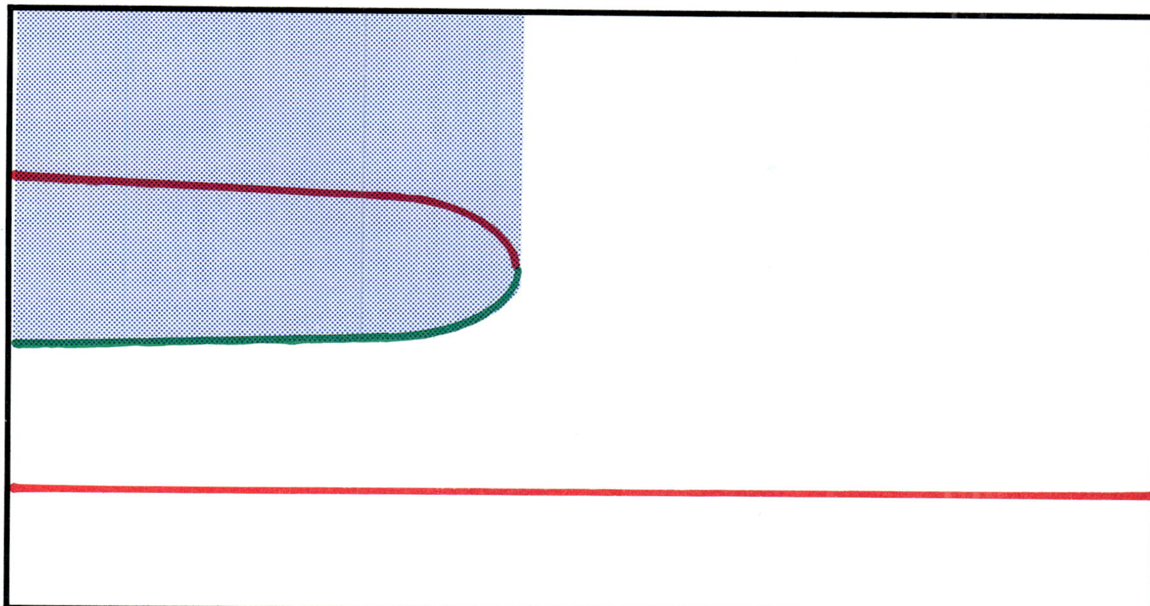

18.3.12. Here is the strobe section plane of the response diagram on its own. Note the similarity to the response diagram of the static fold in 1D, and note that rotating this figure around a circle will recreate the preceding figure.

As in the fold catastrophe in 2D, we are going to illustrate this event in a more realistic global context.

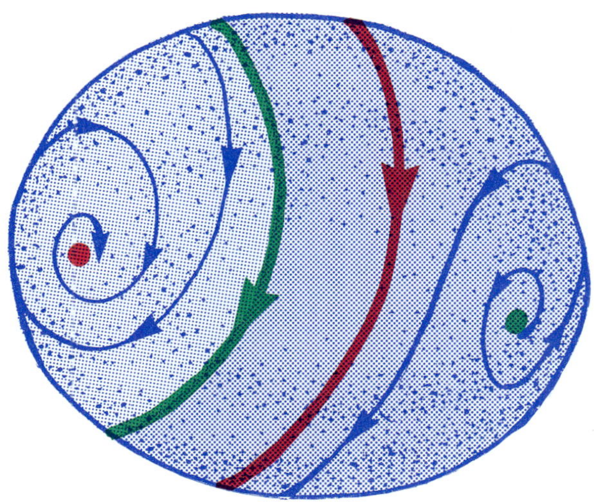

18.3.13. BEFORE: Two limit cycles and two critical points are arranged like this on the sphere. The flow is bistable: there are two basins, "on" and "off."

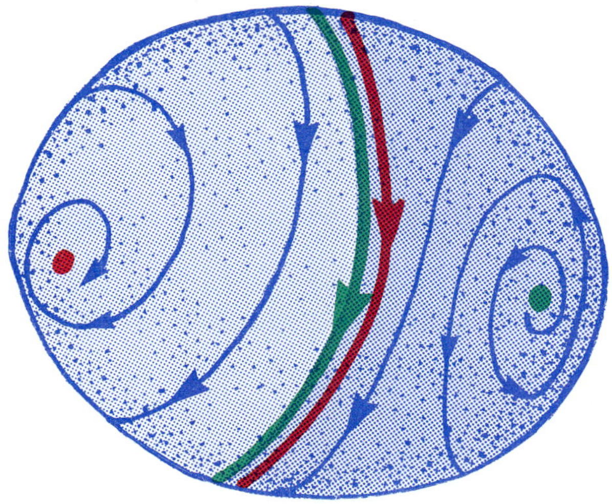

18.3.14. APPROACHING: The "on" attractor and its separatrix drift toward each other.

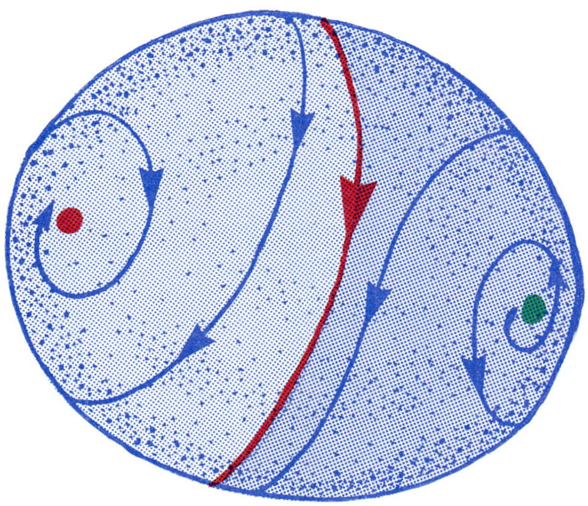

18.3.15. BIFURCATION: The conjunction of the two cycles creates a nonhyperbolic cycle, shown here as a thin red cycle.

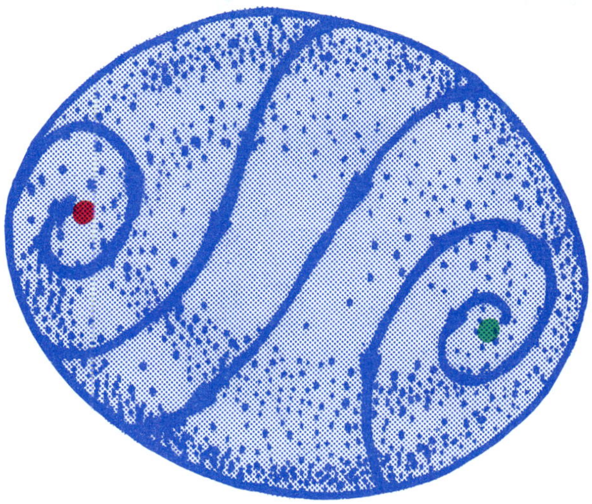

18.3.16. AFTER: The common monostable flow. The blue basin is greatly enlarged.

SUMMARY: This is a new entry for our encyclopedia, but it is very similar to the static fold. We now have five distinct atomic bifurcations on our list, in two categories:

Subtle—first excitation, second excitation, and octave jump.
Catastrophic—static fold and periodic fold.

We continue now with another version of the periodic fold.

WARNING: A strobe section is not exactly revealed by a strobe light blinking periodically. By a strobe section, or Poincaré section, we mean simply a cross-section of the flow.

18.4. Periodic Fold In 3D

This is not a new bifurcation for our list, but just another occurrence of the periodic fold. This introduces not only a more general context for this bifurcation, but also some techniques of visual representation that we will find useful in the sequel.

We need to recall the basic concepts of limit cycles in 3D.

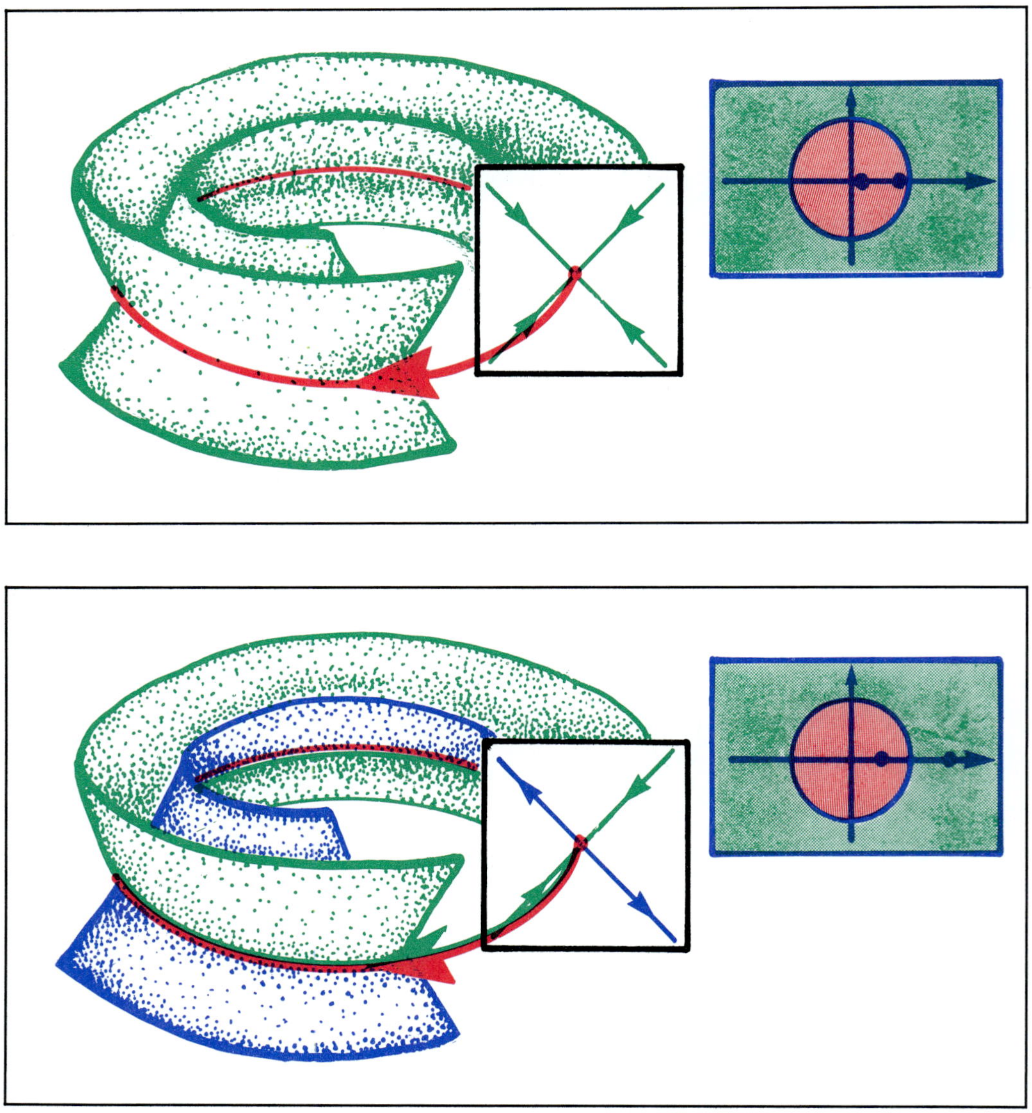

18.4.1. Here are two kinds of elementary limit cycles in 3D, along with their CM's, taken from Figure 7.5.7.

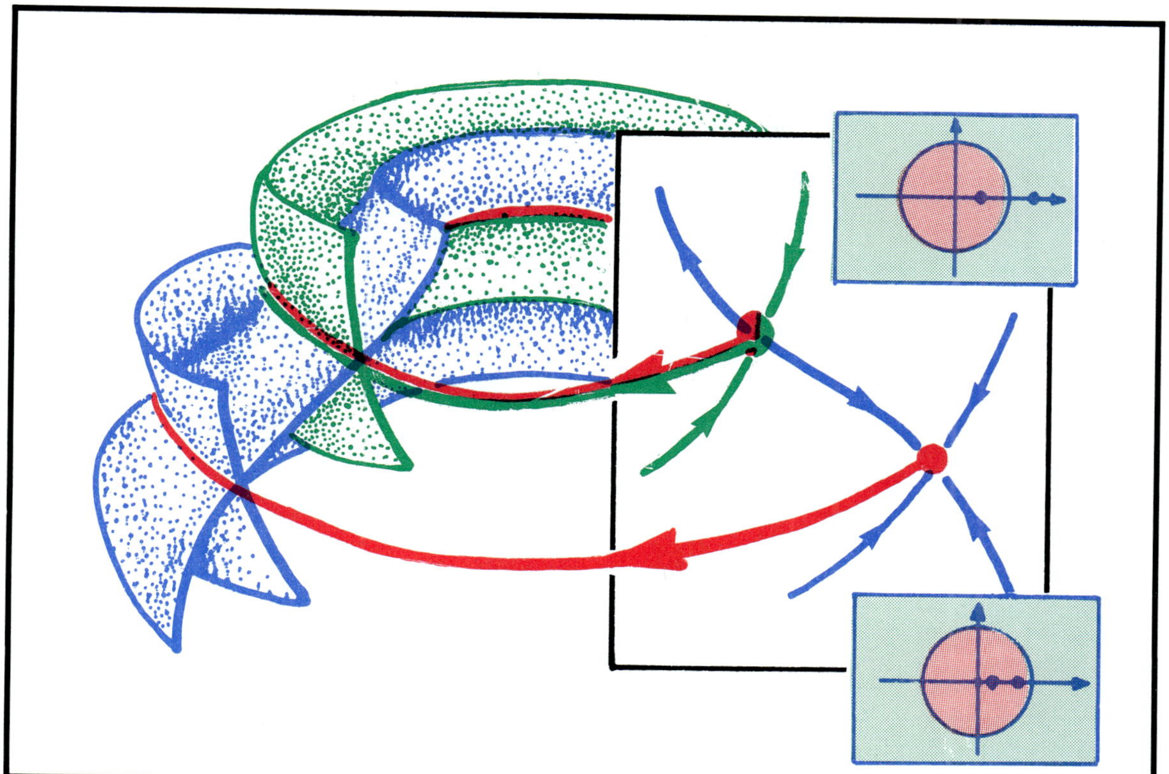

18.4.2. BEFORE: This is a portion of a flow in 3D, showing two elementary limit cycles—a saddle and an attractor—along with their CM's, and some discrete trajectories (the blue points, which belong to a continuously spiraling trajectory, are revealed by the strobe light) in the strobe section.

It will be easier to visualize this event by restricting attention to the strobe planes. But you must keep in mind that the dynamics within the strobe plane is *discrete.* That is, continuous trajectories of the 3D flow appear as a discrete sequence of points in the 2D strobe plane.

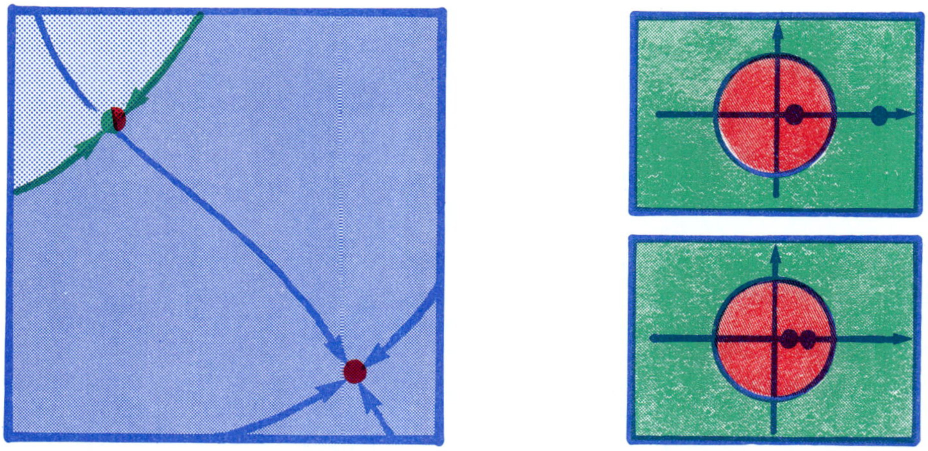

18.4.3. BEFORE, STROBED: Here is the initial configuration in the strobe plane. The green region, the strobed inset of the saddle, is the boundary (within the strobe plane) of the strobed dark blue basin of the strobed periodic attractor (red point). The light blue region is the strobe view of the basin of some other attractor, which is out of sight.

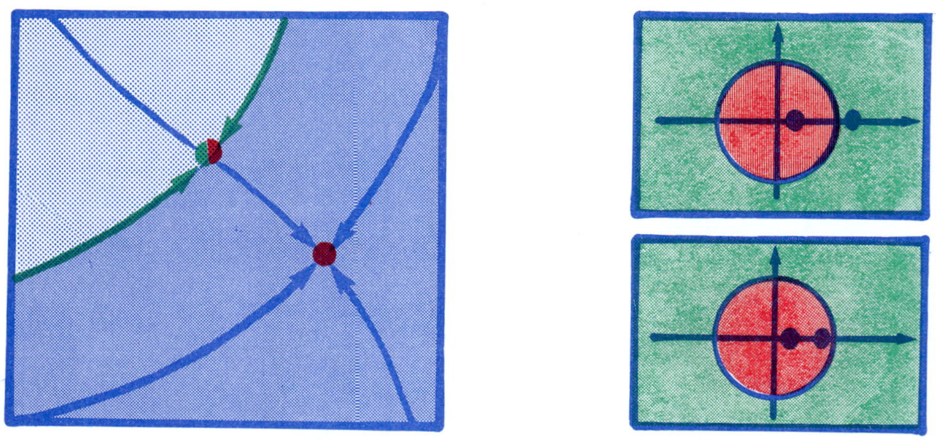

18.4.4. APPROACHING: The attractor and the saddle belonging to its separatrix move toward each other, as the control of the scheme is increased. At the same time, one of the CM's of the attractor (controlling the attraction in the north-south direction) moves outward toward the real number one on the unit circle in the CM plane. This indicates a *weakening* of the strength of attraction in this direction. The other CM of the attractor (controlling east-west attraction) is unaffected. Meanwhile the outer CM of the saddle (controlling north-south repulsion) moves inward toward the unit circle. The other CM of the saddle is unaffected.

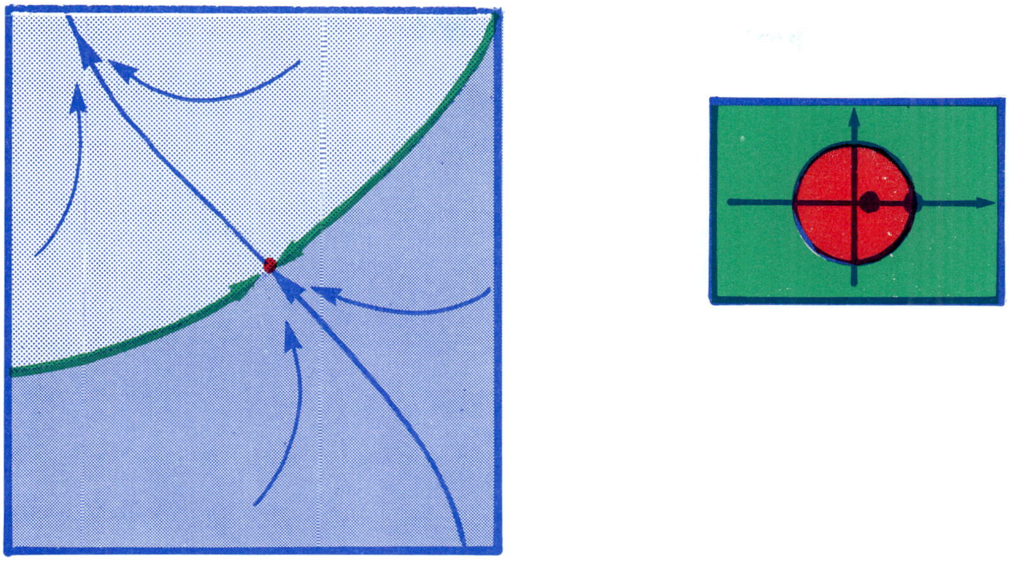

18.4.5. BIFURCATION: At the moment of conjunction, the two points (and the entire limit cycles they represent) coincide. This single limit cycle is nonhyperbolic, as one of its CM's is *one*, on the unit circle.

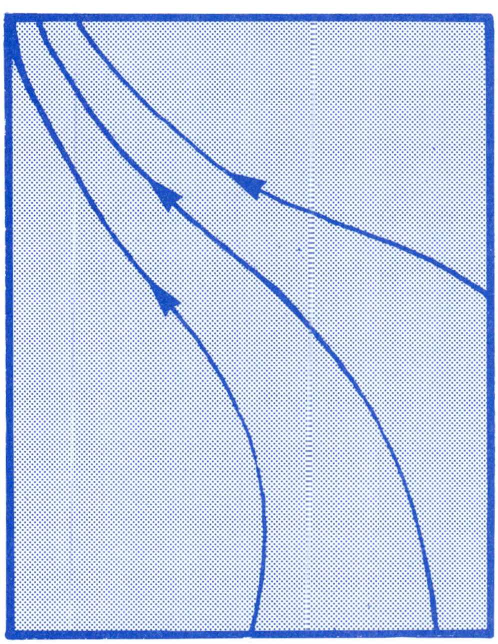

18.4.6. AFTER: No limit cycles. All points in the strobe plane belong to the blue basin.

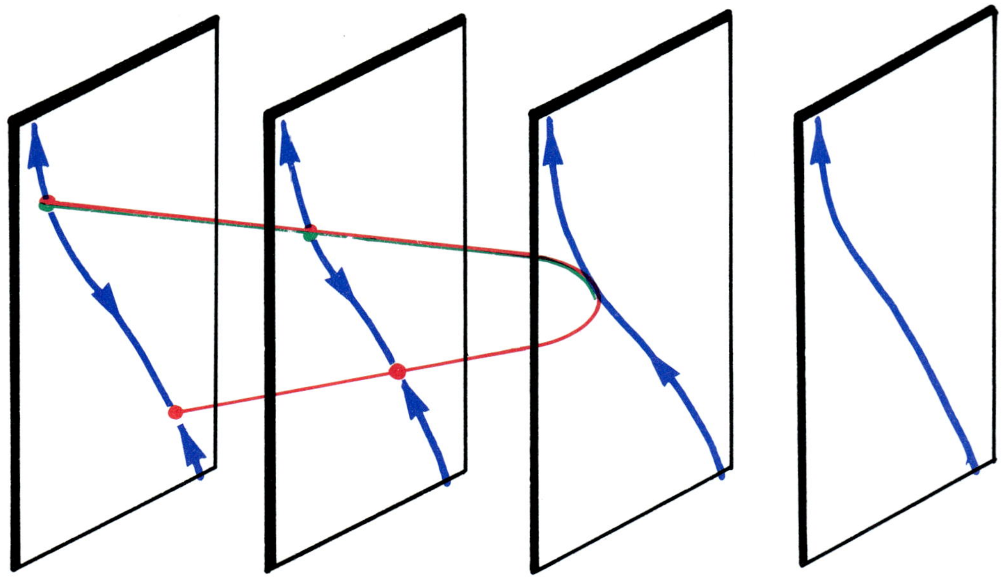

18.4.7. Here are the strobed portraits, erected side-by-side in their proper places within the (strobed) response diagram (recall Figure 18.3.12). Note the parabolic meeting common to all of the fold catastrophes.

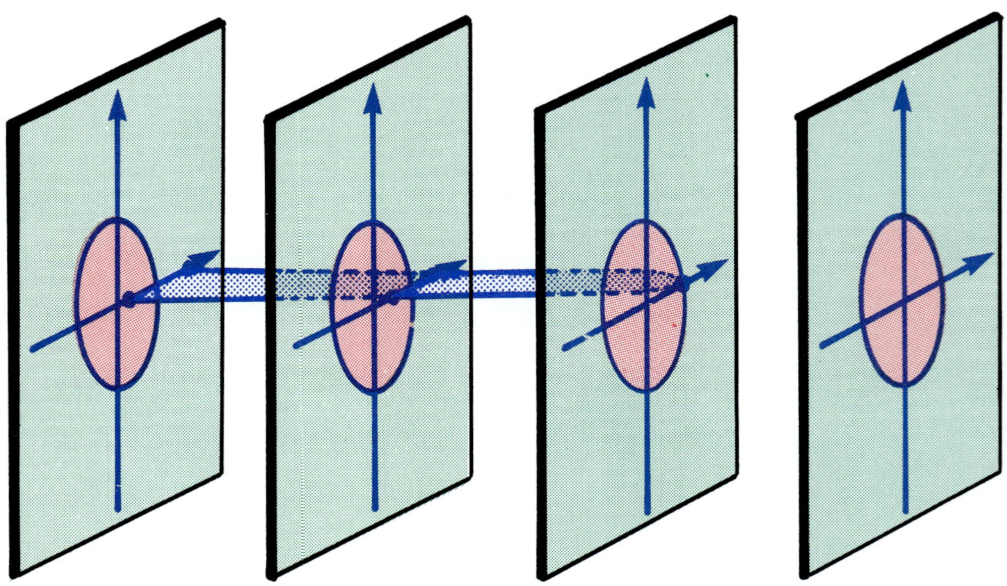

18.4.8. And here is a composite view of the action in the CM plane, showing the affected CM's of each limit cycle only. As the control moves to the right and the two limit cycles move toward each other, the two CM's approach *plus one* on the real axis. (Compare with 18.3.9.)

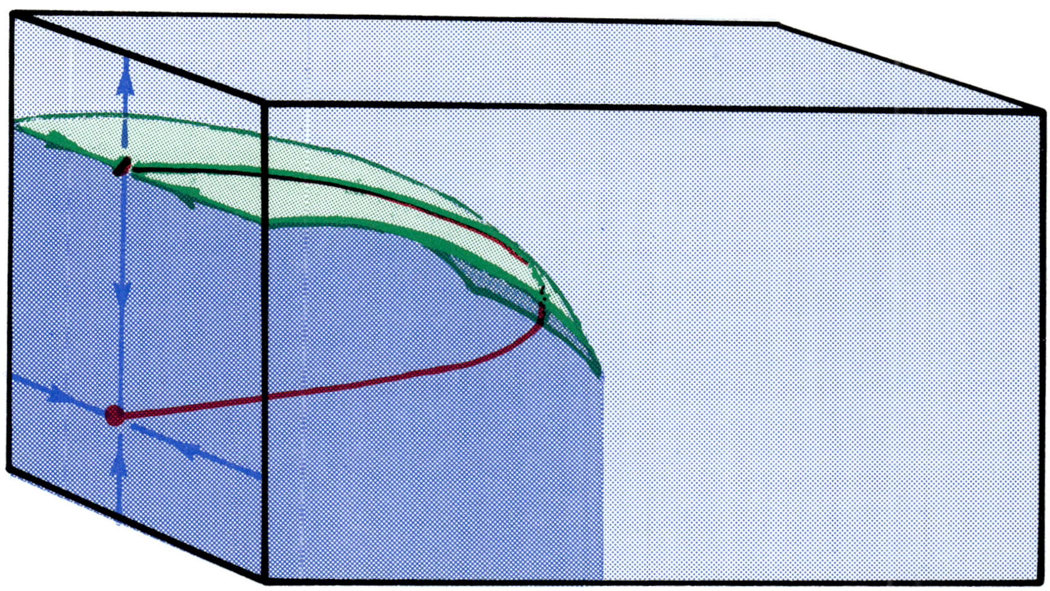

18.4.9. Finally, interpolation of a continuum of strobe planes in the strobed response diagram yields this memorable version (compare 18.2.6).

SUMMARY: This 3D version is harder to visualize than the 2D version of the periodic fold catastrophe. This atomic bifurcation may occur in 4D, and higher dimensions as well. Also, it may occur in much more complicated global phase portraits. It is particularly common in the dynamics of forced oscillators, as we have explained in Section 5.5.

19

Pinch Catastrophes

We now discover some new events for our atlas by reversing the direction of time. Thus, insets become outsets, attractors become repellors, and so on. In this chapter we systematically reverse the four subtle bifurcations of Chapter 17, obtaining a new catastrophe in each case.

19.1. Spiral Pinch In 2D

We begin with the first excitation in 2D, otherwise known as the Hopf bifurcation. What happens when we reverse the direction of time?

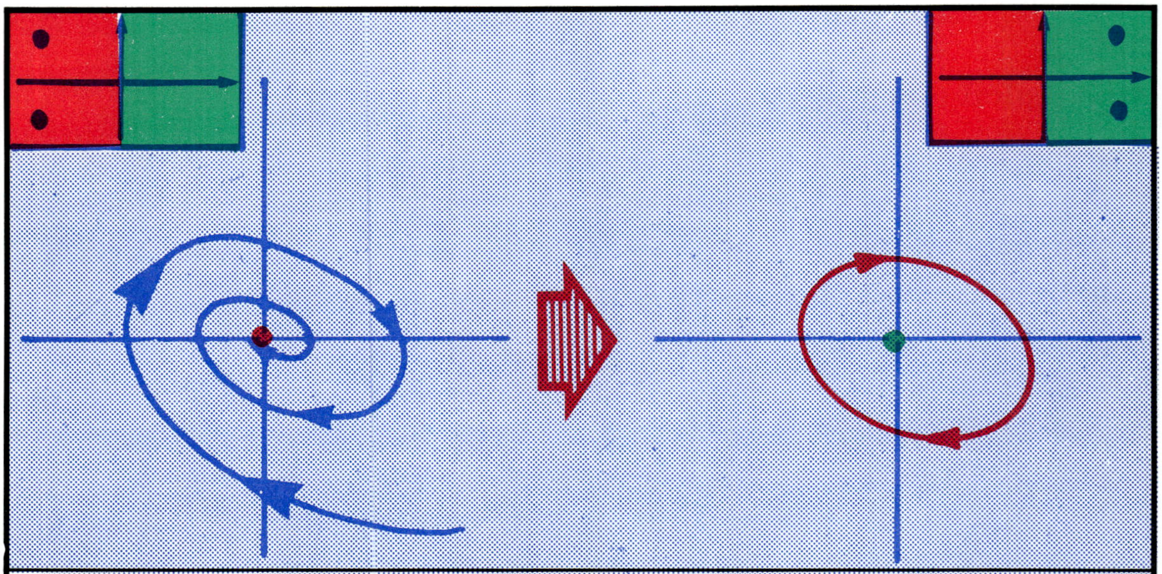

19.1.1. Recall, from Section 17.1, that in this event a point attractor of spiral type expands parabolically into a periodic attractor. The critical point turns into a point repellor of spiral type (a virtual separatrix) as its CE's transit the imaginary axis of the CE plane, from the red to the green region.

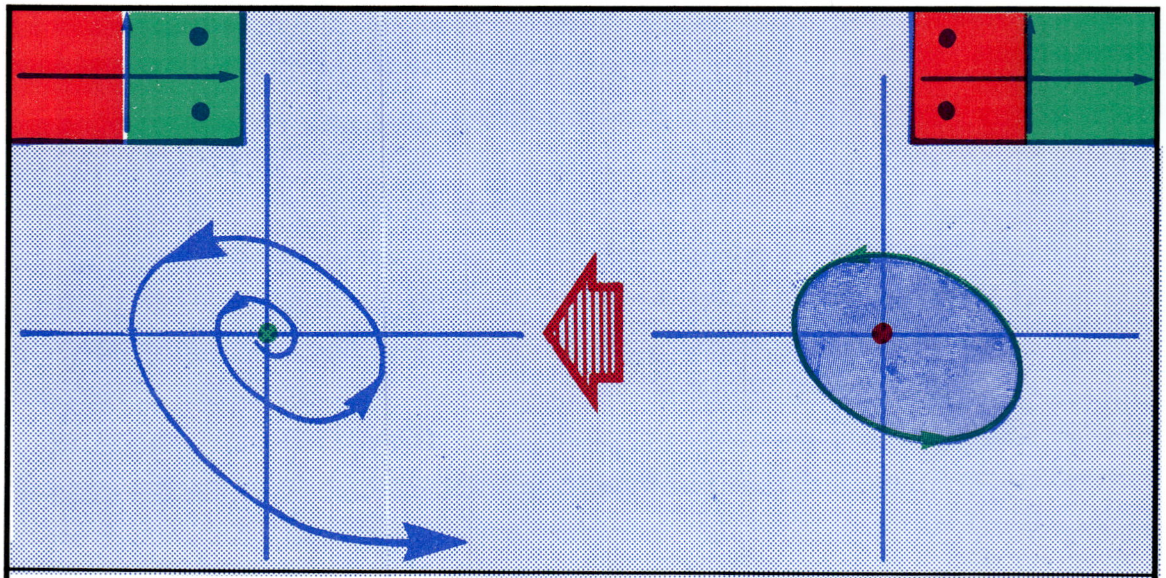

19.1.2. Reversing the direction of flow of all trajectories, we obtain a point repellor of spiral type on the left. We will take this portrait as the final one of the new event. On the right, a point attractor of spiral type is surrounded by a periodic repellor, the actual separatrix of its basin (blue). We will consider this one the initial portrait of this event.

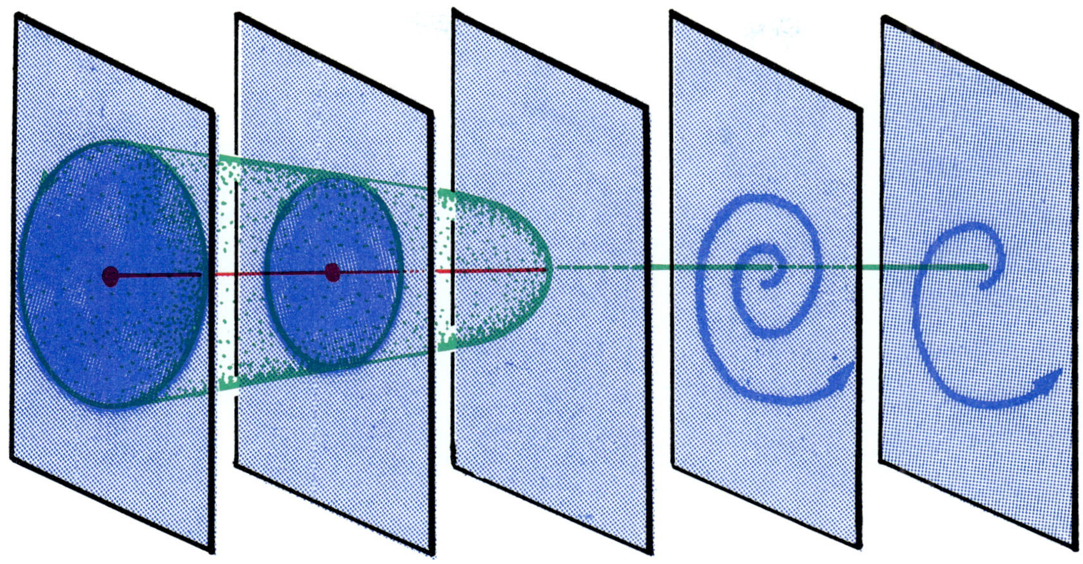

19.1.3. Erecting these portraits in their proper positions in the response diagram and interpolating a few others, we obtain this side-by-side skeleton.

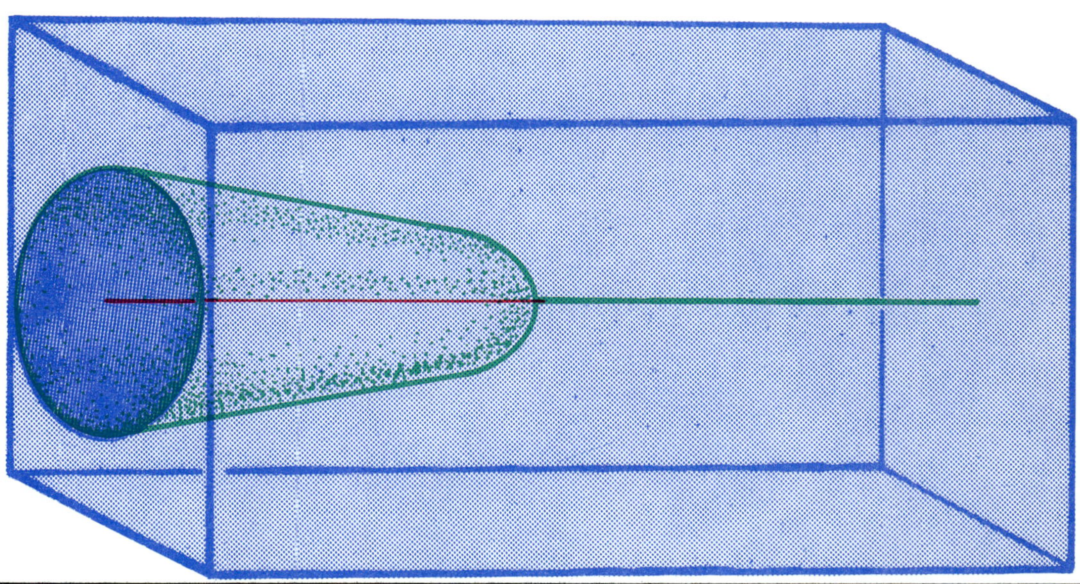

19.1.4. Stripping off the CE details and shading in the continuous locus of attraction, we obtain this image of the spiral pinch event in 2D. Note that as the control is increased to the right. the separatrix (and basin) shrinks down to the point attractor, the strength of which is dwindling as well. At the moment of bifurcation, the actual separatrix becomes a virtual separatrix, replacing the point attractor, which has catastrophically vanished.

SUMMARY: By reversing the direction of flow in a subtle bifurcation, first excitation, we have obtained a new catastrophic bifurcation for our atlas. The spiraling 2D basin shrinks and pinches off its central attractor. Separatrix, basin, and attractor all vanish at once.

19.2. Vortical Pinch In 3D

Reversing direction in first excitation was easy, so we will repeat the operation for the second excitation. Once again, time reversal will create a catastrophe from a subtle bifurcation.

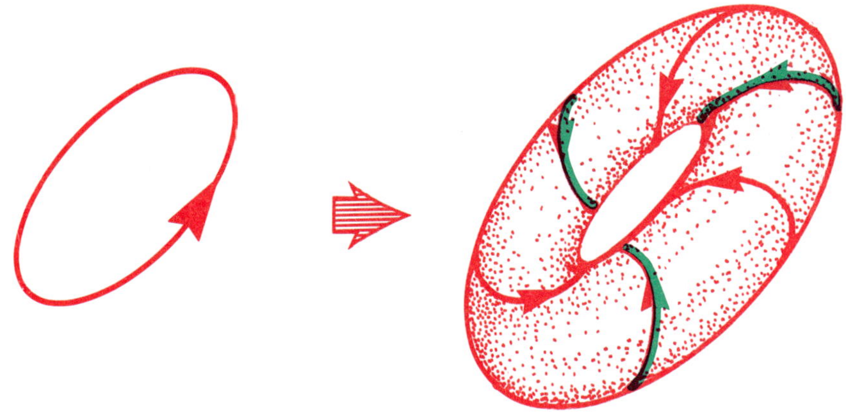

19.2.1. Recall this event in 3D from Section 17.2, in which a periodic attractor is transformed into a periodic repellor within a braided attractive invariant torus or AIT.

Technically, this event is not a single bifurcation. Instead, it consists of a fractal family of bifurcations. As in Section 17.2, we will not dwell on this complication, which concerns the braid dynamics on the AIT, but just concentrate on the central cycle and the AIT.

And now, about face! Attractors become repellors, and so on.

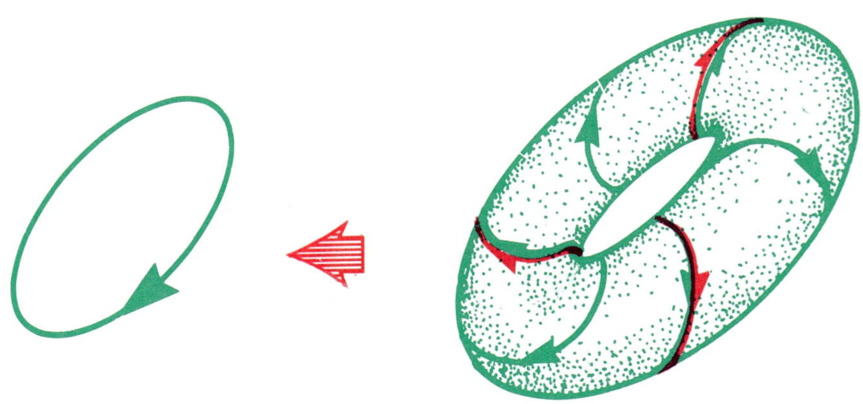

19.2.2. On the left, the periodic attractor of spiral type has become a repellor. And on the right, AIT has become a repulsive invariant torus (RIT). As in the preceding section, we will regard the portrait on the right as the initial configuration.

Now let's strobe these portraits, to cut things down to 2D.

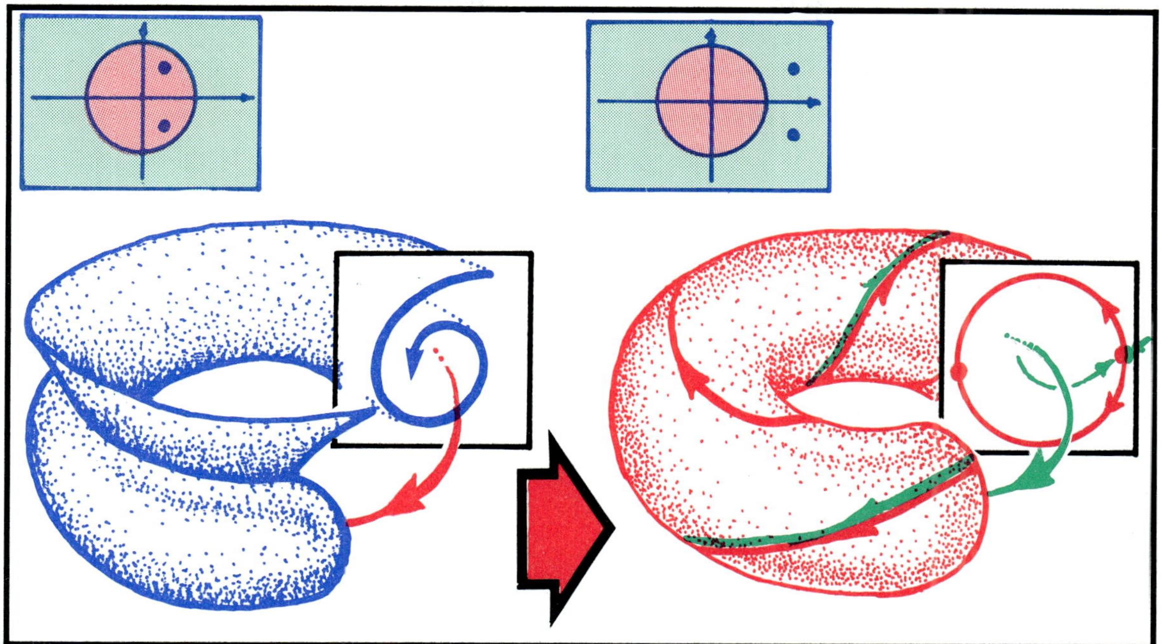

19.2.3. Here again is second excitation, showing the strobe plane, along with the CM's of the central cycle.

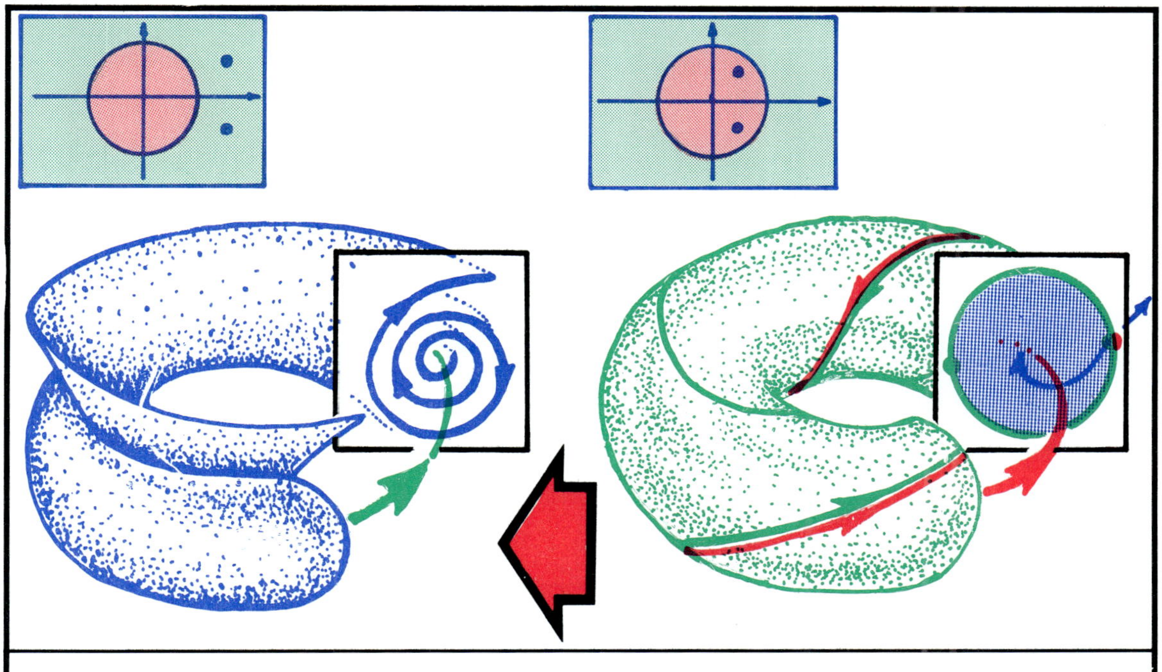

19.2.4. And here is the same portrait, with time reversed, showing similar details.

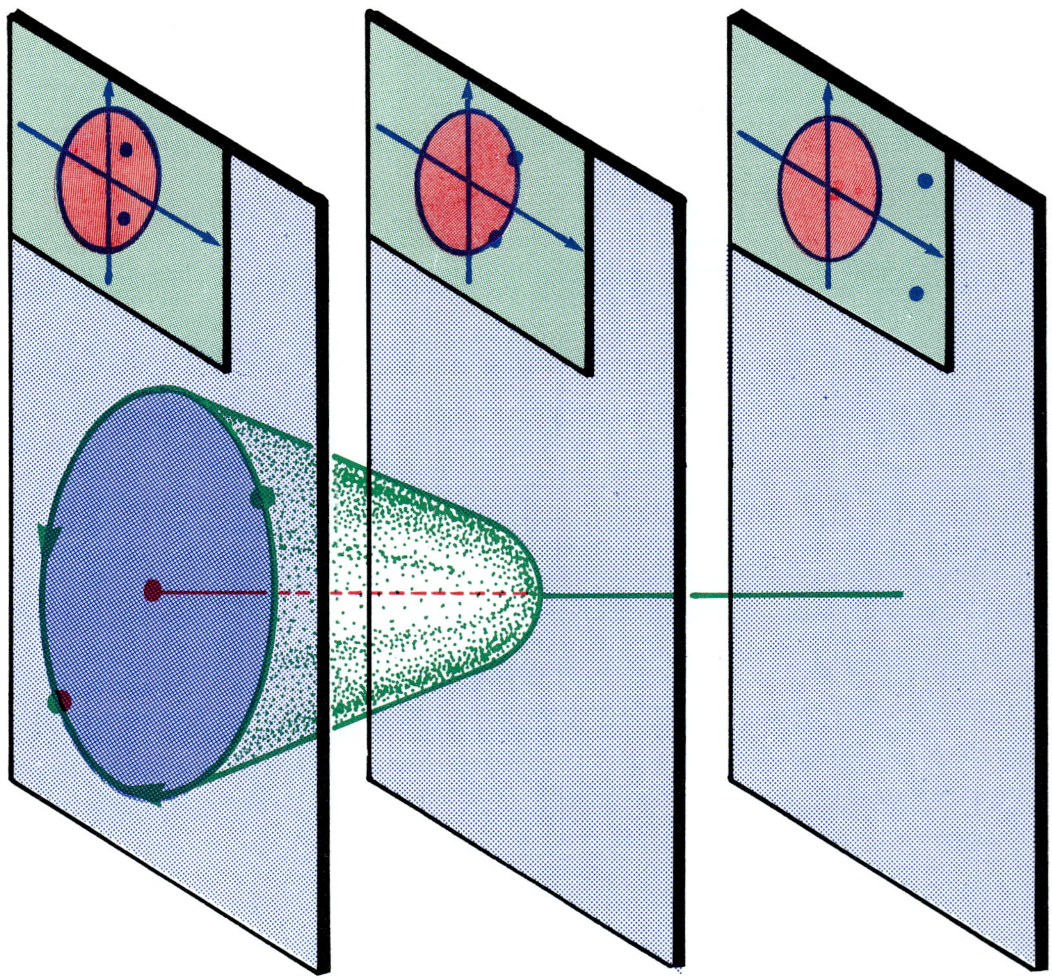

19.2.5. Erecting the two strobe plane portraits in their proper positions in the space of the strobed response diagram, and interpolating a few in-betweens, we have the skeleton of the strobed response diagram.

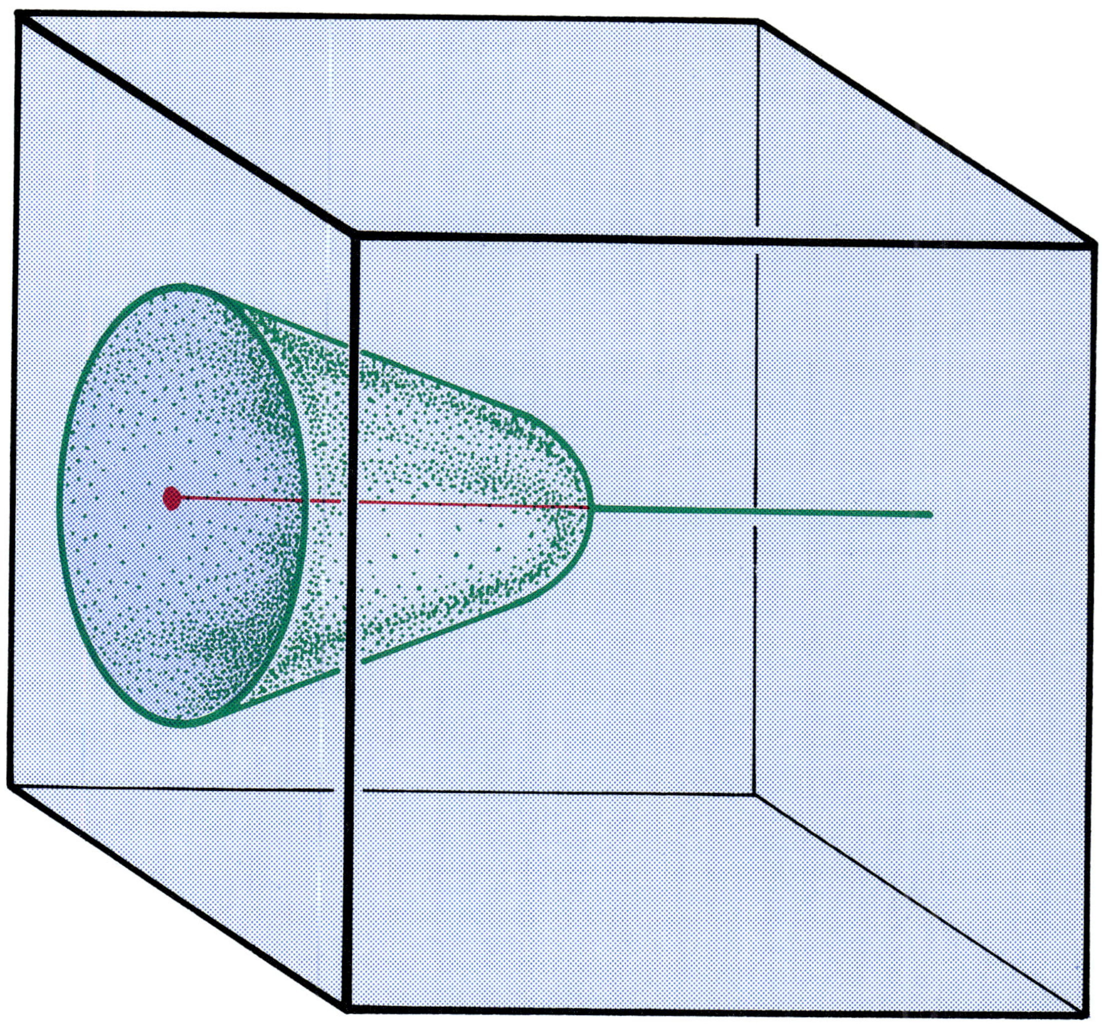

19.2.6. Stripping off the details and filling in the loci of attraction and repulsion, we have this framable image of the vortical basin pinching down to destroy its central attractor.

SUMMARY: Reversal of a subtle bifurcation has again given us a new catastrophe for our atlas of response diagrams of atomic bifurcation events. As long as there is an invariant torus in the portrait, we must expect a fractal set of braid bifurcations on it. But in this case, they do not affect the locus of attraction. The green RIT shrinks, and catastrophically pinches off the central periodic attractor, as the control is moved to the right.

19.3. Octave Pinch In 2D

Recall that the octave jump phenomenon involved a periodic attractor, as one of its CM's transits the unit circle at *minus one.* In the 2D version presented in Section 17.3, the affected limit cycle turns into a periodic repellor, and a period-doubled attractor is emitted. The state space is necessarily a Möbius band.

What happens to the octave jump if the direction of time is reversed?

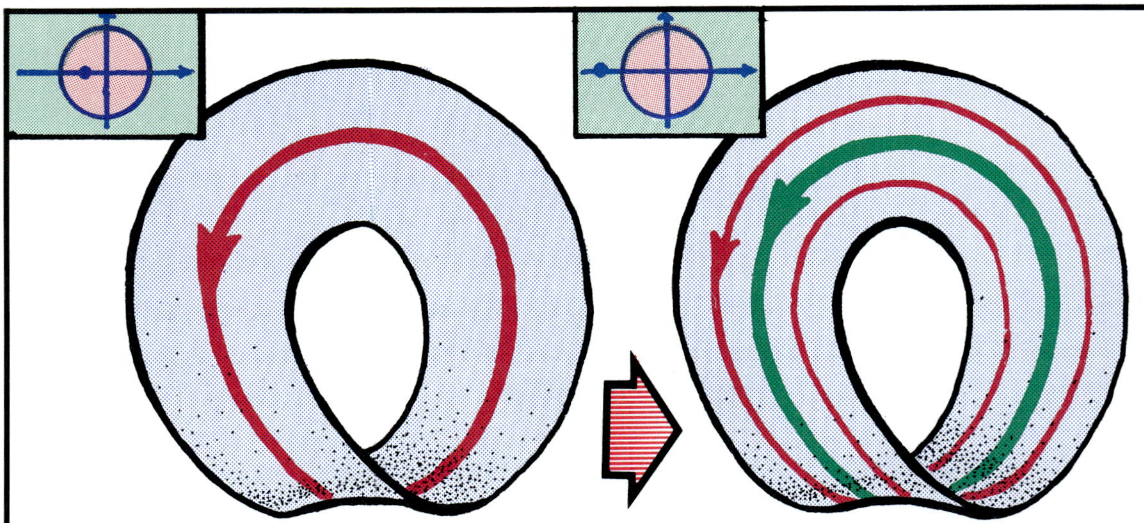

19.3.1. Here are the BEFORE and AFTER portraits, recalled from Section 17.3. Note that the basin is not substantially changed by the bifurcation, it is almost the entire band. After the event, the single-period repellor (green) is a *virtual separatrix.* This means that while technically it is a separatrix (that is, it does not belong to any basin), it does not actually separate two distinct basins. (See Section 1.5 and Section 10.2.)

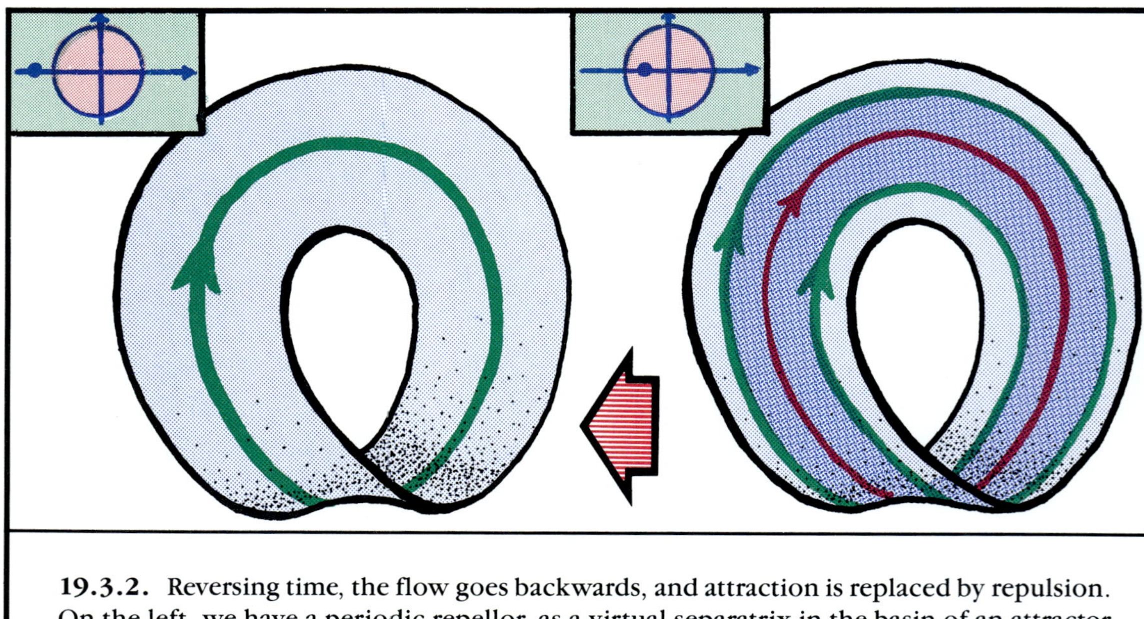

19.3.2. Reversing time, the flow goes backwards, and attraction is replaced by repulsion. On the left, we have a periodic repellor, as a virtual separatrix in the basin of an attractor out of view. On the right, we have a single-period attractor in the dark blue basin bounded by the double-period repellor, an actual separatrix. We will now regard the portrait on the right as the initial configuration.

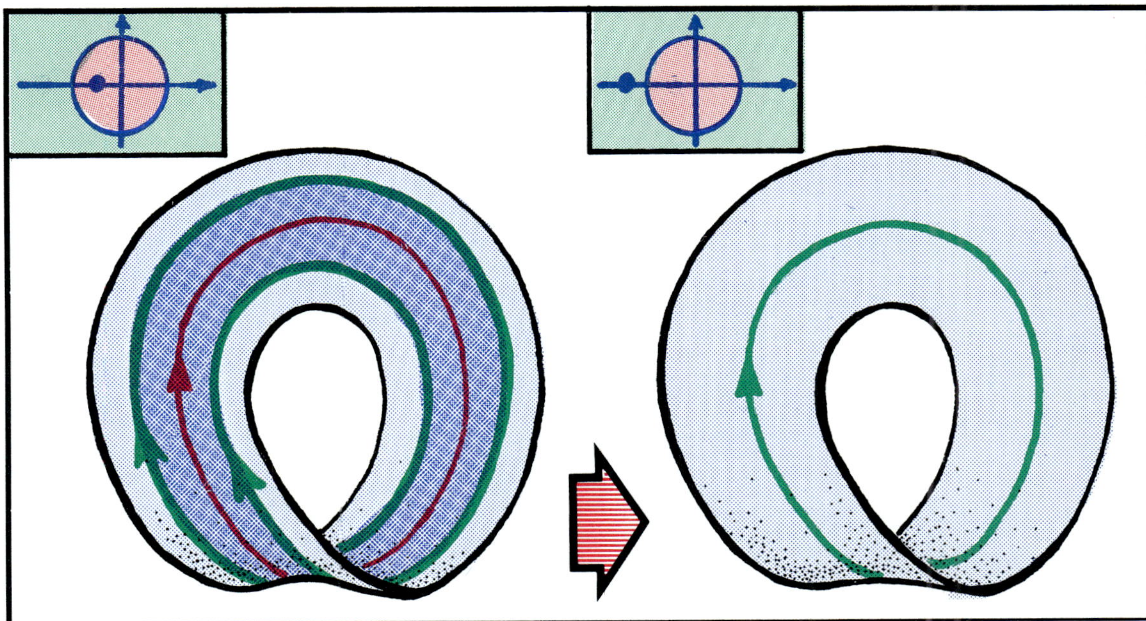

19.3.3. From this perspective, we see a periodic attractor disappear as the control parameter increases. The actual separatrix of the dark blue basin contracts toward its weakening attractor and pinches the entire basin down to a meager repellor and virtual separatrix. The CM's shown are for the central cycle only.

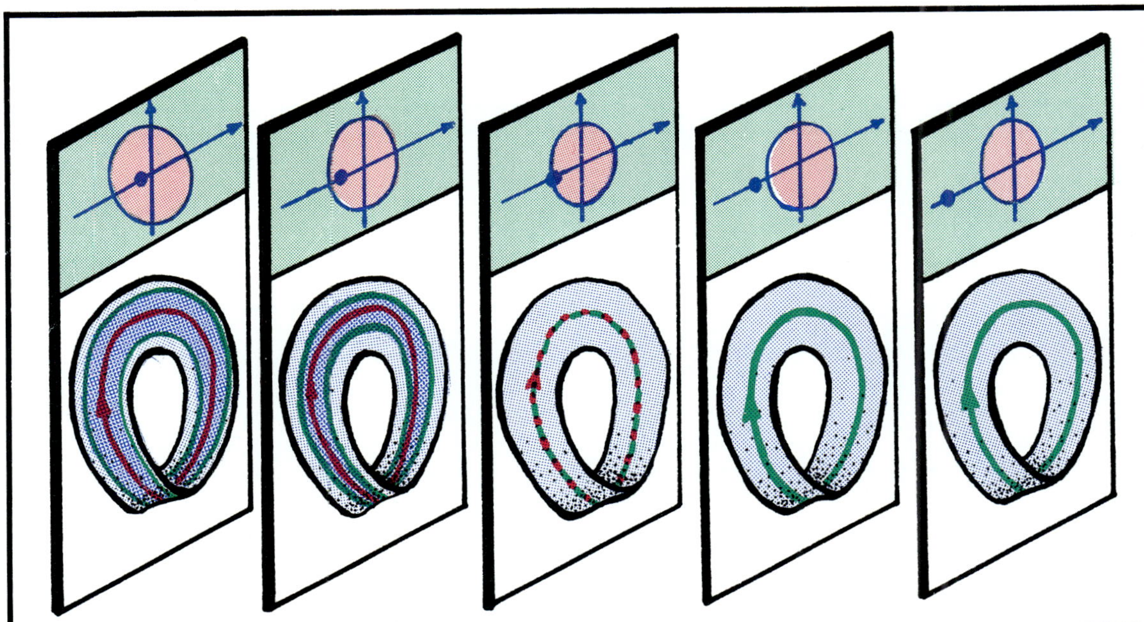

19.3.4. Erecting these two portraits into a side-by-side representation and interpolating three more, we obtain this skeleton of the octave pinch event.

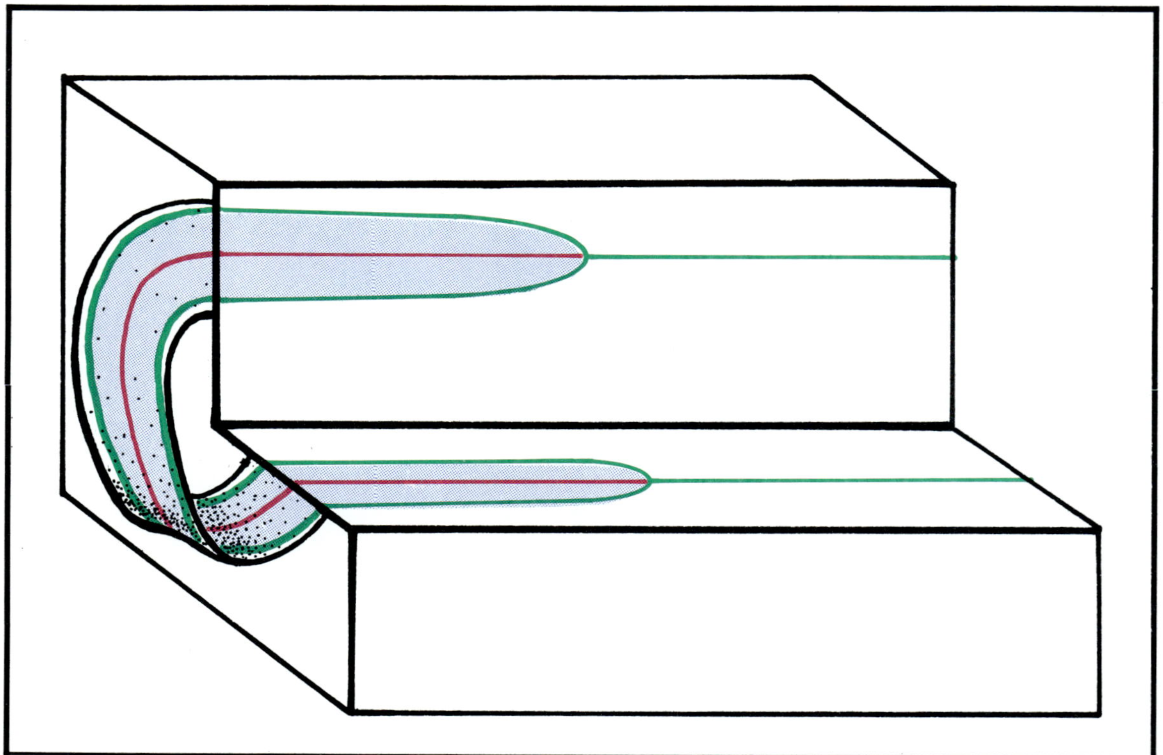

19.3.5. Interpolating a continuum, we fill out the skeleton to create this smoothed and cutaway response diagram. Here we have cut away a segment of the diagram for better viewing and also to suggest the strobed response diagram, in which each full phase portrait is replaced by a strobe section line.

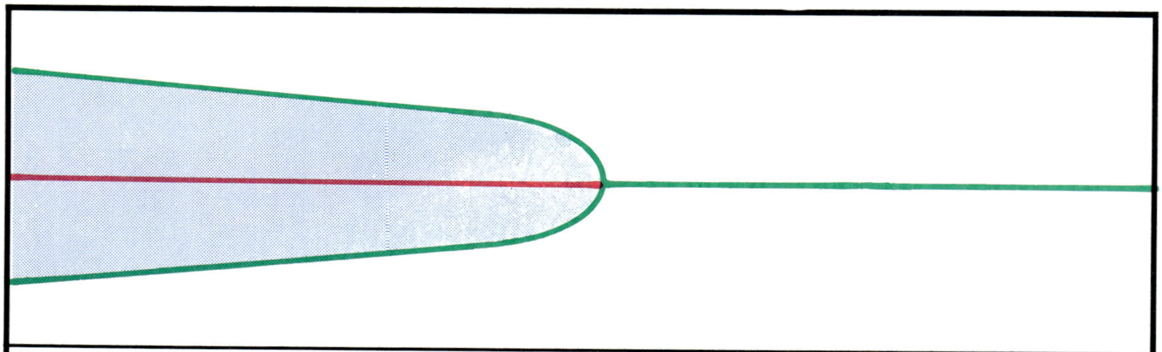

19.3.6. Here is the strobed response diagram. Unlike some similar appearing diagrams shown earlier, the two branches of the green curve both correspond to the same periodic repellor.

SUMMARY: Running the 2D octave jump backwards yields a new entry for our atlas of generic bifurcations with one control. This is our third example of a pinching catastrophe. The basin and separatrix of a periodic attractor shrink, and eventually pinch off the attractor, leaving a green shadow (periodic repellor) in its place.

19.4. Octave Pinch In 3D

As we have seen earlier, an atomic bifurcation event may present itself in contexts of different dimension. Nevertheless, we may regard these presentations as essentially the same event. One way to increase the dimension of a presentation is to embed the phase portraits in a larger state space as an attractive, invariant subspace. We have referred to this previously as the *extension* construction. For example, the octave jump in 3D is the extension of the octave jump in 2D.

Similarly, the octave pinch in 2D may be extended to obtain the octave pinch in 3D.

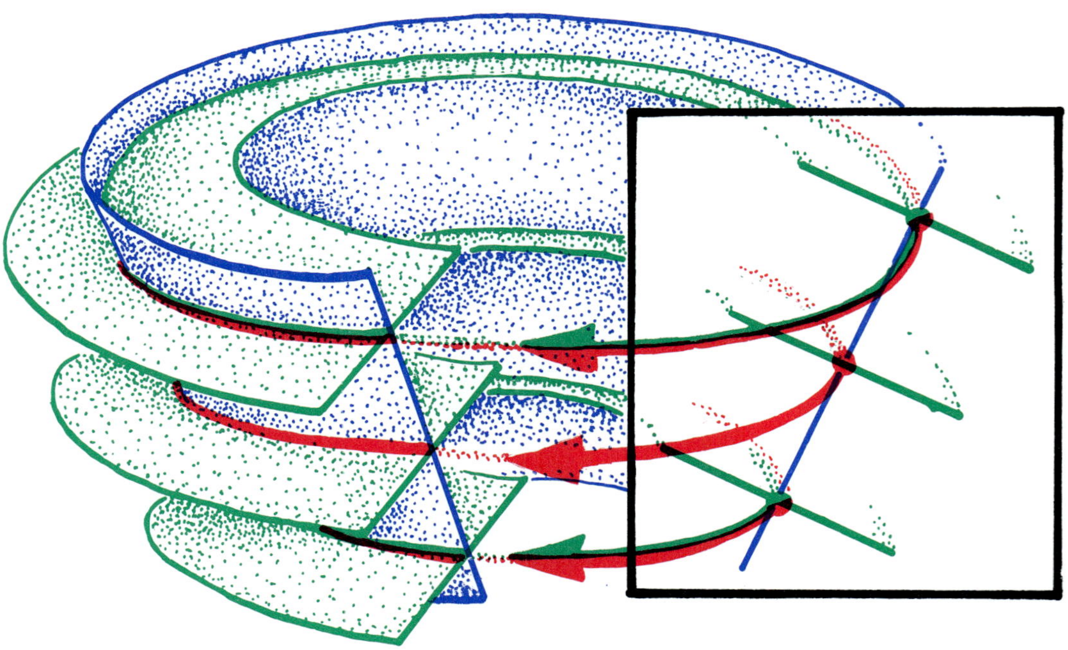

19.4.1. BEFORE: Here is the initial portrait of the 2D octave pinch, from Figure 19.1.3, embedded as an attractive blue band in a 3D flow. The single-periodic attractor within the 2D context becomes a single-periodic 3D attractor after the embedding. But the double-periodic repellor of the 2D band becomes a double-periodic saddle after the embedding, as it is attractive in the north-south direction but repelling in the east-west direction.

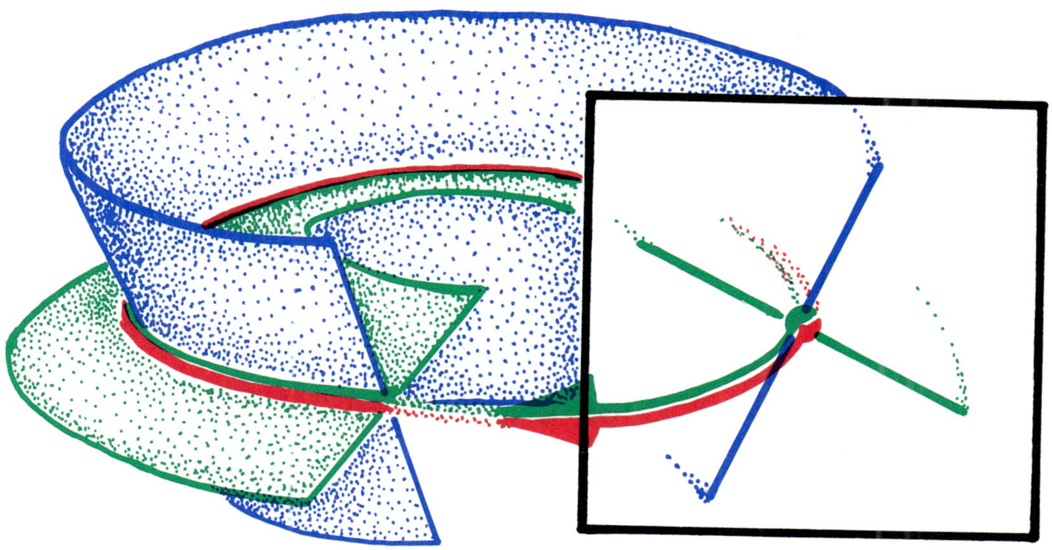

19.4.2. AFTER: And here is the final portrait of the 2D octave pinch, from Figure 19.1.3, as the same blue band. The attraction of the 3D flow to the blue band is unaffected by the control parameter. The single-periodic repellor of the 2D band becomes, after embedding, a single-periodic saddle.

We may create space for the indication of additional detail by extracting the strobe sections.

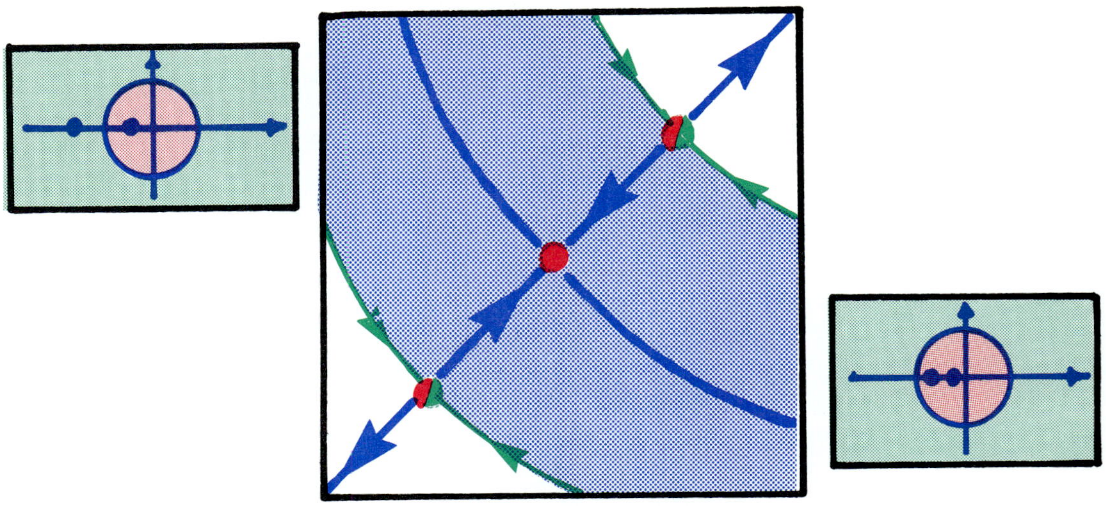

19.4.3. BEFORE: Here is the strobe section plane of the initial 3D flow. The inset band of the saddle (green) cuts the section in two disjoint line segments. These comprise the actual separatrix of the blue basin of the central attractor. The CM's of each limit cycle are shown in their own windows.

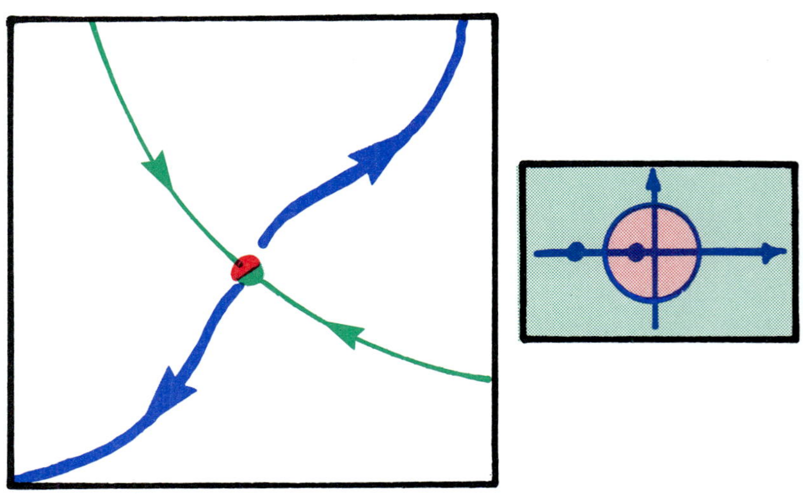

19.4.4. AFTER: Here is the strobe plane of the final portrait, showing the CM's of the solitary limit cycle. The inset is a virtual separatrix, all that remains of the former attractor, basin, and actual separatrix after the pinch.

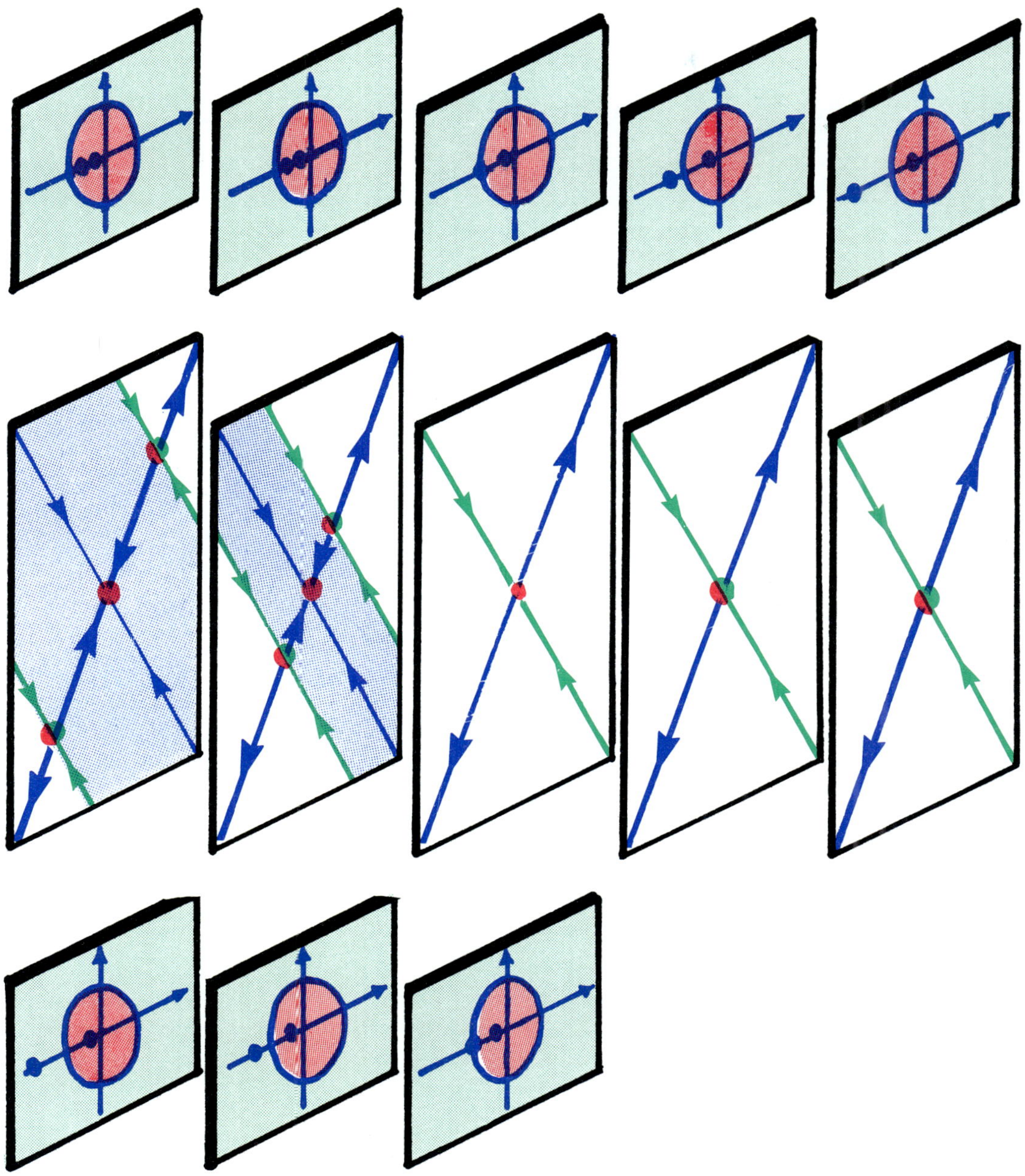

19.4.5. Erecting these portraits with interpolations, we have the side-by-side representation of the response diagram of the 3D octave pinch.

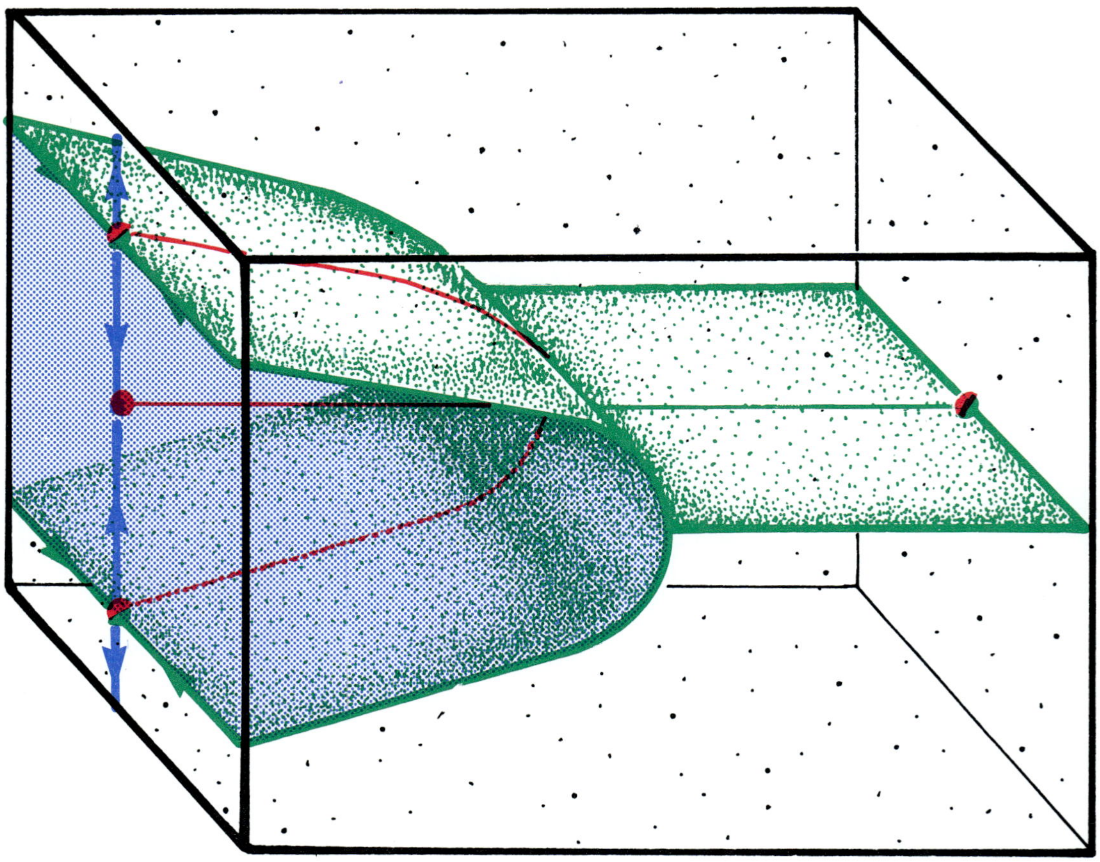

19.4.6. Omitting details and smoothing in the interpolations gives us this cutaway representation for the gallery.

SUMMARY: The 3D octave pinch is not a new entry for the encyclopedia of bifurcations, but simply another presentation of the 2D pinch. Again, a basin pinches down to a virtual separatrix, and its attractor is catastrophically lost.

20

Saddle Connection Bifurcations

This is the third and final chapter on catastrophic bifucations. The fold and pinching bifurcations are rather similar types. In a fold, an attractor drifts toward its separatrix, where it collides with a peer, a similar actor. In a *pinch,* a separatrix squeezes down on an attractor, and the collision involves dissimilar actors. But both types are *local* events: the action takes place in the immediate neighborhood of the affected attractor. We are now going to consider some *global* bifurcations, in which the tangling of insets and outsets create large-scale consequences. The concepts of global behavior from Part Three will be indispensable.

20.1. Basin Bifurcation In 2D

We have encountered this phenomenon previously, under the name *saddle switching,* in Figure 13.1.1. The structural instability, at the moment of bifurcation, is caused by the violation of *generic condition G3* by a *saddle connection* in 2D (see Section 11.1).

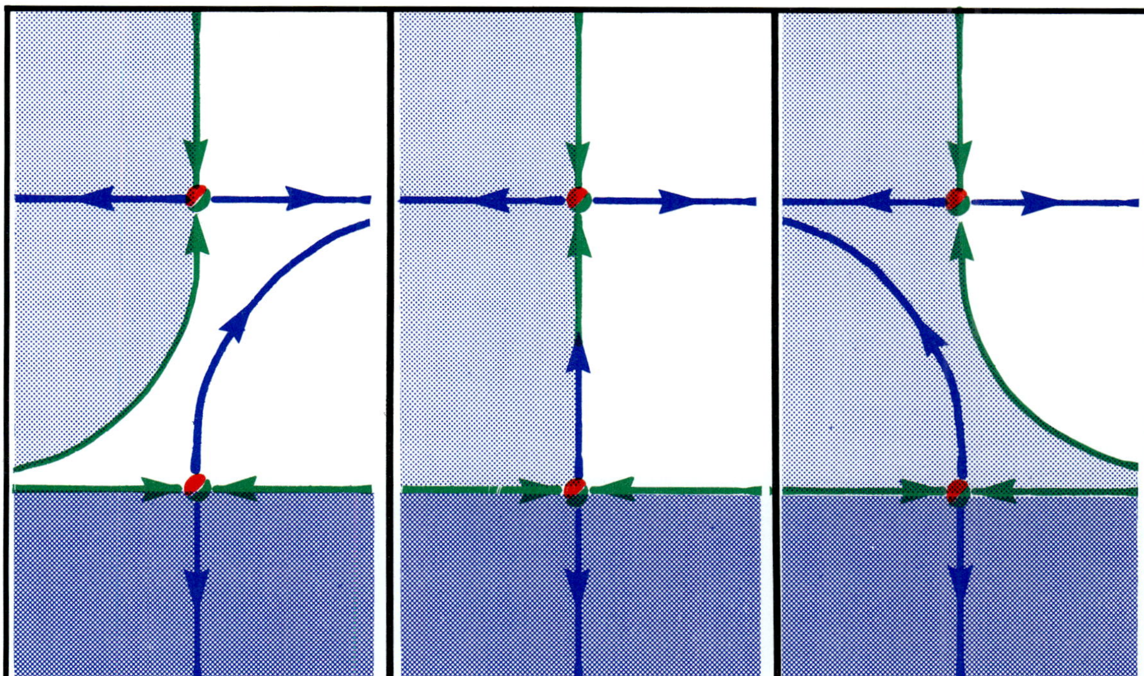

20.1.1. Here are the BEFORE, DURING, and AFTER shots of the event. There is not a single attractor in sight, although three are implied out-of-view. Portions of the three basins of attraction are shown (dark blue, light blue, and white regions). The insets of the saddles (green curves) comprise the separatrices. In this scheme, only one side of one inset and one side of one outset are affected by the control knob.

The effect of the bifurcation (passing through a saddle connection of *heteroclinic* type, that is, connecting two different saddle points) is to radically change the territory claimed by the two competing attractors. As only the basins are affected, and not the attractors, we call this a *basin bifurcation.*

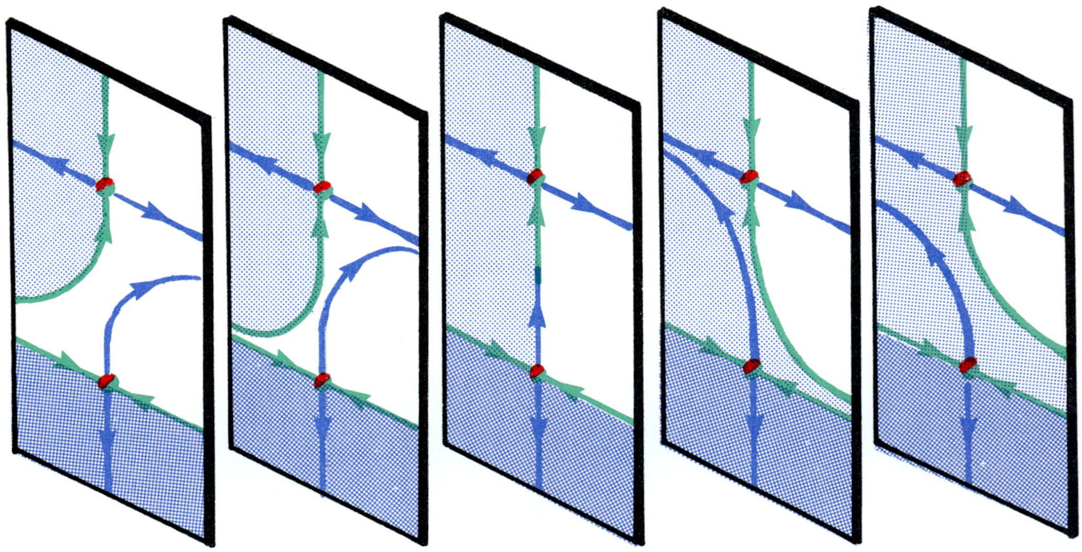

20.1.2. Erecting these portraits, and two in-between interpolations, in their proper positions within the space of the response diagram creates this side-by-side skeleton of the response diagram of this event.

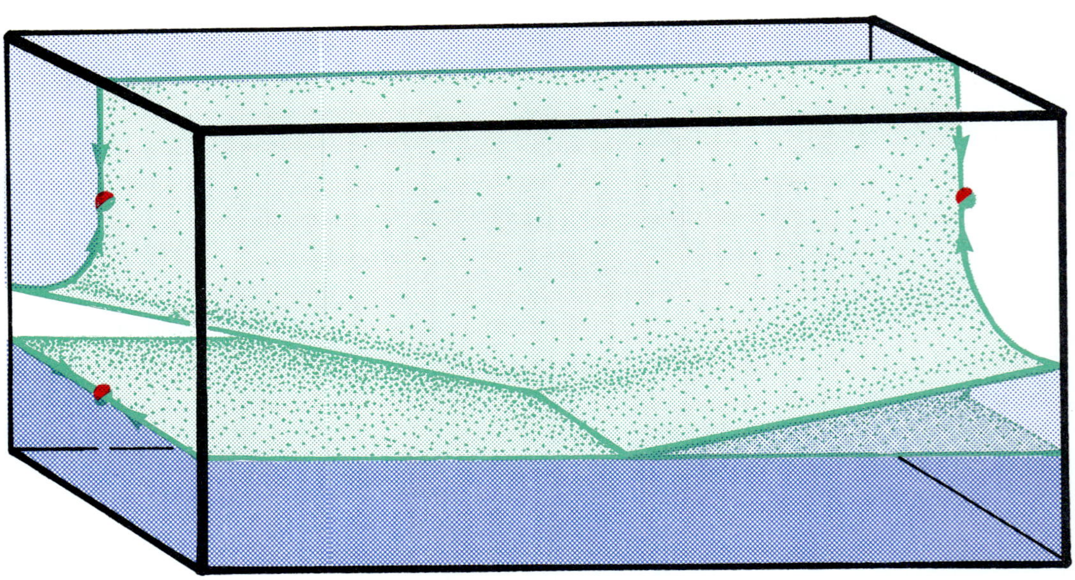

20.1.3. And shading in a continuum of other portraits, we have this cutaway version of the response diagram.

We may understand the effect of this event better in the context of an example. Recall the magnetic pendulum from Part One, Figures 2.1.20 and 2.1.22, remembering that position in the figure indicates position *and speed* of the pendulum.

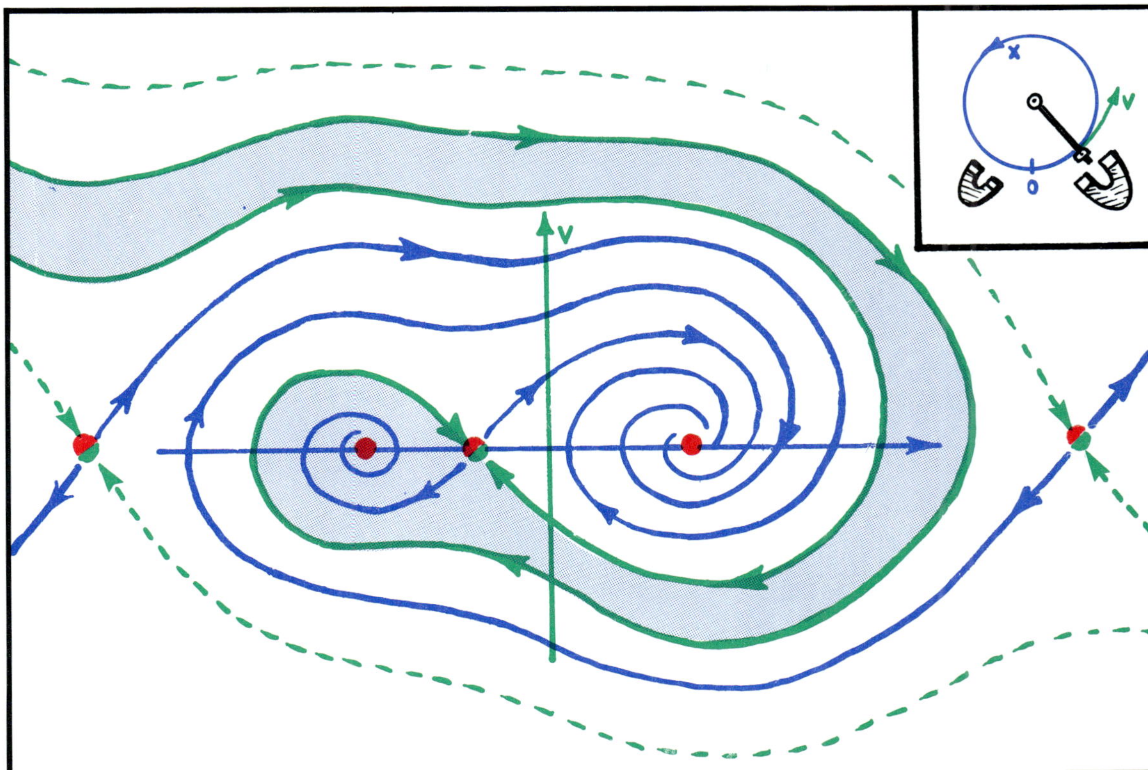

20.1.4. BEFORE: Here is the unrolled phase portrait of the machine in the window. The saddle points indicated on the extreme right and left both repesent the same state, with the pendulum near the top of its swing. We will call this the *top saddle*. The insets, the solid green curves, are actual separatrices, while the dashed green curves represent virtual separatrices. The blue curves are outsets.

Note that the shaded basin winds around the cylinder indefinitely toward the North (upper half-plane).

In this phase portrait, the upper half cylinder (here shown unrolled, hence, the upper half plane) corresponds to counterclockwise (CCW) rotation of the bob. If you want to spin the bob rapidly CCW and to have it fall eventually into the attractor of the smaller magnet on the left, you must start with an initial CCW spin and angle within the shaded basin. As this basin winds around the cylindrical state space (here shown unrolled) indefinitely while tending toward larger and larger CCW rotation rates, there are many good choices (for a given initial position, such as the top) for the initial CCW rate, and for any initial CCW rate, there are good starting portions in the basin. But for CW rates (lower half cylinder) there is only one small portion of the basin, and you must start within this small region to end up at the left equilibrium.

Thus, the probability of throwing the bob CW at random and having it end up at the left equilibrium is much smaller than it is in the CCW case.

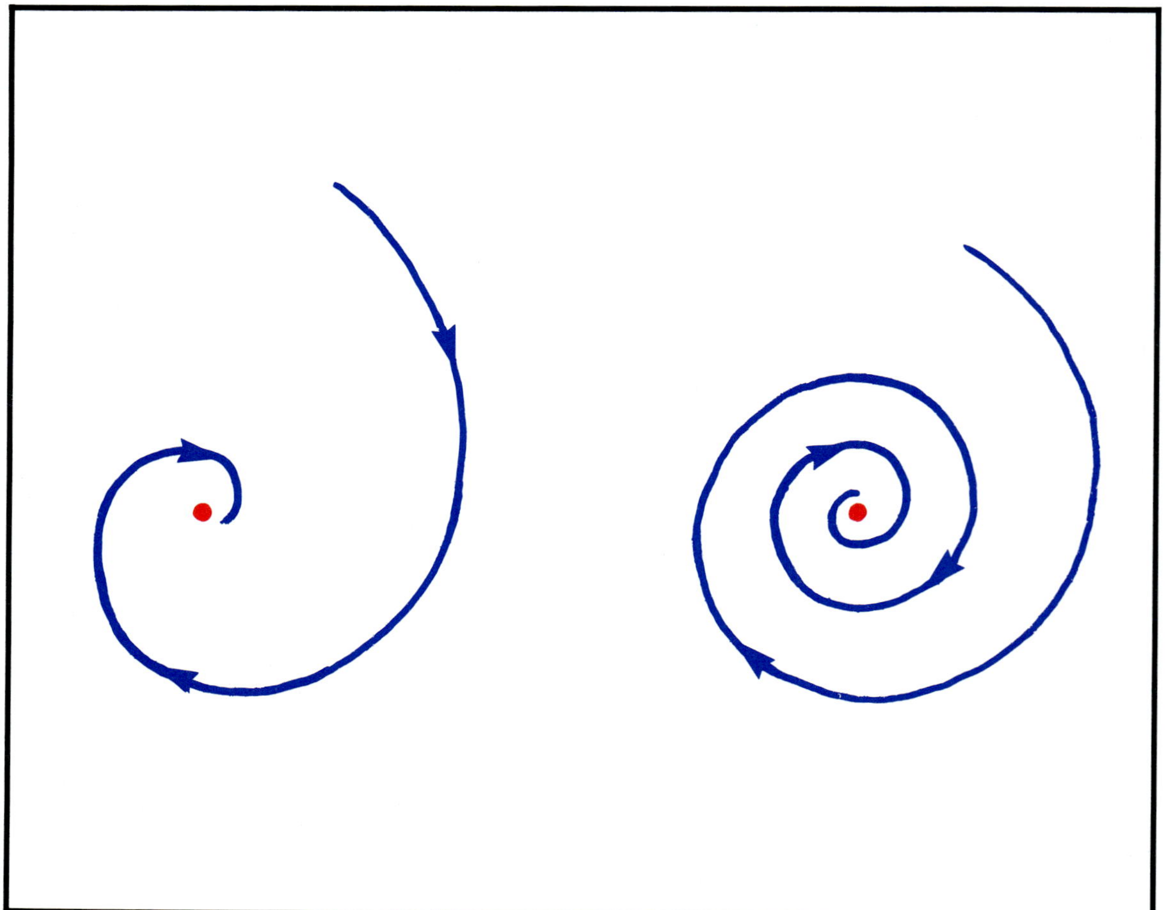

20.1.5. Recall from Part One, Figures 2.1.19, 2.3.18, and 2.3.19 that friction determines the CE's, and thus the rate of decay of each spiraling trajectory, on the way to its attractor. On the left, here, is a closeup of one of the attractors in the case of strong friction. And on the right, the same attractor with weak friction.

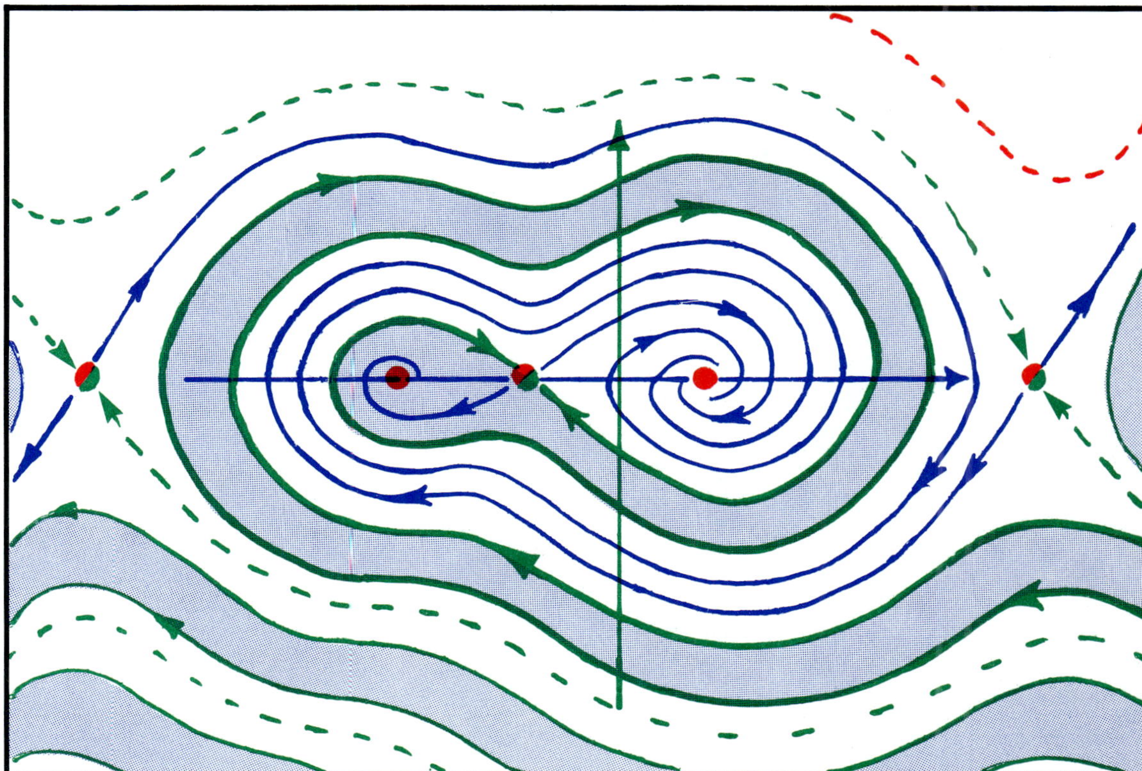

20.1.6. AFTER: Here is the phase portrait of the same pendulum system, with the friction in the hinge substantially reduced by a good lubrication. The key to the difference is the overall rate of spiraling toward the *bottom complex,* the configuration at the bottom of the swing (two point attractors and one saddle point). In fact, we might just pretend for a moment that the entire shaded basin is a single blue trajectory, attracted to this bottom complex. With less friction, more spirals are necessary to make a given amount of progress toward the bottom.

Note that the shaded basin of the equilibrium of the smaller magnet on the left now winds around the cylinder indefinitely toward the south. Surely this makes a difference, especially if you are left-handed. The probability of throwing the bob CW at random and having it end up at the left equilibrium is *much larger* than it is in the CCW case. We will now interpolate from BEFORE to AFTER through four in-betweens. The control parameter in this scheme will be the friction within the support axle of the pendulum, which will *decrease* during the sequence. Two occurrences of the saddle connection bifurcation will be discovered in the process.

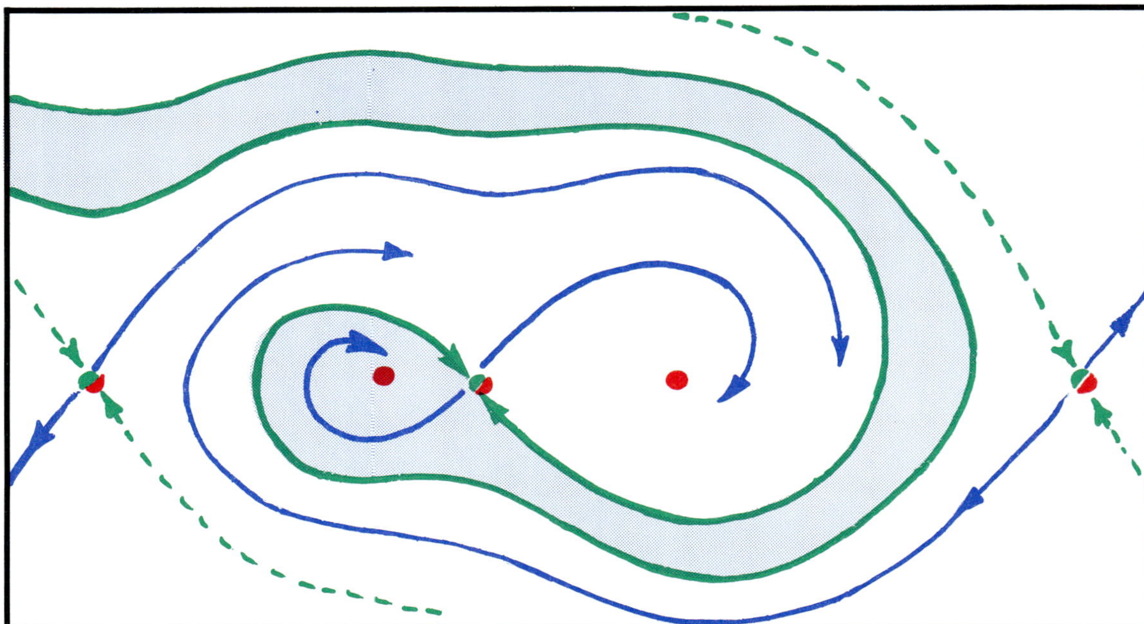

20.1.7. BEFORE: Strong friction. The system is structurally stable, and the tail of the blue snake winds to the North.

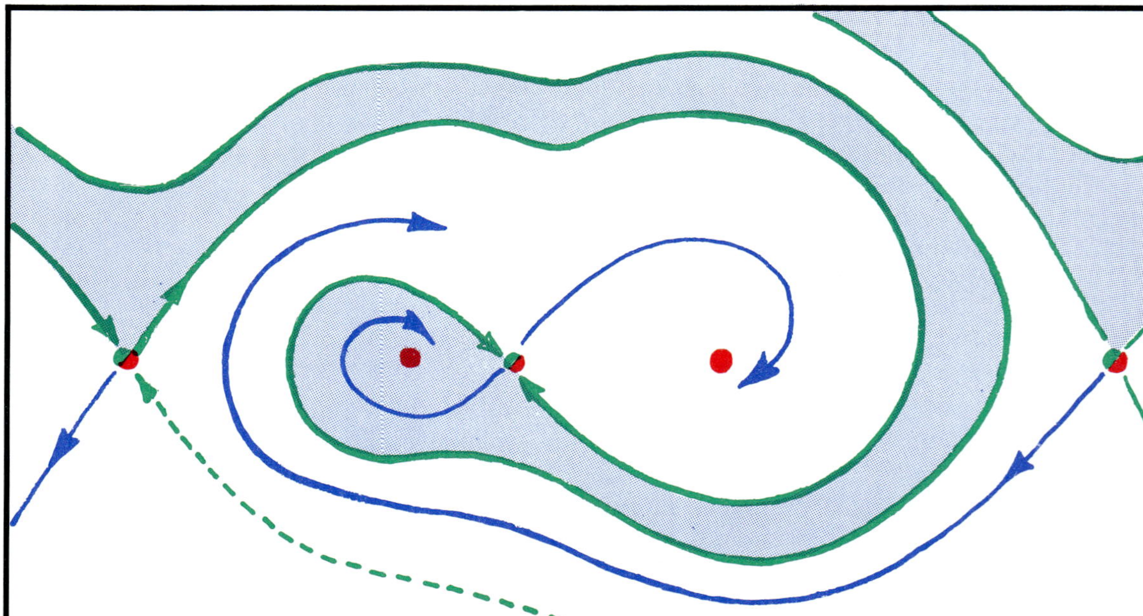

20.1.8. FIRST CONNECTION: After the first lubrication, the lower green boundary of the blue tail (half of the inset of the bottom saddle) makes contact with the blue outset of the top saddle (shown on the left) in a heteroclinic saddle connection. This is the first occurrence of basin bifurcation in this sequence of experiments.

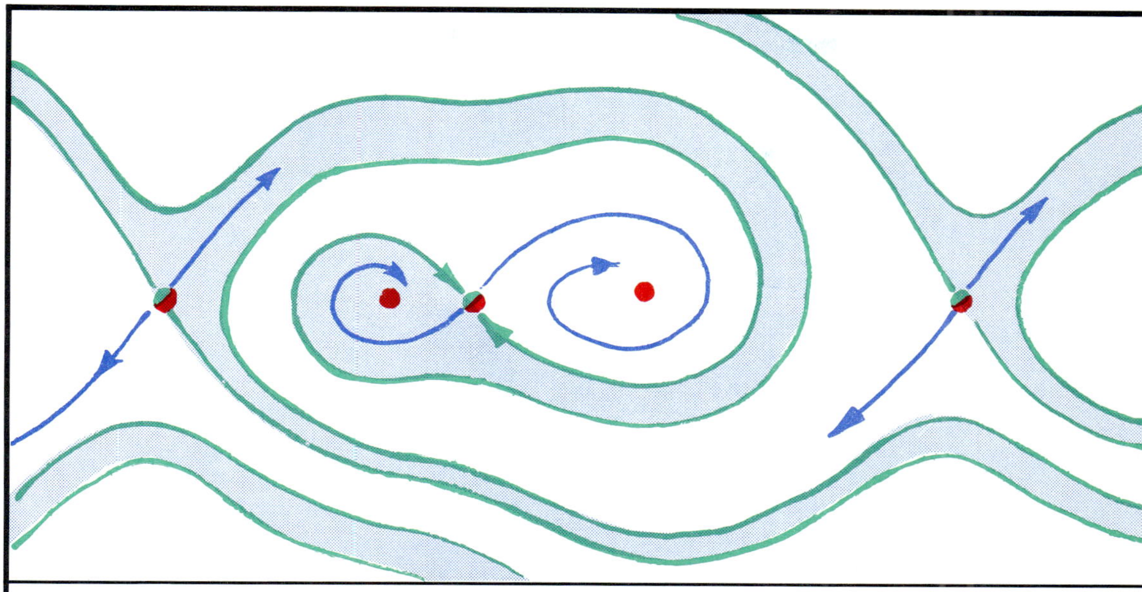

20.1.9. MIDDLE: After the second lubrication, the first basin bifurcation is behind us, the system is again structurally stable, and the blue tail is split into two streamers. There are good chances of hitting the blue basin with either CW or CCW spins. The lower half of the inset of the top saddle (shown on the left) has changed from a virtual to an actual separatrix.

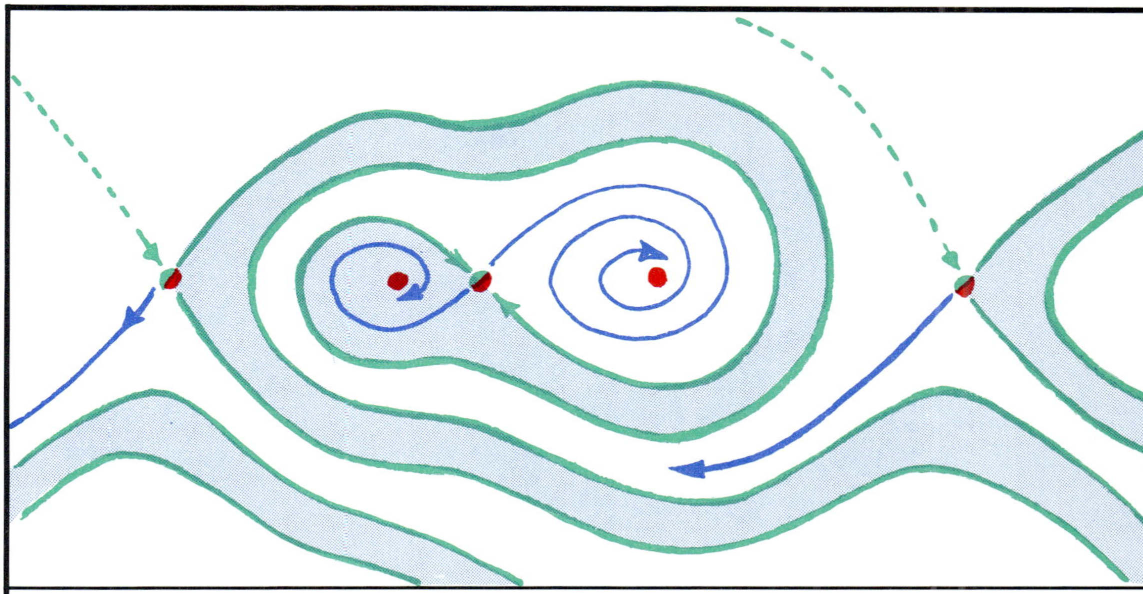

20.1.10. SECOND CONNECTION: After a third lubrication, the upper green boundary of the blue tail conjoins the outset of the top saddle, the upper streamer of the blue basin has been pinched off, and the upper inset of the top saddle has switched from an actual to a virtual separatrix. This is the second occurrence of basin bifurcation in this sequence of experiments.

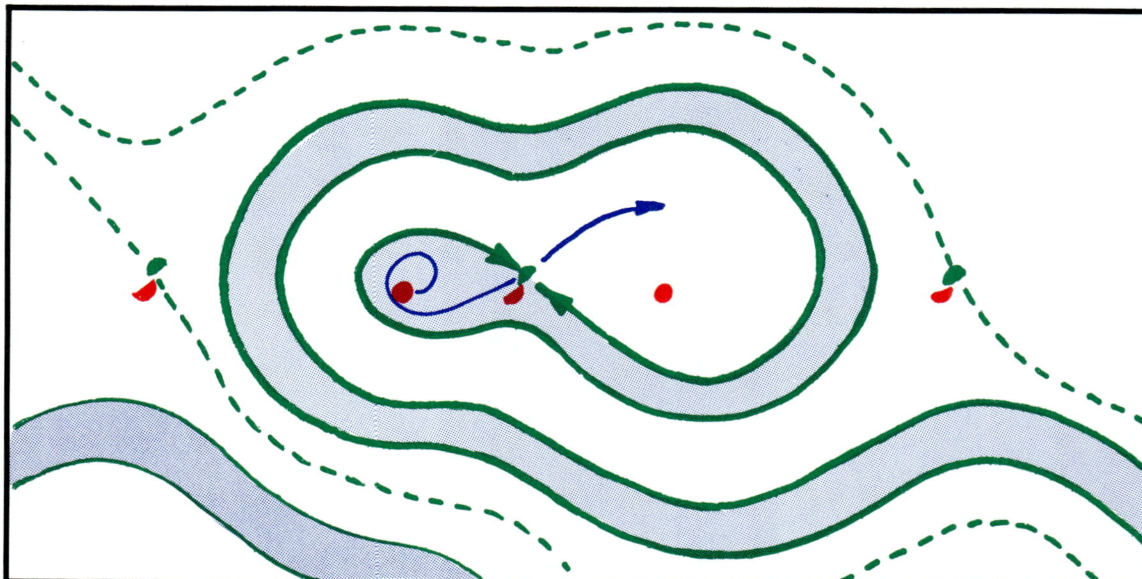

20.1.11. AFTER: After a final lubrication, the friction is reduced, the system is again structurally stable, and the blue tail winds only to the South.

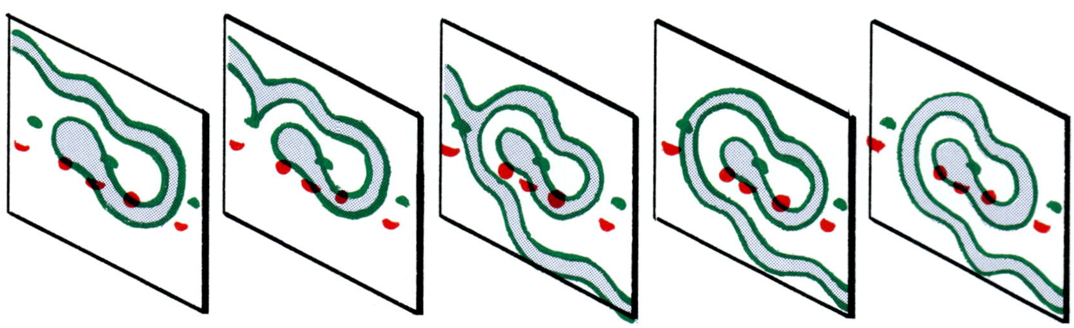

20.1.12. Erecting the five phase portraits in the space of the response diagram of this scheme, we have this side-by-side skeleton sketch.

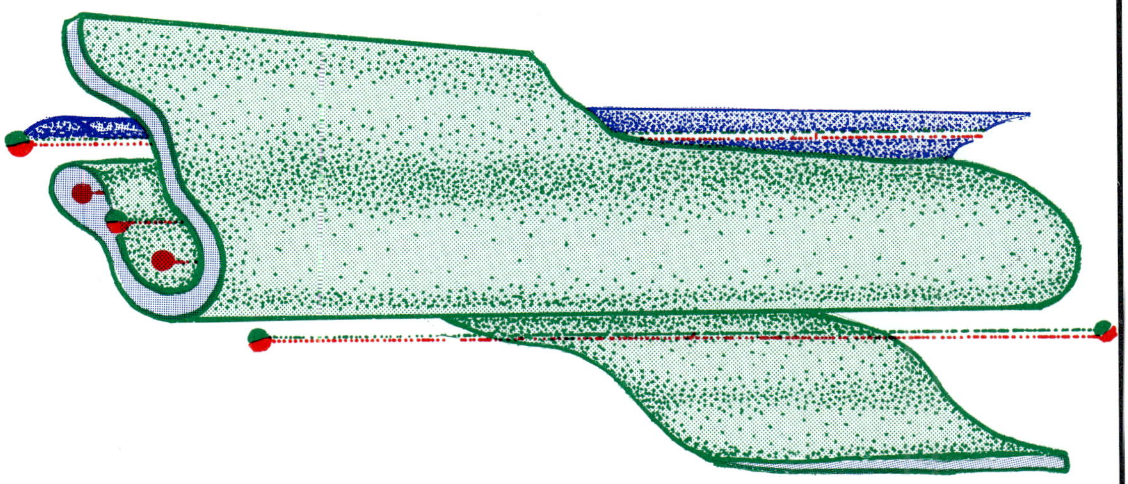

20.1.13. And after continuous interpolation, we have this response diagram, exhibiting two basin bifurcations.

SUMMARY: The saddle connection in 2D is the simplest basin bifurcation. There are numerous other simple examples, in all dimensions. But most saddle connection events in 3D or more entail further complications involving tangles, as we shall see later in this chapter.

20.2. Periodic Blue Sky In 2D

Of saddle connections there are two sorts: heteroclinic and homoclinic. The latter are a much richer source of bifurcation behavior, as we shall now see.

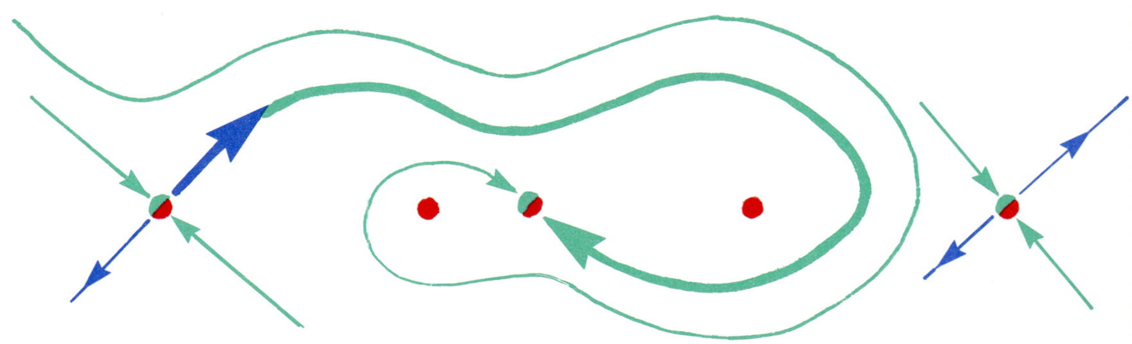

20.2.1. Recall this heteroclinic saddle connection from Figure 20.1.8. What if this connection went from a saddle point to itself?

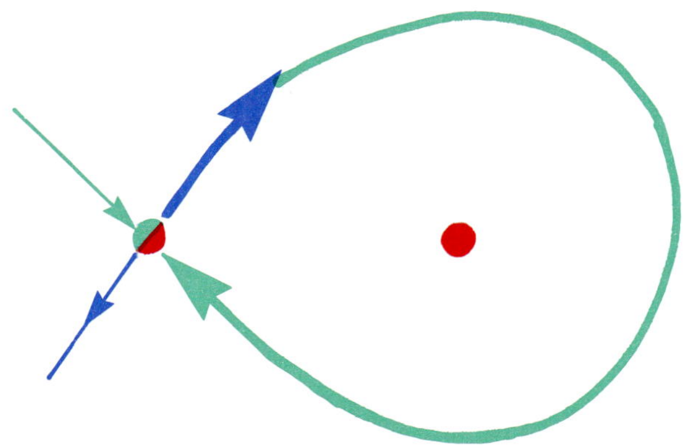

20.2.2. DURING: Here is a homoclinic saddle point in 2D. The outset to the North is conjunct with the inset from the East. This loop is a single trajectory of the flow. It is coming asymptotically from, and going asymptotically to, one and the same critical point. This portrait is not structurally stable. Enclosed within the loop must be, for topological reasons, at least one critical point.

Now let us try to embed this portrait in a bifurcation sequence.

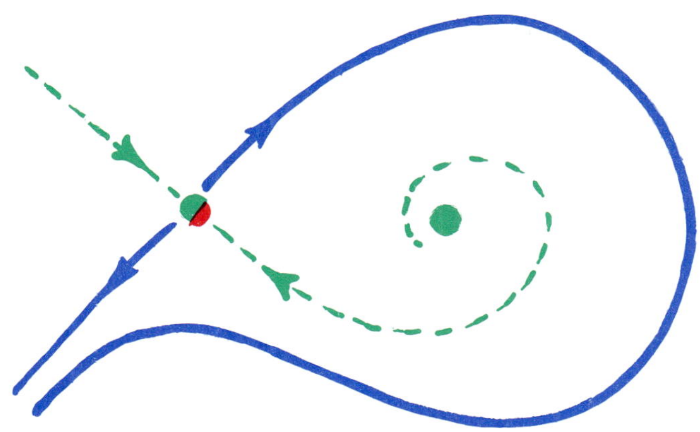

20.2.3. BEFORE: Here is a similar portrait, which is structurally stable. There are two critical points in view, a saddle and a repellor. The inset of the saddle is a virtual separatrix, in the basin of an out-of-sight attractor.

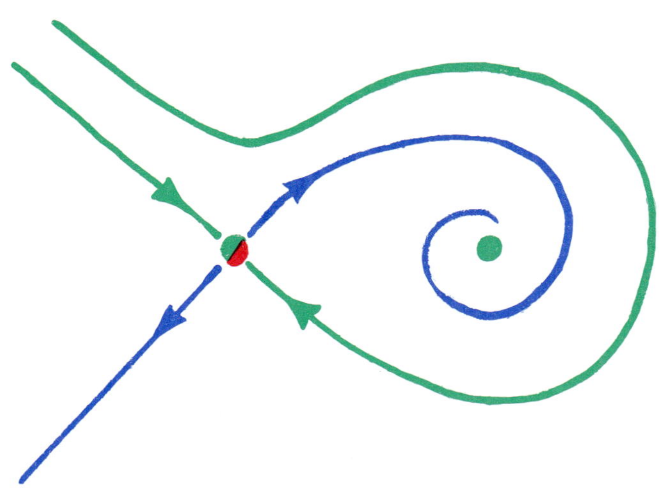

20.2.4. AFTER? If the control parameter steers the outset and inset curves that surround the repellor were to cross, as in Figure 20.2.2, we might expect to end up with this portrait But something is wrong here. The North-Eastern outset goes to a repellor!

Unless the repellor cleverly changed itself into an attractor at the very moment of the homoclinic conjunction, which is not allowed in the one-at-a-time style of a generic bifurcation, this portrait is impossible.

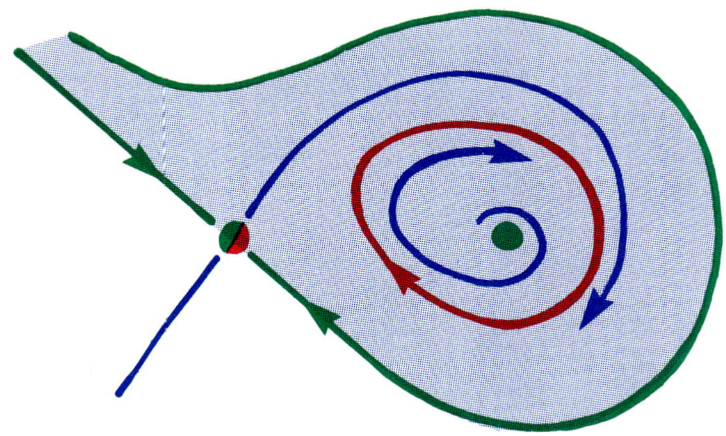

20.2.5. AFTER! Instead, what we find in the generic occurrence of this global bifurcation is the emission of a periodic attractor (of very long period) by the homoclinic conjunction. It simply appears *out of the blue.* The inset of the saddle changes from a virtual to an actual separatrix, and bounds the basin of the new attractor. The repellor is unaffected by the event.

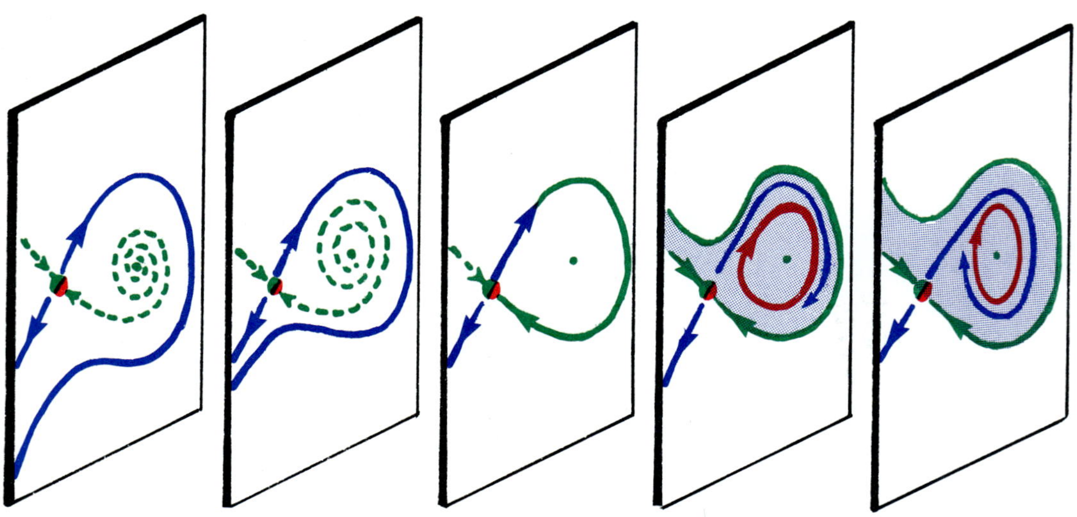

20.2.6. Putting the correct three portraits together with two in-betweens, we obtain this side-by-side skeleton of the response diagram.

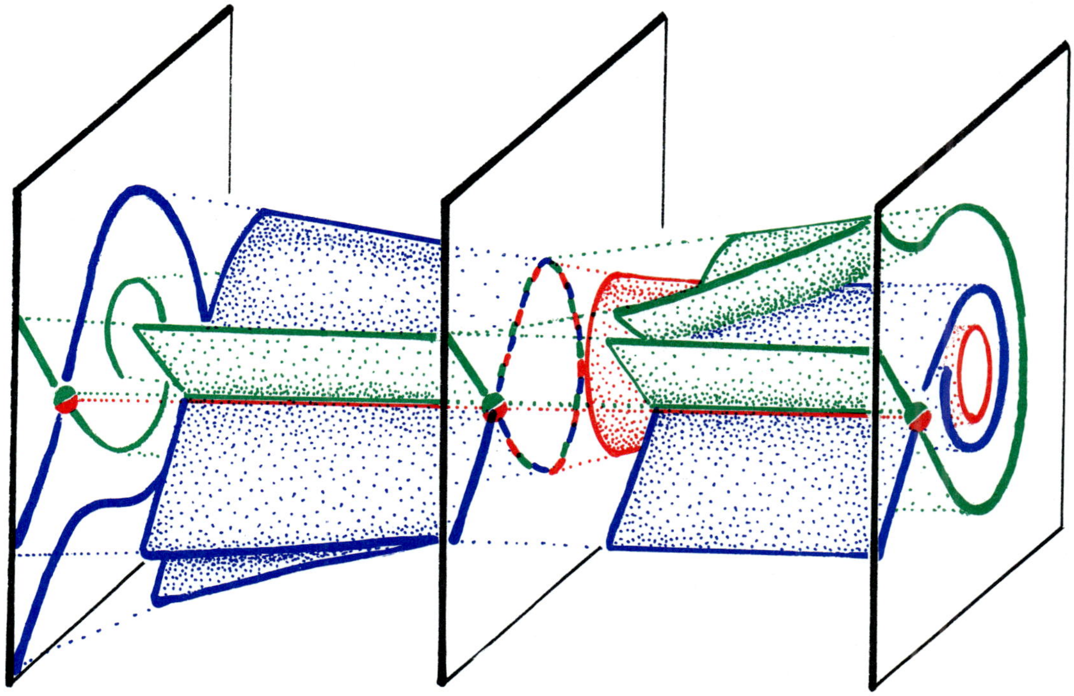

20.2.7. And filling in the loci of insets and outsets smoothly, we obtain this picture of the blue sky catastrophe.

SUMMARY: In this event, the momentary saddle connection is responsible for the appearance out of the blue of a slow, long-periodic attractor and its large basin. This is a basin catastrophe, in that the basin jumps out fully formed, as well as the attractor. As the event is observed in reverse, the periodic attractor moves toward a certain segment of its separatrix, and they vanish into the blue together. As in a fold catastrophe, there is no pinching of the basin.

20.3. Chaotic Blue Sky In 3D

The extension of the periodic blue sky event is not a straightforward construction, because a homoclinic saddle cycle in 3D involves a tangle of inset and outset surfaces (Section 14.1), rather than the simple coincidence of inset and outset curves in 2D that we have seen in the preceding section. A fuller description of this event will be attempted in a later chapter, but here we introduce the basic event: a chaotic attractor (the Birkhoff bagel, see Section 8.2) appears out of the blue.

The initial and final portraits of this event are obtained from those of the 2D periodic blue sky event by simply swinging them around a cycle.

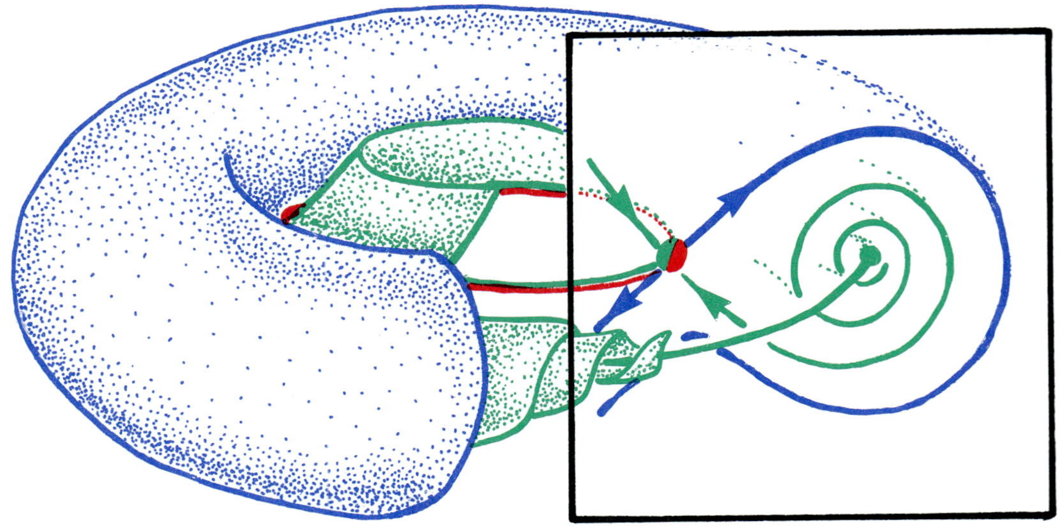

20.3.1. BEFORE: This phase portrait contains two periodic trajectories—a saddle and a repellor of spiral type. The inset scroll of the saddle spirals out from the periodic repellor.

The control parameter affects the positions of the inset and outset scrolls of the saddle cycle, without affecting the repellor.

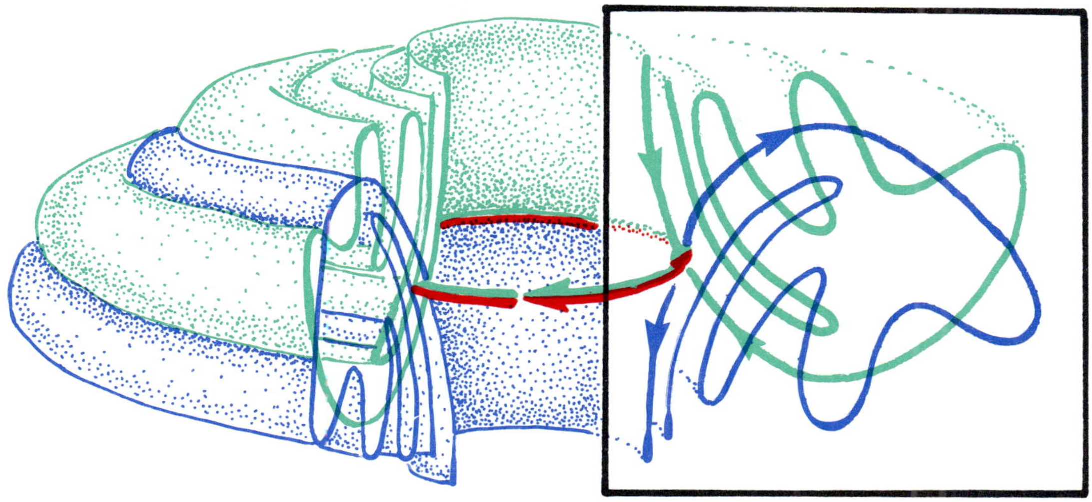

20.3.2. DURING: The homoclinic tangle persists for an interval of control values, not just a single moment of bifurcation. Within this *tangle interval* there is a fractal set of tangency bifurcations, as will be discussed in Section 2.2.4. Ignoring these details for the present, the periodic repellor persists through it all.

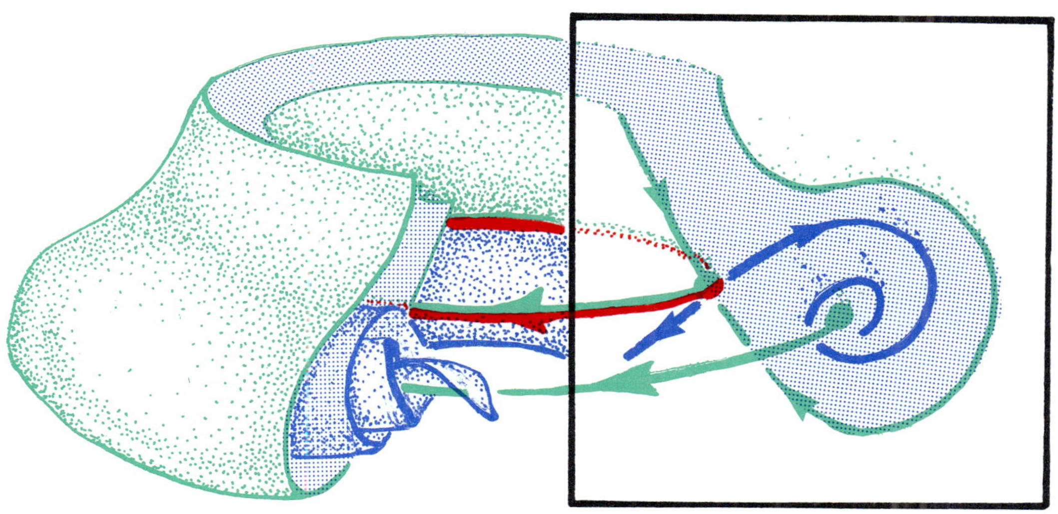

20.3.3. AFTER?: This is the simple *suspension* (prolongation around a cycle) of the impossible Fig. 20.2.4. As the blue outset scroll of the saddle (having passed completely through the green inset scroll as the control parameter moved through the entire tangle interval) rolls up tightly around its limit set, which cannot possibly be the unaffected periodic repellor, there must be a new limit set in the portrait. And there is!

Let's extract the strobe plane for a closer look. Recall that AIT is short for Attractive Invariant Torus.

20.3.4. WELL AFTER! This is the simple suspension of the correct result of the periodic blue sky event, Figure 20.2.5. The blue outset scroll (shown here in strobe section as a blue curve) wraps up around an AIT, a red torus (seen here in strobe section as a red cycle) engulfing the periodic repellor (appearing in strobe section as a green dot).

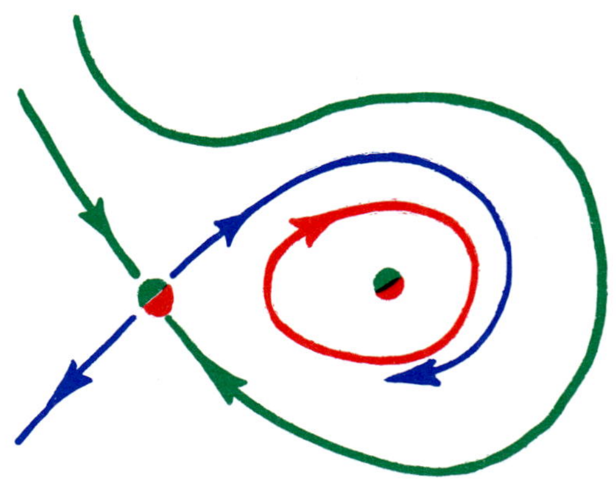

And now, to interpolate some in-betweens, we will run this movie slowly backwards, focusing on the red torus, seen in strobe view.

20.3.5. AFTER. As the control nears the right-hand endpoint of the tangle interval, the section curves of the inset and outset scrolls approach their first tangency bifurcation, the *off-tangency.* Because a tangency, after one turn around the cycle (or one application of the Poincare section map), moves onto another tangency, there must be an infinite number of tangencies created simulteneously. Meanwhile, the red torus expands and develops some bulges, as it is pulled toward the tangle by the blue outset.

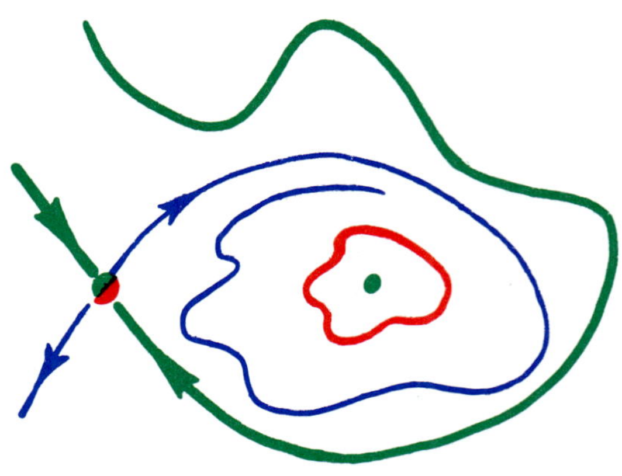

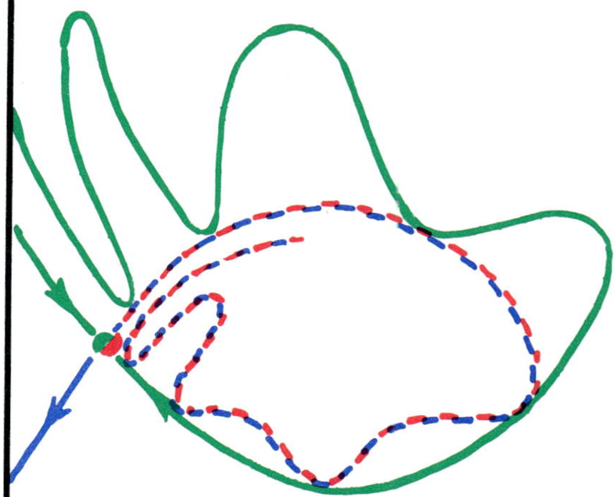

20.3.6. OFF-TANGENCY: Here is the final tangency bifurcation. The inset and outset curves have a one-sided tangle, and the red torus has been pulled into tangency as well, by the infinite sequence of folds of the blue outset.

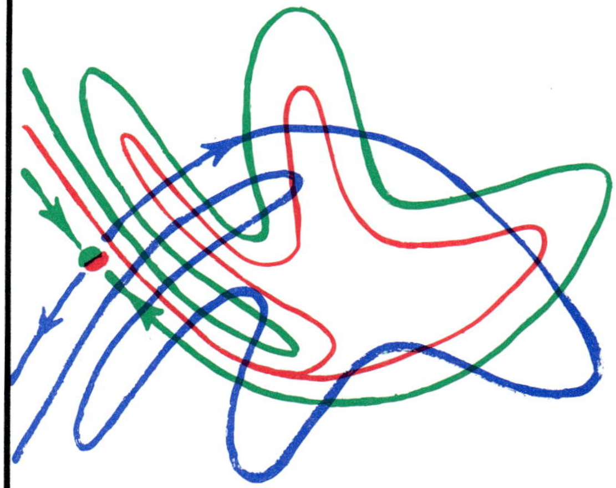

20.3.7. DURING: Within the tangle interval, the red torus becomes a chaotic bagel attractor, as many experiments have shown.[1] There is an infinitude of further tangency bifurcations for control values within this interval.

20.3.8. ON-TANGENCY: Now the green inset has been pulled completely through the blue outset, and the first moment of contact has been attained by this reverse sequence. The folds of the red bagel attractor are tangent to the smooth blue inset, in an infinite sequence of points, as well.

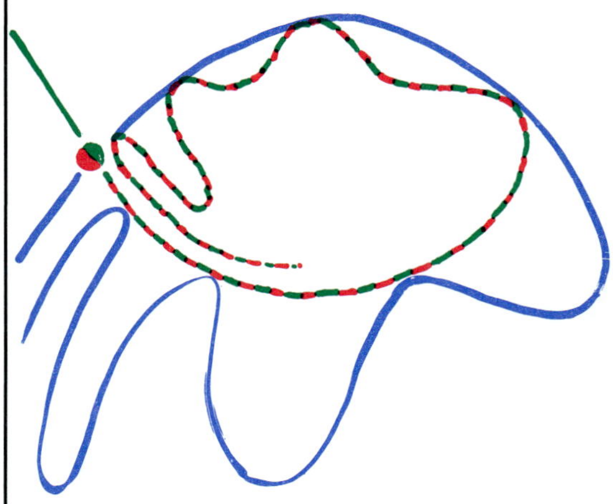

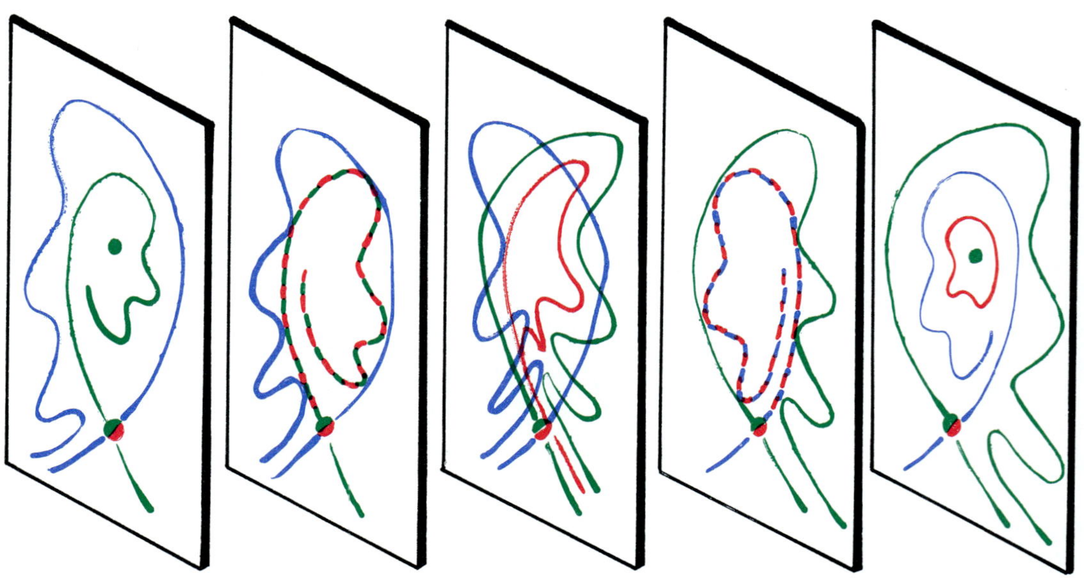

20.3.9. Erecting the strobed portraits in the space of the strobed response diagram, we obtain this side-by-side skeleton. Reading from left to right, we see a chaotic bagel appear out of the blue. This is a catastrophic event involving a chaotic attractor: a *chaotic catastrophe*, or *chaostrophe*.

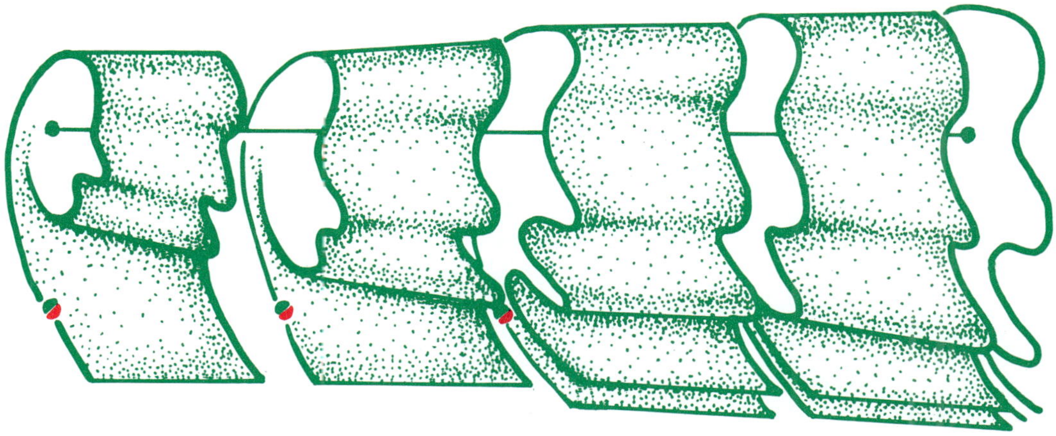

20.3.10. Here we have smoothly interpolated the strobed inset curves, showing their severe folding for control values within the tangle interval.

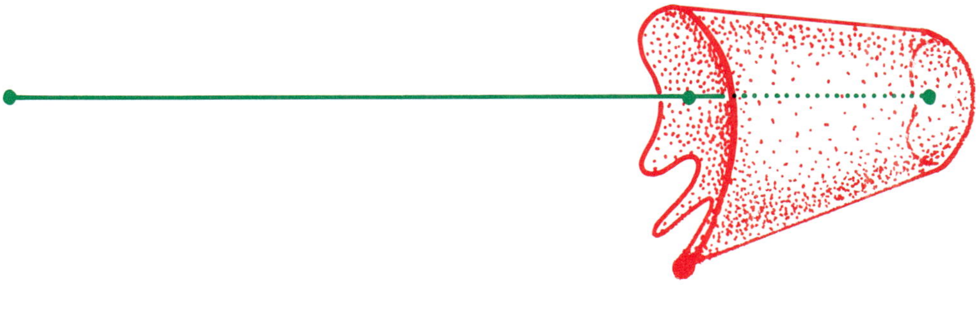

20.3.11. And here, finally, is the cutaway view showing the locus of attraction in the strobed response diagram.

SUMMARY: In this event, an infinitude of microscopic bifurcations within an interval of control values cooperate in the catastrophic creation of a chaotic attractor, which finally settles down to a braid of periodic attractors on an AIT. There are many unanswered questions about this event, which was originally suggested by the extension construction applied to the periodic blue sky event and eventually confirmed in experimental work.

20.4. Rössler's Blue Sky In 3D

A similar event, in which a fully developed chaotic attractor disappeared suddenly, was observed by Rössler early in the history of chaotic bifurcations.[2] We start with a Rössler attractor in a 3D flow (see Section 8.4). In this sequence, it will disappear into the blue.

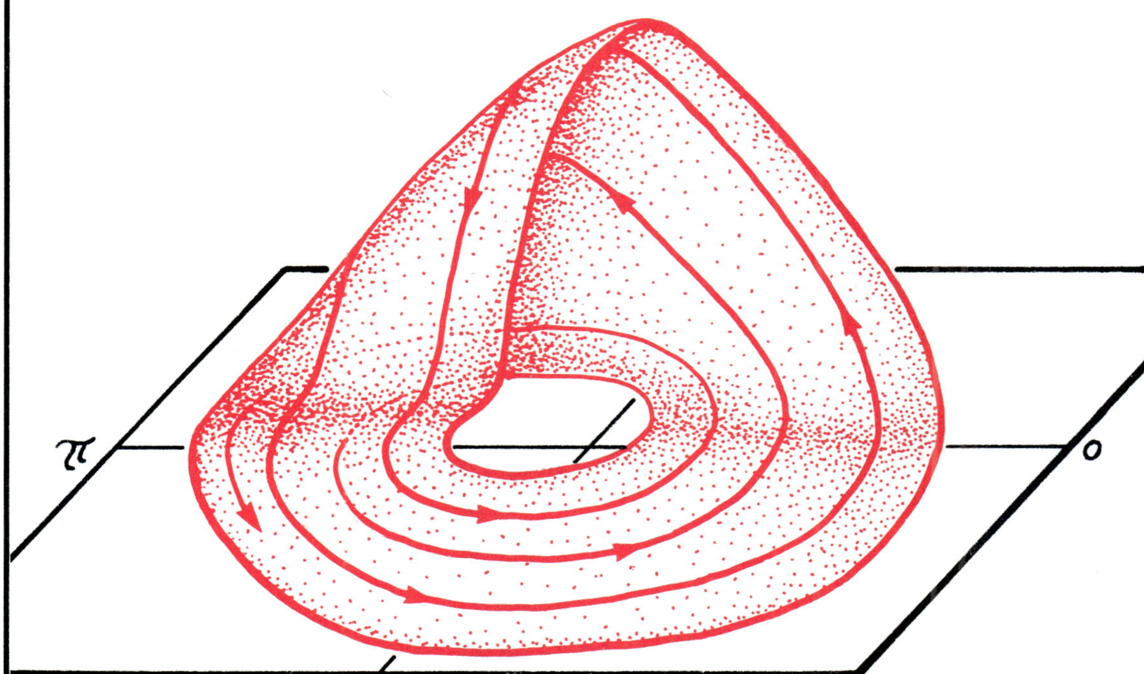

20.4.1. Here is a chaotic attractor that is vaguely periodic. We may cut across it with a strobe section.

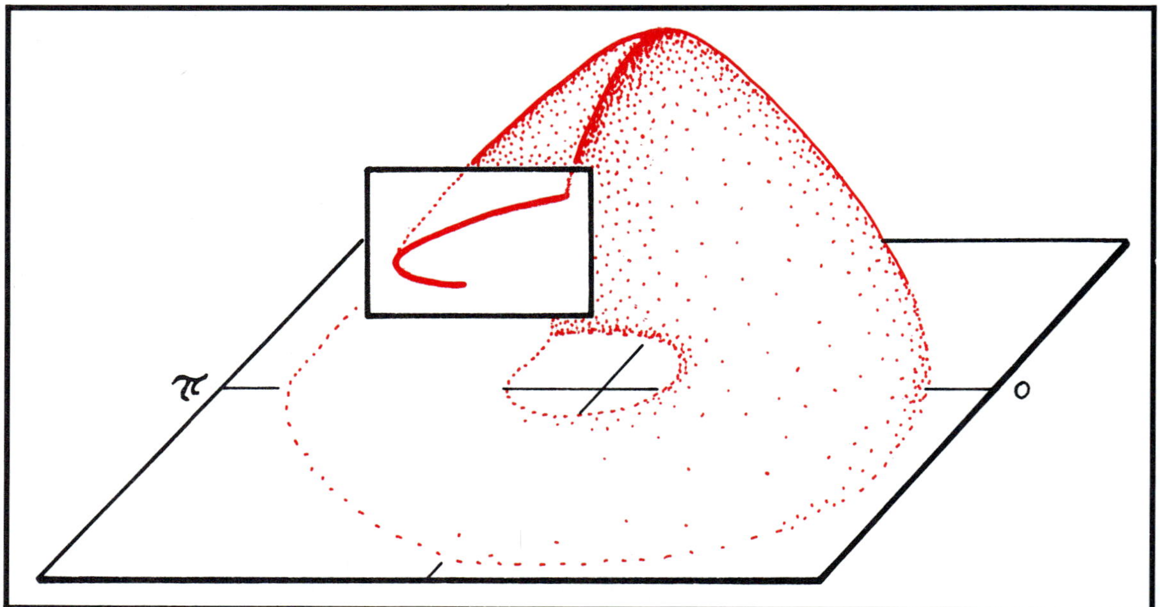

20.4.2. The strobe section reveals the fractal structure of the attractor, and we will describe the blue sky event within this plane.

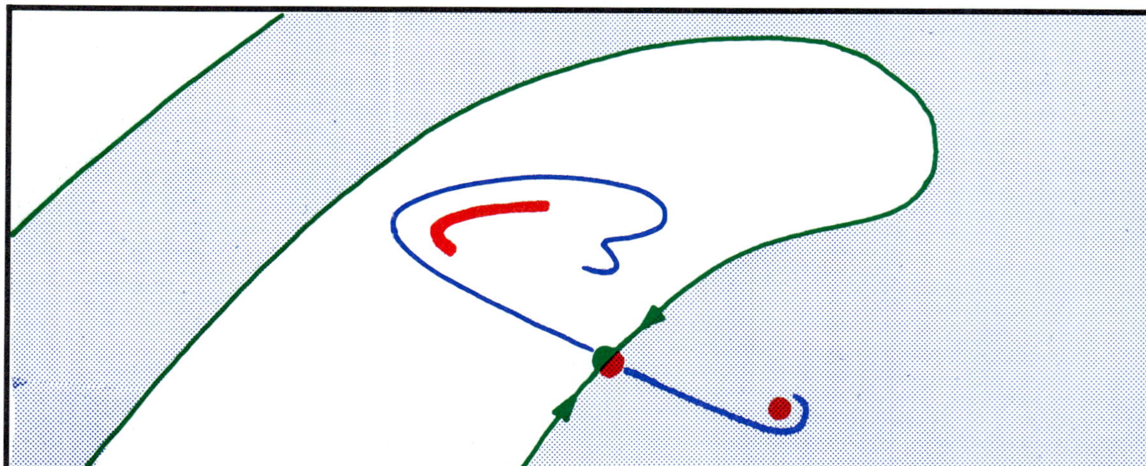

20.4.3. WELL BEFORE: We will start the sequence with this strobe portrait of the Rössler band, the red curve labeled B, within its white basin. There is another attractor, periodic, represented in this strobe portrait by the red point, S, in the lower right. Its basin is shaded blue here. The boundary between the two basins, an actual separatrix, is one-half of the green inset of the saddle point, D. The band attractor is nestled within a curve of this green separatrix, which is getting ready for a homoclinic tangency with one-half of the blue outset of the saddle, D.

Advancing the control parameter, we see the following sequence of events, shown by the computer graphics of Bruce Stewart.[3] We are very grateful to Bruce for his help with this section.

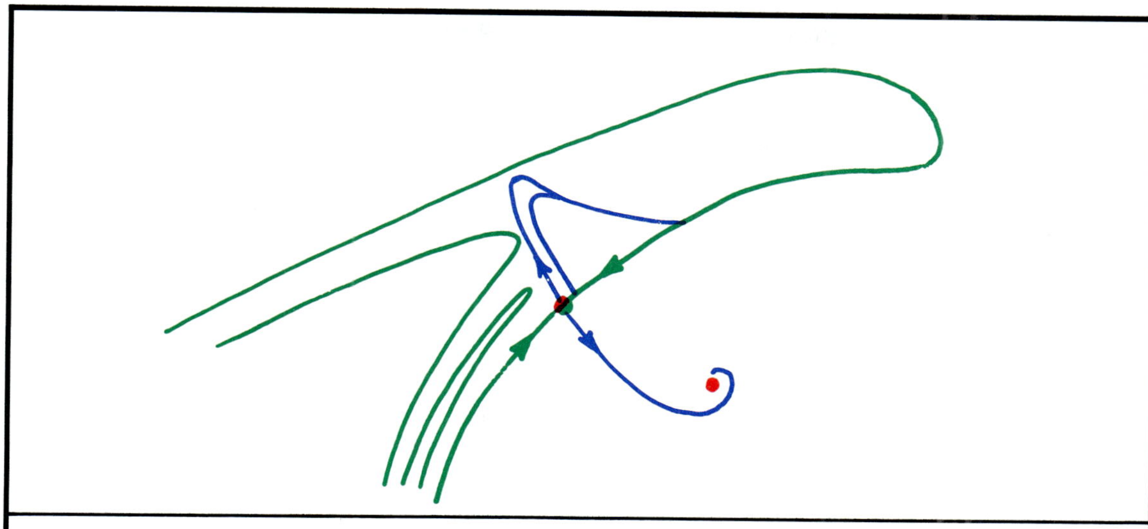

20.4.4. BEFORE: As the control parameter advances, the green inset of the saddle, D, gains a thick set of folds, somewhat like a tangle, and prepares for the onset of homoclinic tangency. The positions of the other actors in the drama are unaffected.

As the control parameter continues its advance, the tangency occurs, and a full-scale homoclinic tangle develops. This is a bifurcation, but it is not the event we wish to describe. It is an essential precursor event.

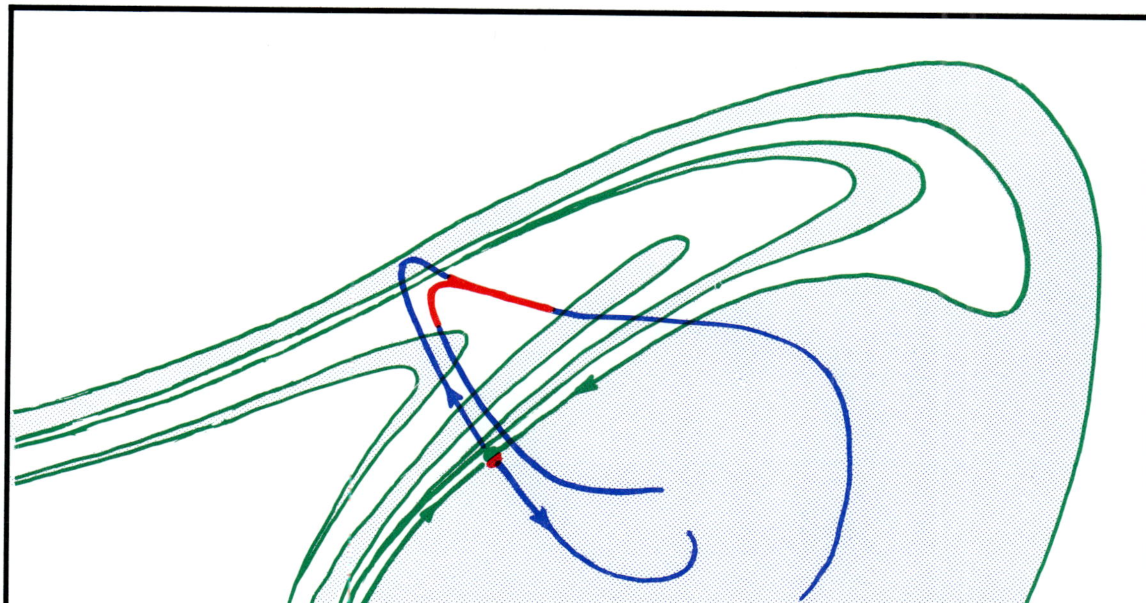

20.4.5. JUST BEFORE: The tangled green inset that separates the blue and white basins has thickened, and is now very close (at two points) to the chaotic band.

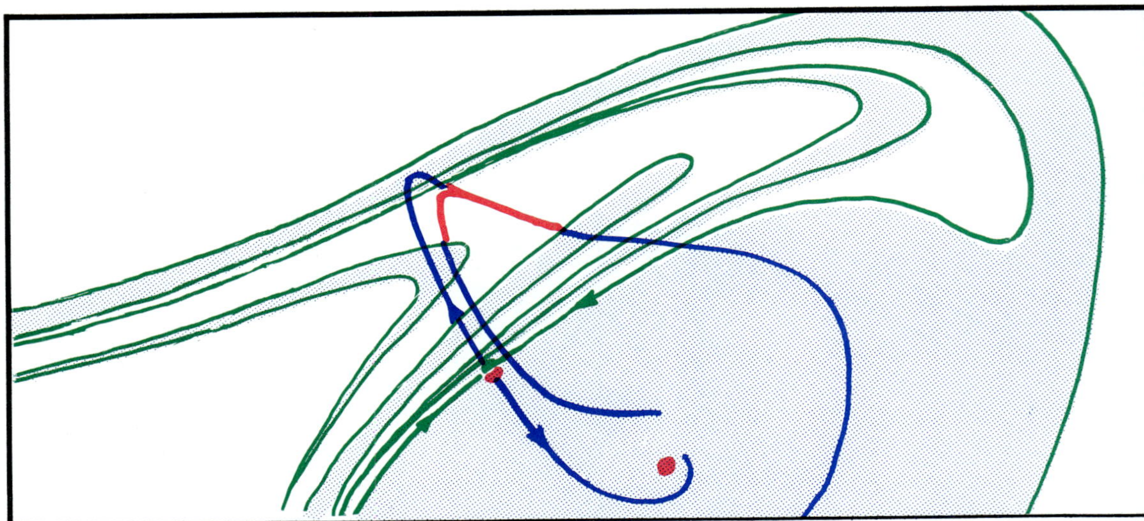

20.4.6. AT CHAOSTROPHE: The tangled, green, basin boundary, thickened further, arrives at the band, touching it at the two extremes with the kiss of death.

The band is actually a tangle, the outset of a homoclinic saddle cycle, which was attractive before the kiss. In strobe section, the saddle cycle appears as a point, I, within the thickened curve, B. The contact with the attractor, B, and its basin boundary is a heteroclinic tangency between the green inset of one saddle, In(D), and the red outset of another saddle, Out(I).

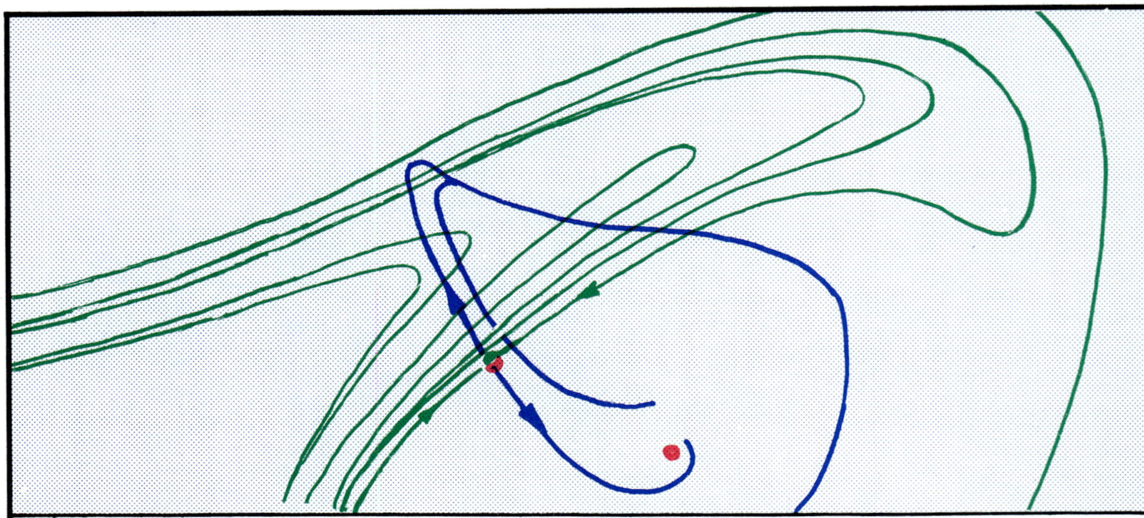

20.4.7. AFTER: Beyond the chaostrophic kiss the tangled band is still here, but is no longer attractive. Experiments now reveal that all orbits eventually leave the tangle and approach the sole remaining attractor S. The blue basin has engulfed the white basin, and the thick separatrix is now virtual.

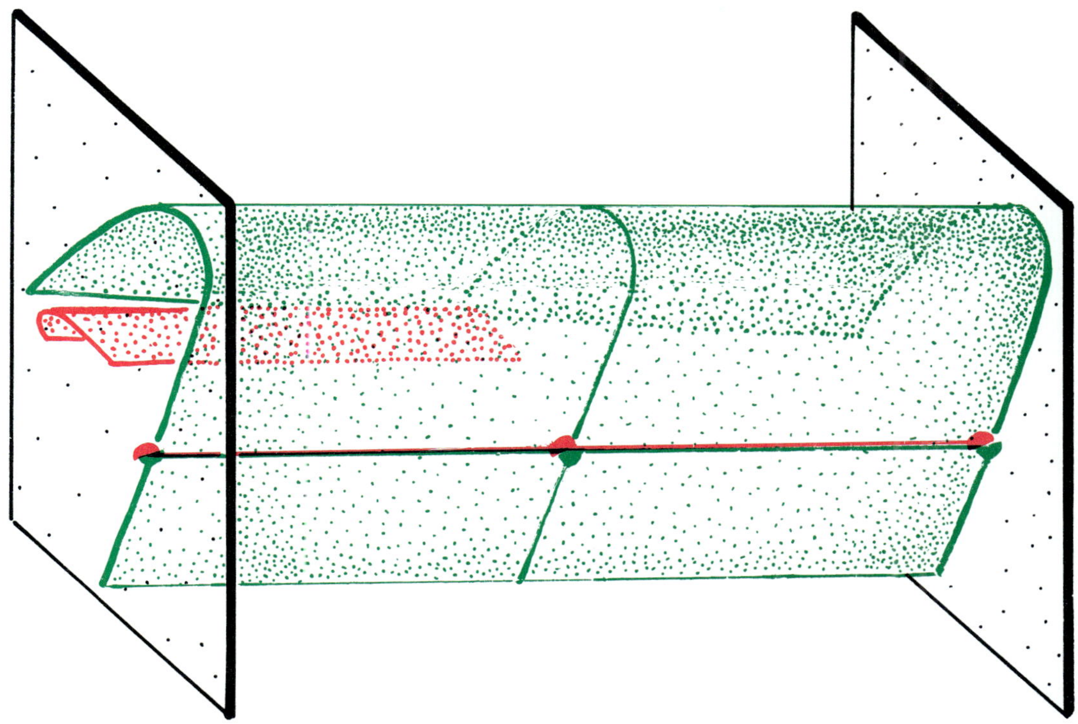

20.4.8. Assembling the strobed portraits side by side, and smoothly interpolating the loci of attraction and separation, we obtain this response diagram of the Rossler event.

SUMMARY: In this event, frequently observed in digital and analog simulation since the early days of chaotic dynamics, a chaotic band attractor disappears catastrophically into the blue. The involvement of the nearby tangle, fully analyzed only recently, is characteristic of many similar events.

21

Explosive Bifurcations

After a chapter on subtle bifurcations and three on catastrophes, you may have forgotten that there are three categories of bifurcation: subtle, catastrophic, and explosive. For years we said that there were only two, subtle and catastrophic. It now seems to us that explosions should be regarded as a separate class, rather than being included among the catastrophes. Explosive bifurcations are discontinuous like catastrophes. But like subtle bifurcations, they lack hysteresis. The name and basic concept were given by Stephen Smale in 1967.[1] Here are some simple examples.

21.1. Blue Loop In 2D

There is another way a periodic attractor can pop up, not out of the blue sky, but from a blue loop associated with a saddle point. This example is due to Zeeman.[2]

Recall this typical portrait on a sphere, from Figure 18.2.7. It has critical points, but no limit cycles.

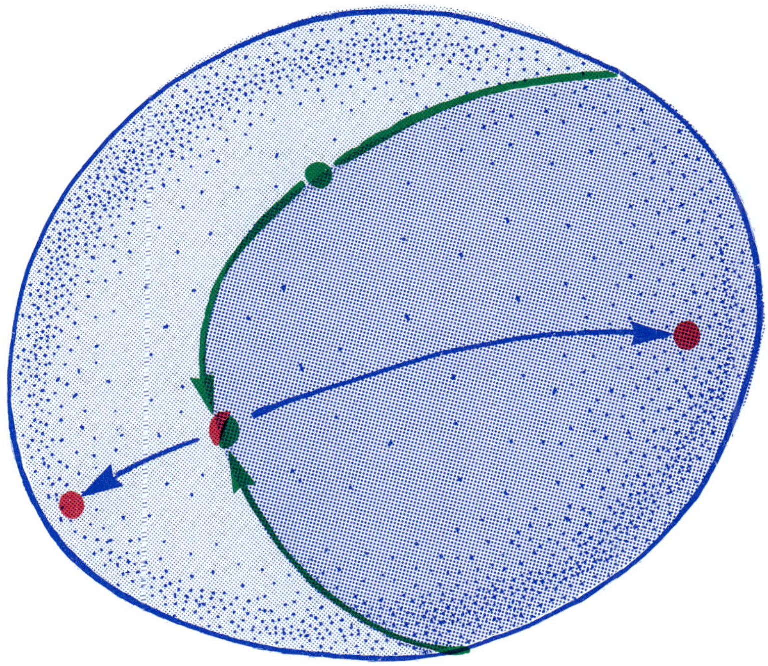

21.1.1. Here, the two outset curves (blue) of the saddle point, which must end up at attractors, belong to *two different* basins. The system is *bistable.* The inset curves (green) actually separate the basins.

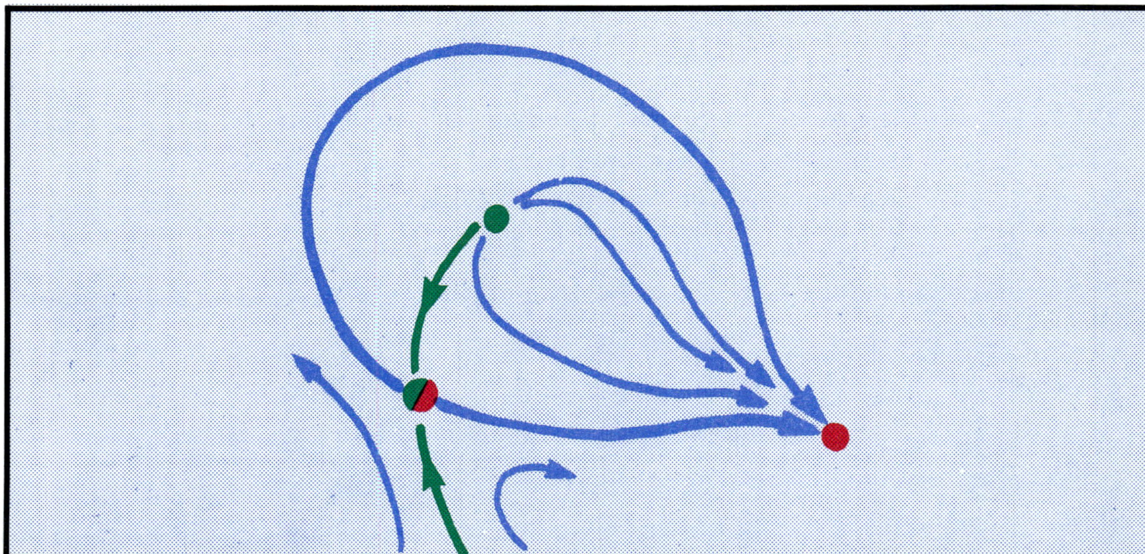

21.1.2. BEFORE: On the other hand, this portrait could arise. Here there is only one attractor and basin. The system is *monostable.* Both outset curves from the saddle point go to the same point attractor. We call this a *blue loop.* The loop encloses a point repellor and touches a point attractor. The two inset curves of the saddle are virtual separatrices.

Now we will perform a static fold catastrophe (Section 18.2) upon the two critical points of the blue loop.

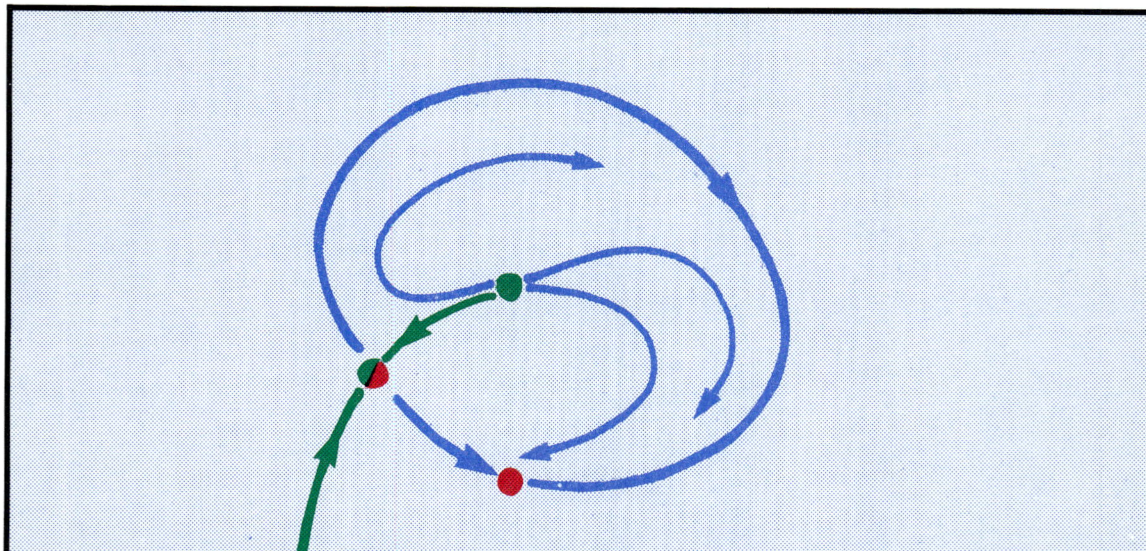

21.1.3. JUST BEFORE: The saddle and attractor points approach conjunction on one of the outset curves of the saddle. One CE of each is approaching *zero.* The angle made by the two outset arcs at the point attractor is increasing toward 180 degrees.

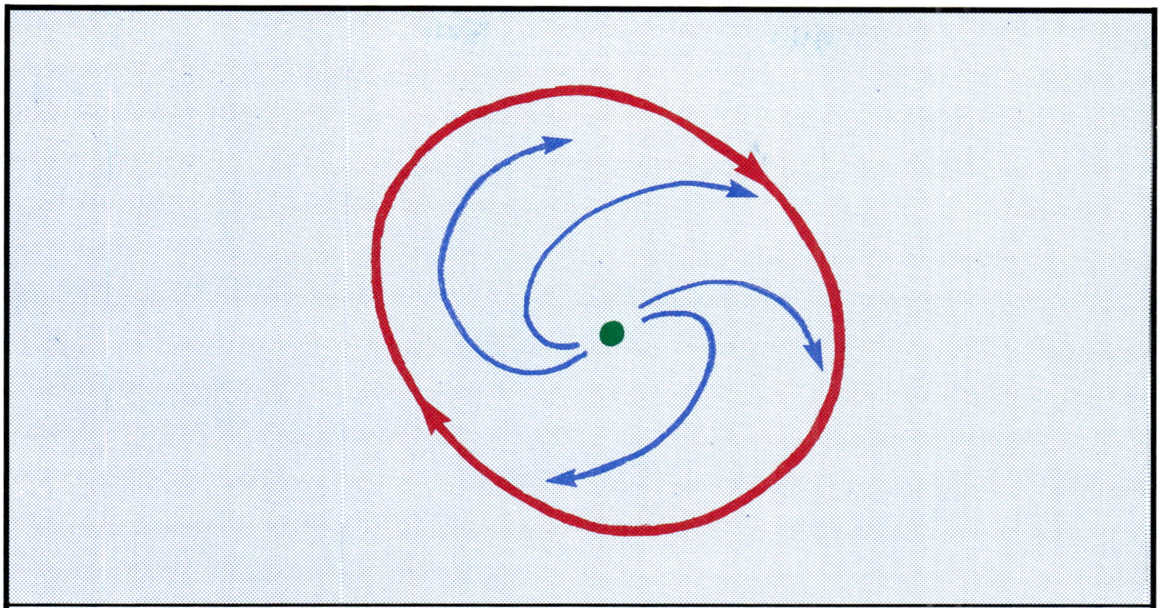

21.1.4. JUST AFTER: The point attractor and the saddle point have mutually annihilated in a static fold catastrophe. *But the blue loop has become a periodic attractor!* The system is still monostable, but the solitary equilibrium state is a slow oscillation with a large amplitude, rather than a rest point.

The point attractor has exploded into a periodic attractor.

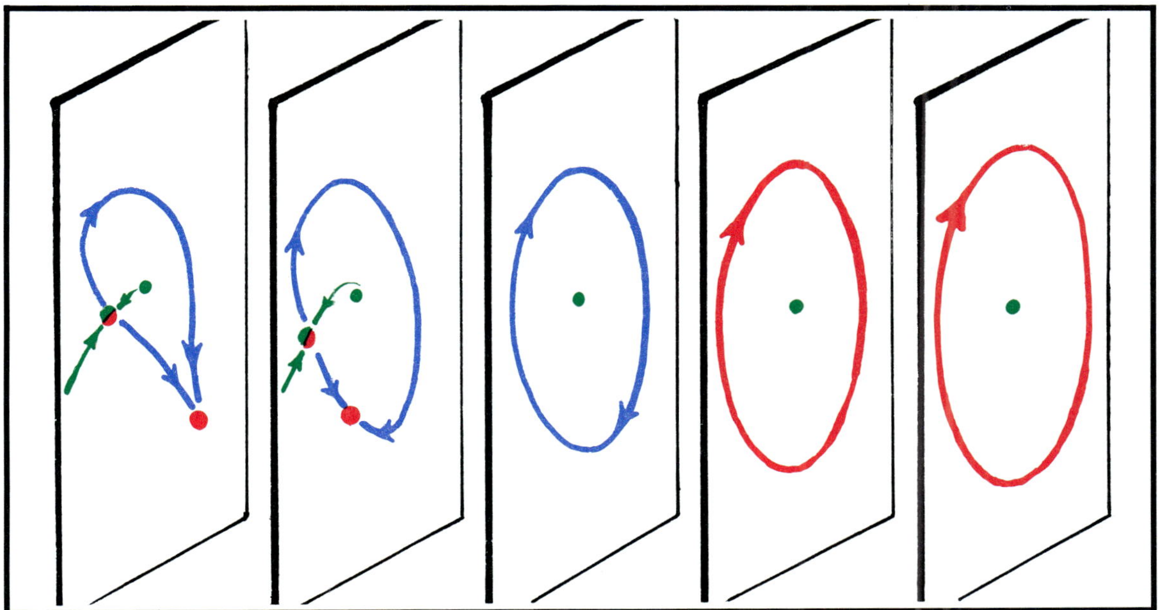

21.1.5. Erecting these phase portraits side-by-side in the control-phase space of the response diagram, we obtain this skeletal representation of the event.

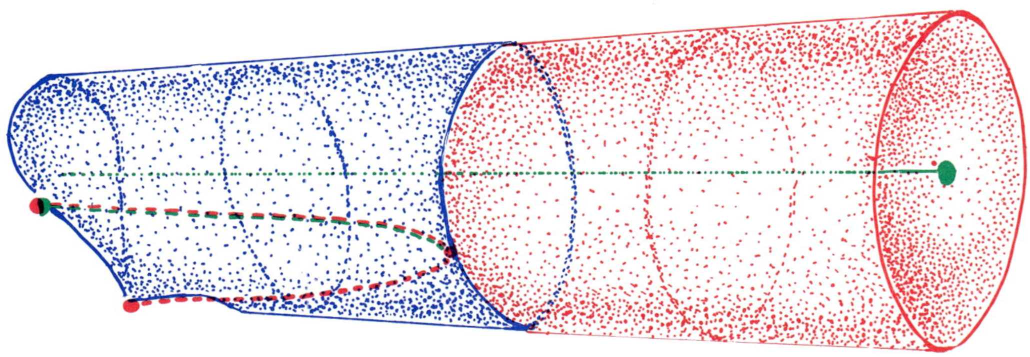

21.1.6. Filling in the continuous red locus of attraction and stripping off some of the green details, we have the response diagram of the blue loop event.

SUMMARY: The static fold catastrophe, in a portrait having a blue loop, results in the explosion of the point attractor into a periodic attractor. The locus of attraction looks like a pot with a handle. The sudden enlargement of an attractor, within its undisturbed basin, is the defining feature of an explosive bifurcation. The event is *reversible* in that the implosion, which takes place as the control is reversed, occurs at the same bifurcation value of the control parameter. The event does not exhibit hysteresis (see Figure 4.3.7). The *local* event involving the fold is identical to that discribed in Section 18.2. But the *global* event involving the blue loop makes this event a new entry in our atlas of bifurcations, in which the loci of attraction within the response diagram have top priority. For this reason, the *response diagram* is not exactly identical to the *bifurcation diagram,* as defined by mathematicians. Bifurcations diagrams indicate bifurcation events living entirely within the separatrices, while response diagrams emphasize the bifurcations affecting the loci of attraction only.

21.2. Blue Loop In 3D

The simple suspension around a cycle of the preceding event gives rise to a 3D version of essentially the same phenomenon.

Within an unperturbed basin, a periodic attractor may explode into a braided torus (AIT).

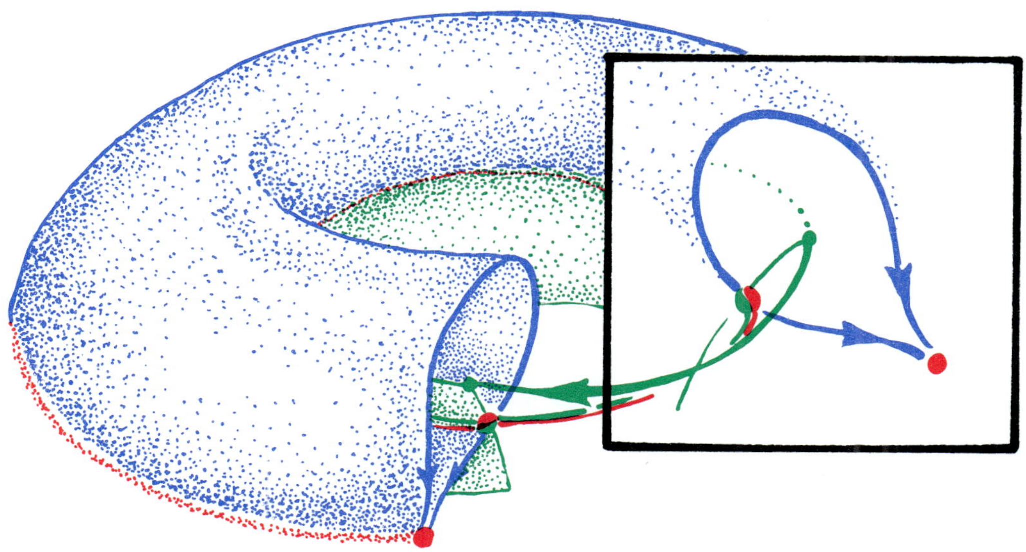

21.2.1. BEFORE: Here is the initial portrait. There are three limit cycles of about the same period going around a ring: attractor, saddle, and repellor. The strobe plane section looks a lot like the initial portrait of the blue loop in 2D. The two leaves of the outset of the saddle (blue surfaces) are both attracted to the same periodic attractor. In strobe section, they form a blue loop. In 3D, they form a blue sleeve, having a crease marked by the periodic attractor (red cycle).

We now increase the control to effect a periodic fold catastrophe, simultaneously annihilating the periodic attractor and saddle.

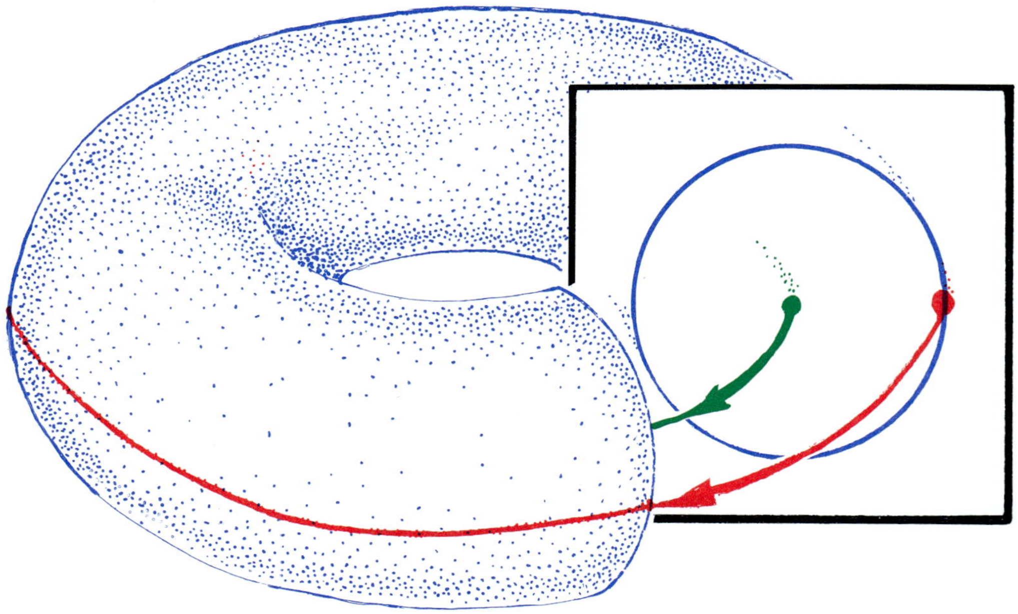

21.2.2. FIRST BIFURCATION: The attractor and saddle are now conjunct, in a honhyperbolic limit cycle. The crease in the blue sleeve has been ironed out flat. The blue loop in the strobe section is an invariant cycle of the first-return map, having rotation number (average rotation) zero. The *rotation number,* which indicates the number of times that trajectories wind around the waist of the AIT between two flashes of the strobe lamp, is about *zero* throughout this example. But any rational number could have been used. The fluctuating braid bifurcations of a fractal bifurcation event, characteristic of toral dynamics, complicate this event. We postpone discussion of this feature to Section 22.2. For the moment, we just need to know that there is an *interval of fluctuation* in the control range, in which there are an infinite number of bifurcations. These affect only the flow on the torus, and involve minor changes in the number of attractor-saddle cycle pairs braided around the AIT. Some changes in the number of braids may be associated with a change of the rotation number. These involve control intervals of irrational solenoidal flow.

For the present, we may simplify this event by regarding the red AIT as the attractor, throughout the interval of fluctuation. (See Section 22.3.)

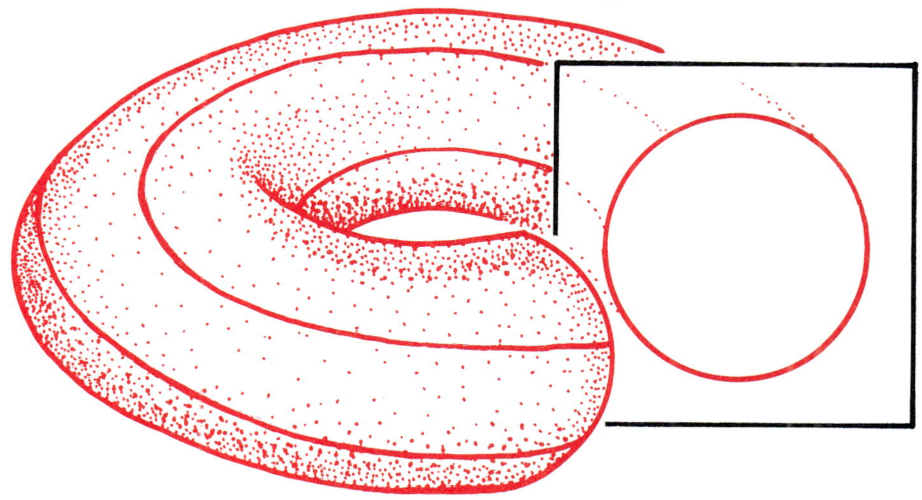

21.2.3. DURING INTERVAL: The blue sleeve has become an attractive invariant torus (AIT), around which trajectories wind only slightly. There are frequent fluctuations in the flow on the torus, between braids and solenoids.

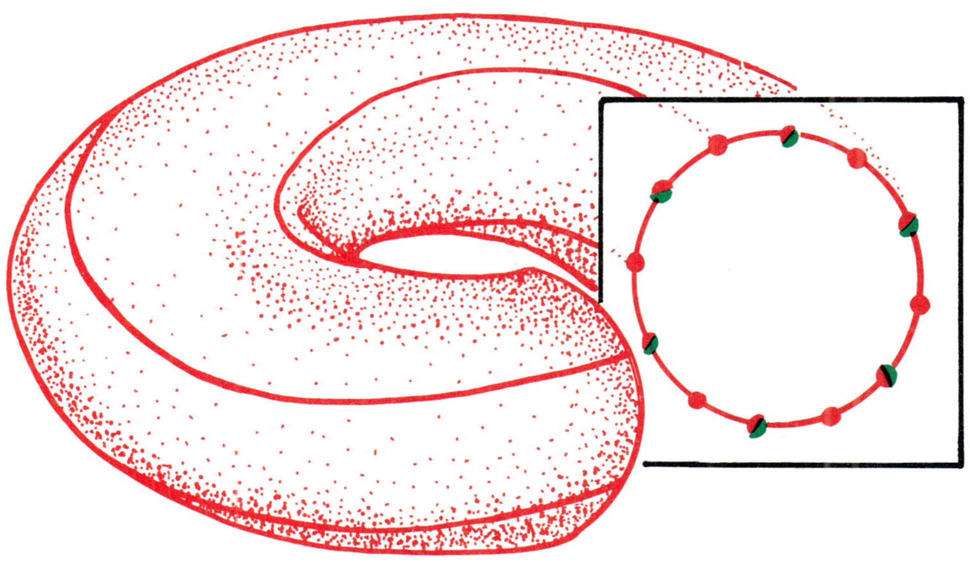

21.2.4. AFTER: The flow on the AIT has settled down to a steady braid, and a few periodic attractors braided on the torus now dominate the portrait.

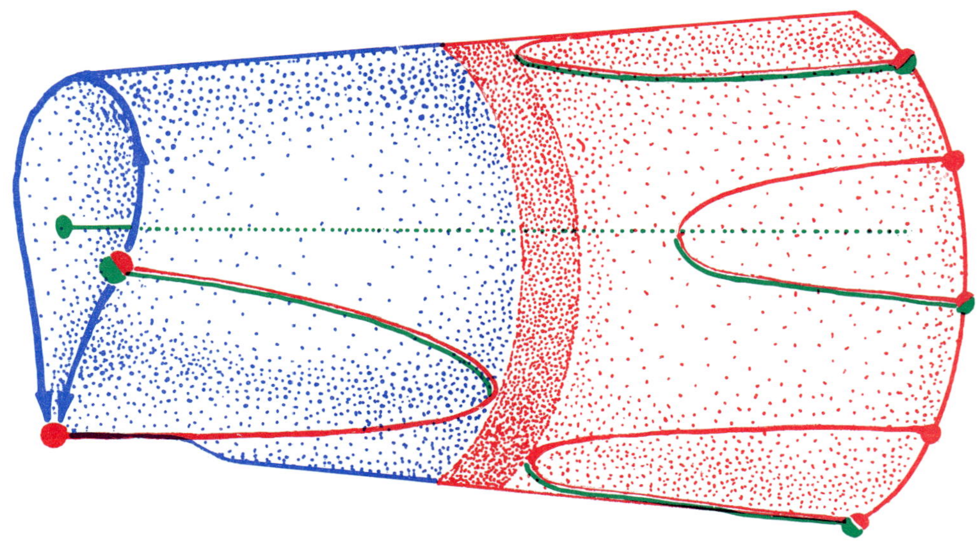

21.2.5. Extracting the strobe planes, erecting them side by side in the control-space, and smoothly interpolating the locus of attraction, we obtain this cutaway representation of the strobed response diagram. The dark ring represents the interval of fluctuation.

SUMMARY: In this event, a direct suspension of the blue loop in 2D to 3D, a local event (periodic fold catastrophe) is complicated by a global feature (blue sleeve) so that an AIT of large amplitude explodes from the vanishing periodic attractor. When the dust settles, there are several new periodic attractors braided around the former blue sleeve. Many other variants of this explosion are known.

21.3. Zeeman's Blue Tangle In 3D

In the early 1960's, Smale constructed his famous horseshoe example, a 3D flow with a tangled, virtual separatrix (see Section 14.3). Eventually, chaotic attractors were observed to explode out of periodic attractors associated with such tangles. In this section we develop a version of this event, following Zeeman.[3] Recall the homoclinic tangle from Figure 14.1.9.

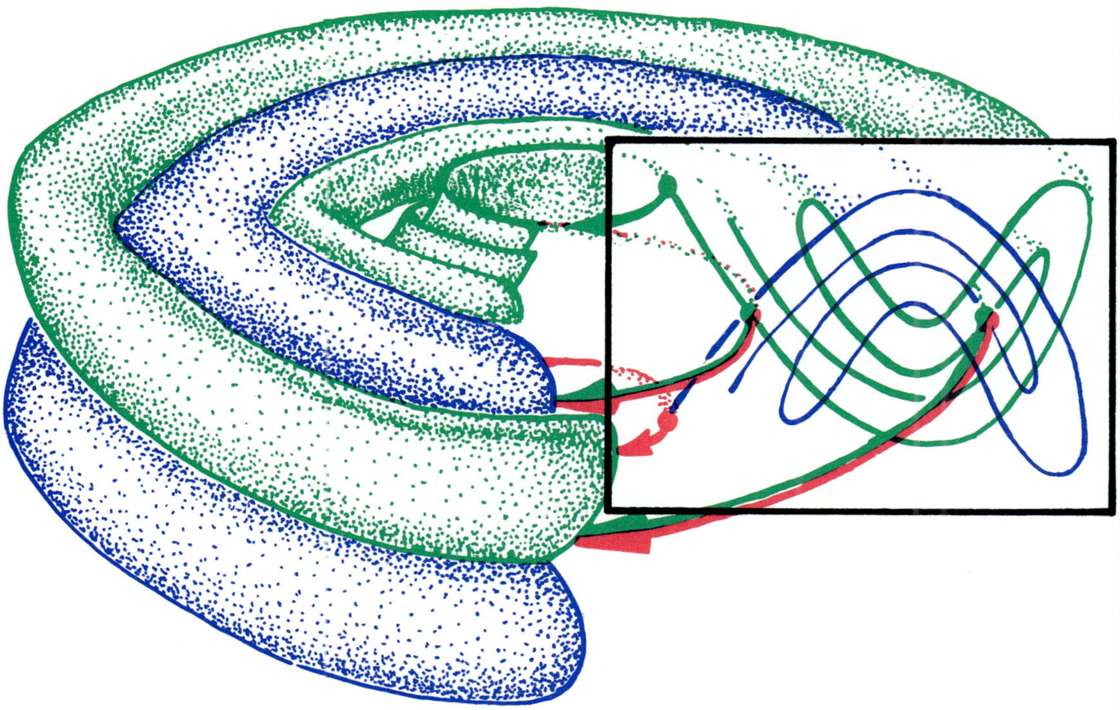

21.3.1. BEFORE: Here we have a global portrait containing four periodic trajectories of about the same period: a repellor, two saddles, and an attractor. (Compare Figure 21.2.1.) The two saddles are individually homoclinic, and heteroclinic to each other. They are involved in the tangle described by Smale's horseshoe. The global system is monostable, with tangled virtual separatrices within a single basin. The inner saddle (passing through the strobe section on the left) has an untangled inset strip arriving from the repellor and an untangled outset strip departing for the attractor.

We now increase the control to effect (locally) a periodic fold catastrophe between the inner saddle and the attractor.

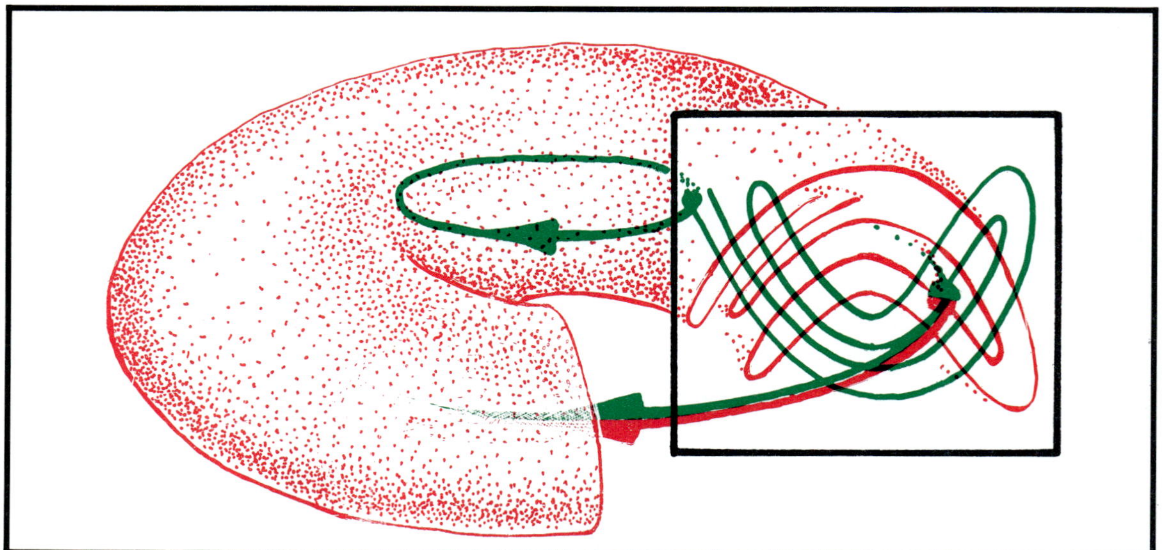

21.3.2. AFTER: The periodic attractor and inner saddle are gone, after mutual annihilation. The remaining saddle is homoclinic as before. But its fractal outset has become the solitary attractor. The global dynamic is still monostable, and the virtual separatrix is only slightly changed.

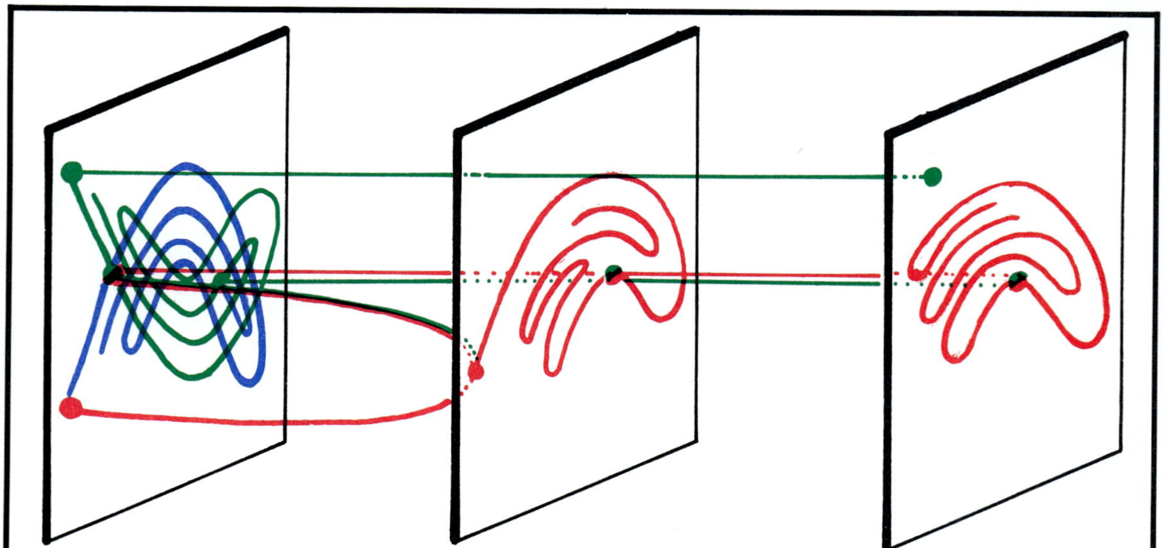

21.3.3. Placing the strobe portraits in their proper places in the space of the strobed response diagram, we interpolate a snapshot of the moment of bifurcation. The inner saddle vanishes along with its blue outset. But the oscillating blue outset of the outer saddle turns red.

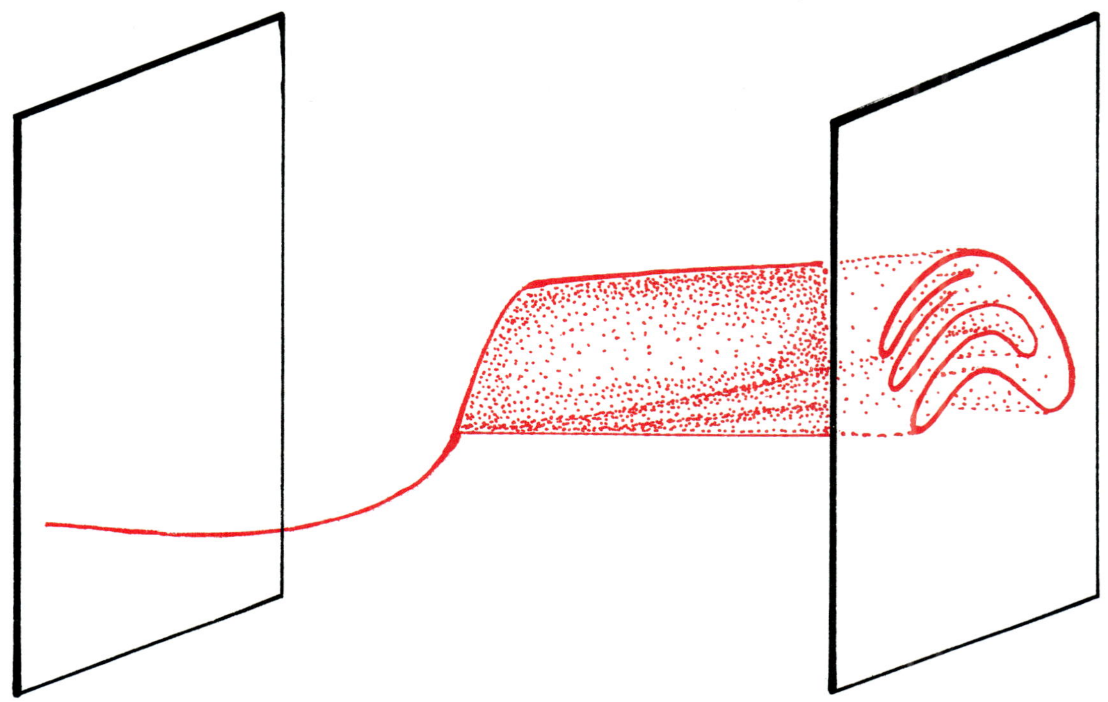

21.3.4. Here is the response diagram, cut away to display the strobed locus of attraction alone.

SUMMARY: In this event, a periodic attractor explodes to a large-scale chaotic attractor. This has been suggested by Zeeman as a model for the onset of turbulence. Without a significant change in the basin, an attractor abruptly changes its type and the volume of the phase space it dominates. Similar explosions can be triggered by either an octave jump or a vertical pinch.[6]

21.4 Ueda's Chaotic Explosion In 3D

So far we have seen explosions from point to cycle, from cycle to braid, and from cycle to chaos. It is also possible to have an explosion from small chaos to large chaos. In 1980, this example was published by Ueda,[4] the first of the great artists of chaos.

Here, translated to four colors, is an early stage on the way to the Rössler band, from Figure 8.4.4. This is obtained from an electronic analog of Duffing's forced pendulum.[5]

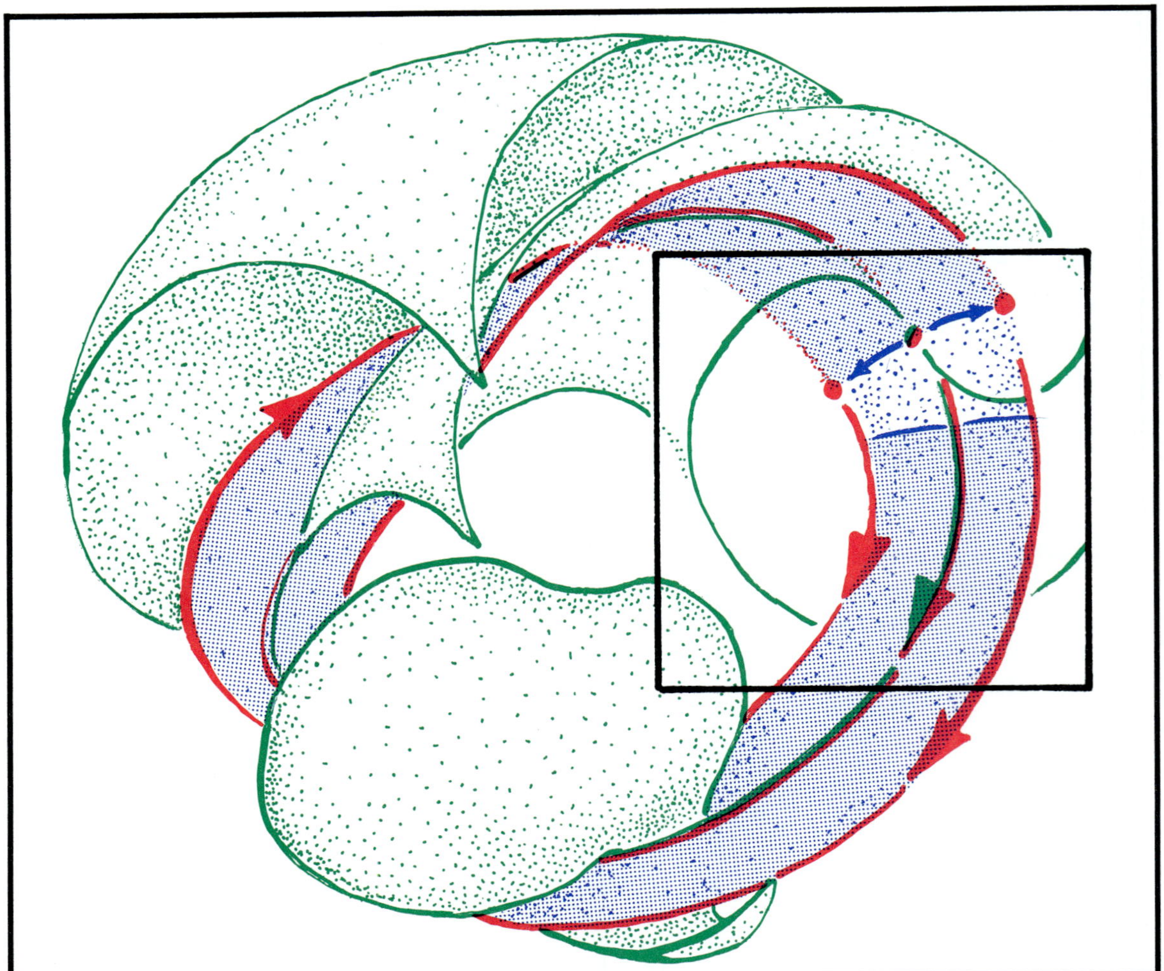

21.4.1. First, we fix the amplitude of the electronic forcing oscillator. Its frequency will be the control knob in this experiment. After a periodic fold catastrophe, a second periodic attractor appears. Here they are, separated by the scrolled green inset of the new periodic saddle. This is a bistable system.

From this point, we will make a few preparatory bifurcations, before beginning the Ueda sequence. These may require changing forcing amplitude as well as frequency, and perhaps other parameters.

We will indicate these preparations in the strobe plane only.

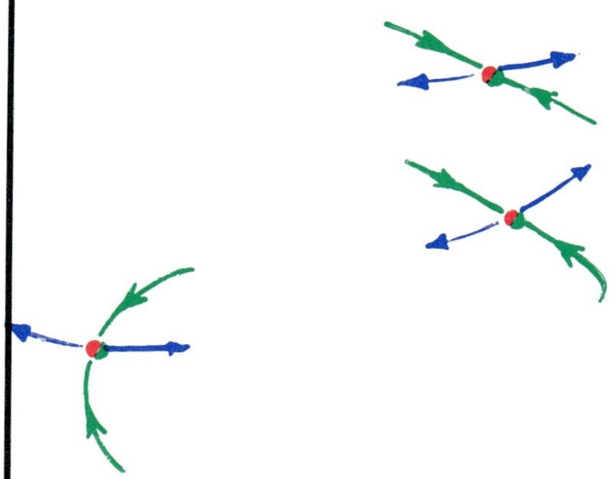

21.4.2. Next, we vary some parameters, and turn these two attractors into saddles. One possibility for accomplishing this would be an octave jump, in which case we imagine the new attractors to have disappeared from view.

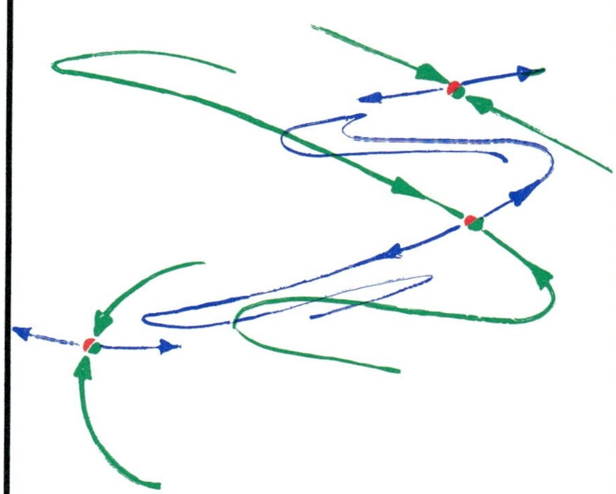

21.4.3. For the next modification, we move the inset and outset of the central saddle, to make two homoclinic tangles. In the window, we show a simplified schematic of the tangles. Note that the loops represent homoclinic tangles (a stable, generic phenomenon), not unstable homoclinic orbits.

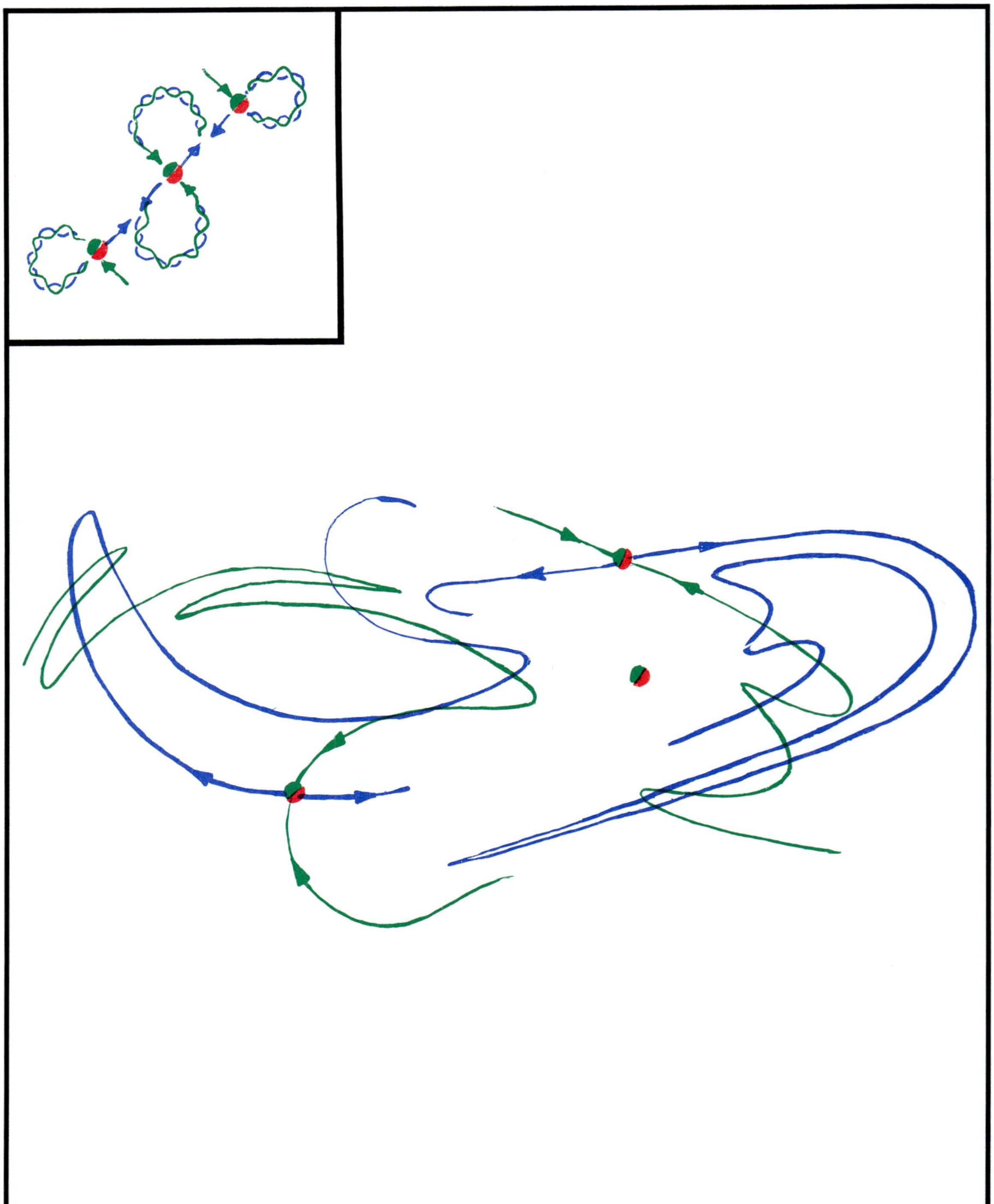

21.4.4. Changing other parameters, we now make both of the outer saddles half-homoclinic, as shown here. The former central tangle is omitted for simplicity, but it has not changed in this modification.

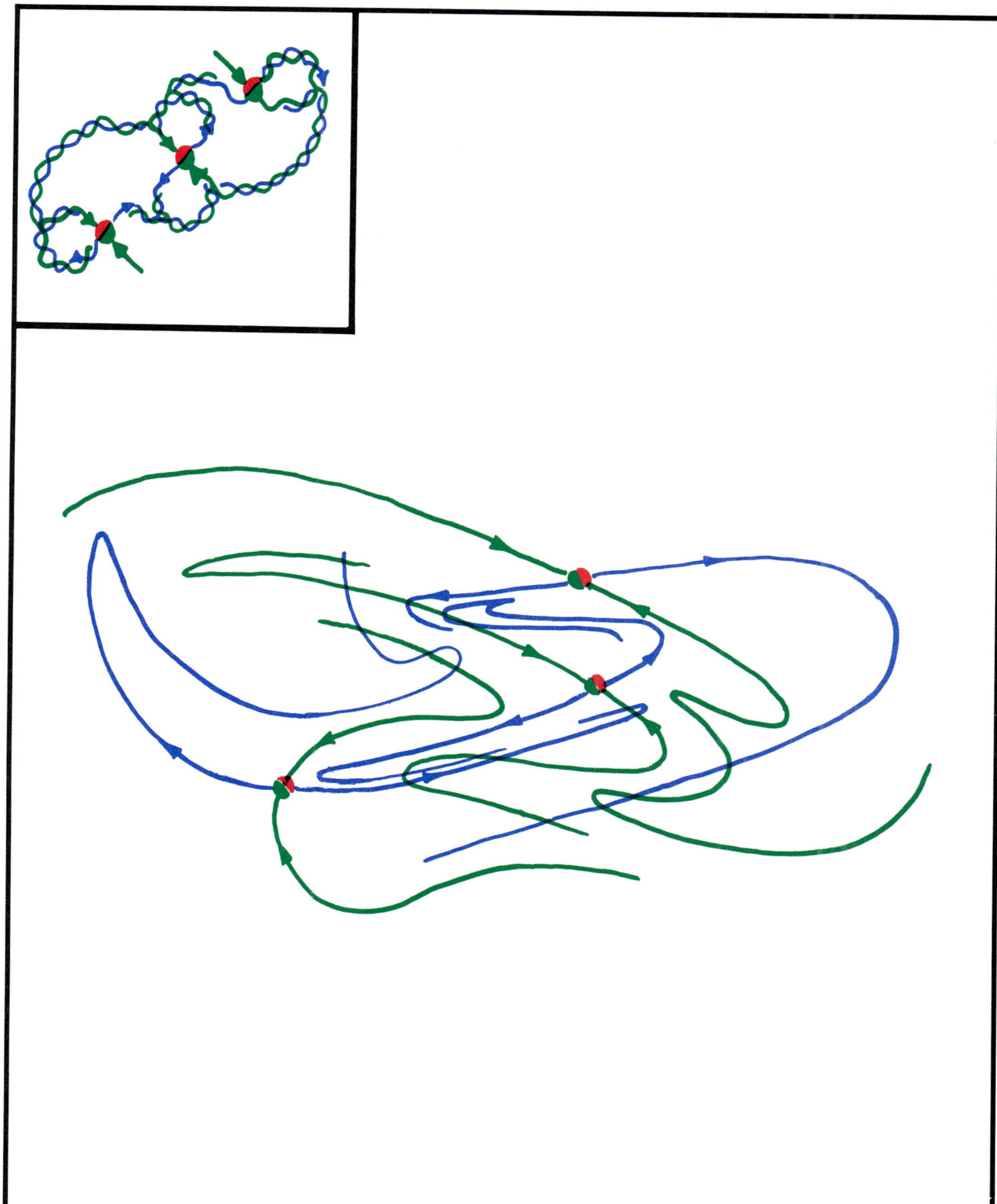

21.4.5. BEFORE: Superimposing the last two strobe portraits, we have this configuration, reproduced from Ueda's computer plot. All tangles are indicated schematically in the window. This is the starting point for our chaotic explosion.

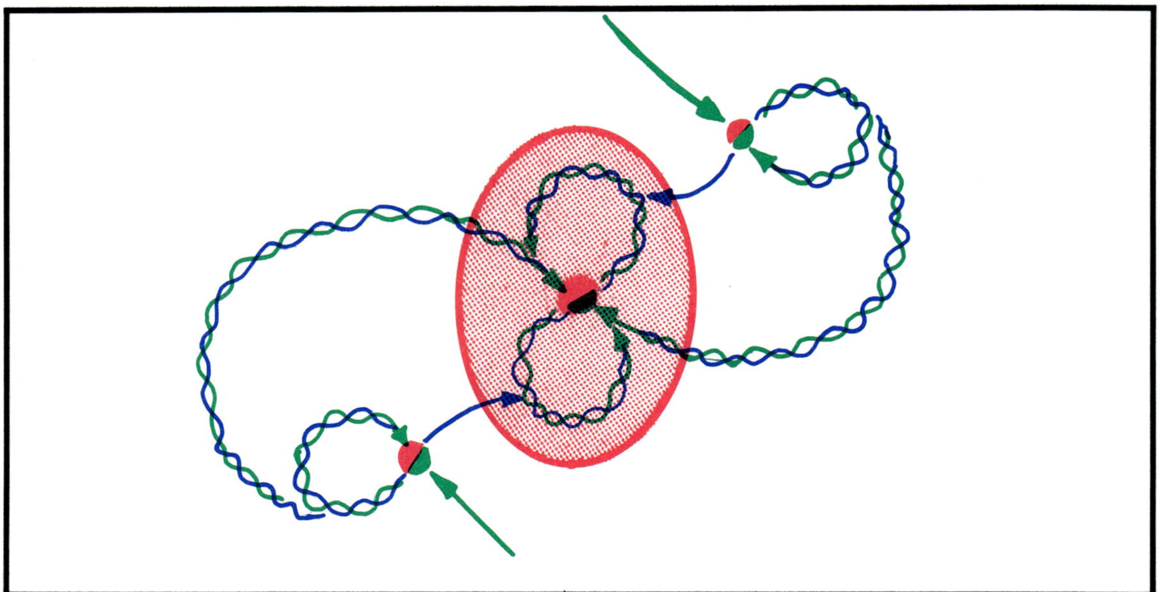

21.4.6. BEFORE: Here is an enlargement of the schematic diagram from the window in the preceding panel. As the outer outsets are attracted to the center tangle, the chaotic attractor in this portrait is the tangled outset of the center saddle. It is contained in the red-shaded region.

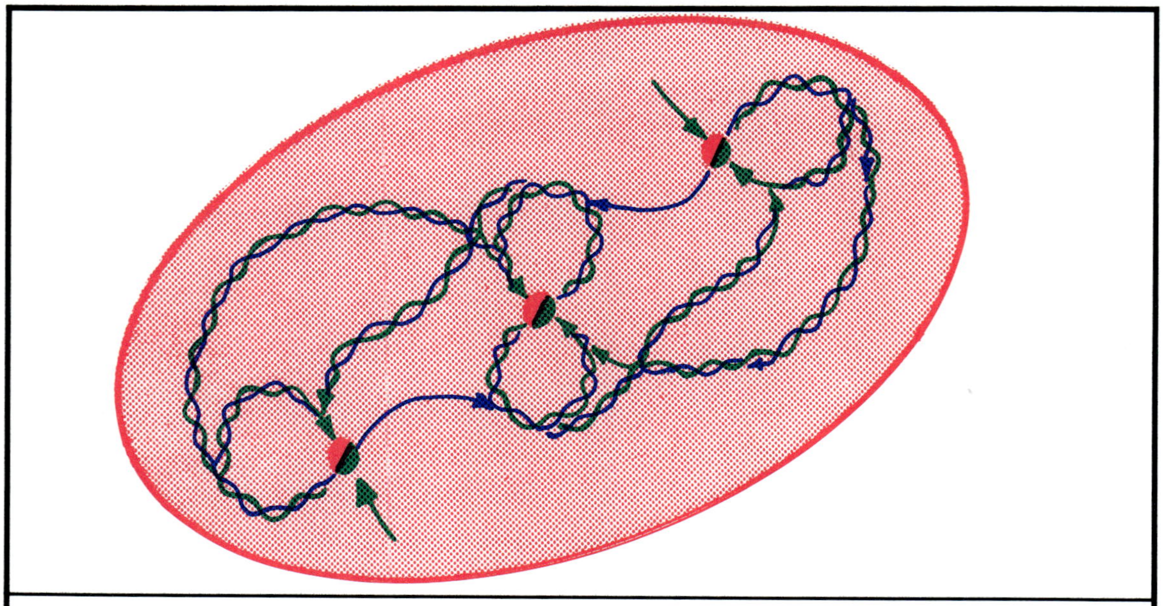

21.4.7. AFTER: The schematic now has new heteroclinic saddle connections, indicated by the bold green arrows here. Now the entire system is tangled: it is a *hypercycle* (see Section 14.4). The chaotic attractor has exploded so as to included the outsets of the outer saddles.

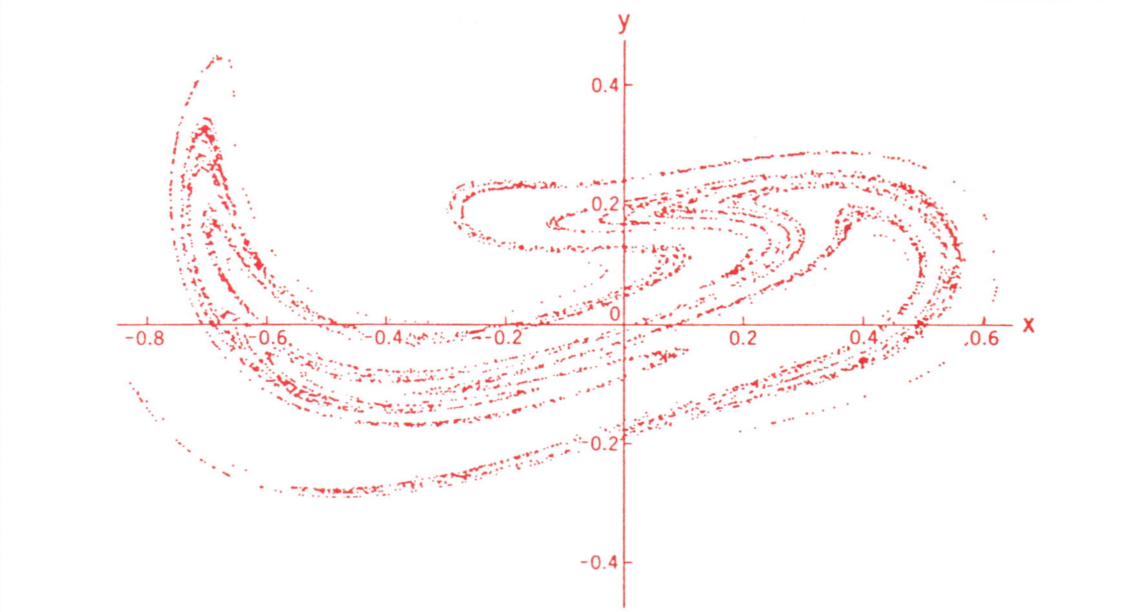

21.4.8. AFTER: Here is the new strobe portrait, corresponding to the new schematic, again reproduced from Ueda.

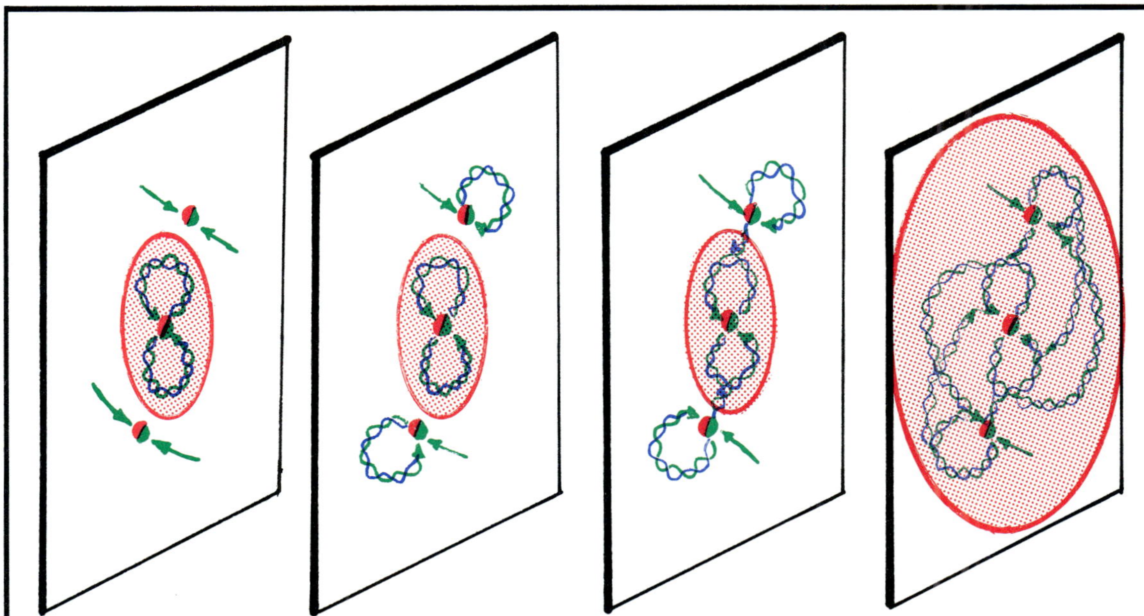

21.4.9. Repeating our usual side-by-side construction in the state/control space of the strobed response diagram, here is the skeleton representation for this chaotic explosion.

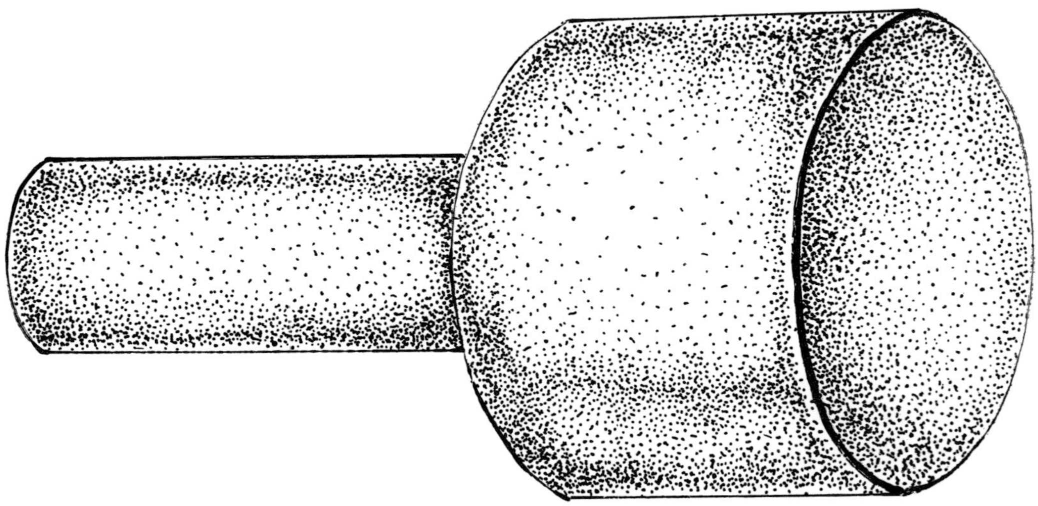

21.4.10. And here is a schematic cutaway portrait of Ueda's explosion.

SUMMARY: In this event, thanks to tangency bifurcations in which structural stability is lost momentarily because of the violation of generic property G3, an attractive tangle is abruptly enlarged through the addition of new foliage to a hypercycle.

22

Fractal Bifurcations

Referring to the *Big Picture,* in which a dynamical scheme is represented as a curve of dynamical systems, we see a bifurcation occurring wherever the curve pierces a hypersurface belonging to the *Bad Set.* An important feature of the Bad Set, which we have not illustrated as yet, is the *accumulation* of an infinite number of these bad sheets in fractal systems. Thus, a curve representing a generic scheme may unavoidably encounter an infinite set of bifurcations. Even though such a configuration involves an infinitude of individual bifurcation events, we may regard it as a single atomic bifurcation event. Thus, we speak of *fractal bifurcation events.* In this chapter we illustrate four examples of these fractal events. Many more may occur.

22.1. Octave Cascade

The simplest fractal bifurcation event is the octave cascade. The infinite set of bifurcation sheets in the big picture are arranged in a simple sequence, like the footprints of the frog jumping halfway to the wall. That is, the sheets get closer and closer to a limit sheet. This configuration is called a *cascade*. Each sheet (individual bifurcation event) in the convergent sequence is the same atomic event: the octave jump in 3D, hence the name, *octave cascade*. (Sometimes this is called the *flip cascade*; see Section 17.4). The limit sheet marks the onset of chaotic behavior.

Recall that the Möbius band is fundamental to the octave jump.

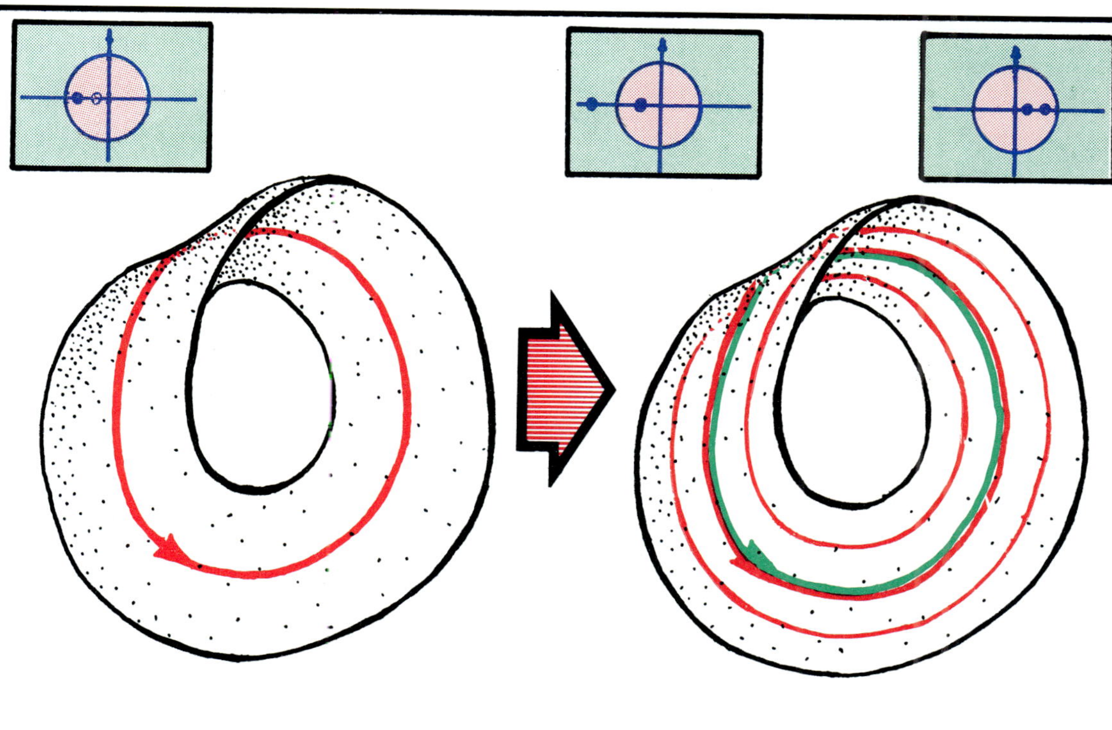

22.1.1. Here is a review of the octave jump in 3D: BEFORE and AFTER. Note that the CM's of the original cycle are both in the left half of the red disk (attractive region of the CM plane) before the event, while one of them (indicating behavior within the twisted band) has moved to the left half of the green (repelling) region after the event. However, the new attractor (corresponding to a tone one octave lower) has one of its CM's on each side of the red disk.

In order to go on with the next step of this cascade, we must adjust the new attractor so that both of its CM's are in the left half of the red disk. This adjustment may be smoothly made if the band is folded in two layers, as indicated in Section 8.1. And if the band has the folded structure of a Rössler attractor, instead of the twisted structure of a Möbius band, then we may adjust the double-period attractor on the band so that it appears to cross itself, but the two branches belong to two different layers of the folded band. The lower branch has been pushed back, over the fold, and down onto the lower sheet. At the same time, the CM's have been pushed to the left of the unit disk.

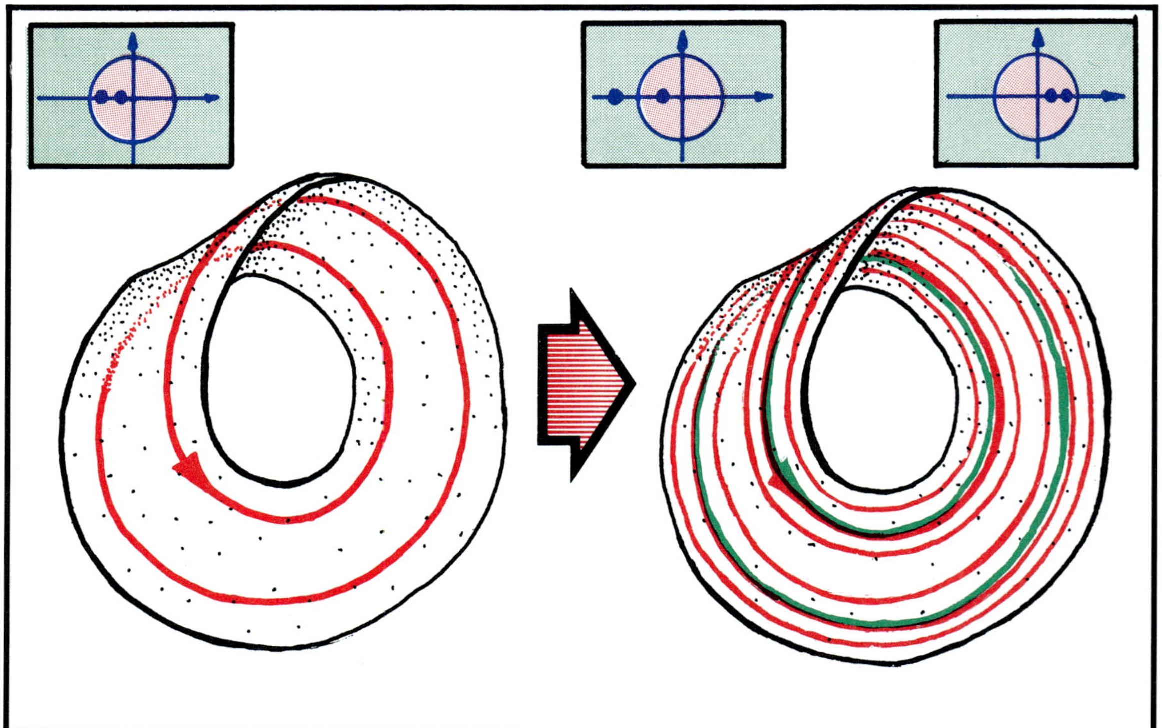

22.1.2. Now we are in a position to repeat the octave jump event. The twice-around attractor becomes a saddle cycle, and a four-times-around attractor takes over the basin.

To continue these events in a infinite cascade, we need more and more layers of the folded band, as are found in the Rössler attractor. And after each octave jump, we need another fold of the underlying band, to adjust the CM's to the left half of the unit disk, in preparation for another jump. Indeed, after the convergence of the cascade of individual octave jumps, we find the attractor has become a chaotic band.

This is one of the first chaotic scenarios to be discovered experimentally, by Rössler in 1976.[1]

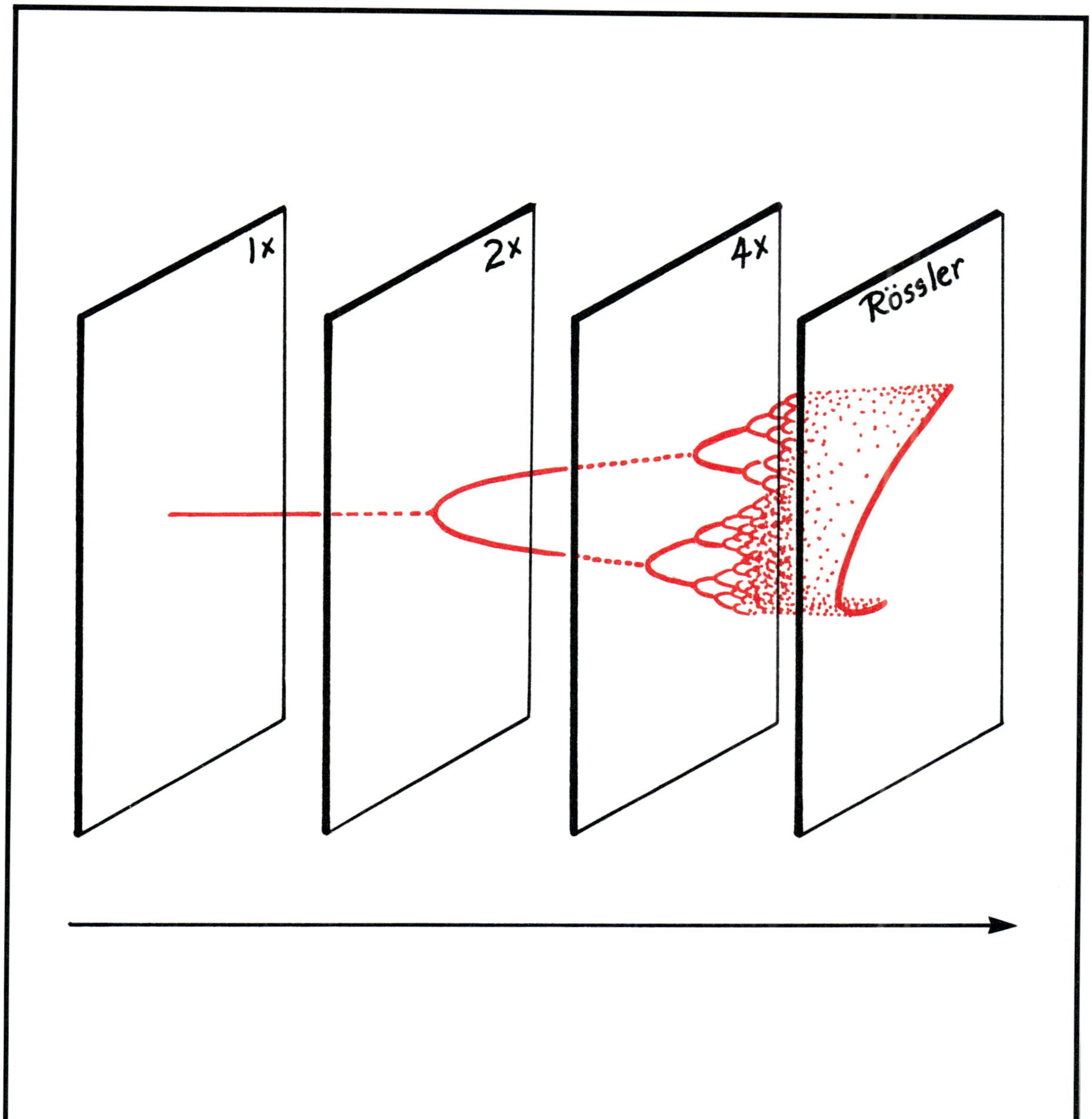

22.1.3. Here, then, is the strobed response diagram of the octave cascade. Each vertical plane corresponds to a strobe plane section through the infinitely folded band. Each octave jump is represented by a fork as explained in Fig. 17.4.5. These accumulate in an infinite sequence of multiplications and converge on a chaotic band. The bifurcation set in the control interval is fractal (see Section 9.4), and the chaotic attractor at the end of the event is fractal as well.

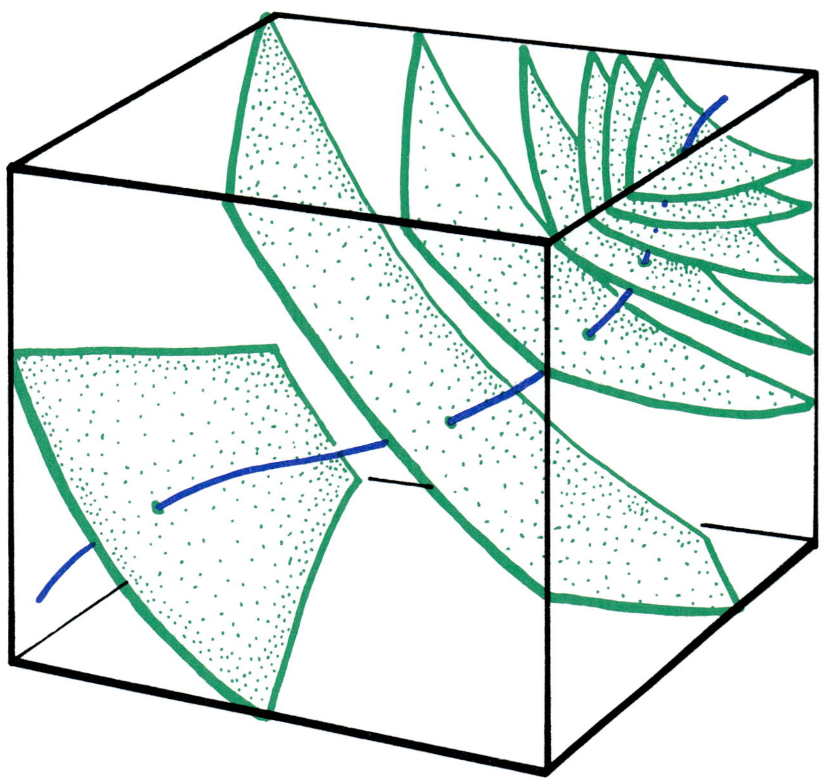

22.1.4. This example, visualized in the Big Picture, reveals a fractal structure in superspace.

SUMMARY: In this fractal bifurcation event, an infinite number of successive octave jumps accumulate on a phantom Rössler band, which eventually becomes the attractor.

NOTE: We have used *fractal* loosely here, as the dimension of the bifurcation set is actually zero.

7.2. The Noisy Cascade

This event may be viewed as a simple modification of the preceding one. First, regard the Rössler attractor as a noisy oscillator, as suggested in Section 9.5. Then replace each oscillator in the octave cascade by a noisy oscillator.

An individual bifurcation that has not been presented earlier in our atlas is the principal actor in the scenario: the chaotic octave jump.

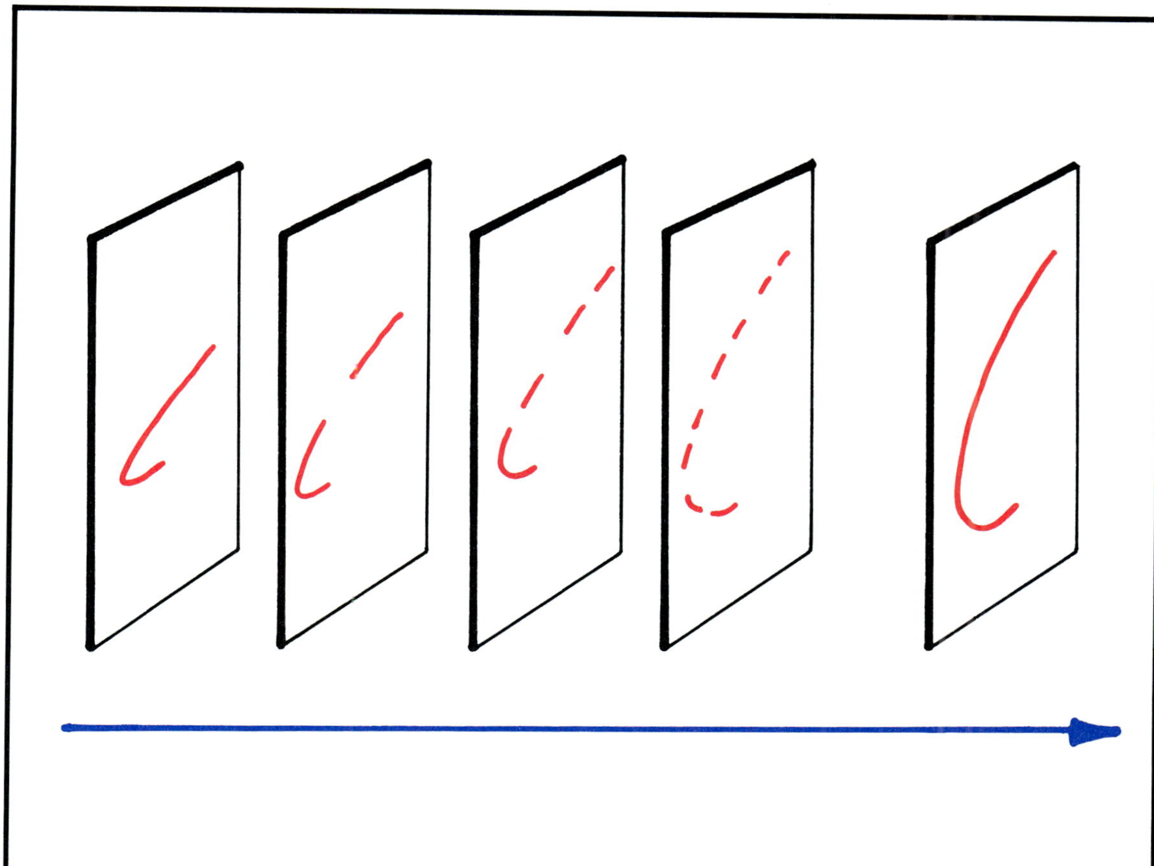

22.2.1. Here is a side-by-side skeleton of the strobed response diagram, showing the entire cascade. From left to right, a Rössler attractor makes an octave jump, turning into a twice-around band, shown here in strobe plane section. This is repeated in a convergent sequence, but at the limiting end, the final configuration on the right is again a once-around band!

In actuality, this cascade occurs on the other side of the limit attained by the octave jump for periodic attractors, as discovered by Lorenz.[2]

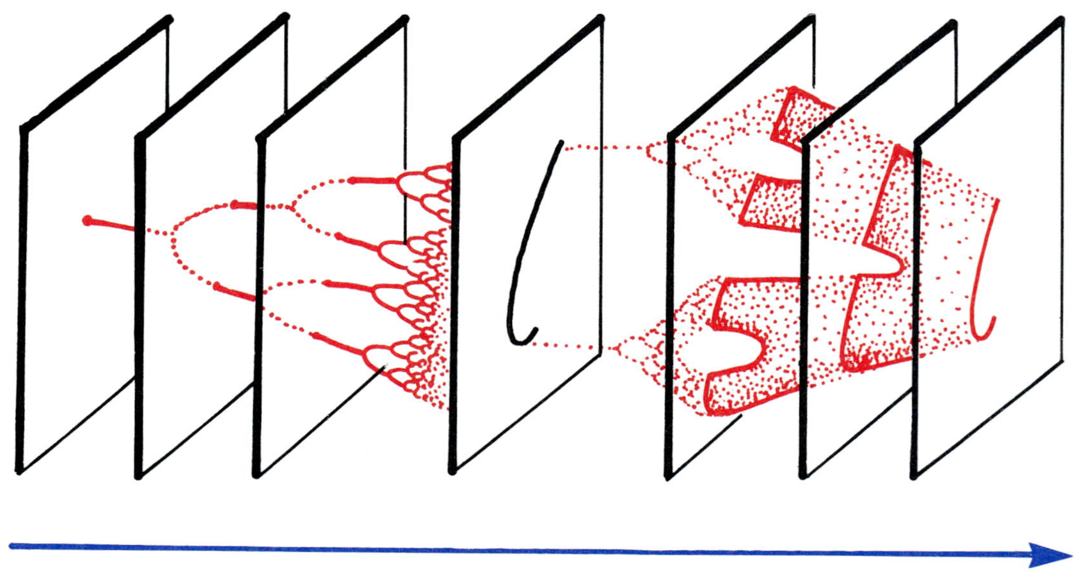

22.2.2. Here is a stripped-down view of the strobed response diagram for the full event, showing a periodic cascade approaching from the left and a noisy one from the right.

SUMMARY: In this event, the bifurcation set in the control interval consists of two convergent sequences, approaching the same limit point from opposite sides. A generic scheme, as a curve in the Big Picture, might encounter a doubly infinite set of sheets of the Bad Set. Unavoidably, this bifurcation event would result.

22.3. Braid Bifurcations

We have referred repeatedly to rearrangements of the braided periodic attractors encountered in toral dynamics (Chapter 5 and Section 8.2). We now review this phenomenon in the Big Picture.

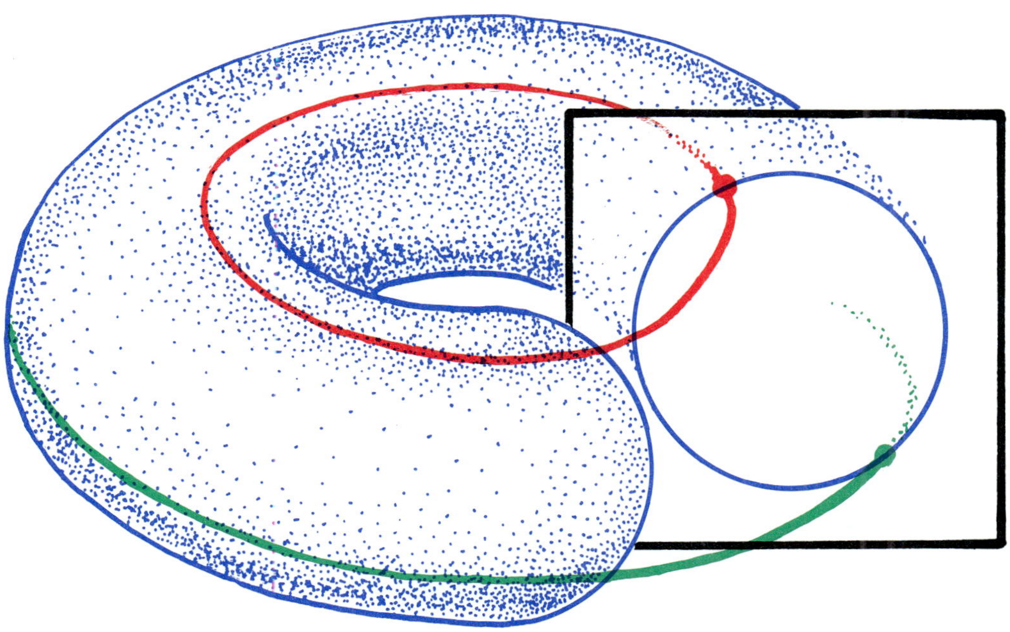

22.3.1. We might begin with the simplest toral flow, with two periodic trajectories: an attractor and a repellor. The rotation number (average rotation of the Poincaré first-return map on the strobe section) in this case is *zero*.

Alternatively, the rotation number could be any number of full circles, if the two periodic trajectories would around the waist of the torus a few times before the first return to the strobe plane. In fact, any rational number may arise as a rotation number in this context. On the other hand, there could be any number of periodic attractors, interspersed with an equal number of periodic repellors on the torus. In this case, the rotation number may still be zero.

However, we now consider the simple case pictured, and ask: what happens if the control parameter changes the rotation number?

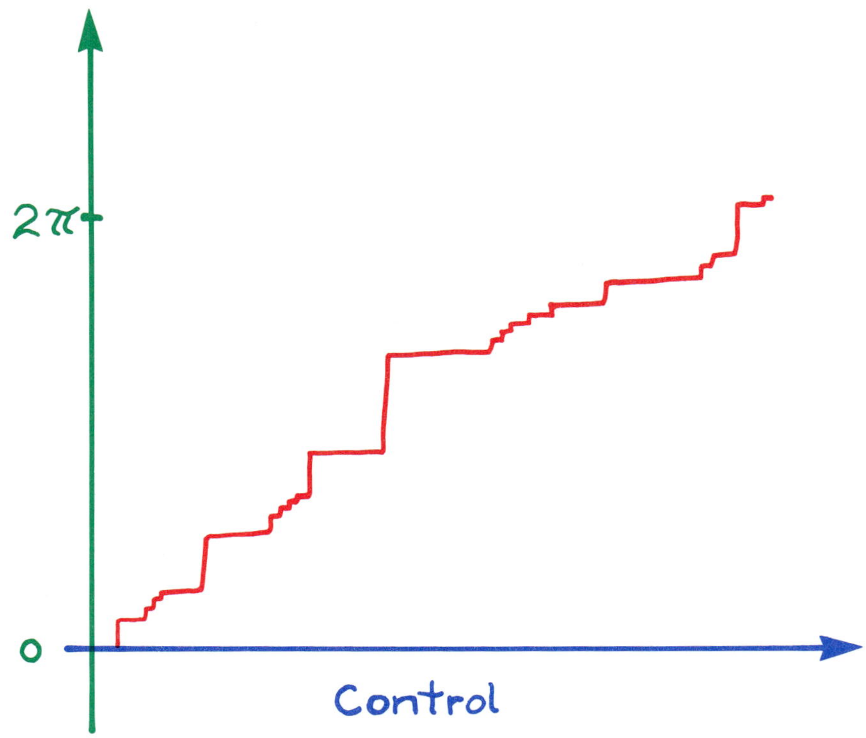

22.3.2. The answer is: this *devil's staircase.* As the control parameter increases to the right in this fictitious example, the rotation number increases from zero to one full revolution, but not smoothly.

But there are an infinite number of rational numbers in this rotation interval. And at each and every one of them, the rotation number tends to dwell awhile, in spite of the still increasing control parameter. This is because toral flows with rational rotations typically are structurally stable (according to Peixoto's theorem, Section 12.2) with a stable braid of periodic attractors and repellors which wind appropriately. The irrational rotations, less likely, always correspond to solenoidal flows (Figure 10.3.13). These may still fill up a set of control parameters having substantial expectation (probability larger than zero). This situation is known as a *fat fractal* bifurcation set. The bifurcation set in the control interval is a Cantor set, or fractal, as there are infinite sequences of control values of solenoidal flows (structurally unstable, hence bifurcations, the risers of the staircase) converging to an infinity of control values at the endpoints of control intervals (the steps of the staircase) corresponding to a stable rational braid. This has been beautifully analyzed by Herman.[3]

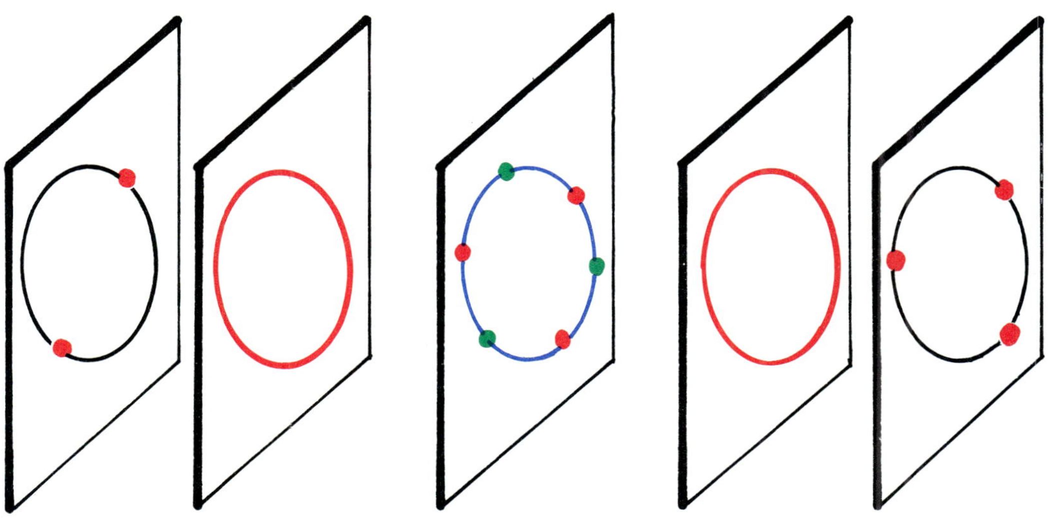

22.3.3. Here are the side-by-side strobe planes for this scheme. The red cycles indicate ephemeral solenoidal flows, while the blue cycles with red and green dots signify the stable braids with rational rotations, which survive for a small interval of control parameter values.

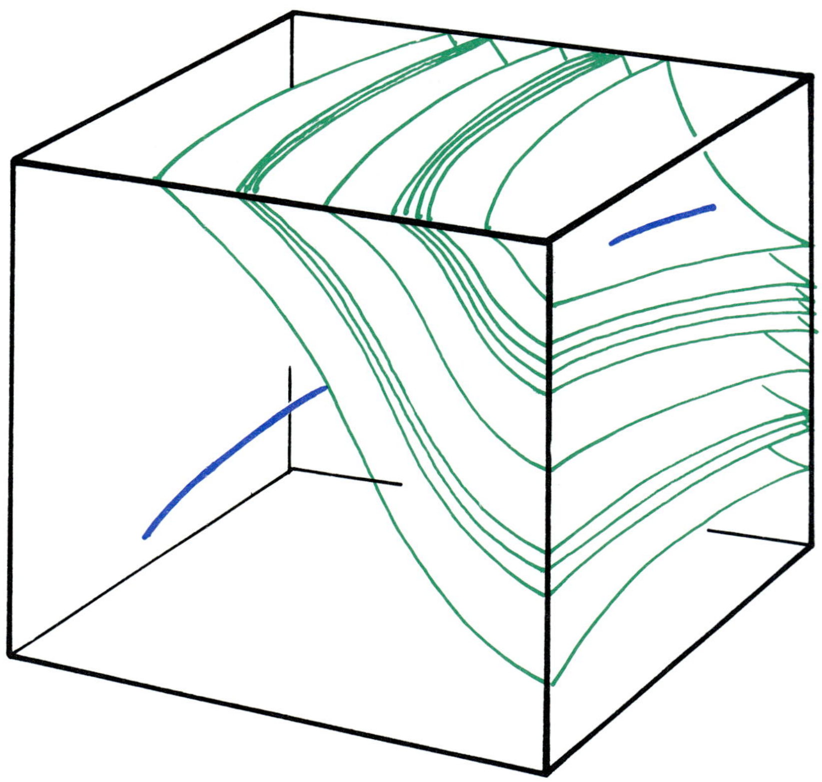

22.3.4. And here is the Big Picture, showing a thick slice of superspace in which the fractal sheets of braid bifurcations predominate.

SUMMARY: All AIT's in a dynamical scheme are troubled by the fluctuating braid bifurcations. Thus, it is pragmatic in many applications to regard an attractive torus as a kind of generalized attractor, and ignore as far as possible the microscopic details of its internal dynamic. As in all the bifurcation events in this chapter, we usually regard the entire fractal set of bifurcations as a single bifurcation event. These are characterized by a *bifurcation interval* enclosing the fractal bifurcation set, rather than a single bifurcation point in the control space.

22.4. Tangle Bifurcations

We have seen, in Chapter 20, that saddle connection bifurcations in 3D or more involve tangles. In the blue bagel, for example, as the control parameter increases and the affected inset and outset strips approach each other, there is a bifurcation point of first contact. The two invariant strips have an infinity of tangencies, which are nongeneric (failing G3) and thus structurally unstable: bifurcations. Later, there was a final bifurcation of tangency, before the two strips have completely passed through one another. These two extreme values define an interval, which we have called the *tangle interval.* For an infinite number of control values (belonging to a fractal set) within the tangle interval, other tangencies may be expected. At these bifurcations, the *signature* of the tangle changes. (See Section 13.5.) Thus, tangles imply an interval of control values that are dominated by a fractal bifurcation set, which may be regarded as a single bifurcation event. These events have been extensively studied by Newhouse and others.[4]

Appendix
Symbolic Expressions

As we have freely admitted in the Preface, our presentation of mathematical ideas by visual and verbal representations alone, without rigorous, symbolic expressions, is incomplete. It is unfair to mathematics and to the mathematically literate reader. But, wanting to give the maximum understanding to readers without experience of symbolic expressions, we decided to collect all the symbols in a closet of their own. This is it.

Our goal is to provide symbolic expressions for the basic concepts of dynamics (section A) and for the main examples of dynamical systems (section B).

A. Basic Concepts

In this section, we complete the representation of the basic concepts from Chapter 1 by adjoining their analytical (formal, symbolic) definitions. This is only a summary. For more details, consult the texts listed in the Bibliography.

State spaces are smooth manifolds. The examples used in the book are:

$R = R^1$, the real number line,
$R^n = R \times \ldots \times R$ (*n times*), Euclidian n-space,
$S^1 = T^1$, the *circle*,
$T^n = S^1 \times \ldots \times S^1$ (*n times*), the n-torus,

and their Cartesian products, the various cylinder spaces.

In such a manifold, M, the space of tangent vectors at a point, $x \varepsilon M$, is naturally identified with Euclidean space, $T_x M = R^n$. Thus, a smooth vectorfield on M is represented by a smooth mapping,

$$V: M \rightarrow R^n;\ x \mapsto (V_1(x), \ldots, V_n(x))$$

and a smooth curve, $c: R \rightarrow M$, *is an integral curve* or *trajectory* of the vectorfield, V, if for all t, $c'(t) = (c(t), V(c(t)))$, where $c'(t)$ is the tangent vector to c at t. In case $M = R^n$, the tangent vector is $c'(t) = (c(t), v(t))$, where $v(t)$ is the velocity vector, $v(t) = \lim [c(t + h) - c(t)]/h$ as h goes to zero.

Thus, the curve, c, is an integral curve if

$$c(t) = (c_1(t), \ldots, c_n(t)),$$

and the component functions satisfy the system of ordinary differential equations of first order,

$$c_1'(t) = V_1(c_1(t), \ldots, c_n(t))$$

$$\vdots$$

$$c_n'(t) = V_n(c_1(t), \ldots, c_n(t))$$

Here c_i' denotes the ordinary derivative.

Two special cases were described in Section 1.3. First, if $V(x) = 0$, then x is a *critical point* of the vectorfield, and the constant curve,

$$c: R \rightarrow M;\ t \rightarrow C(t) = x,$$

is an integral curve.

Second, if c is a nonconstant integral curve and $c(t + T) = c(t)$ for some (smallest positive) real number T, then c is a *periodic trajectory of period T.*

An integral curve is *complete* if it is defined for all real numbers. Suppose c is a complete integral curve at $x = c(0)$. The point x is called the *initial point* of c. The *omega-limit set* of x is the countable intersection,

$$\omega(x) = \cap\{cl(c[n, \infty)) \mid n \varepsilon N\}$$

where N denotes the positive integers, $[n, \ )$ denotes the closed ray, and $cl(A)$ denotes the closure of the set A. Similarly, the *alpha-limit set* of x is the countable intersection,

$$\alpha(x) = \cap\{cl(c(-\infty, -n]) \mid n \varepsilon N\}.$$

Let $L \quad M$ be a subset. Then the *inset of L* is the set

$$In(L) = \{x \varepsilon M \mid \omega(x) \subset L\}$$

and similarly, the *outset of L* is the set

$$Out(L) = \{x \varepsilon M \mid \alpha(x) \subset L\}.$$

If L has an open neighborhood, U, in its inset, $L \subset U \subset In(L) \subset M$, then L is *attractive* (there are many important variants of this definition) and $In(L)$ is its *basin of attraction.* An *attractor* is more than an attractive set. It is a subset, $A \subset M$, which is attractive, and which has *no proper subset* that is attractive. The *separatrix* is the complement of all basins of attraction:

$$Sep = \{x \varepsilon M \mid \omega(x) \textit{ is not contained in an attractor}\}.$$

B. Exemplary Systems

Using the notational conventions described in the preceding section, we give here the formulary of the examples described in the text, in tabular form.

Example: 1.
Section: 1.6.
Type: Negradient system.
Origin: Newton, ca. 1665.
Space: Plane, R^2
System:

$$x' = -x^3 - 2x^2 + 3x$$
$$y' = -y$$

Remarks: Potential function,

$$\mathrm{F}(x) = x^4/4 + 2x^3/3 - 3x^2/2 + y^2/2.$$

Example: 2a.
Section: 2.1.
Type: Simple pendulum.
Origin: Newton, ca. 1665.
Source: Stoker (1950), p. 61.
Space: Cylinder, S^1xR.
Coords: angle of elevation, A,
angular velocity, B.
System:

$$A' = B$$
$$B' = F\sin(A) - cB$$

Remarks: coefficient of viscous damping, c,
weight of pendulum, F.

Example: 2b.
Section: 2.2.
Type: Buckling column.
Origin: Stoker, 1950.
Source: Stoker (1950), p. 54.
Space: Plane, R^2
Coords: displacement, x,
velocity, y.
System:

$$x' = y$$
$$y' = (-1/m)\,[a_3x^3 + a_1x + cy]$$

coefficients,

$$a_1 = A + C - 2P/l$$
$$a_3 = B + D - P/l^3$$

Remarks: coefficient of viscous damping, c,
mass, m,
length or column, $2l$,
vertical force, P,
restoring force of primary (lateral) hard spring, $Ax + Bx^3$,
restoring force of secondary (hinge coil) hard spring, $Cx + Dx^3$.

Example: 2c.
Section: 2.3.
Type: Spring.
Origin: Rayleigh, 1877.
Source: Stoker (1950), p. 15.
Space: Plane, R^2.
Coords: displacement, x,
velocity, y.
System:

$$x' = y$$
$$y' = (-1/m)\,[a_3 x^3\ a_1 x + cy]$$

Remarks: mass, m,
restoring force of spring, $a_1 + a_3x^3$, $a_1 > 0$,
hard spring, $a_3 > 0$,
linear spring, $a_3 = 0$,
soft spring, $a_3 < 0$.

Example: 3.
Section: 2.4.
Type: Predator-prey.
Origin: Volterra, Lotka, 1924.
Source: Hirsch and Smale (1974), p. 259.
Space: Plane, R^2.
Coords: prey population, x,
predator population, y.
System:

$$x' = (A - By)x$$
$$y' = (Cx - D)y$$

Remarks: $A,B,C,D > 0$,
saddle point, (0,0),
center, $(D/C,\ A/B)$.

Example: 4a.
Section: 3.1–3.
Type: Self-sustained oscillation.
Origin: Rayleigh, 1883.
Source: Stoker (1950), p. 119.
Space: Plane, R^2.
Coords: current, x,
voltage, y.
System:

$$x' = y$$
$$y' = (-1/CL)\,[x + By^3 - Ay]$$

Remarks: capacitance, $C > 0$,
inductance, $L > 0$,
characteristic function of vacuum tube, $Bv^3 - Av$, $A, B > 0$.

Example: 2d.
Section: 4.3, 7.4.
Type: Forced spring.
Origin: Duffing, 1908.
Source: Stoker (1950), p. 81.
Space: Ring, $R^2 \times S^1$.
Coords: displacement, x,
velocity, y,
driving phase, θ.
System:

$$x' = y$$
$$y' = (-1/m)\,[a_3x^3 + a_1x + cy] + F\cos(0)$$
$$\theta' = \omega$$

Remarks: mass, m,
restoring force of spring, $a_1x + a_3x^3$, $a_1 > 0$,
hard spring, $a_3 > 0$,
linear spring, $a_3 = 0$,
soft spring, $a_3 < 0$,
coupling strength, F,
driving frequency, ω.

Example: 4b.
Section: 5.4, 7.2.
Type: Forced, self-sustained oscillation.
Origin: Rayleigh, 1888.
Source: Stoker (1950), p. 147.
Space: Ring, $R^2 \times S^1$
Coords: current, x,
voltage, y,
driving phase, θ.

System:

$$x' = y$$
$$y' = (-1/CL)\,[x + By^3 - Ay] + F\cos(\theta)$$
$$\theta = \omega$$

Remarks: capacitance, $C > 0$,
inductance, $L > 0$,
characteristic function of vacuum tube, $By^3 - Ay$, $A, B > 0$,
coupling strength, F,
driving frequency, ω.

Example: 4c.
Section: 7.2.
Origin: Van der Pol, 1922.
System:

$$x' = y$$
$$y' = (-1/CL)\,\{x + (3Bx^2 - A)y\} + F\cos\theta$$
$$\theta' = \omega$$

Remark: version of 4b, obtained by differentiation.

Example: 4d.
Origin: Robert Shaw, 1980.
Section: 7.2.
System:

$$x' = y + F\cos\theta$$
$$y' = (-1/CL)\,\{x + (3Bx^2 - A)y\}$$
$$\theta = \omega$$

Remark: version of 4b, obtained by moving the force to the first equation.

Example: 5.
Section: 7.3.
Type: Polynomial.
Origin: Lorenz, 1962.
Space: Euclidean, R^3.
Coords: x, y, z.
System:

$$x' = 10(y - x)$$
$$y' = x(28 - z) - y$$
$$z' = xy - (8/3)z$$

Example: 6.
Section: 7.4.
Type: Polynomial.
Origin: Rössler, 1968.
Space: Euclidean, R^3.
Coords: x, y, z.
System:

$$x' = -(y + z)$$
$$y' = x + y/5$$
$$z' = 1/5 + z(x - 5.7)$$

Notes

Preface

[1] Abraham and Marsden (1978).
[2] Dynamics, a visual introduction, in F. E. Yates (ed.), *Self-Organizing Systems*, Plenum, 1982.

Hall of Fame

[1] For historical details of this crucial event, see Carl Benjamin Boyer, *The History of the Calculus, and Its Conceptual Development*: Dover, New York, 1959.

Chapter 1

[1] See Zeeman (1977), p. 4.
[2] For an elaborate and carefully considered alternate definition of *attractor*, see David Ruelle, Small random perturbations of dynamical systems and the definition of attractors, *Commun. Math Phys. 82, 137–151 (1981)*.

Chapter 2

[1] See Holmes and Moon, *J. Sound Vibr.* 65(2), 275–296, 1979.
[2] An excellent history of this development is found in M.L. Cartwright, Nonlinear vibrations: a chapter in mathematical history, *Math. Gaz.* 36, 80–88, 1952.
[3] For the history and more discussion of this model, see Rosen (1970). And for the details of the mathematical analysis, see Hirsch and Smale (1974).
[4] See Hirsch and Smale (1974) for details.
[5] See H. I. Freedman, *Deterministic Mathematical Models in Population Ecology*: Decker, New York, 1980.

Chapter 3

[1] See Rosen (1970), Chapter 7, for more discussion of this scheme.

Chapter 4

[1] *The Theory of Sound,* Article 51.
[2] *The Theory of Sound,* Article 42.
[3] If you haven't, this is a good time to begin. See Zeeman (1977), Chapter 9, for seven applications of Duffing's cusp catastrophe to psychological behavior.

[4] An early study of harmonics in the Duffing ring is C. A. Ludeke, *Jour. Appl. Physics* 13, 215–233, 1942.

Chapter 5

[1] For historical details, see M. L. Cartwright, Nonlinear vibrations: a chapter in mathematical history, *Math. Gaz.* 36, 80–88, 1952. Also, see the outstanding text of the subject, Stoker (1950).

Chapter 12

[1] C. Guitierrez (1978) has published some results on structural stability in the nonorientable case.
[2] See Arnol'd (1961) and Herman (1979) for the awesome details on the thickness of the bad set.
[3] See Pugh (1967) and Robinson (1977) for the details of the Closing Lemma.

Chapter 13

[1] See Perello (1980) for an earlier exposition of these tangles.
[2] See Stewart's wonderful film (1985) for a better view of the scrolled insets and outsets.
[3] See Birkhoff and Smith (1950) and Birkhoff (1950) for a simple geometric proof.

Chapter 14

[1] See Birkhoff (1950).
[2] See Smale (1964) for the original horseshoe analysis.
[3] See the cover of Hayashi (1975) for a spectacular drawing of this tangle.
[4] See M. Levi (1981) for the occurrence of horseshoes in the forced Van der Pol system.
[5] Smale (1964).
[6] See Zeeman (1973) for the theory of shoes.
[7] See the original paper of Peixoto (8) for the first use of the quiver.

Chapter 15

[1] See Pugh (1967) for the proof of G4 from the Closing Lemma.
[2] See the cover of Hayashi (175).
[3] See Birkhoff and Smith (1950).
[4] See Birkhoff and Smith (1950).
[5] See the recent papers of R. Mañé (1978) for new results on structural stability in 3D.
[6] See Palis and Smale (1970) for the proof of structural stability under various hypotheses.
[7] See Markus (1961) and Robinson (1973) for the generic consequences of structural stability.
[8] See Smale (1964).

Chapter 16

[1] See Jones (1967) for a thrilling history, and Todhunter (1962), especially Chapters 1–13.
[2] See Anonymous (1941).
[3] See Hagihara (1970), especially the Introduction, and Chandrasekar (1969) Introduction, and Lyttleton (1953), especially Chapters 1 and 2.
[4] Todhunter (1062), p. 181.
[5] Lyttleton (1953), p. 39.
[6] Lyttleton (1953), p. 45.
[7] Lyttleton (1953), p. 45.
[8] Lyttleton (1953), p. 41; Hagihara (1970), p. 2.
[9] Lyttleton, pp. 1–5.

[11] See, for example, Iooss and Joseph (1980).
[12] Abraham, Marsden, and Ratiu (1983), Chapter 8.
[13] Donnelly, et al. (1980). See also Coles (1965).
[14] Thom (1972, 1975, 1983).

Chapter 17

[1] Thom (1972, 1975, 1983) and Prigogine (1980).
[2] Marsden and McCracken (1976).
[3] Hassard, Kazarninoff, and Wan (1981).
[4] Hirsch and Smale (1974) is one of the best for this purpose.

Chapter 18

[1] See Thom (1983), Zeeman (1982), and Poston and Stewart (1978).
[2] Besides the above, see Postle (1980).

Chapter 20

[1] See Abraham and Scott (1985), Abraham and Simo (1986), Abraham and Stewart (1986), and Thompson and Stewart (1986), p. 282.
[2] Rössler (1976), Simo (1979), and Thompson and Stewart (1986), pp. 280–284.
[3] H.B. Stewart, Fig. 6; Thompson and Stewart (1986), p. 282.

Chapter 21

[1] See Smale (1967).
[2] Zeeman (1982), Thompson and Stewart (1986), p. 130.
[3] Zeeman (1982), Thompson and Stewart (1986), p. 143.
[4] Ueda (1980), Simo (1979), Thompson and Stewart (1986), pp. 234, 278.
[5] Actually, this preceded Duffing. See Martienssen (1910).
[6] See Pomeau (1980).

Chapter 22

[1] Rössler (1976), Thompson and Stewart (1986), p. 242.
[2] Lorenz (1980).
[3] See, for example, Herman (1979).
[4] Newhouse (1979), also Guckenheimer and Holmes (1983), p. 331.

Bibliography

Abraham, Ralph H., Is there chaos without noise?, in *Chaos, Fractals and Dynamics* (P. Fischer and W. R. Smith, eds.): Dekker, New York, 1983.

Abraham, Ralph H., and Jerrold E. Marsden, *Foundations of Mechanics,* 2d ed.: Benjamin/Cummings, Reading, Mass., 1978.

Abraham, Ralph H., Jerrold E. Marsden, and Tudor Ratiu, *Manifolds, Tensor Analysis, and Applications:* Addison-Wesley, Reading, Mass., 1983.

Abraham, Ralph H., and Katherine A. Scott, Chaostrophes of forced Van der Pol systems, in *Chaos, Fractals, and Dynamics* (P. Fisher, W. Smith, eds.), pp. 123–134: M. Dekker, New York, 1985.

Abraham, Ralph H., and Carles Simo, Bifurcations and chaos in forced Van der Pol systems, in *Dynamical Systems and Singularities* (S. Pnevmatikos, ed.), pp. 313–323: North-Holland, Amsterdam, 1986.

Abraham, Ralph H., and H. Bruce Stewart, A chaotic blue sky catastrophe in forced relaxation oscillations: *Physica* 21D:394–400, 1986.

Anonymous, Maupertuis and the flattening of the Earth: *Geogr. Jour.* 98:291–293, 1941.

Arnol'd, V. I., On the mappings of a circumference onto itself, *Izv. Akad. Nauk SSSR, Ser. Matem.* 25(1):21–86, 1961.

Arnol'd, V. I., *Ordinary Differential Equations* (transl. by R. A. Silverman): MIT Press, Cambridge, Mass., 1978.

Birkhoff, G. D., Nouvelles recherches sur les systèmes dynamiques, in *Collected Works, vol. 2* 1950.

Birkhoff, G. D., and P. Smith, Surface transformations, in *Collected Works, vol. 2* 1950.

Chandrasekar, S., *Ellipsoidal Figures of Equilibrium*: Yale University Press, New Haven, Conn., 1969.

Coles, C., Transition in circular Couette flow: *J. Fluid Mech.* 21:385–425, 1965.

Crutchfield, J., R. J. Donnelly, D. Farmer, G. Jones, N. Packard, and R. Shaw, Power spectra analysis of a dynamical system: *Phys. Lett.* 76A:1–4, 1980.

Donnelly, R. J., and K. Park, R. Shaw, R. W. Walden, Early nonperiodic transitions in Couette flow, *Phys. Rev. Lett.* 44:987–989, 1980.

Guckenheimer, J., and P. Holmes, *Nonlinear Oscillations, Dynamical Systems, and Bifurcations of Vector Fields*: Springer-Verlag, New York, 1983.

Gurel, Okan, Poincaré's bifurcation analysis: *Ann. N. Y. Acad. Sci.* 316:5–22, 1979.

Gutierrez, C., Structural stability for flows on the torus with a cross-cap, *Trans. American Mathematics Society* 241, 1978.

Hagihara, Yusuke, *Theories of Equilibrium Figures of a Rotating Homogeneous Fluid Mass*: National Aeronautics and Space Administration, Washington, D.C., 1970 [1935].

Hassard, B. D., N. D. Kazarninoff, and Y. H. Wan, *Theory and Applications of Hopf Bifurcation*: Cambridge University Press, Cambridge, U.K., 1981.

Hayashi, Chihiro, *Nonlinear Oscillations in Physical Systems:* McGraw-Hill, New York, 1964.

Hayashi, Chihiro, *Selected Papers on Nonlinear Oscillations*: Kyoto University Press (1975).

Helleman, Robert H. G. (ed.), *Nonlinear Dynamics* (Annals, vol. 357): New York Academy of Sciences, New York, 1980.

Herman, M. R., Sur la conjugaison différentiable des difféomorphismes du cercle à des rotations *Publ. Math. IHES* 49:5–234, 1979.

Hilton, Peter (ed.), *Structural Stability, the Theory of Catastrophes, and Applications in Sciences* (Lecture Notes in Mathematics, 525): Springer-Verlag, New York, 1976.

Hirsch, Morris W., and Stephen Smale, *Differential Equations, Dynamical Systems, and Linear Algebra:* Academic, New York, 1974.

Hoppensteadt, Frank C., *Nonlinear Oscillations in Biology* (Lectures in Applied Mathematics, vol. 17): American Mathematical Society, Providence, R.I., 1979.

Iooss, Gerard, *Bifurcations of Maps and Applications*: North-Holland, Amsterdam, 1979.

Iooss, Gerard, and Daniel D. Joseph, *Elementary Stability and Bifurcation Theory*: Springer-Verlag, New York, 1980.

Irwin, M. C., *Smooth Dynamical Systems:* Academic, New York, 1980.

James, Preston E., and Goethey, J. M., *All Possible Worlds: A History of Geographical Ideas* (2nd ed.): Wiley, New York, 1981.

Jones, Tom B., *The Figure of the Earth*: Coronado Press, Lawrence, Kans., 1967.

Leroi-Gourhan, André, *Treasures of Prehistoric Art*: Abrams, New York, 1967.

Levi, Mark, Qualitative analysis of the periodically forced relaxation oscillations, *Memoires Amer. Math. Soc.* 214:1–147, 1981.

Lichtenberg, A. J., and M. A. Lieberman, *Regular and Stochastic Motion*: Springer-Verlag, New York, 1982.

Lorenz, E. N., Deterministic non-periodic flow, *Journal of Atmospheric Science* 20:130–141, 1963.

Lorenz, E. N., Noisy periodicity and reverse bifurcation, in *Nonlinear Dynamics* (R. H. G. Helleman, ed.): New York Academy of Science, New York, 1980.

Lyttleton, R. A., *The Stability of Rotating Liquid Masses*: Cambridge University Press, Cambridge, U.K., 1953.

Mandelbrot, Benoit, *Fractals*: W. H. Freeman, San Francisco, 1982.

Mañé, R., Contributions to the stability conjecture, *Topology* 17:386–396, 1978.

Markus, Lawrence, *Lectures in Differentiable Dynamics:* American Mathematical Society, Providence, R.I., 1971.

Markus, Lawrence, Structurally stable differential systems: *Ann. Math.* 73:1–19, 1961.

Marsden, Jerrold E., and Marjorie McCracken, *The Hopf Bifurcation and Its Applications*: Springer-Verlag, New York, 1976.

Martienssen, O., Uber neue Resonanzerscheinungen in Wechselstromkreisen: *Physik. Zeitschr.* 11:448–460, 1910.

Newhouse, Sheldon E., The abundance of wild hyperbolic sets and non-smooth stable sets for diffeomorphisms: *IHES Publ. Math.* 50:101–152, 1979.

Nitecki, Z., *Differentiable Dynamics:* American Mathematical Society, Providence, R.I., 1971.

Nitecki, Z., and C. Robinson (eds.), *Global Theory of Dynamical Systems* (Lecture Notes in Mathematics, vol. 819): Springer-Verlag, New York, 1980.

Palis, J., and S. Smale, Structural stability theorems, in *Proceedings of the Symposium on Pure Mathematics, XIV, Global Analysis,* pp. 223–231: American Mathematics Society, Providence, R.I., 1970.

Peixoto, M., Structural stability on two-dimensional manifolds: *Topology* 2:101–121, 1961.

Perello, C., Intertwining invariant manifolds and the Lorenz attractor, in *Global Theory of Dynamical Systems* (Z. Nitecki and C. Robinson, eds.), pp. 375–378: Springer, New York, 1980.

Pomeau, Y., and P. Manneville, Intermittent transition to turbulence in dissipative dynamical systems, *Commun. Math. Phys.* 74:189–197, 1980.

Postle, Denis, *Catastrophe Theory:* Fontana, London, 1980.

Poston, Tim, and Ian Stewart, *Catastrophe Theory and Its Applications*: Pitman, London, 1978.

Prigogine, Ilya, *From Being to Becoming: Time and Complexity in the Physical Sciences:* W. H. Freeman, San Francisco, 1980.

Pugh, C. C., An improved closing lemma and a general density theorem: *Am. J. Math.* 89:1010–1021, 1967.

Rayleigh, Baron, *The Theory of Sound* 2 vols.: Dover, New York, 1945 [1877].

Robinson, R. C., C[1] structural stability implies Kupka-Smale, in *Dynamical Systems* M. Peixoto, ed.), pp. 443–449: Academic, New York, 1973.

Robinson, R. C., Introduction to the closing lemma, in *The Structure of Attractors in Dynamical Systems*: National Science Foundation, Washington, D.C., 1977.

Rosen, Robert, *Dynamical System Theory in Biology*: Wiley-Interscience, New York, 1970.

Rössler, Otto, Different types of chaos in two simple differential equations: *Zeitschr. Naturforsch.* 31a:1664–1670, 1976.

Ruelle, D., Strange attractors: *La Recherche*: 108, 1980.

Shaw, R. S., 1981, Strange attractors, chaotic behavior, and information flow: *Zeitch. Naturforsch.* 36a:80, 1981.

Simo, Carles, On the Henon-Pomeau attractor: *Jour. Stat. Phys.* 21:465–494, 1979.

Smale, Stephen, Diffeomorphisms with many periodic points, in *Differential and Combined Topology Symposium in Honor of Marston Morse*: Princeton University Press, Princeton, N.J., 1964.

Smale, Stephen, Differential dynamical systems: *Bull. Am. Math. Soc.* 73:747–817, 1967.

Sparrow, C., *The Lorenz Equations*: Springer-Verlag, New York, 1982.

Stewart, H. B., *The Lorenz Attractor:* 16mm film, Aerial Press, Santa Cruz, Calif., 1985.

Stewart, I., *Les Fractals*: Belin, Paris, 1982.

Stewart, L. Bruce, A chaotic saddle catastrophe in forced oscillators, *Dynamical Systems, Approaches to Nonlinear Problems in Systems and Circuits* (F. Salam and M. Levi, eds.), 138–149, SIAM, Philadelphia, 1988.

Stoker, J. J., *Nonlinear Vibrations:* Interscience, New York, 1950.

Thom, René, *Mathematical Models of Morphogenesis*: Horwood, Chichester, U.K., 1983.

Thom, René, *Stabilité Structurelle et Morphogénèse: Essai d'une Theorie G*énérale des Models: Benjamin, Reading, Mass., 1972.

Thom, René, *Structural Stability and Morphogenesis* (transl. by D. H. Fowler): Benjamin/Cummings, Reading, Mass., 1975.

Thompson, J. M. T., and H. B. Stewart, *Nonlinear Dynamics and Chaos*: Wiley, New York, 1986.

Todhunter, I., *A History of the Mathematical Theories of Attraction and the Figure of the Earth: From the Time of Newton to that of Laplace*: Dover, New York, 1972 [1873].

Ueda, Yoshi, Explosion of strange attractors exhibited by Duffing's equation in *Nonlinear Dynamics* (R. H. G. Helleman, ed.), pp. 422–434: New York Academy of Science, New York, 1980.

Winfree, Arthur T., *The Geometry of Biological Time* (Biomathematics, vol. 8): Springer-Verlag, New York, 1980.

Zeeman, E. Christopher, Bifurcation, catastrophe, and turbulence, in *New Directions in Applied Mathematics* (Peter J. Hilton and Gail S. Young, eds.): Springer-Verlag, New York, 1982.

Zeeman, E. Christopher, *Catastrophe Theory: Selected Papers, 1972–1977:* Addison-Welsey, Reading, Mass., 1977.

Zeeman, E. Christopher, C^0–Density of stable diffeomorphisms and flows, in *Proceedings of the Dynamic Systems Conference,* Southampton Univ. 1973.

Index

Index numbers refer to panels, rather than pages.